FIVE KINGDOMS

The phyla of life on Earth based on our modification of the Whittaker five-kingdom system and the symbiotic theory of the origin of eukaryotic cells.

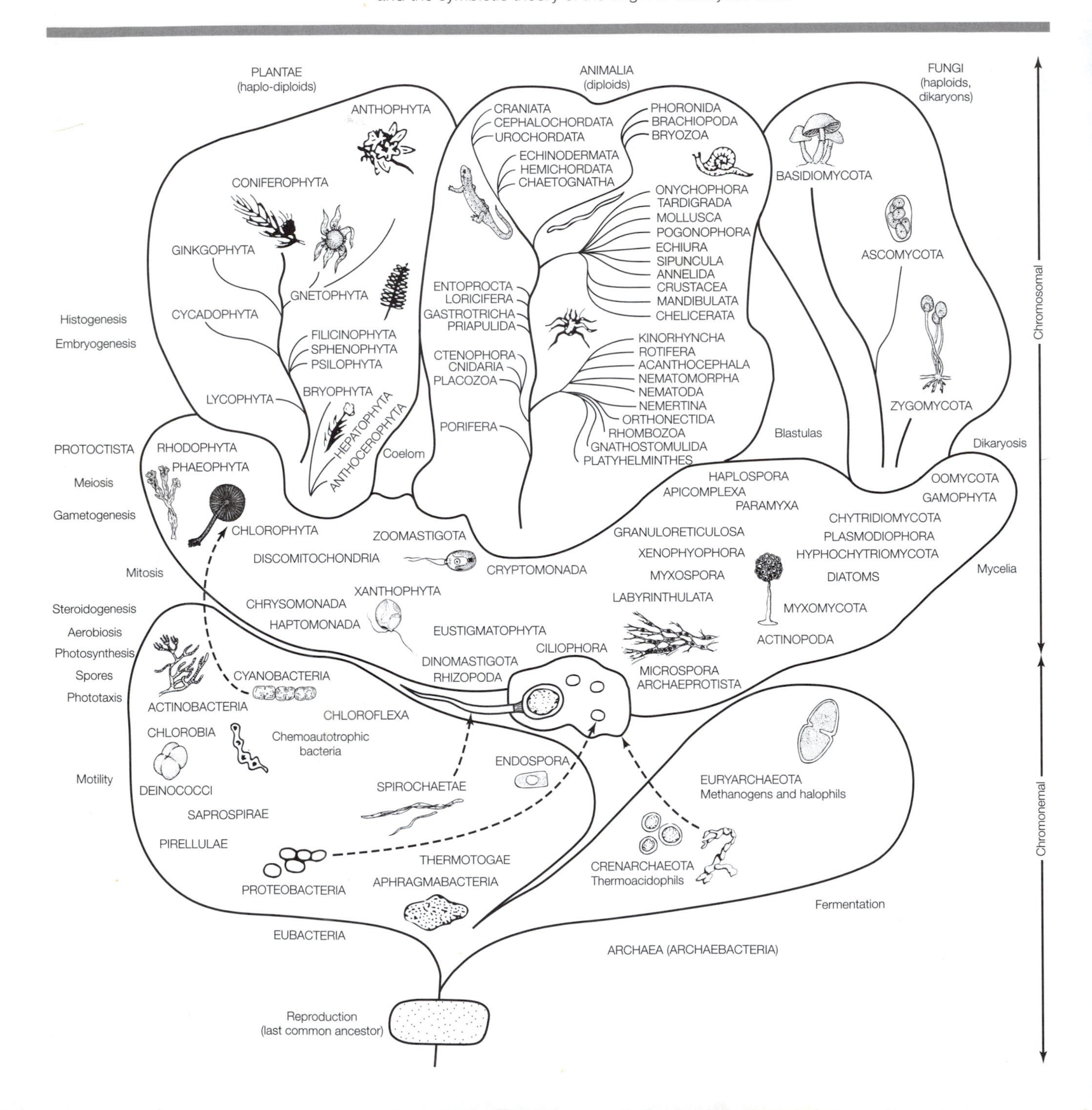

FIVE KINGDOMS

An Illustrated Guide to the Phyla of Life on Earth

THIRD EDITION

Lynn Margulis

University of Massachusetts at Amherst

Karlene V. Schwartz

University of Massachussetts at Boston

W. H. FREEMAN AND COMPANY

New York

Development Editor: Janet Tannenbaum
Project Editor: Georgia Lee Hadler
Cover and Text Designer: Diana Blume
Illustration Coordinator: Susan Wein
Production Coordinator: Maura Studley
Composition: Electronic Publishing Center and Progressive Information Technologies
Manufacturing: The Maple-Vail Manufacturing Group

Library of Congress Cataloging-in-Publication Data

Margulis, Lynn, 1938–
Five Kingdoms: an illustrated guide to the phyla of life on Earth/Lynn Margulis, Karlene V. Schwartz. — 3rd ed.
p. cm.
Includes bibliographical references and index.
ISBN 0-7167-3026-X (hardcover: alk. paper). — ISBN 0-7167-3027-8 (pbk.: alk. paper).—ISBN 0-7167-3183-5 (pbk.: alk paper/ref. booklet).
1. Biology—Classification. I. Schwartz, Karlene V., 1936– . II. Title.
QH83.M36 1998
570′.1′2—dc21 97-21338
CIP

Printed in the United States of America

First Printing, 1997

COVER IMAGE: Classification schemes help us comprehend life on this blue and green planet. But classification schemes are an invention, the human hand attempting to sort, group, and rank the types of life that share Earth with us. Because no person witnessed the more than 3 billion years of the history of life, our kingdoms and phyla, classes, and genera serve only as approximations.

In our metaphor of the hand, the lines within the hand outline separate the kingdoms. The thumb represents the earliest kingdom, Bacteria, which includes all the bacteria. The fingers, more like one another, represent the living forms constructed from nonbacterial cells. The back of the hand and the baby finger are continuous; they form a loosely allied, ancient group of microbes and their descendants: members of Kingdom Protoctista—seaweeds, water molds, ciliates, slime nets, and a multitude of other water dwellers. The ring and middle fingers stand together: The molds and mushrooms of Kingdom Fungi and the green plants of Kingdom Plantae made possible the habitation of the land. Members of Kingdom Animalia, the most recent kingdom to evolve, are on the index finger.

No matter how we care to divide the phenomenon of life, regardless of the names that we choose to give to species or the shapes that we devise for family trees, the multifarious forms of life envelop our planet and, over eons, gradually but profoundly change its surface. In a sense, life and Earth become a unity, each working changes on the other. And it is this interrelation that the image of hand and globe is meant to convey. [Illustration based on a sketch by Dorion Sagan, with design contributions from Donna Reppard.]

This third edition is dedicated to Robert P. Higgins, former head of the Department of Invertebrate Zoology, National Museum of Natural History, Smithsonian Institution, Washington DC, whose research continues to enlighten our perspective of life on Earth. The memories of our author friends David Chase (1933–1986), whose many superb micrographs adorn the book, Kenneth Estep (1954–1995), the first biologist and computer programmer to bring protoctists to the people, and Heinz Lowenstam (1912–1993), marine biogeologist and founder of the field of biomineralization, live on in these pages.

CONTENTS

* Genera illustrated in each phylum are given in italics below the name of the phylum.

SUPERKINGDOM **EUKARYA** 107

CHAPTER TWO
KINGDOM PROTOCTISTA 109

CHAPTER THREE
KINGDOM ANIMALIA 203

FOREWORD

Like bureaucracy, knowledge has an inexorable tendency to ramify as it grows. In the early nineteenth century, the great French zoologist Georges Cuvier classified all "animals"—moving beings, both microscopic and visible—into just four great groups, or phyla. A century earlier, Linnaeus himself, the father of modern taxonomy, had lumped all "simple" animals into the single category "Vermes"—or worms.

Cuvier's four animal phyla have expanded to more than forty, distributed in two kingdoms, the Protoctista (for microscopic forms and their descendants) and the Animalia (for those that develop from embryos)—and remember, we have said nothing of plants, fungi, other protoctists, and bacteria as yet. The very names of these groups are imposing enough—kinorhynchs, priapulids, onychophorans, and gnathostomulids. Some biologists can spit out these names with a certain virtuosity, but most of us know rather few of the animals behind the names. This ignorance arises for two primary reasons: the names are simply now too many, and modern training in zoology is now so full of abstract theory that old-fashioned knowledge of organic diversity has, unfortunately, taken a back seat.

Margulis and Schwartz have generated here that rarest of intellectual treasures—something truly original and useful. If the originality comes before us partly as a "picture book," it should not be downgraded for that reason—for primates are visual animals, and the surest instruction in a myriad of unknown creatures must be a set of figures with concise instruction about their meaning—all done so admirably in this volume. It is remarkable that no one had previously thought of producing such a comprehensive, obvious, and valuable document.

My comments thus far have been disgracefully zoocentric. I have spoken only of animals, almost as if life were a ladder with animals on the top rungs and everything else inconspicuously and unimportantly below. The old taxonomies included two kingdoms (plants and animals, with unicells placed, in procrustean fashion, into one or the other camp), or at most three kingdoms (animals, plants, and unicells). With this work, and its 96 phyla distributed among five kingdoms, we place animals (including ourselves) into proper perspective on the tree of life—we are a branch (albeit a large one) of a massive and ramifying tree. The greatest division is not even between plants and animals, but *within* the once-ignored microorganisms—the prokaryotic Bacteria and the eukaryotic Protoctista. The five kingdoms are arrayed as three great levels of life: the prokaryotes, the eukaryotic microorganisms and their derivatives (Protoctista), and the eukaryotic larger forms (Plantae, Animalia, and Fungi). These last three familiar kingdoms represent the three great ecological strategies for larger organisms: production (plants), absorption (fungi), and ingestion (animals).

Some people dismiss taxonomies and their revisions as mere exercises in abstract ordering—a kind of glorified stamp collecting of no scientific merit and fit only for small minds who need to categorize their results. No view could be more false and more inappropriately arrogant. Taxonomies are reflections of human thought; they express our most fundamental concepts about the objects of our universe. Each taxonomy is a theory about the creatures that it classifies.

The preceding material is a slightly shortened and lightly altered version of the preface that I wrote for the original edition of this book. As I reread my words and consider the remarkable changes in this field during the past 15 years—a growth of knowledge and development of thinking that, for once, justly deserves the overused designation of "revolutionary"—I am particularly struck by the wisdom and discernment of Margulis and Schwartz in their original, and now even more compelling, choice of the Five Kingdom system for ordering the diversity of life.

Molecular sequencing of nucleic acids has provoked the enormous gain in our understanding during the past 15 years. We can now obtain a much more accurate picture of the branching pattern on the tree of life through time by measuring the detailed similarities among organisms for the fine structure of genes held in common by all: as a general rule, the greater the differences between any two kinds of organisms, the longer they have been evolving on separate paths since their divergence from a common ancestor.

The system advocated here—five great kingdoms of life divided into two great domains (the Prokarya with their simple unicellular architecture lacking nuclei and other organelles and forming the kingdom of Bacteria, versus the Eukarya made of more complex cells and including the other four kingdoms of Protoctista, Animalia, Fungi, and Plantae)—might seem to be challenged by the discovery by Carl Woese and others that the genealogical tree of life has only three great branches, including two among the Prokarya (the Archaea and the Eubacteria), with all Eukarya on a third branch, and the three great multicellular kingdoms of plants, animals, and fungi as twigs at the tips of this branch.

But classification must consider more than the timing of branching. Woese's surprising discovery makes excellent sense when we realize that life is at least 3.5 billion years old on Earth, and that only Prokarya lived during the first 2 billion years or so. Since Eukarya arose so much later, they are confined to a single branch on a system that records time of branching alone.

Classifications must also record degree and amount of diversity and complexity (while never violating the primary signal of phylogeny, or order of branching), as well as the timing of branch points. When these criteria are added, the breaking of the enormous eukaryote branch into four kingdoms, and the compression of the two prokaryote branches into one kingdom of Bacteria seems fully justified, if only for our legitimately parochial interest in the astonishing diversity of organisms in our visible range of size and complexity.

Still, as the authors duly and happily note, and from an enlarged and less human-centered perspective, bacteria really are the dominant form of life on Earth—and always have been, and probably always will be. They are more abundant, more indestructible, more diverse in biochemistry (if not in complexity and outward form), and inhabit a greater range of environments than all the other four kingdoms combined. But we cannot grasp this fundamental fact, and so much else about evolution, until we abandon our biased view of life as a linear chain leading to human complexity at a pinnacle, and focus instead upon the rich range of diversity itself as the primary phenomenon of life's spread and meaning. And we cannot grasp life's full diversity without such excellent works as this book, dedicated to presenting the full story of life's vastness—from the "humble" and invisible (to us!) bacteria that really dominate life's history to the arrogant, fragile single species, *Homo sapiens*—a true upstart and weakling, but the Earth's first creature endowed with the great evolutionary invention of language, a device that may only lead to our self-destruction, but that also yields all our distinctive glories, including our ability to understand by classifying.

Stephen Jay Gould
Museum of Comparative Zoology
Harvard University

PREFACE

This book is an illustrated guide to the diversity of life on Earth. It is a comprehensive reference to both microbes and macroscopic organisms, serving as a guidebook to living organisms—what they look like, where they dwell, how they are related to one another, and how scientists group them. Brief essays introduce the broad outlines of kingdoms and phyla, and, if curiosity leads, we provide references to further reading. NASA scientists opened our eyes to the need for an illustrated guide to the diversity of life on Earth to inform their search for possible extraterrestrial life forms; *Five Kingdoms* focuses on diagrams and photographs of whole organisms to enable recognition of life forms in space by understanding life on our home planet. The book is meant for all students of biology at any level of expertise—whether participants in such courses as biology, biodiversity, zoology, botany, mycology, systematics, evolution, ecology, and geomicrobiology—or curious naturalists, geologists, park rangers, space scientists, and armchair explorers. It may serve as a quick reference to learn what characteristics distinguish the members of one phylum or to obtain a broad view of the evolutionary relationships of living organisms.

Readers are encouraged to explore evidence of the history of life firsthand by visits to the fossil sites, prehistoric dioramas, and exhibits listed in the Introduction in the section titled Field Trips through Time. Sites in U.S. and Canadian national parks and monuments, state parks, and science museums, as well as several locations abroad, are included. The extraordinary history of life, best discovered by direct observation of fossil evidence, inspires us to look more appreciatively at present-day life. A list of Internet sites suggests virtual field trips to resources regarding biodiversity.

The Introduction also includes tips for reading this book, which offer tools for understanding the typical habitats illustrated for each phylum. Colophons indicate how we see the organism in each photograph—with the naked eye, a magnifying glass, a light microscope, a scanning microscope, or an electron microscope. A revised chronology of the past four billion years of Earth history is summarized in a chart and a new table outlines our classification scheme.

Our Frontispiece illustrates differences between the archaebacteria and the eubacteria as well as distinctions among the eukaryotes—animals, plants, fungi, and protoctists. These differences and distinctions are summarized in the Introduction and fleshed out in introductory essays for each kingdom. Concepts of kingdoms and phyla originate in the classification proposals of scientists of the twentieth century—including Robert Whittaker and Herbert Copeland—built on earlier attempts of Linnaeus, Jussieu, Cuvier, and Haeckel to order the biota. We have extended these proposals and present a five-kingdom system consistent with both the fossil record and the most recent molecular data.

The molecular data have perhaps most profoundly affected our view of bacteria and protoctists. In accord with changes in bacterial classification that recognize differences in ribosomal RNA molecules, we have classified 16 phyla in Superkingdom Prokarya (Prokaryotae, Monera): two phyla in Subkingdom Archaea and 14 phyla in Subkingdom Eubacteria. We have incorporated protoctist reclassification based on the comprehensive *Handbook of Protoctists* with up-to-date ultrastructural, ecological, and molecular information for the 30 protoctist phyla.

In Kingdom Fungi, the five phyla of our earlier editions are now three. We have incorporated lichens—formerly Phylum Mycophycophyta—into Ascomycota, because the fungal partner in lichen symbiosis is most often an ascomycote that has "enslaved" photosynthetic cyanobacteria and algae. We also include the deuteromycotes—fungi that have lost the ability to differentiate asci or basidia—in Phylum Ascomycota; these fungi previously constituted Phylum Deuteromycota.

Plant classification has changed in accord with recognition of the morphological and biochemical differences between the nonvascular plants indicating that mosses, liverworts, and hornworts may have evolved independently of one another from algal ancestors. Each—formerly a class of Phylum Bryophyta—is now afforded phylum status as Bryophyta (mosses), Hepatophyta (liverworts), and Anthocerophyta (hornworts). Research from the frontiers of botany that has enlightened our understanding of the plant relationships includes the discovery by ethnobotanists of diverse drugs, the tracing of nitrogen pathways in old-growth forest, the elucidation of the evolutionary origin of the seed plants, and advances in understanding modes of plant hybridization.

Our concepts of animal phyla also are changing rapidly. Relationships among the animal phyla continue to be refined as phylogenetic data are correlated based on ultrastructure, developmental biology, morphology, and molecular sequences. Ribosomal RNA sequences and evidence from proteins tell us that the closest relatives of animals are fungi; plants are more distant relatives. As Ernst Haeckel observed in 1874, the Metazoa and a unicellular protoctist share common ancestry; molecular data support his inference. According to molecular evidence, the sponges and comb jellies form a branch within the animal kingdom, the placozoan *Trichoplax* and Cnidaria share a common branch, and the more complex bilateral animal phyla, another, more unresolved lineage.

We now consider the rhombozoans and orthonectids, formerly together in Phylum Mesozoa, to be separate phyla because of their unique characteristics. Having included the pentastome worms as a class within the Crustacea, we have abandoned Phylum Pentastomida. We have moved arthropods into three phyla: Chelicerata, Mandibulata, and Crustacea. Because some recent Burgess shale fossil evidence indicates that arthropods are a good monophyletic group, these three arthropod phyla may eventually be reunited. As fragmentary molecular data become more complete, the arthropod phylogenetic relationships will be refined.

Why do our concepts of classification shift through time? Every taxon—class, order, phylum, kingdom—is artificial but based on the study of relationships. We recognize that only the species is a natural taxon. A case in point is the phylum Loricifera—a group of minute marine organisms first described in 1983. Because loriciferans could not be placed in any previously known phylum without stretching the concept of the phylum, a new phylum was established just for them. Phylum Loricifera was a hypothesis to be tested, as are all new taxa. After more than a dozen years of evidence-gathering regarding the biology of loriciferans, Loricifera persists as a phylum; priapulids appear to be their closest relatives, and the kinorhynchs, more distant relatives. In 1995, a new species, *Symbion pandora,* was reported and a new phylum, Cycliophora, was proposed to accommodate this single species. As is the case for loriciferans, the relationships of *Symbion* to other phyla will be tested—possible relatives are entoprocts and bryozoans. Its hypothetical life cycle may be observed directly and fine details of its structure rigorously scrutinized, with the goal of more accurately learning the classification of *Symbion.* We do not yet accept Cycliophora as a phylum but await firsthand biological evidence, such as the life cycle of *Symbion.*

As in the earlier editions, a handy reference Appendix amplified to about 2000 genera includes all genera mentioned in this book and many others as well. The common (vernacular) names by which these genera are known are given, as is the phylum for each genus. The glossary provides definitions of terms. So that readers can move from glossary to chapter essays, we frequently indicate the phylum and kingdom to which a term applies. New to this edition is an Appendix listing all kingdoms and phyla.

Six sets of 35-mm color transparencies of the five kingdoms including phylogenies are available from Ward's Natural Science Establishment, Inc., 5100 West Henrietta Road, P.O. Box 92912, Rochester NY 14692-9012 (1-800-962-2660). Five of these transparency sets depict a member of each phylum of bacteria, protoctists, fungi, plants, and animals; the sixth set introduces the general features of each kingdom and prokaryote-eukaryote distinctions. A printed teacher's guide describing each slide is included.

Five Kingdoms is available in translation from the following publishers: Spanish: Editorial Labor, Madrid; Japanese: Nikeii Science, Tokyo; and German: Spektrum der Wissenschaft, Heidelberg.

Our readers have generously continued to provide photographs, drawings, manuscripts, publications, and constructive criticism for the complete rewriting of this, our third edition. We thank them for their great kindness and invite them to continue their contributions to us. We welcome suggestions for additional photographs, drawings, and publications. Please send suggestions of prehistoric dioramas and fossil sites accessible to the public to author KVS.

This small book grew out of our own need and that of our students for a single-volume reference to the five kingdoms. Your response tells us that our passion to comprehend the diversity of life is shared. Our list of acknowledgments expresses our appreciation.

ACKNOWLEDGMENTS

The artists' ways of viewing organisms enhance our own appreciation immensely. We thank the artists who gave assistance in the preparation of our earlier editions: Laszlo Meszoly, Michael Lowe, Peter Brady, Christie Lyons, and Robert Golder. Artists who rendered illustrations new for this edition include Christie Lyons, Kathryn Delisle, and Fine Line Studio. We are grateful to Lowell M. Schwartz for creating our chapter-opening background patterns and the rules, strung with tiny organisms, that distinguish the phylum essay pages of the different kingdoms. For their generous provision of photographs and drawings, we thank George Bean, Ray Evert, Robert P. Higgins, Eugene Kozloff, Carl Shuster, Jr., and David K. Smith. David F. Darby of Chase Studios, Cedar Creek, Missouri, generously guided us to locations having paleontological exhibits.

We thank the librarians at the University of Massachusetts in Boston and Duke University for their dedicated labor in obtaining journals and books. We wish to acknowledge the continuing encouragement and critical comments of our colleagues at the University of Massachusetts in Amherst and in Boston. We are grateful to Stephanie Hiebert, Donna Reppard, and especially to Michael Chapman for aid in manuscript preparation.

Since the second edition, help has come from Edward B. Cutler (sipunculans), Charles Cuttress* (invertebrates), Gillian Cooper-Driver (plants), Elizabeth A. Davis (developmental biology), Susan Eichhorn (plants), Gail R. Fleischaker (self-maintenance), Linda Graham (liverworts), Rolf Haugerud (pentastomes), Lawrence Kaplan (economic botany), Les Kaufman (reefs), Julian Monge-Nájera (onychophorans), Karl Niklas (ephedra), James W. Nybakken (molluscs), Jennifer Purcell (ctenophores), and Peter Del Tredici (ginkgo).

* Deceased.

One of us (LM) acknowledges the support of the Biology Department, University of Massachusetts at Amherst. The other (KVS) acknowledges the support of the Biology Department, University of Massachusetts at Boston, for her sabbatical at Duke University (1994) with the inspiring science writer, Steven Vogel. The enthusiastic encouragement of the Dean of the Sciences, Christine Arnett-Kibel, University of Massachusetts, Boston, provided critical impetus.

For urging us forward in countless ways, we thank our families: Lowell and Jonathan Schwartz; Dorion, Jeremy, and Tonio Sagan; Jennifer and Zachary Margulis.

We extend our thanks to the instructors and many others who provided reviews during our preparation of this third edition: Chris Anderson, Barbara H. Bowman, Lawrence Brand, Michael Chapman, Gillian Cooper-Driver, John Corliss, Peter Del Tredici, Michael Dolan, Niles Eldredge, Peter Gogarten, James Grant, Ricardo Guerrero, P. T. Handford, Margo Haygood, Robert P. Higgins, John Holt, Louis L. Jacobs, Jeri Jewett-Smith, Heather McKhann, Kenneth H. Nealson, James W. Nybakken, Lorraine Olendzenski, Kris A. Pirozynski, Peter H. Raven, J. John Sepkoski, Jr., Joseph W. Spatafora, Diana B. Stein, and Paul Whitehead.

Our heartfelt gratitude to Janet Tannenbaum and Georgia Lee Hadler, our editors at W. H. Freeman and Company, and Patricia Zimmerman, our manuscript editor, for their probing questions and superb professional contributions to this third edition of our work.

FIVE KINGDOMS

INTRODUCTION

Great contributors to our concepts of kingdoms and phyla

Carolus Linnaeus (Carl von Linné, Swedish) originated the concept of binomial nomenclature and a comprehensive scheme for all nature.
Antoine-Laurent de Jussieu (French) established the major subdivisions of Kingdom Plantae.
Georges Leopold Cuvier (French) established the major "embranchements" (phyla) of Kingdom Animalia.
Ernst Haeckel (German) was the innovator of Kingdom Monera for many microorganisms.
Herbert F. Copeland (American, Sacramento City College, California) reclassified all the microorganisms, recognizing Hogg's (1860) Kingdom Protoctista for all the nucleated organisms, all eukaryotes except plants and animals.
Robert H. Whittaker (American, Cornell University, New York) founded the five-kingdom system, recognizing Kingdom Fungi.

Carolus Linnaeus
1707–1778
[The Bettmann Archive.]

Antoine-Laurent de Jussieu
1748–1836
[The Bettmann Archive.]

Georges Léopold Cuvier
1769–1832
[Bibl. Museum Hist. Nat. Paris.]

Ernst Haeckel
1834–1919
[The Bettmann Archive.]

Herbert F. Copeland
1902–1968

Robert H. Whittaker
1924–1980
[Courtesy of R. Geyer.]

INTRODUCTION

CLASSIFICATION SYSTEMS

This book is about the biota, the living surface of the planet Earth. A catalogue of life's diversity and virtuosity, *Five Kingdoms* gives the reader a manageable system of ordering living beings. We present here an internally consistent, complete classification system, one that we judge to be valid and up to date, given the varying, fragmented, and often inconsistent professional literature from which our information is drawn.

Biologists, whether in the field or in the laboratory, study individual organisms or parts of populations, communities, or ecosystems. These organisms are classified—on the basis of body form, genetic similarity, metabolism (body chemistry), developmental pattern, behavior, and (in principle) all their characteristics—together with similar organisms in a group called a species. Scientists estimate that at least 3 million and perhaps 30 million species of living organisms now exist. An even greater number have become extinct. Our book can only touch on the greatest diversity of all: that of past life as documented in the fossil record. Only living groups are represented.

The effort to discern order in this incredible variety has given rise to systematics, the classification of the living world. Modern systematists group closely related species into genera (singular: genus), genera into families, families into orders, orders into classes, classes into phyla (singular: phylum), and phyla into kingdoms (Table I-1). This conceptual hierarchy grew gradually, in the course of about a century, from a solid base established by the Swedish botanist Carolus Linnaeus (1707–1778), who began the modern practice of binomial nomenclature. Every known organism is given a unique two-part name, Latin in form. The first part of the name is the same for all organisms in the same genus; the second part is the species within the genus. For example, *Acer saccharum, Acer nigrum,* and *Acer rubrum* are the scientific names of the sugar maple, the black maple, and the red maple, respectively.

Table I-1 The Classification of Two Organisms

Taxonomic level	Humans	Garlic
Kingdom	Animalia	Plantae
Phylum (Division)*	Chordata	Anthophyta
Subphylum†	Vertebrata	
Class	Mammalia	Monocotyledoneae
Order	Primates	Liliales
Family	Hominoidea	Liliaceae
Genus	*Homo*	*Allium*
Species	*sapiens*	*sativum*

* Botanists use the term "division" instead of "phylum."
† Intermediate taxonomic levels can be created by adding the prefixes "sub" or "super" to the name of any taxonomic level.

Groups of all sizes, from species on up to kingdom, are called taxa (singular: taxon); taxonomy is the analysis of an organism's characteristics for the purpose of assigning the organism to a taxon. Since the time of Linnaeus, the growth of biological knowledge has greatly extended the range of characteristics used in taxonomy. Linnaeus based his classification on the visible structures of living organisms. Later, extinct organisms and their traces—fossils—were named and classified. In the nineteenth century, the discoveries of paleontologists and Charles Darwin's revelation of evolution by natural selection encouraged systematists to believe that their classifications reflected the history of life—classifications were converted into phylogenies, family trees of species or higher taxa. To this day, very few lineages from fossil organisms to living ones have actually been traced, yet the truest classification is still held to be the one that best reflects the evidence for relationship by common ancestry.

In the twentieth century, advances in developmental biology and biochemistry have given the taxonomist new tools. For example, phylogenies can now be based on patterns of larval development, on the linear sequence of amino acids that compose proteins, or on gene sequences—the sequences of nucleotides in nucleic acids. Techniques of electron microscopy and optical microscopy have greatly improved in recent years, enabling scientists to study the internal structure of the smallest life forms and of the constituent cells of large forms in unprecedented detail. Computers that can handle massive quantities of sequence data allow scientists to measure the relatedness of organisms by comparing their gene sequences—the procedure that underlies molecular systematics.

MOLECULAR SYSTEMATICS

A new field of biological science, molecular systematics, or molecular evolution as it is often called, stems from two innovations. The first is the deciphering of the linear, genetically determined sequence of component monomers (amino acid residues or nucleotides) in macromolecules such as proteins and nucleic acids (DNA, RNA). The second is the computer-based handling of these immense quantities of data. Each cell of every live being has from 500 to 50,000 genes, and the average size of a gene exceeds 1000 nucleotides. Hence the molecular data for only the linear order of the protein and nucleic acid components in a single cell can range from half a million (0.5×10^6) to 50 million pieces of relevant information. Only high-speed computers, properly programmed, can organize and meaningfully compare these quantities of sequence data.

Fine structure, genetics, metabolism, behavior, development, and natural history—the biology of organisms—are refractory to quantitative study and resist the uniform descriptions needed for statistical measures of relatedness; thus molecular systematics, which compares molecules of constant function broadly distributed among many species, has revolutionized our understanding of evolution, especially of microbial evolution. All life, because it evolved from common ancestors, must at all times make and use certain long-chain molecules to maintain cell metabolism. The number and order of components (amino acids in proteins and nucleotides in DNA and RNA) in these long-chain macromolecules are called their sequences. Molecular sequence comparisons, for example, ribosomal RNA or ATPase (enzymes essential for all life that, retaining their identical functions and overall structure, have changed slowly through time), make a useful standard that contributes to whole-organism systematics. In the classification system presented in this book, we make extensive use of the insights that have emerged from molecular systematics. In all cases, however, the molecular data have been integrated with the biology of the whole organism to provide a system consistent with all the information available.

THE KINGDOMS OF LIFE

From Aristotle's time to the middle of the twentieth century, nearly everyone classified members of the living world into two kingdoms, plant or animal. Since the middle of the nineteenth century, however, many scientists have noted that certain organisms, such as bacteria and slime molds, differ from plants and animals more than plants and animals differ from each other. Third and fourth kingdoms to accommodate these anomalous organisms were proposed from time to time. Ernst Haeckel (1834–1919), the German proponent and popularizer of Darwin's theory of evolution, for example, made several proposals for a third kingdom of organisms. The boundaries of Haeckel's new kingdom, Protista, fluctuated in the course of his long career, but his consistent aim was to set the most primitive and ambiguous organisms apart from the plants and animals, with the implication that the larger organisms evolved from protist ancestors. Haeckel recognized the bacteria and blue-green algae as a major group—the Monera, distinguished by their lack of a cell nucleus—within the protist kingdom. However, most biologists either ignored proposals for additional kingdoms beyond plants and animals or considered them unimportant curiosities, the special pleading of eccentrics.

The climate of opinion regarding the kingdoms of life began to change in the 1960s, largely because of the knowledge gained by new biochemical and electron-microscopic techniques. These techniques revealed fundamental affinities and differences on the subcellular level that encouraged a spate of new proposals for multiple-kingdom systems. Among these proposals, a system of five kingdoms (plants, animals, fungi, protoctists, and bacteria), first advanced by Robert Whittaker in 1959 and greatly indebted to the earlier and highly original four-kingdom (plants, animals, protoctists, and bacteria) work of Herbert Copeland, has steadily gained support for more than three decades. With some modifications necessitated by more recent data, the Whittaker system is used in this book. Briefly, our five kingdoms are Bacteria (with its two subkingdoms, Archaea and Eubacteria), Protoctista (algae, protozoa, slime molds, and other less-known aquatic and parasitic organisms), Animalia (animals with or without backbones), Fungi (mushrooms, molds, and yeasts), and Plantae (mosses, ferns, and other spore- and seed-bearing plants). With regard to the plant kingdom, we are following the suggestion of James Walker (University of Massachusetts, Amherst; personal communication) for distributing the 12 plant phyla among two broad groups: we use Bryata for all the nonvascular plants (mosses, liverworts, and hornworts) and Tracheata for all other plants—that is, the vascular plants. Although Walker uses "Anthocerophyta" for the nonvascular group, in keeping with our book's policy of name simplification when possible, we instead call only the hornworts Anthocerophyta. We group our five kingdoms into two superkingdoms: (1) Prokarya, containing the sole prokaryote kingdom, Bacteria, and (2) Eukarya, containing the other four kingdoms, which encompass all the eukaryotes. We recognize that sociopolitical terms such as kingdom, class, order, and family are anachronisms that eventually will be replaced. Yet their current widespread use makes it convenient for us to continue using them in our classification of all life on Earth.

The only serious challenge to any of the five-kingdom schemes is the three-domain system of the microbiologists led by Carl Woese of the University of Illinois. Using molecular criteria, especially ribosomal RNA nucleotide sequences, these microbiologists argue for three major groups: two domains (Archaea and Bacteria) consisting of prokaryotic cells and one domain (Eukarya) containing all other organisms (Figure I-1). Fungi, plants,

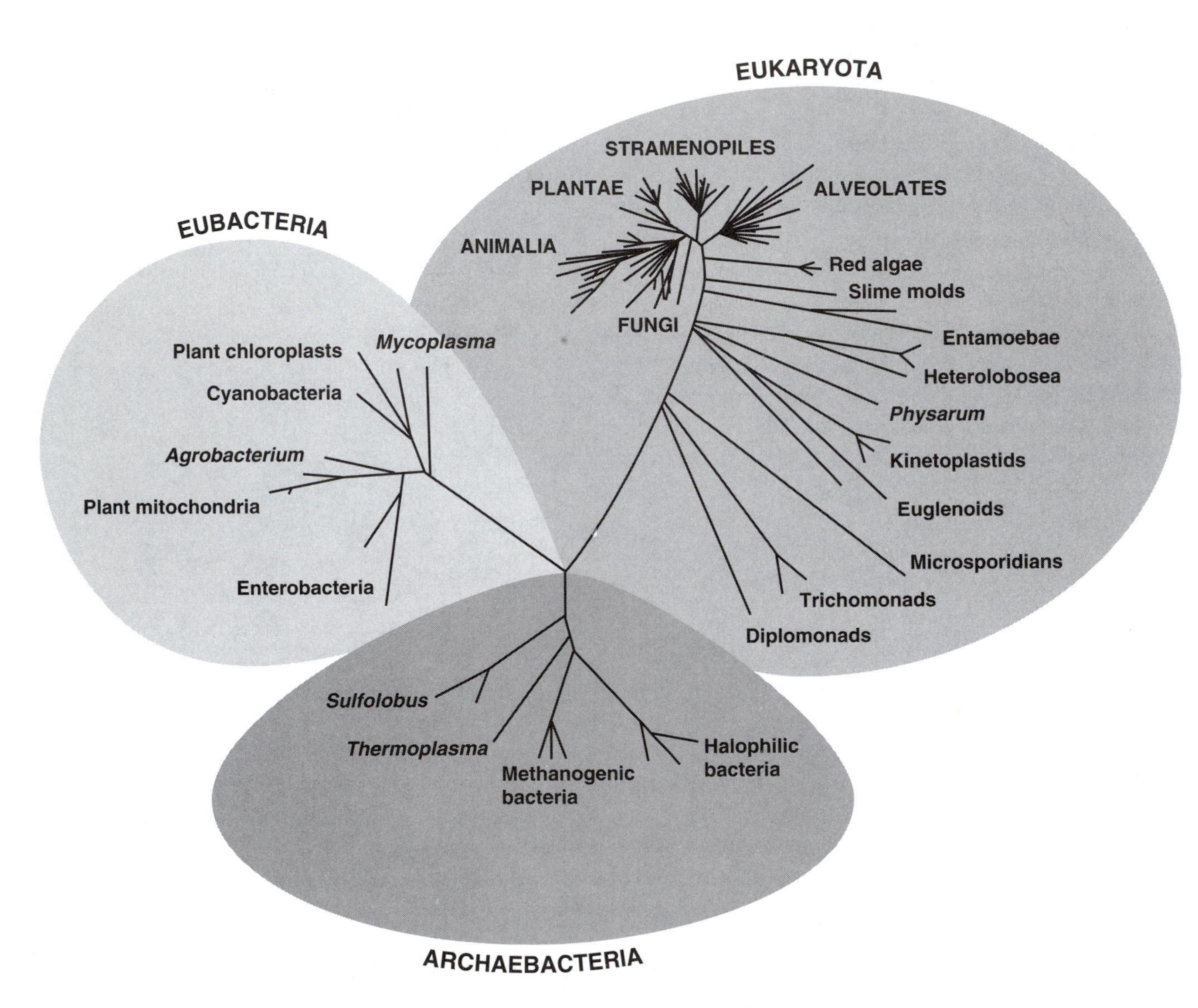

Figure I-1 Relations between eukaryotic higher taxa based on a single important criterion: nucleotide sequences in the genes for small-subunit ribosomal RNAs. The lengths of the lines are proportional to the number of differences in the nucleotide sequences. The "crown group" (Fungi, Animalia, Plantae, Stramenopiles) is envisioned to be those more recently evolved eukaryotes most closely related to large organisms. The difference between this scheme, based solely on molecular biology criteria, and ours is that we try to take into account all the biology of the living organisms. [Modified from G. Hinkle and M. Sogin, unpublished, with permission.]

and animals are three of the kingdoms of the Eukarya domain, as they are in our five-kingdom scheme. However, within each of the three domains are numerous additional kingdoms—many corresponding to phyla in the five-kingdom scheme.

Although we are deeply indebted to Carl Woese (University of Illinois), Mitchell Sogin (Marine Biological Laboratory at Woods Hole), and other molecular-sequence analysts for their unprecedented contributions to the reorganization of the living world, we reject the bacteriocentric three-domain scheme on biological and pedagogical grounds. Biologically, this trifurcation fails to recognize cell symbiogenesis (fusion of former bacteria) as the major source of innovation in the evolution of eukaryotes. Furthermore, its three domains and multiple kingdoms are established solely by the criteria of molecular sequence comparisons, whereas each kingdom in our five-kingdom scheme can be uniquely defined by using all features of the whole organism—molecular, morphological, and developmental. Pedagogically, proliferation of so many kingdoms in the three-domain system defeats the purpose of manageable classification of the fundamental diversity of our planetmates into a system from which information can be retrieved by teachers, naturalists, and other nonspecialists. For these reasons, although we have made extensive use of molecular sequence data in our classification, we reject the scheme that has made these data its sole criterion.

HISTORY OF THE PROKARYOTE-EUKARYOTE DISTINCTION

In 1937, the French marine biologist Edouard Chatton (1883–1947) wrote a short paper for an obscure journal published in Sète, in southern France, suggesting that the term *procariotique* (from Greek *pro,* meaning before, and *karyon,* meaning seed, kernel, or nucleus) be used to describe bacteria and blue-green algae, organisms that lack a nucleus, and that the term *eucariotique* (from Greek *eu,* meaning true) be used to describe animal and plant cells. In the past four decades, Chatton's insight into the nature of cells has been abundantly verified. Virtually all biologists now agree that

> this basic divergence in cellular structure which separates the bacteria and the blue green algae from all other cellular organisms, probably represents the greatest single evolutionary discontinuity to be found in the present-day world.*

This fundamental distinction is retained at the superkingdom level: Superkingdom Prokarya contains prokaryotes and only prokaryotes, the great diversity of organisms with bacterial cell organization. Members of the other four kingdoms are all eukaryotes.

Both in structure and in biochemistry, eukaryotes and prokaryotes differ by far more than the presence or absence of a cell nucleus (Figure I-2; Table I-2 summarizes their major differences). As you can see by comparing the illustrations in Chapter 1 with those in Chapter 2, prokaryotic cells are usually simpler in structure (but not necessarily in chemistry) and smaller than eukaryotic cells. The distinction between prokaryote and eukaryote is immediate and definitive from electron micrographs, which show multiple structures within eukaryotic cells. Any visible structure inside a cell is an organelle. Prokaryotes have them (carboxysomes or gas vacuoles, for example), but eukaryotic cells contain unique heritable organelles, some of them separated by their own membranes from the cytoplasm (see Figure I-2). Mitochondria (singular: mitochondrion), ovoid organelles that specialize in producing energy by enzymatic oxidation of simple organic compounds, are found in nearly all eukaryotes. Prokaryotes lack mitochondria, but enzymes bound to their membranes may catalyze the oxidations.

Green plant cells and algal cells contain one or several plastids—membrane-enclosed bodies in which complex structures made of membranes, chlorophyll, and other biochemicals photosynthesize. In photosynthetic bacteria, on the other hand, chlorophyll and other photosynthetic chemicals are found as granules in and on membranes.

Many eukaryotic cells—plant sperm, those of most protoctists and animal sperm—at some stage in their life history have flexible, long, protruding organelles called undulipodia (singular: undulipodium). All cilia, eukaryotic flagella, and most "sperm tails" are examples of undulipodia. All

---

* R. Y. Stanier, E. A. Adelberg, and M. Doudoroff, *The microbial world,* 3d ed., Prentice-Hall, Englewood Cliffs, NJ; 1963.

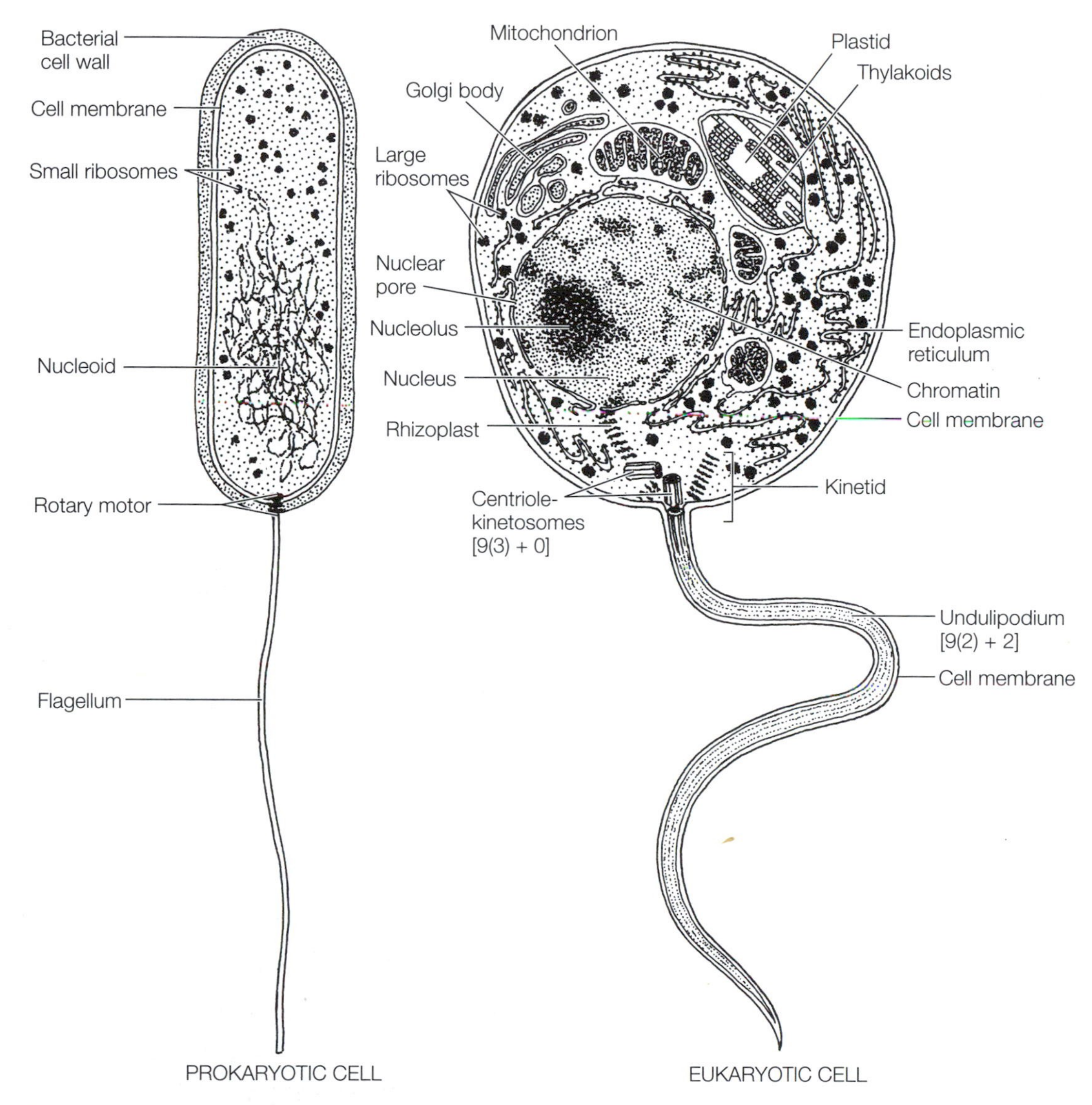

Figure I-2 Typical organism cells, based on electron microscopy. Not all prokaryotic or eukaryotic organisms have every feature shown here. "[9(3)+0]" and "[9(2)+2]" refer to the microtubule arrangement in cross section of kinetosomes and undulipodia, respectively (see Figure I-3). [Drawing by K. Delisle.]

Table I-2 Comparisons between Prokaryotes and Eukaryotes

Prokaryotes	Eukaryotes
DNA (The genophore[1]) seen in electron micrographs as nucleoid (not membrane bounded). Genophores, plasmids, not chromosomes. DNA not coated with protein.	Membrane-bounded nucleus containing chromosomes made of DNA and proteins.
Evolved by duplication and mutation of DNA.	Evolved by duplication and mutation of DNA and by symbiogenesis: permanent associations between at least two different kinds of prokaryotes.
Cell division direct, by binary or multiple fission or budding. Centriole-kinetosomes, mitotic spindle, microtubules lacking. In sexual recombination, genetic material (as plasmid, virus, genophore, or other replicon) is transferred from donor to recipient. Reversible transfer of genes (DNA) without any cytoplasmic fusion.	Cell division by various forms of mitosis; formation of mitotic spindles (or at least some arrangement of microtubules). Sexual systems in which two partners (often male and female) participate in fertilization. Alternation of diploid and haploid forms by fertilization and meiosis. Reversible formation of hybrid (fused) nuclei (karyogamy) at fertilization by mating. Gamontogamy,[2] syngamy,[3] conjugation, etc., usually include cytoplasmic fusion, called cytogamy.
All lack tissue development.	Some have extensive development of tissues and organs.
Strict anaerobes (which are killed by oxygen), facultative anaerobes, microaerophiles, aerotolerant and aerobic organisms.	Almost all are aerobes, needing oxygen to live; exceptions are either archaeprotists (Phylum Pr-1) or organisms evolved from aerobes.
Highly diverse modes of metabolism: vary in sources of energy (light, organic or inorganic molecules), sources of electrons (organic or inorganic molecules), and sources of carbon (organic or CO, CO_2, CH_4).	Same metabolic patterns of oxidation within the group (Embden-Meyerhof glucose metabolism, Krebs-cycle oxidations, cytochrome-electron transport chains). They are either organoheterochemotrophs[4] (most) or photolithoautotrophs[5] (most plants and algae).
Bacterial flagella, composed of flagellin protein. Flagella rotate.	Complex [9(2) + 2] undulipodia composed of tubulin and hundreds of other proteins (see Figure I-3).
Most are small cells (1–10 μm). All are microbes; many, if not most, are multicellular (colonial) in nature.	Most are large cells (10–100 μm). or composed of large cells. Some are microbes; most are large multicellular organisms or colonial.
Mitochondria absent; Cofactors and enzymes for oxidation of organic molecules, if present, are bound to cell membranes (not packaged separately).	Enzymes for oxidation of three-carbon organic acids are packaged in mitochondria [except for archaeprotists (Phylum Pr-1)].
In photosynthetic species, enzymes for photosynthesis are bound as chromatophores to cell membranes, not packaged separately. Various patterns of anaerobic and aerobic photosynthesis, including the formation of end products such as sulfur, sulfate, and oxygen.	In photosynthetic species, enzymes for photosynthesis are packaged in membrane-bounded plastids. All photosynthesizers produce oxygen.

[1] "Bacterial chromosome," a term used by molecular biologists to refer to the genophore, is confusing and its use should be avoided.

[2] Gamontogamy: in protoctists, such as foraminiferans, fusion of two or more gamonts (reproducing cells or organisms) followed by gametogamy (fusion of gametes). Examples: Copulation, conjugation.

[3] Syngamy: process by which two haploid cells fuse to form a diploid zygote; fertilization.

[4] Organoheterochemotroph: refers to metabolic mode in which organism uses organic compounds (from other organisms, living or dead) as sources of energy, carbon, and electrons (e.g., *Escherichia coli*).

[5] Photolithoautotroph: refers to metabolic mode in which organism uses light as a source of energy and uses inorganic compounds (such as CO_2 and H_2S) as sources of carbon and electrons (e.g., *Chlorobium*).

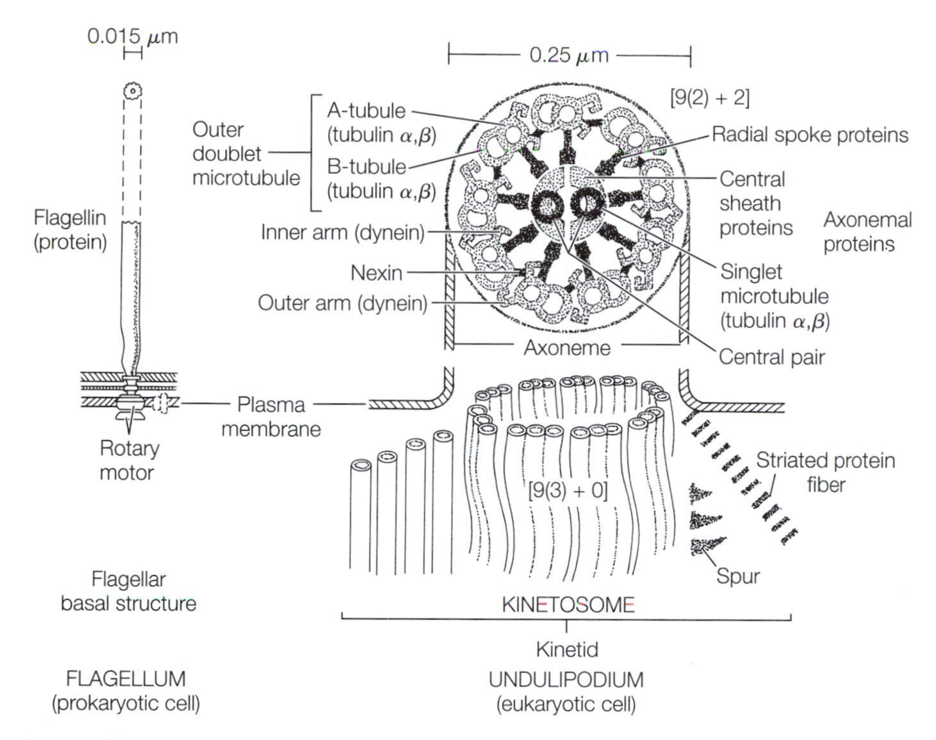

Figure I-3 A bacterial flagellum (left) compared with the undulipodium of eukaryotes (right). Kinetosomes, which always underlie axonemes, are associated with fibers, tubules, and possibly other structures. The organelle system, the kinetosome with its associated structures, whatever the detail, is called the kinetid. nm, nanometer; μm, micrometer. [Drawing by K. Delisle.]

undulipodia are composed of microtubules in bundles that, in cross section, show a characteristic ninefold symmetry (Figure I-3). The undulipodium is enclosed in an undulipodial membrane, which is simply an extension of the cell membrane. The ninefold symmetry characteristic of undulipodia is also found in the centriole-kinetosomes (often called eukaryotic basal bodies) from which the undulipodial shafts grow. The centrioles by themselves are small barrel-shaped bodies that, in many cells, appear at the poles of the spindle during mitosis. Undulipodia are composed of more than 500 different proteins. The main type of protein, which makes up the walls of the microtubules, is called microtubule protein, or tubulin.

Swimming prokaryotic cells bear long thin moveable extensions, called flagella (singular: flagellum; see Figure I-2). However, these extensions are not undulipodia—flagella are far smaller, are not composed of microtubules, and have an 11-fold symmetry. A flagellar shaft is an extracellular structure that protrudes through the cell wall and is composed of a single globular protein belonging to a class of proteins called flagellins. The beating motion of an undulipodium is caused by the transformation of chemical into mechanical energy by "motor" proteins arrayed along the full length of the microtubules in this organelle. The flagellum's motion results from the rotation of its "rotary motor," a basal attachment embedded in the cell wall and membrane.

SEX AND REPRODUCTION: PROKARYA AND EUKARYA

Reproduction is the increase in number of cells or organisms whether unicellular or multicellular. Growth is increase in size. All species of organisms grow and reproduce, although the details of how they do it vary. Even though fusion of parental gametes accompanies reproduction in humans and in the animals we best know, biologically, sex is entirely distinguishable from reproduction. Sex is best defined as the formation of an organism having genes from more than a single individual. Sex, the recombining of genes from two or more individuals, does occur in prokaryotes, but prokaryotic sex is not directly required for reproduction.

Bacterial cells—prokaryotes—never fuse. Rather, genes from the fluid medium, from other bacteria, from viruses, or from elsewhere enter bacterial cells on their own. A bacterium carrying some of its original genes and some new genes is called a recombinant. This propensity for gene uptake, along with the lack of a nucleus and all the other features listed in Table I-2, defines one of the two highest taxa, or superkingdoms: Prokarya, all organisms composed of bacterial cells. All other organisms are Eukarya, organisms composed of eukaryotic, nucleated cells, with the features listed in Table I-2.

Eukaryotic cells reproduce by mitosis. They form chromosomes—tightly coiled gene packages bound together by proteins and attached to the inner membrane of the nucleus. At least two chromosomes are located in the nucleus of every eukaryotic cell; some protoctists have more than 16,000 chromosomes in a single nucleus at certain stages. Although all cells and species of organisms made of cells must either reproduce or die, the way that eukaryotes make more eukaryotic cells or organisms made of cells is highly peculiar to each of the eukaryotic kingdoms and forms the basis of our biological classification.

All animals reproduce by fertilization of an egg by a smaller sperm to form the fertile egg—zygote—that divides by mitosis to make a rudimentary embryo. The first embryonic stage is called the blastula (see Figure 3-1).

All plants form spores that, by themselves, grow into one of two kinds of gametophyte (plant that forms gametes): either male and sperm forming or female and egg forming. The egg stays on the mother plant and is fertilized by fusion of egg with sperm nuclei. The fertilized egg—with a chromosome set from each parent—then develops into a plant embryo, a young multicellular stage common to all plant groups. The embryo becomes the adult plant, capable again of making spores with only one set of chromosomes (see Figure 5-3).

Fungi reproduce by means of fungal spores, propagules capable of generating—from a single parent, asexually—the entire fungal organism again. Some fungi enter their sexual phase only when the environment no longer favors asexual reproduction. At such times, genetically novel spores, but no embryos, are produced. Fungal spores are usually more resistant to water loss, starvation, and other adverse conditions than is the growing fungus. Fungi lack undulipodia.

Protoctists display a huge range of variation in life cycle features—but none fits the description of animal, plant, or fungus. The protoctist kingdom includes the microbial (few- or single-celled) eukaryotes and their immediate multicellular descendants. Because protoctists are grouped together as the microbial symbiotic complexes from which animals, plants, and fungi were removed, it is not too surprising that their life cycles are extraordinarily varied. "Protist" refers to the smaller protoctists, but some people use the term for all of them.

The differences between members of the five kingdoms are summarized in Table I-3 and will be further described and explained in the opening sections of the five chapters that follow this one.

Table I-3 Kingdom Summary

Superkingdom PROKARYA (PROKARYOTAE)
Nonnucleated (prokaryotic) cells. Chromonemal genetic organization ultrastructurally visible as nucleoids. Cell-to-cell transfer of genophores, i.e., of the chromoneme (large replicons) and of plasmids (and other small replicons). Ether- (isoprenoid-derivative) or ester-linked membrane lipids, without steroids, cytoplasmic fusion absent. Flagellar rotary motor motility. Concept of ploidy inapplicable.

Kingdom BACTERIA (PROKARYOTAE, PROCARYOTAE, MONERA)
Bacterial cell organization.

Subkingdom ARCHAEA
Methanogens, thermoacidophiles, halophiles, and probably some Gram-positive bacteria.

Subkingdom EUBACTERIA
Gram-negative and most other bacteria. Variable metabolic modes.

Superkingdom EUKARYA (EUKARYOTAE)
Nucleated (eukaryotic) cells, all evolved from integrated bacterial symbioses. Membrane-bounded organelles. Chromosomal genetic organization. Intracellular, microfilament- and microtubule-based motility (actin, myosin, tubulin-dynein-kinesin). Microtubule organizing centers. Whole-cell fusion (karyogamy). Flexible steroid-containing (e.g., cholesterol, ergosterol) membranes. Meiosis and fertilization cycles underlie Mendelian genetic systems. Levels of ploidy vary.

Kingdom PROTOCTISTA (Hogg, 1860)
Mitotic organisms capable of internal cell motility (i.e., cyclosis, phagocytosis, pinocytosis). Many motile by undulipodia. Binary or multiple fusion. Meiosis and fertilization cycles absent or details unique to the group. Photoautotrophs, ingestive and absorptive heterotrophs.

Kingdom ANIMALIA
Embryo called a blastula (diploid) formed after fertilization of egg by sperm (fusion of haploid anisogametes—karyogamic cells that differ in size). Females deliver mitochondria to the zygote in cytogamy. Meiosis produces gametes. Diploids. Most are ingestive heterotrophs; some are absorptive heterotrophs.

Kingdom FUNGI
Hyphal or cell fusion. Zygotic meiosis to form resistant propagules (spores). Lack undulipodia at all stages. Haploids. Absorptive heterotrophs.

Kingdom PLANTAE
Maternally retained diploid embryo formed from fusion of mitotically produced gamete nuclei. Sporogenic meiosis produces male (antheridium; sperm-producing haploid plant) or female (archegonium; egg-producing haploid plant). Gametes formed in antheridium and archegonium and fertilized in archegonium. Alternating generations of haploid and diploid organisms. Most are photoautotrophs.

NOTE: For the major higher taxa (Prokarya, Eukarya, Bacteria, Protoctista, Fungi, Animalia, and Plantae), brief technical descriptions accompany the introduction of each of their sections.

VIRUSES

Antony van Leeuwenhoek (1632–1723), the discoverer of the microbial world, called the microorganisms that he found everywhere in vast numbers "very many little animalcules." For more than a century after his discoveries, it was commonly held that these little animals arose spontaneously from inanimate matter. The chemist Louis Pasteur (1822–1895) and the physicist John Tyndall (1820–1893) showed conclusively that, like large organisms, microbes are produced only by other microbes.

Thus, all organisms in the five kingdoms either are cells or are composed of cells. The arguably living forms that do not fit this description are the viruses (Figure I-4). Composed of DNA (deoxyribonucleic acid) or RNA (ribonucleic acid) enclosed in a protein coat, viruses are much smaller than cells. Although viruses replicate, they can do so only by entering a cell and using its living machinery. Outside the cell, viruses cannot reproduce, feed, or grow. Some viruses can even be crystallized, like minerals. In this state, viruses can survive for years unchanged—until they contact the specific living tissues they require.

Viruses are probably more closely related to the cells in which they replicate than to one another. They may have originated as replicating nucleic acids that escaped from cells—they must always return to living tissue to use the complex chemicals and structures they require for replication. Thus, the polio and flu viruses are probably more closely related to people, and the tobacco mosaic virus (TMV) to tobacco, than polio virus and TMV are to each other.

EARTH HISTORY

The planet Earth is almost 5 billion years old. The oldest fossils yet discovered, bacterium-like filaments from rocks of the Pilbara gold fields near North Pole in Western Australia, are 3.5 billion years old. Bacterium-like spheroids also have been found in rocks of that age from the Swaziland rock system of southern Africa. Thus, for most of its history, Earth has supported

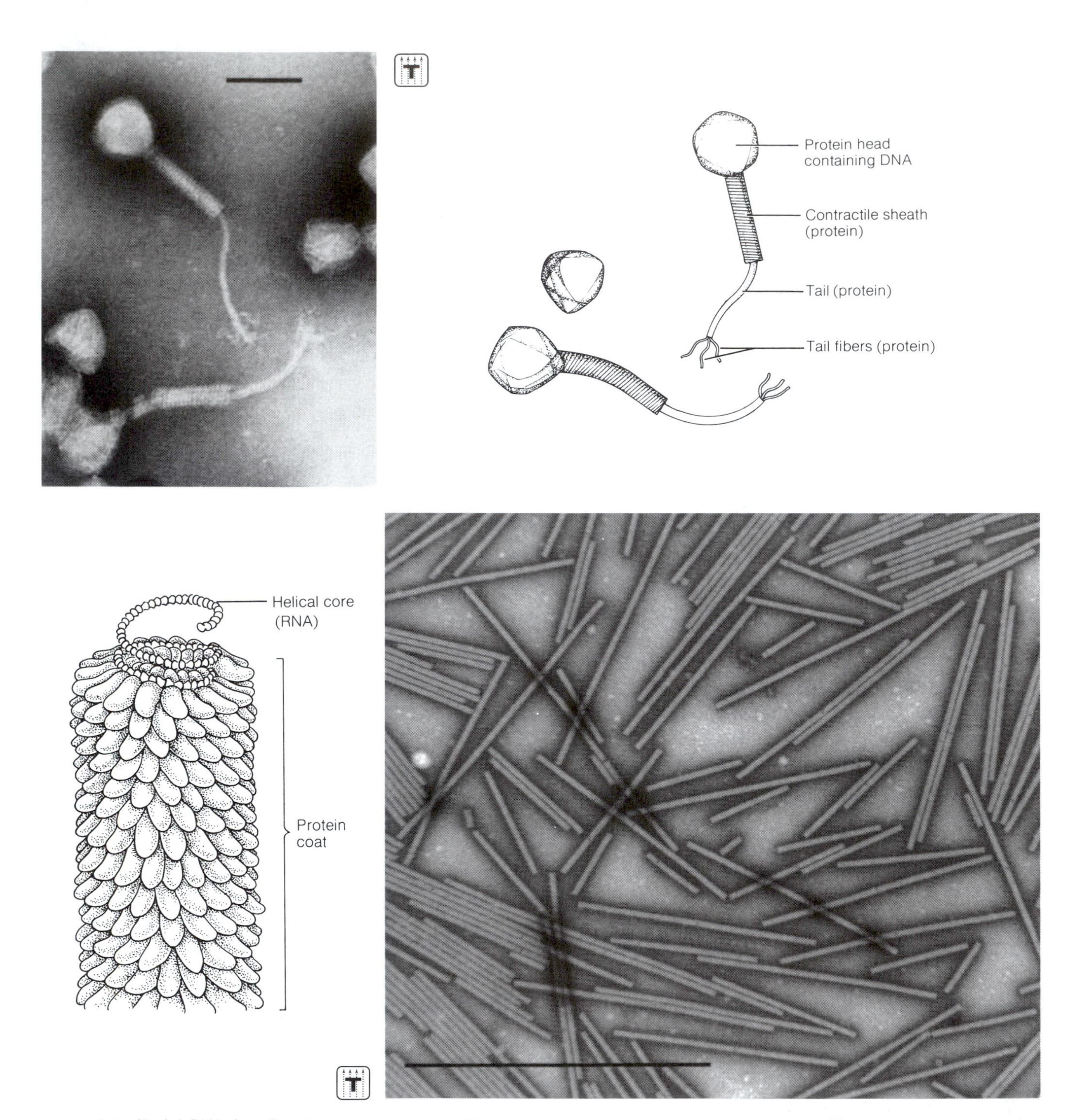

Figure I-4 (Top) A DNA virus, *Botulinum* ϕ, which attacks *Clostridium botulinum;* TEM, bar = 0.1 μm. (Bottom) An RNA virus, tobacco mosaic (TMV), which causes a blight of tobacco plants; TEM, bar = 1 μm. [Photographs courtesy of E. Boatman; botulinum drawing by R. Golder; TMV drawing by M. Lowe.]

life. Almost everything that we know about the history of the planet and of life comes from evidence in its rocks. Geologists have made a chronology of this history from the composition of rocks, the order of their formation, and the fossils in them (Figure I-5). The upper layers of an undisturbed sequence of sediments are younger than the lower ones. Rock layers at different places on Earth's surface are matched by examining the fossils in them, whereas absolute ages are determined by radioactive-dating methods.

The longest divisions of geological time are called eons. The sequence of rocks in the latest, the Phanerozoic, is known in such detail that this eon is divided into eras, eras into periods, and the periods into epochs (not shown). Of the other eons, only the Proterozoic is known well enough to allow generally accepted subdivisions (eras, epochs). The subdivisions of the Phanerozoic are so well known because of the worldwide abundance of its fossils. In fact, the fossil record of eukaryotic organisms is Phanerozoic, except for microfossils of protoctist cysts and many Proterozoic trails, burrows, and body fossils of unknown origin. Such evidence of pre-Phanerozoic eukaryotes ("Edicara biota") is known from more than 20 localities.

The most abundant fossils from before the Phanerozoic eon are rocks called stromatolites. A typical stromatolite is a column or dome of rock a few centimeters wide made of thin horizontal layers. The layers are apparently the remains of sediment trapped or precipitated and bound by growing communities of bacteria, primarily cyanobacteria. Fossil stromatolites may extend over hundreds of meters laterally and several meters in height. Comparable bacterial communities exist today, but in only a few isolated, extreme environments, such as salt ponds, do they harden into stromatolites. During the Archaean and Proterozoic eons, more than half of Earth's existence, the planet was the uncontested territory of Kingdom Bacteria. The stromatolite communities began to withdraw to their modern places of refuge only after the rise of protoctists and animals—when some of the new organisms must have grazed voraciously in the lush bacterial pastures.

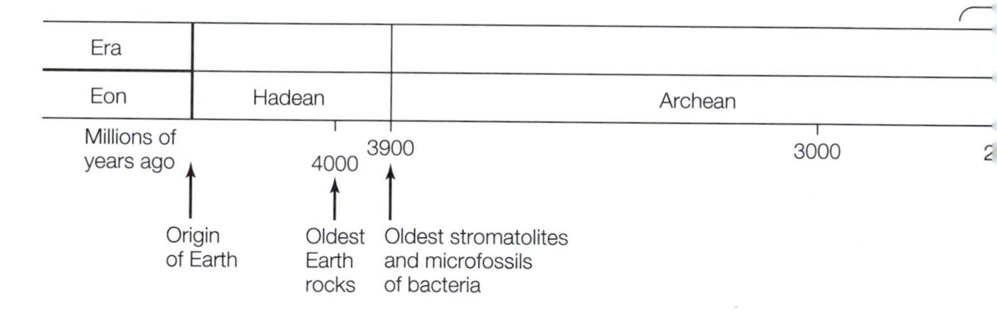

Figure I-5 Time line of Earth history. Eons (time-rock divisions) in which unambiguous fossils first appear: bacteria — early Archean; protoctista — middle Proterozoic; animals — late Proterozoic (Vendian era); plants and fungi — early Phanerozoic (Paleozoic era, Ordovician period).

READING THIS BOOK

We recognize and describe two superkingdoms: Prokarya (all bacteria) and Eukarya (all nucleated organisms). In these two highest, most inclusive taxa are the kingdoms. Superkingdom Prokarya contains Kingdom Bacteria (equivalent to Prokaryotae, Procaryotae, or Monera in other schemes) with its two subkingdoms, Archaea and Eubacteria. The subkingdom Archaea contains two phyla: B-1, the Euryarchaeota, or the methanogens and

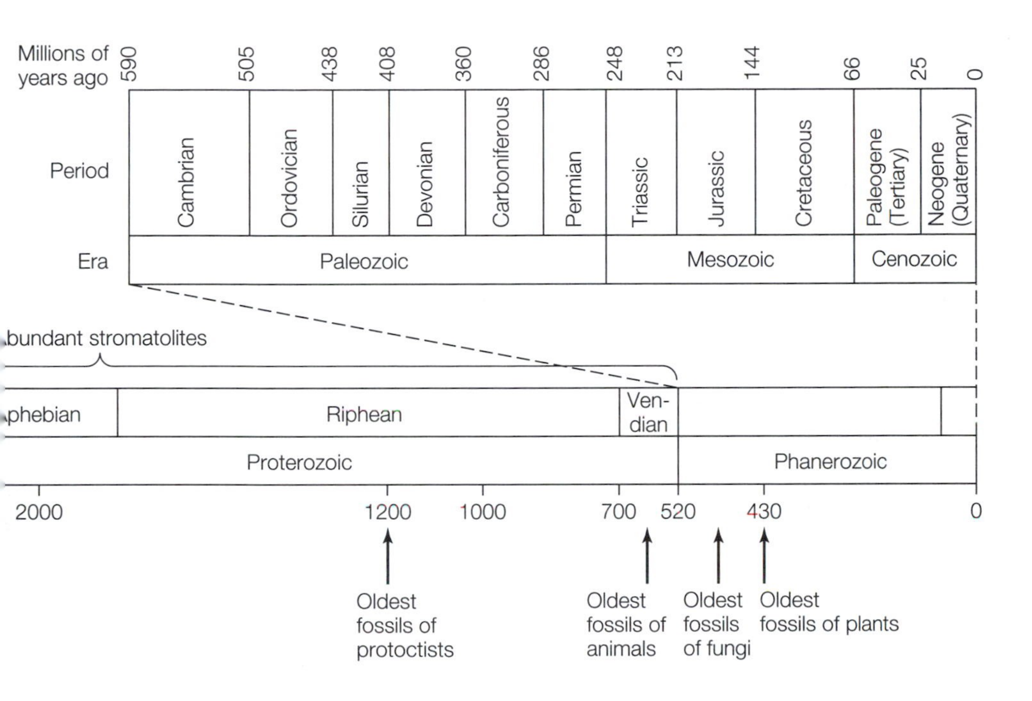

extreme halophils, and B-2, the Crenarchaeota, or extreme thermophils. The subkingdom Eubacteria contains 12 phyla. In the Eukarya kingdoms (Protoctista, Animalia, Fungi, and Plantae) are 30 protoctist phyla, 37 animal phyla, 3 fungal phyla, and 12 plant phyla. Within each Eukarya kingdom, the phyla are arranged in an approximate order from the simplest (presumably the earliest to have evolved) to the more complex (presumably more recent) forms. As a rule, the highest taxa within a kingdom, the phyla and classes, represent the most ancient evolutionary divergences; the lowest taxa, the genera and species, represent the most recent. However, this is not an absolute rule, because the evolutionary relationships of many groups are unknown. Organisms are grouped in the same taxon for now only because they have some clearly distinguishable trait in common (for example, the rays of the actinopods), whether or not their common ancestry has been documented.

We begin each chapter with a phylogeny—a diagram showing the likely evolutionary relationships among the phyla of that kingdom. Each phylogeny is a branching structure evolving through time from a single ancestral group to the extant phyla. A time line along one axis shows relevant geological eras. Solid lines indicate accepted lineages; dashed lines are lineages that are provisional. Each extant phylum is illustrated with a thumbnail sketch of a member organism.

The introduction to each chapter defines and describes the general features of the entire kingdom, and it is followed by essays, each describing one phylum in that kingdom. Phyla differ enormously in size. Some have a single species; others have millions. Each phylum description begins with a list of examples of its genera, not all of which are mentioned in the text; we have selected some genera that are significant research models, some genera of vast economic import, and some new genera. The lists give experienced students a firm idea of each phylum and are clues to further reading.

At the top of the right-hand page of each phylum essay is a scene with one or more arrows pointing to the typical habitats of the members of the phylum. Figure I-6 shows seven different environments and their habitats: temperate seashore—rocky, sandy, and muddy; temperate forests, lakes, and rivers; deserts and high mountains; tropical forest; tropical seas, reefs, continental shelves and slopes, and seashore; tectonically active anoxic environment; and the abyss at a tectonically active ocean rift zone. Tectonically (Greek *tekton,* carpenter) active pertains to changes in the structure of Earth's crust. The tropical seashore, which is not shown as a separate environment, will be indicated by arrows pointing toward the top of the tropical seas scene when appropriate. Our depictions are certainly not complete, because members of a phylum may be found in habitats not illustrated, yet our scenes include habitats in which life frequently and abundantly abides.

Temperate seashore — rocky, sandy, and muddy; sunlit surface

Temperate forests, lakes, and rivers; sunlit surface

Deserts and high mountains; sunlit surface

Tropical forest; sunlit canopy

Tropical seas, reefs, continental shelves and slopes, and seashore; sunlit to depths of 200 or fewer meters

Tectonically active anoxic environment; sunlit surface

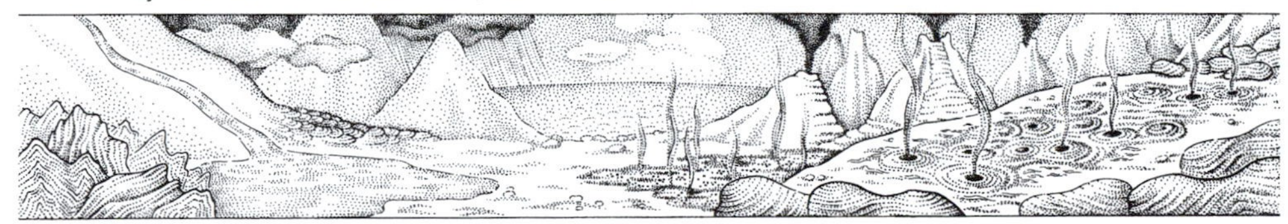

Abyss at tectonically active ocean rift zone; sunlight absent

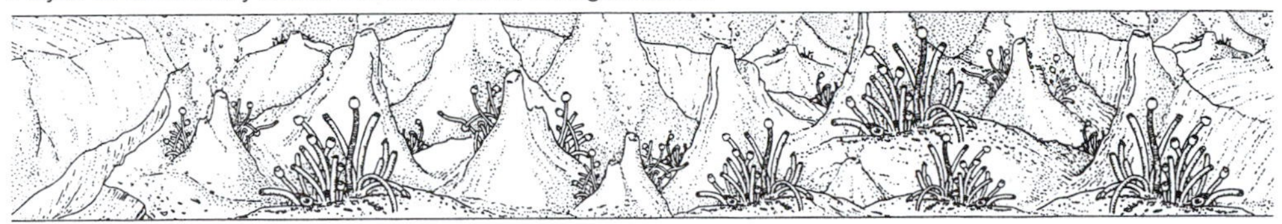

Figure I-6 Environments: the seven scenes used to designate typical habitats.

A gray strip embedded with tiny organisms runs across the top of the text on pages that describe phyla. Distinctive organisms for each kingdom serve to indicate the kingdom to which the phylum belongs: budding bacteria for Bacteria, euglena for Protoctista, salamanders for Animalia, mushrooms for Fungi, and tulip tree leaves for Plantae. The text is accompanied by photographs and drawings of representative species; those species illustrated are listed in the Contents. In most phyla, the main photograph is interpreted in a labeled anatomical drawing. Unfamiliar terms are defined in the Glossary.

Most organisms were photographed alive. The chief exceptions are those visualized by electron microscopy. Transmission electron microscopy requires dead samples that have been chemically fixed, embedded in a transparent matrix, and cut extremely thin. Even scanning electron microscopy, which allows a more three-dimensional view of samples, usually requires that organisms be treated with deadly fixation techniques and viewed in a vacuum.

The legend of each photograph states the type, if any, of microscopy used to take the photograph: LM stands for light (optical) microscopy, TEM and SEM for transmission and scanning electron microscopy, respectively. The legend also gives the organism's length represented by the scale bar in the photograph. A colophon with each photograph indicates the kind of optical equipment needed to see the subject of the photograph (Figure I-7).

COLOPHON	 [eye icon]	 [hand lens icon]	 [microscope icon]	 S	 T
OPTICAL EQUIPMENT	Naked eye	Hand lens	Light microscope	Scanning electron microscope	Transmission electron microscope
SIZE OF SUBJECT IN METERS (approximate)	10^{-3}–10^{1}	10^{-4}–10^{-2}	10^{-6}–10^{-4}	10^{-8}–10^{-2}	10^{-9}–10^{-5}

Figure I-7 Key to photograph colophons.

Bibliography: Introduction

Bengtson, S., ed., *Early life on Earth.* Columbia University Press; New York; 1994.

Calder, N., *Timescale: An atlas of the fourth dimension.* Viking Press; New York; 1983.

Copeland, H. F., *The classification of lower organisms.* Pacific Books; Palo Alto, CA; 1956.

Corliss, J. O., "The protista kingdom and its 45 phyla." *Biosystems* 17:87–126; 1984.

DeDuve, C., "The birth of complex cells." *Scientific American* 274(4):56–57; April 1996.

Eldredge, N., *Fossils.* Abrams; New York; 1993.

Erwin, D. H., *The great Paleozoic crisis: Life and death in the Permian.* Columbia University Press; New York; 1993.

Gould, S. J., *Wonderful life: The Burgess shale and the nature of history.* Norton; New York; 1989.

Hillis, D. M., and C. Moritz, eds., *Molecular systematics.* Sinauer; Sunderland, MA; 1990.

Hogg, J., "On the distinctions of a plant and an animal, and on a fourth kingdom of nature." *Edinburgh New Philosophical Journal* 12:216–225; 1860.

Johnson, K. R., and R. K. Stucky, *Prehistoric journey: A history of life on Earth.* Denver Museum of Natural History (Roberts Rinehart Publishers); Boulder, CO; 1995. A superb vicarious field trip through time.

Lipps, J., ed., *Fossil prokaryotes and protists.* Blackwell; New York; 1995.

Margulis, L., *Symbiosis in cell evolution,* 2d ed. W. H. Freeman and Company; New York; 1993.

Margulis, L., "Archaeal-eubacterial mergers in the origin of Eukarya." *Proceedings of the National Academy of Sciences, USA* 93:1071–1076; 1996.

Margulis, L., and D. Sagan, *Microcosmos: Four billion years of evolution from our microbial ancestors.* Simon and Schuster; New York; 1986.

Margulis, L., and D. Sagan, *Origins of sex: Three billion years of genetic recombination.* Yale University Press; New Haven and London; 1991.

Margulis, L., and D. Sagan, *What is life?* Simon and Schuster; New York; 1995.

Margulis, L., and D. Sagan, *What is sex?* Simon and Schuster; New York; 1997.

McMenamin, M. A., and D. L. S. McMenamin, *Emergence of animals.* Columbia University Press; New York; 1990.

Raff, R. A., *The shape of life: Genes, development, and the evolution of animal form.* University of Chicago Press; Chicago; 1996.

Schopf, J. W., *Major events in the history of life.* Jones and Bartlett; Sudbury, MA; 1992.

Sogin, M. L., H. G. Morrison, G. Hinkle, and J. D. Silberman. "Ancestral relationships of the major eukaryotic lineages." *Microbiologia SEM* 12:17–28, 1996.

Taylor, T. N., and E. L. Taylor. *Biology and evolution of fossil plants.* Prentice-Hall; Englewood Cliffs, NJ; 1993.

Ward, P. D., *On Methuselah's trail: Living fossils and the great extinctions.* W. H. Freeman and Company; New York; 1992.

Whittaker, R. H., "On the broad classification of organisms." *Quarterly Review of Biology* 34:210–226; 1959.

Whittington, H. B., *The Burgess shale.* Yale University Press; New Haven, CT; 1985.

Woese, C. "Microbiology in transition." *Proceedings of the National Academy of Sciences, USA* 91:1601–1603; 1994.

Woese, C. R., O. Kandler, and M. L. Wheelis, "Towards a natural system of organisms: Proposal for the domains Archaea, Bacteria and Eucarya." *Proceedings of the National Academy of Sciences, USA* 87:4576–4579; 1990.

Selected recommendations for further reading appear at the end of each kingdom chapter. The Appendix contains a list of all kingdoms and an alphabetical list of genera with the phylum and common names (if any) of each genus.

FIELD TRIPS THROUGH TIME
Dioramas/Dinosaurs/Fossils

Visit the museums and parks listed here to journey through prehistoric life on Earth. Encounter animals, plants, seaweeds, and other ancient ancestors in their habitats. Most of the museums display the fossils on which their dioramas are based; and many of the parks offer interpretive walks through deposits of real fossils. Some dioramas feature animated models of extinct creatures and sounds of antiquity.

We have attempted to verify the information in this list. However, because telephone numbers, hours, fees, and exhibits on display change, we recommend that you contact well in advance any location that you plan to visit.

UNITED STATES—BY REGION

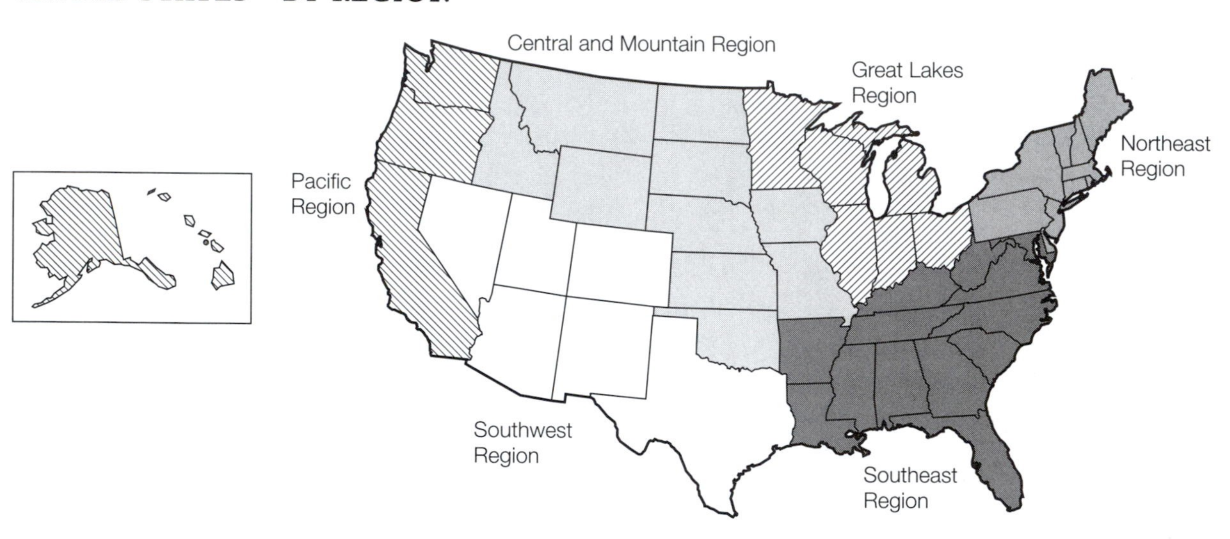

NORTHEAST REGION

Connecticut

Dinosaur State Park
West Street, Rocky Hill CT 06067
Telephone (860) 529-8423

Exhibit center with:
- 200-million-year-old Jurassic dinosaur tracks
- Two large Triassic and Jurassic dioramas with full-scale animals and plants
- Arboretum of plant families originating in the Age of Dinosaurs

Road-cut exhibit of fossil tracks from Connecticut River valley
Guided tours by reservation

New York

Paleontological Research Institution (PRI)
1259 Trumansburg Road, Ithaca NY 14850
Telephone (607) 273-6623
Internet URL: http://www.englib.cornell.edu/pri

One of the largest American collections of fossils, many invertebrates—tiny shells of single-celled organisms, dinosaur bones, and woolly mammoth teeth are displayed. Visitors see what paleontologists do and how they do it.

Ithaca's Sciencenter:
- Exhibit tour
- Fossil hunt
- Exhibits: (1) What are fossils?, (2) How fossils are used to

determine the age of rocks, (3) How fossils tell us about evolution, and the history of the places in which we find fossils, (4) How vast geologic time is and how the geologic time scale was constructed and is used, (5) How to find fossils and classify them, (6) The difference between paleontology, the study of fossils, and archeology, the study of human artifacts
- Fossils of the Ithaca area: New York 380 million years ago shows fossils from the Devonian period (the "Age of Fishes") from the Cayuga Lake area, history of the area, and a reconstruction of the Devonian sea floor
- Fossil invertebrates from around the world (echinoderms, molluscs, arthropods, brachiopods, sponges, corals); a large fossil sponge, collected in Ithaca at the Llenroc stone quarry, is among the largest of its kind known
- The Geologic Time Scale answers the question "How do you know how old it is?"
- A Fossil and Its Living Relative: a life-size cast of a fossil ammonite and a photograph of a living chambered nautilus suggest how paleontologists reconstruct the soft parts and other aspects of the biology of long-dead organisms

Snee Hall houses the Department of Geological Sciences and fossils from the Paleontological Research Institute in the first floor atrium
The Earth Window location at the web site allows viewing of some of the collection.

Pennsylvania

Carnegie Museum of Natural History
4400 Forbes Avenue, Pittsburgh PA 15213-4080
Telephone (412) 622-3247

Dioramas in the paleontological galleries:
- Pennsylvanian coal swamp/forest, natural size, depicting land life, including large amphibians, in Pittsburgh area
- Eight scale dioramas of life in the seas over this area: Cambrian (arthropods); Ordovician (brachiopods, nautilids); Silurian (with eurypterids—chelicerate arthropods—from western New York); Devonian; Upper Devonian glass sponges of the Catskill Delta; Carboniferous sea; Permian sea; Permian land including tetrapods—below each diorama are actual fossils
- Scale diorama of Cretaceous duckbill dinosaur *Corythosaurus casuarius*
- Scale diorama during the Pleistocene at the La Brea tar pits—located in present-day Hancock Park, Wilshire Boulevard, Los Angeles, California

Philadelphia Academy of Natural Sciences
1900 Benjamin Franklin Parkway, Philadelphia PA 19103-1195
Telephone (215) 299-1000

- Cretaceous botanical models
- Discovering dinosaurs: *Tyranosaurus rex* and 14 skeletal mounts
- Full mount of Argentinian dinosaur *Giganotosaurus,* largest carnivorous dinosaur in the world
- Largest *Tylosaurus* ever found
- Fossil preparation lab
- Hands-on dig experience

SOUTHEAST REGION

District of Columbia

Smithsonian Institution, National Museum of Natural History
10th Street and Constitution Avenue NW, Washington DC 20560
Telephones (202) 357-1300 (general information);
(202) 786-2178 (guided tours)

- Diorama of life in a Precambrian (Ediacaran) sea
- Burgess shale diorama
- Ordovician marine diorama
- Upper Silurian eurypterid diorama
- 9-ft *Pterygotus* model
- Devonian marine diorama with marine invertebrates
- Early terrestrial diorama with *Cooksonia* and eurypterids
- Restoration of Permian reef from west Texas
- Three Mesozoic dioramas—Triassic, Jurassic, Cretaceous
- 40 Mesozoic fish, bivalves and cephalopod mollusc models
- Horse evolution exhibit

North Carolina

Schiele Museum of Natural History
1500 East Garrison Boulevard, Gastonia NC 28054-5199
Telephone (704) 866-6900

Four habitat groups depicting marine life in four geologic periods:
- Life-size coal forest diorama
- Life-size *Coelophysis* dinosaur, mastodon, and saber-toothed cat
- Cretaceous mammal diorama
- Footprints of hominids from east Africa (casts)
- Lucy skeletal reproduction—3.2-million-year-old hominid
- Four scale model dioramas of human evolution

GREAT LAKES REGION

Illinois

Burpee Museum of Natural History
737 North Main Street, Rockford IL 61103
Telephone (815) 965-3132

- Geology hall paleoexhibits depict ancient life in Illinois including many fossils found in the region

Field Museum of Natural History
Roosevelt Road at Lake Shore Drive, Chicago IL 60605-2496
Telephone (312) 922-9410

Life over Time—the entire history of life on Earth, from the first organic molecules to the present day:
- Walk-through Carboniferous forest (Middle Pennsylvanian) coal swamp and levee vegetation including 41 trees
- Dioramas depicting invertebrate animals and fish from Silurian and Carboniferous
- 12 dinosaur miniature dioramas (Triassic, Jurassic, Cretaceous)

Indiana

Falls of the Ohio State Park
201 West Riverside Drive, Clarksville IN 47129
(Mail: PO Box 1327, Jeffersonville IN 47131-1327)
Telephone (812) 280-9970

Interpretive visitor center with exhibits including:
- Silurian-Devonian marine diorama and fossils
- Devonian forest with *Ichthyostega*
- *Dunkleosteus* fish and skull cast (Devonian armored fish)
- *Cladoselache* shark
- Marine aquaria comparing modern to ancient coral reef environments and fish
- Ice Age exhibit with mammoth
- Early paleontologists at the Falls with fossils
- Film including recreations of a Devonian patch reef

Indiana State Museum
202 North Alabama Street, Indianapolis IN 46204
Telephone (317) 232-1641

- Mounted cast of giant short-faced bear from Indiana's Ice Age
- Numerous bones and teeth from Indiana's Ice Age
- Paleozoic undersea Indiana mural with real fossils
- Fossil Mississippian Indiana crinoids (sea lillies)
- Coal forest mural with model trees

Ohio

Caesar Creek Lake Visitor Center
4020 North Clarkesville Road, Waynesville OH 45068
Telephone (513) 897-1050

- Painting of prehistoric underwater life
- Real Ordovician fossils in the visitor center and in the field: trilobites, horn coral, cephalopods, gastropods, brachiopods, and crinoids (sea lillies)
- Free collecting permits for visitors
- Activities and programs all year round

Cincinnati Museum Center, Cincinnati Museum of Natural History and Science
1301 Western Avenue, Cincinnati OH 45203
Telephone (513) 287-7000

- Immersion environment of the Cincinnati Pleistocene
- Walk-through Pleistocene (Quaternary 19,000 years ago) diorama with botanical models
- Life-size glacial cave with ice sounds and dripping water, life-size glacier and glacial plain with kettles and other glacial features, musk oxen, Dire wolves, giant beaver, sabre-toothed cat, Cervalces, mastodon, and *Megalonyx*
- Quaternary fossil site Big Bone Lick, Kentucky
- Interactive exhibits

Highbanks Metro Park Nature Center
9466 Columbus Pike (US 23 North), Lewis Center OH
(Mail: Columbus Metro Park, PO Box 29169, Columbus OH 43229)
Telephone (614) 846-9962

- 12-ft *Dunkleosteus* model (a large Devonian armored fish)
- Life-sized *Dunkleosteus* skull replica
- Geology exhibit including brachiopod fossils in limestone
- Fauna at the edge of the Ice Age glacier including woolly mammoth
- Fossil jawbone of *Dinichthys,* Devonian fish found at Highbanks
- Hikes and talks all year

Wisconsin

Milwaukee Public Museum
800 West Wells Street, Milwaukee WI 53233-1478
Telephone (414) 278-2700

- Precambrian fossils: stromatolites
- Silurian Reef diorama (408–438-million-year-old corals,

bryozoans, brachiopods, cephalopod molluscs, trilobites, sponges, algae)
- Carboniferous coal forest diorama and fossils (giant club mosses, *Calamites*—horsetail, tree ferns, seed ferns, giant cockroach, giant dragonflies, early reptiles, amphibians, and lobe-finned fishes)
- Mesozoic era—Triassic period fossils (cycads, ammonoid cephalopods, crinoids), Jurassic period fossils (*Ichthyosaurus*, brachiopods, bony fishes), Cretaceous period fossils (*Metasequoia*, dawn redwood; cephalopod molluscs)
- Dinosaur dioramas showing tree ferns

CENTRAL AND MOUNTAIN REGION

Idaho

Hagerman Fossil Beds National Monument
221 North State Street, PO Box 570, Hagerman ID 83332-0570
Telephone (208) 837-4793

- World's richest known fossil deposit from 3.5-million-year-old ecosystem (late Pliocene); famous for early zebralike horse, *Equus simplicidens*, sabre-tooth cats, camels, peccaries, beavers, muskrats, antelope, hyenalike dogs, mastodon fossils; 140 animal species of vertebrates and 35 plant species have been exposed, are displayed, and interpreted
- Visitor center with fossil exhibits, slide program, and videos
- Ranger-guided tours to horse quarry on summer weekends
- Self-guided driving tour with boardwalk to view of Hagerman Fossil Beds and interpretive exhibits

Iowa

University of Iowa Museum of Natural History
University of Iowa, Iowa City IA 52242
Telephone (319) 335-0481

- Marine diorama depicting a Devonian coral reef with profuse undersea life including a reconstruction of a 20-ft *Dunkleosteus*, a predatory armored fish

Kansas

University of Kansas Natural History Museum
Dyche Hall, Lawrence KS 66045
Telephone (913) 864-4450

- Bird evolution exhibit—includes an *Archaeopteryx* cast
- Fossils of the Kansas sea—a mural with real fossils
- Horse evolution exhibit
- La Brea tar pit exhibit
- Fossils of mastodon, mammoth, other mammals, dinosaurs, fish, invertebrate animals, and plants

Missouri

St. Louis Science Center
5050 Oakland Avenue, St. Louis MO 63110
Telephone (314) 289-4400

- Walk-through Pennsylvanian coal forest
- 20-ft Mississippian crinoid bank diorama
- Cretaceous habitat group with a three-story mural
- *Triceratops* skull

St. Louis Zoo
Forest Park, St. Louis MO 63110
Telephone (314) 781-0900

- The Living World: a mural of Miocene landscape with *Indricotherium* (giant hornless rhinoceros) fossil life-size leg; life-size *Pteranodon* model
- Life-size hologram of *Tyrannosaurus rex*
- Introduction to the Animals: real fossils of many animals
- Introduction to Ecology: real fossils and casts of stromatolites, extinct invertebrates, and vertebrates through mammals including *Archeopteryx* and *Smilodon*

Montana

Museum of the Rockies
Montana State University, 600 West Kagy,
Bozeman MT 59717-2730
Telephone (406) 994-2251
Internet URL: http://www.montana.edu/wwwmor/

Landforms/Lifeforms dioramas:
- Cambrian: 530-million-year-old Burgess shale fauna showing algal reef in what is now Yoho National Park, BC, Canada
- Devonian (Jefferson): 370-million-year-old reef with corals and cephalopods
- (Carboniferous) Mississippian (Bear Gulch): 320-million-year-old ray- and lobe-finned marine fishes including coelacanths
- Cretaceous diorama with 25-ft tall trees
- Pangea Theater: plate tectonics—dynamic earth polarmotion model showing continental drift

Nebraska

AGATE FOSSIL BEDS NATIONAL MONUMENT
PO Box 427, Gering NE 69341
Telephone (308) 436-4340
Internet URL: http://www.nps.gov/agfo/

Visitor center with exhibits including a diorama of Miocene waterhole/bonebed with life-size *Moropus* (relative of horse with clawed feet), *Dinohyus* (terrible hog), and *Daphoenodon* (carnivorous bear dog) casts; *Stenomylus* (early camel) fossil; film on Agate Fossil Beds paleontology

- 1.6-km paved trail to fossil beds in western Nebraska
- 0.8-km trail to *Daemonelix* fossil burrows of giant land beaver
- Miocene mammal fossils concentrated in a 60-acre quarry

ASHFALL FOSSIL BEDS STATE HISTORICAL PARK
PO Box 66, Royal NE 68773
Telephone (402) 893-2000

Visitor center with displays and interpretive talks—evolution, migration, and extinction on the Great Plains
Rhino barn covering a fossil-bearing ash bed of mid- to late Miocene, 10-million years ago featuring:

- A mounted skeleton of a barrel-bodied rhino calf *Teleoceras*
- A barrel-bodied rhino calf and cow skeletons embedded in ash
- Five early horse species skeletons
- 17 additional fossil species, including three birds, camels, giant tortoises, and sabre-toothed deer
- A mural depicting the subtropical landscape before volcanic ashfall trapped the animals around a watering hole

UNIVERSITY OF NEBRASKA STATE MUSEUM
307 Morrill Hall, Corner of 14th and U Street, University of Nebraska, Lincoln NE 68588-0332
Telephone (402) 472-2642
Internet URL: http://www.museum.unl.edu/

- Dioramas depict underwater marine life: Precambrian, Silurian, Devonian (Age of Fishes), Mississippian, Permian reef
- Dioramas depict prehistoric terrestrial environments: Pennsylvanian coal swamp, Cretaceous landscape, real fossil displays of prehistoric elephants found in Nebraska including touchable fossil tusks and teeth, prehistoric land beaver corkscrew burrow
- Computer interactive Mesozoic sites with videoclips of fossil sites in Montana (Egg Mountain), Texas, Nebraska, and Colorado
- Mesozoic sea mural
- Age of Dinosaurs gallery with fossil mosasaur, a carnivorous lizard that evolved flippers

South Dakota

BADLANDS NATIONAL PARK
PO Box 6, Interior SD 57750
Telephone (605) 433-5361

- A site containing Eocene and Oligocene epoch vertebrates, such as oreodonts (sheep-like animals), giant pigs, tiny horses, aquatic rhinos, elephant-sized titanotheres, and extinct ancestors of the camel, cat, dog, mouse, beaver, marmot, rabbits, owls and eagles
- Visitor center open year round with exhibits, park movie, book sales, and a children's touch room
- Fossil Exhibit Trial is accessible to those with mobility impairments and includes casts of several fossils found in the park
- Fossil talks and other naturalist programs

MAMMOTH SITE OF HOT SPRINGS
PO Box 692, Hot Springs SD 57747-0606
Telephone (605) 745-6017
Internet URL: http://www.mammothsite.com/

Visitor center with walkway situated over sinkhole from the Ice Age, 26,000 years ago:

- Large collection of woolly mammoth and Columbian mammoth bones and tusks; other Ice Age animals
- Cast of standing Columbian mammoth
- Giant short-faced bear skull and bones
- Guided tours
- Watch paleontologists dig
- Junior paleontologist excavations by reservation

Wyoming

FOSSIL BUTTE NATIONAL MONUMENT
PO Box 529, Kemmerer WY 83101
Telephone (307) 877-4455

Visitor center featuring:

- Fossils of rare fish, evidence of aquatic life 48–52 million years ago in this presently semi-arid area of southwest Wyoming
- Models depicting stages in the fossilization process, including one Eocene Green River diorama

4-km and 2.4-km interpretive trails

SOUTHWEST REGION

Arizona

Petrified Forest National Park
PO Box 2217, Petrified Forest National Park AZ 86028
Telephone (520) 524-6228

- Painted Desert visitor center with exhibits and film and Rainbow Forest Museum—geology and paleontology exhibits
- 225-million-year-old late Triassic early conifers and cycads from a period when northern Arizona was near a sea—one of the largest exposed concentrations of petrified wood in the world
- Interpretive walks and talks available seasonally
- Fossil evidence of plants, reptiles, amphibians, early dinosaurs, insects, freshwater ecosystem including fish, crabs, snails, and clams

Colorado

Denver Museum of Natural History
2001 Colorado Boulevard, Denver CO 80205-5798
Telephone (303) 370-6387

Dioramas (Prehistoric Journey exhibits based on actual fossil sites):
- Ediacaran (late Precambrian, 600 million years ago) ancient sea floor: possible primitive metazoans, marine lichens, immense protoctists—first macroscopic life forms, based on Ediacara site, Australia
- Cambrian (525 million years ago) sea floor: early arthropods, sponges, anomalocarids, based on Burgess shale, British Columbia, Canada
- Silurian (425 million years ago) sea lily reef: crinoid meadow, coral, and stromatoporid reef, based on fossil reef, Racine, Wisconsin
- Early Devonian (395 million years ago) colonization of land: lungfish, placoderms, psilophyton- and lycopsid plants, based on Beartooth Butte, Wyoming
- Late Pennsylvanian (Carboniferous, 295 million years ago) Kansas coastline: lycopod trees, early conifers, protomammals (sphenopsids), based on quarry in Flint Hills at Hamilton, Kansas
- Cretaceous (66 million years ago) creek bed: forest, late dinosaurs, evolution of flowering plants, *Stygimoloch* pachycephalosaurs, and a *Triceratops* skull, based on Hell Creek Formation, North Dakota
- Early Eocene (50 million years ago) rainforest: *Notharctus* primate, treetop tropical ecosystem in present-day Wyoming, based on Wind River Formation, Wyoming
- Cenozoic (20 million years ago) Nebraska woodland: rhino, camel *Stenomylus,* entelodont *Dinohyus,* based on Hamilton Formation, Agate Springs quarry, Nebraska

Late Triassic (225 million years ago) mural: early conifers, sphenopsids (horse tails), cycads, dinosaurs, phytosaurs (crocodile-like reptiles) based on Petrified Forest National Park, Arizona
Time Travel theater

Dinosaur National Monument
4545 East Highway 40, Dinosaur CO 81610 (also in Utah)
Telephone (970) 374-3000

Dinosaur quarry visitor center:
- Interpretive programs
- Displays of 10 kinds of dinosaur fossils ranging from juveniles to adult brontosaurs and other Jurassic animals
- Nine 1/12-scale dinosaurs

Quarry contains bones of large reptiles exposed in sandstone from an ancient river—an exhibit-in-place—with a museum over it

Florissant Fossil Beds National Monument
PO Box 185, Florissant CO 80816-0185
Telephone (719) 748-3253

- Visitor center with exhibits of 34-million-year-old Oligocene fossil butterflies and other insects, leaves, Sequoia cones, pine cones, and seeds of ancient Lake Florissant
- Natural history talks and interpretive walks
- Two self-guided trails: 1.6-km petrified forest trail around the largest standing petrified Sequoia stumps in the world; 0.8-km walk-through-time trail to fossil-bearing shales

Nevada

Berlin-Ichthyosaur State Park
HC 61, Box 61200, Austin NV 89310
Telephone (702) 964-2440

- Visitor center
- Remains of the largest known ichthyosaur (50 ft) in actual fossil quarry
- Interpretive walks
- Skeleton cast of ichthyosaur

Texas

PERMIAN BASIN PETROLEUM MUSEUM
1500 Interstate 20 West, Midland TX 79701
Telephone (915) 683-4403

- Permian marine diorama depicting life on an ancient reef

PACIFIC REGION

Alaska

BERING LAND BRIDGE NATURAL PRESERVE
National Park Service, Box 220, Nome AK 99762
Telephone (907) 443-2522

- Woolly mammoth remnants and other fauna and flora sealed in permafrost
- Imuruk lava flow—northernmost lava flow in the United States

California

BERKELEY MUSEUM OF PALEONTOLOGY
1101 Valley Life Sciences Building
University of California, Berkeley CA 94720-4780
Telephone (510) 642-1821

- Complete skeleton of *Tyrannosaurus rex*
- Three large dinosaur skulls
- California fossils

CALIFORNIA ACADEMY OF SCIENCES
Golden Gate Park, San Francisco CA 94118-4599
Telephone (415) 221-5100
Internet URL: http://www.calacademy.org/

Life through Time: Evidence for Evolution walk-through dioramas:
- Carboniferous coal forest: 300-million-year-old giant tree ferns, club mosses, seed ferns, early conifers, horsetails, large dragonflies, coelacanth fish, fossils, animated giant scorpion and millipedelike *Arthropleura*
- Cretaceous diorama featuring a hunting pack of *Deinonychus* also includes an animated display of crocodilian young hatching from eggs in their nest and vocalizing; overhead flies a life-size *Quetzalcoatlus* model
- Mesozoic sea diorama with fossil and model ammonites along with paintings and/or fossil skeleton casts of ichthyosaurs, mosasaurs, and plesiosaurs (marine reptiles)
- Minidiorama of early land plants: *Rhynia, Psilophyton, Zosterophyllum*
- Rise of the amniotes exhibit
- Eurypterid model (chelicerate)

Additional exhibits:
- Late Silurian/Early Devonian display with lifesize *Cladoselache* shark model and a placoderm skull fossil cast
- Aquaria housing living horseshoe crabs, lungfish, and horned shark
- 48-million-year-old *Diatryma* adult birds defending young from raiding *Pachyaena*
- Modern archaelogical dig showing *Triceratops* being unearthed
- Earliest prairie habitat

UNIVERSITY OF CALIFORNIA MUSEUM AT BLACKHAWK
3700 Blackhawk Plaza Circle, Danville CA 94506
Telephone (510) 736-2280

- Video interviews with University of California scientists entitled "What a Paleontologist Does"
- Sabre-tooth cats, mastodons, camels, horses, and plants from 9 to 10 million years ago
- Geological time scale
- Upright cast of Lucy *(Australopithecus afarensis)*
- Diorama of Burgess shale (Cambrian) marine organisms

Hawaii

BISHOP MUSEUM
1525 Bernice Street, Honolulu HI 96817
Telephone (808) 847-3511
Internet URL: http://www.bishop.hawaii.org

- Hall of Hawaiian Natural History contains exhibits depicting Hawaii's geological origins and biological heritage

Oregon

JOHN DAY FOSSIL BEDS NATIONAL MONUMENT
HCR 82 Box 126, Kimberly OR 97848-9701
Telephone (541) 987-2333

- Visitor center with exhibits spanning 40 million years of the Age of Mammals (Cenozoic)
- Interpretive talks, ranger-led hikes and special programs
- Animal and plant fossils from five epochs from Eocene to the close of the Pleistocene

Three separate locations:
- Sheep Rock Unit (visitor center with film and fossil museum; five trails, two self-guiding, including a walk along fossil beds; visitors can view a fossil preparation laboratory)

- Painted Hills Unit (four trails, two with exhibits: geology and paleontology; interpretive talks)
- Clarno Unit (two trails—one self-guiding; exhibits: geology and paleontology; interpretive talks)

Washington

Ginkgo Petrified Forest State Park
PO Box 1203, Vantage WA 98950
Telephone (509) 856-2700

Interpretive center including:
- Audiovisual program describing formation of petrified wood
- Many kinds of fossilized wood including ginkgo
- Displays of Miocene forest

Self-guiding interpretive trail through a ginkgo petrified forest and live ginkgos
Group tours by appointment

OUTSIDE THE UNITED STATES

AUSTRALIA

Australian Museum
6–8 College Street, Sydney South, New South Wales 2000
Telephone 02-9320-6000

- Dioramas: Precambrian (Ediacara) and Cambrian (Burgess shale)

South Australian Museum
North Terrace, Adelaide, South Australia 5000
Telephone 61-8-8207-7431

- Cretaceous marine vertebrates: *Diprotodon, Genyornis* (giant bird), *Protemnodon* (giant wallaby), *Allosaurus fragilis* (carnivorous lizard-hipped dinosaur)
- Numerous Australian fossils

CANADA

Joggins Fossil Centre
30 Main Street, Joggins, Nova Scotia B0L 1A0
Telephone (902) 251-2618

Open June through September; by appointment only in May and October:
- Carboniferous fossils including fish, clams, snails, worms, ferns, seeds, cones, leaves, tree stumps, and bark; early amphibians
- Footprints and travelways made by prehistoric arthropods, amphibians, and reptiles
- Interpretive walks of Fossil Cliffs

Royal Tyrrell Museum of Paleontology
Drumheller, Alberta T0J 0Y0
Telephone (403) 823-7707
Internet URL: http://tyrrell.magtech.ab.ca/tour/

- Exhibits on all aspects of fossil history from stromatolites to hominids
- Walk-through diorama of Cambrian Burgess shale fauna and algae
- Diorama showing Paleozoic plants and animals in Silurian, Devonian, Carboniferous, and Permian.
- Walk-around Devonian Reef diorama
- 35 full dinosaur skeletons
- Paleoconservatory features living plants that grew in Alberta during the Cretaceous: *Agathis robusta, Araucaria araucana, Platanus occidentalis, Taxodium mucronatum, Hamamelis mollis, Sabal palmetto, Gunnera manicata, Dicksonia antarctica, Tectaria cicutaria*
- Opportunities to participate in work on dinosaur fossil quarries.
- Prehistoric mammal skeletons include woolly mammoth, mastodon, giant sloth, sabre-toothed cat, brontotheres, and smaller animals
- Interpretive hikes

University of Alberta
Department of Geology, Edmonton, Alberta T6G 2E3

- Three exhibits depicting life in the Cambrian, Devonian, and Cretaceous.

University of Waterloo Earth Sciences Museum
Waterloo, Ontario N2L 3G1
Telephone (519) 885-1211, ext. 2469
Internet URL: http://www.science.uwaterloo.ca/earth/museum/rgarden/rgarden.html and http://www.science.uwaterloo.ca/earth/museum/museum.html

- Exhibits: Devonian bioherm (carbonate rock in the form of an ancient reef) and Cambrian shelf
- Ontario rocks: stromatolitic marble, glacial scratches
- Dinosaur eggs, skeletons, life-size replicas including *Albertosaurus, Parasaurolophus, Tyrannosaurus,* and others, film "64,000,000 Years"

- Evolution of Earth from Precambrian to present—slide presentation
- 3-D film, interactive exhibits, computer programs
- Special programs by appointment

Yoho National Park
Box 99, Field, British Columbia V0A 1G0
Telephone (604) 343-6324

- Trilobite fossil bed
- Exhibit at the actual Burgess shale site (a Cambrian algal reef)
- Guided tours by reservation
- Fossils and video at Field visitor center

ENGLAND

Royal Botanic Gardens
Kew, Richmond, Surrey TW9 3AB
Telephone 44-81-332-5543

- Evolution House—models of Prephanerozoic lower Paleozoic life including 12-m-tall trees: Precambrian, Silurian, Devonian, Carboniferous

FRANCE

Músee National d'Histoire Naturelle
57 rue Cuvier in the Jardin des Plantes, 75005 Paris
Telephone 40-79-30-00

Paleontology gallery showing the history of life including:
- Enlarged models of Cambrian organisms
- Paleobotany
- Fossil insects
- Mastodon and dinosaur skeletons
- Evolution of vertebrate skeleton with fossils

GERMANY

Museum Mensch und Natur
Schloss Nymphenburg, 8000 Munich 19
Telephone 49-89-171-382

- Dioramas: Precambrian, Cambrian, Silurian, Upper Devonian forest with *Ichthyostega* (full-scale), Carboniferous coal forest, Cretaceous, Eocene, Miocene
- Mural: Upper Permian landscape
- Prehistoric insect models

Naturkunde Museum Coburg
Park 6, 96450 Coburg
Telephone 449-9561-8081-0

- Dioramas: Late Precambrian—Ediacaran, Middle Cambrian (Burgess shale), Silurian reef

Naturkunde Museum Ostbayern
Am Prebrunntor 4, 8400 Regensburg 93047
Telephone 49-9415073443

- Carboniferous dragonfly model

SWEDEN

Swedish Museum of Natural History
Frescativagen 40, Stockholm
(Mail: Box 50007, S10405, Stockholm)
Telephone 46-8666-4000

- Dioramas: Cambrian and Silurian

SWITZERLAND

Museum d'Histoire Naturelle
Case Postale 6434, Route de Malagnou, CH-1221 Geneve 6
Telephone 41-22-418-6300

- Diorama depicting Cambrian marine life
- Skeletons of dinosaurs and of early mammals
- Evolution of life on Earth
- Evolution of humankind

WALES

National Museum and Gallery of Wales, Cardiff
Cathays Park, Cardiff CF1 3NP
Telephone 01222 397951

- Silurian reef diorama

VIRTUAL FIELD TRIPS
Internet Resources

Several Internet sites provide further information on organisms described in this book. For each site, we list (1) the name, (2) the address (the URL) by which the site can be called, (3) a brief description of the type of information offered, and, in some cases, specific features within the site. Because most sites here offer links to others with related topics, our list is not comprehensive. (Please note that the URLs and other details of these sites were verified in June 1997 but may have changed since then.)

GENERAL REFERENCE

Site name: UNIVERSITY OF CALIFORNIA MUSEUM OF PALEONTOLOGY
URL: **http://www.ucmp.berkeley.edu**

- Click on Subway for links to many phylogenetic sites
- From Subway, click on Phylogenetics, then click on Databases, then click on UCMP Phylogeny Exhibit to reach The Phylogeny of Life—the ancestor-descendant relationships connecting all organisms that ever lived

Site name: THE TREE OF LIFE
URL: **http://phylogeny.arizona.edu/tree/phylogeny.html**

- A phylogenetic navigator for the internet

Site name: CALIFORNIA ACADEMY OF SCIENCES (CAS)
URL: **http://www.calacademy.org/**

On the Home Page in the area entitled "What's in the California Academy of Sciences Web Server"

- Click on African Frogs, then click on Arabuko-Sokoke Forest to reach photos of forest habitat, amphibians, forest predators, and sounds of Kenya; click on Discovery of a New Serengeti Tree Frog for photos of newly discovered frog, its plains habitat, plains mammals, and the ecosystem in Serengeti, Tanzania.
- Click on CAS Echinoderm Home Page for descriptions, photos, habitat, and classification of many echinoderms
- Click on CAS Diatom Collection for information and photos of marine and freshwater diatoms
- Click on CAS Species of Fishes Catalog for catalog of 53,500 species of fishes; no photos

Site name: MICROBE ZOO
URL: **http://35.8.189.77/CTLProjects/dlc-me/zoo/**

- Microbes and their relationships to all kingdoms

Site name: SMITHSONIAN INSTITUTION RESEARCH INFORMATION SYSTEM
URL: **http://www.siris.si.edu/**

- Bibliography of the natural history collection of the Smithsonian Institution

MICROBES

Site name: AMERICAN SOCIETY FOR MICROBIOLOGY
URL: **http://www.asmusa.org/others.htm**

- Many links to other sites, principally about bacteria and protoctists. In particular, click on Public Education, then Bugs in the News.

Site name: PROTIST IMAGE DATA
URL: **http://megasun.bch.umontreal.ca/protists/protists.html**

- Pictures and information on selected genera of protoctists

Site name: CELLS ALIVE!
URL: **http://www.cellsalive.com**

- Selected topics on microbes and other cells

Site name: MICROBIOLOGY VIDEO LIBRARY
URL: **http://www-micro.msb.le.ac.uk/Video/Video.html**

- Source of video clips of bacteria in motion

PLANTS

Site name: AUSTRALIAN NATIONAL BOTANIC GARDENS IN CANBERRA
URL: **http://155.187.10.12/anbg/index.html**

- Photographs of native Australian plants and information about aboriginal uses of plants

Site name: **Internet Directory for Botany**
URL: **http://www.helsinki.fi/kmus/botmus.html**

- Links to botanical museums, herbaria, natural history museums—sources of information on plants and fungi

Site name: **TreeBASE, a Database of Phylogenetic Knowledge**
URL: **http://herbaria.harvard.edu/treebase**

- A database of phylogenetic information about green plants, especially angiosperms

ANIMALS

Site name: **Marine Biological Laboratory**
URL: **http://www.mbl.edu**

- Source of information on marine animals and algae. Click on Marine Resources to reach the Aquatic Resources Division page, then click on Marine Specimens to reach the Live Marine Specimens page, then click on the Database by phylum

Site name: **National Zoo**
URL: **http://www.si.edu/organiza/museums/zoo/nzphome.htm**

For photographs of animals (mammals, birds, reptiles, amphibians, and invertebrates), click on Animal Photos, then click on Animal Photo Library-Enhanced

- Click on Zoo Views for information on Animal Research and on plant-animal relationships

Site name: **Florida Museum of Natural History**
URL: **http://www.flmnh.ufl.edu**

- Click on Fossil Horses in Cyberspace for an exhibit of the paleontology and evolution of horses, including a time line and pictures of skulls and toes
- Click on Photo Galleries for photos of crocodiles, alligators, fish and birds
- Click on Surf the Web for Related Sites for links to many other natural history museums, botanical gardens, zoos, aquariums and biological parks

Site name: **Cephalopoda**
URL: **http://www.soest.hawaii.edu/tree/cephalopoda/cephalopoda.html**

- Information on cephalopods

Site name: **Academy of Natural Sciences, Philadelphia**
URL: **http://www.acnatsci.org**

- Click on Dinofest '98 for information on the World's Fair of Dinosaurs scheduled for April 1998
- Click on Dinolinks for links to other dinosaur-related web sites

Site name: **World's First Dinosaur Skeleton**
URL: **http://www.levins.com/dinosaur.html**

- List of dinosaur-related web sites
- List of museums that feature dinosaur exhibits

SUPERKINGDOM PROKARYA

Single-membrane–bounded genetic systems: nucleoids, protein synthesis on small ribosomes, DNA-level recombination only. No cell fusion; lack of nuclear and cytoplasmic fusion (that is, fertilization) implies absence of Mendelian genetics. Display unidirectional gene transfer and, in microscopic observation, lack visible intracellular motility. Reproduction by binary fission, budding, budding of filaments, fission of stalked sessile parent to produce flagellated offspring, polar (end-to-end) growth, or multiple fission.

B-5 *Anabaena* [Courtesy of N. J. Lang.]

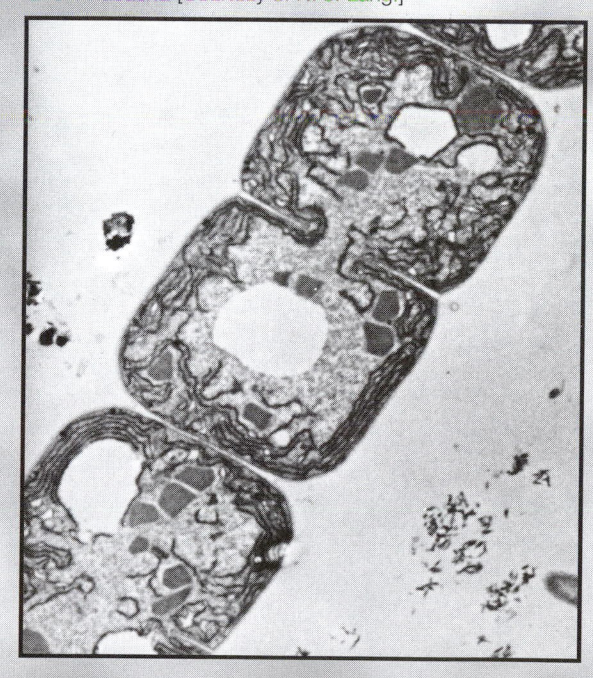

Superkingdom Prokarya
- Kingdom Bacteria
  - Subkingdom Archaea (Archaebacteria)
  - Subkingdom Eubacteria

CHAPTER ONE

BACTERIA

(Prokaryotae, Procaryotae, Monera)

B-3 *Nitrobacter winogradskyi* [Courtesy of S. W. Watson, *International Journal of Systematic Bacteriology 21*, 1971.]

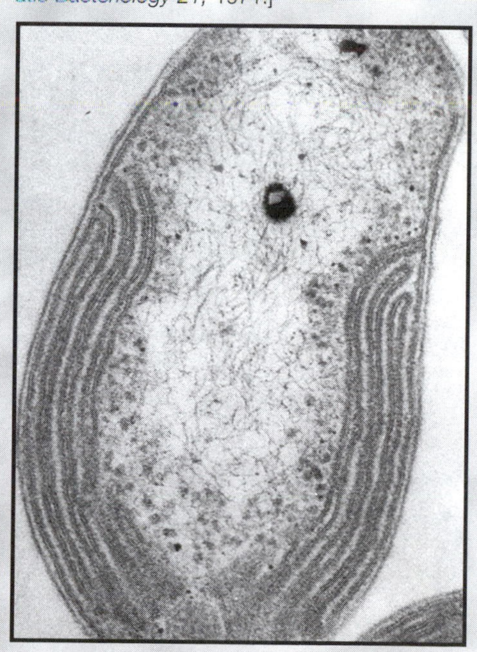

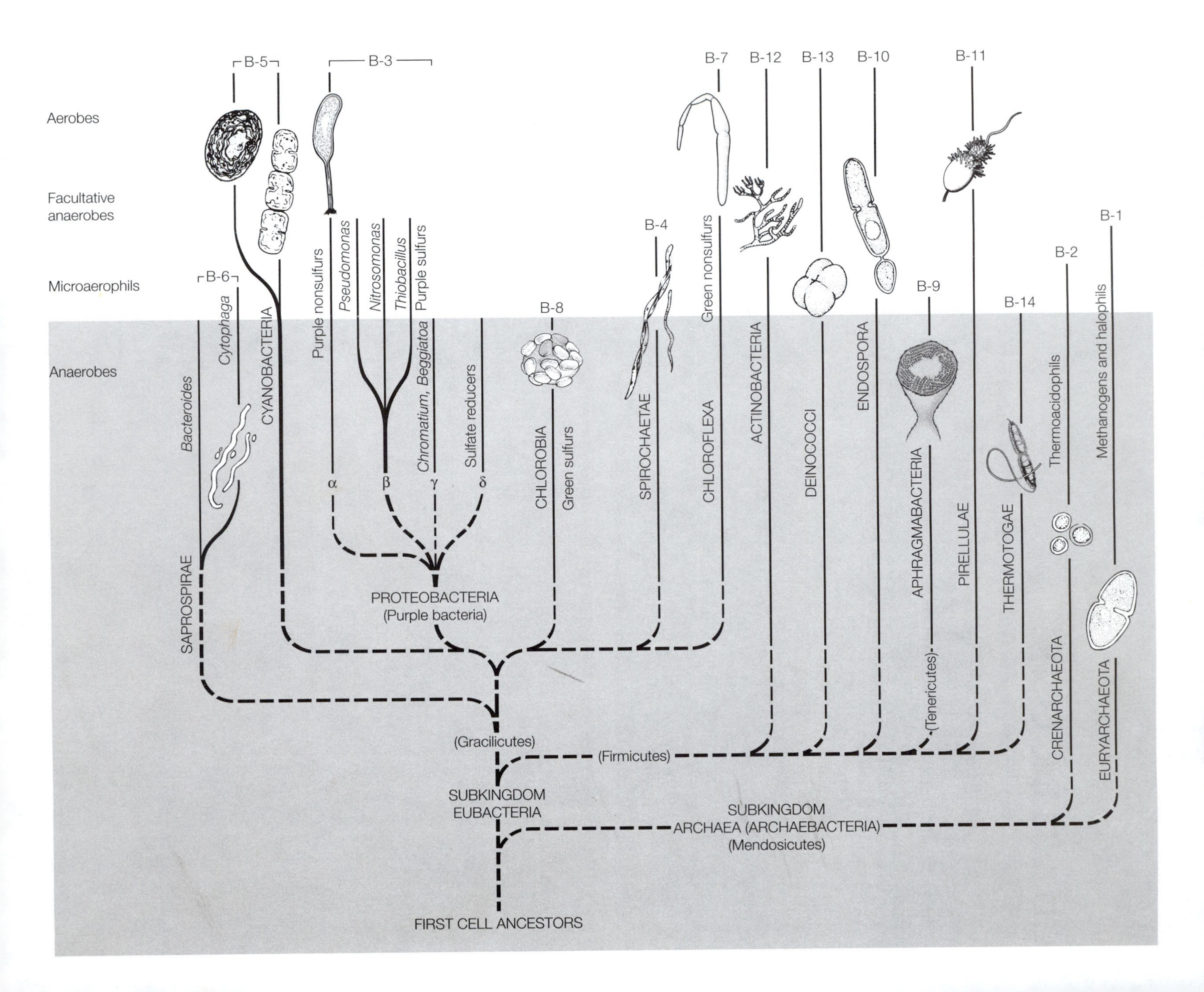
Aerobes
Facultative anaerobes
Microaerophils
Anaerobes
B-5
B-3
B-7
B-12
B-13
B-10
B-11
B-4
B-1
B-6
B-2
B-8
B-9
B-14
Bacteroides
Cytophaga
CYANOBACTERIA
Purple nonsulfurs
Pseudomonas
Nitrosomonas
Thiobacillus
Purple sulfurs
Chromatium, Beggiatoa
Sulfate reducers
α
β
γ
δ
CHLOROBIA
Green sulfurs
SPIROCHAETAE
CHLOROFLEXA
Green nonsulfurs
ACTINOBACTERIA
DEINOCOCCI
ENDOSPORA
APHRAGMABACTERIA
(Tenericutes)
PIRELLULAE
THERMOTOGAE
Thermoacidophils
Methanogens and halophils
CRENARCHAEOTA
EURYARCHAEOTA
SAPROSPIRAE
PROTEOBACTERIA
(Purple bacteria)
(Gracilicutes)
(Firmicutes)
SUBKINGDOM
EUBACTERIA
SUBKINGDOM
ARCHAEA (ARCHAEBACTERIA)
(Mendosicutes)
FIRST CELL ANCESTORS

BACTERIA
(Prokaryotae, Procaryotae, Monera)

Greek *pro,* before; *karyon,* seed, kernel, nucleus
Greek *moneres,* single, solitary
Greek *bakterion,* little stick (diminutive of *baktron,* stick, staff, rod)

Bacterial cell structure, all prokaryotes. Sexuality produces genetic recombinants temporally and spatially independent of reproduction. Single or multicellular, single or branched filaments. Terminally and cyclically differentiated cells (for example, heterocysts, endospores). All major modes of metabolism represented in the group (see Table 1-1): maximum metabolic diversity, lithospheric (geologic) and atmospheric interaction relative to eukaryotes. Fossil record of bacteria organized in communities extends from the lower Archean eon (see Figure I-4) to the present. Comparison of gene sequence in rRNA molecules useful for distinguishing modern lineages.

Kingdom Bacteria comprises all organisms with prokaryotic cell structure: they have small ribosomes surrounding their nucleoids (see Figure B-1A), but they lack membrane-bounded nuclei. In activity and potential for rapid unchecked growth, bacteria are unrivaled among living organisms. About 10,000 different forms have been described as "species;" most of them are cyanobacteria (Phylum B-5). Their genes are easily passed from one to another; the set of genes, or genophore, is organized into thin (25 Å) fibrils that, when visible as a light region in electron micrographs, are called the nucleoid. The distinguishing traits of all the prokaryotes (and therefore of bacteria) are listed and compared with those of eukaryotes in Table I-2. Prokaryotes, unlike all members of the superkingdom Eukarya, lack pore-studded nuclei that contain chromosomes. Prokaryotes also lack membrane-bounded organelles such as mitochondria and plastids; they did not evolve by cell symbiosis.

In this third edition, we have changed the kingdom name (from Monera or Prokaryotae) to Kingdom Bacteria. *Bergey's Manual of Systematic Bacteriology* provides information for identifying and classifying bacteria. It unites all bacteria on the basis of their prokaryotic nature under the name Prokaryotae. Although we support the bacteriologists who bring together all living organisms that lack a nuclear membrane under the term Prokaryotae (*pro,* before; *karyon,* seed, nucleus), we know that our readers prefer the term Bacteria, because it is more familiar. Today's kingdom of prokaryotes corresponds to the "Moneres" of the German biologist Ernst Haeckel. Haeckel used the term "Monera" (from the Greek *moneres* for single, solitary) to refer to bacteria as a group within his "Protista" kingdom in his three-kingdom scheme. Our acceptance of bacteria as the only kingdom of life that did not evolve by symbiosis supports the use of the terms Bacteria and Prokarya rather thān the older Monera.

Bacteria are the most hardy of living beings. Some can survive very low temperatures, well below freezing, for years; others thrive in boiling hot springs; and still others even grow in very hot acid or live by deriving hydrogen and carbon dioxide from rocks. By forming propagules such as spores—traveling particles of life that contain at least one copy of all the genes of an organism—many tolerate boiling water or total desiccation. Bacteria are the first to invade and populate new habitats: land that has been burned, volcanic soils, or newly emerged islands.

We dimly recognize the activities of thriving communities of bacteria (usually supported by photosynthesizers) as "scum," "slime," "gloop," "microbial mats," "floc," and other derogatory terms. Prokaryotic communities of different kinds of bacteria living together survive in an extraordinary range of habitats inhospitable to protoctists, plants, animals, and fungi. The absolute requirements for growth of all of them are liquid water and sources of energy and matter (carbon, hydrogen, nitrogen, sulfur, phosphorus, oxygen, magnesium, sodium, potassium, zinc, and a few others) in the appropriate form and amounts. Some bacteria survive and grow at great oceanic depths or even inside granites or carbonate rocks. Others have been captured in nets of stratospheric airplanes from above the atmosphere. Yet no organism—not even the hardiest of bacteria—is known to complete its life history suspended in the air or any other gas.

Some other activities of bacteria are still only poorly known. The incorporation of soluble metal ions such as those of manganese and iron into rocks—nodules on lake and ocean floors—are accelerated by bacterial action. Layered chalk deposits called stromatolites (Figure 1-1) are produced at the seashore by the trapping and binding of calcium carbonate–rich sediment by growing communities of bacteria, especially by cyanobacteria. Gold in South African mines is found with rocks rich in organic carbon, associated with fossil bacteria and probably of microbial origin. In Witwatersrand, the miners find the gold, deposited apparently more than 2500 million years ago, by following the "carbon leader" that leads people to the gold. Copper, zinc, lead, iron, silver, manganese, and sulfur all seem to have been concentrated into ore deposits by biogeochemical processes that include bacterial growth and metabolism. Bacteria are the only organisms

A

B

C

Figure 1-1 (A) Live stromatolites, Shark Bay, Australia. [Photograph by E. S. Barghoorn.] (B) Fossil stromatolites from the Kuuvik Formation of the Kilohigok Basin, Victoria Island, Northwest Territories, Canada. These domed ancient stromatolites are about 2 billion years old and more than 2 m high. [Photograph courtesy of Fred Campbell, Geologic Survey of Canada.] (C) Microbial mat, Matanzas, Cuba. Each block on the scale marker represents 10 cm. [Photograph by R. Horodyski.]

that, in a process called nitrogen fixation, convert the air's major gas, nitrogen (N_2), into usable organic nitrogen.

Because of their limited morphology and the paucity of their fossil record, bacteria have evolutionary relationships that have been exceedingly difficult to ascertain. However, in recent years, advances in molecular biology, the result of detailed studies of macromolecules, have enhanced our understanding of the evolutionary relationships among the tiny but highly diverse prokaryotes. Great insights have emerged from comparative studies of the long-chain ribonucleic acid (RNA) molecules that are components of the ribosomes of all organisms. The assumption, probably valid, on which this work is based is that changes in the sequence of units (nucleotides) in RNA molecules reveal evolutionary histories of the modern bacteria. Because ribosomes are universally found in all organisms and are crucial for the same cell function, ribosomal RNA (rRNA) molecules are thought to have changed very slowly through evolutionary time. Under this assumption, Carl Woese, of the University of Illinois, and Otto Kandler and Wolfram Zillig, of the University of Munich, and many other colleagues have concluded that bacteria assort into two fundamentally different, and therefore ancient, groups: the Archaebacteria (Archaea) and the Eubacteria. We agree and therefore recognize these groups as subkingdoms. However, we cannot accept the trifurcation of all life into three "primary kingdoms" [(1) Archaebacteria = Archaea, (2) Eubacteria = Bacteria, (3) Eukarya], as these authors urge, because eukaryotes, as composite beings that evolved through symbiogenesis, are different in principle from eubacteria and archaebacteria, both of which are prokaryotes.

Small though they are, bacteria, which are extremely numerous and fast growing, are crucial to the health of digestive systems, soil maintenance in agriculture and forestry, and the very existence of the air that we breathe. Modern food processing began with an awareness of the nature of bacteria. Canning, preserving, drying, salting, and pasteurization are techniques that prevent entry by even a single bacterium or growth of the few that remain. The success of these techniques is remarkable in view of the ubiquity of bacteria. Every spoonful of garden soil contains some 10^{10} bacteria; a small scraping of film from your gums might reveal some 10^9 bacteria per square centimeter of film—the total number in your mouth is greater than the number of people who have ever lived. Bacteria make up some 10 percent of the dry weight of mammals. They normally cover our skin, especially on damp surfaces such as under the arms and between the toes. They line nasal, ear, and mouth passages and live in pockets in the gums and between the teeth. Most pack the digestive tract, especially the large intestine. Most bacteria are never pathogenic.

Pathogens are simply bacteria (or, occasionally, protoctists or fungi) capable of causing infectious diseases in animals or plants. The word "germ," like the word "microbe," has no precise or specific meaning. A germ is a small living organism capable of growth at the expense of another organism; a microbe, or microorganism, is an organism so small that one needs a microscope to see it [for example, cyanobacteria (Phylum B-5) and myxobacteria (Phylum B-3) are microbes]; thus the smaller fungi, most protoctists, and all but the largest bacteria are also called microbes. Bacteria can cure as well as cause disease. Many of our most useful antibiotics (a kind of allelochemical, a compound made by one form of life that inhibits the growth of a different, usually microbial, life form) come from microbes. Of the best-known antibiotics, streptomycin, erythromycin, chloromycetin, and kanamycin come from bacteria, whereas penicillin and ampicillin come from fungi.

Bacteria are rather simple morphologically: spherical (cocci), rod shaped, or spiral shaped. The most complex undergo developmental changes in form: single bacteria may reproduce, producing populations that metamorphose into stalked structures, grow long, branched filaments, or form tall bodies that release resistant sporelike microcysts. Some produce highly motile (swimming) colonies. Knowledge of bacterial structure, unless complex, is seldom a source of insight into function. In this respect, bacteria are very different from protoctists, animals, fungi, and plants. Because their differences lie chiefly in their metabolism (their internal chemistry), many kinds of bacteria can be distinguished only by the chemical transformations that they cause. Bacteria are easily grouped by cell wall properties that are distinguished by a color-staining procedure. A universally applied diagnostic test of bacteria is whether they stain purple or pink with the Gram test, a staining method developed by the Danish physician Hans Christian Gram (1853–1938). Gram-positive organisms (which stain deep purple) differ from Gram-negative ones (which stain light pink). The chemistry of their cell walls—the presence of an extra membrane in Gram-negative bacteria—is the basis for classification (see page 63).

Bacteria, especially in highly ordered but flexible communities, can effect a large number of different chemical transformations. A summary of metabolic patterns of strains of bacteria growing in pure culture is presented in Table 1-1. The range of metabolic capabilities is far greater than that of all the eukaryotes. Metabolically speaking, bacteria represent the extreme range of biological diversity. The full extent of bacterial diversity has yet to be appreciated.

Although some very complicated molecules are made by certain plants and fungi, the biosynthetic and degradative patterns—the chemistry of food use and energy generation—in all plant and fungal cells are remarkably similar. Animals and protoctists exhibit even less variation in their chemical repertoires. In short, the metabolism of eukaryotes is rather uniform; its patterns of photosynthesis, respiration, glucose breakdown, and synthesis of nucleic acids and proteins are fundamentally the same in all eukaryotic organisms. Bacteria, on the other hand, are not only metabolically different from eukaryotes, but also from each other.

The work of most microbiologists concerns the role of bacteria in health and disease. Bacterial activities in our environment have been studied much less, but they are even more significant. Bacteria release to Earth's atmosphere and remove from it all the major reactive gases: nitrogen, nitrous oxide, oxygen, carbon dioxide, carbon monoxide, several sulfur-containing gases, hydrogen, methane, and ammonia, among others. Protoctists and plants also make substantial contributions to atmospheric gases, such as oxygen, and ruminant animals contribute methane, but few, if any, that differ from those of bacteria, whereas many important reactions are limited to the prokaryote repertoire.

The soil of Earth and the regolith—the loose, rocky covering of any planet—on the surface of Mars and of the Moon differ enormously. Mars and the Moon are very dry and lack atmosphere relative to Earth, but the differences extend far beyond just moisture content. The surface of Earth—its regolith, sediments, and waters—is rich not only in living bacteria, small animals, protists, yeasts, and other fungi, but also in the complex organic (carbon plus hydrogen) compounds that they produce. The less tractable substrates, such as tannic acids, lignin, and cellulose, tend to accumulate, whereas much more actively metabolized organics, such as sugars, starches,

Table 1-1 Metabolic Modes in Prokarya

Energy	Nutrition: Electrons (Hydrogen)*	Nutrition: Carbon	Examples†	Phylum
Light	Inorganic compounds and compounds with one carbon atom	Carbon dioxide	Photoautotrophs:	
			Chlorobium (H_2S)	B-8
			Chromatium (H_2S)	B-3
			Cyanobacteria (H_2O)	B-5
			Rhodospirillum (H_2)	B-3
	Organic compounds	Acetate, lactate, pyruvate	Photoheterotrophs:	
			Chromatium (some)	B-3
			Chloroflexus	B-7
			Halobacterium	B-1
			Heliobacterium	B-10
			Rhodomicrobium (some)	B-3
			Rhodospirillum (some)	B-3
Chemical compounds	Inorganic compounds	Carbon dioxide	Methanogens:	
			Methanococcus (H_2)	B-1
			Hydrogen-oxidizing pseudomonads (H_2)	B-3
			Nitrosomonas (NH_3)	B-3
			Methylotrophs:	
			Methylosinus (CH_4)	B-3
		Organic compounds	Manganese oxidizers (Mn^{2+})	B-10
			Sulfide oxidizers:	
			Beggiatoa (H_2S)	B-3
			Desulfovibrio (SO_4^{2-})	B-3
	Organic compounds	Organic compounds	Heterotrophs:	
			Escherichia coli	B-3
			Bacillus	B-10

* Source of electrons for reduction of carbon to synthesize cell material.

† Examples include genera that have more than a single possible physiological mode. In parentheses is the most common source of electrons for the example in question.

organic phosphorous compounds, and proteins, are produced and removed more rapidly. All these organic compounds are—directly or indirectly—the products of chemosynthesis or of photosynthesis, processes that use chemical energy or sunlight, respectively, to convert the carbon dioxide of the air into the organic compounds of the biosphere and, ultimately, into the organic-rich sediments from which we obtain oil, gas, and coal. In fact, the soil and rocks of Earth contain about 100,000 times as much carbon as Earth's living forms do.

Chemosynthesis is limited to certain groups of bacteria. Photosynthesis, which often is incorrectly attributed only to algae and plants, is carried out by many groups of bacteria. Chemosynthesis and photosynthesis are often, but not always, correlated with processes that use inorganic chemicals or light, respectively, to generate energy to make organic compounds. Both types of synthesis are forms of strict autotrophic nutrition, the synthesis of all food and derivation of energy exclusively from inorganic sources. Heterotrophy, the alternative mode of nutrition, is the deriving of food and energy from preformed organic compounds—from either live or dead sources. Like algae and plants, most photosynthetic bacteria convert atmospheric carbon dioxide and water into organic matter and oxygen; unlike them, many bacteria are also capable of very different modes of photosynthesis—for example, the use of hydrogen sulfide instead of water and the elimination of sulfur but not oxygen. Bacterial photosynthesis and chemosynthesis are essential for cycling the elements and compounds on which the biosphere, including ourselves, depends. Bacteriologists refer to phototrophy as the mode of nutrition for organisms (at the top of Table 1-1) that nourish themselves by light reactions using O_2. Photosynthesis refers to any process of living tissue in which light energy is used to build organic matter.

Probably the most important evolutionary innovation on Earth, if not in the solar system and the galaxy, was photosynthesis, the transformation of the energy of sunlight into usable form: the chemical energy of food or energy-storage molecules (that is, carbohydrates, lipids, and proteins). Photosynthesis, the process, began in anaerobic bacteria more than 3 billion years ago. Bacteria that derive their energy from sunlight, their carbon from CO_2 of the air, and their electrons from H_2, H_2S, H_2O, or other inorganic sources are called photolithoautotrophic bacteria. They "feed" on sunlight.

Photosynthesis is an essentially anaerobic process, and none of the proteobacteria (Phylum B-3) can carry it out when they are exposed to oxygen. Except for the Chloroflexa (Phylum B-7), green nonsulfur bacteria are hypersensitive to oxygen; they can grow only photosynthetically and in the total absence of oxygen. Some purple nonsulfur bacteria (Phylum B-3) can grow microaerophilically—that is, under oxygen concentrations less than the modern norm—or even aerobically, but only in total darkness. In that case, they derive their energy not from photosynthesis, but from the breakdown of food, as do respiring bacteria.

Oxygen release is not an essential property of photosynthesis, even though it is characteristic of photosynthesis in plants, algae, and cyanobacteria. The essential properties are the incorporation of carbon dioxide (CO_2) from the air into organic compounds needed for growth of the photosynthesizer and the conversion of the energy of visible light into chemical energy in a form useful to cells. The conversion of light energy requires chlorophyll and other pigment molecules, although not always precisely the same chlorophyll molecules. The chemical energy currency produced is adenosine triphosphate (ATP), a nucleotide used in transformation reactions of all cells. Although the details of the enzymatic pathways of photosynthesis are still being worked out, it is clear that the five types of photosynthetic bacteria (purple sulfur, purple nonsulfur, green nonsulfur, green sulfur, and oxygenic) differ in details of metabolism, in their source of electrons for CO_2 reduction, and in other ways.

To reduce the CO_2 in the air to organic compounds, cells need a source of electrons, which, as a rule, are carried by hydrogen atoms. The source of these electrons varies with the organism. In green sulfur bacteria (Chlorobia, Phylum B-8), the electrons come from hydrogen sulfide (H_2S), although they may also come from hydrogen gas (H_2). The purple sulfur bacteria (Phylum B-3) also use H_2S as the hydrogen donor. In purple nonsulfur bacteria, such as *Rhodospirillum* and *Rhodopseudomonas* (Phylum B-3), the hydrogen donor is a small organic molecule such as lactate, pyruvate, or ethanol. Thus, the general photosynthetic equation can be written as

$$2n\ H_2X + n\ CO_2 \xrightarrow{\text{light}} n\ H_2O + n\ CH_2O + 2n\ X$$

in which X varies according to species. The molecule H_2X is the hydrogen donor. In proteobacteria (Phylum B-3), the hydrogen donor is never water; thus, oxygen is not a by-product of their photosynthesis. In cyanobacteria (Phylum B-5), algae (Phyla Pr-7, Pr-10, Pr-11, Pr-13 through Pr-17, Pr-25, Pr-26, and Pr-28), and plants (Phyla Pl-1 through Pl-12), on the other hand, because water is the hydrogen donor, oxygen is released. When H_2S is the hydrogen donor, the by-product of photosynthesis is sulfur, which may be excreted, stored as elemental sulfur, or further oxidized as sulfur compounds, such as sulfate, and then excreted. No gas such as oxygen is released; rather, the form of the sulfur product depends on conditions. When buried in sulfur-rich muds full of high concentrations of sodium sulfide, for example, intracellular sulfur globules are made even by some cyanobacteria (Phylum B-5). These same cyanobacteria generate oxygen from water in aerated, sulfur-poor conditions.

Genera of photosynthesizers that are placed in bacterial phyla include the archaean *Halobacterium* (Phylum B-1), which uses rhodopsin rather than chlorophyll in its processing of light energy, and several groups that contain some kind of bacterial chlorophyll (that is, cyanobacteria and some proteobacteria). A new type of photosynthesizer, *Heliobacterium,* which by 16S rRNA criteria is related to Gram-positive low-GC (guanosine plus cytosine) eubacteria (Phylum B-10), grows either heterotrophically or phototrophically; this option, rare in prokaryotes, is common in algae (Kingdom Protoctista). *Heliobacterium* uses bacteriochlorophyll G, whereas the green sulfur (Phylum B-8) and nonsulfur (Phylum B-7) bacteria use still other chlorophylls to generate cell energy in the form of ATP. All cyanobacteria (Phylum B-5) use at least chlorophyll *a*. Photosynthesis evidently evolved many times separately (polyphyletically) in prokaryotes, although members of only one group, cyanobacteria, were ancestral to the plastids of protoctists and plants.

The notion that the food web starts with the plants, followed by the herbivores, and ends with carnivorous animals is shortsighted. Zooplankton of the seas feed on protoctists; nonphotosynthesizing protoctists feed on bacteria; bacteria (and fungi and animal scavengers) break down the carcasses of animals, plants, and algae, releasing back into solution such elements as nitrogen and phosphorus required by the phytoplankton.

Because phyto means plants and because no plants float in the open ocean, we prefer the term photoplankton to refer to floating bacteria and algae that photosynthesize at sea and in lakes. Bacteria, because they are eaten by others, facilitate entire food webs. The ways in which we and other forms of life depend on bacteria, and evolved from them, will be explained in the descriptions of the phyla. Life on Earth would die out far faster if organisms in the kingdom of Bacteria became extinct than if any of the other kingdoms disappeared. We believe that bacterial life on our planet thrived long before the large organisms that evolved by symbioses from communities of bacteria ever appeared.

Bacteria have an ancient and noble history. They were probably the first living organisms and, with respect to everything but size, have dominated life on Earth throughout the ages. The oldest fossil evidence for bacteria dates to about 3400 million years ago, whereas the oldest evidence for organisms belonging to the eukaryotic kingdoms is about 1200 million years.

Biologists and geologists agree that, some 2000 million years ago, the cyanobacteria (the oxygen-releasing photosynthetic prokaryotes that used to be called "blue-green algae"; Phylum B-5) began one of the greatest changes known in the history of this planet: the increase in concentration of atmospheric oxygen from far less than 1 part per thousand (<0.001) to about 200 parts per thousand, or 20 percent. Without this high concentration of oxygen, plants, people, and other animals would not have evolved.

All bacteria reproduce nonsexually by binary fission (for example, one cell divides, giving rise to two identical offspring cells) or budding. Although bacteria participate in sexual donation of DNA from one (the donor) to another (the recipient) in the process of conjugation, this bacterial sexuality is not associated in time or space with reproduction. The extent of bacterial sexuality in nature is not well known, partly because—as with most other organisms—the "sex life" of bacteria is elusive. Any process that leads to the formation of a bacterium with genes from more than a single parental source is bacterial sexuality. If the source is a second bacterium in contact with the first, the bacterial sexual process is conjugation. If the second source of genes is a plasmid (circular fragment of DNA), virus, or linear piece of DNA carrying bacterial genes, the bacterial sexual process has another name: transformation. When genes are transferred to recipient bacteria by viruses that infect bacteria, the process (a special case of transformation) is called transduction. Many bacteria excrete DNA. In the laboratory and most likely in nature, DNA excreted by one bacterium is taken up and incorporated by another to form genetically recombinant organisms.

No location anywhere on Earth lacks bacteria, but only a few places today are dominated by them. Some exclusively bacterial habitats, most often found in intemperate climates, are the bare rocks of cliffs, the interior of certain carbonate rocks, and muds lacking oxygen. Perhaps the most spectacular are the boiling hot springs and muds such as those in Yellowstone National Park, in Wyoming, or the brightly colored salt flats and shallow embayments of tropical and subtropical areas. Many such thermal springs, flats, and bays are dominated by microbial mats—cohesive, domed or flat structures on soil, air, or in shallow water that are caused by the growth and metabolism of bacteria, primarily filamentous cyanobacteria. By entrapping bits of sand, carbonate, and other sediment, such microbial communities grow to be quite conspicuous manifestations of biological activity.

The habitat scenes are notably arbitrary in this chapter because so many bacteria can be found in protoctist, animal, fungal, and plant hosts, as well as in soil, air, and water samples, in vastly different habitats and locations. Bacterial communities in the intestines of animals (Figure 1-2) have been studied disproportionately.

Except for those rather extreme environments where microbial mats or thermal springs abound, eukaryotes seem to dominate our landscape. However, microscopic examination of a sample from any forest, tide pool, riverbed, chaparral, or other habitat reveals bacteria in abundance. When a specific type of bacterium is removed from nature for growth on its own (pure culture) or with other microbes (mixed culture), that type, called a strain, is given a name or identifying number. When environmentalists mourn the destruction of habitats by pollution, they are usually thinking of the loss of fish, fowl, and fellow mammals, and not strains of bacteria. If our sympathies were with the cyano- and other bacteria instead, we would recognize the pollution of green scummy lakes, for example, as a sign of flourishing life.

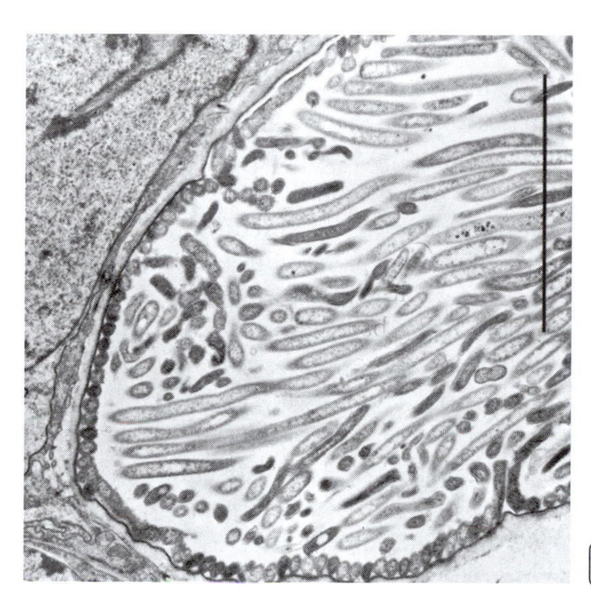

Figure 1-2 An intact bacterial community from a pocket in the hindgut wall of the Sonoran desert termite *Pterotermes occidentis* (Phylum A-20). More than 10 billion bacteria per milliliter have been counted in these hindgut communities. Many are of unknown species. All survive anoxia. In our studies, from 28 to 30 strains isolated were facultative aerobes that metabolize oxygen when available. Most are motile, Gram-negative heterotrophs and thus most likely proteobacteria. Notice that some of the bacteria line the wall of the gut, whereas others float freely in the lumen. TEM, Bar = 5 μm. [Courtesy of D. Chase.]

Much of bacterial nomenclature is in dispute; there is no consensus among scientists on how to name and group the thousands of strains or how to relate them to bacteria in nature or in the literature. Most bacteria are still not identified; microbiologists, who study bacteria (as well as the smaller fungi and anaerobic protoctists), assert that the vast majority of bacteria have not yet been carefully studied and described. Microbiologists lack standard nomenclatural and taxonomic practices. Relative to the strict rules of those who study animals (zoologists) or plants (botanists), the terminology and taxonomic practices of microbiologists are inconsistent and not directly comparable. Thus, inevitably, several of our groupings differ from those found in *Bergey's Manual of Systematic Bacteriology* or the four-volume work called *The Prokaryotes* (see Balows et al. in the Bibliography).

We aim to make the taxonomic level of phylum conceptually comparable throughout all life and to avoid confusion and contradiction. We recognize 14 major prokaryotic phyla, which group the bacteria by clearly distinguishable morphological and metabolic traits, but, unlike the phyla in other kingdoms, these prokaryotic phyla are not all-inclusive. Many small groups of bacteria that are difficult to classify have been omitted from the descriptions of members of phyla for reasons of clarity and lack of space. As the amount of molecular systematic data increases and 16S ribosomal RNA and other detailed information becomes available, we employ it, keeping members of natural groups together as well as we can.

Because bacteria that differ in nearly every measurable trait can receive and permanently incorporate any number of genes from each other or from the environment, Sorin Sonea (University of Montreal) and others have argued that bacteria form a single worldwide web of relations. Although strain names are easily applied, the concept of "species," applicable to named eukaryotes, seems inappropriate for the Prokarya. Because prokaryotes can change their genetic properties so quickly and easily, we agree with Sonea and Panisset's analysis (see Bibliography). We doubt that bacteria ever evolved to form permanent species characteristic of all Eukarya. A certain flexibility must be tolerated in any bacterial taxonomic scheme; one of our goals is to be maximally informative and useful in our treatment of these fascinating planetmates.

SUBKINGDOM ARCHAEA

The distinguishing characteristics of archaebacteria concern primarily the gene (DNA) sequences that determine the small-subunit ribosomal RNA sequences. A 16S rRNA, about 1540 nucleotides long, from small ribosomal subunits is comparable in function in all prokaryotes. On the basis of the nucleotide sequence in both 16S and 5S RNA (constituents of the 30S rRNA small subunit shown in Figure 1-3), archaebacteria are more closely related to each other than they are to eubacteria. Archaebacterial ribosomes are more similar in ultrastructure to eukaryotic ribosomes than to eubacterial ribosomes (Figure 1-3). The major lipids of archaebacteria are ether linked with phytanol side chains (C_{20}). In other bacteria and in eukaryotes, the major lipids are ester linked. Archaebacteria lack the peptidoglycan layer characteristic of cell walls of eubacteria. A single DNA-dependent RNA polymerase (an enzyme) with complex subunit structure—more than six subunits—is present in archaebacteria. Archaebacteria include methanogenic, halophilic, and thermoacidophilic bacteria. All other—that is, most—bacteria are eubacteria.

We emphasize the ancient environments in which members of two archaebacterial phyla—Euryarchaeota and Crenarchaeota—tend to be found: although the habitats of archaebacteria were thought to be typical of those on the surface of Earth during the Archean eon more than 3 billion years ago (that is, tectonically active environments), recent studies have shown archaebacteria to be widespread in seawater, lakes, soils, and other environments not subject to extremes. The distribution of archaebacteria in nature is under intense investigation. Here we depict them thriving as methanogenic (methane-producing), halophilic (salt-loving), and thermoacidophilic (heat- and acid-loving) bacteria in settings of oxygen-depleted muds and soils (conditions of anoxia) or comparable places: hot springs, sites where sea vents spew oxygen-depleted gases, salty seashores, boiling muds, or places where ash-ejecting volcanoes abound. The extreme environments that dominated the early Earth certainly harbor archaebacteria, but not exclusively. The direct investigation of the distinctive 16S rRNA sequences in nature in microbes that cannot be grown in the laboratory reveal an astounding diversity and emphasize the depth of our ignorance concerning prokaryotic life on Earth, especially archaebacteria with unknown morphology and physiology.

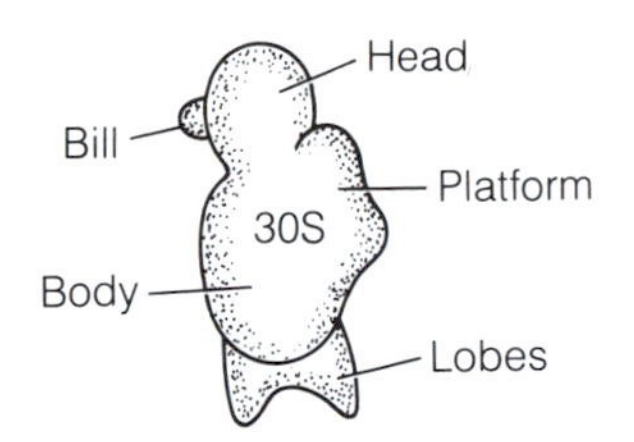

Figure 1-3 A small subunit (called the 30S subunit)) of the ribosome, a universal cell structure. A comparison of this subunit in eukaryotes, archaebacteria, and eubacteria shows systematic differences in morphology. [Drawing after J. A. Lake, "Ribosomal evolution: The structural basis for protein synthesis in archaebacteria, eubacteria, and eukaryotes," *Cell* 33:318–319 (1983).]

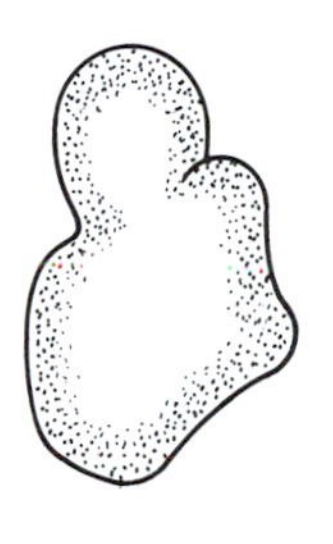

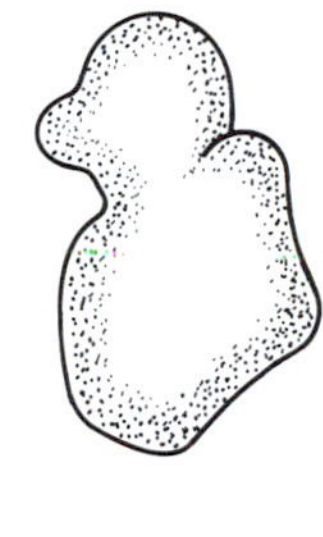

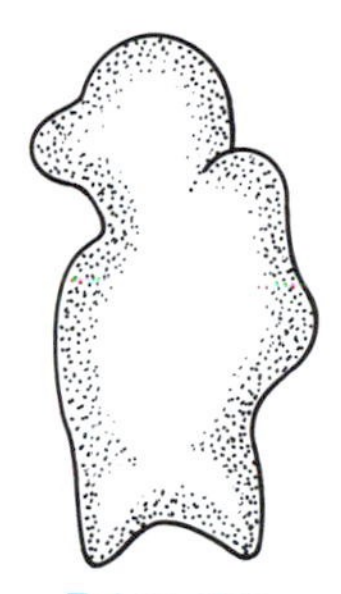

Division: Mendosicutes

B-1 Euryarchaeota
(Methanogens and halophils)

Greek *eury*, broad, wide; *archae*, old; *-otes*, quality

EXAMPLES OF GENERA

Haloarcula
Halobacter
Halococcus
Haloferax
Methanobacillus
Methanobacterium
Methanobrevibacter
Methanococcoides
Methanococcus
Methanocorpusculum
Methanoculleus
Methanogenium
Methanohalophilus
Methanolobus
Methanomicrobium
Methanoplanus
Methanopyrus
Methanosaeta
Methanosarcina
Methanosphaera
Methanospirillum
Methanothermus
Methanothrix
Natronobacterium
Natronococcus

The phylum Euryarchaeota comprises two very strange and very different groups of archaebacteria, methanogens and halophils. They are grouped together on the basis of rRNA sequence similarities. Methanogens can be Gram-positive or -negative, motile (by means of flagella) or immotile. All three classic bacterial shapes—rod, spirillum, and coccus—are represented: methanogens vary from short rods and irregular cocci to spirilla, large cocci in packets, and filaments. So far, no methanogens with branched filaments or internal (periplasmic) flagella have been discovered. Most species can tolerate moderate or high temperatures. Methanogens are found worldwide in sewage, in marine and freshwater sediments, and in the intestinal tracts of animals—both ruminants and cellulose-eating insects. Formally recognized genera are distinguished by morphology and physiology. All have names prefixed by "methano-": *Methanosarcina, Methanobacterium, Methanococcus, Methanobacillus,* and *Methanothrix.*

Halophils live in salty or alkaline extreme environments and do not form spores. Before the use of molecular phylogenetic techniques, the close relationship between halophils and methanogens was unknown.

Methanogenic bacteria cannot use sugars, proteins, or carbohydrates as sources of carbon and energy. In fact, most can use only three compounds as carbon sources: formate, methanol, and acetate. Methanogens are characterized by their extraordinary way of gaining energy: they form methane (CH_4) by reducing carbon dioxide (CO_2) and oxidizing hydrogen (H_2), a scarce commodity. Because they obtain both these gases from the air around them and they cannot tolerate oxygen, their distribution is limited. In addition to using H_2, some use formate, methanol, or acetate as a source of electrons for reducing CO_2. The characteristic overall metabolic reaction is

$$CO_2 + 4\ H_2 \rightarrow CH_4 + 2\ H_2O$$

Methanobacillus omelianski, which was thought for many years to be a methane-producing bacterium requiring ethanol, is now recognized as a symbiosis between two similarly shaped but metabolically distinct bacteria. One, known simply as "organism S," is a eubacterium: an anaerobic, Gram-negative fermenting bacterium that produces hydrogen from ethanol. It forms no CH_4 whatsoever. At a certain concentration, the H_2 produced by *M. omelianski* is toxic to the fermenting bacterium. The other bacterium, called "strain MOH," is a methanogen that combines the H_2 provided by its colleague with atmospheric CO_2 to form CH_4 in the preceding autotrophic reaction.

Methanogenic bacteria are the source of "marsh gas" and, indeed, of most natural gas. More than 90 percent of natural gas is methane. Methane is produced in swamps, estuaries, bogs, and sewage treatment plants by these archaebacteria. If methanogenic bacteria did not move organic carbon from the sediments—in the form of acetate, formate, methanol, and CO_2 derived by other organisms from sugars, starch, and other photosynthate—to the atmosphere in its gaseous form (as CH_4), the carbon produced by photosynthesis would be irretrievable; it would remain buried in the ground. Methane may be made from CO_2 and H_2 or formal-

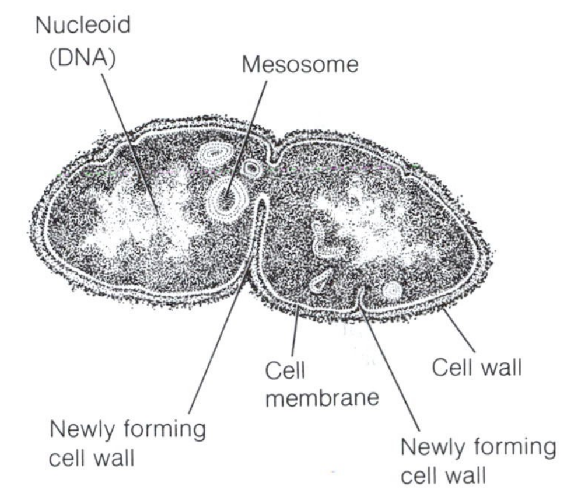

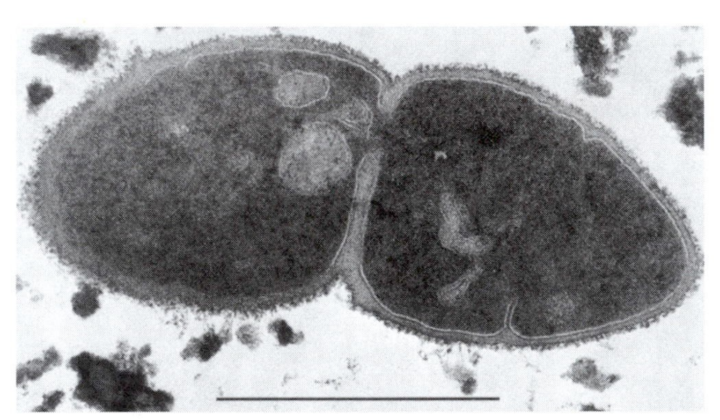

A *Methanobacterium ruminantium,* a methanogenic bacterium taken from a cow rumen. The bacterium has nearly finished dividing: a new cell wall is almost complete. Notice that a second new cell wall is beginning to form in the right-hand cell. TEM, bar = 1 μm. [Photograph courtesy of J. G. Zeikus and V. G. Bowen, "Comparative ultrastructure of methanogenic bacteria." Reproduced by permission of J. G. Zeikus and the National Research Council of Canada from the *Canadian Journal of Microbiology* 21:121–129 (1975). Drawing by I. Atema.]

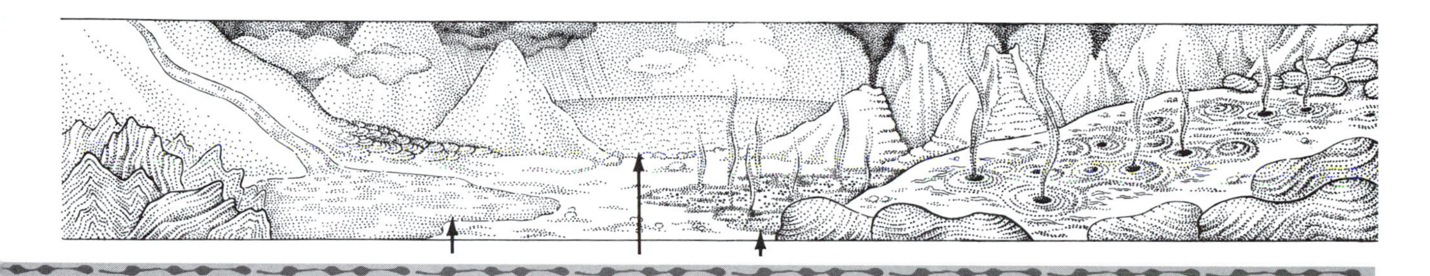

dehyde (H_2CO) (as in *Methanobacterium* or the sulfur-reducing methanogen *Methanothermus*), or it may be made from acetate (as in *Methanothrix* or *Methanosaeta*).

For every atom of carbon buried as carbohydrate or other reduced compound produced by oxygenic photosynthesis, a molecule of oxygen (O_2) has been released. Without the intervention of methanogens, the amount of buried carbon and the amount of O_2 in the atmosphere would increase. An excess of atmospheric O_2 would lead to spontaneous fires that would threaten the entire biosphere. However, most methane released by methanogens reacts with O_2 spontaneously or, in methylotrophic eubacteria (Phylum B-3), to form cell material or CO_2. Balance is thereby maintained.

Methanogenic bacteria produce some 2 billion tons of methane per year—a quantity equivalent to several percent of the total annual production of photosynthesis on the entire Earth. Thus carbon is spared from burial and returned to the atmosphere, where it can be used again by plants and other life forms. All methane in Earth's atmosphere, amounting to more than one part per million, is produced by methanogenic bacteria. Although some bacteria emit methane as an end product of carbohydrate fermentation, no eukaryotes or prokaryotes other than methanogenic bacteria are capable of formation of methane from CO_2 and H_2. A good part of the world's methanogenesis, perhaps some 30 percent, comes from animal fermentation tanks on four legs that we recognize as cows, elephants, and other mammals that have a diet high in cellulose (Figure A). Wood-eating termites, roaches, beetles, and other cellulose eaters also harbor methanogens. The rumen (a specialized part of the cow stomach) could not function, just as our sewage treatment plants could not function, without methanogenesis.

Enzymes connected in series, including the autofluorescing flavin derivative coenzyme F_{420} (so named because it absorbs light at 420 nm), are essential for generating methane and therefore are always present in cells that are actively methanogenic. When ultraviolet light is shone on these often ordinary-looking bacteria, they fluoresce blue green, betraying their methanogenic identity. Another coenzyme, which is unique to methanogenic bacteria and catalyzes the last step of the pathway before methane is vented to the air, is a yellow tetrapyrrole (like chlorophyll and heme) that (unlike iron-containing heme) contains a nickel atom in its carbon ring system. Called coenzyme F_{430}, this compound does not fluoresce. The presence of additional unique coenzymes, such as methanofuran, tetrahydromethanopterin, and coenzyme M, further supports the idea that these anaerobes are unique.

Halophils are aerobes that respire oxygen, unlike oxygen-intolerant methanogens. The salt-requiring bacteria are incapable of methane production, but, like methanogens, they are adapted to extreme environmental conditions and none form spores. Halophils abide in a variety of newly discovered hypersaline environments, taking in as food a variety of carbon compounds. Some examples include rod-shaped bacteria such as *Halobacterium salinarium,* which lives on salted fish; *H. sodomense,* which requires high concentrations of magnesium; and the salt-tolerant sugar eater *H. saccharovorum.* Flattened disc- or cup-shaped cells are members of the genus *Haloferax,* whereas triangular and rectangular halophils are usually assigned to the genus *Haloarcula.* Most halophils have obligate aerobic metabolism and are motile.

Halophils live in environments of high ionic strength (Figure B). Notably, most studied halophils abide in saturated solutions of sodium chloride in salt works and brine all over the world. They produce bright pink carotenoids and can even be spotted from airplanes and orbiting satellites as pink scum on salt flats. Many of their proteins have modifications that allow them to function only at high salt concentrations. Their cell walls are quite different from those of other bacteria in that they lack derivatives of diaminopimelic and muramic acids. Ordinary lipoprotein membranes burst or fall apart at high salt concentrations, but the halobacters' special lipids include derivatives of glycerol diether that stabilize the membranes under high salt concentrations.

Natronobacterium (including *N. gregori, N. magadii,* and *N. pharaonis*) has been found in extremely alkaline "soda lakes." All are rod-shaped organisms growing best at pH 9.5. The related coccus, *Natronococcus occultus,* also has been isolated from a highly alkaline soda lake.

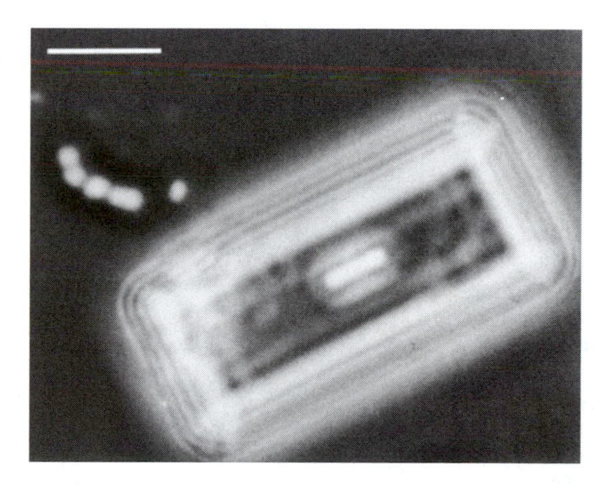

B Halophilic bacteria in saturated salt solution. A string of five spherical bacteria (*Halococcus* sp.) are shown near a salt (sodium chloride) crystal. A rod-shaped bacterium (probably *Halobacter* sp.) is on the surface of the crystal. These salt-loving archaebacteria are tiny; the fuzzy rings around the three-dimensional salt crystal are due to the microscopic imaging. LM, bar = 5 μm.

Division: Mendosicutes

B-2 Crenarchaeota (Thermoacidophils)

Greek *krene,* spring; *archae,* old; *-otes,* quality

EXAMPLES OF GENERA

Acidianus
Desulfurococcus
Pyrodictium
Pyrolobus
Sulfolobus
Thermococcus
Thermofilum
Thermoplasma
Thermoproteus

The archaebacteria in Crenarchaeota are encountered in sulfurous hot springs all over the world. They inhabit the geothermal sources of Iceland, the geysers of Yellowstone National Park, submarine volcanic eruption fluids, and other habitats with conditions far too hot, too acidic, and too sulfur rich and oxygen poor for the far more familiar eubacteria. Thermoacidophils studied, such as *Thermoproteus* and *Sulfolobus,* have strong, acid-resistant cell walls composed of a glycoprotein material arranged in a hexagonal subunit pattern. *Thermoplasma,* like the mycoplasmas of Phylum B-9, to which it is not related, lack walls entirely.

Cells of the genus *Sulfolobus,* which thrive in waters at 90°C and pH values of 1 to 2 (the acidity of concentrated sulfuric acid), were first isolated in culture in 1972. Some die of freezing at temperatures below 55°C. Growing in environments ranging in pH from 0.9 to 5.8, preferring acid waters at a pH ranging from 2 to 3, *Sulfolobus acidocaldarius* is well named (Figure A). These archaebacteria have cell walls that lack peptidoglycan, and they are facultatively autotrophic. They use elemental sulfur as their energy source and fix CO_2, but they may also use glutamate, yeast extract, ribose, and other organic compounds. *Sulfolobus* generally is an aerobe or a microaerophil (colonizer of low-oxygen habitats) that oxidizes organic matter. Some strains can aerobically reduce iron (Fe^{3+} to Fe^{2+}). Live *Sulfolobus* cells can be seen tightly adhering to the surface of elemental sulfur crystals when they are viewed by fluorescence microscopy.

Thermoplasma is a genus with one well-known species—*Thermoplasma acidophilum*—and about seven newly isolated strains (Figure B). The new *Thermoplasma* isolates are under investigation

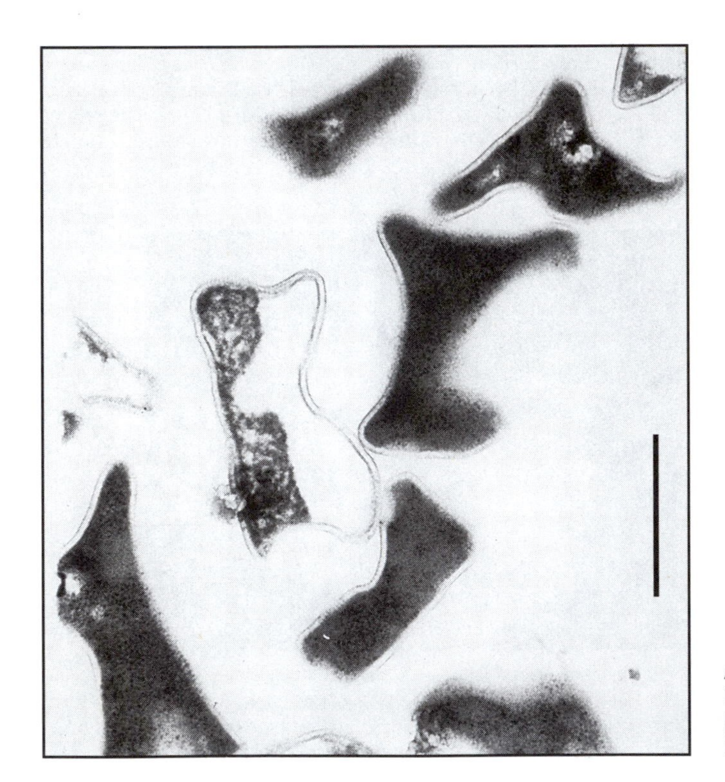

A *Sulfolobus acidocaldarius,* although pleiomorphic like *Thermoplasma,* has well-bounded cells. TEM (negative stain), bar = 1 μm. [Photograph courtesy of D. W. Grogan.]

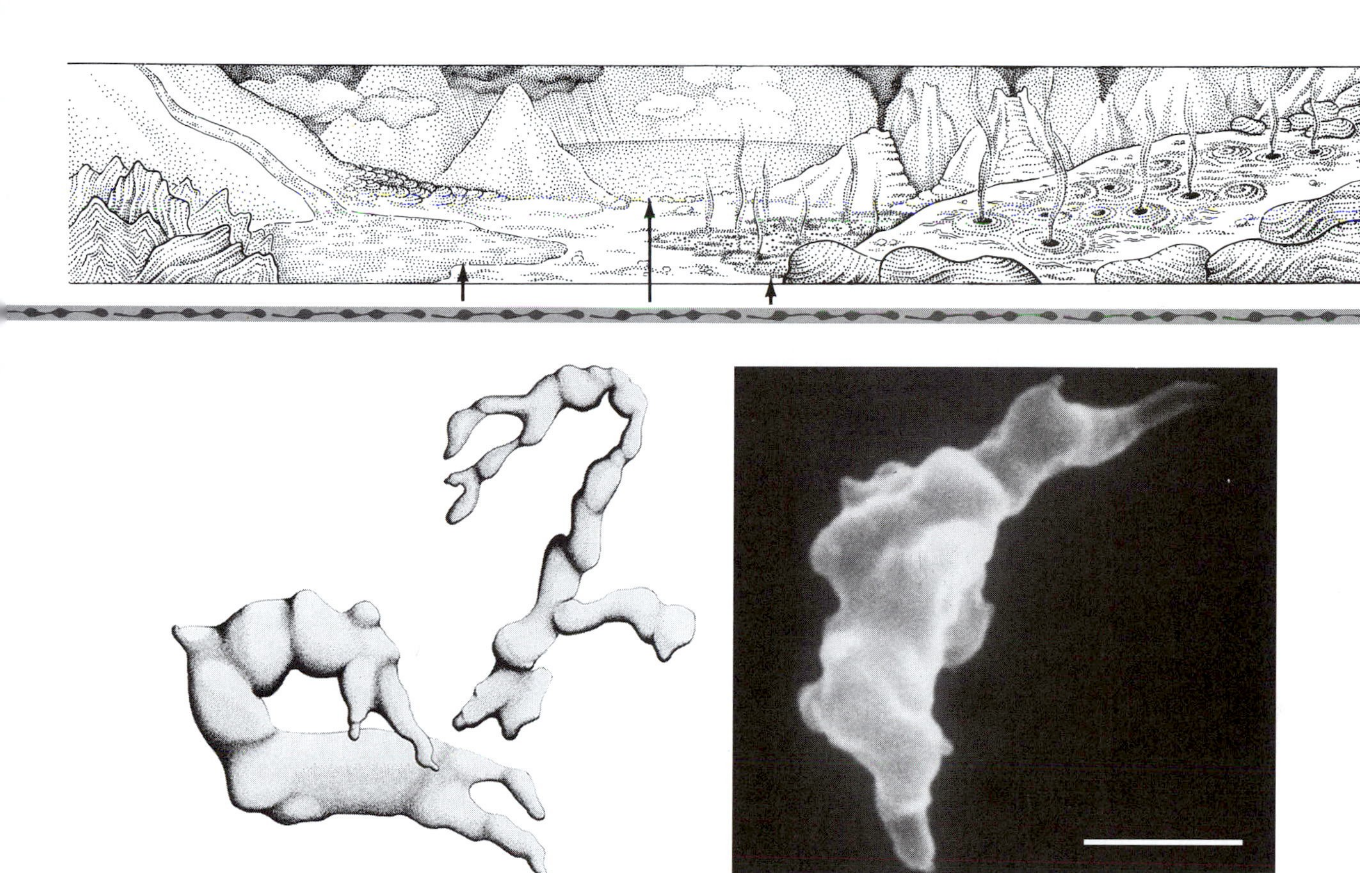

B *Thermoplasma acidophilum* from a culture at high temperature, less than 50 percent oxygen, and low pH. Scanning electron microscopy reveals a great variety of morphologies in a single culture of *Thermoplasma.* When these same organisms are grown with particles of elemental sulfur, they flatten and adhere. SEM, bar = 0.5 μm. [Photograph courtesy of D. G. Searcy; drawing, based on serial photographs, by C. Lyons.]

in Germany and Japan. The best-studied *T. acidophilum* come from hot coal refuse piles (waste tailings from coal mines) or from hot springs in Yellowstone National Park. First described in the 1970s, thermoplasmas are ecologically distinctive because of the extremely hot and acidic conditions under which they live, thriving at nearly 60°C and pH values of from 1 to 2. Having no competition under such conditions—because no other organisms tolerate the hot acid so dangerous to their DNA and proteins—thermoplasmas can easily be grown in pure culture (that is, the extreme conditions favored by thermoplasmas exclude potential contaminating bacteria from the thermoplasma laboratory culture). However, observation of these live cells is very difficult because, at 37°C (human body temperature) or cooler and at pH 3 or greater, these thermoplasmas die. (Microscopes are ill equipped to maintain samples at high temperatures.)

Thermoplasmas are the only prokaryotes known to contain DNA coated with basic proteins similar to histones, the chromosomal proteins of most eukaryotes. The protein coating is believed to protect their DNA from destruction in hot acid. *Methanothermus,* a methanogen, also has this protein coating on its DNA. *Thermoplasma* may be related to the ancestor of the nucleocytoplasm of eukaryotes.

In recent years, submarine vents have yielded new genera of thermoacidophils such as *Pyrodictium,* a strict anaerobe that forms an anchorlike structure, a flagellin-like fibrous network for attachment. Growing at temperatures as high as 110°C, it rivals the bacterium growing at the hottest temperature known (113°C): *Pyrolobus.*

SUBKINGDOM
EUBACTERIA

All bacteria that are not archaebacteria are eubacteria. This latter group, comprising the vast majority of bacteria described in the literature, is exceedingly diverse in morphology and metabolism. All multicellular bacteria with complex structures or capacity for cell differentiation belong to this group, as do all bacteria that derive energy directly from sunlight and carbon from the air (that is, phototrophic bacteria). All cells, including all bacteria, must maintain an intact cell membrane at all times and the cell wall is always outside this intact cell (or plasma) membrane (which is also called the cell envelope); thus prokaryotes are grouped by bacteriologists according to their walls. Classification based on cell walls is not necessarily consistent with that based on 16S rRNA; the comprehensive and consistent classification in this book has reconciled the views and data published by as many biologists as possible in a form feasible for students to follow.

The various classifications for cell walls of eubacteria are: Gram-negative walls (Phyla B-3 through B-8), no walls (Phylum B-9), and Gram-positive walls (Phyla B-10 through B-14). The Gram-negative stain reaction correlates with the presence of an outer lipoprotein layer of the cell wall and a thin inner peptidoglycan (peptide units attached to nitrogenous sugars) layer bounded on the inside by the plasma membrane. Gram-positive walls are associated with certain wall-bound proteins.

Bergey's Manual recognizes three major groups of eubacteria on the basis of their cell walls. The archaebacterial division Mendosicutes (Latin *mendosus,* having faults; *cutis,* skin), by contrast, lack conventional peptidoglycan walls and show a great diversity in wall structure, which means that they may or may not take up the Gram stain. The three eubacterial divisions are Firmicutes, Gracilicutes, and Tenericutes.

Division Firmicutes (Latin *firmus,* strong, durable) includes all the Gram-positive eubacteria, which tend to have thick conspicuous peptidoglycan walls that lack the outer lipoprotein layer. They may be spherical, rod-shaped, filamentous, or composed of branching filaments. No Gram-positive eubacteria are phototrophic, but many genera form resistant propagules called spores.

Division Gracilicutes (Latin *gracilis,* slender, thin) includes all Gram-negative eubacteria that, even though they have an outer lipoprotein membrane, have generally thinner cell walls. Very few spore-forming eubacteria are Gram-negative, but some, such as *Sporomusa,* do exist.

Division Tenericutes (Latin *tener,* soft, tender) comprises eubacteria that lack cell walls. They correspond to Aphragmabacteria (Phylum B-9) and are genetically incapable of synthesizing precursors of peptidoglycan. On the basis of their RNA molecules and other features, biologists believe that most wall-less eubacteria evolved from members of the Firmicutes, the Gram-positive eubacteria.

Both subkingdoms Archaebacteria (Mendosicutes: Phyla B-1 and B-2) and Eubacteria (Gracilicutes: Phyla B-3 through B-8; Tenericutes: Phylum B-9; and Firmicutes: Phyla B-10 through B-14) are diagrammed on the phylogenetic tree at the beginning of this chapter.

Division: Gracilicutes

B-3 Proteobacteria

(Purple bacteria)

EXAMPLES OF GENERA

Acetobacter
Actinobacillus
Aerobacter
Aeromonas
Agrobacterium
Alcaligenes (Achromobacter)
Alginobacter
Amoebobacter
Ancalomicrobium
Aquaspirillum
Archangium
Asticcacaulis
Azotobacter
Bdellovibrio
Beggiatoa
Beijerinckia
Beneckea
Bradyrhizobium
Caedibacter

(continued)

The members of this great group of eubacteria are classified by 16S rRNA data, by morphology, and by metabolism rather than by only one method of subdivision. Because these ways of grouping the proteobacteria are mutually exclusive, we have tried to summarize the basic biology of the members of the group by briefly pointing out distinguishing characteristics.

Comparisons of the nucleotide sequences of 16S rRNA of thousands of bacterial isolates have led to reorganization of this vast group of bacteria into four nameless major lineages referred to by Greek letters: α, β, γ, and δ (see the table on the facing page). These groups correspond to no consistent morphology or metabolism. Only representative genera are listed in this book. New genera are discovered frequently.

Large, slime-producing, multicellular gliding bacteria (for example, *Archangium, Myxococcus, Chondromyces,* and *Stigmatella*) traditionally grouped together as myxobacteria have been placed with other genera in the δ group of proteobacteria.

Of the metabolic variations displayed in this enormous and extremely diverse phylum, respiration is aerobic in the presence of oxygen for many genera, reducing O_2 to H_2O. In the absence of oxygen, they do not stop growing, as obligate aerobes must; rather, as facultative aerobes, they continue to respire, using compounds and ions such as sulfate (SO_4^{2-}), nitrate (NO_3^-), or nitrogen (N_2) as the terminal electron acceptor and reducing them to sulfide (S^{2-}) or elemental sulfur (S^0), nitrite (NO_2^-), and nitrous oxide (N_2O), respectively. Cytochrome electron-transport pathways are used in these reductions that are called respiration. The pathways used with oxygen are the same as those used with NO_3^-. In facultatively aerobic species, two respiration products can be excreted by the same eubacterium, depending on physiological and ecological conditions. Many members of the phylum are chemoheterotrophic; that is, they require reduced organic compounds both for energy and for growth. But at least two genera of oxygen-respiring proteobacteria, *Bdellovibrio* and *Daptobacter,* are predaceous. They attack and live off other members, *Chromatium,* by reproducing in the cytoplasm of their prey.

Morphologically, these organisms range from solitary, simple unicells, on the one hand, to several classes of complex morphological types, such as stalked, budding, and aggregated bacteria, on the other.

Enterics, Gram-negative eubacteria that inhabit intestines, have long been associated with human, plant, and other animal diseases; many have been isolated from intestinal tissue or from diseased plants. The enterics include many rod-shaped microbes (Figure A), most of which have flagella distributed all around the cell (peritrichous): *Escherichia, Edwardsiella, Citrobacter, Salmonella, Shigella, Klebsiella, Enterobacter, Serratia, Proteus, Yersinia (Pasteurella),* and *Erwinia.*

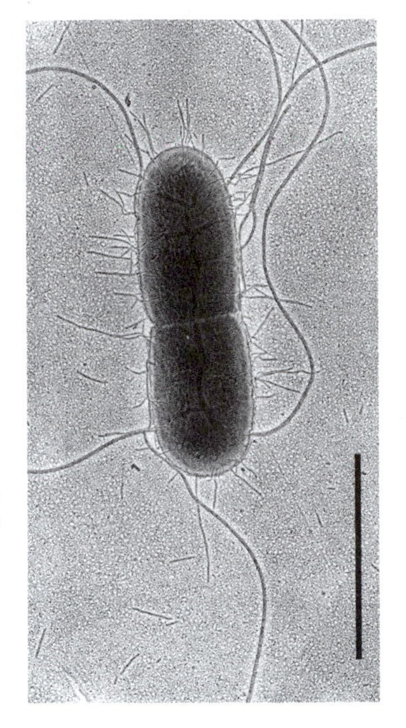

A Peritrichously (uniformly distributed) flagellated *Escherichia coli*. A new cell wall has formed and the bacterium is about to divide. The smaller appendages, called pili, are known to make contact with other cells in bacterial conjugation. However, even many strains that do not conjugate have pili. TEM (shadowed with platinum), bar = 1 μm. [Courtesy of D. Chase.]

The enterics are distinguished from one another by the carbohydrates that they use (lactose, glutamic acid, arabinose, sugar, alcohols, citrate, tartrate, and other fairly small organic compounds) and by chemical abilities. Some hydrolyze urea, produce gas from glucose, or break down gelatin. They are also distinguished by their sensitivity to specific bacteriophages, by their surface antigens, by their attraction to certain hosts, by their pathogenicity, and by morphological traits, such as mode of motility and distribution of flagella.

These Gram-negative rod-shaped unicellular heterotrophs grow rapidly and well. Although none produce spores, they seem to have remarkable persistence, as do some vibrios, waiting things out under conditions of adversity and vigorously taking advantage of new food sources. Water samples everywhere yield enterics when incubated under the proper conditions of growth. It is not too extreme to assert that most life on Earth takes the form of facultatively aerobic Gram-negative unicellular rod-shaped bacteria. At the base of microbial food webs, they provide food for innumerable protoctists and other organisms. As the chief object of study by

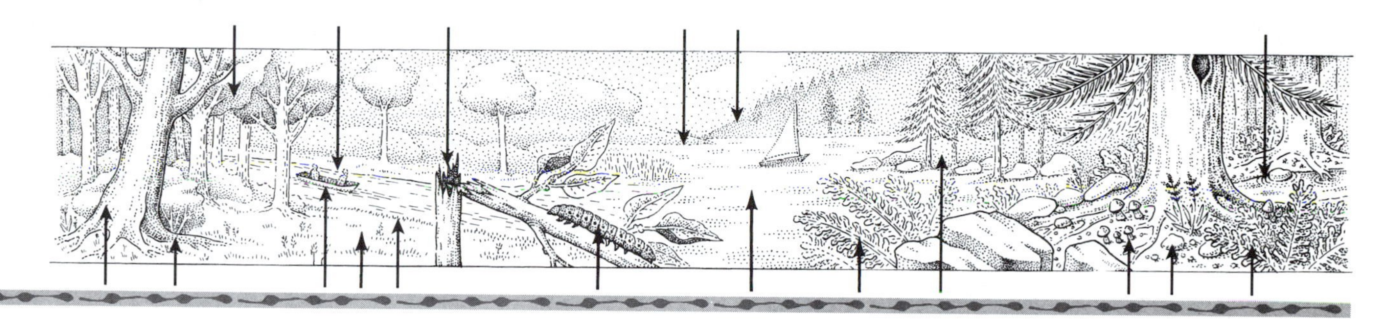

Best-Known Proteobacteria

Group	Genus	Phenotypes: Morphology	Phenotypes: Metabolism
Alpha (α) group	*Agrobacterium*	Motile rod*	Crown gall former, heterotroph
	Aquaspirillum	Motile spirillum-shaped bacterium	Heterotroph
	Beijerinckia	Motile rod	Nitrogen fixer
	Hyphomicrobium	Budding bacterium	Heterotroph
	Nitrobacter	Short budding rod	Ammonia oxidizer, nitrate oxidizer (marine)
	Paracoccus	Motile rod	Heterotroph
	Pseudomonas†	Motile rod, polar flagella	Heterotroph
	Rhizobium	Symbiotic with root nodules	Nitrogen fixer
	Rhodobacter	Flagellated rod	Purple nonsulfur phototroph
	Rhodomicrobium	Budding bacterium	Purple nonsulfur phototroph
	Rhodopseudomonas	Flagellated rod	Purple nonsulfur phototroph
	Rhodospirillum	Spirillum-shaped bacterium	Purple nonsulfur phototroph
Beta (β) group	*Alcaligenes*	Motile rod	Hydrogen oxidizer
	Neisseria	Nonmotile coccus	Heterotroph
	Nitrosomonas	Rod	Ammonia oxidizer, nitrite oxidizer
	Pseudomonas†	Motile rod, polar flagella	Heterotroph
	Sphaerotilus	Filament	Iron oxidizer
	Spirillum	Motile spirillum-shaped bacterium	Heterotroph
	Thiobacillus	Motile rod	Sulfide oxidizer
Gamma (γ) group	*Azotobacter*	Rod	Nitrogen fixer
	Beggiatoa	Filaments form rosettes	Sulfide oxidizer
	Chromatium	Motile rod, sulfur globules in cells	Purple sulfur phototroph
	Escherichia	Motile rod, peritrichous flagella	Heterotroph
	Legionella	Motile rod	Heterotroph
	Leucothrix	Filament	Sulfide oxidizer
	Pseudomonas† (fluorescent)	Motile rod, polar flagella	Heterotroph
	Salmonella	Motile rod	Heterotroph
	Shigella	Motile rod	Heterotroph
	Thiocapsa	Gelatinous colonies of cocci	Purple sulfur phototroph
	Thiospirillum	Motile spirillum-shaped bacterium	Purple sulfur phototroph
	Vibrio	Motile comma-shaped bacterium	Heterotroph
Delta (δ) group	*Archangium*	Myxospores stalkless, rod cell ends tapered	Aerobic heterotroph
	Bdellovibrio	Motile, tiny curved rod	Predator
	Chondromyces	Single branched myxospore-bearing stalks	Aerobic heterotroph
	Desulfotomaculatum	Spore-forming rod	Sulfate reducer
	Desulfovibrio	Motile rod, polar flagella	Sulfate reducer
	Desulfuromonas	Small rod	Sulfate reducer, obligate anaerobe, obligate sulfur reducer
	Myxococcus	Glider	Heterotroph
	Polyangium	Multiple myxospore-bearing stalks	Aerobic heterotroph
	Stigmatella	Myxospores stalkless, rod cell ends rounded	Aerobic heterotroph

* Motility may be by flagella (swimming) or by unknown means, slow movement in contact with a surface (gliding).

† Pseudomonads are aerobic, straight or slightly curved rod-shaped cells that swim by polar flagella. They are incapable of fermentation but oxidize a wide range of organic substances. Analysis of their 16S rRNA genes shows that any given strain may belong to any one of the three groups of Proteobacteria: α, β, and γ.

B-3 Proteobacteria
(continued)

Calymmatobacterium
Cardiobacterium
Caulobacter
Chloropseudomonas
Chondromyces
Chromatium
Chromobacterium
Citrobacter
Coxiella
Daptobacter
Desulfacinum
Desulfonema
Desulfotomaculum
Desulfovibrio
Desulfuromonas
Edwardsiella
Enterobacter
Erwinia
Escherichia
Ferrobacillus
Flavobacterium
Gluconobacter
Haemophilus
Hyphomicrobium

molecular biologists, *Escherichia coli* is better known than any other single organism on Earth.

A second group of enterics comprises mainly comma-shaped organisms, most of which have single polar flagella. Bacteria of this shape, called vibrios, include at least seven genera: *Vibrio, Beneckea, Aeromonas, Pleisiomonas, Photobacterium, Xenorhabditis,* and *Zymomonas.* Members of the genus *Vibrio* are associated with cholera. They ferment carbohydrates into mixed products, including acids, but do not give off carbon dioxide and hydrogen. *Beneckea,* a marine vibrio that requires salt, is capable of fermentative or aerobic respiring growth on a broad range of carbon sources. Some vibrios, *Photobacterium* and *Xenorhabditis,* are bioluminescent. Photobacteria associate with certain tropical, marine fish, which culture the bacteria in special pockets called light organs. The growth requirements of *Photobacterium* are generally far more restricted than those of the free-living marine genus *Beneckea. Xenorhabditis* is known from its associations with nematodes and nematode-eating insects.

Members of the genus *Aeromonas,* informally called aeromonads, are common in ponds, lakes, and soils. They are coccoids or straight rods with rounded ends; most are motile by means of polar flagella. When growing anaerobically, they reduce nitrate to nitrite. Most of them contain cytochrome *c,* an electron-transporting protein whose activity is due to its inclusion of a porphyrin molecule, and catalase, an enzyme that decomposes hydrogen peroxide into water and oxygen. Very eclectic in their tastes, they use a wide variety of food sources, especially plant materials such as starch, casein, gelatin, dextrin, glucose, fructose, and maltose.

Similar unicellular Gram-negative bacteria include *Chromobacterium, Haemophilus, Actinobacillus, Cardiobacterium, Streptobacillus, Calymmatobacterium,* and several symbionts of ciliates in Phylum Pr-8. Certain paramecium symbionts, originally called kappa, lambda, sigma, and mu particles, were revealed to be proteobacteria with complex requirements for growth. Once thought to be cytoplasmic genes of the ciliate itself, they are now known to be members of the genus *Caedibacter.*

Many enterics produce colorful pigments: the violet, ethanol-soluble violacein of *Chromobacterium;* the red prodigiosin of *Serratia* and *Beneckea;* and the red yellow, orange, and brown pigments, some of which are carotenoids and some not, of *Flavobacterium* and *Aeromonas.* No one yet understands the functions of these pigments.

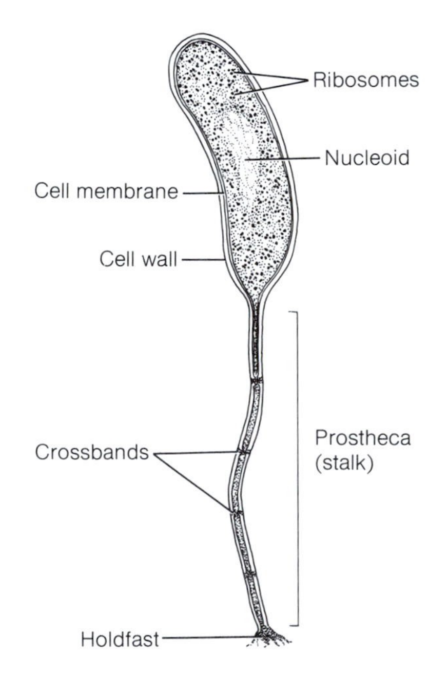

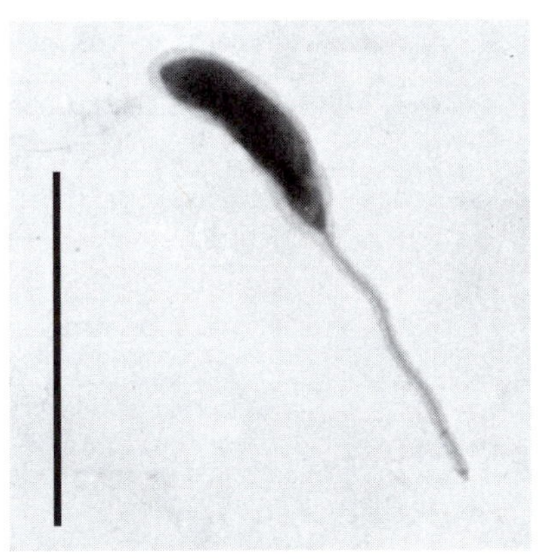

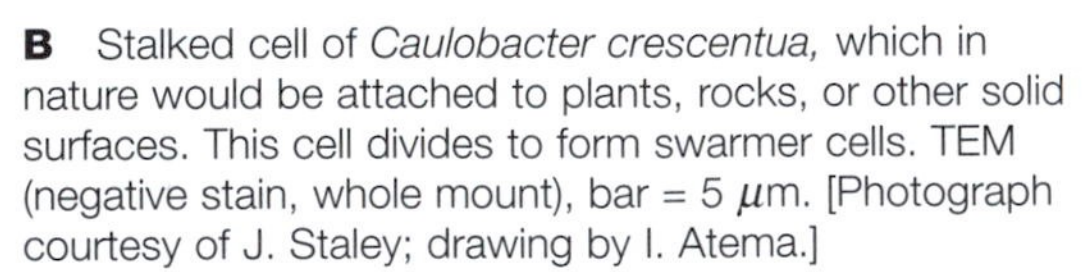
B Stalked cell of *Caulobacter crescentua,* which in nature would be attached to plants, rocks, or other solid surfaces. This cell divides to form swarmer cells. TEM (negative stain, whole mount), bar = 5 μm. [Photograph courtesy of J. Staley; drawing by I. Atema.]

Klebsiella	*Neisseria*	*Nitrosomonas*	*Proteus*	*Rickettsia*	*Stigmatella*
Legionella	*Neorickettsia*	*Nitrosospira*	*Pseudomonas*	*Rickettsiella*	*Streptobacillus*
Leucothrix	*Nitrobacter*	*Nitrospira*	*Rhizobium*	*Rubrivivax*	*Thermodesulfor-habditis*
Macromonas	*Nitrococcus*	*Oceanospirillum*	*Rhodobacter*	*Salmonella*	*Thiobacillus*
Methylococcus	*Nitrocystis*	*Paracoccus*	*Rhodoferax*	*Serratia*	*Thiobacterium*
Methylomonas	*Nitrosococcus*	*Photobacterium*	*Rhodomicrobium*	*Shigella*	*Thiocapsa*
Methylosinus	*Nitrosogloea*	*Pleisiomonas*	*Rhodopseudomonas*	*Sphaerotilus*	
Myxococcus	*Nitrosolobus*	*Polyangium*	*Rhodospirillum*	*Spirillum*	

(continued)

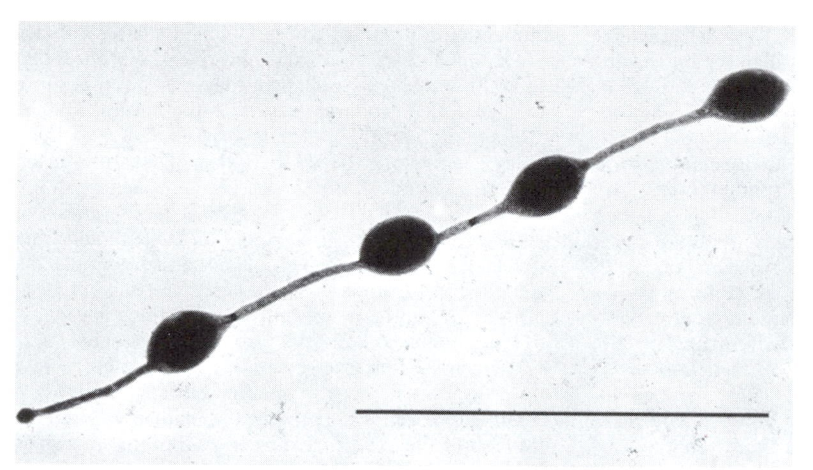

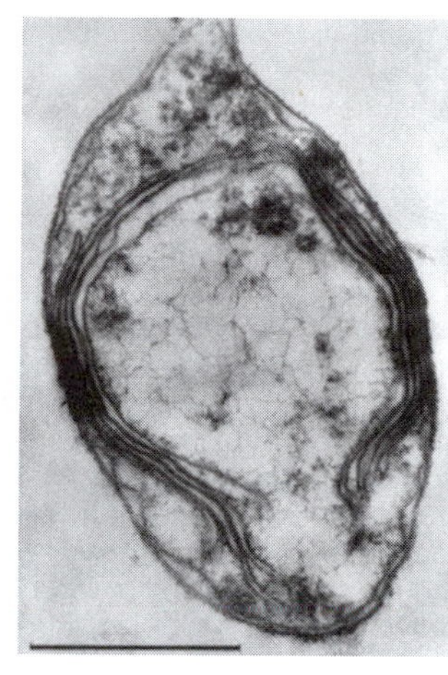

C *Rhodomicrobium vannielii,* a phototrophic purple nonsulfur bacterium that lives in ponds and grows by budding. (Left) A new bud is forming at lower left. TEM, bar = 1 μm. (Right) Layers of thylakoids (photosynthetic membranes) are visible around the periphery of this *R. vannielii* cell. TEM, bar = 0.5 μm. [Courtesy of E. Boatman.]

Other proteobacteria are extremely heterogeneous. In one group, the prosthecate bacteria, appendages called prosthecae (stalks) made of living material, protrude from the cells. *Caulobacter* and *Asticcacaulis* have single polar or subpolar prosthecae (Figure B). Their life histories superficially resemble those of some marine animals: a stalked sessile form divides to produce a motile form, which swims away; its offspring, in turn, is a sessile form.

The budding bacteria, or hyphomicrobia, reproduce by the outgrowth of buds that eventually swell to parent size. Hyphomicrobial colonies may form quite complex networks that resemble the mycelia of fungi. This sort of budding is found both in photosynthetic microbes, such as *Rhodomicrobium* (Figures C and D), and in heterotrophs, such as *Hyphomicrobium*.

The aggregated bacteria, which include *Sphaerotilus*, can be recognized in the light microscope by their distinctive metal-rich clumps of cells. These microbes oxidize iron or manganese, which they then deposit around themselves as manganese and iron oxides. The bacteria are thought to gain some energy from these oxidations; however, there is still no proof that any of them are obligate lithotrophs. All prefer fixed nitrogen and carbon compounds and grow faster if supplied with organic food.

Nonflagellated bacteria of the genus *Neisseria* are infamous as the cause of gonorrhea and one form of bacterial meningitis. They can grow on their own.

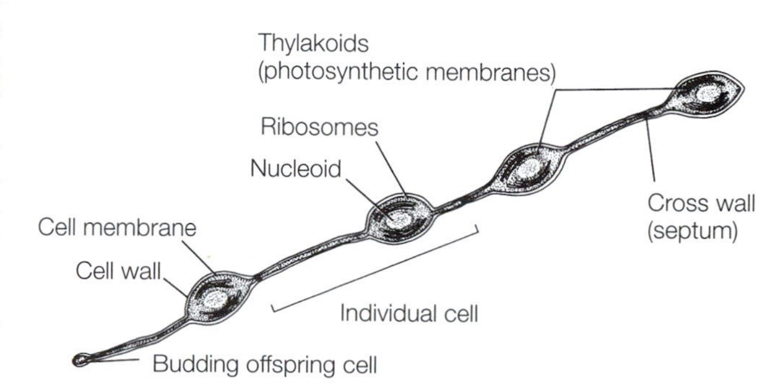

D *Rhodomicrobium vannielii*. [Drawing by I. Atema.]

B-3 Proteobacteria
(continued)

Thiocystis
Thiodictyon
Thiomicrospira
Thiopedia
Thiosarcina
Thiospira
Thiospirillum
Thiothece
Thiovulum
Vibrio
Xenorhabditis
Yersinia (Pasteurella)
Zymomonas

The acetic acid bacteria, *Gluconobacter* and *Acetobacter*, oxidize ethanol to acetic acid—wine to vinegar. They form sheaths having a rectangular cross section.

The rickettsias, unable ever to grow on their own, have an obligate intracellular existence in animals. They have residual Gram-negative walls, cell walls reduced in size, and their metabolism is limited. Several strains cause Rocky Mountain spotted fever.

The best-known chemolithoautotrophs belong to the α, β, or γ group of Proteobacteria. They probably evolved from phototrophic ancestors. Chemolithoautotrophy is metabolism that functions without sunlight and without preformed organic compounds—not a single vitamin, sugar, or amino acid. Thus, chemolithoautotrophic bacteria represent the pinnacle of metabolic achievement. They live on air, salts, water, and an inorganic source of energy. Provided with nitrogenous salts, oxygen, carbon dioxide (CO_2), and an appropriate reduced compound to use as an energy source, they make all of their own nucleic acids, proteins, and carbohydrates and derive their energy from oxidation of the reduced compound. Some are capable of using organic compounds as food, but all can do without them. Chemolithoautotrophic bacteria are crucial to the cycling of nitrogen, carbon, and sulfur throughout the world because they convert gases and salts unusable by animals and plants into usable organic compounds. The maintenance of the biosphere depends on such metabolic virtuosity, yet chemolithoautotrophy is strictly limited to bacteria.

Chemolithoautotrophic bacteria are grouped by the compounds that they oxidize to gain energy. At least three types are distinguished: oxidizers of nitrogen compounds, of sulfur compounds, and of CH_4.

Chemolithoautotrophs that oxidize reduced nitrogen compounds include morphologically distinct organisms that oxidize nitrite (NO_2^-) to nitrate (NO_3^-): *Nitrobacter, Nitrospira, Nitrocystis,* and *Nitrococcus*. Nitrobacters are short rods, many of them pear shaped or wedge shaped (Figure E). Elaborate internal membranes extend along the periphery of one end of the cell. Old cultures of *Nitrobacter winogradskyi,* a widely distributed soil microbe, form a flocculent sediment made of gelatinous sheaths produced by the bacteria. Organic compounds and even ammonium salts inhibit the growth of this nitrobacter. *Nitrospira* are long slender rods that lack elaborate internal membranes. They are marine bacteria, strict aerobes, and strict chemolithoautotrophs. Nitrococci are spherical cells containing distinctive internal membranes that form a branched or tubular network in the cytoplasm.

The other group of nitrogen-compound oxidizers contains the chemolithoautotrophs that oxidize ammonia (NH_3) to nitrite (NO_2^-) for energy: *Nitrosomonas, Nitrosospira, Nitrosococcus,* and *Nitrosolobus*. They live in environments containing both oxygen and ammonia, such as at the edges of the anaerobic zone at the sedimentation interface where the solid surface contacts seawater or fresh water of soil or of lakes and rivers. *Nitrosomonas* species are either ellipsoidal or rod shaped; they may be single, in pairs, or in short chains. They are rich in cytochromes, which impart a yellowish or reddish hue to laboratory cultures. Internal membranes extend along the cell periphery. They grow at temperatures between 5°C and 30°C. *Nitrosospira* is a genus of spiral-shaped freshwater microbes that lack internal membranes. *Nitrosococcus* cells are spheres; they grow singly or in pairs and often form an extracellular slime. Aggregates of cells attach to surfaces or become suspended in liquid. *Nitrosolobus* cells are variously shaped, lobed cells that are motile by means of peritrichous flagella. They divide by binary fission (cell division producing two equal offspring cells).

There are at least five genera of organisms currently recognized that grow by oxidizing inorganic sulfur compounds. Their cells contain sulfur globules (products of oxidizing sulfide to elemental sulfur), and they live in high concentrations of hydrogen sulfide or other oxidizable sulfur compounds. The genera are of four distinct morphological types: nonmotile rods embedded in a gelatinous matrix (*Thiobacterium*), cylindrical cells having polar flagella (*Macromonas*), ovoid cells having peritrichous flagella (*Thiovulum*), and spiral cells having polar flagella (*Thiospira* and *Thiobacillus*).

Thiobacillus is the best-known genus of sulfur oxidizers and has been grown in culture. Most are Gram-negative rods and are motile by means of a single polar flagellum. They derive energy from the oxidation of sulfur or its compounds, such as sulfide (S^{2-}), thiosulfate ($S_2O_3^{2-}$), polythionate, and sulfite (SO_3^{2-}). The final oxidation product is sulfate (SO_4^{2-}), but other sulfur compounds may accumulate under certain conditions. One species, *Thiobacillus ferrooxidans,* also oxidizes ferrous compounds. Thiobacilli will grow on strictly inorganic media, fixing CO_2 to produce cell material.

Proteobacteria that oxidize the reduced single-carbon compounds CH_4 or methanol (CH_3OH) are methylomonads. Two genera may be distinguished by morphology: *Methylomonas* (various Gram-negative rods) and *Methylococcus* (spherical cells usually appearing in pairs). Methylomonads cannot grow on complex organic compounds; rather, they use CH_4 or CH_3OH as their sole source of both energy and carbon. In fact, the growth of many is inhibited by the presence of organic matter.

Besides these chemolithoautotrophs are organic carbon–requiring chemotrophs that respire sulfur compounds. The sulfate reducers of the δ group of proteobacteria are obligate anaerobes quickly poisoned by exposure to oxygen. They require sulfate (SO_4^{2-}), just as we require oxygen, for respiration. In this energy-yielding process, electrons from food molecules are transferred to inorganic compounds; by this transfer, the food molecules are oxidized and the inorganic compounds are reduced in oxidation state.

E *Nitrobacter winogradskyi*. This specimen is young and thus lacks a prominent sheath. Carboxysomes are bodies in which are concentrated enzymes for fixing atmospheric CO_2. This species is named for the Russian Sergius Winogradsky, who pioneered the field of microbial ecology. TEM, bar = 0.5 μm. [Photograph courtesy of S. W. Watson, *International Journal of Systematic Bacteriology* 21:261 (1971); drawing by I. Atema.]

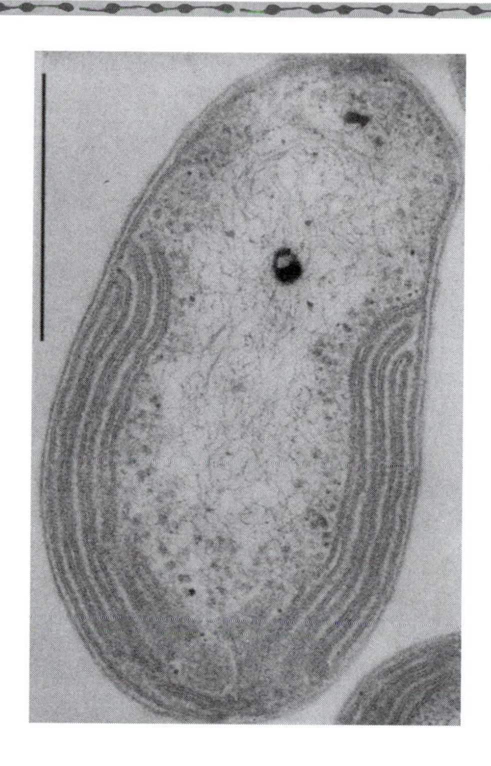

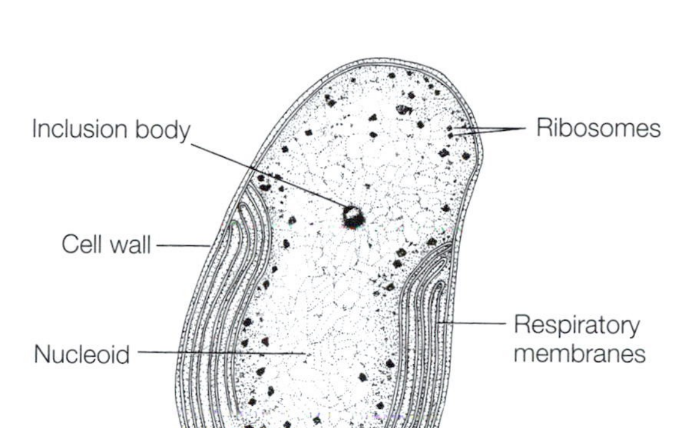

All reduce SO_4^{2-} to some other sulfur compound, such as elemental sulfur or hydrogen sulfide (H_2S), and synthesize cytochromes. Although these chemotrophs obtain energy from SO_4^{2-}, they also take in organic compounds, usually the three-carbon compound lactate or pyruvate, as a source of carbon, electrons, and energy. Thus, they are not autotrophs.

More than 18 genera of sulfate-reducing bacteria are known. Some, such as *Desulfovibrio* and the motile rod *Desulfotomaculum,* cannot oxidize acetate but can use other carbon sources for energy, such as lactate or pyruvate. Others, strict anaerobes, oxidize acetate to CO_2 by using this carbon compound as an electron source. Genera of acetate-using sulfate reducers include the large filament *Desulfonema* and the thermophils *Desulfacinum* and *Thermodesulforhabdus.*

The δ group of sulfate reducers—members of the best-known genus, *Desulfovibrio*—are unicellular bacteria widely distributed in marine muds, estuarine brines, and freshwater muds. Either single polar flagella or bundles of them (lophotrichous flagella) provide motility. The genera that require sodium chloride for growth are considered marine bacteria. Sulfate reducers contain cytochrome c_3 and a pigment called desulfoviridin, which gives them a characteristic red fluorescence. Many also synthesize hydrogenases, enzymes that generate hydrogen, which protects the organisms from the hostile aerobic world.

Sulfate reducers release gaseous sulfur compounds, including H_2S, into the sediments, thus playing a crucial role in cycling sulfur—a constituent of all proteins—throughout the world. In iron-rich water, the H_2S formed by these bacteria reacts with iron, leading to the deposition of pyrite (iron sulfide, also known as fool's gold). It is thought that Archean and Proterozoic iron deposits may be due, at least partly, to the activity of sulfate-reducing bacteria. No symbiotic forms of sulfate-reducing bacteria have been reported; the group is free living.

Desulfotomaculum is a genus of unicellular, straight or curved rod bacteria that are motile by means of peritrichous flagella. The genus is distinguished by the formation of resistant endospores and is commonly found in marine and freshwater muds, in the soil of geothermal regions, in the intestines of insects and bovine animals, and in certain spoiled foods.

Also in the δ group of proteobacteria are large, heterotrophic, multicellular bacteria: the myxobacteria. The myxobacteria represent, with the cyanobacteria (Phylum B-5), the acme of morphological complexity among the prokaryotes that form upright, propagule-dispersing, multicellular bodies. Individual myxobacteria are obligately aerobic, unicellular, Gram-negative rods that may be as long as 5 μm. Some aggregate into complex colonies that show distinctive behavior and form. The cells are typically embedded in slime consisting of polysaccharides of the cells' own making.

When soil nutrients or water are depleted, members of certain genera of myxobacteria (for example, *Stigmatella* and *Chondromyces*) aggregate and form upright structures composed of extracellular excretion products and many cells (Figures F and G). Bacterial cells within these reproductive bodies enter a resting stage; the resting cells are called myxospores. In some taxa, these resting cells may become encapsulated, thick walled, and shiny; in others, they seem to be quite like growing bacteria. Some, such as *Polyangium violaceum,* form brightly colored reproductive structures. Others form branched stalks; these tiny "trees" may be barely visible to the unaided eye.

Some myxobacteria form thick-walled, darkly colored, spore-filled cysts, called sporangioles, that open when wetted to release huge numbers of individual gliding bacteria; the gliding cells move together to form migrating colonies. The entire history of these myxobacteria is uncannily analogous to that of the slime molds (Phylum Pr-6).

Phototrophic proteobacteria are morphologically diverse. Many are single cells and motile or immotile. Some grow as packets or as stalked budding structures, extensive filaments, or sheets of cells in which the spaces between the cells are filled with coverings, called sheaths, composed of mucous material. Some contain gas vacuoles, giving them buoyancy and a sparkling appearance. In anoxic environments, most purple sulfur bacteria convert H_2S into elemental sulfur, which they deposit inside their cells in tiny but visible granules; the presence, distribution, and shape of these granules can be used to distinguish them.

The phototrophic bacteria are delightfully colored in an astonishing range of pinks and greens, although in the bright sunlight in the top layers of anaerobic muds they become very dark, nearly black. Because each species has an optimum growth at a given acidity, oxygen tension, sulfide and salt concentration, moisture content, and so on, they often grow in layers—each in its appropriate niche. Well-lit anoxic sediments become layered communities of phototrophic and other bacteria.

Although many of the major photosynthetic bacteria belong here in the proteobacteria—for example, the purple nonsulfur bacteria such as *Rhodospirillum, Rhodomicrobium, Rhodoferax,* and *Rubrivivax* are in either the α or the β lineage, and the purple sulfur bacteria such as *Chromatium* or *Amoebobacter* are genera of photosynthesizers grouped together as δ purples—many others do not.

Among the many different species of phototrophs, some are tolerant of extremely high or extremely low temperatures or salinities. In each group, some kinds are capable of fixing atmospheric nitrogen. The ability to fix atmospheric nitrogen, the conversion of N_2 into organic compounds, such as amino acids, that include nitrogen in their structure, is conspicuously present in many members of this huge phylum. This important process is entirely limited to bacteria. Most notable are the free-living aerobic (oxygen-respiring) soil nitrogen fixers, *Azotobacter* (Figure H) and *Beijerinckia* among them. Other close relatives of these free-living bacteria include some soil bacteria that can also live as plant root symbionts, such as the motile rods *Rhizobium* and *Bradyrhizobium.* All atmospheric N_2 fixers contain nitrogenase, a large enzyme complex. Nitrogenase is composed of azo- and molybdoferredoxins, iron and molybdenum-containing proteins absolutely necessary for the reduction of N_2 to organic nitrogen compounds such as glutamine. The genes for the entire process may be borne on plasmids, relatively small pieces of DNA, that transfer from one to another kind of bacteria. For this reason, organisms are not classified according to this ability.

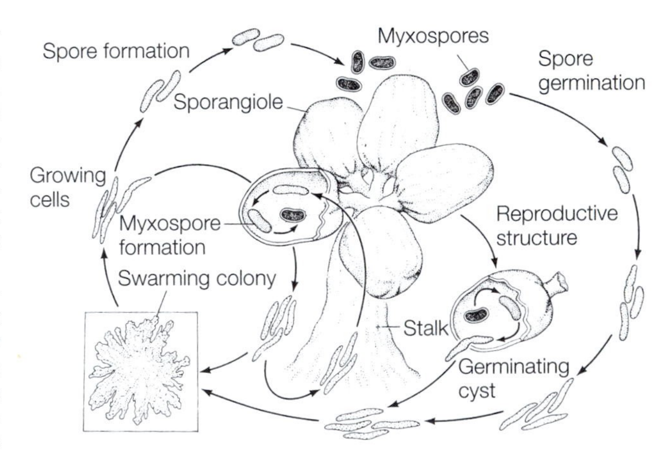

F Life cycle of *Stigmatella aurantiaca.* [Drawing by L. Meszoly; labeled by M. Dworkin.]

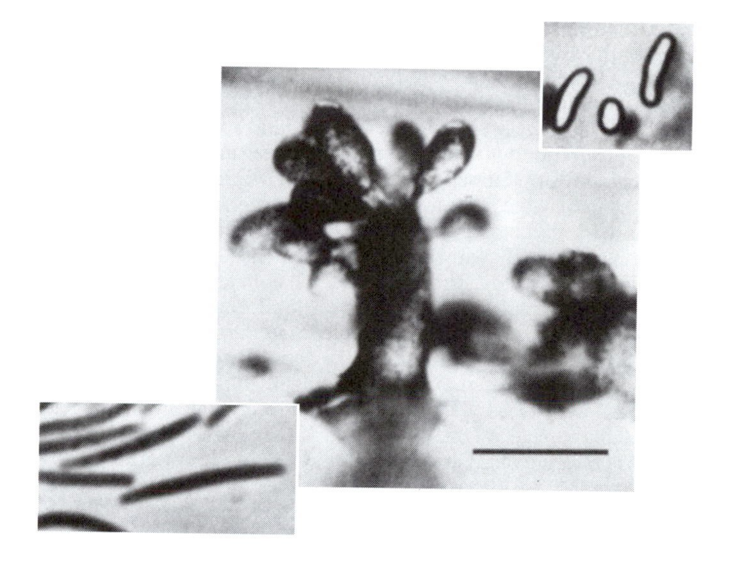

G The reproductive body of *Stigmatella aurantiaca,* which grows on remains of vegetation in soil. LM, bar = 100 μm. (Inset, bottom left) Growing cells, which glide in contact with solid surfaces. (Inset, top right) Myxospores. [Photographs courtesy of H. Reichenbach and M. Dworkin, in *The prokaryotes,* M. Starr et al., eds. (Springer-Verlag; New York; 1981).]

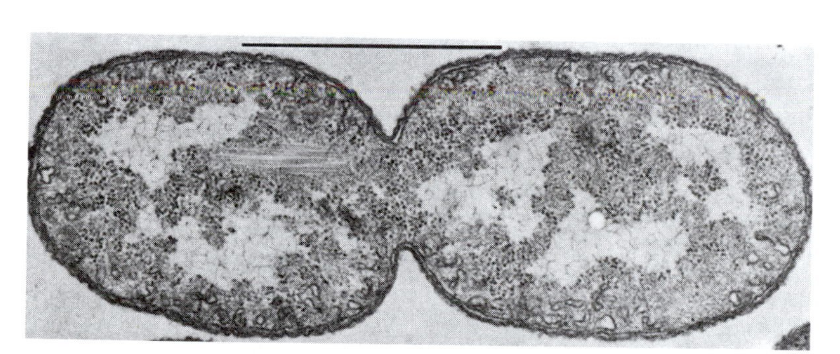

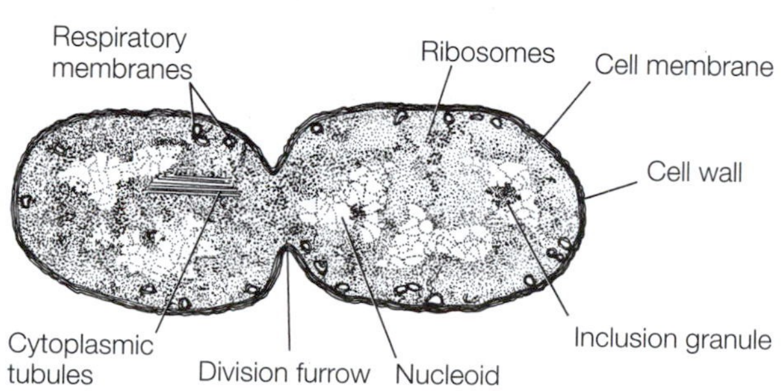

H *Azotobacter vinelandii,* commonly found in garden soils. In this photograph, division into two cells is nearly complete. TEM, bar =1 μm. [Photograph courtesy of W. J. Brill; drawing by I. Atema.]

Division: Gracilicutes

B-4 Spirochaetae

Latin *spira*, coil; Greek *khaite*, long hair

ALL GENERA
Borrelia
Clevelandina
Cristispira
Diplocalyx
Hollandina
Leptonema
Leptospira
Mobilifilum
Pillotina
Spirochaeta
Spirosymplokos
Treponema

Spirochetes look like tightly coiled snakes. Unlike other motile bacteria, they have from 2 to more than 200 internal flagella (axial filaments or endoflagella) in the space between the inner (plasma) membrane and the outer cell membrane of the Gram-negative cell wall; that is, the flagella are in the cell wall. The range of numbers is specific to the genus. Spirochetes are found in marine and fresh waters, deep muddy sediments, the gastrointestinal tracts of several different kinds of animals, and elsewhere. Their long, slender, corkscrew shape enables them to move flexibly through thick, viscous liquids with great speed and ease. In more dilute environments, many swim quickly with complex movements—rotation, torsion, flexion, and quivering. The unusual flagellar arrangement, probably responsible for their corkscrew shape and characteristic movements, distinguishes spirochetes from other bacteria. Each flagellum originates near an end of the cell and extends along the body. Thus, flagella anchored at opposite ends of the cell often overlap, like the fingers of loosely folded hands. Presumably, internal rotation of these flagella is responsible for the motility of spirochetes, just as is external flagellar rotation for other motile bacteria.

Major groups of spirochetes include leptospires (*Leptospira* and *Leptonema*), spirochaetas (*Spirochaeta*), and pillotinas (*Pillotina, Hollandina,* and *Diplocalyx*). Leptospires require gaseous oxygen, whereas most other spirochetes are quickly poisoned by the slightest trace. Certain leptospires (members of *Leptospira*) live in the kidney tubules of mammals. Often, they are carried with the urine into water supplies and can enter the human bloodstream through cuts in the skin, causing the disease leptospirosis. Because leptospires are so thin and hard to see in the microscope, the disease is often misdiagnosed.

The spirochaetas include at least three genera—*Borrelia, Treponema,* and *Spirochaeta*—some species of which are internal parasites of animals. Infamous as the cause of syphilis, *Treponema pallidum* is also responsible for yaws, a debilitating and unsightly tropical eye disease. *Treponema* have from one to four flagella at each end of the cell. The genus *Spirochaeta* (Figure A) contains free-living marine and freshwater spirochetes that are less than a micrometer wide and have a small number of overlapping flagella, resembling *Treponema.*

The pillotinas, all symbionts of animals, are much larger than other spirochetes (Figures B and C). Some may be 3 μm wide and hundreds of micrometers long. All have been found in symbiotic relations with animals. Members of the genus *Cristispira* inhabit the crystalline style of clams and oysters; the style is an organ that helps these molluscs (Phylum A-26) grind their algal food. The presence of the cristispires is contingent on environment and does not seem to influence the mollusc or its style. The other pillotina spirochetes—*Pillotina, Hollandina, Diplocalyx,* and *Clevelandina*—are found associated with other bacteria and protoctists in the

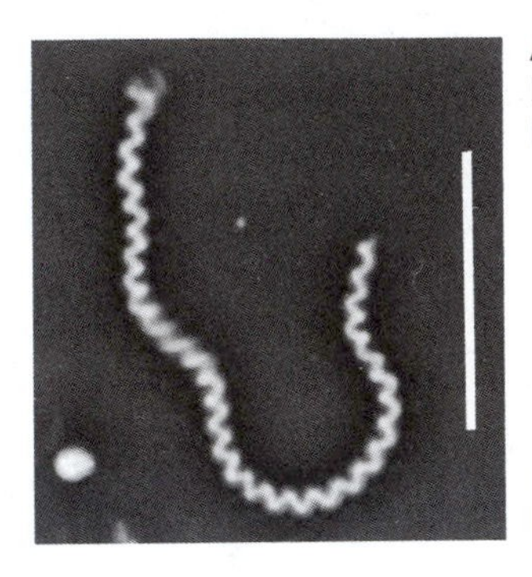

A *Spirochaeta plicatilis* from the Fens, Boston. LM, bar = 10 μm. [Courtesy of W. Ormerod.]

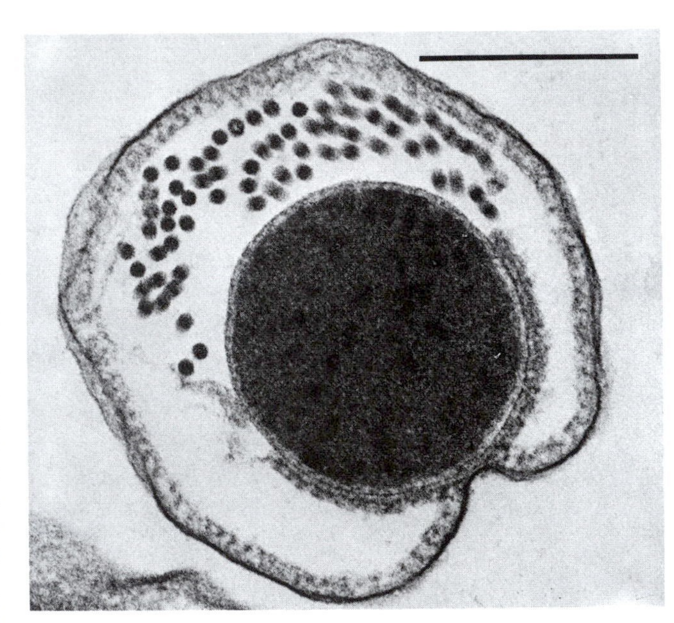

B *Diplocalyx* sp., in cross section. These large spirochetes, which belong to the family Pillotaceae (the pillotinas), have many flagella. The several genera of Pillotaceae all live in the hindguts of wood-eating cockroaches and termites. This specimen was found in the common North American subterranean termite *Reticulitermes flavipes* (Phylum A-20). TEM, bar = 1 μm. [Courtesy of H. S. Pankratz and J. Breznak.]

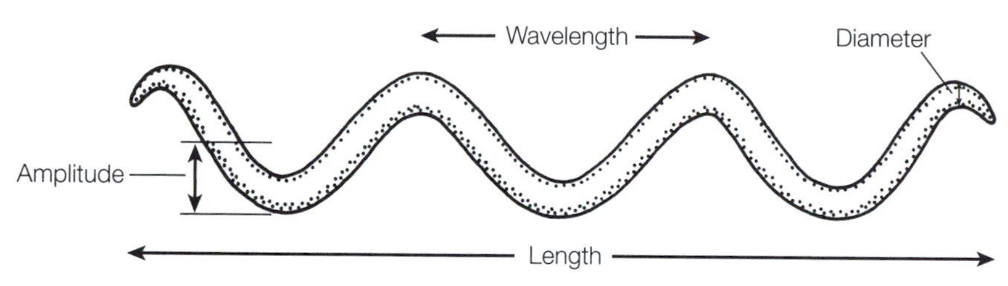

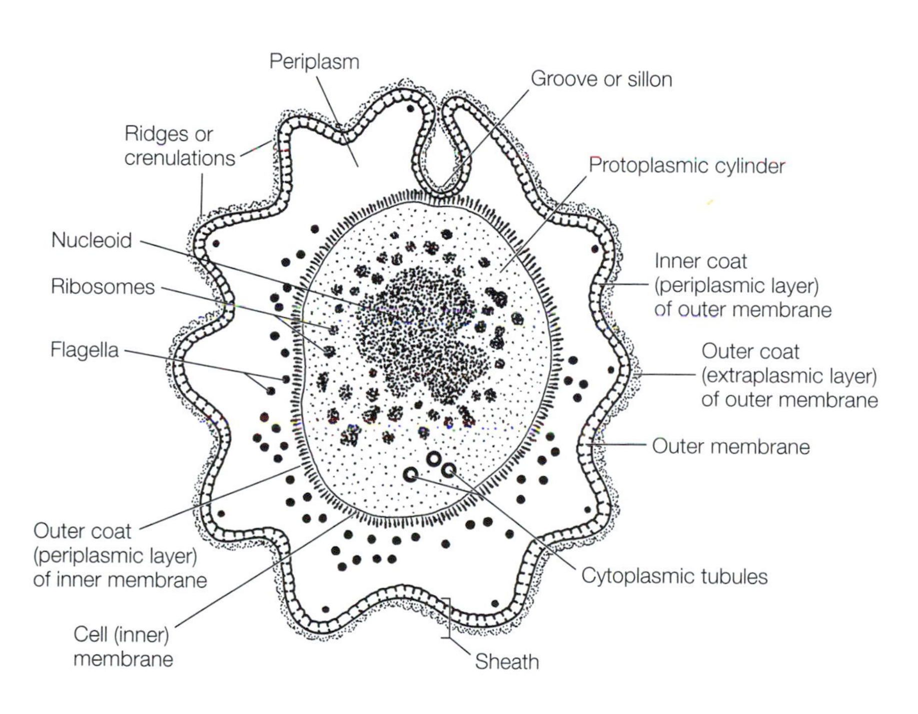

C (Top) Features, in principle, measurable in all spirochetes. (Bottom) Cross section of a generalized pillotina spirochete. No single member of the group has all these features. [Drawing by K. Delisle.]

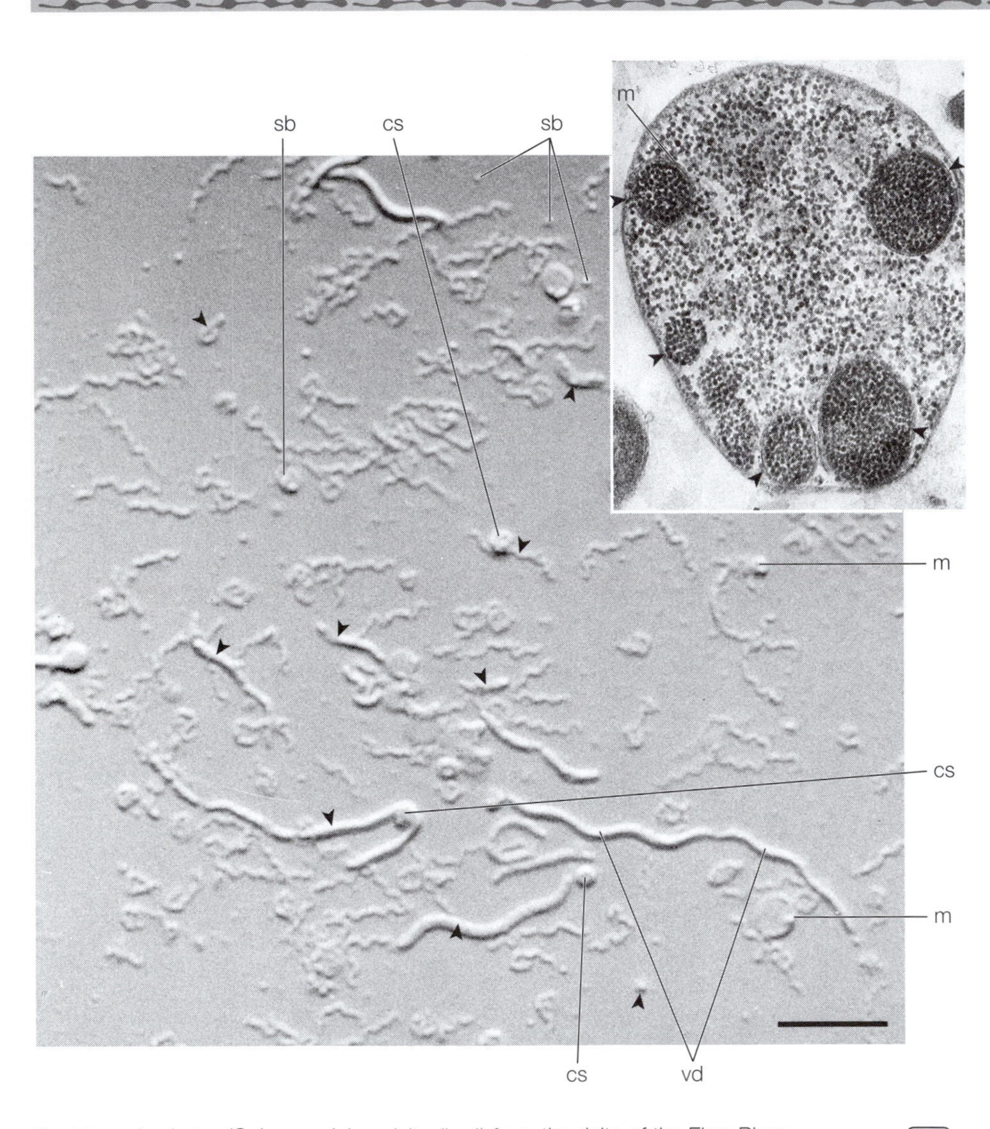

D Live spirochetes (*Spirosymplokos deltaeiberi*) from the delta of the Ebro River, northeastern Spain. Variable diameter (vd), spherical bodies (sb), internal membranous structures (m), and probably composite structure (cs) can be inferred. TEM, bar = 10 μm. (Inset) Transverse section of internal development of composite structure as the membranes form around the internal offspring (arrows). TEM, bar = 1 μm.

hindgut of wood-eating cockroaches and dry-wood, damp-wood, and subterranean termites. The animals ingest wood, but the microbes inhabiting their intestines digest it. One of these microbes, *Spirosymplokos deltaeiberi* from microbial mats, is unique. Not only is its diameter tapered (variable), but it seems to produce "babies" by releasing small spirochete and membranous bodies that may function as spores (Figure D).

The spirochetes themselves probably lack cellulases, enzymes that initiate the breakdown of wood, but do have enzymes for digesting the products of the initial breakdown. In fact, the termite spirochetes are often in intimate contact with parabasalid protoctists (Phylum Pr-1) that contain cellulases. Many spirochetes, mostly unidentified, are observed in environments where active breakdown of algal or plant cellulose is taking place.

Most spirochetes are difficult to study in the laboratory. Some require nutritional media containing large, complex fatty acids. Requirements in general are not known, because only a few spirochetes have been cultured and none of them are pillotinas.

Division: Gracilicutes

B-5 Cyanobacteria

(Blue-green bacteria and chloroxybacteria, grass green)

Greek *kyanos,* dark blue

EXAMPLES OF GENERA

Anabaena	*Gloeocapsa*	*Prochlorococcus*
Anacystis	*Gloeothece*	*Prochloron*
Chamaesiphon	*Hyella*	*Prochlorothrix*
Chroococcus	*Lyngbya*	*Spirulina*
Dermocarpa	*Microcystis*	*Stigonema*
Entophysalis	*Nostoc*	*Synechococcus*
Fischerella	*Oscillatoria*	*Synechocystis*

These two groups of oxygenic photosynthetic bacteria differ in color and pigmentation. By far the largest group, with perhaps a thousand genera, are the blue-greens. Only a few recently described genera of the green, or chloroxybacteria, photosynthesizers are known: *Prochlorococcus, Prochloron,* and *Prochlorothrix.* Because studies of 16S rRNA show these chloroxybacteria to be more closely related to certain cyanobacteria than they are to each other, the group has been united. They are called "chloroxybacteria" for their colors and oxygenic habits.

Until less than two decades ago, cyanobacteria were called blue-green algae or cyanophyta and were considered plants. Physiologically, cyanobacteria are remarkably similar to algae and plants. As the ultimate producers, these three kinds of life feed and energize all of the others—with the minor exception of some obscure chemolithoautotrophic bacteria (see Table 1-1). Together with anoxygenic phototrophic bacteria, algae, and plants, cyanobacteria sustain other life on Earth by converting solar energy and carbon dioxide (CO_2) into organic matter that provides food and energy for the rest. Algae, plants, and cyanobacteria photosynthesize according to the same rules, using water as the hydrogen donor to reduce CO_2:

$$12\,H_2O + 6\,CO_2 \xrightarrow[\text{chlorophyll}]{\text{light}} C_6H_{12}O_6 + 6\,O_2 + 6\,H_2O$$

Like algae and plants, cyanobacteria and their grass-green relatives respire the oxygen that they produce by photosynthesis. Furthermore, like certain anoxygenic phototrophs (Phylum B-3), many cyanobacteria can use hydrogen sulfide (H_2S) instead of water as the hydrogen donor. Thus, under high sulfide conditions, their photosynthetic metabolism deposits sulfur instead of releasing oxygen:

$$2\,H_2S + CO_2 \xrightarrow{\text{light}} CH_2O + 2\,S + H_2O$$

Stained preparations viewed in the light microscope, confirmed by detailed electron microscopic observations, show that the blue-greens and chloroxybacteria have Gram-negative walls (see page 63). The similarity between cyanobacteria, on the one hand, and nucleated algae and plants, on the other, is now understood to apply only to the plastids (photosynthetic organelles and their nonphotosynthetic derivatives) of algal and plant cells: the plastid evolved when cyanobacteria became permanent symbionts.

Many cyanobacteria in high sulfide conditions photosynthesize as the green and purple sulfur bacteria do. They deposit sulfur in their cells. Yet all can produce oxygen gas. The production of oxygen gas distinguishes the cyanobacteria and their grass-green relatives from other photosynthetic bacteria.

Cyanobacteria, like many other bacteria, contain organelles: distinct subcellular structures visible with the light microscope in live cells. Many have nucleoids and carboxysomes. The former are fibrils of DNA; the latter stores the most abundant enzyme in the biosphere: ribulose bisphosphate carboxylase, the universal CO_2-fixing enzyme of cyanobacteria. Like most other photosynthetic organisms, cyanobacteria contain membranes called thylakoids, often at the best-lit periphery of the cells. In most blue-greens, thylakoids are associated with spherical structures called phycobilisomes. Photosynthetic pigments called phycobilins are bound to proteins in phycobiliprotein complexes that are embedded in the phycobilisomes. Chlorophyll *a* and two similar lipid-soluble pigments, phycocyanin and allophycocyanin, give cyanobacteria their bluish tinge. Many have an additional reddish pigment called phycoerythrin. These pigments, which resemble nitrogen-containing porphyrin whose ring has been opened up to form a chain, are biosynthesized by the same reactions that produce porphyrins. Bile pigments of animals also contain porphyrin rings. The grass-green oxygenic phototrophic bacteria (*Prochloron, Prochlorothrix,* and *Prochlorococcus*) lack phycobiliprotein and phycobilisomes but contain a chlorophyll: chlorophyll *b.*

Unobtrusive though they are, there are thousands of living types of cyanobacteria; in the ancient past, they dominated the landscape. The Proterozoic eon, from about 2500 million until about 600 million years ago, was the golden age of cyanobacteria. Remains of their ancient communities include trace fossils called stromatolites, layered sedimentary rocks produced by the metabolic activities of microorganisms, especially of filamentous cyanobacteria. Certain stromatolitic communities still live in salt flats and shallow embayments of the Persian Gulf, the west coast of Mexico, the Bahamas, western Australia, and even under the ice in Antarctica. In the Proterozoic, however, such communities extended to all the continents. Stromatolites formed in fairly deep open waters at the lowest end of the photic zone, the surface waters into which light penetrates (about 200 m in clear seawater). Cyanobacteria occupied the kinds of environments that coral reefs do today.

Prochlorococcus, a marine bacterium resembling *Synechococcus* but with green pigmentation, may be one of the most common bacteria in the world. Thriving at the base of the photic zone all over the ocean, for years it was called "unidentified green coccoid."

There are two great classes of cyanobacteria:

Class Coccogoneae: coccoid (spherical) cyanobacteria

- Order Chroococcales: coccoids that reproduce by binary fission. Some genera are *Gloeocapsa, Chroococcus, Anacystis, Prochloron, Synechocystis, Synechococcus,* and the stromatolite-building coccoid *Entophysalis.*
- Order Chamaesiphonales: coccoids that reproduce by releasing exospores. Unlike bacillus spores, exospores are not necessarily resistant to heat and desiccation. *Chamaesiphon* and *Dermocarpa* are two genera in this group.
- Order Pleurocapsales: coccoids that reproduce by forming propagules (baeocytes). The parent organism disintegrates when the propagules are released. Unlike the endospores of

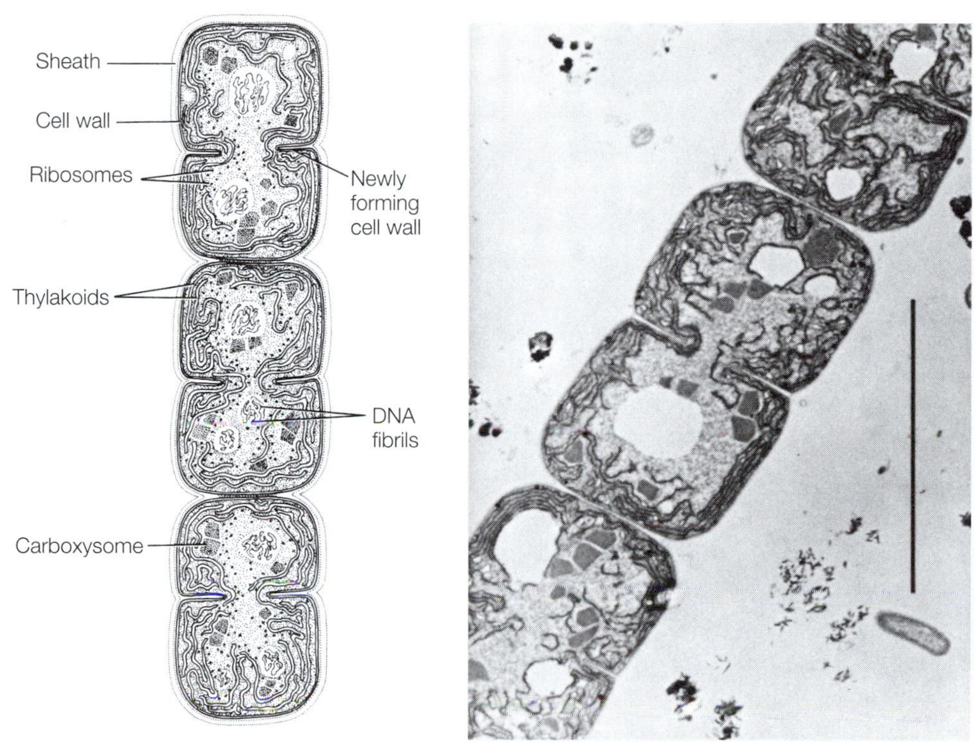

A *Anabaena.* This common filamentous cyanobacterium grows in freshwater ponds and lakes. Within the sheath, the cells divide by forming cross walls. TEM, bar = 5 μm. [Photograph courtesy of N. J. Lang; drawing by R. Golder.]

other bacteria, these propagules are not resistant to desiccation and high heat.

Class Hormogoneae: filamentous cyanobacteria

In many members of this class, filament fragments, called hormogonia, containing as many as several dozen cells break off, glide away, and begin new growth.

Order Nostocales: filaments that either do not branch or exhibit false branching—that is, a branch formed not by only a single growing cell but by slippage of a row of cells. Examples include *Anabaena* (Figure A), *Nostoc, Oscillatoria,* and *Prochlorothrix.*

Order Stigonematales: filaments that exhibit true branching. A single cell dividing by binary fission in a multicellular body may have two places of growth on it, leading to the formation of two new filaments from the same cell. These cyanobacteria are among the morphologically most complex of the prokaryotes. Examples include *Fischerella* and *Stigonema.* The *Stigonema* shown in Figure B is from the Nufunen Pass at about 2500 m in the Austrian Alps, where, like moss (Phylum Pl-1), it grows on granite as ground cover as a terrestrial cyanobacterium.

Many cyanobacteria take in nitrogen from air and incorporate (fix) it into organic compounds. Some of the species fix nitrogen in cells that specialize by losing their chlorophyll-containing thylakoids and turning colorless. The photosynthetic cell changes into a generally larger and more spherical nonphotosynthetic cell called a heterocyst. Several genera of nitrogen-fixing cyanobacteria that lack heterocysts or any other morphological clue to their ability

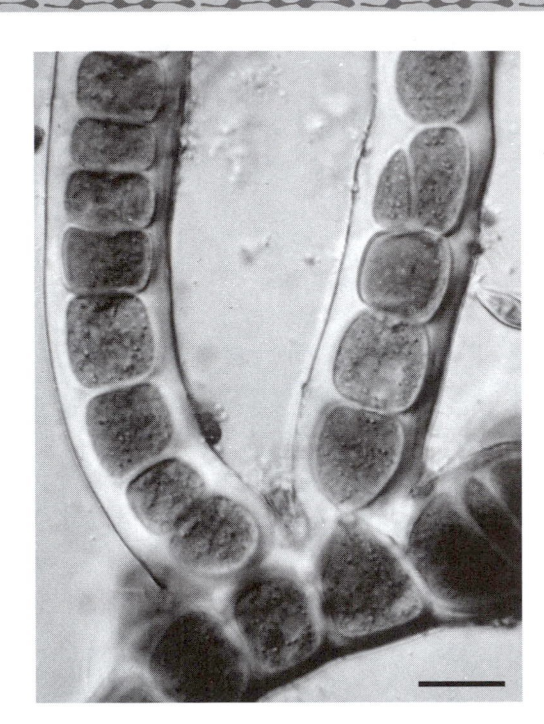

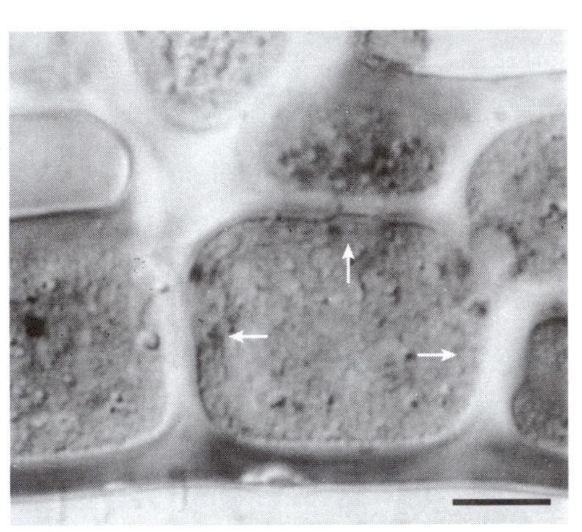

B (Left) *Stigonema informe,* a multicellular, terrestrial cyanobacterium that grows luxuriantly in the high Alps, showing true branching. (Right) Close-up view of true branching, showing three growth points (arrows) on a single cell. LMs, bars = 10 μm. [Courtesy of S. Golubic.]

have been described. Early claims that cyanobacteria do not inhabit the open ocean are incorrect. The tiny marine cyanobacterium *Synechococcus,* abundant in seawater, was ignored for years. *Oscillatoria* species capable of nitrogen fixation but lacking heterocysts also have been reported.

Prochloron was seen before it was scientifically described in the late 1960s, but it was assumed to be a green alga. When they were carefully studied with the electron microscope, however, it became clear that these green coccoids are oxygenic phototrophic prokaryotes. Morphologically, they are cyanobacteria. They were named "prochlorophytes" to indicate their relationship to green plants (Greek *phyton,* plant). They do resemble the plastids of plants and green algae.

Several strains, simple nonmotile coccoids primarily from the South Pacific, belong to the genus *Prochloron.* All form associations with tropical and subtropical marine animals, primarily tunicates (Phylum A-35). *Prochloron* has been found on two species of *Didemnum*—as gray or white surface colonies on *Didemnum carneolentum* and as internal but not intracellular cloacal colonies in *Didemnum ternatanum.* They also reside as colonies in the cloaca of *Diplosoma virens* (Figure C), *Lissoclinum molle, Lissoclinum patella* (Figure D), and *Trididemnum cyclops.* Why *Prochloron* grows only on the surface or in the cloaca of these tunicate hosts is a mystery.

Although they lack the phycobiliproteins of other cyanobacteria, prochlorons contain carotenoid pigments. A large percentage of the total carotenoid is beta-carotene, the pigment in carrots, green algae, and plants.

After *Prochloron,* a second filamentous green oxygenic photosynthesizer was found free living in the Loosdrecht Lakes, the Netherlands, in 1984. It resembled blue-greens such as *Oscillatoria* but had other features of *Prochloron.* Named *Prochlorothrix hollandica,* after its form and place of discovery, it is the only grass-green cyanobacterium that grows in pure culture.

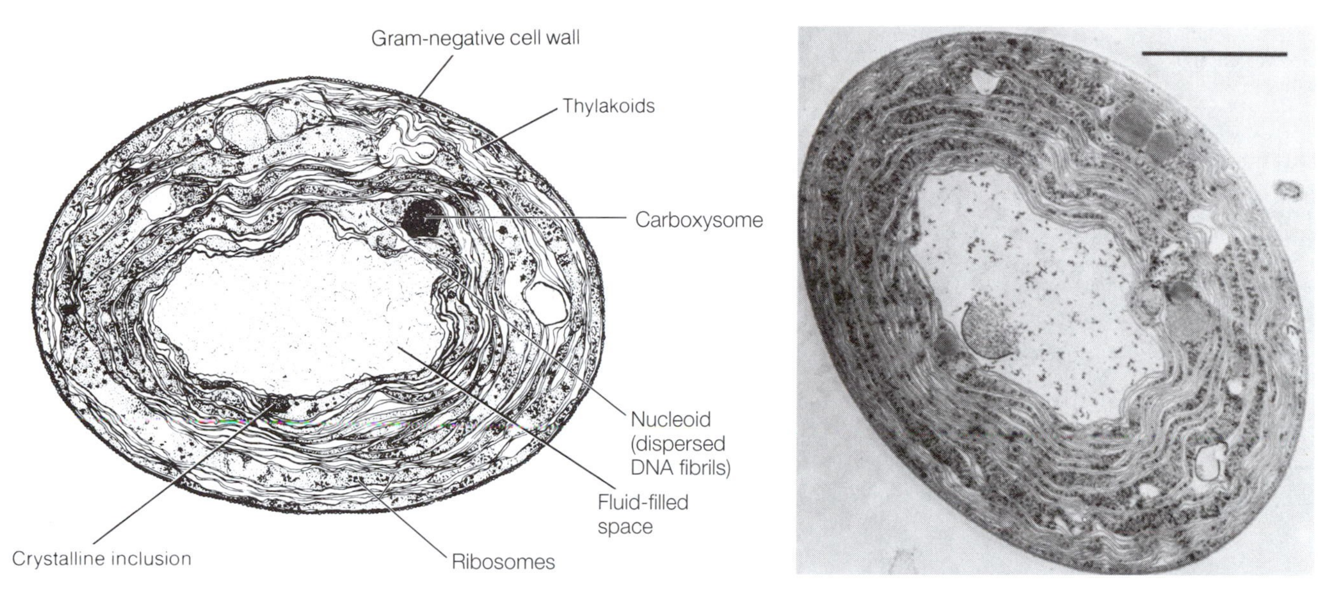

C Thin section of *Prochloron* from the tunicate *Diplosoma virens* (Phylum A-35). TEM, bar = 2 μm. [Photograph courtesy of J. Whatley and R. A. Lewin, *New Phytologist* 79:309–313 (1977); drawing by E. Hoffman.]

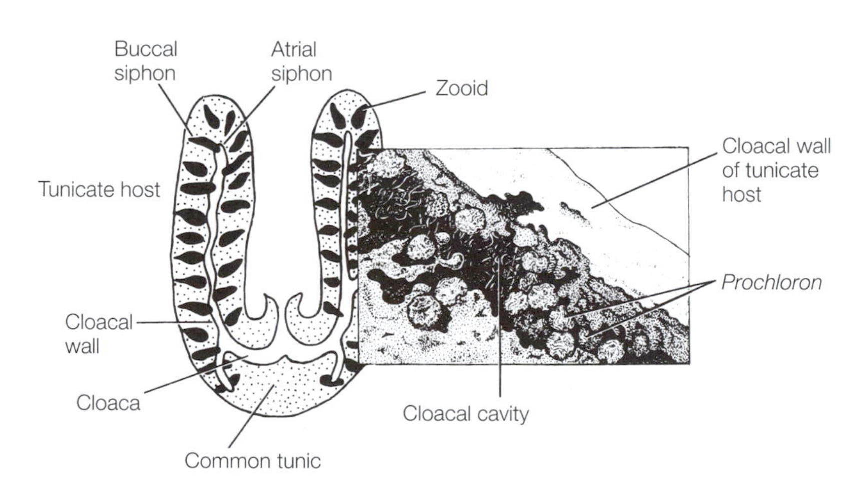

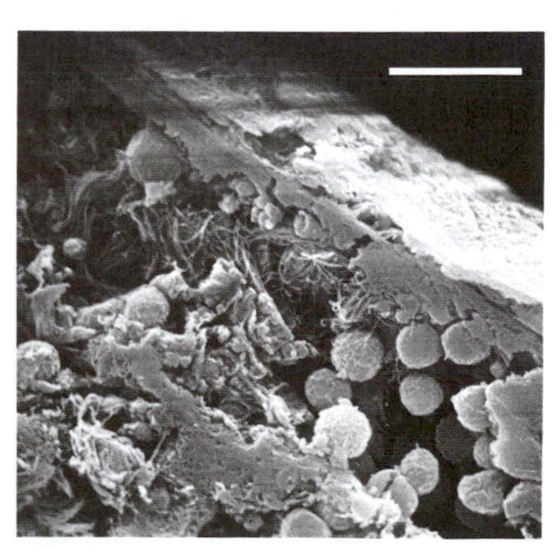

D Cloacal wall of *Lissoclinum patella* (Phylum A-35) with embedded small spheres of *Prochloron.* The tunicate *L. patella* is native to the South Pacific. SEM, bar = 20 μm. [Photograph courtesy of J. Whatley; drawing by E. Hoffman.]

Division: Gracilicutes

B-6 Saprospirae
(Fermenting gliders)

Greek *sapros,* rotten; *spira,* coil, twist

EXAMPLES OF GENERA

Alysiella	*Flexibacter*	*Microscilla*
Bacteroides	*Flexithrix*	*Saprospira*
Capnocytophaga	*Fusobacterium*	*Simonsiella*
Cytophaga	*Herpetosiphon*	*Sporocytophaga*
Flavobacterium		

Gene sequencing of the 16S small subunit rRNA molecule has revealed this lineage of Gram-negative eubacteria. *Bacteroides* and its relatives, anaerobic fermenters, form one subgroup (Figure A); the other, the flavobacterium subgroup, unites a set of oxygen respirers that move by gliding motility (Figure B): *Capnocytophaga, Cytophaga, Flexibacter, Microscilla, Saprospira,* and *Sporocytophaga.* There is another group of gliders, not related to Saprospirae by 16S rRNA criteria, that belong to Proteobacteria (Phylum B-3).

Members of the *Bacteroides* group, like us, are nutritionally organoheterochemotrophs: their sources of energy, carbon, and electrons are all organic (food) compounds. Unlike us, however, they are anaerobic fermenters restricted to anoxic environments. Members of the genus *Bacteroides* inhabit our intestinal tracts in large numbers.

Fermentation is a metabolic process that uses organic compounds to produce energy; the end products are a different set of organic compounds from which some chemical energy has been extracted. Fermenting bacteria are distinguished by their inability to synthesize the metal- and nitrogen-containing ring compounds porphyrins, which all photosynthetic and respiring organisms are capable of synthesizing.

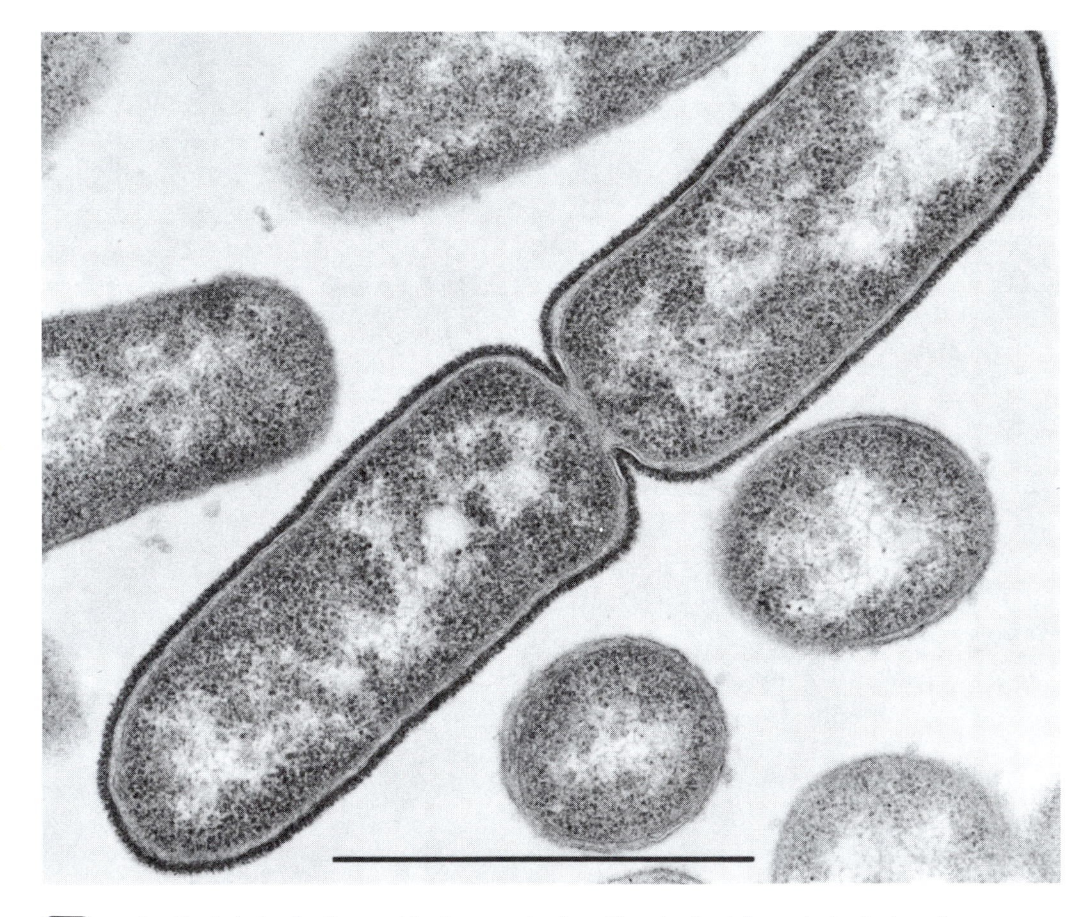

A *Bacteriodes fragilis,* an obligate anaerobe found in animal gut tissue, just prior to cell division. TEM, bar =1 μm. [Photograph courtesy of D. Chase; drawing by L. Meszoly.]

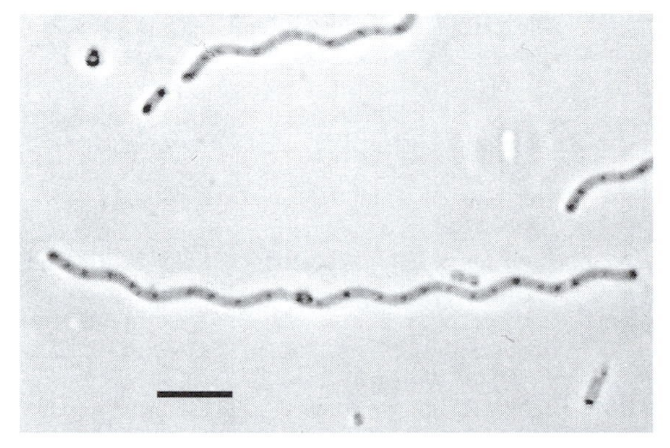

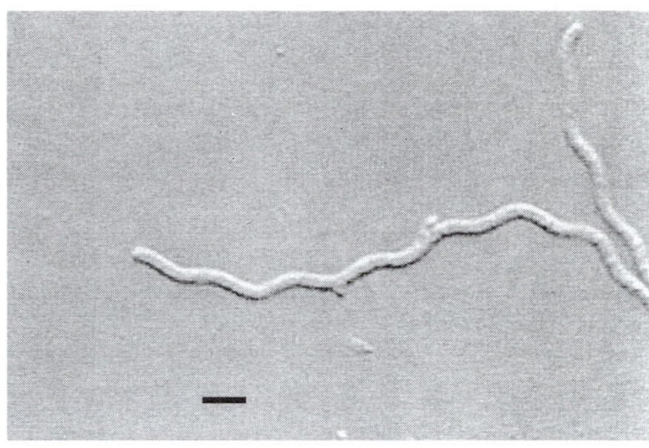

B *Saprospira* sp., live from a microbial mat from Laguna Figueroa, Mexico. (Left) Internal polyphosphate granules (dark spots) are visible in this gliding cell. LM (phase contrast), bar = 5 μm. (Right) The surface of these helical rigid gliders, as seen by using Nomarski phase-contrast optics. LM, bar = 5 μm.

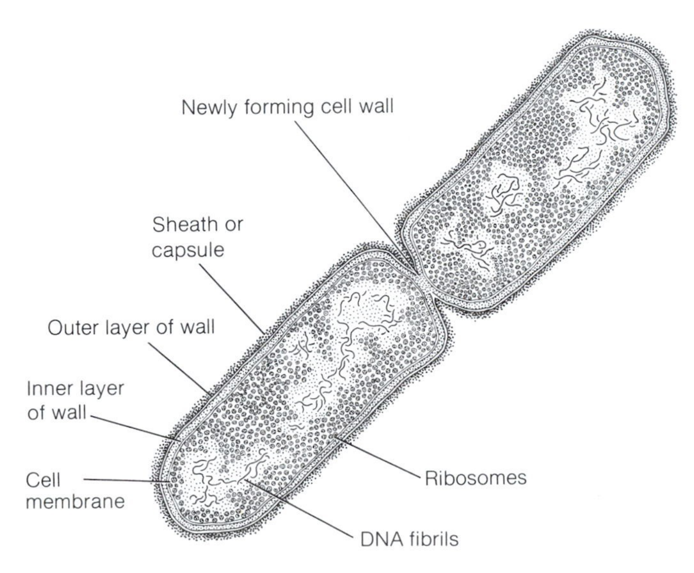

Saprospira and its relatives are oxygen-respiring aerobes that tend to live in organic-rich environments: decaying vegetation, rotting seaweed, and the like. Members of the flavobacterium subgroup are unified by their ability to glide at some stage in their life history and by heterotrophic (parasitic or saprobic) growth.

Glider cells—characteristic of these organisms but also of the δ group of Proteobacteria (Phylum B-3)—typically embedded in slime (polysaccharides of their own making), require contact with a solid surface in order to move; they cannot swim. Although they unmistakably move by gliding along surfaces, the part of the bacterium in contact with the surface shows no organelles of motility, such as flagella. Numerous tiny intracellular fibrils seen in electron micrographs of some species have been correlated with motility, but the means of motility in these organisms is really not understood.

Cytophaga and its relatives (*Capnocytophaga, Saprospira,* and *Sporocytophaga*) constitute a natural group of straight rods or rigid helical filaments, gliding forms that produce orange pigments (carotenoids). *Cytophaga* breaks down agar, cellulose, or chitin; *Flexibacter* metabolizes less-tough carbohydrates, such as starch and glycogen; and *Herpetosiphon,* a filamentous member of this group with a sheath around its cells, breaks down cellulose, but not agar or chitin. Three filamentous genera are *Flexithrix* and helical *Saprospira,* neither of which forms myxocysts (cases from which myxobacteria emerge), and *Sporocytophaga,* which forms small, resistant cells called microcysts.

Division: Gracilicutes

B-7 Chloroflexa

(Green nonsulfur phototrophs)

Greek *chloro,* green; Latin *flexus,* bend

ALL KNOWN GENERA
Chloroflexus
Heliothrix
Oscillochloris

New knowledge, especially of the sequence of nucleotides in their 16S rRNA genes, has led to the separation of these phototrophic bacteria in their entirety from other green sulfur phototrophic bacteria, with which they had been grouped. Since they were discovered in the 1960s in hot springs and microbial mats, there have been hints that *Chloroflexus* were not the standard green sulfur phototrophic bacteria, members of Chlorobia (Phylum B-8), that they superficially resembled. These hints include tolerance for oxygen gas (O_2) and lack—at all times—of a need for a sulfur-rich environment. In contrast, Chlorobia are obligate phototrophs and obligate anaerobes that are extraordinarily intolerant of free oxygen and, because they require sulfur in their photosynthetic metabolism, dwell only in sulfide-rich, anoxic environments.

Chloroflexus is grouped now not on the basis of its phototrophic metabolism but with only two other genera, *Heliothrix* and *Oscillochloris.* Like *Heliothrix* and *Oscillochloris, Chloroflexus* is a filamentous, gliding, nonsulfur bacterium (Figure A). *Oscillochloris* is such a large filament, with cells as large as 5 μm in diameter and a conspicuous holdfast that anchors it to surfaces, that it superficially resembles cyanobacteria such as *Oscillatoria* (Phylum B-5). [In contrast, the gliding filaments of one of only two recognized species of *Chloroflexus* (*C. aurantiacus*) are 0.8 μm in diameter.] However, both the physiology and the 16S rRNA gene sequence of *Oscillochloris* group it here with the Phylum Chloroflexa.

Chloroflexus, a typical bacterial cell with tendencies to form filaments, can be grown in quantity in the laboratory. The presence of chlorosomes, cigar-shaped membrane-bounded structures (Figure B), rather than the flat-membraned thylakoids, further distinguishes them. The lack of CO_2 fixation through the Calvin-Benson cycle (the ribulosediphosphocarboxylase cycle) will probably be observed in the heretofore unstudied members of this small phylum, with its four species (two of *Chloroflexus* and one each of

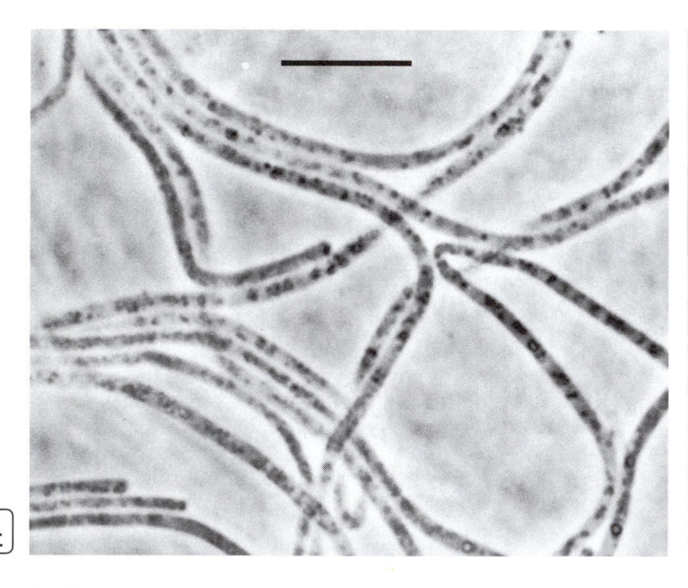

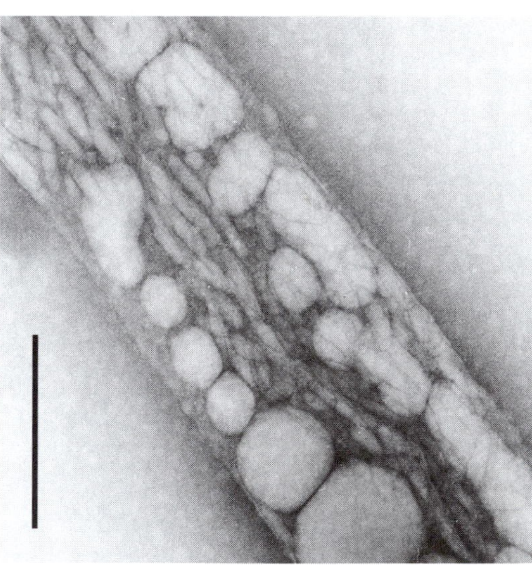

A (Left) Live photosynthetic gliding filamentous cells, 1 μm in diameter, of *Chloroflexus* from hot springs at Kahneeta, Oregon. LM (phase contrast), bar = 5 μm. [Courtesy of B. Pierson and R. Castenholz. *Arch. Microbiol.* 100:5–24 (1975).] (Right) Magnified view showing the typical membranous phototrophic vesicles that contain the enzymes and pigments for photosynthesis. EM (negative stain), bar = 1 μm. [Courtesy of R. Castenholz.]

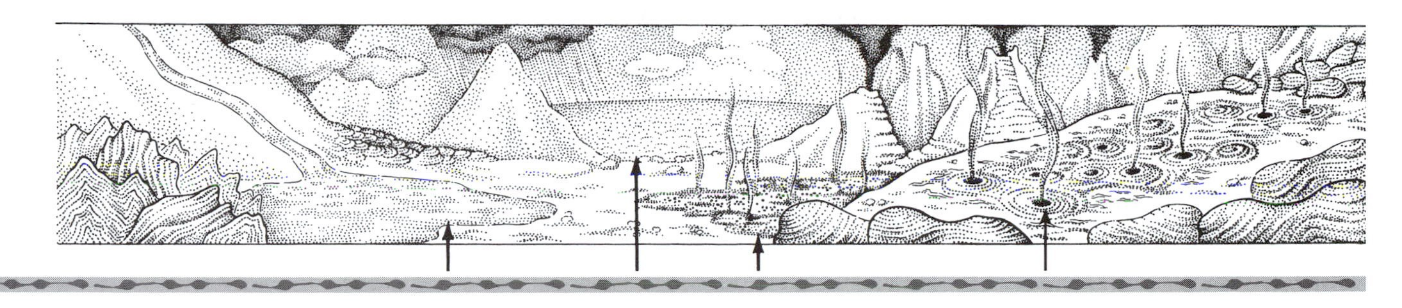

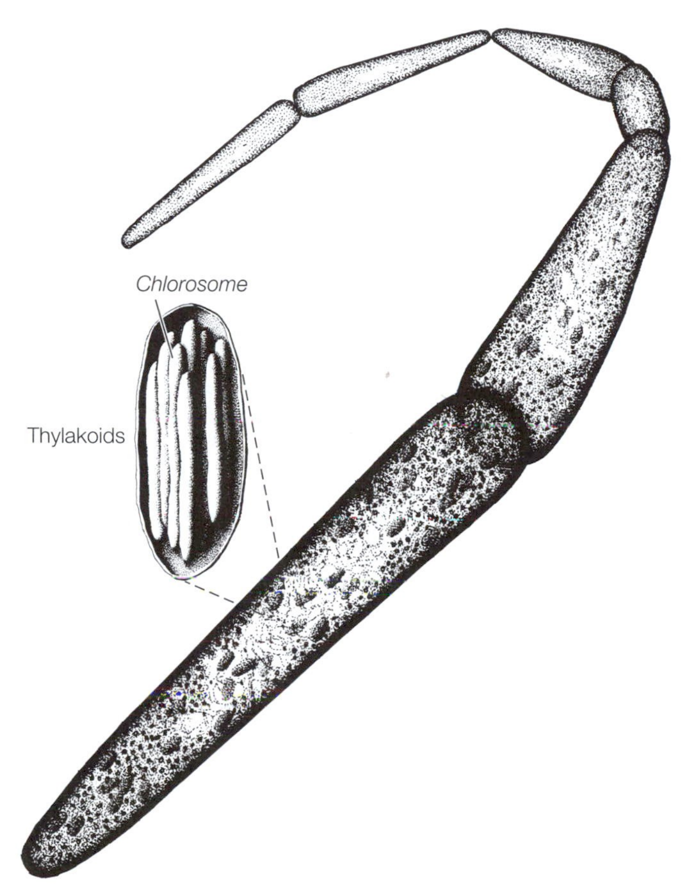

B *Chloroflexus aurantiacus.* Filamentous, thin photosynthesizers showing distribution of their chlorosomes as seen by light microscopy. (Inset) The entire chlorosome as reconstructed from electron micrographs. The membranous plates are the sites of the bacterial chlorophylls and their bound proteins. [Drawings by C. Lyons.]

the other two genera). On 16S rRNA phylogenies, the Chloroflexa green nonsulfur bacteria are closer to the spirochetes (Phylum B-4) and the Saprospirae (in the rather isolated *Bacteroides-Flavobacterium* group, Phylum B-6) than they are to the green sulfur phototrophs (Chlorobia, Phylum B-8).

Because of its cultivability, nearly all the detailed information on members of this phylum comes from studies of *Chloroflexus.* Although large populations of cells of *Chloroflexus,* like those of *Chlorobium,* appear green, the photosynthetic apparatus of *Chloroflexus* is organized into chlorosomes. Chlorosomes, like thylakoids of cyanobacteria, algae, and plants, are repositories of the chlorophylls and their binding proteins. The single (nonunit) membrane is composed of light-harvesting bacteriochlorophylls and the straight-chain (aliphatic) carotenoids alpha- and beta-carotene (Figure C). (These carotenoids resemble those of cyanobacteria, algae, and plants, although oxygenic photosynthesis does not take place in Chloroflexa.)

Chloroflexus may grow well photoautotrophically with CO_2 as its carbon source and with hydrogen (H_2) or hydrogen sulfide (H_2S) as its hydrogen donor. These properties, including the assumption that the complex and demanding ability to photosynthesize is fundamental, led to the classification of this green sulfur phototroph with the Chlorobia (Phylum B-8). But the 16S rRNA gene sequence data are not the only evidence that leads us to place *Chloroflexus* in its own phylum. Another characteristic that distinguishes *Chloroflexus* is that, atypically for other green phototrophs, it grows well heterotrophically. *Chloroflexus* thrives in the dark on organic foods, including a wide variety of sugars, amino acids, or other small organic acids, as sources of carbon.

The electron acceptor-reduction potential of *Chloroflexus* during phototrophic growth resembles that of the phototrophic purple bacteria (Phylum B-3), which are more tolerant of oxygen than *Chloroflexus* and *Chlorobium* are, much more than it does that of Chlorobia (Phylum B-8). This measurable reduction potential, a property that predicts the conditions in which cells absorb light and generate chemical energy for cell reactions, is useful for understanding both the habitat distribution and the physiological limitations of photosynthetic and respiring microbes.

Chloroflexus and many purple bacteria enjoy a reduction potential of about -0.15 V (compared with the more negative and thus more reduced reduction potential of -0.5 in the Chlorobia). Like all the purple and the other green phototrophs, *Chloroflexus* is not oxygenic; nevertheless, in spite of its oxygen tolerance and pigment composition (green color and plantlike carotenoids), it is not like purple bacteria or cyanobacteria. *Chloroflexus* is also unique in the way that it handles carbon dioxide. Two atmospheric CO_2 molecules, one at a time, are bound to acetyl coenzyme A, making hydroxypropionyl CoA molecules. Twice-carboxylated acetyl CoA yields methylmalonyl CoA. Rearranged, this molecule forms acetyl CoA and glyoxylate. Glyoxylate, probably through an amino acid pathway (serine or glycine), is converted into cell material in a series of reactions in which the acetyl CoA is reused.

Thus, neither the ribulosediphosphocarboxylase cycle (Calvin-Benson cycle, in which CO_2 is fixed by a five-carbon ribulose 1,5-bisphosphate molecule and ultimately yields two 3-carbon glyceraldehyde 3-phosphate molecules) nor the reverse Krebs cycle (TCA, or tricarboxylic acid, cycle) of Chlorobia reduces carbon dioxide into cell material in *Chloroflexus.* No other group of organisms uses such a pathway for carbon dioxide removal from the air and incorporation into the cell material of the biosphere. The *Chloroflexus* hydroxypropionate pathway is one of about half a dozen distinct and independently evolved CO_2-fixing metabolic schemes in all of life. From this perspective, cyanobacteria, algae, and plants, all of which use only the Calvin-Benson cycle, are extremely uniform.

Chloroflexus has been discovered associated with heat-tolerant filamentous cyanobacteria in hot springs at temperatures between 40°C and 70°C. *Heliothrix* was found at the surface of microbial mats replete with *Chloroflexus. Heliothrix* requires high light intensities for growth but is not a *Chloroflexus,* because it lacks chlorosomes and contains only a single bacteriochlorophyll (bacteriochlorophyll *a*). Its gliding filamentous structure, its 16S rRNA sequences, and the fact that it contains carotenoids nearly identical with those in *Chloroflexus* support its placement in this phylum.

C *Chloroflexus aurantiacus.* Antenna pigment (secondary, light-harvesting pigment) bacteriochlorophyll *c*–containing chlorosomes apposed to cytoplasmic membrane, which contains the reaction center bacteriochlorophyll *a*. TEM (freeze fracture), bar = 1 μm. [Courtesy of C. Fuller.]

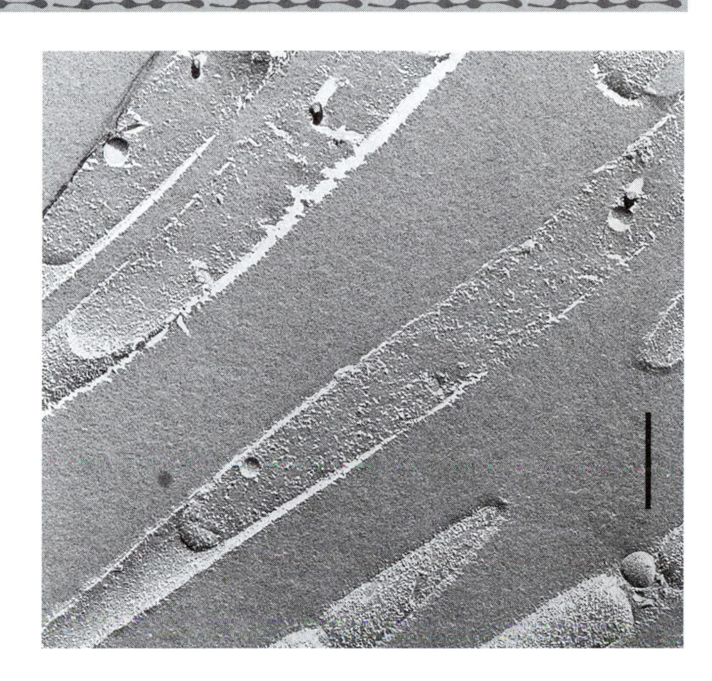

Division: Gracilicutes

B-8 Chlorobia

(Anoxygenic green sulfur bacteria)

Greek *chloro,* green; *bios,* life

EXAMPLES OF GENERA

Ancalochloris	*Chloronema*
Chlorobium	*Pelodictyon*
*Chlorochromatium**	*Prosthecochloris*
Chloroherpeton	

*Only the phototrophic member of the symbiotic complex.

Not all photosynthetic organisms are phototrophic. Any life form that uses light for the synthesis of cell material is photosynthetic, but only those that do not need fixed organic compounds in the medium are phototrophs. The most independently productive photosynthesizers are the photolithoautotrophs: their energy comes from visible light, their carbon comes from atmospheric CO_2, and their electrons for the reduction of CO_2 to cell material comes from inorganic compounds: hydrogen sulfide (H_2S), sodium sulfide (Na_2S), hydrogen gas (H_2), or water (H_2O).

To reduce the CO_2 in the air to organic compounds, any phototroph needs a source of electrons; members of *Chlorobia* depend on sulfur for this source. These light-requiring eubacteria dwell in sulfide-rich, sunlit habitats. In such obligately anaerobic green sulfur bacteria, photosynthetic pigments are organized on chlorosomes, as they are in *Chlorochromatium* here and in *Chloroflexus* (see Phylum B-7, Figure B). When H_2S or Na_2S is the hydrogen donor, the by-product of photosynthesis is sulfur, which may be stored or excreted by cells as elemental sulfur or as more oxidized sulfur compounds.

Chlorobia are inevitably found in anaerobic muds—those dark, sulfur-smelling beach areas at the edge of the ocean or in ponds and lakes over sulfur-rich rock. In today's oxygenated world, the greenish, brown, or yellow Chlorobia are banished beneath the blue and green cyanobacteria (Phylum B-5), the red, pink, and purple phototrophs (Phylum B-3), and the Chloroflexa (Phylum B-7). Chlorobia tend to be at the bottom of the photic zone. So oxygen-intolerant and light-requiring on their own, Chlorobia escape the need to dwell at the bottom of the photic zone by establishing associations with facultatively oxygen-respiring heterotrophic bacteria (see figure).

All these phototrophic bacteria, delightfully colored in an astonishing range of yellows, pinks, and greens, are cosmopolitan in

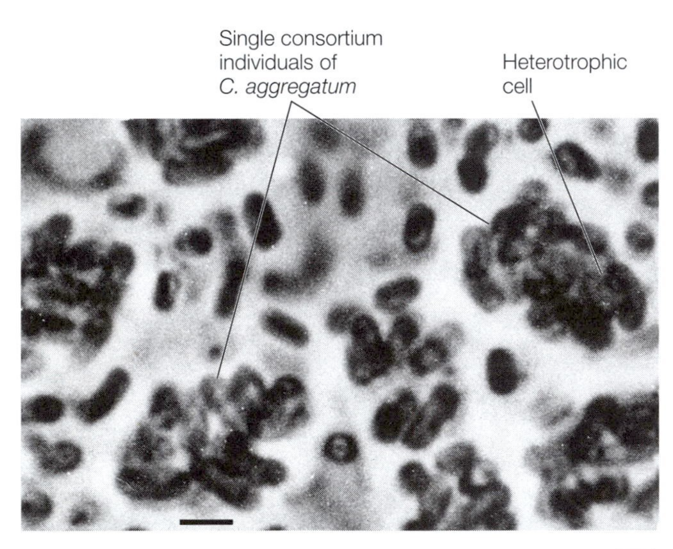

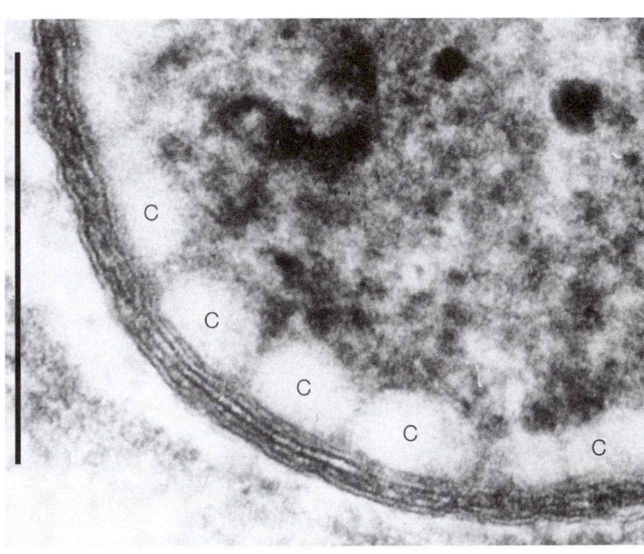

Chlorochromatium aggregatum. Transmission electron micrograph (above, left; bar = 1 μm) of this consortium bacterium, in which a single heterotroph (facing page, left; bar = 1 μm) is surrounded by the several pigmented phototrophs with their chlorosomes (c), seen here as peripheral vesicles (above, right; bar = 0.5 μm). From Lake Washington, near Seattle. The

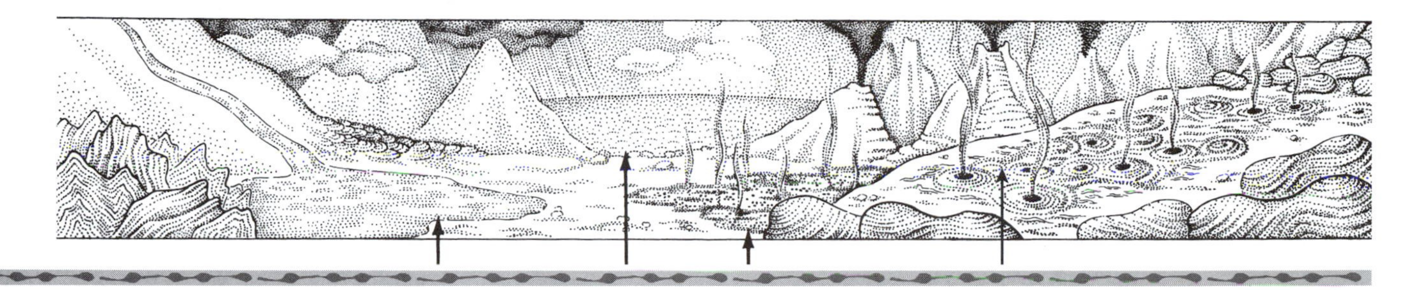

sunlit muds, scums, and most limestones. In the bright sunlight in the top layers of anaerobic muds, absorbing all incident light, they become very dark, nearly black. Because each species has an optimum growth at a given acidity, oxygen tension, sulfide and salt concentration, moisture content, and so on, they grow in layers—each in its appropriate niche—which we see as bands of differing color of sediment. Layered communities of phototrophs support with their productive mode of life many other types of bacteria; these form sedimentary structures such as microbial mats and living stromatolites (see Figure 1-1).

Chlorobium is a large genus; some strains are tolerant of extremely high or extremely low temperatures or salinities. They tend to be small nonmotile coccoids deeply embedded in their anoxic communities. Most of these green photosynthesizers are refractory to propagation in pure culture under laboratory conditions. Members of the genus *Chlorobium* lack gas vesicles, membranous intracellular inclusions used in floating. Less well known genera of Chlorobia, anaerobic phototrophs, include only one other genus that lacks gas vesicles: *Prosthecochloris.* Gas vesicles in all the other genera of nonflagellated Chlorobia aid in regulation of position in the water column. Both of the two known species of *Prosthecochloris* form cell protrusions from their ellipsoid or spherical cells. As wall appendages that are not filled with cytoplasm, the protrusions are called prosthecae.

The remaining four gas-vacuolate genera of Chlorobia are less familiar to us. They can be morphologically distinguished. *Pelodictyon* cells, which are branched, nonmotile rods, form a loose, irregular network. The spherical cells of *Ancalochloris* bear prosthecae. Whereas *Chloroherpeton* is a gliding rod, *Chloronema* forms large-diameter (2.0–2.5 μm) gliding filaments. No doubt a great variety of other Chlorobia reside in nature, awaiting students of anoxic, sulfurous, well-lit habitats to discover them.

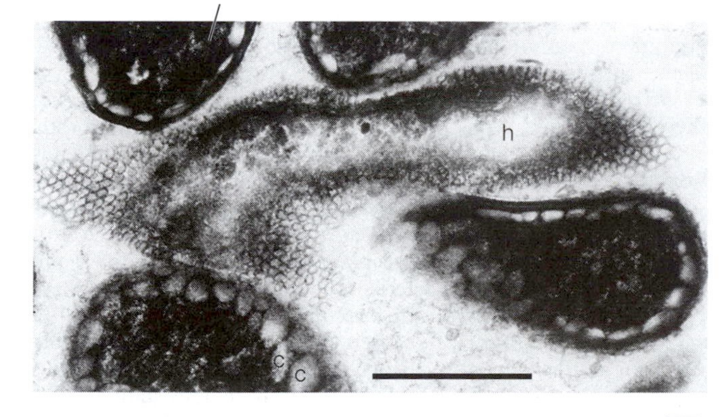

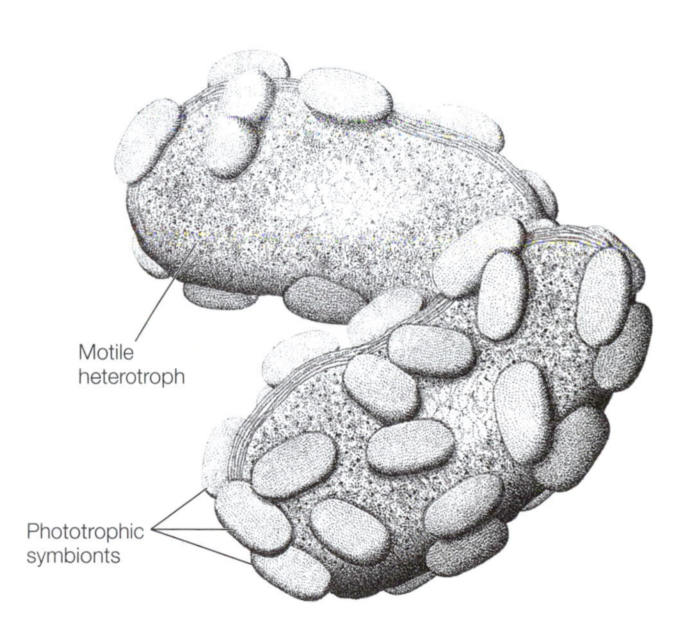

photosynthetic cells responsible for the productivity of the consortium are *Chlorobium,* whereas the motility needed to approach the light but flee from oxygen gas is due to the central heterotroph (h). [Drawing by C. Lyons, based on these three micrographs by Douglas Caldwell and others.]

Division: Tenericutes

B-9 Aphragmabacteria
(Mycoplasmas)

Greek *a,* without; *phragma,* fence

ALL KNOWN GENERA

Acholeplasma	*Ehrlichia*
Anaplasma	*Mycoplasma*
Bartonella	*Spiroplasma*
Cowdria	*Wolbachia*

All cells, including those of wall-less eubacteria, the mycoplasmas, are bounded by a cell membrane of their own making that permits passage to water, salt ions, and small organic compounds. Outside the ubiquitous lipid-protein bilayer (double looking in electron micrographs), called the unit membrane in most other bacteria, cells are bounded by a rigid cell wall that, although varied in composition, contains polysaccharides attached to short polypeptide molecules. This peptidoglycan wall is absent in all aphragmabacteria, although extramembranous materials that, under the microscope, look like fuzz may coat the cells. Aphragmabacteria are bounded by the single simple or decorated unit membrane because they are incapable of synthesizing certain polysaccharides (for example, diaminopimelic and muramic acids) that form the finished walls of most other bacteria. Lacking cell walls, they are resistant to penicillin and other drugs that inhibit wall growth.

Typically, bacteria range from 0.5 to 5.0 μm along their longest axis. Because the diameter of many aphragmabacteria is less than 0.2 μm, they are invisible even with the best light microscopes. Their shapes vary: irregular blobs, filaments, and even branched structures reminiscent of tiny fungal hyphae.

How wall-less bacteria reproduce is not clear. In some, tiny coccoid structures appear to form inside the cells, emerging when the "parent" organism breaks down. Others seem to form buds that become new organisms. Some apparently reproduce by binary fission (division of the cell into two roughly equal parts). On agar plates, the best-known mycoplasmas typically form tiny colonies, with a dark center and lighter periphery, resembling the shape of a fried egg.

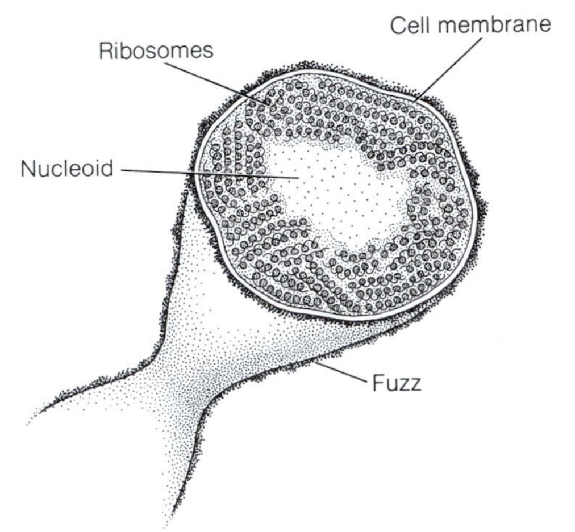

A A generalized mycoplasma. [Drawing by L. Meszoly.]

Most well-known kinds are studied because they are seen in large numbers in diseased mammals and birds. They live in profusion inside cells of animal tissues, where as symbiotrophs they derive their carbon, energy, and electrons from their hosts. Some, such as *Ehrlichia,* cause tick-borne diseases, but the presence of many, probably most, is invisible without high-power electron microscopy. Few have been cultivated outside the animal in which they reside. Those that have, primarily *Mycoplasma,* require very complicated growth media that include steroids, such as cholesterol. These lipid compounds, produced by most and required by all eukaryotes, are seldom, if ever, found in prokaryotes. However, in aphragmabacteria of the genus *Mycoplasma,* cholesterol constitutes more than 35 percent of the membrane's lipid content. For a bacterium, this is an extremely large fraction and may be the legacy of a long biological association between *Mycoplasma* and animal tissue rich in lipids. Through evolutionary time, long association of these bacteria with eukaryotes' lipids led to dependency. All strains so far cultured require long-chain fatty acids (a kind of lipid) for growth, and most ferment either glucose (a sugar) or arginine (an amino acid). The fermentation products are usually lactic acid and some pyruvic acid.

Mycoplasma are of economic and social importance because they cause certain types of pneumonia in humans and domestic animals (Figures A and B). Under conditions of host cell debilitation, these common symbiotrophs that are normally benignly present can be responsible for the death of cells in animals or in laboratory tissue cultures. Widespread in insects (Phylum A-20), vertebrates (Phylum A-37), and plant tissues, many are too small to identify as aphragmabacteria—to see whether cell walls are present—even with electron microscopes.

Bacteriologists have often used the word mycoplasma informally for all members of the genus *Mycoplasma* and other wall-less organisms without formally raising them to phylum status. Other well-known genera include *Acholeplasma,* which do not require steroids, and *Spiroplasma* and others, which do require them (as *Mycoplasma* does).

Members of *Spiroplasma* were isolated from the leaves of citrus plants affected with a disease called "stubborn." Whether the spiro-

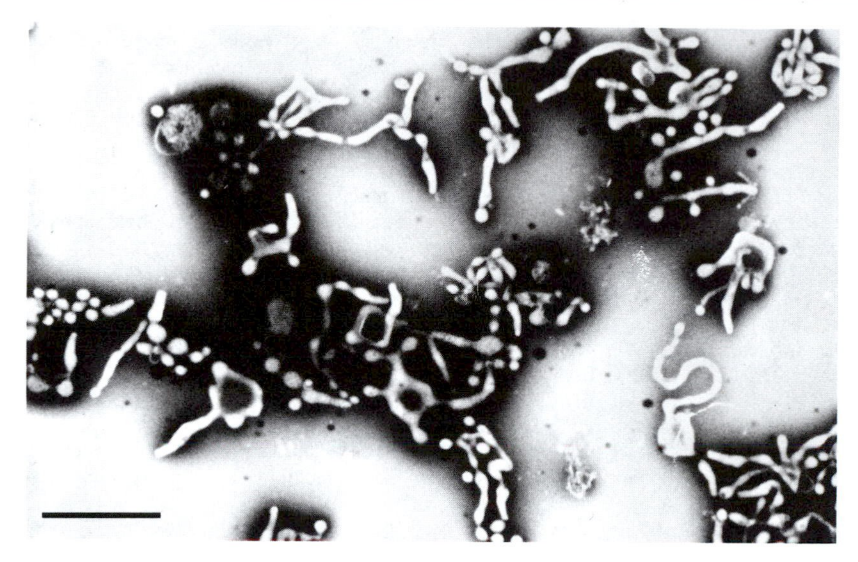

B *Mycoplasma pneumoniae,* which lives in human cells and causes a type of pneumonia. TEM (negative stain), bar = 1 μm. [Courtesy of E. Boatman.]

plasmas actually cause the disease has not been determined with certainty. Like other aphragmabacteria, the spiroplasmas are variable in form, lack cell walls, and form colonies that look like fried eggs. *Spiroplasma* cells are helical in shape and motile, showing a rapid screwing motion or a slower waving movement, yet they lack flagella or any other obvious organelles of motility. How do they move? Nothing is known.

An old hypothesis that the various aphragmabacteria separately (convergently) evolved from different Gram-positive bacteria by loss of walls is borne out by modern observations. Some *Mycoplasma* probably do represent minimal life on this planet and are truly primitive in the sense that their ancestors never had walls. Among them are the smallest organisms known. Most bacteria have from 3000 to 6000 genes. The DNA of the strain *Mycoplasma genitalium* has only about 4.5×10^8 daltons of DNA, which is ten times less than most bacteria. The complete sequence reveals this organism to have fewer than 500 genes and to make fewer than 500 proteins. Figure C shows a similar mycoplasma.

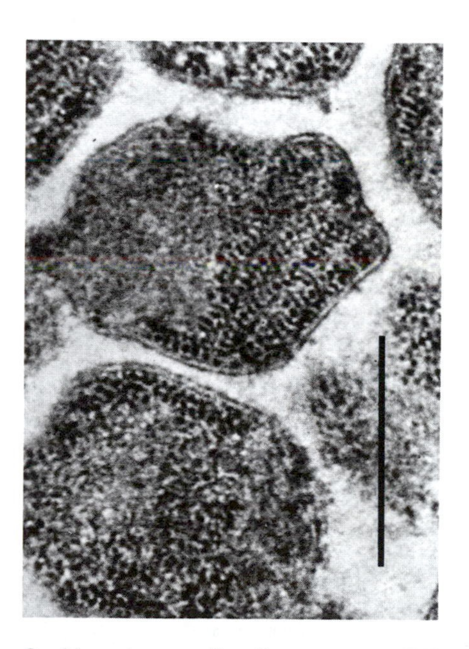

C *Mycoplasma gallisepticum,* symbiotroph in the cells of chickens. TEM, bar = 0.5 μm. [Courtesy of J. Maniloff, *Journal of Cell Biology* 25: 139–150, 1965.]

Division: Firmicutes

B-10 Endospora

(Endospore-forming and related low–G + C Gram-positive bacteria)

Greek *endos,* within; *spora,* seed

EXAMPLES OF GENERA

Bacillus	*Leuconostoc*	*Sporolactobacillus*
Clostridium	*Lineola*	*Sporomusa*
Coleomitus	*Peptococcus*	*Sporosarcina*
Heliobacterium	*Peptostreptococcus*	*Streptococcus*
Lactobacillus	*Ruminococcus*	

Endospora include fermenting obligate anaerobes and obligate or facultative aerobes, most of which stain as Gram-positive walled cells and all of which belong to the 16S rRNA low-GC group of eubacteria. They are heterotrophs, many of which form endospores within their cells. Endospores, specialized propagules formed within the parent ("mother") cell, are resistant to conditions such as heat and desiccation that harm the trophic (growing) cells.

Most bacteria belonging to this phylum are motile by means of polar, peritrichous or laterally inserted flagella. At least three genera are capable of oxygen respiration: *Bacillus,* a huge and important genus; *Sporosarcina;* and *Sporolactobacillus.* Only members of the genera *Clostridium* and *Sporomusa* are obligate anaerobes. Whereas *Clostridium* and all the other thousands of bacterial strains in this phylum are Gram-positive, *Sporomusa*—insect symbionts—surprisingly have Gram-negative cell walls.

Endospora comprise thousands of strains distributed in a multitude of habitats all over the world. Growing cells are rod or sphere shaped, as a rule, and their spores either elliptical or spherical. Each parent cell produces only a single, water- or airborne spore, which may land anywhere. A special compound, calcium dipicolinate, constitutes from 5 to 15 percent of the dry weight of the spore. The core of the spore contains one copy of the parent bacterium's genetic material (the nucleoid in the figure). It is enclosed in a cortex made of peptidoglycan cell-wall material and surrounded by an outer layer, called the spore coat. Spore development is quite complicated. In this dormant stage, endospora cells survive for years without water and nutrients. If nutrients and moisture are plentiful, the spore germinates and growth ensues until nutrient or water is depleted. As conditions become unfavorable for growth, sporulation sets in; the spore's position in the cell may be either central or terminal. Finally, the parent cell shrivels or disintegrates, releasing the spore.

Many strains of endospora produce antibiotics during the active growth stage. Most species produce acid, and all can metabolize glucose. Several produce gas or acetone as products of glucose catabolism. Many species can hydrolyze starch to glucose, and some also produce pigments such as the reddish brown or orange pulcherrinin and the brown or black melanin.

Some endospore formers break down tough plant substances such as pectin, polysaccharides, and even the lignin and cellulose of wood. The ability to degrade lignin and cellulose is unusual in the living world, which accounts in part for the persistence of forest litter, petroleum, peat, coal, and other organic-rich sediments. Lignin- and cellulose-degrading bacteria tend to associate in communities. Because no population of a single bacterial species totally degrades any of these refractory materials—lignin and cellulose—all the way to CO_2 and H_2O in days or weeks, in nature the breakdown requires communities of differently metabolizing bacteria.

The requirements for growth vary widely among endospora. Some have a high tolerance or even a requirement for salt; others do not. Vitamins and other complex growth factors are needed by certain strains, but not by others. Some clostridia can fix molecular nitrogen from the air, but most have more complex organic nitrogen requirements. Some may grow at a temperature as low as –5°C; some strains, called psychrophils, grow optimally at –3°C. Some endospore formers, even members of *Bacillus,* are thermophilic; found in hot springs, they grow at temperatures above 45°C.

The genus *Bacillus* has so many members that information about it requires entire books. *Arthromitus,* first described as a "plant" (because it was not an animal) rooted to the intestines of beetles, termites, and other animals, is a symbiotic bacillus. Joseph Leidy (1823–1891), founder of the Philadelphia Academy of Natural Sciences, and his successors recognized the ubiquity of animal-associated large filamentous spore-forming bacteria before the bacteriologists had developed a vocabulary for their description. Hence, many animal-associated bacilli, endospore-forming filaments, have obsolete names such as *Lineola longa, Coleomitus,* and *Arthromitus.* Some of these animal symbionts are morphologically indistinguishable from *Bacillus anthracis* (some strains of which are associated with the disease anthrax); they deserve further study.

Entire colonies of *Bacillus circulans* are motile, rotating as a unit for unknown reasons and by unknown mechanisms. Many *Bacillus* grow as filaments or chains and form colonies characterized by distinctive growth patterns.

Sporosarcina ureae are composed of spherical cells 1 to 2 μm in diameter, form tetrads (packets of four), and occasionally grow in distinctive cubical bundles that give the genus its name (Latin *sarcina,* packet). Urea, the metabolic product excreted by many animals, is converted by *S. ureae* into ammonium carbonate.

In appearance and metabolism, *Sporolactobacillus* resembles *Lactobacillus*—a well-known lactic acid–producing bacterium that grows vigorously on milk products—except that *Sporolactobacillus* produces endospores and respires in the presence of oxygen.

Miscellaneous important Gram-positive bacteria, many grouping with lactic acid bacteria (*Peptococcus, Leuconostoc*), including some obligate anaerobes, are discussed here, even though they do not form spores. Their classification is in flux. Lactic acid bacteria do not use oxygen in their metabolism, and many do not tolerate oxygen at all: it inhibits their growth or quickly kills them. Nutritionally, they are heterotrophs of the kind called chemoorganotrophs—they require a mixed set of organic compounds to grow and reproduce.

Lactic acid bacteria, such as *Lactobacillus, Streptococcus* (the bacterium associated with strep throat infections), and *Leuconostoc,* are rod-shaped organisms famous for their ability to ferment sugar, in particular that in milk, and to produce lactic acid as well as acetate,

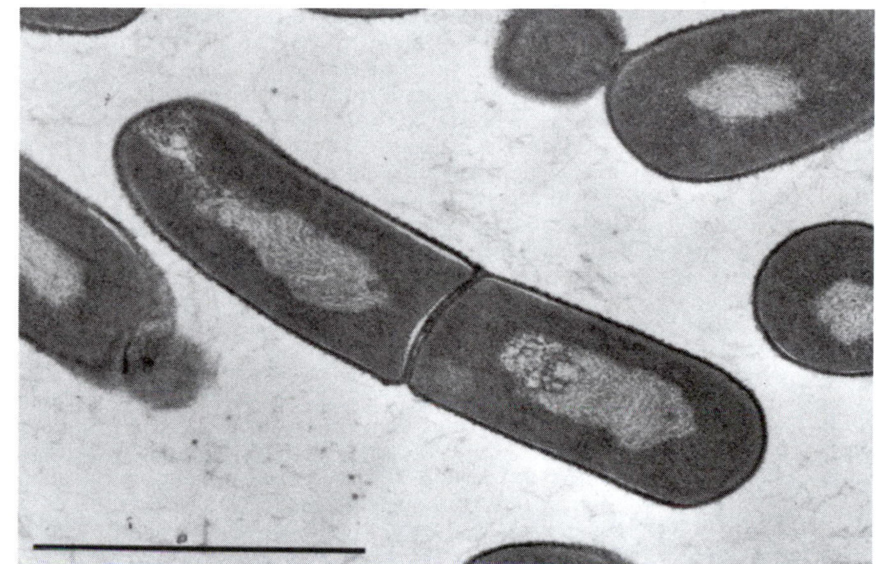

This unidentified *Bacillus* has just completed division into two offspring cells. Such spore-forming rods are common both in water and on land. TEM, bar = 1 μm. [Photograph courtesy of E. Boatman; drawings by I. Atema.]

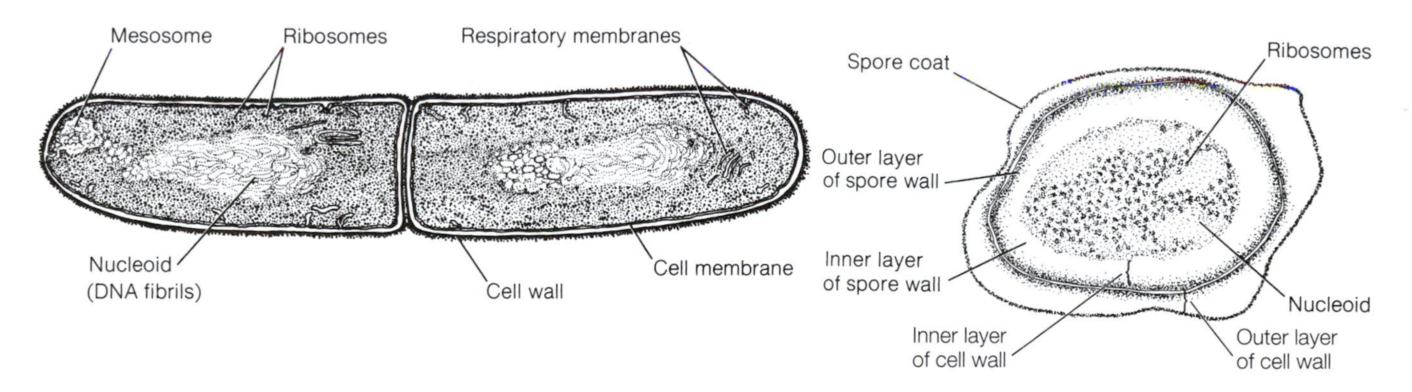

formate, succinate, CO_2, and ethanol. Their complex nutritional requirements include amino acids, vitamins, fatty acids, and other compounds depending on species and strain. A 5 to 10 percent solution of CO_2 enhances their growth. They cannot form spores. Some may tolerate oxygen, but none use it for metabolism.

Peptococcaceae, spherical bacteria that lack flagella and spore formation, range in diameter from 0.5 to 2.5 μm. They are found singly, in pairs, in irregular masses, as three-dimensional packets, and as long and short chains of cocci. Many produce gases such as CO_2 and H_2 when they ferment carbohydrates, amino acids, and other organics. Like most lactic acid bacteria, they have complex growth requirements. Many have been isolated from the mouths or intestines of animals. Three genera are recognized: *Peptococcus; Peptostreptococcus,* which uses protein or its breakdown products for energy; and *Ruminococcus,* which can break down cellulose.

Division: Firmicutes

B-11 Pirellulae

(Proteinaceous-walled bacteria and their relatives)

Named after the genus *Pirellula* (Latin *pirus,* pear)

EXAMPLES OF GENERA

Blastobacter
Chlamydia
Gemmata
Isosphaera
Pirellula (formerly *Pirella*)
Planctomyces

This new diverse group was revealed by 16S rRNA sequence information. At least two of these genera—*Pirellula* and *Planctomyces* (and probably *Gemmata*)—have unique proteinaceous, nonpeptidoglycan cell walls. They do have bacterial cell walls, and, although they superficially resemble the rickettsias, from sequence data *Chlamydia* are classified on their own branch, far closer to the pirellas than to the rickettsias. *Pirellula* and *Planctomyces* contain large quantities of two amino acids: proline and the sulfur-rich amino acid cysteine. Their lipids—palmitic, oleic, and palmitoleic—are strange for bacteria, although common in eukaryotes. Their aberrant walls render them insensitive, as are eukaryotes, to cell–wall-inhibiting antibiotics such as cycloserine, cephalosporin, and penicillin.

Obligate aerobes, *Gemmata, Pirellula,* and *Planctomyces* are found in fresh water. They are heterotrophs that grow in extremely dilute solutions of salts and food. Some attach to surfaces of rock or vegetation by a kind of stalk (holdfast). Those that have prosthecae (appendages) differ from the ordinary prosthecate bacteria of Phylum B-3 (Proteobacteria—for example, *Ancalochloris,* a green phototroph, or the heterotrophs *Asticcacaulus* and *Ancalomicrobium*) by the nature of their cell walls and RNA sequences.

Pirellula, Blastobacter, and *Planctomyces* form buds in their reproductive process: a small cell appears as a protrusion on its parent (Figures A and B). These bacteria grow by polar growth: a new cell wall is formed from a single point instead of the typical (nonpolar) growth by intercalation of new peptidoglycan units all over the surface of the cell. In polar growth, internal membranous and other organelles, such as carboxysomes or gas vesicles, do not take part in cell division, freeing up the parent cell for internal elaborations. This is most conspicuous and fascinating in *Gemmata obscuriglobus* (Figures C and D), which, totally independently of eukaryotes, has evolved a nucleoid membrane. The double-layered unit membrane entirely encloses its bacteria-type DNA, producing an organelle that superficially resembles true nuclei. It lacks the pores, chromatin attachment sites, microtubules, and other characteristics of nuclear membranes. Nor in any other way is *Gemmata* like a eukaryotic organism.

Planctomyces grow as pear-shaped or globular (large spherical) cells with long stalks. Because stalks contain no cytoplasm within the cell wall, they are not prosthecae. Although these stalks may be long and actually accompany flagella as protrusions on the surfaces of the cells, they are simply proteinaceous extensions.

Thought to be fungi when they were first described, *Planctomyces* still bear this misleading name (meaning "floating fungus"). Because they do float on lakes (they are planktonic organisms), they should be renamed *Planctobacter* or something else more fitting of their prokaryotic nature.

Members of the genus *Chlamydia,* as obligate symbiotrophs refractory for study in pure culture, are enigmatic—well known from their medical context only. These bacteria, too, lack conventional peptidoglycan walls; hence, although they stain as Gram-negative bacteria, like all eukaryotes that also stain the telltale pink of the Gram-negative cell, they are better thought of as neither Gram-positive nor Gram-negative. *C. psittaci* is correlated with a

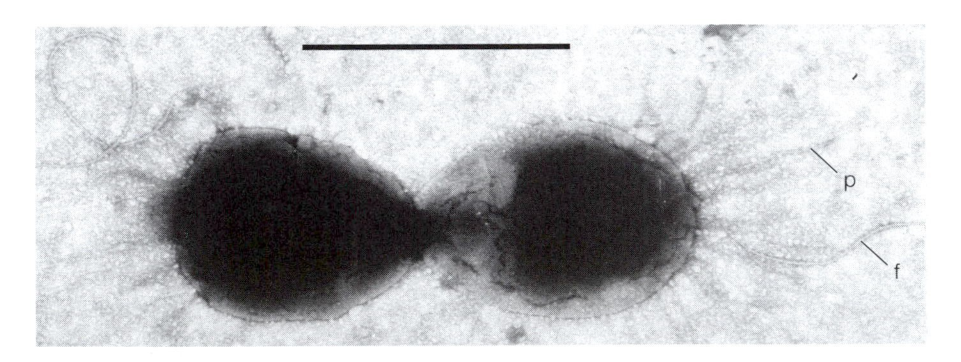

A Dividing cells of *Pirellula staleyi* still attached to one another. Note pili (adhesive fibers; p) and polar flagella (f). TEM (negative stain, whole mount), bar = 1 μm. [Courtesy of J. Staley.]

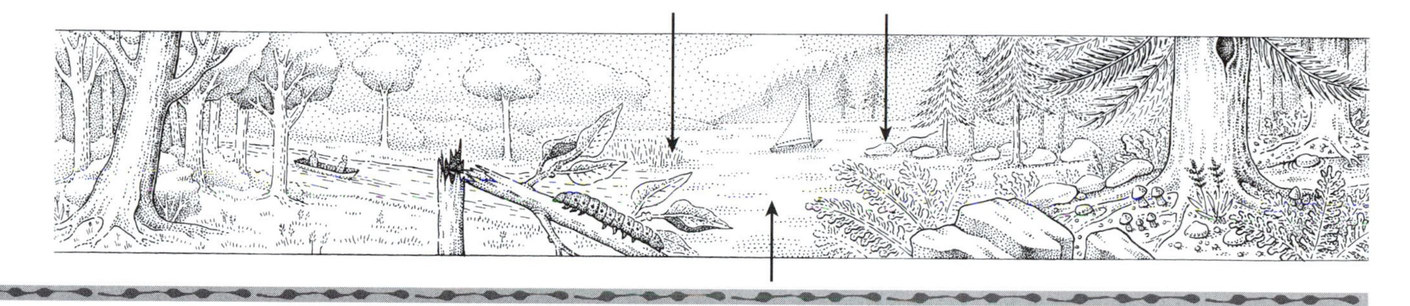

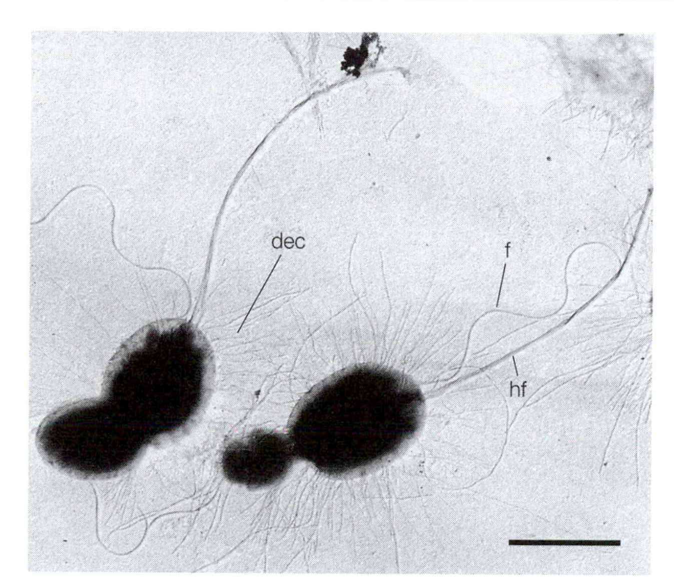

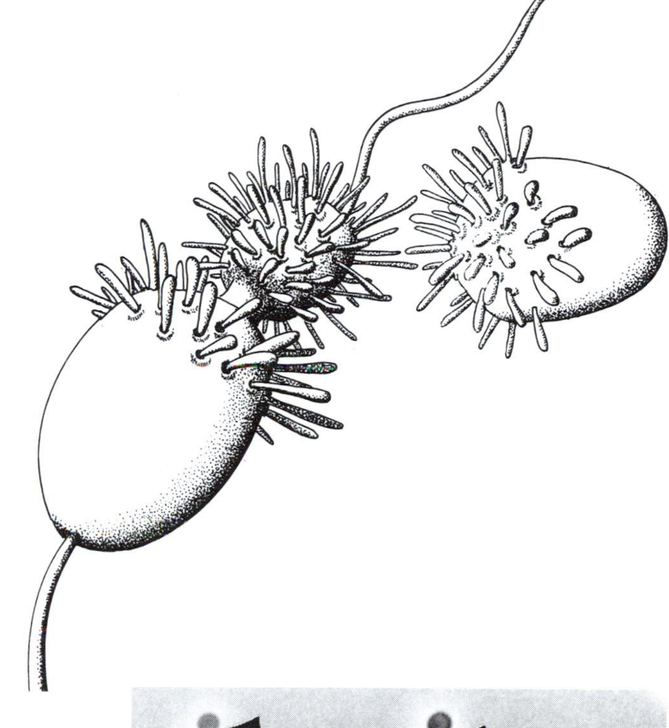

B Two budding cells of *Planctomyces maris,* anterior buds. Polar flagella (f), holdfasts (hf), and unidentified decorations (dec) are clearly visible. The two cells use their holdfasts to attach to the diatom test. TEM (negative stain, whole mount), bar = 1 μm. [Photograph courtesy of J. Staley; drawing by C. Lyons.]

C *Gemmata obscuriglobus.* Budding globular cells (arrowheads) as seen in a growing population. LM, bar = 10 μm. [Courtesy of J. Fuerst.]

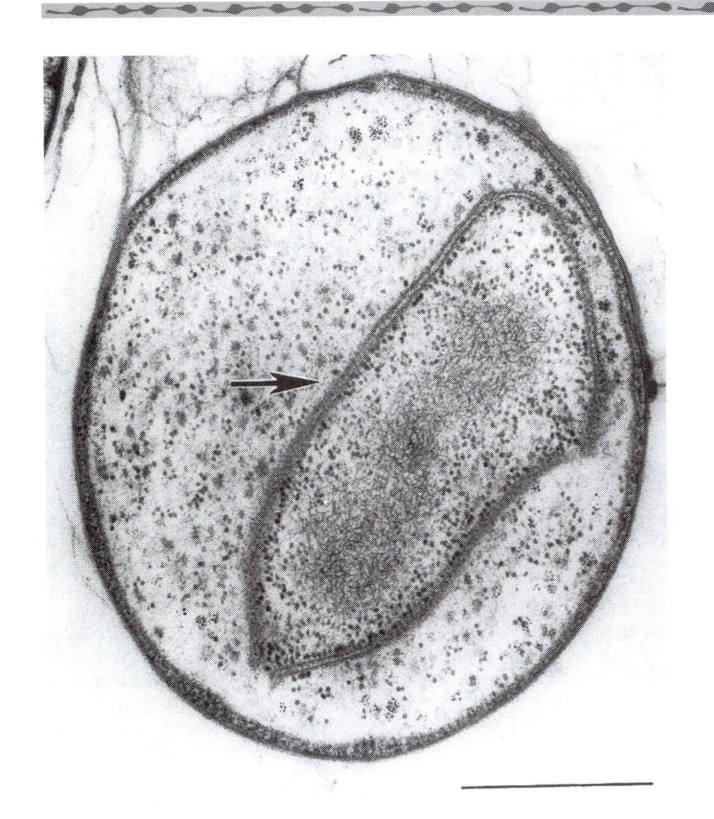

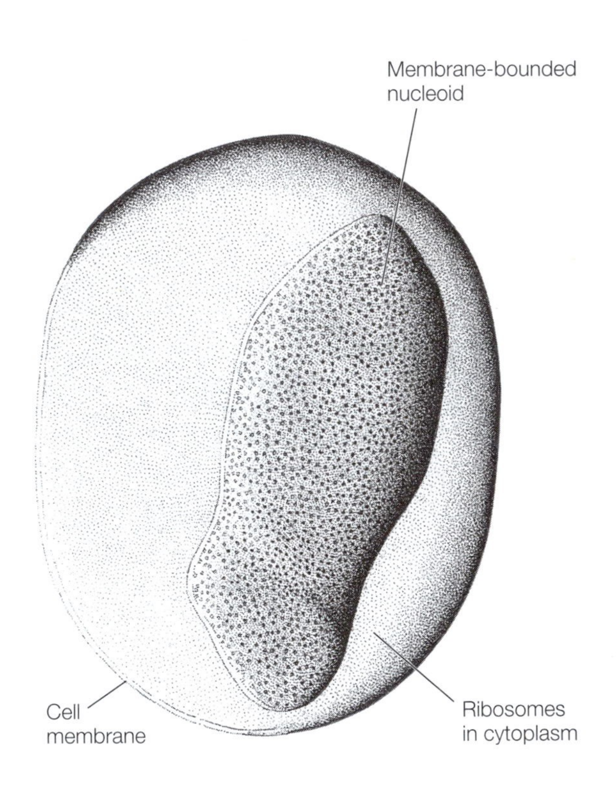

D *Gemmata obscuriglobus.* Equatorial thin section of a single cell, showing the unique membrane-bounded nucleoid (arrow). TEM, bar = 0.5 μm. [Photograph courtesy of J. Fuerst; drawing by C. Lyons.]

parrot-transmitted disease of humans, and *C. trachomatis* is associated with the trachoma type of blindness. All dwell obligately inside animal cells (Figure E), and most work on these organisms has been directed at eliminating them from human cells.

Although severely reduced in size relative to other bacteria by their obligate symbiotrophy, *Chlamydia* form propagules that are small, dense, and relatively resistant to desiccation. These structures, called elementary bodies, convert into reticulate structures capable of multiplication inside the animal cell. No known biochemistry for the metabolic production of energy molecules has been detected in purified preparations of *Chlamydia*, leading to the conclusion that they are energy parasites, dependent on ATP from the animals cells in which they reside.

The DNA content of these tiny propagules is only twice that of large viruses such as *Vaccinia*, but at least some of their RNA is standard ribosomal, suggesting that, in principle, they synthesize proteins and evolved from free-living bacteria.

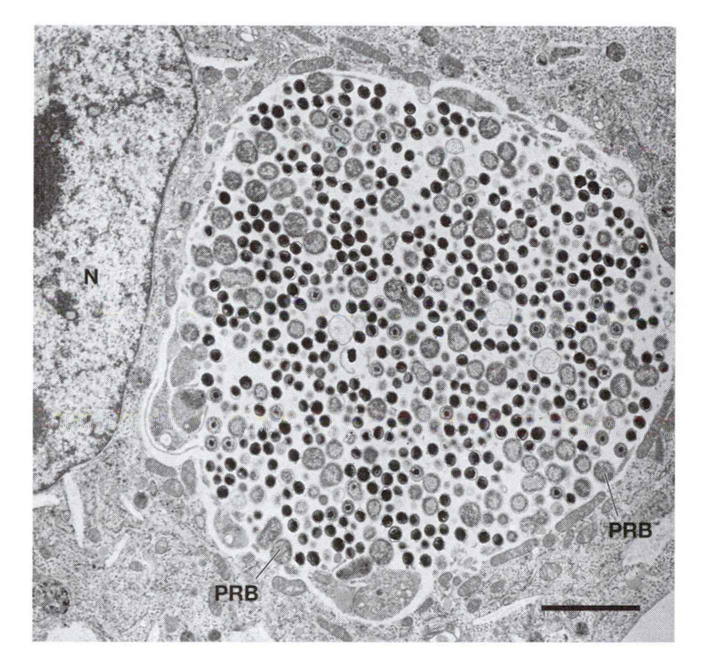

E *Chlamydia psittaci.* Elementary bodies (dark small spheres) and progeny reticulate body (PRB) of *Chlamydia* in mammalian cells in tissue culture. The nucleus (N) of the animal cell is at left. TEM, bar = 1 μm. [Courtesy of P. B. Wyrick.]

Division: Firmicutes

B-12 Actinobacteria

(Actinomycetes, actinomycota; and related high–G + C Gram-positive bacteria)

Greek *aktis,* ray; *bakterion,* little stick

EXAMPLES OF GENERA

Actinomyces	*Dermatophilus*	*Nocardia*
Actinoplanes	*Frankia*	*Propionibacterium*
Arthrobacter	*Micromonospora*	*Streptomyces*
Cellulomonas	*Mycobacterium*	*Thermoactinomyces*
Corynebacterium	*Mycococcus*	*Thermomonospora*

This phylum, uniting similar bacteria as determined by morphological, physiological, and 16S rRNA criteria, includes the coryneform bacteria and filamentous actinobacteria. The coryneforms are unicellular Gram-positive organisms, straight or slightly curved rods with a tendency to form club-shaped swellings. Corynebacterial genera include *Corynebacterium* and the long or spiny *Arthrobacter;* members of the latter genus may absorb food from nearly pure water through as many as 20 spines (cytoplasmic projections) per cell. This phylum also contains the cellulose-attacking *Cellulomonas* and *Propionibacterium,* whose species produce propionic or acetic acids as products of sugar metabolism. The offspring cells of many fissioned coryneform bacteria or actinobacteria typically remain attached in a Y- or V-shaped configuration or remain attached to form hyphaelike structures.

The multicellular actinobacteria include filamentous prokaryotes that were originally mistaken for fungi. Unfortunately, even though they are prokaryotic in all of their features, they are still sometimes called "actinomycetes." Actinobacteria are distinguished by their stringy threads, superficially like fungal hyphae, and by the production at their tips of resistant propagules, actinospores. Actinobacteria probably evolved the "fungal" habit of growing hyphae (long strings) from cells that did not separate after binary fission. Hyphae in profusion form a visible mass of filaments, called a mycelium. The actinobacterial mycelium probably evolved long before that of the true fungi, all of which are eukaryotes.

This large and diverse group of eubacteria includes some that grow septate or nonseptate multicellular filaments. In *Mycobacterium,* the tuberculosis microbe, the filaments are short, whereas *Actinoplanes* and *Streptomyces* produce long and complex filaments that form and release actinospores.

Actinospores are sometimes inaccurately called conidia, but true conidia are the eukaryotic, haploid-propagating spores of the basidio- and ascomycotous fungi (Phyla F-2 and F-3). Actinospores also differ from bacterial endospores, which are formed within a parent cell and then released (see Phylum B-10). In the development of all actinospores, the entire cell converts into a thick-walled, resistant spore. Thus, actinospores, endospores, and fungal conidia are all convergent structures—they represent a response to similar environmental pressures.

At least six of the actinobacteria groups form true actinospores, but only two enclose them in external structures. In these two (*Frankia* and *Actinoplanes*), actinospores are borne inside structures called sporangia, by analogy with fungi. The many members of the genus *Frankia* are symbiotic in plants, where, like *Rhizobium* (Phylum B-3), they induce nodules that fix atmospheric nitrogen. The complex of threads of *Actinoplanes* forms dense mycelia that are still easily mistaken for fungal hyphae.

In *Dermatophilus,* the cells of the filaments of the mycelium divide transversely and in at least two longitudinal planes to form coccoid, motile bacteria that swim away and then lose flagellar motility, settle, and develop again into filaments. Some species form pathogenic lesions on human skin; others have been isolated from soil. The motile bacteria have been miscalled zoospores, a name restricted to motile eukaryotic propagules that swim by undulipodia and are capable of continued growth. Zoospores, for example, are commonly produced by oomycotes (Phylum Pr-20), chytridiomycotes (Phylum Pr-29), and plasmodiophorans (Phylum Pr-19). The term "zoospore" ought always, as it is in this book, to be restricted to eukaryotes. A multicellular mycelial habit alternating with motile stages (for example, flagellated bacteria or undulipodiated zoospores), in which there is no sexual encounter or fusion in either stage, has evolved independently in aquatic organisms: in several types of prokaryotes [these actinobacteria and *Caulobacter* (see Phylum B-3, Figure B), for example] and in several distinctly different groups of eukaryotic microorganisms. In microbes living on land, especially in soil, the tendency to form and release aerial propagules to be scattered by wind or by other organisms has evolved on repeated occasions: in myxobacteria (Phylum B-3), in actinobacteria here, and in several protoctist and most fungal groups.

The family Nocardiaceae includes the widely distributed genus *Nocardia.* Nocardias typically form mycelial filaments that fragment, yielding single nonmotile bacteria. They tend, especially in old cultures, to be Gram-variable (some Gram-negative cells) rather than clearly Gram-positive. Nocardias are very tenacious, and many survive, but do not grow, under noxious conditions of acidity, high salt, dryness, and other environmental extremes. If their entire developmental pattern is not observed, they can easily be mistaken for unicellular bacteria. Eventually, however, they betray their nocardial nature by forming filaments, mycelia, Y- and V-shaped cell groups, and actinospores. Some pathogenic strains and at least one capable of nitrogen fixation are known.

Members of the huge group constituting *Streptomyces* (more than 500 species have been described for this genus) form mycelia that tend to remain intact (Figure A). From the mycelia grow quite remarkable and well-developed chains of aerial actinospore-bearing structures (Figures B and C). Some are easily confused, at least superficially, with the smaller fungi. This group is justly well known for its synthetic versatility in producing streptomycin and other antibiotics.

The two genera grouped with *Micromonospora*—*Thermomonospora* and *Thermoactinomyces*—are found in soil, usually on plant debris. They form spores singly, in pairs, or in short chains on either aerial or subsurface mycelia. The mycelia are

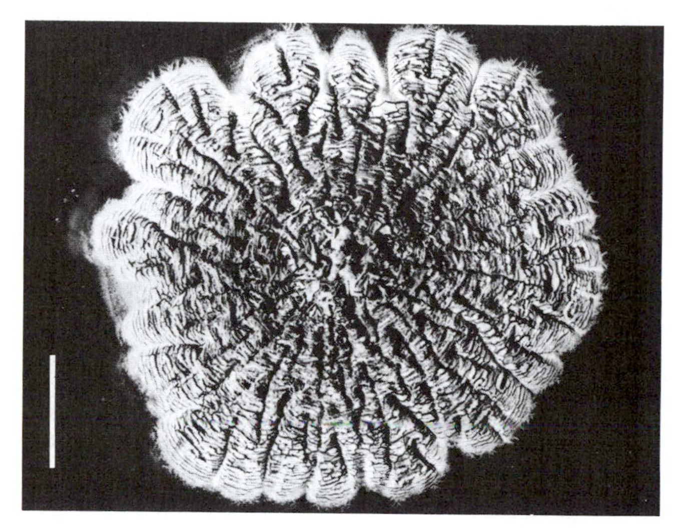

A Colony of *Streptomyces rimosus* after a few days of growth on nutrient agar in petri plates. Bar = 10 μm. [Courtesy of L. H. Huang, Pfizer, Inc.]

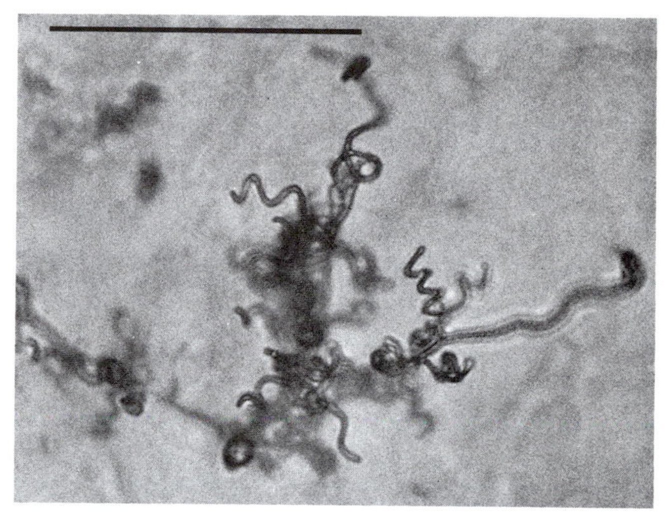

B Aerial trichomes (filaments) bearing actinospores of *Streptomyces*. LM, bar 50 = μm. [Courtesy of L. H. Huang, Pfizer, Inc.]

branched and septate, and the propagules are often brown. Some strains tolerate high temperatures.

Analysis of gene sequences of 16S rRNA confirm the validity of grouping together all the actinobacteria. This tight cluster of thousands of strains shows a range in guanosine-cytosine ratios from 51 to 79 and is called the high–G + C Gram-positive eubacteria. These diverse actinobacteria are distinguished from the genus *Bacillus* and its relative, which are low–G + C Gram-positive eubacteria (Phylum B-10). However, nature fails to cooperate with our tidy classifications: at least one strain of *Micromonospora* produces typical endospores instead of actinospores.

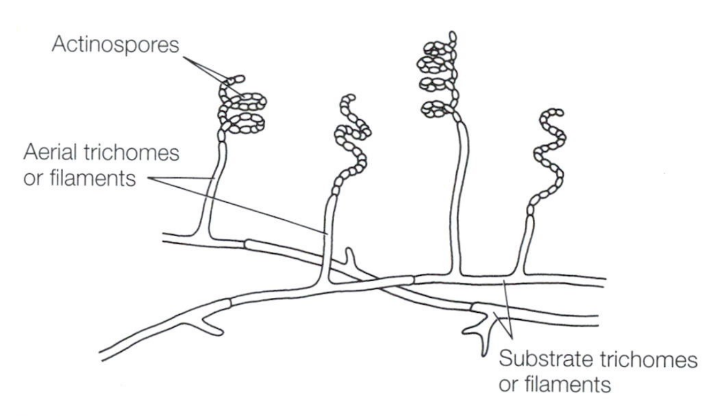

C Part of a mycelium of *Streptomyces*. [Drawing by R. Golder.]

Division: Firmicutes

B-13 Deinococci

(Radiation-resistant or heat-resistant Gram-positive bacteria)

Greek *deino,* terrible, whirling; Latin *coccus,* berry

EXAMPLES OF GENERA
Deinococcus
Thermus

Deinococci, a new phylum based primarily on uniqueness as determined by 16S rRNA gene sequences, comprises Gram-positive, highly resistant, heterotrophic bacteria that require oxygen for growth. All deinococci and many other aerobic spheres that do not form spores or other distinctive propagules were called "micrococci" in the literature before the 1990s. Although at present only two well-studied genera (*Deinococcus* and *Thermus*) are formally assigned to this phylum, the colonization of radiation- and heat-resistant surfaces by coccoid bacteria suggests that more will be revealed by further study.

The spherical cells in *Deinococcus* are found singly or grouped in pairs and characteristically divide by binary fission in more than one plane to produce tetrads, irregular cubical packets of four cells (Figures A through C). Those micrococci that morphologically, but not by 16S rRNA criteria, resemble *Deinococcus* have been removed from this phylum. *Thermus* is heat resistant (60°–80°C), whereas *Deinococcus* is radiation resistant.

Deinococci are strictly or facultatively aerobic—some can ferment, but all respire, using oxygen as the terminal electron acceptor. Because they tolerate as much as 3,000,000 rads of ionizing radiation, they are easily isolated by bombardment with radiation, a procedure that kills everything else. (The human lethal dose is about 500 rads, and other bacteria can be killed by 100 rads.) Deinococci synthesize the respiratory pigments (cytochromes) and a class of quinones, also participants in respiration, called menaquinones. Most species metabolize sugars: glucose, for example, is oxidized either to acetate or completely to carbon dioxide and water. They metabolize glucose by the hexose monophosphate pathway, rather than by the Embden-Meyerhof pathway used by eukaryotes and many heterotrophic bacteria. Some species also oxidize smaller organic compounds, such as pyruvate, acetate, lactate, succinate, and glutamate, by the citric acid, or Krebs, cycle, characteristic of mitochondria. Some deinococci can grow in hypersaline environments, such as 5 percent NaCl in water (seawater is about 3.4 percent total salts). They break down hydrogen peroxide by using the enzyme catalase. *Deinococcus radiodurans* distinctively groups with *Thermus* by 16S rRNA analysis. Both *Deinococcus* and *Thermus* have tough cell walls of strange composition. The diaminopimelic acid of the peptidoglycan is absent in *Deinococcus;* instead, ornithine, a nonprotein amino acid, is a wall component. Unlike other Gram-positive bacteria, an extra outer-membrane layer surrounds the wall. The composition of the outer membrane in *Deinococcus* differs from that of Gram-negative bacteria, which also have an outer, as well as an inner (plasma) membrane.

Deinococcus cells are protected by many-layered walls. Most *Deinococcus* strains are colored pink or red, owing to the high concentration of carotenoids in their cells. Besides having resistance to ionizing radiation, many resist desiccation and ultraviolet light treatment even better than do endospores. The remarkable resistance of *Deinococcus* to mutagenic chemicals, radiation, and other treatments that generate genetic change in most bacteria has led to failure to induce stable mutations. The genetics of these organisms is poorly understood, although it is clear that excellent enzyme repair systems for DNA lead to highly efficient recovery from attempts to induce heritable damage.

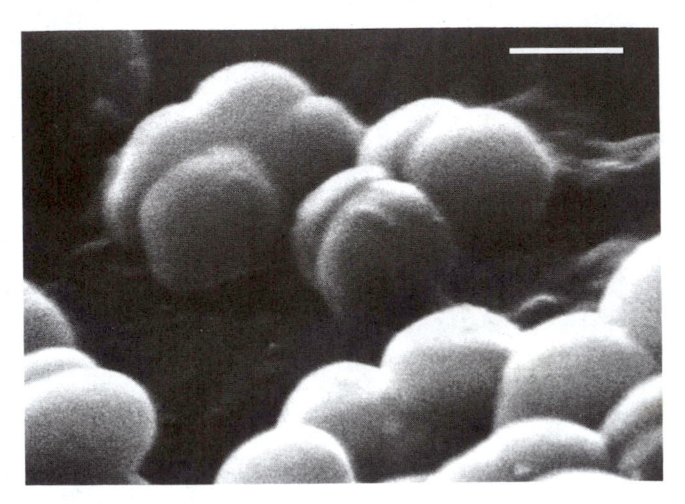

A *Deinococcus radiodurans.* SEM (whole mount), bar = 1 μm. [Courtesy of J. Troughton.]

Thermus aquaticus, discovered by Thomas Brock in studies of natural hot springs bacteria of Yellowstone National Park, Wyoming, was catapulted to fame when it became the source of the Taq polymerase. This enzyme, a DNA polymerase that is so heat resistant that it survives the temperature cycling required for the PCR (polymerase chain reaction) is now in everyday use in thousands of laboratories that sequence DNA for scientific, legal, and industrial applications. The PCR technique permits small amounts of specific pieces of DNA to be replicated (amplified).

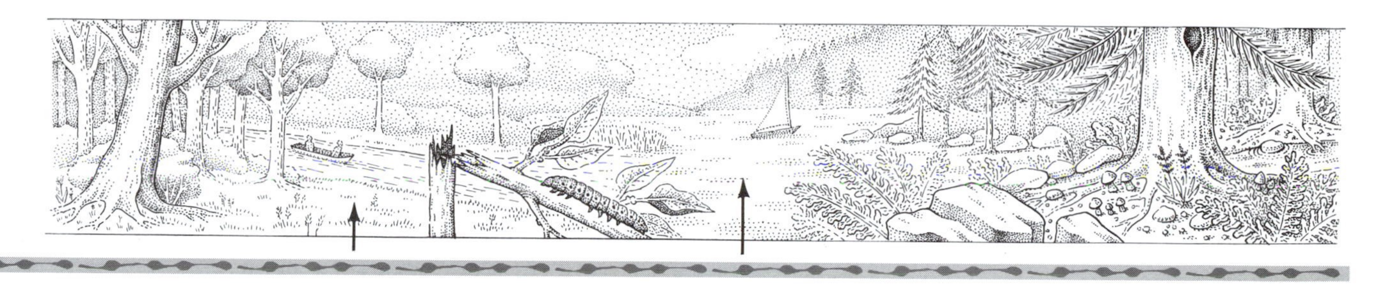

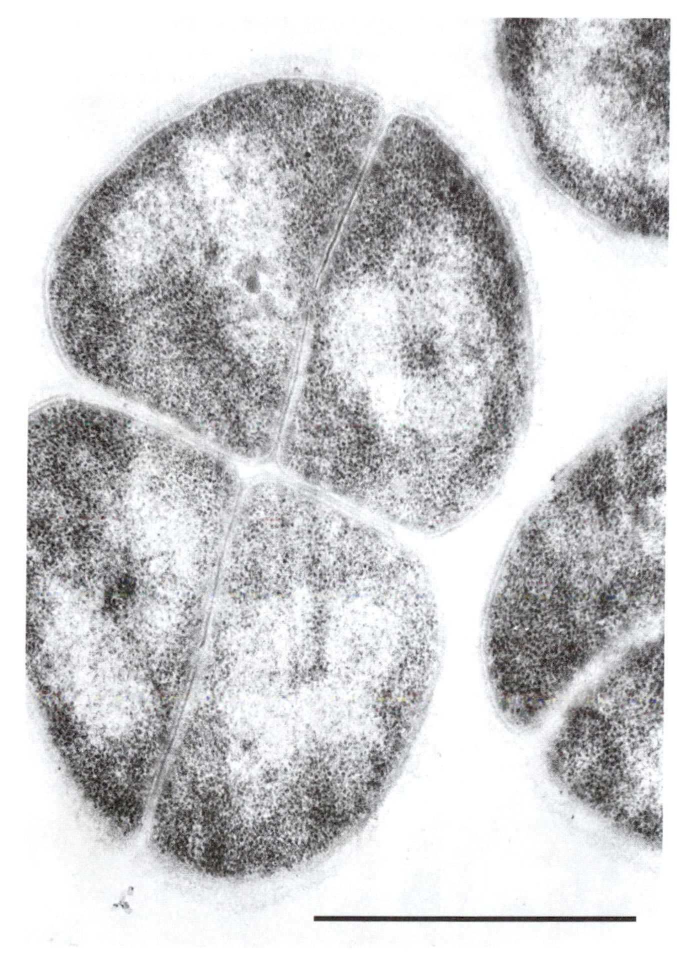

B Transverse section of packet of four radiation-resistant *Deinococcus radiodurans* cells. TEM, bar = 1 μm. [Courtesy of A. D. Burrell and D. M. Parry.]

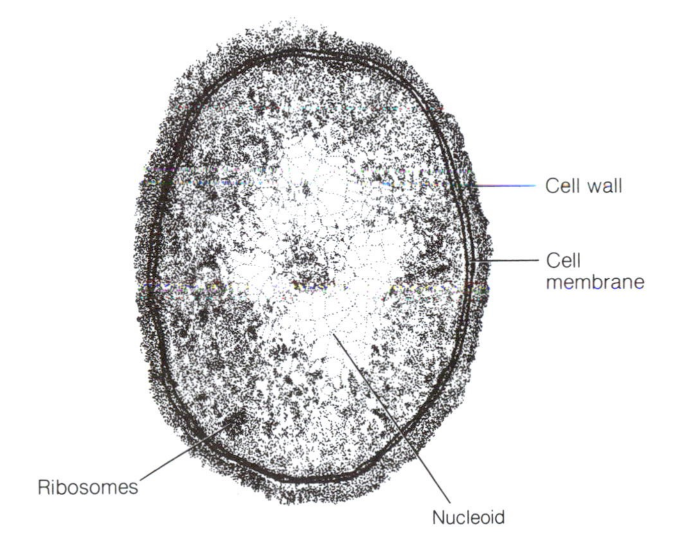

C One cell from a tetrad of *Deinococcus radiodurans.* [Drawing by I. Atema.]

Division: Firmicutes

B-14 Thermotogae

(Thermophilic fermenters)

Named after the genus *Thermotoga* (Greek *thermo,* heat; Latin, *toga,* Roman citizen garment)

EXAMPLES OF GENERA
Aquifex
Fervidobacterium
Thermosipho
Thermotoga

These newly discovered eubacteria are so strikingly different in their rRNA sequence from all others that they warrant a new phylum. Four genera have been described; no doubt many others exist. Both *Thermotoga* and *Thermosipho* are inhabitants of the hot vents at the bottom of the sea, where, as the continents separate, their spreading centers produce hot water and gases at temperatures that exceed 90°C. The differences between them are mainly morphological. *Thermotoga* is conspicuously covered by its "toga," several times thicker than and outside of its cell wall (Figures A and B). The guanosine-cytosine content of *Thermosipho* is lower than that of *Thermotoga,* another feature by which the isolates can be distinguished.

Thermotoga is exceedingly tolerant of high heat—one of the most heat-resistant eubacteria known. The cells have been cultured in the laboratory at temperatures ranging from 50°C to 90°C and grow optimally at 80°C. Although not an archaebacterium, *Thermotoga* produces unique lipidlike, extremely long chain fatty acids. Nevertheless, like bacteria, these organisms do have peptidoglycan cell walls. As anaerobes, these cells have a strict requirement to avoid oxygen. Lacking a respiratory apparatus for any reduced inorganic compounds, they are obligate fermenters of sugars and other organic compounds, including complex protein digests.

The hot deep-sea marine sediments have been under scientific study only since 1977, when the "dark gardens" replete with tubeworms, sulfur-rich clams, and other unique fauna were first discovered. Microbiologists require growth in "pure" culture (single bacterium) for the introduction of new bacterial names into the literature. Because high-temperature, high-pressure, anoxic conditions are not typically obtained in the laboratory, we believe that it is highly likely that only a few thermotogas have been described; many other relatives of thermotogas lie waiting to be discovered in the hottest nether regions of the planet.

A new one, *Thermotoga subterranea,* another strict anaerobe, was isolated as recently as 1995 from a deep, continental oil reservoir in the East Paris Basin (France). It grew at temperatures between 50°C and 75°C, with optimum growth at 70°C. This *Thermotoga* was inhibited by elemental sulfur, but reduced cystine and thiosulfate to hydrogen sulfide. The guanosine-cytosine content, the presence of a lipid structure unique to the genus *Thermotoga,* and the 16S rRNA sequence put it with the other thermotogas but not as one of the recognized species: *T. maritima, T. neapolitana,* and *T. thermarum.*

Although *Aquifex*—another hot springs, submarine, volcanic eubacterium—belongs to this group, it lacks the ability to grow on sugar or protein. Rather, it oxidizes inorganic compounds—hydrogen (H_2), elemental sulfur (S^0), or thiosulfate ($S_2O_3^{2-}$)—by using either O_2 (under low oxygen tensions, hence microaerophilically) or NO^{3-} (anaerobically).

Fervidobacterium isolated from Iceland hot springs (hence *F. islandicum*) is also very new to science. As extreme environments come under scrutiny, many new strains of bacteria are expected to surprise us.

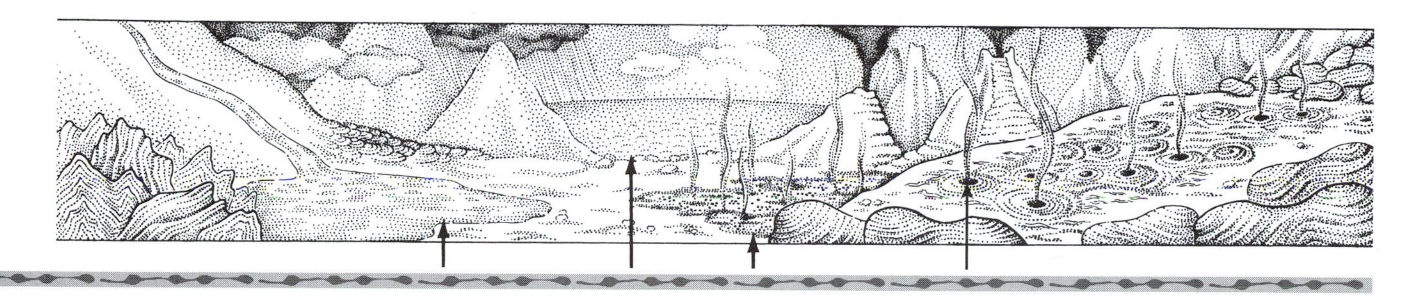

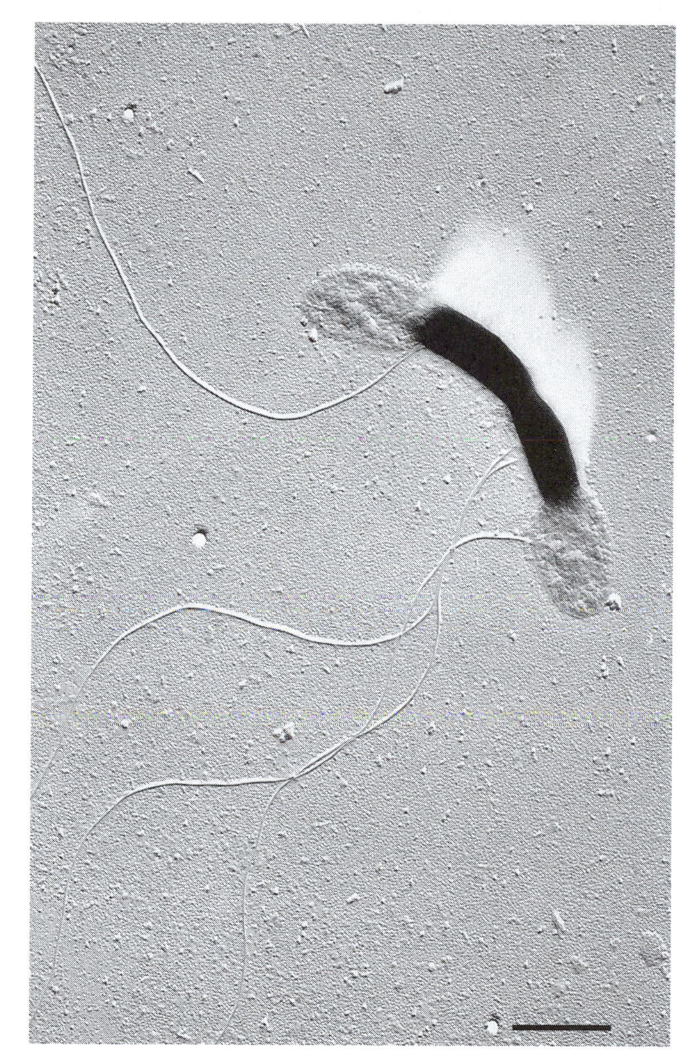

A *Thermotoga thermarum.* The two cells are in division inside the thick toga. Here, the toga extensions can be seen by shadowcasting. TEM (negative stain), bar = 1 μm. [Courtesy of K. O. Stetter and R. Huber.]

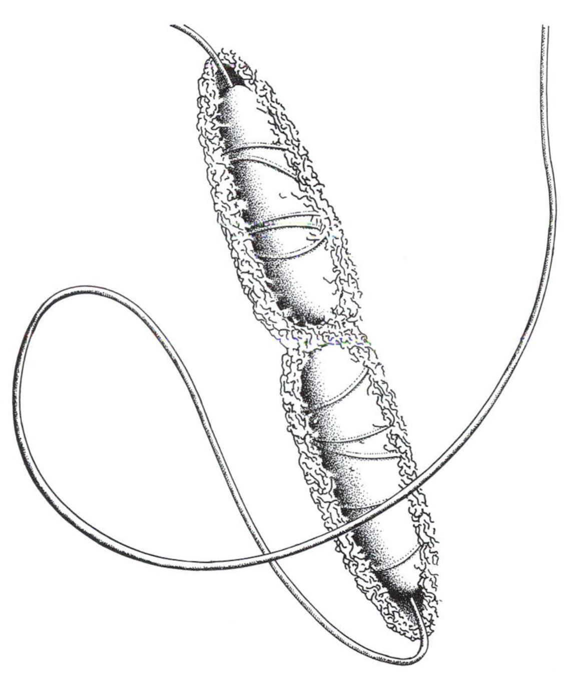

B *Thermotoga* cell in division, entirely surrounded by the toga. The composition and function of the toga that surrounds the cell and the nature of the cell projections are not known. [Drawing, based on thin-section TEM images, by C. Lyons.]

Bibliography: Bacteria

General

Balows, A., H. G. Trüper, M. Dworkin, W. Harder, and K.-H. Schleifer, eds., *The prokaryotes: A handbook on the biology of bacteria: Ecophysiology, isolation, identification, applications,* 4 vols., 2d ed. Springer-Verlag; New York; 1991.

Bengtson, S., ed., *Early life on Earth.* Columbia University Press; New York; 1994.

Krieg, N. R., and J. G. Holt, eds., *Bergey's manual of systematic bacteriology.* Williams & Wilkins; Baltimore and London; 1984 – 1989. Vol. 1, *Gram-negative bacteria of medical and commercial importance:* spirochetes, spiral and curved bacteria, Gram-negative aerobic and facultatively aerobic rods, Gram-negative obligate anaerobes, Gram-negative aerobic and anaerobic cocci, sulfate and sulfur-reducing bacteria, rickettsias and chlamydias, mycoplasmas. Vol. 2, *Gram-positive bacteria of medical and commercial importance:* Gram-positive cocci, Gram-positive endospore-forming and non-spore-forming rods, mycobacteria, nonfilamentous actinomycetes. Vol. 3, *Remaining Gram-negative bacteria:* phototrophic, gliding, sheathed, budding, and appendaged bacteria, cyanobacteria, lithotrophic bacteria, and the archaeobacteria. Vol. 4, *Filamentous actinomycetes and related bacteria.*

Dixon, B., *Power unseen: How microbes rule the world.* W. H. Freeman and Company; New York; 1994.

Ferry, J. G., *Methanogenesis: Ecology, physiology, biochemistry and genetics.* Chapman & Hall; New York; 1993.

Gillies, R. R., and T. C. Dodds, *Bacteriology illustrated,* 4th ed. Churchill; New York; 1976.

Ingraham, J., and K. Ingraham, *Introduction to microbiology.* Wadsworth; Belmont, CA; 1995.

Madigan, M. T., *Brock's biology of microorganisms,* 8th ed. Prentice-Hall; Englewood Cliffs, NJ; 1996.

Margulis, L., *Early life.* Jones and Bartlett; Boston, MA; 1982.

Sagan, D., and L. Margulis, *Garden of microbial delights: A practical guide to the subvisible world.* Kendall-Hunt; Dubuque, IA; 1993.

Schleifer, K. H., and E. Stackebrandt, eds. *Evolution of prokaryotes.* Academic Press; London; 1985.

Sneath, P. H. A., ed., *International code of nomenclature of bacteria.* American Society for Microbiology; Washington, DC; 1992.

Sonea, S., and M. Panisset, *A new bacteriology.* Jones and Bartlett; Boston; 1982.

Starr, M. P., H. Stolp, H. G. Trüper, A. Balows, and H. G. Schlegel, eds., *The prokaryotes: A handbook on habitats, isolation and identification of bacteria,* 4 vols., 2d ed. Springer-Verlag; Heidelberg and New York; 1991.

Woese, C. R., O. Kandler, and M. L. Wheelis, "Towards a natural system of organisms: Proposal for the domains Archaea, Bacteria, and Eucarya." *Proceedings of the National Academy of Sciences, USA* 87:4576–4579; 1990.

Woese, C. R., and R. S. Wolfe, *The bacteria: A treatise on structure and function,* Vol. 3: *Archaebacteria.* Academic Press; New York; 1985.

Wolfe, S., *The biology of the cell,* 4th ed. Wadsworth; Belmont, CA; 1995.

B-1 Euryarchaeota

Bult, C. J., et al., "Complete genome sequence of the methanogenic archaeon, *Methanococcus jannaschii.*" *Science* 273:1058–1073; 1996.

Madigan, M. T., *Brock's biology of microorganisms,* 8th ed. Prentice-Hall; Englewood Cliffs, NJ; 1996.

Mah, R. A., D. M. Ward, L. Baresi, and L. Glass, "Biogenesis of methane." *Annual Review of Microbiology* 31:309–341; 1977.

Morell, V., "Life's last domain." *Science* 273:1043–1045; 1996.

Woese, C. R., "Archaebacteria." *Scientific American* 244(6): 98–122; June 1981.

B-2 Crenarchaeota

Kandler, O., "Evolution of the systematics of bacteria." In K. H. Schleifer and E. Stackebrandt, eds., *Evolution of prokaryotes.* FEMS Symposium 29. Academic Press; London; pp. 335–361; 1985.

Lanyi, J. K., "Physical chemistry and evolution of salt tolerance in halobacteria." In C. Ponnamperuma and L. Margulis, eds., *Limits of life.* D. Reidel; Dordrecht, the Netherlands; pp. 61–68; 1980.

B-3 Proteobacteria

Barton, L. L., ed., *Sulfate-reducing bacteria.* Plenum Press; New York; 1995.
Bermudes, D., L. Margulis, and G. Hinkle, "Do prokaryotes have microtubules?" *Microbiological Reviews* 58:387–400, 1994.
Blankenship, R. E., M. T. Madigan, and C. E. Bauer, eds., *Anoxygenic photosynthetic bacteria.* Kluwer Academic; Dordrecht, the Netherlands; 1995.
De Ley, J., "Proteobacteria (purple bacteria)." In A. Balows et al., eds., *The prokaryotes,* Vol. 2, 2d ed. Springer-Verlag; New York; pp. 2111– 2140; 1992.
Frederickson, J. K., and T. C. Onsott, "Microbes deep inside the earth." *Scientific American* 275(4):68–73; October 1996.
Schlegel, H. G., and B. Bowien, eds., *Autotrophic bacteria.* Springer-Verlag; Heidelberg; 1989.
Schlegel, H. G., and H. W. Jannasch, "Prokaryotes and their habitats." In A. Balows et al., eds., *The prokaryotes,* Vol. 1, 2d ed. Springer-Verlag; New York; pp. 75–125; 1992.
Shapiro, J. A., "Bacteria as multicellular organisms." *Scientific American* 258(6):82–89; June 1988.

B-4 Spirochaetae

Bermudes, D., D. Chase, and L. Margulis, "Morphology as a basis for taxonomy of large spirochetes symbiotic in wood-eating cockroaches and termites: *Pillotina* gen. nov., nom. rev.; *Pillotina calotermitidis* sp. nov., nom. rev.; *Diplocalyx* gen. nov., nom. rev.; *Diplocalyx calotermitidis* sp. nov., nom. rev.; *Hollandina* gen. nov., nom. rev.; *Hollandina pterotermitidis* sp. nov., nom. rev.; and *Clevelandina reticulitermitidis* gen. nov., sp. nov." *International Journal of Systematic Bacteriology* 38:291–302; 1988.
Breznak, J. A., "Genus II. *Cristispira* Gross, 1910, 44." In N. R. Krieg and J. G. Holt, eds., *Bergey's manual of systematic bacteriology,* Vol. 1. Williams & Wilkins; Baltimore; pp. 46–49; 1984.
Canale-Parola, E., "The genus *Spirochaeta.*" In A. Balows et al., eds., *The prokaryotes,* Vol. 4, 2d ed. Springer-Verlag; New York; pp. 3524– 3536; 1992.
Holt, S. C., "Anatomy and chemistry of spirochetes." *Bacteriological Reviews* 42:114–160; 1978.
Margulis, L., and G. Hinkle, "Large symbiotic spirochetes: *Clevelandina, Cristispira, Diplocalyx, Hollandina,* and *Pillotina.*" In A. Balows et al., eds., *The prokaryotes,* Vol. 4, 2d ed. Springer-Verlag; New York; pp. 3965–3978; 1992.

B-5 Cyanobacteria

Carr, N., and B. Whitton, eds., *The biology of the blue-green algae.* Blackwell; Oxford, England; 1973.
Cohen, Y., and E. Rosenberg, eds., *Microbial mats.* American Society for Microbiology; Washington, DC; 1989.
Golubic, S., "Microbial mats of Abu Dhabi." In L. Margulis and L. Olendzenski, eds., *Environmental evolution.* MIT Press; Cambridge, MA; pp. 103–130; 1992.
Golubic, S., "Stromatolites of Shark Bay." In L. Margulis and L. Olendzenski, eds., *Environmental evolution.* MIT Press; Cambridge, MA; pp. 131–148; 1992.
Golubic, S., M. Hernandez-Marine, and L. Hoffman, "Developmental aspects of branching in filamentous Cyanophyta/Cyanobacteria." *Algological Studies* 83:303–329; 1996.

B-6 Saprospirae

Reichenbach, H., and M. Dworkin, "Introduction to the gliding bacteria." In M. P. Starr, ed., *The prokaryotes.* Springer-Verlag; Berlin; pp. 315–327; 1981.

B-7 Chloroflexa

Fuller, C., ed. *Phototrophic bacteria.* Proceedings of the second international meeting. ASM Press; Washington, DC; 1991.
Stolz, J. F., ed., *Structure of photosynthetic bacteria.* CRC Press; Boca Raton, FL; 1990.

B-8 Chlorobia

Pearson, B., and R. Castenholz, "The family Chloroflexaceae."

In A. Balows et al., eds., *The prokaryotes,* Vol. 4, 2d ed. Springer-Verlag; New York; pp. 3754–3774; 1992.
Stolz, J. F., ed., *Structure of photosynthetic bacteria.* CRC Press; Boca Raton, FL; 1990.

B-9 Aphragmabacteria

Fraser, C. M., et al., "The minimal complement of *Mycoplasma genitalium.*" *Science* 270:397–403; 1995.
Gordon, R. E., W. C. Haynes, and C. H. N. Pang, *The genus* Bacillus. Handbook No. 427, U.S. Department of Agriculture; Washington, DC; 1973.
Razin, S., and E. A. Freundt, "Mycoplasmas." In N. R. Krieg and J. G. Holt, eds., *Bergey's manual of systematic bacteriology,* Vol. 3. Williams & Wilkins; Baltimore; pp. 742–775; 1986.
Woese, C. R., E. Stackebrandt, and W. Ludwig, "What are mycoplasmas?: The relationship of tempo and mode in bacterial evolution." *Journal of Molecular Evolution* 21:305–316; 1985.

B-10 Endospora

Breznak, J. A., J. M. Switzer, and H.-J. Seitz, "*Sporomusa termitide* sp. nov., an H_2/CO_2-utilizing acetogen isolated from termites. *Archives of Microbiology* 150:282–288; 1988.
Clements, K. D., and S. Bullivant, "An unusual symbiont from the gut of surgeonfishes may be the largest known prokaryote." *Journal of Bacteriology* 173:5359–5362; 1991.
Slepecky, R. A., and H. E. Hemphill, "The genus *Bacillus*—nonmedical." In A. Balows et al., eds., *The prokaryotes,* Vol. 2, 2d ed. Springer-Verlag; New York; pp. 1663–1696; 1992.
Sonenschein, A. L., J. A. Hoch, and R. Losick, Bacillus subtilis *and other gram positive bacteria: Biochemistry, physiology and molecular genetics.* American Society of Microbiology; Washington, DC; 1993.

B-11 Pirellulae

Franzmann, P. D., and V. B. D. Skerman, "*Gemmata obscuriglobus,* a new genus and species of the budding bacteria." *Antonie van Leeuwenhoek* 50:261–268; 1984.
Fuerst, J. A., "The planctomycetes: Emerging models for microbial ecology, evolution and cell biology." *Microbiology* 141:1493–1506; 1995.
Fuerst, J. A., and R. I. Webb, "Membrane-bounded nucleoid in the eubacterium *Gemmata obscuriglobus,*" *Proceedings of the National Academy of Sciences, USA* 88:8184–8188; 1991.
Staley, J. T., J. A. Fuerst, S. Giovannoni, and H. Schlesner, "The order Planctomycetales and the genera *Planctomyces, Pirellula, Gemmata,* and *Isosphaera.*" In A. Balows et al., eds., *The prokaryotes,* Vol. 4, 2d ed. Springer-Verlag; New York; pp. 3710–3730; 1992.

B-12 Actinobacteria

Callwell, R., and D. Callwell, *Actinomycetes and streptomycetes.* American Society of Microbiology; Washington, DC; 1996.
Ensign, J., "Actinomycetes." In A. Balows et al., eds., *The prokaryotes,* Vol. 1, 2d ed. Springer-Verlag; New York, pp. 811–815; 1992.

B-13 Deinococci

Kocur, M., W. E. Kloos, and K.-H. Schleifer, "The genus *Micrococcus.*" In A. Balows, et al., eds., *The prokaryotes,* Vol. 2, 2d ed. Springer-Verlag: New York; pp. 1300–1311; 1992.

B-14 Thermotogae

Jeanthon, C., A. L. Reysenbach, S. L'Haridon, A. Gambacorta, N. R. Pace, P. Glenat, D. Prieur, "*Thermotoga subterranea* sp. nov., a new thermophilic bacterium isolated from a continental oil reservoir." *Archives of Microbiology* 164(2):91–97; 1995.
Kristjansson, J. K., ed., *Thermophilic bacteria.* CRC Press; Boca Raton, FL; 1992.
Schliefer, B., ed., *Deep-sea bacteria.* CRC Press; Boca Raton, FL; 1996.

SUPERKINGDOM
EUKARYA

Organisms composed of cells containing more than two chromosomes per cell that reproduce by mitosis. In mitosis, pore-studded membrane-bounded nuclei totally or partly dissolve and re-form as two offspring nuclei. Display viable cytoplasmic and nuclear fusion and their reciprocal processes (for example, cytoplasmic and chromosomal doubling with subsequent meiotic or equivalent reduction of cytoplasmic and nuclear content; hence, Mendelian genetics). Cells with tubulin-actin cytoskeletons capable of intracellular motility. Lack unidirectional gene (naked DNA) transfer.

A-6 *Bolinopsis infundibulum* [Courtesy of M. S. Laverack.]

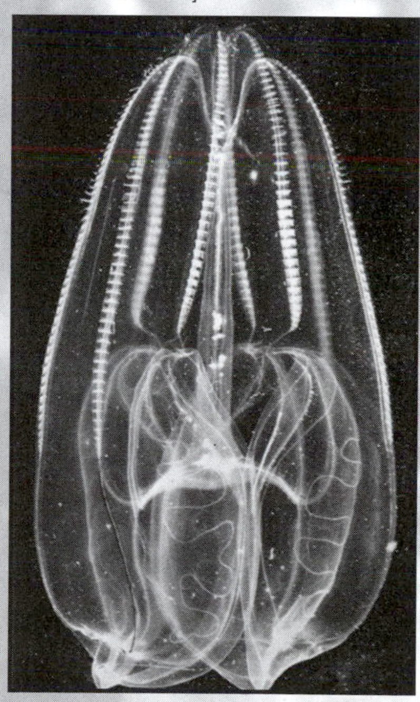

Superkingdom Eukarya
- Kingdom Protoctista
- Kingdom Animalia
- Kingdom Fungi
- Kingdom Plantae

CHAPTER TWO

PROTOCTISTA

PR-16 *Diploneis smithii*

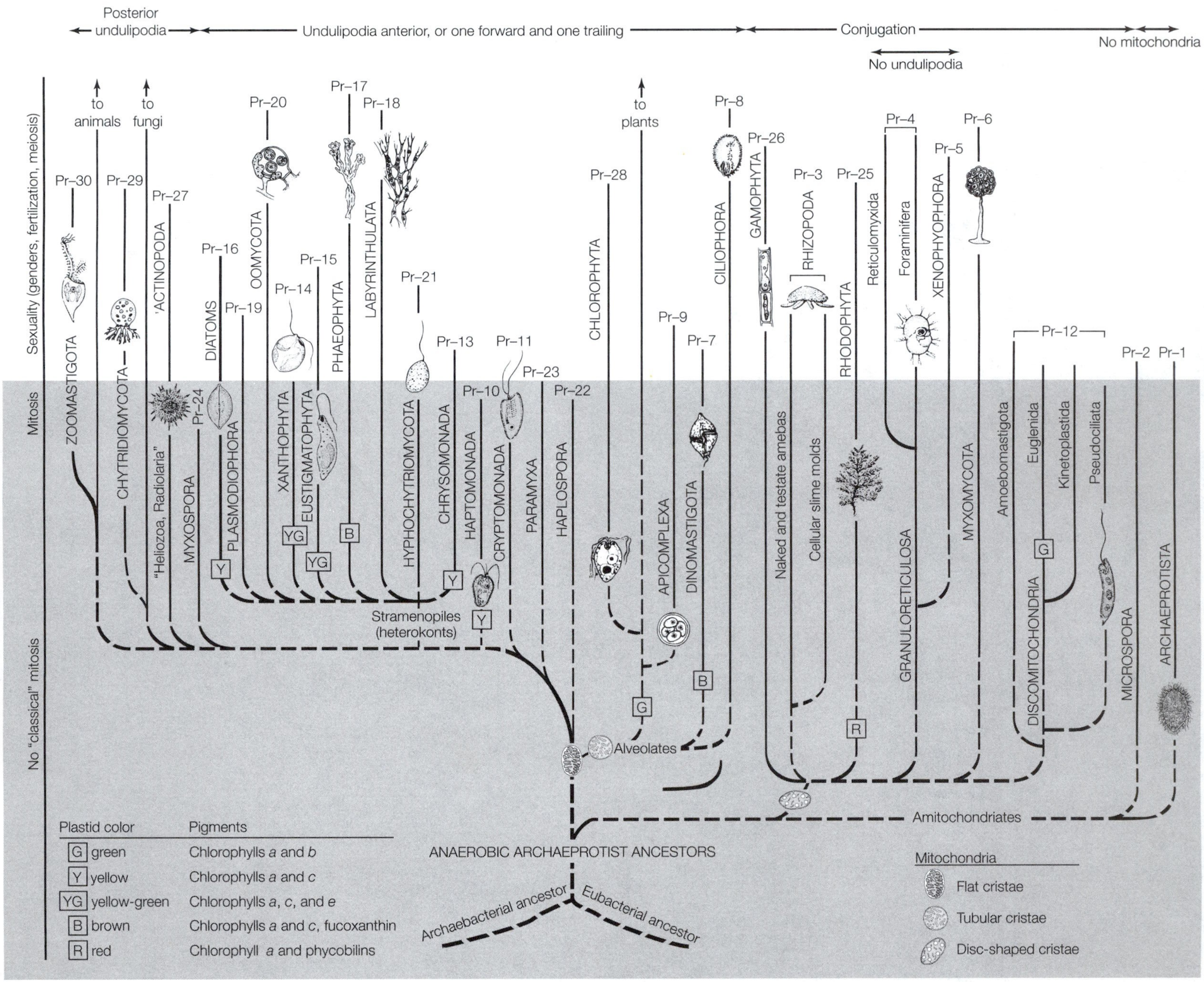

Posterior undulipodia
Undulipodia anterior, or one forward and one trailing
Conjugation
No undulipodia
No mitochondria

Sexuality (genders, fertilization, meiosis)
Mitosis
No "classical" mitosis

Pr-30 ZOOMASTIGOTA
to animals
Pr-29 CHYTRIDIOMYCOTA
to fungi
Pr-27 "ACTINOPODA
"Heliozoa, Radiolaria"
Pr-24 MYXOSPORA
Pr-16 DIATOMS
Pr-19 PLASMODIOPHORA
Pr-20 OOMYCOTA
Pr-14 XANTHOPHYTA
Pr-15 EUSTIGMATOPHYTA
Pr-17 PHAEOPHYTA
Pr-18 LABYRINTHULATA
Pr-21 HYPHOCHYTRIOMYCOTA
Pr-13 CHRYSOMONADA
Pr-10 HAPTOMONADA
Pr-11 CRYPTOMONADA
Pr-23 PARAMYXA
Pr-22 HAPLOSPORA
Stramenopiles (heterokonts)
Pr-28 CHLOROPHYTA
to plants
Pr-9 APICOMPLEXA
Pr-7 DINOMASTIGOTA
Pr-8 CILIOPHORA
Alveolates
Pr-26 GAMOPHYTA
Pr-3 RHIZOPODA
Naked and testate amebas
Cellular slime molds
Pr-25 RHODOPHYTA
Reticulomyxida
Pr-4 Foraminifera
GRANULORETICULOSA
Pr-5 XENOPHYOPHORA
Pr-6 MYXOMYCOTA
Amoebomastigota
Pr-12 Euglenida
Kinetoplastida
Pseudociliata
DISCOMITOCHONDRIA
Pr-2 MICROSPORA
Pr-1 ARCHAEPROTISTA
Amitochondriates

ANAEROBIC ARCHAEPROTIST ANCESTORS
Archaebacterial ancestor
Eubacterial ancestor

Plastid color	Pigments
G green	Chlorophylls *a* and *b*
Y yellow	Chlorophylls *a* and *c*
YG yellow-green	Chlorophylls *a*, *c*, and *e*
B brown	Chlorophylls *a* and *c*, fucoxanthin
R red	Chlorophyll *a* and phycobilins

Mitochondria
Flat cristae
Tubular cristae
Disc-shaped cristae

PROTOCTISTA

Greek *protos,* very first; *ktistos,* to establish

Nucleated microorganisms and their descendants, exclusive of fungi, animals, and plants; evolved by integration of former microbial symbionts. Nonmeiotic or meiotic with variations in the meiosis-fertilization cycle. Fossil record extends from the Lower Middle Proterozoic era (about 1.2 billion years ago) to the present.

Kingdom Protoctista comprises the eukaryotic microorganisms and their immediate descendants: all algae, including the seaweeds, undulipodiated (flagellated) water molds, the slime molds and slime nets, the traditional protozoa, and other even more obscure aquatic organisms. Its members are not animals (which develop from a blastula), plants (which develop from an embryo), or fungi (which lack undulipodia and develop from spores). Nor are protoctists prokaryotes. All protoctist cells have nuclei and other characteristically eukaryotic features. Many photosynthesize (have plastids), most are aerobes (have mitochondria), and most have [9(2)+2] undulipodia with their kinetosome bases (see Figure I-2) at some stage of the life cycle. All protoctists evolved from symbioses between at least two different kinds of bacteria—in some cases, between many more than two. As the symbionts integrated, a new level of individuality appeared.

Many different combinations of ancient bacteria into symbiotic consortia did not pass the test of natural selection. But those that survived gave rise to modern-day protoctist lineages (see the figure on the facing page), which may be grouped according to their organelle structure. In the mitochondrion, for example, the most essential (and therefore slowly evolving) membranous structures are the cristae. These structures may be flat [as in the stramenopiles (Phyla Pr-13 through Pr-21), chytrids (Phylum Pr-29), and zoomastigotes (Phylum Pr-30)]; tubular [as in the alveolates (Phyla Pr-7 through Pr-9)]; discoid [as in the amebas (Phyla Pr-3 through Pr-6), slime molds, and discomitochondriates (Phylum Pr-12)]; or altogether absent [as in the archaeprotists (Phylum Pr-1) and microsporans (Phylum Pr-2)]. Photosynthetic pigment profiles, essential to chloroplast function, are major criteria similarly employed by taxonomists to resolve the bewildering diversity of Kingdom Protoctista.

Undulipodia and their insertions, the kinetosomes always embedded in kinetids (see Figure I-3), are crucial to an understanding of protoctists. Undulipodia were present in common ancestors to all the phyla even before mitochondria, given that the anaerobic archaeprotists bear them. Their behavior as they move and reproduce is related to mitotic cell division. In some phyla, all members bear undulipodia; in other phyla, they are absent; but most protoctists produce and retract them as a function of their life histories. Although the importance of the undulipodia that develop from kinetosomes is emphasized by all who study protoctists—algologists, invertebrate zoologists, microbiologists, mycologists, parasitologists, protozoologists, and others—some feel that the use of the term "flagella" should be retained. But "flagella" are entirely unrelated rotary structures of bacteria (see Figure I-3), and hence the word, when applied to cilia, sperm tails, and other undulipodia, is confusing.

Why "protoctist" rather than "protist"? Since the nineteenth century, the word protist, whether used informally or formally, has come to connote a single-celled organism. In the past two decades, however, the basis for classifying single-celled organisms separately from multicellular ones has weakened. Multicellularity evolved many times in unicellular organisms—many multicellular beings are far more closely related to certain unicells than they are to other multicellular organisms. For example, the ciliates (Phylum Pr-8, Ciliophora), most of which are unicellular microbes, include at least one species that forms a sorocarp, a multicellular cyst-bearing structure. Euglenids (Phylum Pr-12, Discomitochondriates), chrysomonads (Phylum Pr-13, planktonic), and diatoms (Phylum Pr-16) also evolved multicellular descendants.

Here we adopt the concept of protoctist propounded in modern times by American botanist Herbert F. Copeland in 1956. The word had been introduced by English naturalist John Hogg in 1861 to designate "all the lower creatures, or the primary organic beings;—both *Protophyta*, . . . having more the nature of plants; and *Protozoa* . . . having rather the nature of animals." Copeland recognized, as had several scholars in the nineteenth century, the absurdity of referring to giant kelp by the word "protist," a term that had come to imply unicellularity and, thus, smallness. He proposed an amply defined Kingdom Protoctista to accommodate certain multicellular organisms as well as the unicells that may resemble their ancestors—for example, kelp as well as the tiny brownish cryptomonad alga *Nephroselmis.* The protoctist kingdom thus defined also solves the problem of blurred boundaries that arises if the unicellular organisms are assigned to the intrinsically multicellular kingdoms.

Attempting to reconcile ultrastructural and genetic information with newly acquired molecular data, we here propose 30 protoctist phyla. This number is more a matter of taste than tradition, because no rules for defining protoctist phyla are enforced. Our groupings are debatable; for example, some argue that the cellular and plasmodial slime molds (Phyla Pr-23 and Pr-6, Paramyxa and Myxomycota, respectively) should be united. Some believe that the oomycotes, hyphochytrids, and chytrids—which we place in Phyla Pr-20, Pr-21, and Pr-29, respectively—are really fungi and that chlorophytes (Phylum Pr-28) are plants. Some insist that chaetophorales and prasinophytes, which here are within Chlorophyta (Phylum Pr-28), ought to be raised to phylum status. Most would reunite conjugating green algae (Phylum Pr-26, Gamophyta) with the others in Chlorophyta (Phylum Pr-28). There are arguments for and against these views.* Our system has the advantage of limiting the number of highest taxa and precisely defining the three kingdoms of large organisms. Although it has the disadvantage that these eukaryotes have little in common with one another, grouping together xenophyophores, cercomonads, water molds, and the others as the single kingdom Protoctista is superior to the tradition of ignoring them entirely.

Protoctists are aquatic: some marine, some freshwater, some terrestrial in moist soil, and some parasitic or symbiotic in moist tissues of others. Nearly all animals, fungi, and plants—perhaps all—have protoctist associates. Phyla such as Microspora (Phylum Pr-2) and Apicomplexa (Phylum Pr-9) include myriad species, all of which live in the tissues of others.

No one knows the number of protoctist species. Although 40,000 extinct foraminifera alone are documented in the paleontological literature and more than 10,000 live protoctists are described in the biological literature, Georges Merinfeld (Dalhousie University, Halifax, Nova Scotia) esti-

---

* References to alternative kingdoms and phyla are listed in the bibliography.

mates that there are more than 65,000 extant species and John Corliss (University of Maryland) suggests that there are more than 250,000. Water molds and plant parasites are described in the literature on fungi, parasitic protozoa in the medical literature, algae by botanists, and free-living protozoa by zoologists. Contradictory practices of describing and naming species have led to confusion that this book attempts to dispel. Another problem is that much protoctist diversity is in tropical regions, where scientists are scarce. Furthermore, the documentation of new species of protoctists often requires time-consuming life-cycle and ultrastructural study. Most funding is limited to those temperate-zone protoctists that are sources of food, industrial products, or disease.

Remarkable variation in cell organization, patterns of cell division, and life cycle is evident in this diverse group of eukaryotic microbes and their relatives. Whereas the algae are oxygenic phototrophs, the others are heterotrophs that ingest or absorb their food. In many, the type of nutrition varies with condition: they photosynthesize when light is plentiful and feed in the dark. Although protoctists are more diverse in life style and nutrition than are animals, fungi, or plants, metabolically they are far less diverse than bacteria.

Increasing knowledge about the ultrastructure, genetics, life cycle, developmental patterns, chromosomal organization, physiology, metabolism, fossil history, and especially the molecular systematics of protoctists has revealed many differences between them and animals, fungi, and plants. The major protoctist groups, described here as phyla or groups of phyla, are so distinct as to deserve kingdom status in the minds of some authors, as explained in the *Handbook of Protoctista* (Margulis, Corliss, Melkonian, and Chapman, editors) and the *Illustrated Glossary of Protoctista* (Margulis, McKhann, and Olendzenski, editors). The *Glossary* contains seven taxonomic tables including the classes, informal names, and summaries of the technical criteria for distinguishing these groups. New molecular biological data relating the protoctist taxa are described by Mitchell Sogin, and the book contains an organism glossary (after the general glossary), where hundreds of taxonomic categories and names in current use are defined and depicted. Because no single person or group can master all the biological details of the protoctists, we expect years of animated discussion about their optimal taxonomy. With awe for protoctist diversity, a recognition of their common eukaryotic heritage, and a sense of humility toward both their complexity and our ignorance, we present our 30 protoctist phyla.

Pr-1 Archaeprotista

Greek *karyon,* nucleus, kernel; *blastos,* bud, sprout

EXAMPLES OF GENERA
Barbulanympha
Blastocrithidia
Calonympha
Chilomastix
Coronympha
Deltotrichonympha
Devescovina
Dinenympha
Foaina
Giardia
Herpetomonas
Hexamastix
Hexamita
Histomonas
Holomastigotoides
Joenia
Mastigamoeba
Mastigina
Metadevescovina
Microthopalodia
(continued)

Although the vast majority of nucleated organisms—in fact, all of them except those in this phylum and one other (Phylum Pr-2)—are aerobes with a mandate to take in atmospheric oxygen, the ancestors of eukaryotes were originally anaerobes, killed by oxygen. Living in habitats that recall ancient anoxic environments, members of this phylum are relicts of those early days of eukaryotic life: all are anaerobes. Molecular phylogenies show that these anaerobes branch off very early from all other eukaryotes, and no evidence that they ever did have mitochondria exists. New knowledge, especially from coupling ultrastructure, physiology, and sequence studies of nucleotides in ribosomal RNA, has led to impressive new insights about our ultimate eukaryotic ancestors. Because every member of every taxon here lacks mitochondria, this new information leads biologists to agree that these organisms were without mitochondria from the onset of their evolution. (It is more logical to argue that mitochondria were never in the ancestors than that mitochondria disappeared independently in many aerobic species without leaving a trace of their former presence.) The former phylum Zoomastigina has therefore been dissolved, with members lacking mitochondria now placed in this phylum and those with mitochondria in Phylum Pr-12 (Discomitochondriates).

Phylum Archaeprotista and a second phylum, Microspora (Phylum Pr-2), are best grouped as the amitochondriates; other names for them in the literature are hypochondriates and Archaeozoa.

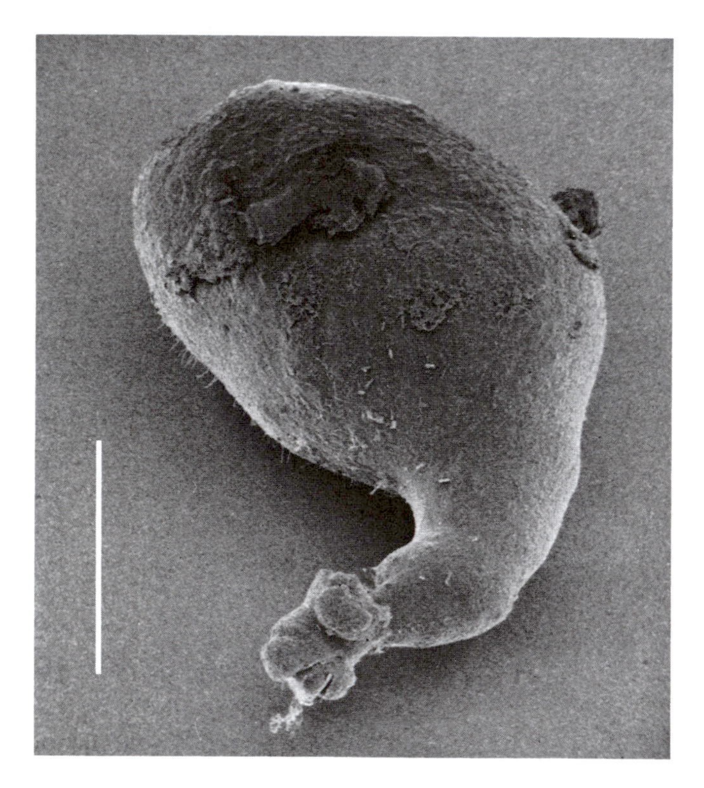

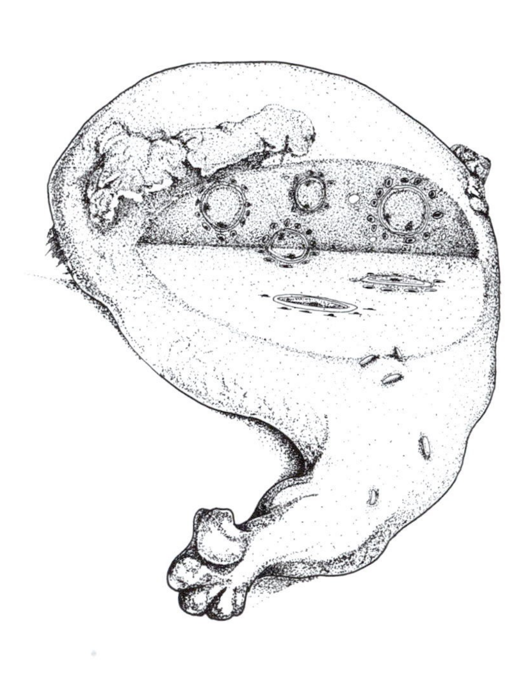

A *Pelomyxa palustris.* SEM, bar = 100 μm. [Photograph courtesy of E. W. Daniels, in *The biology of amoeba,* K. W. Jeon, ed. (Academic Press, New York, 1973); drawings by R. Golder.]

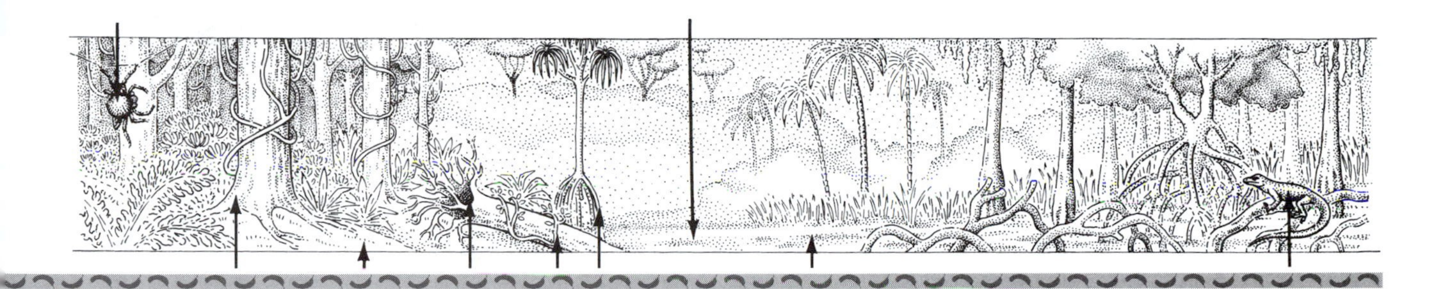

Members of Phylum Microspora are easily unified as one of the old "sporozoan" groups: tiny amitochondriate intracellular parasites of animals. The grouping of organisms that creates Phylum Archaeprotista, however, is new.

Archaeprotists were traditionally ignored or poorly known as tiny "microflagellates" or "animal parasites." Despite that neglect, the work of some master biologists—Harold Kirby (1900–1952), Lemuel R. Cleveland (1892–1969), Andre Hollande (1910–1994), Guy Brugerolle, J.-P. Mignot, and others—now allows three classes to be well delineated: Archamoebae, Metamonada, and Parabasalia.

Class Archamoebae includes free-living freshwater and marine organisms grouped into two subclasses: (1) Pelobiontae (or Caryoblastea) consists of the anaerobic amebas that lack undulipodia and swimming stages and (2) Mastigamoebae unites the anaerobic ameboids that bear undulipodia at some stage in their life histories, conferring the ability to swim.

Subclass Pelobiontae contains only one genus, *Pelomyxa* (Figure A). These giant cells, visible to the naked eye, may be relicts of the earliest living eukaryotes. They are classified as eukaryotes by definition—they have membrane-bounded nuclei. However, they

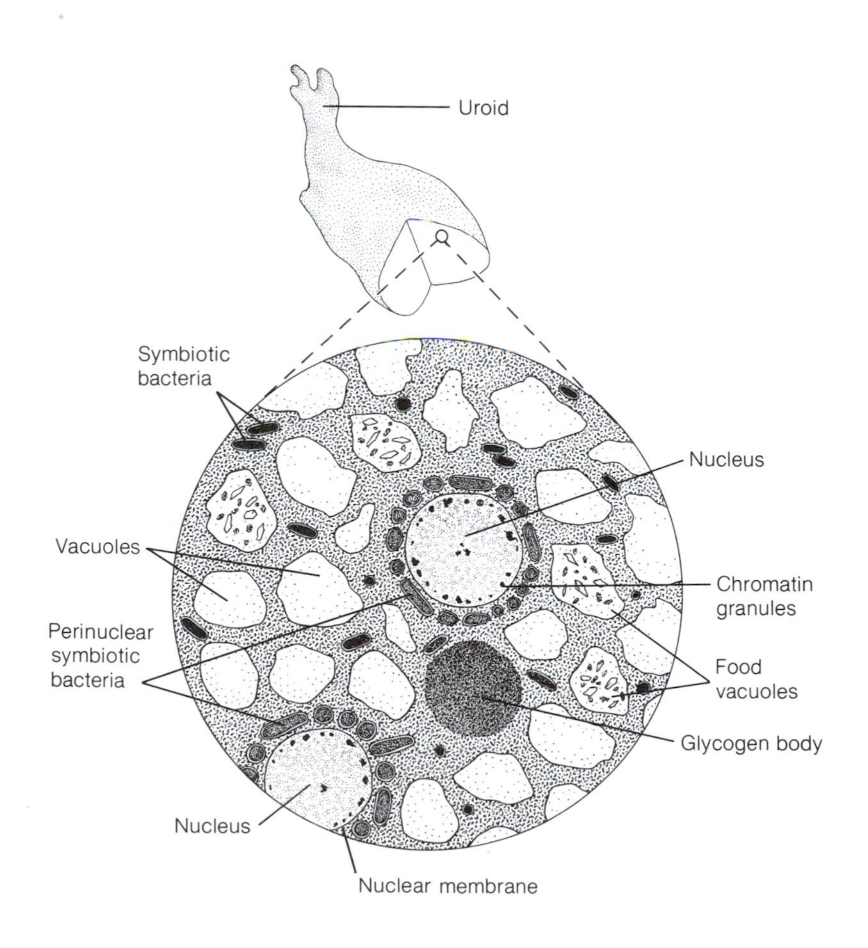

Monocercomonas
Notila
Oxymonas
Pelomyxa
Pseudotrichonympha
Pyrsonympha
Retortamonas
Saccinobaculus
Snyderella
Spironympha
Spirotrichonympha
Staurojoenina
Stephanonympha
Trichomitus
Trichomonas
Trichonympha

lack nearly every other cell inclusion characteristic of eukaryotes; they have no endoplasmic reticula, Golgi bodies, mitochondria, chromosomes, or centrioles. *Pelomyxa palustris,* a large ameba with many nuclei, is the only well-documented species described in recent literature. Its nuclei do not divide by standard mitosis. Like bacterial nucleoids, the nucleus divides directly; new membranes form and two nuclei appear where there had been only one.

In the late 1970s, some submembraneous microtubules were seen in thin section, as was an intracellular, apparently nonfunctional undulipodium, suggesting that *Pelomyxa* evolved from undulipodiated ancestors. The microtubules and undulipodium do not take part in nuclear or cell division. No gametes are formed, and sexuality is absent.

Although it lacks mitochondria, *Pelomyxa* is microaerophilic, requiring lower concentrations of oxygen than do most eukaryotes. This giant ameba has three types of bacterial endosymbionts, functional analogues of mitochondria. At least two of the three endosymbiont types are methanogenic—that is, they produce methane gas instead of carbon dioxide. The endosymbionts of one type lie in a regular ring around each nucleus and are therefore called perinuclear bacteria. The others lie scattered in the cytoplasm; they are different species of bacterium having their own characteristic wall structures. *Pelomyxa* dies when treated with antibiotics to which its endosymbiotic bacteria are sensitive. Before the ameba dies, lactic acid and other metabolites accumulate in the cytoplasm; it is thought that the healthy symbiotic bacteria remove and metabolize the lactic acid from the cytoplasm. In any case, the ever-present bacterial partners seem to be required. *Pelomyxa* has vacuoles and stores glycogen, a polysaccharide that is also stored by many types of animal cells.

Pelomyxa palustris has been discovered in only one habitat: mud on the bottom of freshwater ponds in Europe, the United States, and, probably, North Africa. In fact, most studies were made on organisms taken from the former Elephant Pond at Oxford University, so named because, in the nineteenth century, the discarded carcasses of elephants were thrown there by taxidermists preparing museum exhibits. At the muddy bottom of such small ponds, including some in Massachusetts and Illinois, *Pelomyxa* feeds on algae and bacteria; there it grows and divides. It survives severe winters; *Pelomyxa* is best detected in autumn, when one ameba looks like a tiny glistening droplet of water on dead leaves or submerged bark. Despite its hardiness, no scientist has been able to grow it in the laboratory. Those who study *P. palustris* must collect it from ponds.

Members of the Mastigamoeba subclass of Class Archamoebae are unicellular, with each cell bearing at least one undulipodium.

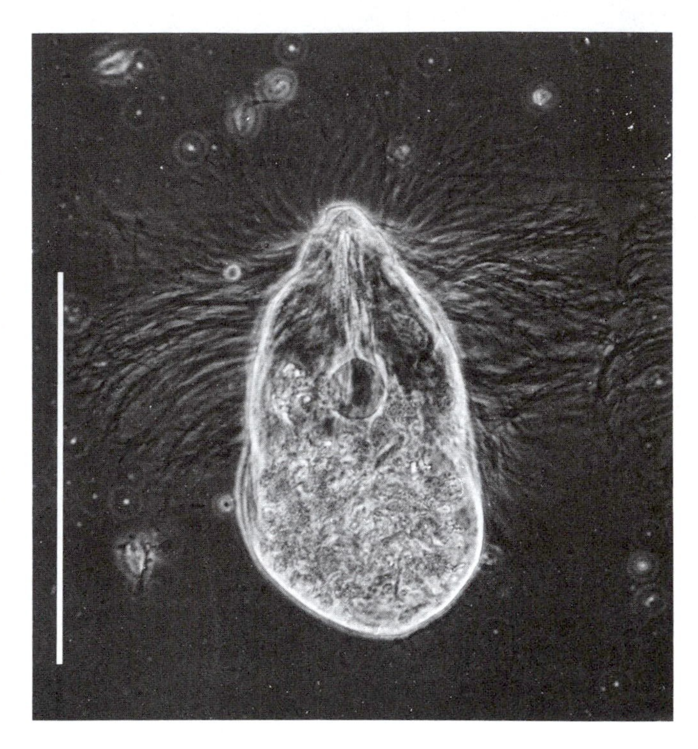

B *Staurojoenina* sp., a wood-digesting hypermastigote from the hindgut of the dry-wood termite *Incisitermes* (*Kalotermes*) *minor* (Phylum A-20, Mandibulata). LM (stained preparation), bar = 50 μm.

They may be either free living or parasitic, and all species known so far are asexual. They are heterotrophs that lack plastids. They are osmotrophs or phagotrophs. Two genera, *Mastigina* and *Mastigamoeba* (*mastig* means "whip" in Greek), are single-celled heterotrophic protists known to protozoologists for the entire twentieth century.

The second class of Phylum Archaeprotista, Metamonada, consists of cells in which the nuclei are attached to the undulipodia by thin fibers called nuclear connectors, or rhizoplasts. (The nucleus with associated fibers is called a karyomastigont.) The metamonads

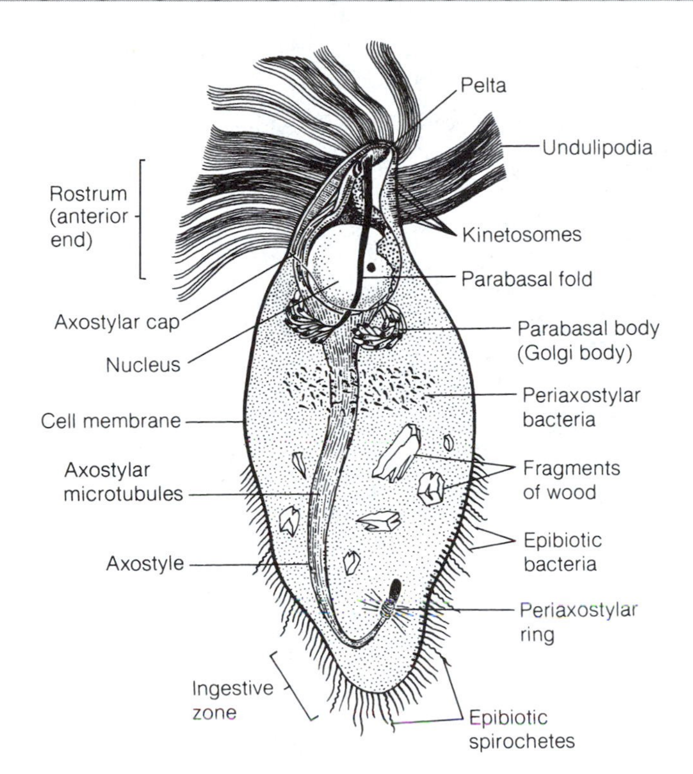

C *Joenia annectens,* a hypermastigote that lives in the hindgut of a European dry-wood termite. *Joenia* is closely related to *Staurojoenina.* [Drawing by R. Golder.]

comprise three subclasses: Diplomonadida, Retortamonadida, and Oxymonadida—all referred to as "polymonads" (cells with a small number of undulipodia) in the old literature. As long ago as the 1940s, Kirby warned about the artificiality of using numbers of reproducing organelles, such as the number of nuclei and associated undulipodia, as a basis for taxonomy. Kinetid structure (see Figure I-2), the details of which are shared by close relatives but differ markedly in organisms only distantly related, makes it abundantly clear that cells with the same number of undulipodia are not necessarily related to each other. The number of kinetosomes per kinetid—usually one or two—is the same for thousands of unrelated organisms, but the elaborate fine structure of the kinetosome arrangement, the kinetid, is common only to members of the same lower taxa (for example, families, genera). The approach of grouping organisms on the basis of the number of intrinsically reproducing organelles (for example, "polymonads" in one class) has been abandoned as electron microscopic techniques increasingly enable our understanding of kinetids to deepen. *Giardia,* the most notorious of all metamonads, belongs to the subclass Diplomonadida. Like others in this subclass, *Giardia* has two karyomastigonts, as well as a Velcro-like adhesive ventral pad, allowing it to stick to our intestines and share our food. Members of the subclass Retortamonadida are small mastigotes with twisted cell bodies; this small mastigote has a trailing undulipodium that propels food into its mouth (cytostome). Most retortamonads (*Retortamonas, Chilomastix*) live as symbiotrophs in the digestive tracts of animals. *Pyrsonympha,* like all other members of the subclass Oxymonadida, live as parasites in the intestines of wood-eating cockroaches and termites. They have ribbon-shaped organelles called axostyles. These pulsating axostyles are composed of hundreds of microtubules, which are sometimes connected to each other by bridges and arranged in elaborate patterns. The other oxymonad genera are *Dinenympha* (the group has also been called Dinenymphida or Pyrsonymphida), *Notila, Oxymonas, Saccinobaculus,* and the one multinucleate genus, *Microthopalodia.*

Members of the third class of Phylum Archaeprotista, Parabasalia, also are parasitic or symbiotic in the intestines of insects. Apparently, these microbes digest cellulose, from which they derive sugars both for themselves and for their hosts. Particles of wood are taken up through a sensitive posterior zone. In various stages of digestion, wood is often seen in the cytoplasm. A parabasalid bears at least four undulipodia, an axostyle, and conspicuous parabasal bodies. Well-defined homologues of the Golgi apparatus (dictyosomes) of animal cells and plant cells, parabasal bodies take part in the synthesis, storage, and transport of proteins. The presence of the membranous and granular parabasal body—often many parabasal bodies—distinguishes parabasalids from oxymonads. Sexuality is known, but, because parabasalids are limited to the intestines of insects, no detailed study of it has been possible. In some species, the entire haploid adult is seen to transform into a sexually receptive gamete. Two gametes fuse into a gametocyst in which, it is thought, meiosis takes place—meiotic products, haploid adults, emerge from the cyst.

The class Parabasalia comprises two orders: Trichomonadida and Hypermastigida. Trichomonads typically bear from 4 to about

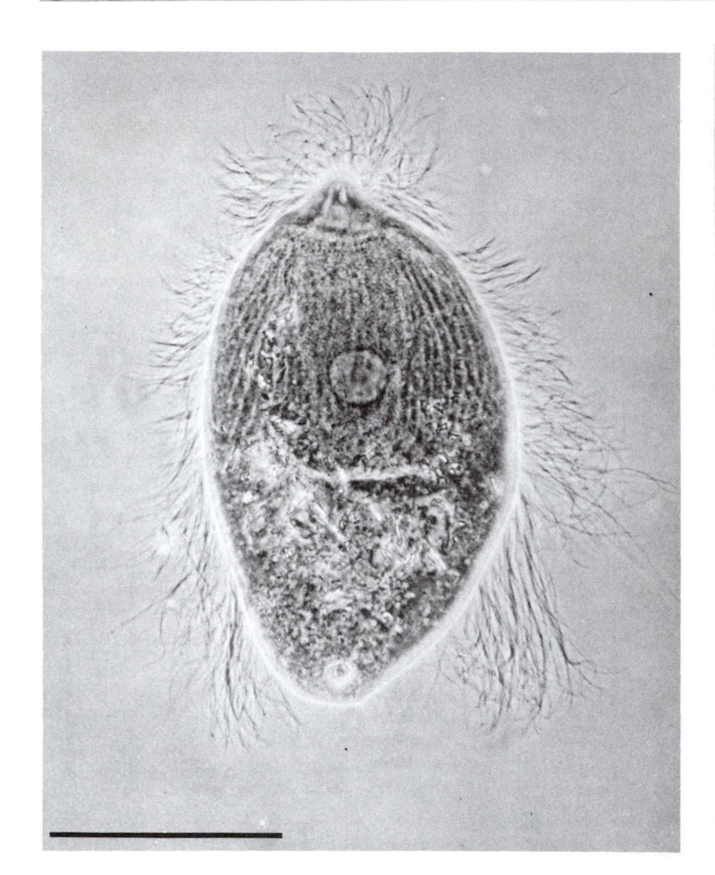

D The hypermastigote *Trichonympha ampla* from the Sonoran desert dry-wood termite *Pterotermes occidentis* (Phylum A-20, Mandibulata). LM, bar = 100 μm. [Courtesy of D. Chase.]

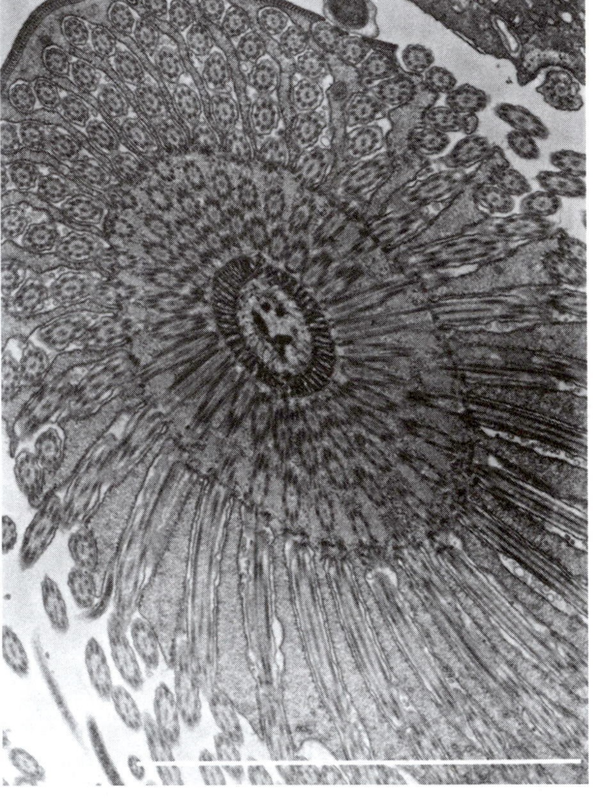

E Transverse section through the rostrum of a *Trichonympha* sp. from the termite *Incisitermes* (*Kalotermes*) *minor* from near San Diego, California, showing the attachment of undulipodia. TEM, bar = 5 μm. [Courtesy of D. Chase.]

16 undulipodia. The undulipodia are often associated with supernumerary nuclei, two undulipodia per nucleus. On the basis of distinctive cell structure, in particular the manner of insertion of the undulipodia and its fibers in the cortical cell layer (just beneath the plasma membrane), four families of trichomonads have been described: Monocercomonadidae (*Hexamastix, Monocercomonas, Histomonas*), Devescovinidae (*Devescovina, Metadevescovina*), Trichomonadidae (*Trichomonas, Trichomitus*), and Calonymphidae (*Calonympha, Snyderella*). Some members of the family Calonymphidae have more than a thousand nuclei. The order Hypermastigida, informally called hypermastigotes, have hundreds and even hundreds of thousands of undulipodia attached to special bands (Figures B through E) but generally only one or a few nuclei. The mitotic spindle, which grows out from these bands, is external to the nuclear membrane. Members of this order include *Staurojoenina, Joenia,* and *Trichonympha.*

Pr-2 Microspora
(Microsporida)

Greek *mikros*, small; Latin *spora*, spore

EXAMPLES OF GENERA
Encephalitozoon
Glugea
Ichthyosporidium
Nosema
Vairamorpha

Microsporans, the second phylum of amitochondriates, are heterotrophic microbes that lack mitochondria and produce some sort of polar filament or thread. In the older literature, the microsporans were classified with the organisms in our Phylum Apicomplexa (Phylum Pr-9) as "sporozoan parasites." However, microsporans differ from apicomplexans and the other symbiotrophs in many ways. Apicomplexans have intricate sexual lives, but whether microsporans engage in sexual activity at all is still not known. The microsporan spores are a resting stage in the life cycle and a modification for safe dissemination; they are not infective. Some microsporans have small ribosomes rather like those of prokaryotes.

All microsporans are intracellular parasites, often of vertebrate hosts. They have a great reproductive capacity. A thick-walled chitinous spore contains the conspicuous polar filament and an infec-

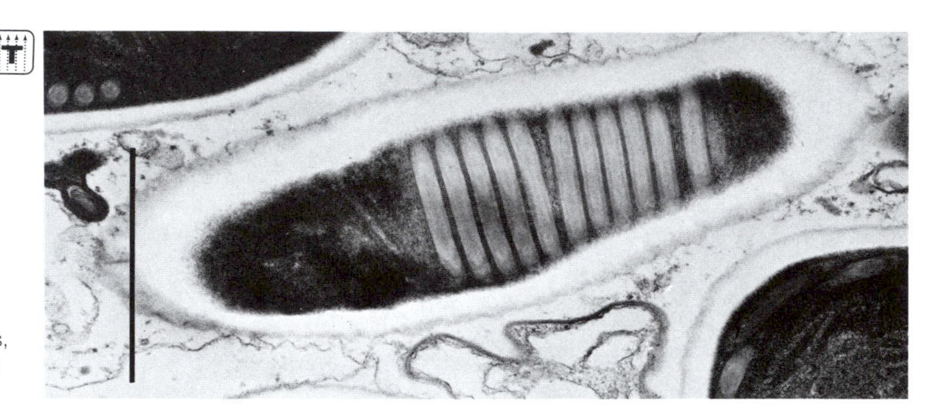

Glugea stephani, a microsporan parasite in the starry flounder, *Platiclothus stellatus.* (Phylum A-37). Ultrastructure of a mature spore. TEM, bar = 1 μm. [Photograph courtesy of H. M. Jensen and S. R. Wellings, *Journal of Protozoology* 19:297–305 (1972); drawings by R. Golder.]

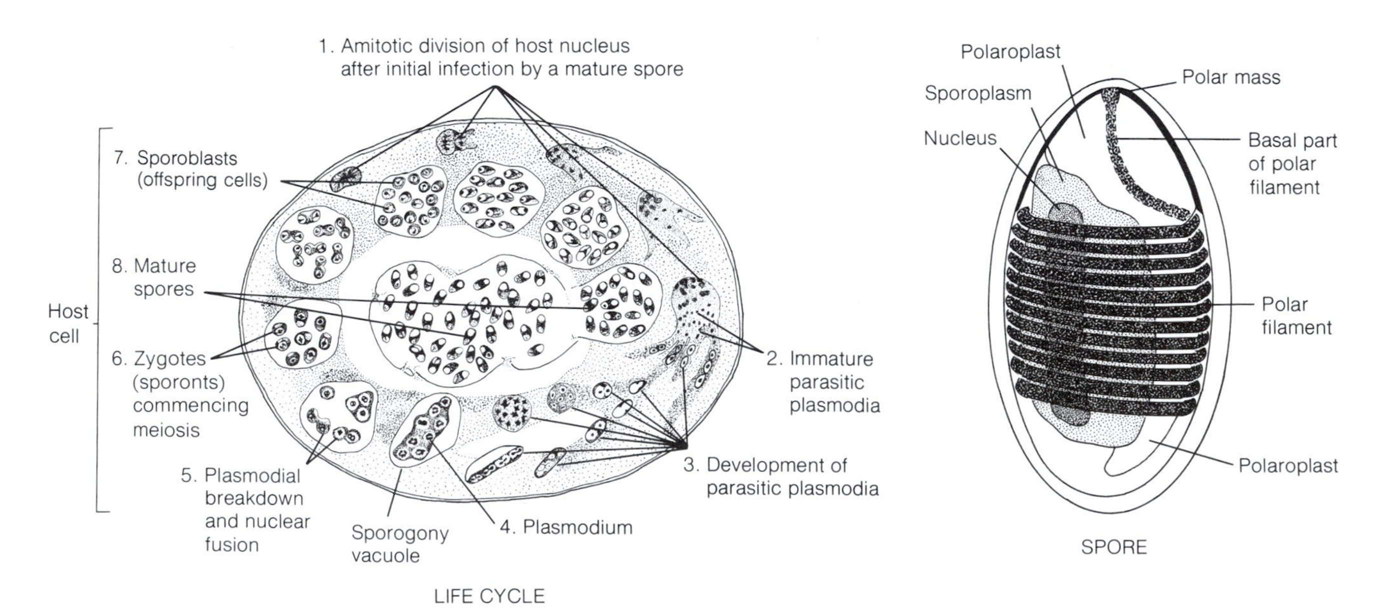

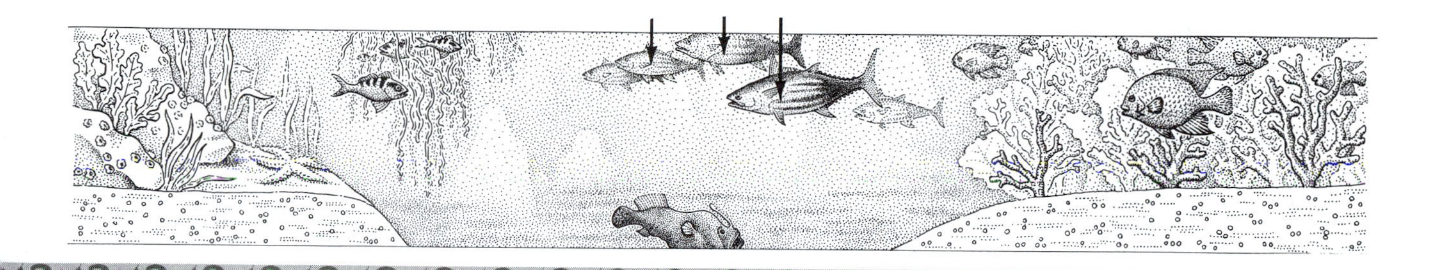

tive body called the sporoplasm. When penetrating the host, the sporoplasm everts from the spore; it is squeezed through a narrow hollow tube (derived from the polar filament) and forced into the host cell. Thus, the microsporans have independently evolved the injection needle.

Glugea stephani, illustrated here, is a microsporan parasite of the starry flounder (Phylum A-37, Craniata). Many microsporans form large single-cell tumors in the tissue of their hosts. The least well-known microsporans are extremely well integrated into their hosts and do not cause them any harm. Others are severe pathogens; *Glugea,* for example, forms single-cell tumors on fish. *Ichthyosporidium* also grows on fish. Members of the genus *Encephalitozoon* parasitize warm-blooded vertebrates. *Nosema* has caused devastating damage to the silk industry, because members of the genus are agents of pebrine, a disease of silkworm larvae.

The sexual life cycles that have been reported of microsporans are not well documented, because, situated deep in tissue, these organisms are difficult to study. Microsporans reproduce asexually inside the host cell by single or multiple fission. Their 1- to 20-μm spores are complex in structure, containing the anchoring disc, extrusion apparatus, and infective sporoplasm. Microsporan tissue may contain one or two nuclei per cell. The fish parasite *Glugea* develops a multinucleate plasmodium. The nuclei of the plasmodium fuse—without having formed undulipodiated gametes. The resulting diploid zygotes, called sporonts, undergo meiosis, the eventual products of which are filamented spores.

Pr-3 Rhizopoda

(Amastigote amebas and cellular slime molds)

Greek *rhiza,* root; *pous,* foot

EXAMPLES OF GENERA

Acanthamoeba
Acrasia
Acytostelium
Amoeba
Arcella
Centropyxis
Coenonia
Dictyostelium
Difflugia
Entamoeba
Guttulina
Guttulinopsis
Hartmannella
Hyalodiscus
Mayorella
Minakatella
Paramoeba
Pocheina
Polysphondylium
Thecamoeba

As defined here, members of Phylum Rhizopoda—amastigote amebas—are amebas that have mitochondria and lack undulipodia at all stages in their life histories. There are two classes. All the amebas in the first class, Lobosea, are single celled, either naked or with shells called tests. In the second class, Acrasiomycota (from the Greek *acrasia,* bad mixture, and *mykes,* fungus), are cellular slime molds—multicellular, land-dwelling derivatives of members of the first class. Even though amastigote amebas lack undulipodia and the [9(3)+0] centrioles from which kinetosomes derive undulipodia, they are motile. Defining features for phylum members are pseudopods (false feet; Figures A through D), flowing cytoplasmic processes used for forward locomotion and to surround and engulf food particles. Where studied, nonmuscle forms of actomyosin proteins have been found to underlie pseudopodial movements. Like contraction of muscle, such movement is sensitive to variations in the concentration of calcium ion (Ca^{2+}).

Members of the first class (Lobosea, naked and testate amebas) are distributed worldwide in both fresh and marine waters, and they are especially common in soil. Many are symbiotrophic in animals; they may pass from host to host or from the soil or fodder to host. Although morphologically these amebas are among the most simple of the protoctists, from a molecular-evolution viewpoint, they are also among the most diverse. They are not monophyletic. All are microscopic, yet some are very large for single cells, hundreds of micrometers long. Lacking meiosis and any sort of sexuality, these amebas reproduce by direct division into two offspring cells of equal volume. They have mitotic spindle microtubules and nuclear chromatin granules, which in some species form chromosomes. In these species, metaphase through telophase stages of cell division have been observed. The nuclear membrane persists well into the later stages of mitotic division; in some amebas, the nuclear membrane does not disperse at all during division.

The first subclass (naked amebas; subclass Gymnamoebia) in this phylum contains a single order (Amoebida) and five suborders: Tubulina, Thecina, Flabellina, Conopodina, and Acanthopodina.

The Tubulina are uninucleate, cylindrical, naked amebas. They are grouped into three families. The Amoebidae, which include the well-known *Amoeba proteus,* tend to be polypodial—an ameba has many feeding, changing, flowing pseudopods at one time. The Hartmannellidae, on the other hand, are monopodial—this ameba forms one pseudopod at a time. Some form desiccation-tolerant resting cysts; the cell inside each cyst is binucleate. The Entamoebidae also are monopodial. The rhizopod nucleolus, which contains the ribosomal precursors, is organized into a conspicuous organelle (or several) called an endosome. These amebas are probably the most ancient to have endosomes, which are also found in the nuclei of other protoctists (for example, euglenids, Phylum Pr-12). That the Entamoebidae form cysts is of great importance because nearly all of them live in animals. Some, such as *Entamoeba histolytica,* are responsible for amebic dysenteries. The cysts enable the amebas to resist animal digestive enzymes. The amebal nuclei can divide inside the cysts without accompanying cytoplasmic division; this leads to four, eight, or even more nuclei per encysted cell. The cysts germinate in the animal digestive tract or they are transported to the soil in the host's feces.

The Thecina ameba seems to roll its wrinkled surface as it moves. Thecina amebas form a rather obscure group of free-living forms having various mitotic patterns.

The Flabellina form spatula-shaped pseudopods in which flowing endoplasm seems to erupt. Some, such as *Hyalodiscus,* are fan shaped.

The Conopodina are shaped like fingers. When they move, they are longer than they are wide; some float on water, where they extend slender radiating pseudopods. *Mayorella* (see Figures A and B) and *Paramoeba,* both marine genera, belong to the family Paramoebidae, the only family in the suborder Conopodina. *Paramoeba eilhardi* contains two distinctive bodies called nebenkörper, which are packages of benign, omnipresent bacteria-like symbionts. *Paramoeba eilhardi* can be attacked and killed by certain other marine bacteria that are able to grow and divide only in its nucleus.

Members of the suborder Acanthopodina have finely tipped subpseudopodia; that is, each pseudopod extends smaller pseu-

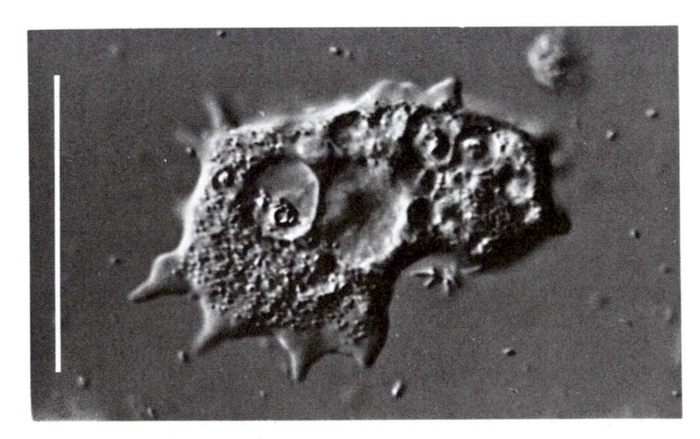

A *Mayorella penardi,* a living naked ameba from the Atlantic Ocean. LM (differential interference contrast microscopy), bar = 50 μm. [Courtesy of F. C. Page, in "An illustrated key to freshwater and soil amoebae," *Freshwater Biological Association Scientific Publication* 34 (1976).]

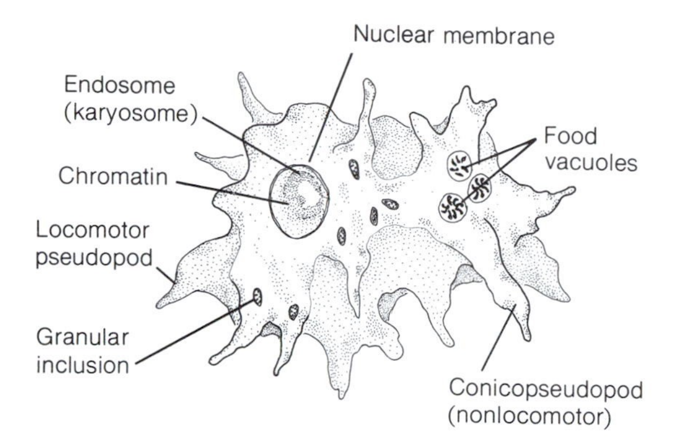

B Structure of *Mayorella penardi* seen from above. [Drawing by E. Hoffman.]

dopods of its own. The cell as a whole may be disc shaped. The suborder contains two families: Acanthamoebidae and Echinoamoebidae. *Acanthamoeba* forms a polyhedral or thickly biconvex cyst having a wall that contains cellulose. The many ubiquitous acanthamebas aggressively devour bacteria, other amebas, and ciliates. Their populations achieve huge numbers, and tough cysts permit prolonged survival in soil—even very dry soil. The Echinoamoebidae, members of the other family, are more or less flattened when they move; they have tiny, finely pointed pseudopods that look like spines.

Although naked amebas constantly change shape, the range of shapes that each takes on is genetically limited and species specific. Many naked amebas correspond to shelled forms (of which Testacealobosa, the second subclass within Class Lobosea, consists) that are thought to have derived from them. To construct their tests, amebas glue together sand grains, bits of carbonate particles, and other inorganic detritus, depending on what is available. Some of these tests, such as those of *Arcella* (see Figures C and D), are distinctive enough to be recognized in the fossil record. Such tests give the testate members of Rhizopoda a fossil record that extends well into the Paleozoic era. Some of the pre-Phanerozoic microfossils called acritarchs have been interpreted as tests of shelled amebas as well.

The organisms in the second class of this phylum, Acrasiomycota—cellular slime molds—are multicellular, land-dwelling, heterotrophic protoctists found in fresh water, in damp soil, and on rotting vegetation, especially on fallen logs. They enjoy a fascinating "dispersed" life history. In the course of their life history, independently feeding and dividing amebas aggregate into a slimy mass or slug that eventually transforms itself into a spore-forming reproductive body; the scattered spores germinate into amebas. Sexuality is rare or absent.

Because slime molds have features commonly taken to be animal (they move, they ingest whole food by phagocytosis, and they metamorphose), plant (they form spores on upright reproductive bodies), or fungal (their spores have tough cell walls and germinate into colorless cells with absorptive nutrition—they live on dung and decaying plant material), the taxonomy of slime molds has always been contested. The zoologists have called them "mycetozoa" (slime animals) and classified them with protozoa; the mycologists call them "myxomycetes." In some classifications, three of our phyla—Pr-6 (Myxomycota), Pr-18 (Labyrinthulata), and Pr-19 (Plasmodiophora)—have been classified together with these acrasiomycotes as a single phylum—Gymnomycota or Gymnomyxa (naked fungi)—in Kingdom Fungi. As early as 1868, Ernst Haeckel considered them neither plants nor animals but primitive forms that had not yet evolved to be members of either of the two great kingdoms. Haeckel erected his new Kingdom Protoctista to accommodate such unruly organisms.

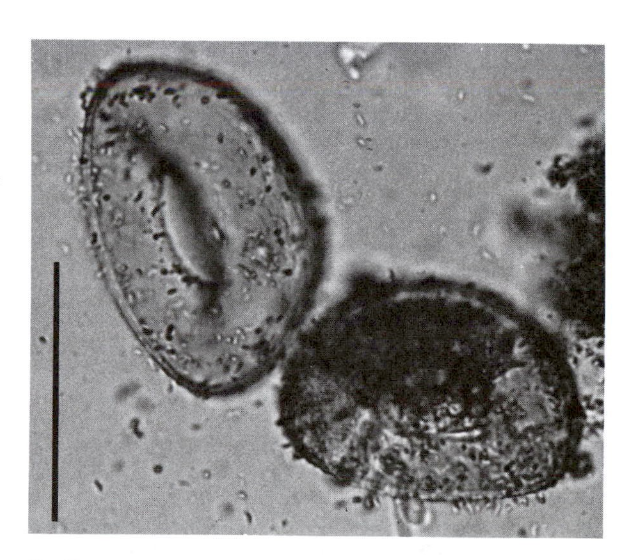

C Two empty tests (shells) of the freshwater ameba *Arcella polypora*. LM, bar = 10 μm. [Courtesy of F. C. Page.]

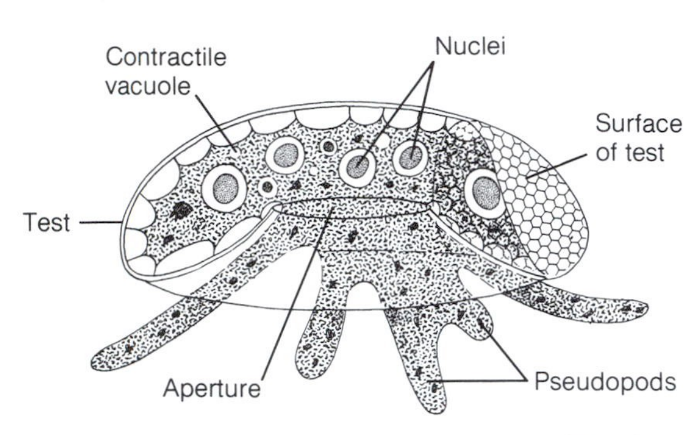

D Structure of *Arcella polypora*, cutaway view. [Drawing by R. Golder.]

The class Acrasiomycota contains two subclasses, Acrasea and Dictyostelia. The members of both subclasses pass through a unicellular stage of ameboid cells that feed on bacteria. Later they form a multicellular, stalked reproductive structure, the sorocarp, that produces spores borne in a swelling called the sorus, which lies at the top or just below the top of the stalk. In passing from the first stage to the second, the ameboid cells aggregate to form a pseudoplasmodium (slug). A true plasmodium, or syncytium, is a mass of protoplasm containing many nuclei formed by mitotic divisions but not separated by cell membranes. The acrasiomycote structure is called a pseudoplasmodium because it is made of mononucleate constituent cells that retain their cell membranes. It only superficially resembles the plasmodium of the true plasmodial slime molds (Phylum Pr-6).

Most acrasiomycotes will begin to aggregate if food is depleted and light is present. However, the exposure to light must be followed by a minimum period of darkness before development can continue.

The two subclasses differ in many ways and may not be directly related. In Acrasea, the stalk of the sorocarp consists of live cells that are capable of germination and lack cellulosic walls, whereas the stalk in Dictyostelia consists of a tube of cellulosic walls of dead cells. Dictyostelid ameboid cells are aggregated by their attraction to cyclic adenosine monophosphate (cAMP); the acrasids do not respond to cAMP.

In Subclass Acrasea, the feeding stage consists of ameboid cells having broad, rounded pseudopods. The families of acrasids are distinguished primarily by the structure of the sorocarp. In some, the spore cells are different from the stalk cells; in others, all the cells are alike.

Acrasids, like all other rhizopods as grouped here, lack undulipodia. However, members of one genus of Subclass Acrasea, *Pocheina*, exhibit an undulipodiated form. Once enough molecular and cellular details are reported to allow its reclassification, the presence of this swimming form suggests that *Pocheina* will be moved to another phylum. When spores of *Pocheina* germinate, they divide into motile, swimming cells, each bearing two undulipodia of equal length. Still poorly known, pocheinas have been reported from the former Soviet Union (Kazan) and North America; they live on conifer bark and lichenized dead wood.

Members of Subclass Dictyostelia are far better known than members of Subclass Acrasea. Four dictyostelid genera have been described: *Acytostelium, Dictyostelium, Polysphondylium,* and *Coenonia*. Each has a number of species, at least 16 in *Dictyostelium*. The ameboid cells of dictyostelids are usually uninucleate and haploid. However, cells having more than one nucleus and aneuploids, cells having an uneven number of chromosomes, have been reported. Some strains consist of stable diploid cells.

E The development of a reproductive body from a slug of *Dictyostelium discoideum*. Bar = 1 mm. [Courtesy of J. T. Bonner, from *The cellular slime molds,* © 1959, rev. ed. © 1967 by Princeton University Press; Plate III reprinted with permission.]

The typical dictyostelid life cycle is illustrated here for *Dictyostelium discoideum* (Figures E and F). The amebas have thin pseudopods and feed mainly on live bacteria. After the food supply is exhausted and the amebal population has reached a certain density, the cells cease feeding and dividing. Because of a pheromone (chemical attractant) called acrasin, cAMP (cyclic adenosine monophosphate), secreted by the amebas themselves, they begin to aggregate, streaming toward aggregation centers. The dispersed feeding stage of the life history terminates when the pseudoplasmodium forms. A thin slime sheath is produced around the mass of cells, forming a pseudoplasmodium that takes on the form of a slug. The slug begins to wander, leaving behind a slime track. As conditions become drier, the migration stops and the differentiation of the reproductive structure begins. In a complicated developmental sequence including differentiation, but not cell division, the sporophore (also called the sorocarp) with its cellulosic stalk, forms. *Dictyostelium discoideum* is valued in developmental biology research because it grows rapidly and the separation in time of its trophic stage (feeding and growing stage of the ameba population) from its differentiation into stalk and propagule (spore) facilitates manipulation of its developmental stages.

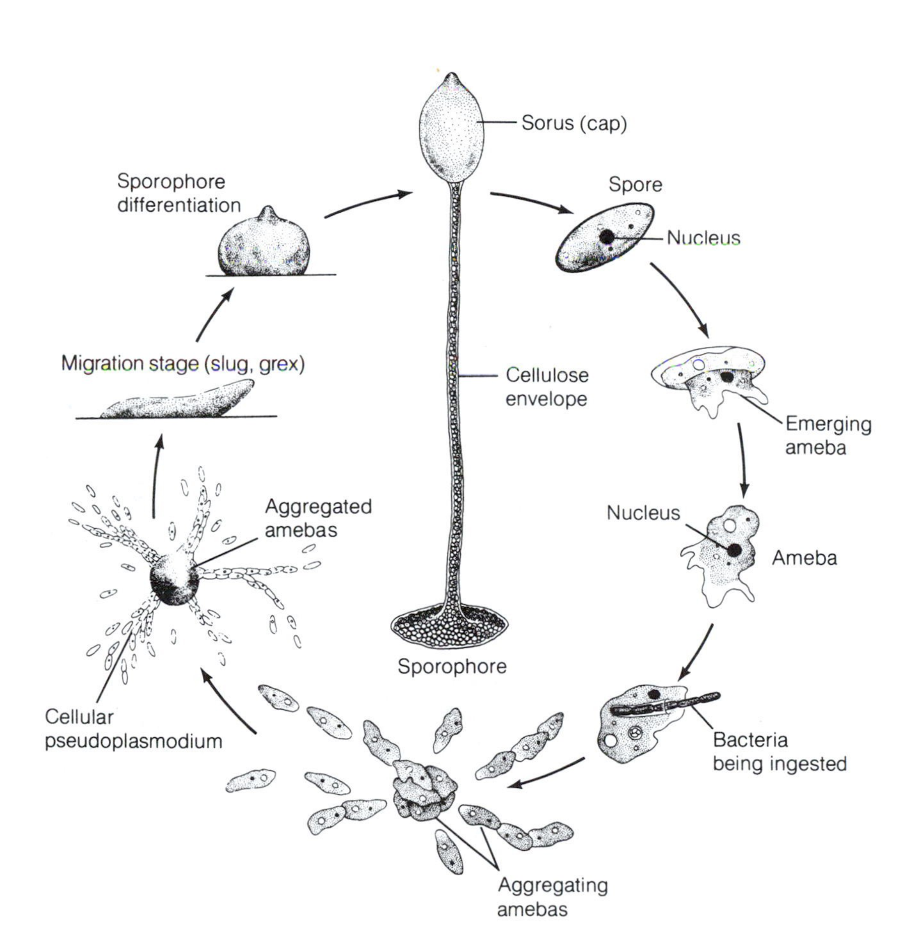

F Life cycle of the cellular slime mold *Dictyostelium discoideum.* [Drawing by R. Golder.]

Pr-4 Granuloreticulosa

Latin *granulum,* small grain, seed; *reticulum,* network

EXAMPLES OF GENERA

Allogromia	*Glabratella*	*Nodosaria*
Discorbis	*Globigerina*	*Reticulomyxa*
Elphidium	*Iridia*	*Rotaliella*
Fusulina	*Miliola*	*Textularia*

Granuloreticulosans are easily defined: these organisms bear reticulopods, cells that fuse to form networks in which bidirectional (two-way) streaming can be seen. Two classes make up this phylum: Reticulomyxida and Foraminifera (from the Latin *foramen,* little hole, perforation, and *ferre,* to bear). By far the better-known class, Foraminifera—affectionately known as forams—have pore-studded shells, or tests. In contrast, reticulomyxids are "snot shaped" slimy nets of messy, bactivorous soft masses that lack shells. Very few have been studied. These "naked forams," reticulomyxids, are the presumed ancestral group; therefore, with their discovery, the former Phylum Foraminifera was renamed.

Forams are exclusively marine organisms. The smallest are some 10 μm in diameter, and the largest ones, visible to the naked eye, grow to several centimeters in diameter. The majority are tiny and live in sand or mud or attached to rocks, algae, or other organisms. Two families of free-swimming modern planktonic forams (Globigerinidae and Globorotalidae) are very important in the economy of the sea as food for many marine animals.

The pore-studded tests of forams are composed of organic materials, often reinforced with minerals. Some are made of sand grains; most are neatly cemented granules of calcium carbonate deposited from sea water. Some forams, by mechanisms that are still unknown, choose echinoderm plates (Phylum A-34) or sponge spicules (Phylum A-2) to construct their tests. The test and the organism itself may be brilliantly colored—salmon, red, or yellow brown. Whereas the simplest forams have single-chambered tests, most are multichambered. A typical test looks like a clump of blobs of partial spheres (Figure A). Pores in the test permit thin cytoplasmic projections, the microtubule-reinforced filopodia, to emerge. Anastomosing (linked-up) filopodia form nets called reticulopodia. The filopodia are used for feeding, swimming, and gathering materials for tests. Forams are omnivorous: they eat algae, ciliates (Phylum Pr-8), actinopods (Phylum Pr-27), and even nematodes (Phylum A-10) and crustacean larvae (Phylum A-21). Many forams, probably most that live in shallow water, harbor photosynthetic symbionts—dinomastigotes (Phylum Pr-7), chrysomonads (Phylum Pr-13, planktonic), and diatoms (Phylum Pr-16).

Although some foram genera (for example, *Textularia*) have been seen reproducing only by asexual budding into multiple offspring, others that have been well studied—some dozen species—show a remarkably complex life cycle. The known cycles are variations on the theme of *Rotaliella* (Figure B). Meiosis takes place in the agamont, a fully adult diploid organism that produces and releases smaller haploid forms called agametes. These agametes disperse and grow by mitotic cell divisions into a second kind of adult, called gamonts. The gamonts reproduce sexually, by fusion of haploid nuclei, to produce diploid offspring, which are agamonts.

The alternation of the diploid agamont and haploid gamont generations is obligatory in the forams that have been studied, just as alternation of generations is obligatory in plants, such as mosses (Phylum Pl-1) and ferns (Phylum Pl-7). In fact, forams are the only heterotrophic protoctists that alternate morphologically distinct free-living adult generations. What complicates matters is that, unlike other organisms except ciliates (Phylum Pr-8), forams show a striking nuclear dimorphism. The agamonts of *Rotaliella roscoffensis,* for example, contain four diploid nuclei. Three of these nuclei, the generative nuclei, reside in a chamber separate from that in which the larger somatic nucleus remains. The somatic nucleus never undergoes meiosis; it eventually becomes pycnotic (it stains heavily) and disintegrates. The three generative nuclei give rise to 12 haploid products by meiosis. These products become the nuclei of the small haploid agametes. Later, in the gamonts, pairs of haploid nuclei, apparently of opposite sex, fuse to form diploid zygotes. In effect, these organisms show programmed cell death (selective "death" of the somatic nucleus), and each gamont fertilizes itself, although neither egg nor sperm is formed.

Foram tests have contributed greatly to the sediment on the bottom of marine basins, especially since the Triassic period. There are fossil giant forams of great fame. Some, such as *Lepidocyclina*

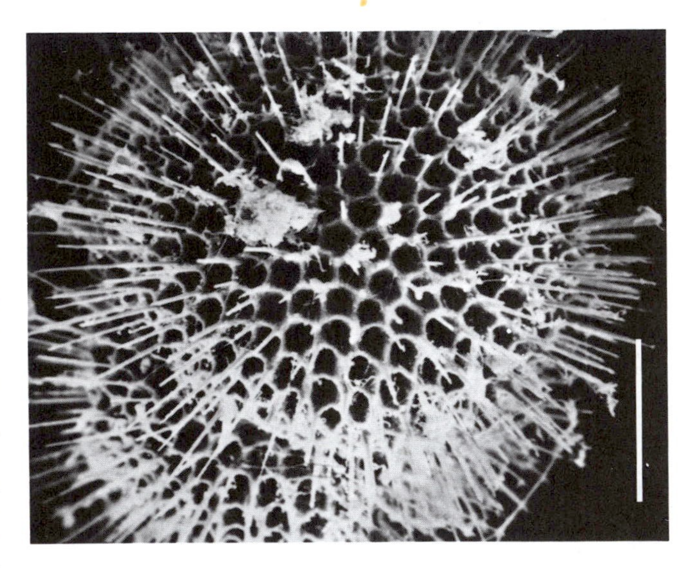

A Adult agamont stage of *Globigerina* sp., an Atlantic foraminiferan. SEM, bar = 10 μm. [Courtesy of G. Small.]

elephantina, had tests as thick as 1.5 cm. *Camerina laevigata* (also known as *Nummulites,* the "coin stone") was a large (10 cm wide) foram that lived in warm shallow waters during the Cenozoic era from the Eocene to the Miocene epoch (some 38 million to 7 million years ago). Rocks bearing Miocene forams, many of them easily visible to the naked eye, abound on the shores of the Mediterranean. It is from such "nummulitic" limestone that the pyramids of Egypt were constructed.

The abundance of foram tests and their detailed architecture (the earliest ones appeared in the Cambrian) make them excellent stratigraphic markers. Geologists use the 40,000 or so fossil species to identify geographically separate sediment layers of the same age. Because the tests are often found in strata that cover oil deposits, recognition of foram morphology and knowledge of their distribution is helpful in petroleum exploration.

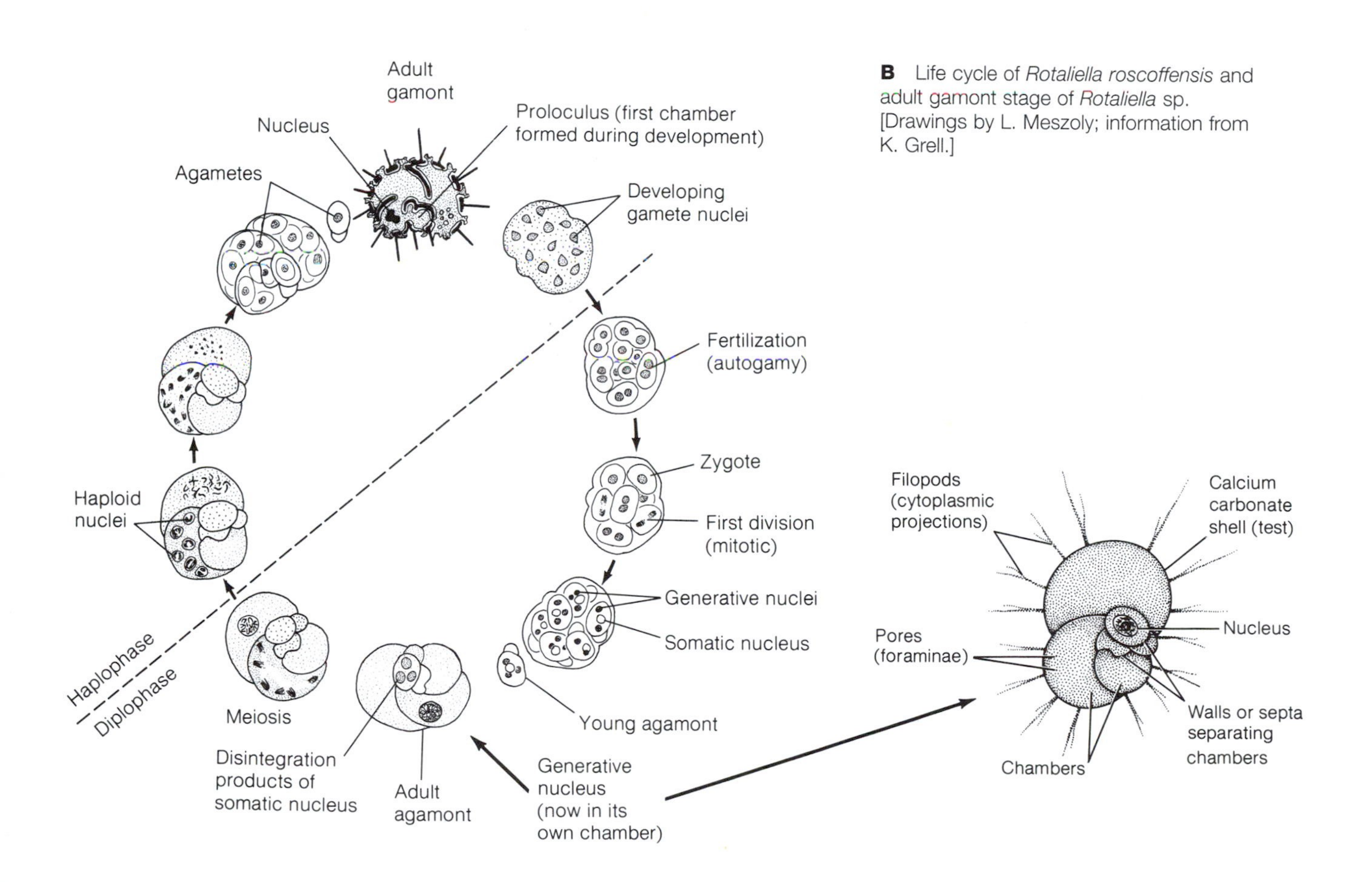

B Life cycle of *Rotaliella roscoffensis* and adult gamont stage of *Rotaliella* sp. [Drawings by L. Meszoly; information from K. Grell.]

Pr-5 Xenophyophora

Greek *xenos*, stranger; *phyein*, to bring forth, beget; *pherein*, to bear

ALL GENERA
Aschemonella
Cerelasma
Cerelpemma
Galatheammina
Maudammina
Ocultammina
Psammetta
Psammina
Reticulammina
Semipsammina
Stannoma
Stannophyllum
Syringammina

Large, benthic (seafloor) and even abyssal dwellers, these enigmatic protoctists are objects of great curiosity. Ernst Haeckel first identified them as a kind of sponge in 1889. Later, in 1907, they were taken to be gigantic marine rhizopods (testate amebas) and given the apt name Xenophyophora by F. E. Schulze, an investigator on the German Valdivia expedition. They are not sponges or indeed animals of any kind. Nor are they testate amebas (Phylum Pr-3) or agglutinating foraminifera (Phylum Pr-4), both of which they somewhat superficially resemble in size and appearance as amorphous, incongruent, sediment-particle bearers. Some are huge; lumpy adults 7 cm in maximum dimension have been described. A flat *Stannophyllum* specimen measuring 25 cm in diameter but only 1 mm thick is the largest reported xenophyophore.

Xenophyophores seem to feed phagotrophically (for example, by engulfment) on bacteria and protists on the sediment bottom by stretching out pseudopods. The details of these cell extensions have not been described, but further study may demonstrate that they resemble the filopodia of forams or the pseudopods of rhizopod amebas. Even though these testate heterotrophic protoctists contain large concentrations of barium sulfate, their contribution to sedimentary barite is deemed insignificant.

The life history stages of xenophyophores are unknown; xenophyophores are generally damaged upon collection, and cannot be cultured, so knowledge about them is extremely limited. In their dark abyssal locations, their populations may be numerous and their presence obvious, yet very few investigators have even seen them alive. One would love to know if they form resistant stages or sexual propagules.

Xenophyophores are cosmopolitan in distribution. There is a clearly greater species diversity at the sea bottom in localities with nutrient-rich overlying sea surfaces. Evidence from photographs of the seafloor and samples collected at the same locations shows that xenophyophores prevail in certain deep-sea communities, where hundreds of specimens per 100 square meters can be detected. The xenophyophoran pseudopodial system, which moves and removes things around the sea bottom, must greatly affect the distribution of organic particles and organisms from bacteria to small animals.

Xenophyophores are plasmodia (multinucleate masses of cytoplasm) enclosed by a branched tube system, a complex called a granellare. Fecal pellets called stercomes hang outside the granellare in dark strings or masses called stercomares. The granellare, which is made of a transparent, cementlike organic substance, forms a system of anastomosing (fusing) branches with diameters from about 30 to 90 mm. The organic tube wall itself is very thin, less than 0.5 mm. Plasmodial pseudopodia may extend through the free ends of the granellare branches. The most obvious bodies in the strings of plasmodial cytoplasm running through the tubes are numerous nuclei and huge numbers of barite crystals called granellae. Xenophyophores may be heterokaryotic, with a differentiation of nuclei into somatic and generative. The nuclei, evenly distributed

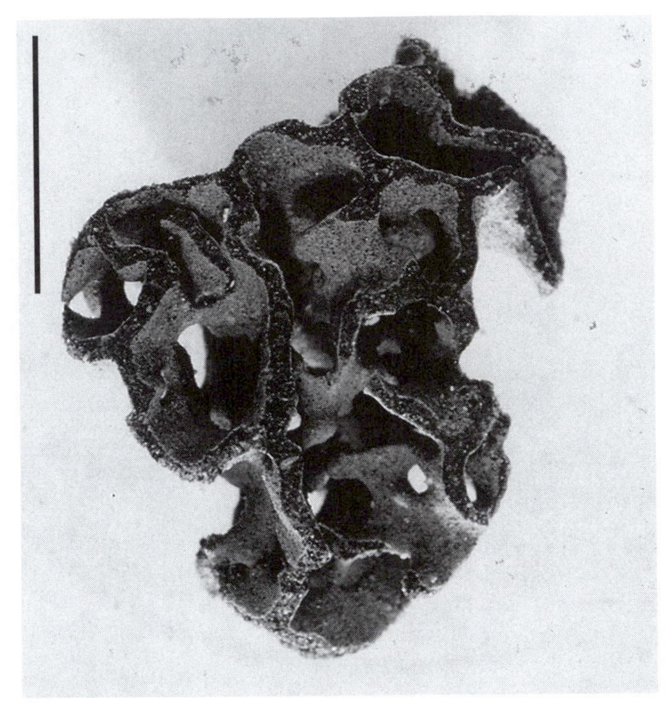

A *Psammetta globosa* Schulze, 1906. "John Murray Expedition" St. 119. The specimen measures about 20 mm in diameter. Bar = 1 cm. [Photograph by O. S. Tendal.]

B *Galatheammina tetraedra* Tendal, 1972. "Galathea Expedition" St. 192. Greatest dimension from tip of arm to tip of arm is 18 mm. Bar = 2 cm. [Photograph by O. S. Tendal.]

throughout the cytoplasm, are spherical or ellipsoidal and measure from 2 to 10 mm in diameter, usually from 3 to 4 mm. The granellae are generally the size of large bacteria (2–5 mm).

The tests of xenophyophores consist of foreign matter (xenophyae, "stranger particles")—whole or parts of foram tests, radiolarian skeletons, sponge spicules, and mineral grains. These are bound by patches of a cementlike substance. The test surrounds the granellare and stercomare. In one class of xenophyophores, the test also contains a slimy mass of extracellular, proteinaceous, 2- to 3-mm-thick fibers (linellae). The test can be hard, brittle, or more or less flexible, depending not only on the quantity of xenophyae and cement, but also on the absence or presence of linellae. Color varies with the kind of agglutinated foreign particles. Selectivity for certain kinds of particles has been shown in some species.

Recognition of xenophyophores is not difficult for those interested in protoctists. The presence of granellare and stercomare, visible with a binocular microscope as yellowish and black strings, respectively, permits a presumptive identification for marine abyssal benthic organisms. Microscopic investigation, showing the multinucleate cytoplasm with granellare, as well as the stercomes, permits definitive recognition of these amorphous beings. All life activities and processes, however, must be inferred from morphological analysis of preserved material and in situ observation.

Single specimens are said to reproduce by gametogamy and have several gamete-producing rounds during the life history. The gametes bear two undulipodia or may be ameboid. Extreme morphological changes are suspected to take place in the course of development, but live material will be needed to verify these speculations.

The 42 known species of xenophyophores are placed in 13 genera. The current classification system recognizes a total of five families of xenophyophores organized into two classes, Psamminida and Stannomida. We include a complete list of genera in our enumeration of these groupings, as follows.

In Class Psamminida, which consists of four families, the xenophyae are generally organized in some kind of order, and linellae are absent. Family Psammettidae has xenophyae cemented into the massive, lumpy body: *Maudammina, Psammetta* (Figure A). Family Psamminidae has xenophyae cemented into one or more layers; the body form varies from lumpy to branched or reticulate: *Galatheammina* (Figure B), *Reticulammina* (Figure C), *Psammina, Semipsammina, Cerelpemma.* Family Syringamminidae has xenophyae cemented together to form the walls of tubes; the body is composed of a number of such tubes: *Syringammina* (Figure D), *Ocultammina, Aschemonella.* Family Cerelasmidae has xenophyae cemented together with great amounts of cement in no obvious order: *Cerelasma.*

In Class Stannomida, which contains only one family, xenophyae are organized poorly or not at all, and linnellae are present. Family Stannomidae has a branched, treelike or flat body: *Stannoma, Stannophyllum.*

Pr-5 Xenophyophora
(continued)

C *Reticulammina lamellata* Tendal, 1972. NZOI "Taranui Expedition" St. F 881. Greatest dimension is about 30 mm. Bar = 1 cm. [Photograph by O. S. Tendal.]

D *Syringammina fragillissima* Brady, 1883. "Triton Expedition" St. 11. Greatest dimension is about 40 mm. Bar = 1 cm. [Photograph by O. S. Tendal.]

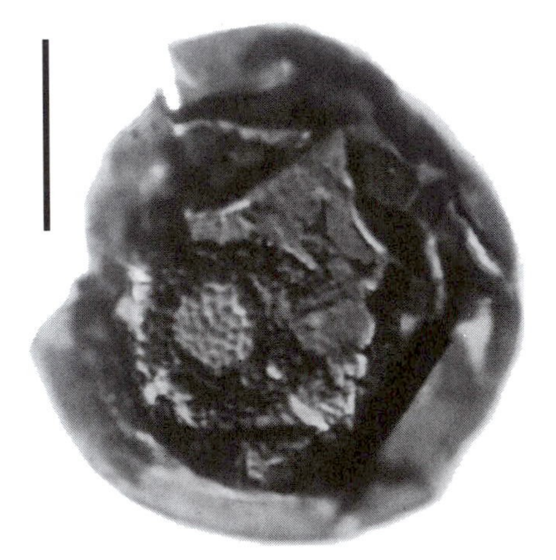

E Acritarch with excystment structure. (Neoproterozoic from approximately 800 to 700 million years ago, Chuar Group, Arizona.) LM, bar = 20 μm. The size and complexity of the surface structures convince us that this, although unidentified, was a fossil eukaryote. [Photograph by G. Vidal.]

Paleodictyon, a well-known fossil with a history dating from at least the Ordovician, may be xenophyophoran in origin, but the limited evidence available permits no firm conclusion at this time. The presence of varying and sometimes abundant fossils indicates that enigmatic organisms frequently produced sea-bottom lumps, tracks, and tunnels. Furthermore, the fossil record is replete with complex spherical or nearly spherical hard-walled structures called acritarchs (Figure E). Most paleontologists suspect that these structures are sexual or resistant stages of protoctists. Because so many kinds of acritarchs have been described, these fossils probably have many different origins from various groups of protoctists. Some, when alive, may even have been xenophyophores.

Pr-6 Myxomycota

(Myxogastria, plasmodial slime molds)

Greek *myxa*, mucus; *mykes*, fungus

EXAMPLES OF GENERA

Arcyria
Barbeyella
Cavostelium
Ceratiomyxa
Cercomonas
Clastoderma
Comatricha
Dictydium
Diderma
Didymium
Echinostelium
Fuligo
Hemitrichia
Leocarpus
Licea
Lycogala
Metatrichia
Perichaena
Physarella
Physarum
Protostelium
Sappinia
Stemonitis

The myxomycotes enjoy many names, including myxomycetes, mycetozoa, plasmodial slime molds, true slime molds, Myxomycotina, and others. Like the cellular slime molds—the acrasiomycotes (Phylum Pr-3)—these plasmodial slime molds form an ameboid stage that lacks cell walls and feeds on bacteria by engulfing them with pseudopodia—for example, phagocytosis. Also like the acrasiomycotes, the plasmodia can differentiate into reproductive structures that are stalked and funguslike in appearance. Unlike the acrasiomycotes, however, the myxomycotes are overtly sexual.

Like foraminifera and plants, some species of myxomycotes alternate haploid and diploid generations. The haploid cell in this phylum bears two anterior undulipodia of unequal lengths. These cells convert into ameboid cells called myxamebas. The myxamebas reconvert just as readily into undulipodiated cells (mastigotes). Both the undulipodiated cells and the myxamebas can differentiate into complementary mating types. Either two amebas or two mastigotes fuse to form a diploid zygote. This zygote divides repeatedly by mitosis to form a large mass of multinucleated cytoplasm, the plasmodium. Unlike comparable structures in acrasiomycotes (Phylum Pr-3), the plasmodium of myxomycotes is not cellular, because the diploid nuclei are not separated by cell membranes. The overall form of the plasmodium varies between classes, but, in some, the plasmodium takes on a veined or reticulated structure.

Plasmodia are found as a slimy wet scum on fallen logs, bark, and other surfaces. The plasmodia, although usually pigmented orange or yellow, do not photosynthesize. They are phagotrophs feeding on bacteria (and certain small protists) whose abundant populations develop on decaying vegetation. The size and shape of these slime molds is in no way predetermined; bits taken from a plasmodium form new individuals that can feed and grow independently. The plasmodium moves only by differential growth. Its innards pulsate back and forth in a movement that is obvious, under the microscope, as an incessant intraplasmodial flow. The movement distributes metabolites and oxygen evenly.

If its surroundings become drier, the plasmodium may mature to the fruiting stage. Portions become concentrated into mounds, from which stalked spore-producing organs (sporangia, also called sporocarps) grow. Meiosis takes place inside the maturing spores contained in the sporocarp. In some cases, three of the meiotic products degenerate and the fourth develops into a mature spore, which germinates into either a haploid ameba or a haploid mastigote. Most of the life cycle of these organisms, then, is spent in the diploid stage. The development of haploid amebas and mastigotes into plasmodia directly, without fusion to form a diploid zygote, has also been reported, so the ploidy of any slime-mold mass must be ascertained in each case.

Some 400 or 500 species of myxomycotes are documented, with larger members of the phylum being best known. The color, shape, and size of the reproductive structure, the presence of a stalk, the presence of a sterile structure (the columella) at the top of the stalk, the presence of calcium carbonate ($CaCO_3$) granules or crystals in or on the reproductive body, and the structure of the spores distinguish species. Inside the young sporangium, a system of threads develops, which are sterile in that they do not give rise to spores. This thread system, called the capillitium, differs among various myxomycote groups and is used as a taxonomic marker.

Although the mastigote ameba *Cercomonas* does not form plasmodia, we group it here with the myxomycotes as representative genus of the ancestral class, Cercomonadida. Insight into the many mastigotes of the subvisible world leads biologists to believe that these diverse, tiny, ubiquitous swimmers (classified in Phyla Pr-1, Pr-6, Pr-11, Pr-12, or Pr-30, depending on detailed structure) are ancestral to several lineages of larger heterotrophic protoctists (such as slime molds and water molds). *Cercomonas,* with its orthogonally arranged kinetosomes and cone of microtubules, kinetid structure, cyst formation (akin to myxomycote spore), its ameboid posterior, and other features inspires us to classify it here at the base of Phylum Myxomycota. The other two classes, both of which do form plasmodia, are Protostelida and Myxomycetes.

The major subgroups, here subclasses within Class Myxomycetes, are Echinosteliales, Trichiales, Liceales, Stemonitales, and Physarales.

The spores of members of the first three subclasses—Echinosteliales, Trichiales, and Liceales—are pale, as a rule, and do not deposit $CaCO_3$. The sporocarps of members of Subclass Echinosteliales are tiny (less than 1 mm high) and contain capillitia. The diploid feeding stage is a protoplasmodium—a microscopic ameba-like plasmodium that lacks veins or reticulations until it matures in preparation to fruit. Subclass Echinosteliales comprises two families: Echinosteliidae, having only the genus *Echinostelium* (illustrated here), and Clastodermidae, three tiny species belonging to the genera *Clastoderma* and *Barbeyella.* The sporocarps of members of Subclass Trichiales contain sculptured capillitia. The diploid feeding plasmodium is midway between an aphanoplasmodium (a thin, inconspicuous plasmodium consisting of a reticulum, or network, of veins having fan-shaped leading fronts) and a phaneroplasmodium (a thick, more conspicuous structure whose veins and leading fronts are visible to the unaided eye). The genera include *Perichaena, Arcyria,* and *Metatrichia.* Members of Subclass Liceales may be either proto- or phaneroplasmodial. Their sporocarps are of diverse shapes and lack capillitia. Genera include *Licea, Lycogala,* and *Dictydium.*

Members of Subclass Stemonitales, which typically form aphanoplasmodia, and of Subclass Physarales, which typically form phaneroplasmodia, bear dark purplish to brown or black spores. They are the largest and best-known slime molds. Members of Stemonitales, which include *Stemonitis* and *Comatricha,* form dark fingerlike upright sporocarps. Members of Subclass Physarales often have conspicuous $CaCO_3$ deposits in the capillitium and other parts of the sporocarp. Some 85 species of the genus *Physarum* are known. Their very active protoplasmic streaming and the ease with which some myxomycotes can be grown in the laboratory have made them useful in the study of proteins engaged in cell motility. The yellow plasmodia of *Physarum polycephalum* have nonmuscle actin and myosin proteins homologous to the actomyosin complexes of animal muscle.

Sporophore and life cycle of the plasmodial slime mold *Echinostelium minutum.* LM, bar = 0.1 mm. [Photograph courtesy of E. F. Haskins; drawing by L. Meszoly.]

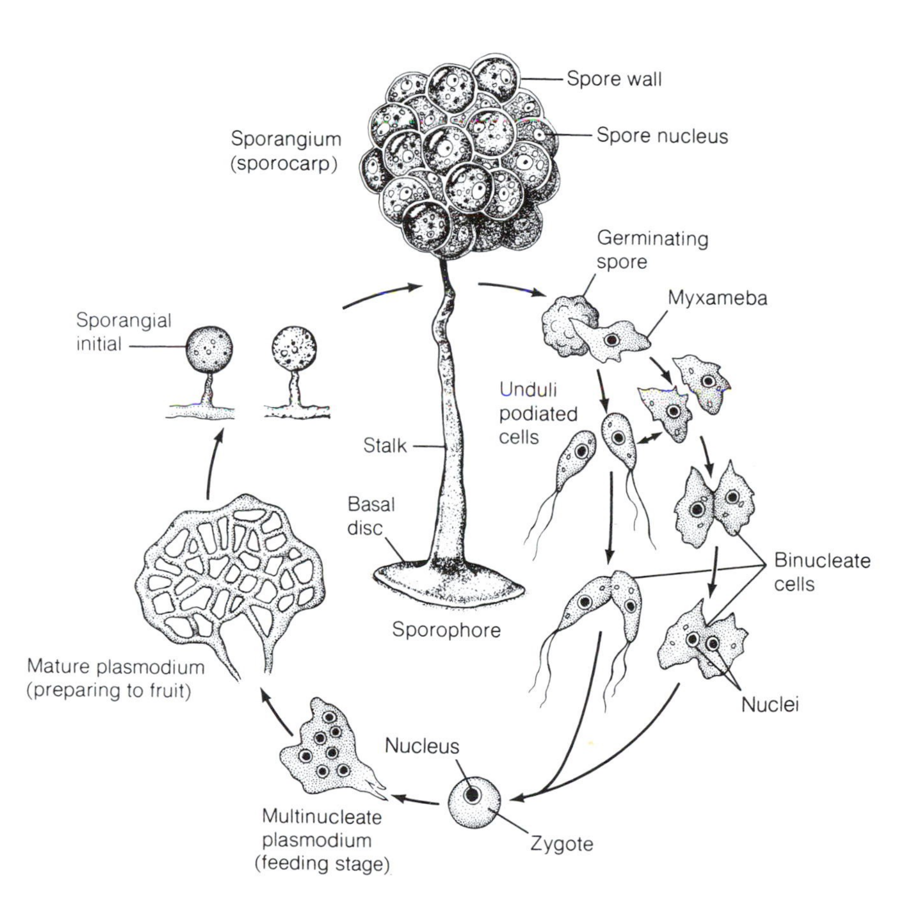

Pr-7 Dinomastigota

(Dinoflagellata, Dinophyta)

Greek *dinos,* whirling, rotation, eddy;
Greek *mastigio,* whip

EXAMPLES OF GENERA

Amphidinium	*Gonyaulax*	*Pfeisteria*
Ceratium	*Gymnodinium*	*Polykrikos*
Cystodinium	*Nematodinium*	*Prorocentrum*
Dinothrix	*Noctiluca*	*Protopsis*
Erythropsidium	*Peridinium*	*Warnowia*

Dinomastigota, Ciliophora (Phylum Pr-8), and Apicomplexa (Phylum Pr-9) are grouped together as alveolates because all have both surface alveoli, or pits, and common ribosomal RNA sequences. Further investigation may lead to their placement together as three classes in a new phylum, "Alveolata."

The dinomastigote, often called "dino," has two undulipodia. One of them rests in a characteristic groove, the girdle (cingulum), encircling the cell, and the other in a longitudinal groove, the sulcus. The undulipodia lie at a right angle to each other. When they both move, the cell whirls. Most dinomastigotes are enclosed in rigid walls, called tests; the test, consisting of an upper half called an epicone and a lower half called a hypocone, is made of plates embedded in the cytoplasmic side of the plasma membrane. The plates tend to be composed of cellulose and are encrusted with silica. Many, the "naked dinos," lack walls. Specialized vacuole-like organelles (pusules), usually two per cell, open by canals into the kinetosomes and from there to the exterior of the cell. The pusules respond to changes in pressure and are presumably osmoregulatory in function. Stinging organelles (trichocysts) are found underlying the cell membrane of many dinomastigotes; a kind of extrusome, they are capable of sudden discharge to sting prey.

Of more than 4000 known species of dinomastigotes in 550 genera, most swim as members of marine plankton and are especially abundant in warm seas. Many genera have freshwater representatives. Most are single celled, but some form colonies.

Some dinomastigotes produce powerful toxins that are accumulated by fish and marine invertebrates. The sometimes toxic red tides are colorful blooms of marine microbes, many of which are dinomastigotes such as *Gonyaulax tamarensis* (illustrated here). Many dinomastigotes are bioluminescent. Some of these bioluminescent species, such as *Noctiluca miliaris* ("a thousand night lights") cause the twinkling of lights in the waves of the open ocean at night.

An occasional species of dinomastigote is symbiotrophic or epibiotic on marine animals or seaweed. Others, such as the complex predator *Noctiluca* that grows to be 2000 μm in diameter, are bactivorous. *Noctiluca miliaris* is a large, carnivorous cell bearing an immense feeding tentacle with which it sweeps in various microorganisms. About half of the species of dinomastigotes are photosynthetic, crucial symbionts in marine corals and sea anemones (Phylum A-3) and clams (Phylum A-26). The most common intracellular photosynthetic symbiont in the reef communities of the world is the dinomastigote *Gymnodinium microadriaticum.* Photosynthetic dinos contain brownish plastids. The pigments of these plastids generally include chlorophylls *a* and c_2 (sometimes c_1 as well), beta- and gamma-carotenes, several xanthins (diadinoxanthin, diatoxanthin, dinoxanthin, fucoxanthin, neofucoxanthin, and neoperidinoxanthin), and a carotenoid peculiar to dinomastigotes, called peridinin. Photosynthetic dinomastigotes store starch.

Although dinomastigotes are undoubtedly eukaryotic, their nuclear organization is so idiosyncratic that they have been called mesokaryotic (between prokaryotic and eukaryotic). The DNA of nearly all other eukaryotes is organized into a complex with histone proteins to form fibrils that are 10 nm wide. The DNA of dinomastigotes is complexed with very tiny quantities of a peculiar alkaline protein rather than with the four or five common eukaryotic histones to form fibrils that are only 2.5 nm wide. In this sense, then, dinomastigote chromatin is organized like that of prokaryotes.

Furthermore, rather than condensing only during mitosis, the chromatin of dinomastigotes is always condensed into chromosomes. This is particularly strange, because the condensed-chromosome stages in animal and plant cells are just those in which the genome is turned off: RNAs and proteins are not synthesized while chromatin is condensed. However, the genes in the condensed chromatin of dinomastigotes are not turned off—the genes in these condensed chromosomes continue to be expressed. The typical mitotic stages (interphase, prophase, metaphase, anaphase, and telophase) are absent in dinomastigotes. In some, microtubules penetrate the nucleus during division; the kinetochores, which in plants and animals are attached directly to chromosomes, in dinos are embedded in the nuclear membrane, which remains intact during the division

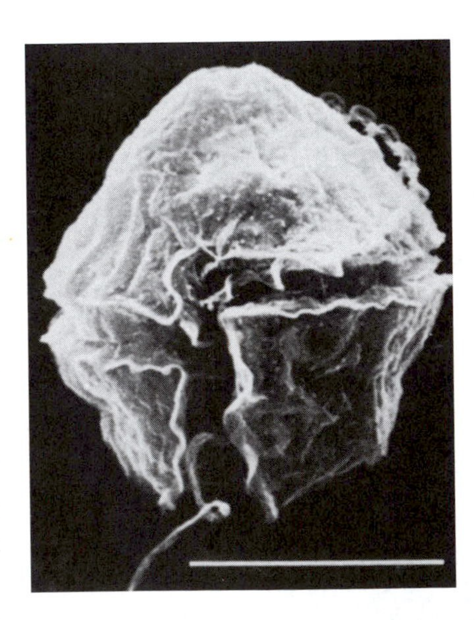

Gonyaulax tamarensis, a dinomastigote from the Pacific Ocean. SEM, bar = 50 μm. [Photograph courtesy of F. J. R. Taylor; drawings by R. Golder (external view) and E. Hoffman (cutaway view); information from L. Leoblich (external view) and F. J. R. Taylor (cutaway view).]

cycle. Chromatin is segregated to offspring cells by membrane attachment. The pattern of cell division differs from one dinomastigote species to another as if, within the phylum, mitosis evolved in its own peculiar fashion. Many dinos also form hard, resistant cysts. Before cyst formation, they engage in fusion, which seems to entail mating and gene exchange of some kind.

Dinomastigotes have speciated in some remarkable ways. An eyespot consisting of a layer of light-sensitive bodies containing carotenoid pigments overlain by a clear zone is characteristic of *Protopsis, Warnowia,* and *Nematodinium.* In the sedentary species *Erythropsidium pavillardii,* a complex cell eye, the ocellus, evolved. The ocellus, which detects the approach of prey, includes a lens and a fluid-filled chamber underlaid by a light-sensitive pigment cup. The lens changes shape, and the pigment cup moves freely about the lens. The whole ocellus protrudes from the cell to point in different directions. The development of *E. pavillardii* reveals that the pigment cup evolved from a plastid.

Dinomastigotes are masters of transformation. Members of a single genus may display different morphologies, such as undulipodiated swimming forms and test-covered amebas, as a function of life history stage. Perhaps the most impressive case of dino duplicity is documented for the astounding, recently characterized species *Pfeisteria piscicida,* which is of great—if negative—economic importance to the fishing industry.

Pfeisteria piscicida are the virtuosic fish killers promised by the species name. They are capable of rapid development from one stage to another. These phantoms form terrible toxins, forcing many fish that feed on them—such as Atlantic menhaden and southern flounder—to turn belly up in a few minutes, before they settle into the sediment and transiently disappear. In the continued presence of fish flesh, the typical dinomastigote swimmer cell transforms into a stellate test-covered ameba, one of more than a dozen guises. In the absence of food, these shelled amebas—dinos in disguise—sink to become invisible benthic cysts lying in wait. The volatile toxins, inhaled or absorbed through the skin, cause illness in the scientists who have sleuthed the multiple rapid transformations of dinomastigotes.

Many dinomastigotes make complex tests. Each has its characteristic size, apical pore, epicone, and hypocone. Hystrichospheres, fossils known to micropaleontologists, are actually resistant, robust dinomastigote tests. The fossil record of these organisms extends at least to the base of the Cambrian period; there is some evidence that they existed even earlier, in the late Proterozoic eon.

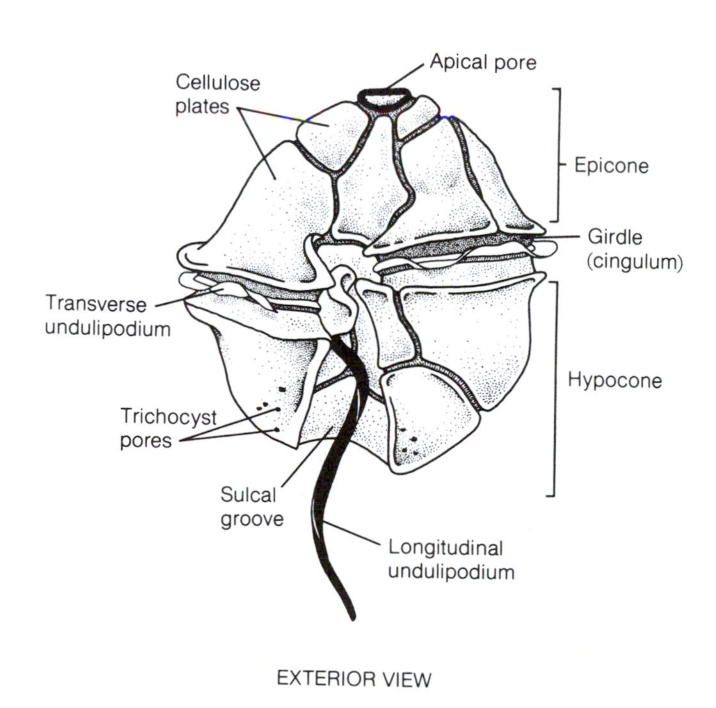

EXTERIOR VIEW

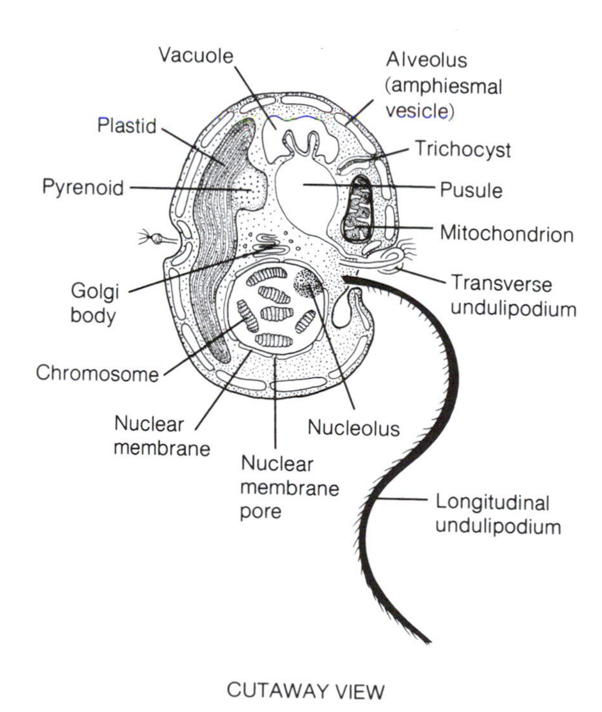

CUTAWAY VIEW

Pr-8 Ciliophora

(Ciliates)

Latin *cilium*, eyelash, lower eyelid; Greek *pherein*, to bear

EXAMPLES OF GENERA

Balantidium	*Gastrostyla*	*Spirostomum*
Blepharisma	*Halteria*	*Stentor*
Carchesium	*Nyctotherus*	*Stylonychia*
Colpoda	*Oxytricha*	*Tetrahymena*
Didinium	*Paramecium*	*Tokophrya*
Dileptus	*Pleurotricha*	*Uroleptus*
Ephelota	*Prorodon*	*Vorticella*
Euplotes	*Sorogena*	

Most ciliates, which are among the best-known protoctists, are bactivorous single cells. Ciliates are characteristically covered with cilia—short undulipodia with kinetosomes embedded in a tough, fibrillar outer cortex (proteinaceous cell layer) of the cell. Like the dinomastigotes (Phylum Pr-7) and apicomplexans (Phylum Pr-9), ciliates are alveolates, with pits embedded in their cortices. They possess two different types of nuclei, small micronuclei and larger macronuclei, usually more than one of each kind. Nearly 10,000 freshwater and marine species have been described in biology literature. Probably many more exist in nature. Nearly all are phagotrophic, eating bacteria, tissue, or other protists, or they are osmotrophs, utilizing dissolved nutrients in rich waters.

Although ciliates overwhelmingly are single-celled, multicellular slime-mold-like ciliates are known. *Sorogena* forms a stalked structure from aggregates of cells. It develops propagules, sporelike spheres that germinate into free swimmers, each with its macronucleus and micronucleus.

The cilia of ciliates, like other undulipodia, including sperm tails, have the same ultrastructure, the ninefold symmetrical array of microtubules (the axoneme), with a kinetosome at its base (see Figure I-2). Cilia are modified to perform specialized locomotory and feeding functions. The most usual modification is the grouping of cilia and their underlying kinetosomes into cirri (bundles) or membranelles (sheets). Cirri or membranelles function as mouths, paddles, teeth, or feet. The ciliate undulipodia are embedded in an outer proteinaceous cell layer (the cortex) about 1 μm thick containing rows of kinetids (the kineties) comprising complex fibrous connections between them. Associated with each kinetid of the ciliate cortex is a parasomal sac, a small invagination of the plasma membrane of unknown function.

Of the two types of nuclei in each ciliate, only the micronuclei, which apparently contain standard chromosomes, divide by mitosis. The huge macronuclei, which develop from precursor micronuclei in a series of complex steps, do not contain typical chromosomes. Instead, the DNA is broken into a great number of little chromatin bodies; each body contains hundreds or even thousands of copies of only one or two genes. Macronuclei are always required for growth and reproduction. They divide by elongating and constricting—not by standard mitosis. They take part in cellular functions such as the production of messenger RNA to direct protein synthesis. The micronuclei, not required for growth or reproduction, are dispensable, essential only for sexual processes unique to ciliates.

Most ciliates reproduce by transverse binary fission, dividing across the short axis of the cell to form two equal offspring. The anterior new cell is called the proter and the posterior one is the opisthe. However, certain stalked and sessile species, such as some suctorians, asexually bud off "larval" offspring. These offspring are "born": small rounded offspring, covered with cilia, emerge through "birth pores" of their entirely different looking, stalked "mother."

Most ciliates undergo a sexual process called conjugation. The conjugants, two cells of compatible mating types ("sexes") remain attached to each other for as long as many hours. Each conjugant retains some micronuclei and donates others to its partner. A series of nuclear fusions, divisions, and disintegrations follows, resulting in the two conjugants becoming "identical twins," as far as their micronuclei are concerned. The conjugants eventually separate and undergo a complex sequence of maturation steps. Although the micronuclei of the two exconjugants are now genetically identical (each conjugant having contributed equally), each new cell retains the cytoplasm and cortex of only one of the original conjugants. Because cytoplasmic and cortical inheritance in ciliates can be definitively distinguished from nuclear inheritance, these organisms are used in cell genetic analysis.

Ciliate classification has been revised dramatically in the past two decades because of the new information derived from riboso-

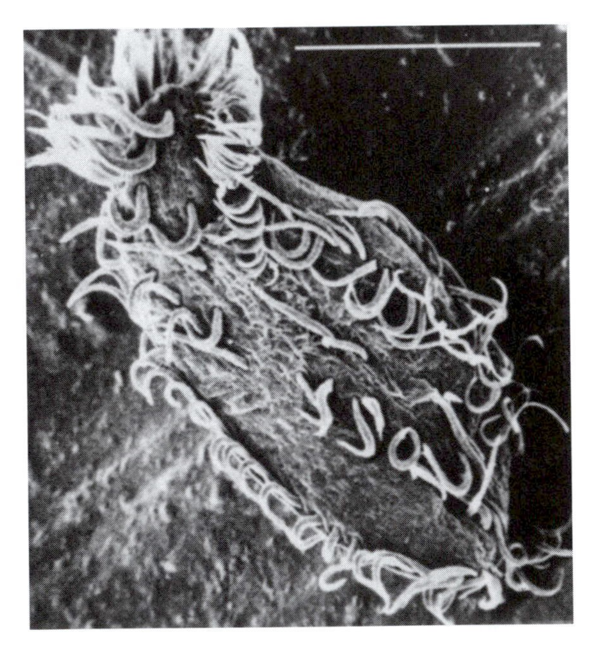

Gastrostyla steinii, a heterotrichous ciliate. The adoral membranelles, consisting of rows of cilia, are used in feeding. They sweep particulate food (bacteria and small ciliates) into the gullet. SEM, bar = 100 μm. [Photograph courtesy of J. Grim, *Journal of Protozoology* 19:113–126 (1972); drawings by L. Meszoly.]

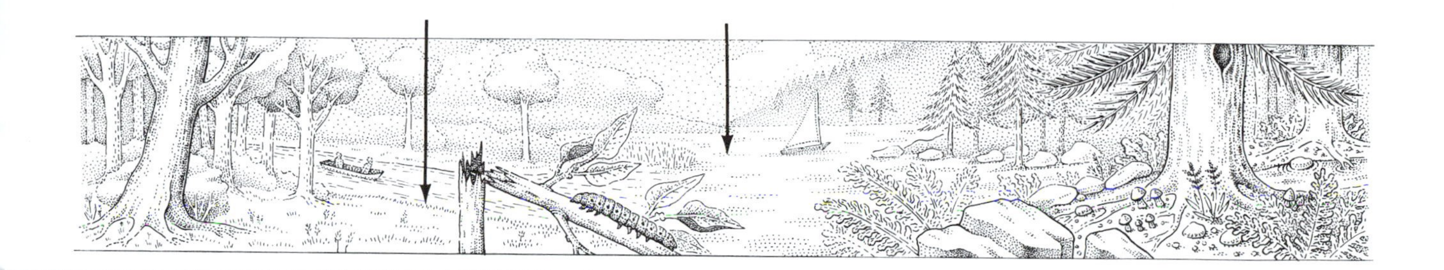

mal RNA sequencing studies and correlated with electron microscopy. "Holotrichs," ciliates with cilia over the entire surface, are not necessarily related—this formal name has been abandoned. Groups thought to be only very distantly related or unrelated, such as karyorelictans and stentors, are now known to be related, whereas organisms resembling each other superficially, such as *Euplotes* and *Stylonychia,* are more distantly related.

The most useful structure for the comparison of ciliates and the reconstruction of evolutionary history is the ciliate kinetid, the structure consisting of one or more kinetosomes and their undulipodia, as well as the ribbons of microtubules and filaments, including kinetodesmal fibers, that surround them. Kinetids are universal units of structure in all protoctist, animal, and plant cells that bear undulipodia. A kinetid with a single kinetosome is called a monokinetid, that with two kinetosomes is a dikinetid, and the rarer kinetid with many is called a polykinetid.

On the basis of kinetid structure and organization, the ciliates are divisible into three subphyla: Postciliodesmatophora, a group of ciliates with dikinetids, contains two classes: Karyorelictea and Spirotrichea. Heterotrichs, including *Stentor* and *Gastrostyla* (illustrated here) belong in this subphylum. The subphylum Rhabdophora, ciliates that have kinetids with short kinetodesmal fibers and tangential transverse ribbons of microtubules, contains two classes: Prostomea and Litostomea. Among many others, entodiniomorphs, bizarre-looking ciliates living as symbionts in the mammalian rumen, are classified in Litostomea. The third subphylum, Cyrtophora, contains four classes: Nassophorea, Phyllopharyngea, Colpodea, and Oligohymenophora. The cyrtophoran, unlike the rhadophoran, disassembles its complex oral ciliature and makes two new ones in the process of cell division. Nearly all the well-known ciliates—*Colpoda* (Class Colpodea: Subclass Colpodida), *Tetrahymena* (Class Oligohymenophora: Subclass Scuticociliatida), *Vorticella* (Class Oligohymenophora: Subclass Peritrichia), *Paramecium* (Class Nassophorea: Subclass Peniculida), and *Stylonychia, Oxytricha,* and *Pleurotricha* (Class Nassophorea: Subclass Hypotrichia), as well as the subclass Suctoria (Class Phyllopharyngea), belong in this great subphylum of ciliates.

Although many form spherical, resistant cysts, most ciliates lack hard parts and therefore do not fossilize. The tintinnids, heterotrichs in Subphylum Postciliodesmatophora, are exceptional marine ciliates that make shell-like structures from sand and organic cements. Their ancestors left evidence in the fossil record that they evolved before the Cretaceous period, some 100 million years ago.

Because of their various ciliary modifications, their rapid and controllable growth rates, and the ease with which they can be handled in the laboratory, ciliates are valuable for anatomical, genetic, and neurophysiological studies of single cells. Except for the parasite *Balantidium,* which occasionally grows in the human gut, ciliates cause no disease and are of little immediate economic importance.

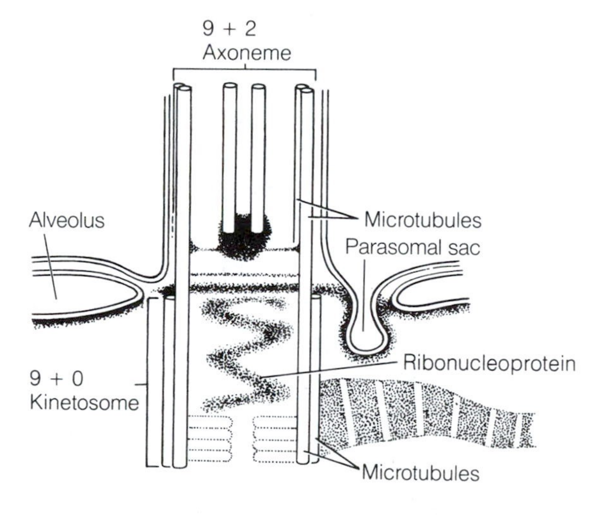

LONGITUDINAL SECTION OF KINETID

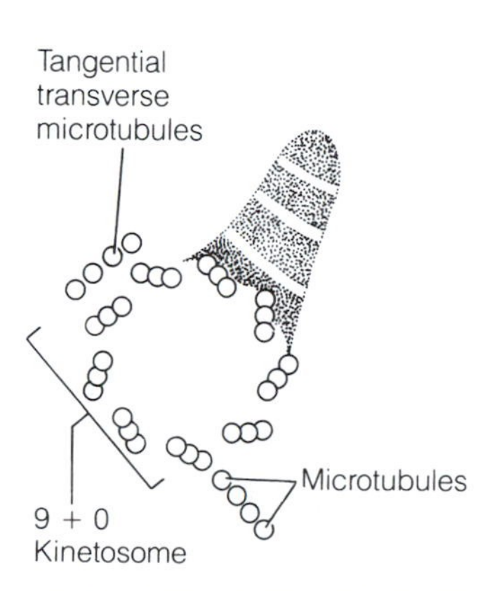

TRANSVERSE SECTION OF KINETID

Pr-9 Apicomplexa
(Sporozoa, Telosporidea)

Latin *apex,* summit; *complexus,* an embrace, enfolding

EXAMPLES OF GENERA

Babesia	*Gregarina*	*Plasmodium*
Coccidia	*Haemoproteus*	*Schizocystis*
Coelospora	*Haplosporidium*	*Selenidium*
Eimeria	*Isospora*	*Toxoplasma*

The Apicomplexa are single-celled symbiotrophs modified to penetrate tissue and obtain food from animals. All form spores. Unlike bacterial spores, apicomplexan spores are not heat- and desiccation-resistant cells; rather they are compact infective bodies that permit dissemination and transmission of the species from host to host. Along with Phyla Pr-22, Pr-23, and Pr-24, apicomplexans have traditionally been grouped together as "sporozoa" because of their common habitat—animal tissue. Detailed studies of their structure, nucleic acid sequences, and proteins have firmly established the great differences that justify their separation into distinct phyla.

This phylum of alveolates is named for the "apical complex," a distinctive arrangement of fibrils, microtubules, vacuoles, and other cell organelles at one end of each cell. The apicomplexan group is probably monophyletic. Three classes are recognized: Class Gregarinia, the gregarines, which includes *Gregarina;* Class Coccidia, the coccidians, which includes *Eimeria* and *Isospora;* and Class Hematozoa, the hemosporans and piroplasmids, which includes *Plasmodium, Haemoproteus,* and *Babesia.*

Apicomplexans reproduce sexually, with alternation of haploid and diploid generations. Both diploids and haploids can also undergo schizogony, a series of rapid cell divisions by mitosis that does not alternate with cell growth. Schizogony produces the small infective spores.

In fertilization, the undulipodiated male gamete (the microgamete) fertilizes a larger female gamete (the macrogamete) to produce a zygote (Figures A and B). The formation of the zygote is followed by the formation of a thick-walled oocyst (Figure C). The oocyst, rather than the infective spore, is the desiccation-, heat-, and radiation-resistant stage. The oocysts serve to transmit the microbes to new hosts. These cysts develop further by sporogony—rapid meiotic divisions inside the cyst produce infective haploid cells called sporozoites (Figures D and E).

The life cycles of apicomplexans may be complex and require several very different species of host. Many are bloodstream parasites. Many cause hypertrophy (gigantism) of the host cells in which they divide. The infection leads to duplication of host chromatin, causing a striking increase in the amount of host DNA, probably by polyploidization.

The coccidians are perhaps the best-known group of apicomplexans, because many of them cause serious and even fatal diseases of their animal hosts. *Isospora hominis* is the only coccidian that

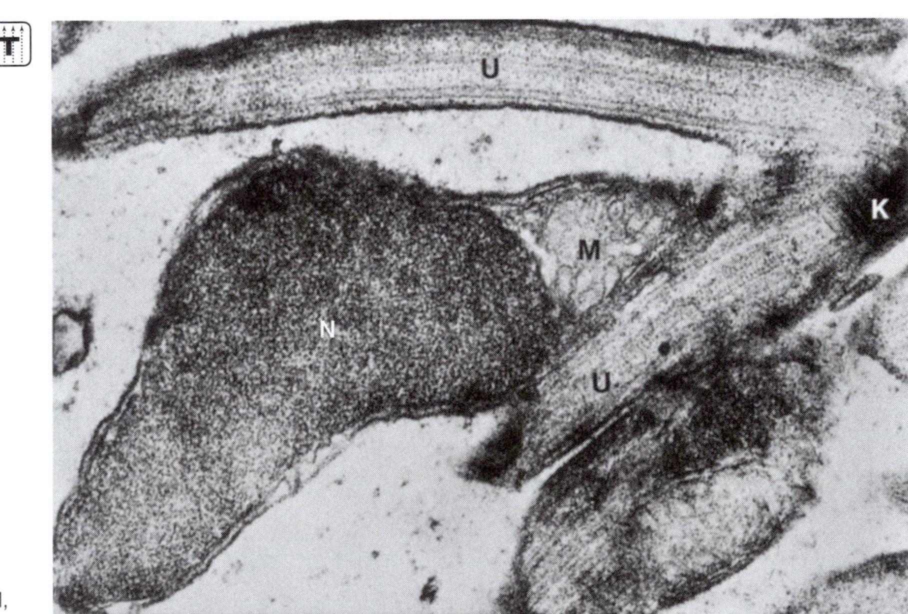

A Microgamete ("sperm") of *Eimeria labbeana,* an intracellular parasite of pigeons (Phylum A-37). N = nucleus; M = mitochondria; U = undulipodium; K = kinetosome. The structures above the nucleus are part of the apical complex. TEM, bar = 1 μm. [Courtesy of T. Varghese.]

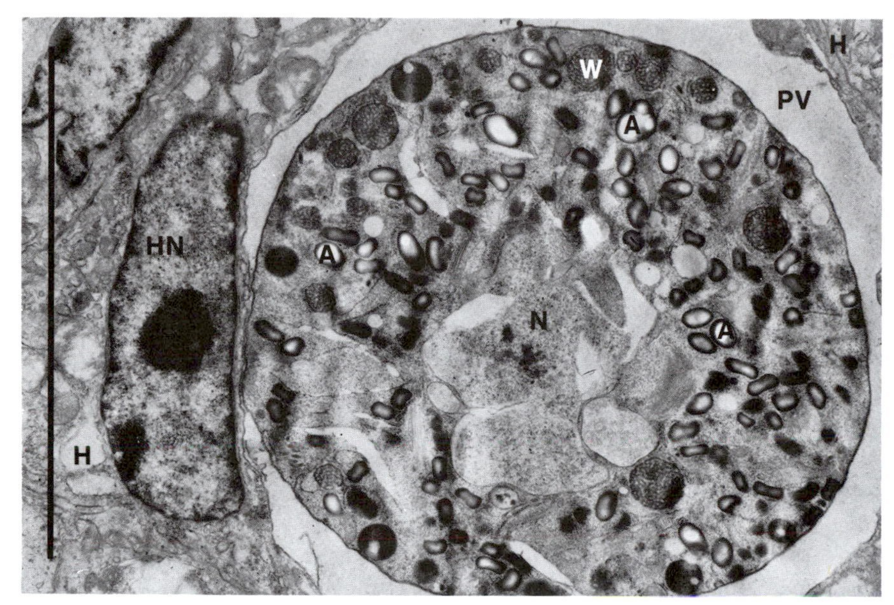

B Macrogamete ("egg") of *Eimeria labbeana.* H = host cell; HN = host nucleus; PV = parasite vacuole in host cell; N = macrogamete nucleus; A = amylopectin granule; W = wall-forming bodies, which later coalesce to form the wall of the oocyst. TEM, bar = 5 μm. [Courtesy of T. Varghese.]

parasitizes humans, but others, such as *Eimeria,* affect livestock and fowl. Because these apicomplexans are generally acquired with food and thus find their way into the digestive tract, the major symptoms of coccidian disease are diarrhea and dysentery.

An *Eimeria* infection begins when an oocyst is eaten. The oocyst germinates and produces sporozoites, which escape from the oocyst (Figures E and F) and enter the epithelial cells of the host, typically those of the gut lining, where they multiply by mitosis. Within the host cells, they develop into various forms called trophozoites, merozoites, and schizonts. The merozoites escape to infect more host cells. This cycle, called the schizogony cycle, can be repeated many times. Eventually, some of the merozoites within host cells develop into microgametes, which escape from the host cell. Other merozoites develop into macrogametes within host cells, where they remain. Fertilization takes place inside host cells. The resulting diploid zygotes undergo meiosis and develop into oocysts that typically exit from the anus with the feces (Figure G). Oocysts survive in soil until they are eaten by other potential hosts.

The most infamous apicomplexans important to human history are the malaria parasites, *Plasmodium* species. They are transmitted to humans by the female *Anopheles* mosquito (Phylum A-20, Mandibulata). Fertilization of *Plasmodium* takes place in the gut of the mosquito. The undulipodiated zygote embeds itself in the gut wall, where it transforms into a resistant oocyst. Within the oocyst, infective cells are formed by meiosis and sporogony (multiple fission). The sporozoites migrate to the salivary glands of the mosquito. With the bite of the mosquito, the sporozoites are injected into a human bloodstream, where they infect red blood cells. Inside a blood cell, a sporozoite develops further to become a feeding stage, the trophozoite, which grows at our expense—the sporozoan diet requires iron obtained from human hemoglobin. The trophozoite eventually undergoes schizogony to produce merozoites, small infective cells that escape from the ruined blood cell into the bloodstream. The merozoites attack and penetrate more blood cells, develop into trophozoites, and divide into more merozoites. After several such cycles, the merozoites differentiate into male and female gametes. These gametes must be taken up from the blood into the *Anopheles* mosquito gut—as they are when the mosquito draws in blood during its bite—to complete the fertilization stage of the *Plasmodium* life cycle. All the merozoites in a human host are produced and released more or less simultaneously—the pulse of formation and release of successive generations causes the characteristic periodic attacks of malarial fever.

There is a baffling variety of sexual life cycles and attack strategies in this group of protoctists; an enormous and contradictory terminology has made their study an arcane delight for specialists—especially for those interested in veterinary medicine.

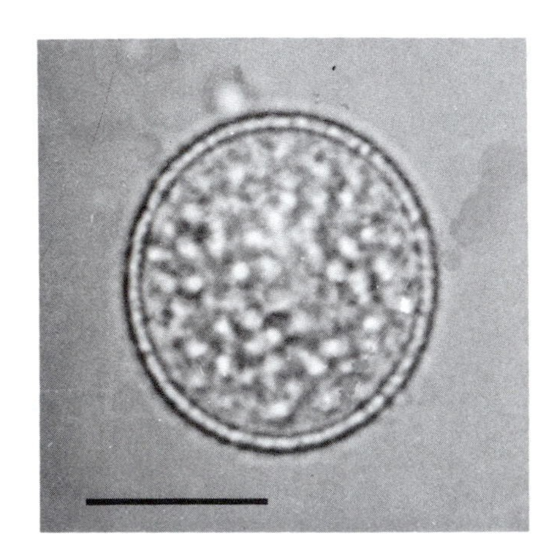

C Unsporulated oocyst of *Eimeria falciformes*. LM, bar = 10 μm. [Courtesy of T. Joseph, *Journal of Protozoology* 21:12–15 (1974).]

D Four sporocysts of *Eimeria nieschulzi* in sporulated oocyst. LM, bar = 10 μm. [Courtesy of D. W. Duszynski.]

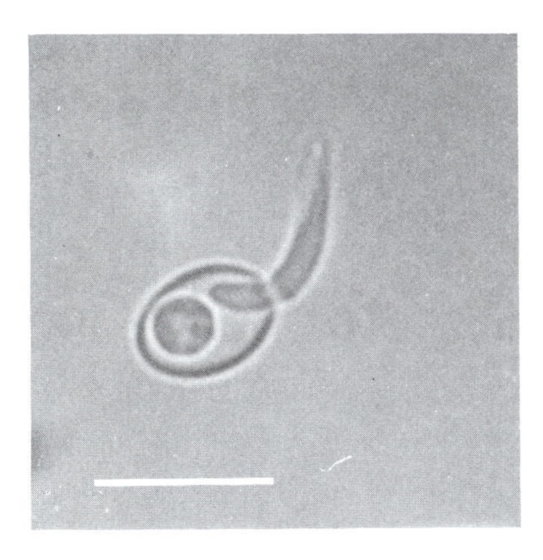

E Sporozoite of *Eimeria indianensis* excysting from oocyst. LM, bar = 10 μm. [Courtesy of T. Joseph, *Journal of Protozoology* 21:12–15 (1974).]

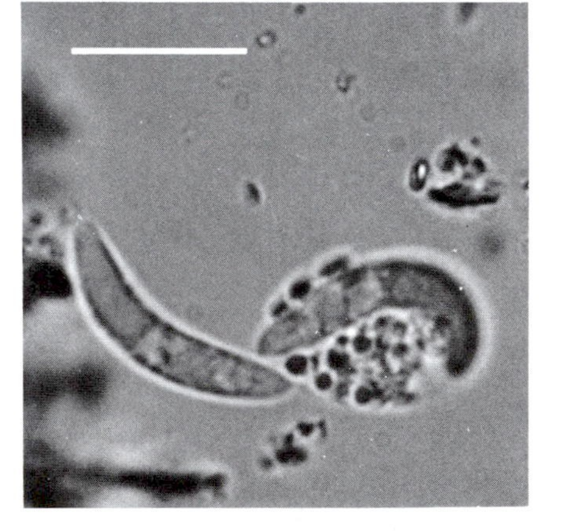

F Free sporozoites of *Eimeria falciformes*. LM, bar = 10 μm. [Courtesy of D. W. Duszynski.]

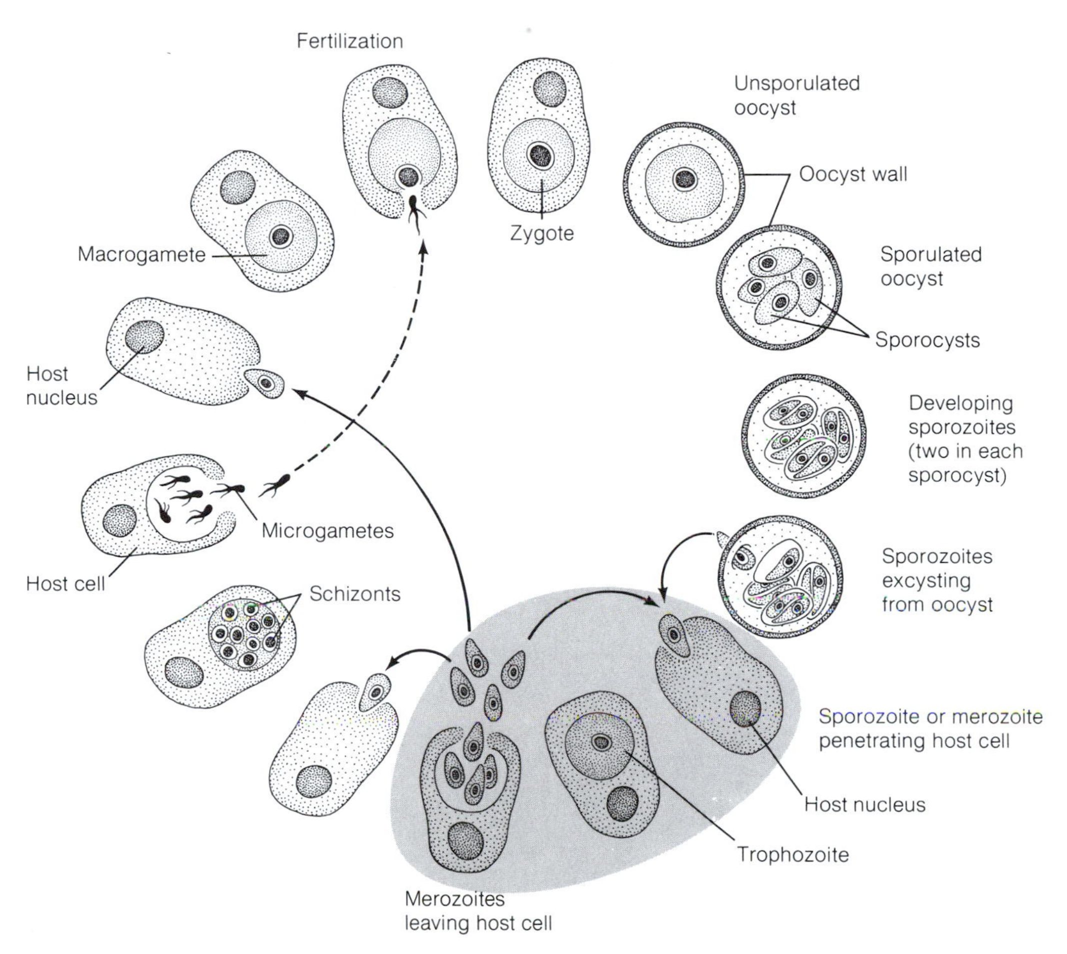

G The life cycle of *Eimeria* sp. The shaded part of the diagram represents the schizogony cycle, which may repeat itself many times before some of the merozoites differentiate into gametes. [Drawing by L. Meszoly.]

Pr-10 Haptomonada

(Prymnesiophyta, Haptophyta, coccolithophorids)

Greek *haptein,* fasten; *phyton,* plant

EXAMPLES OF GENERA

Calcidiscus	*Discosphaera*	*Pontosphaera*
Calciosolenia	*Emiliania*	*Prymnesium*
Calyptrosphaera	*Gephyrocapsa*	*Rhabdosphaera*
Chrysochromulina	*Hymenomonas*	*Syracosphaera*
Coccolithus	*Phaeocystis*	

The tiny, planktonic haptomonads have been viewed by marine biologists and paleontologists as two different types of organisms: (1) golden, motile algae (Figure A) resembling planktonic chrysomonads (Phylum Pr-13) and (2) coccolith-covered coccolithophorids (Figure B). Coccoliths, the microscopic disclike calcium carbonate structures of renown to paleontologists, are produced by coccolithophorids, which bear them as packed surface plates. Ultrastructural, developmental, and molecular biological studies have united the two views: the golden chrysomonad-like alga and the coccolithophorid are the same organisms at different stages in their life histories.

Haptomonads are primarily marine organisms, although some freshwater genera are known. Although they have golden yellow plastids in common with the chrysomonads, their cell structure differs enough from that of chrysomonads to justify placing haptomonads in a different phylum. Unlike the chrysomonads, haptomonads have only a very weak tendency to become multicellular. Sexual stages are unknown.

Haptomonads are distinguished by their haptonemes, scales, and, in some, their coccoliths. The haptoneme is a thread, often coiled, that may be used as a holdfast to anchor the free-swimming protoctist to a stable object (Figure C). Each cell has one haptoneme, generally at its anterior end. The haptoneme, which may be a specialized modification of the ubiquitous [9(2)+2] undulipodium (see Figure I-2), is a microtubular structure consisting of six doublets of microtubules arranged in a circle whose center is occupied by a single microtubule or none at all. With the haptoneme at the anterior end of the cell are two standard undulipodia and, generally, a membranous Golgi body.

B *Emiliania huxleyi,* a coccolithophorid from the Atlantic. That coccolithophorids are resting stages of haptomonads has been realized only in the past decade. SEM, bar = 1 μm. [Courtesy of S. Honjo.]

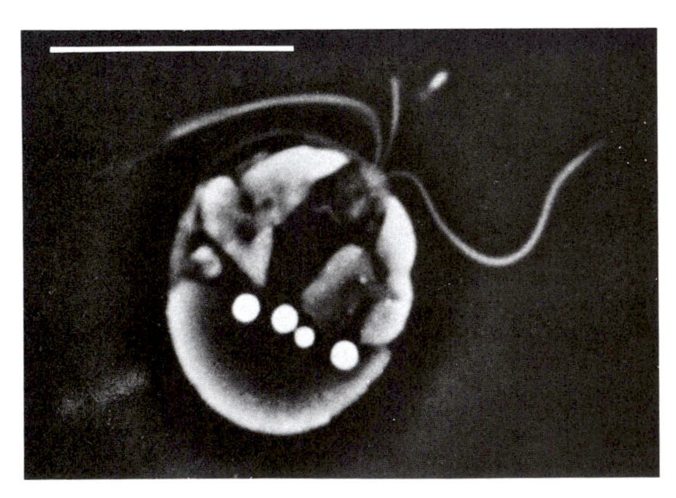

A *Prymnesium parvum,* a living marine haptomonad, showing undulipodia and haptoneme. LM, bar = 10 μm. [Courtesy of I. Manton and G. F. Leedale, *Archiv für Microbiologie* 45:285–303 (1963).]

In the transformation of the free-swimming stage into the resting, coccolithophorid stage, coccoliths and scales develop inside the cell: calcium carbonate crystals precipitate on conspicuous scales, which are made of organic polymers inside the Golgi apparatus. The gradually assembling coccoliths are transported to the edge of the cell by microtubule-mediated processes. They are deposited, in some cases with exquisite regularity, on the cell surface. The scales and the coccoliths that form on them bear intricate patterns that are species or genus specific. These coccolithophorid stages are often resistant, permitting tolerance of conditions that would be prohibitive to the swimming, growing forms of haptomonads.

Coccolithophorids have produced great quantities of particulate calcitic carbonate continuously for 100 million years or so

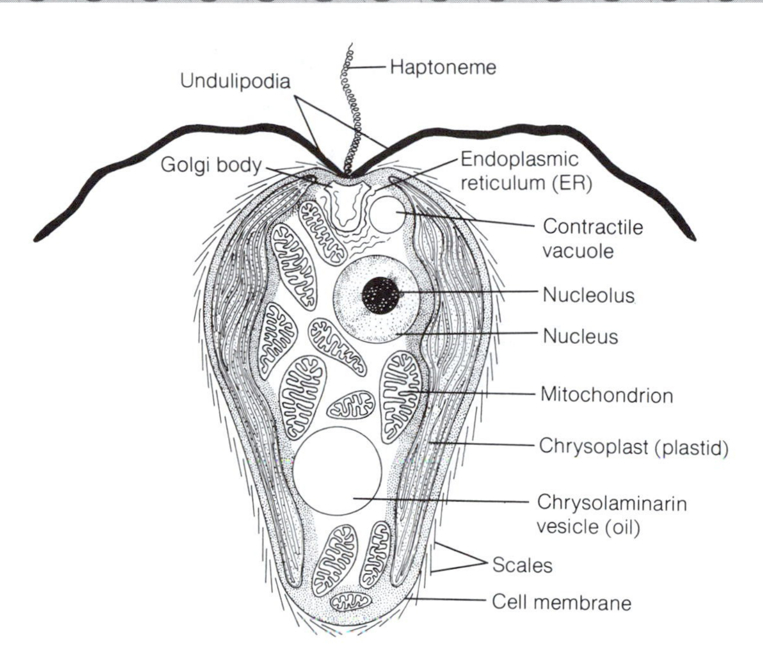

C *Prymnesium parvum,* the free-swimming haptonemid stage of a haptomonad. The surface scales shown here are not those on which coccoliths form. [Drawing by R. Golder.]

since the Cretaceous period; they have contributed significantly to the chalk deposits of the world. Because coccolithophorids are distinctive, they serve as stratigraphic markers; several hundred morphotypes or fossil species have been studied by geologists. For haptomonads, more is thus known about the morphology of fossils than of extant forms. The correlation between the haptonemid and the coccolithophorid stages has still not been made for many species. More work on haptomonad life cycles, including the development of calcium carbonate skeletal patterns, is needed in the context of the biology of the living cells.

Haptomonads typically have two golden yellow plastids (chrysoplasts) surrounded by a plastid endoplasmic reticulum that is continuous with the nuclear membrane. The plastids contain chlorophylls a, c_1, and c_2 but lack chlorophylls b and e. In addition to beta-carotene, they have alpha- and gamma-carotenes. They have fucoxanthin, an oxidized isoprenoid derivative that is also found in diatoms (Phylum Pr-16) and brown algae (Phylum Pr-17). Fucoxanthin is probably the most important determinant of the brownish yellow color. Haptomonads do not store starch; rather, like euglenids (Phylum Pr-12), they form a glucose polymer having the β-1–3 linkage of the monosaccharides. This white storage material, called paramylon, is stored within pyrenoids (proteinaceous structures) between the thylakoids (photosynthetic membranes) of the plastids.

Populations of the naked haptomonad *Phaeocystis poucheti,* which lack coccoliths, are responsible for the production of large quantities of the atmospheric gas dimethyl sulfide.

Pr-11 Cryptomonada
(Cryptophyta)

Greek *kryptos,* hidden; Latin *monas,* unit; Greek *phyton,* plant

EXAMPLES OF GENERA

Chilomonas
Chroomonas
Copromonas
Cryptomonas
Cyanomonas
Cyanophora
Cyathomonas
Hemiselmis
Hillea
Nephroselmis

Cryptomonads are flattened, elliptical swimming cells. Both heterotrophic and photosynthetic, they are found all over the world in moist places. Some commonly form blooms on beaches, whereas others have been found as intestinal parasites in domesticated animals. Palmelloid colonies (for example, nonmotile cells) embedded in gel of their own making, are known as well. These widely differing habitats have led differently trained scientists—such as marine botanists and parasitologists—to study them. Confusions in terminology and in taxonomy and general ignorance of their existence abounded, especially before electron microscopy revealed them to be a clearly delineated group.

Like the euglenids of Phylum Pr-12, cryptomonads may be pigmentless animal-like "protozoa" or brightly pigmented and photosynthetic plantlike algae. Found primarily as free-living single cells, commonly in fresh water, they are unlike euglenids in details of cell structure and division. Their photosynthetic pigmentation, if present, also is unique.

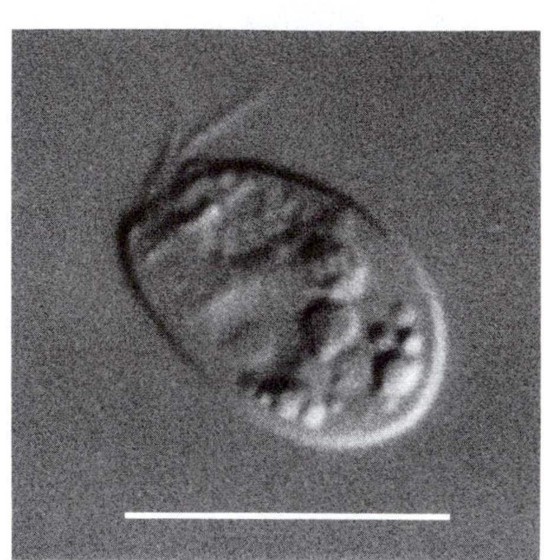

B *Cyathomonas truncata,* live cell. LM, bar = 5 μm. [Courtesy of F. L. Schuster.]

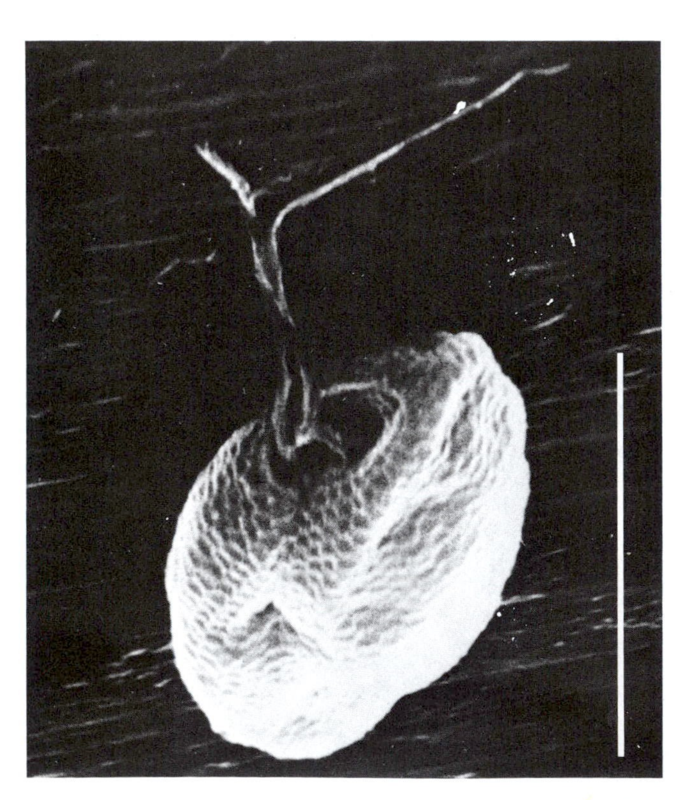

A *Cyathomonas truncata,* a freshwater cryptomonad. SEM, bar = 5 μm.

The cryptomonad bears two anterior undulipodia inserted in a characteristic way along the gullet, also called the crypt. The colorless genus *Cyathomonas,* for example, ingests particulate food through its gullet (Figures A through C). In the phagotrophic members of the group, which eat bacteria or other protoctists, the crypt is typically lined with trichocysts and bacteria-like bodies. Trichocysts expel poisons, which subdue and kill the microbial prey. Most members of the photosynthetic genera also have trichocysts.

Pigmented cryptomonads as a rule contain in their plastids chlorophyll c_2 in addition to chlorophyll *a.* Members of photosynthetic genera, such as *Cryptomonas* and *Chroomonas,* contain unique protein-pigment complexes called phycobiliproteins. Unlike most algae, they lack beta-carotene and zeaxanthin, but they contain alpha-carotene, cryptoxanthin, and alloxanthin. Many cryptomonads with those pigments are green or yellowish green. Others also contain phycocyanin or phycoerythrin and so tend to be deeper blue or deeper red. In general, phycocyanin pigments are strictly limited in nature: they are found in most cyanobacteria (Phylum B-5) and in the rhodoplasts of red algae (Rhodophyta, Phylum Pr-25), as well as in miscellaneous anomalous algae such as *Cyano-*

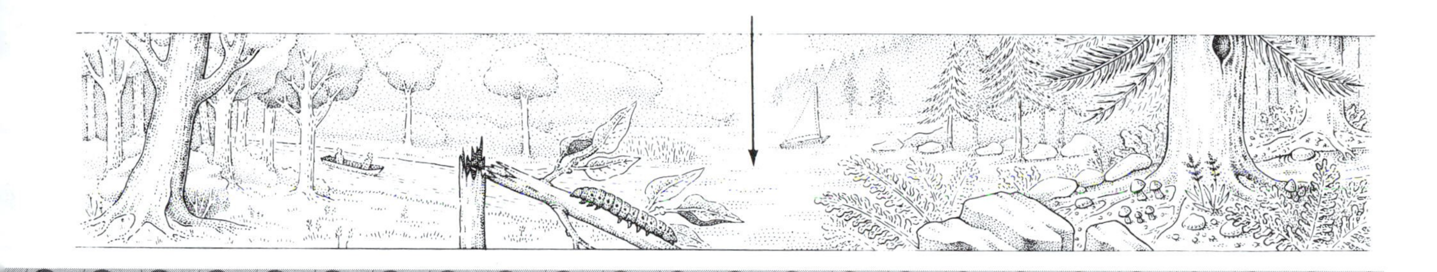

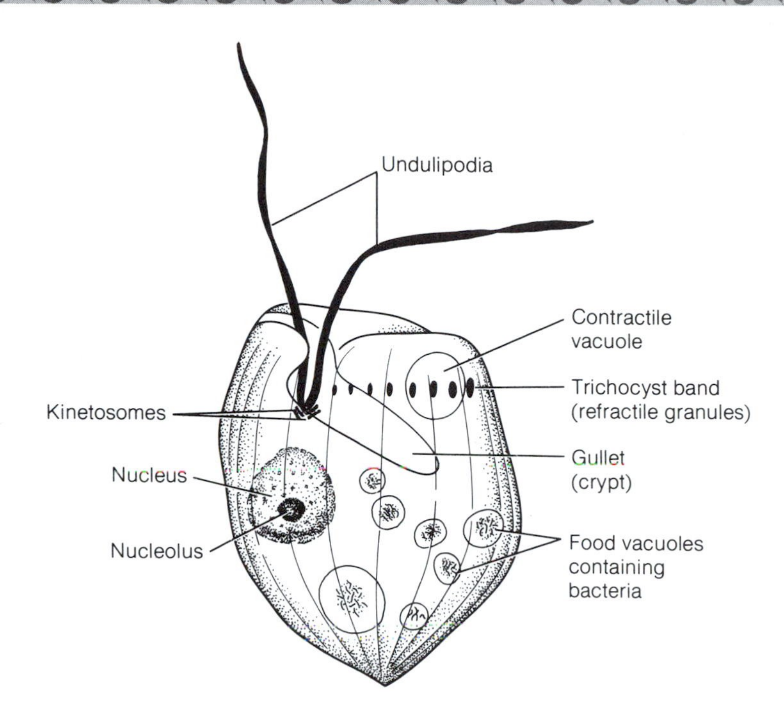

C *Cyathomonas truncata.* [Drawing by M. Lowe.]

phora paradoxa, organisms that are thought to harbor blue-green remnant-walled cyanobacteria symbionts instead of plastids. Because of the peculiar cryptomonad cell structure combined with plastids of various colors, cryptomonads probably became photosynthetic secondarily. More than once in the course of their evolution, they acquired phototrophic prokaryotic or eukaryotic symbionts that became chloroplasts or rhodoplasts. *Chilomonas* has a reduced plastid and may therefore have acquired photosynthesis and lost it secondarily.

Meiotic sexuality and gametogenesis is unknown in cryptomonads. Many have been grown and observed in the laboratory. They simply divide into two offspring cells. Just before cell division, new kinetosomes and undulipodia appear with a new crypt in proximity to the old one. The new oral structure then rotates, migrating to the opposite end of the cell. In the meantime, chromatin inside the closed nuclear membrane forms small knobby chromosomes that segregate into two bundles at opposite sides of the nucleus. The nucleus divides, cytokinesis ensues, and two offspring cells with a plane of mirror symmetry between them separate. This type of reproduction by binary fission distinguishes cryptomonads, regardless of their nutritional mode, from other protoctists. It was first documented, beautifully, by Karl Bělăr in 1926.

Pr-12 Discomitochondria

(Flagellates, zoomastigotes, zooflagellates)

Greek *zoion,* animal; *mastix,* whip

EXAMPLES OF GENERA

Astasia
Blastocrithidia
Bodo
Cephalothamnium
Colacium
Crithidia
Dinema
Distigma
Euglena
Herpetomonas
Heteronema
Leishmania
Leptomonas
Naegleria
Oikomonas
Paranema
Paratetramitus
Phacus
Phytomonas
Stephanopogon
Trachelomonas
Trypanosoma
Vahlkampfia

Until the details of its relation to the rest of the protoctist world are discovered, this new group conveniently holds four well-defined subgroups of motile, bactivorous or osmotrophic, swimming, mastigote unicells and their colonial derivatives. They have in common mitochondria with discoid cristae and the lack of meiotic sexual fertilization cycles. Their mitochondria bar them from inclusion in other phyla that house similar organisms: Phylum Pr-2 (no mitochondria) and Phylum Pr-20 (tubular cristae in mitochondria). In two-kingdom classifications, all these organisms were in the class Flagellata in the phylum Protozoa in the animal kingdom. Their ultrastructure, as revealed by the electron microscope, reveals a great diversity, and changes in their classification are expected in the future.

Four classes are recognized. Members of the first class, Amoebomastigota, or Schizopyrenida, transform from amebas into undulipodiated swimmers. The second and third classes are Kinetoplastida and Euglenida. Although it would have come as no surprise to an observant biologist at the turn of the twentieth century, the evolutionary connection between euglenids (then considered in the botanical realm) and kinetoplastids (in the province of the zoologists) has been confirmed by molecular biology. Both share ribosomal RNA sequences, and we thus put these two groups of early evolved eukaryotes together as classes in the same phylum. Members of the fourth class, Pseudociliata, are free-living cells covered with undulipodia. These organisms, once thought to be ciliates, are all members of the genus *Stephanopogon.*

The amebomastigotes, such as *Paratetramitus* (Figure A), are freshwater or symbiotrophic microbes distinguished by their ability to change from an undulipodiated to an ameboid stage and back again. This transformation, induced by the depletion of nutrients, is best studied in *Naegleria* because this genus can be cultured. When *Naegleria* amebas are suspended in distilled water, they develop kinetosomes, grow [9(2)+2] axonemes from them, and soon elongate into a mastigote form. They quickly swim in search of food bacteria. After they find it, they lose their undulipodia and return to an ameboid lifestyle.

Kinetoplastids include the bodos, mainly free-living biundulipodiated cells common in stagnant water, and the infamous trypanosomes. Prerequisite for inclusion in this group is the presence of a special large mitochondrion called the kinetoplast. Meiotic sexuality and fertilization, although widely sought, have never been found in any member of the class. Many genera, such as *Trypanosoma* and *Crithidia,* contain important pathogens and parasites affecting humans and domestic animals. Sleeping sickness and Chagas disease are caused by kinetoplastids. Marked changes—for example, the elongation or near disappearance of the kinetoplast—take place when the cells move from vertebrate blood to insect salivary gland and again when they move back. A few multicellular derivatives are known. For example, *Cephalothamnium cyclopum,* a bodo having a characteristically backward-directed undulipodium, forms colonies of several dozen cells attached to a common stalk.

There are about 800 species of euglenids (Figure B), most of which are unicellular and live in fresh or stagnant water. Six orders

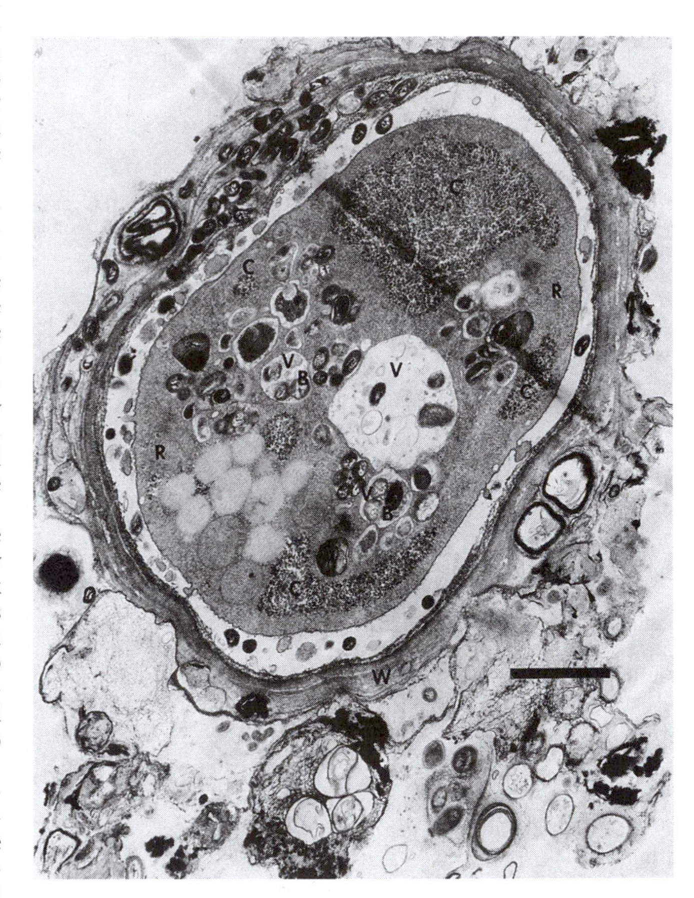

A *Paratetramitus jugosus,* an amebomastigote that grows rampantly in microbial mats. From Baja California Norte, Laguna Figueroa, Mexico, these cysts and amebas are found with *Thiocapsa* (Phylum B-3) and other phototrophic bacteria. W = cyst wall; R = ribosome-studded cytoplasm; B = bacteria being digested in vacuoles (V); C = well-developed chromatin. TEM, bar = 10 μm.

are recognized. Some genera are marine, some are colonial, and some are symbiotrophic. Most euglenids are photosynthetic, but many lack chloroplasts and hence are limited to heterotrophy. Even the photosynthetic euglenids under certain conditions eat dissolved or particulate food.

Euglenids are thought to have evolved from free-living bodos. The most likely explanation of the genesis of phototrophic green plastid–bearing euglenids is that an aggressive free-living kinetoplastid, such as *Bodo,* acquired a plastid secondarily by feeding. This scenario is bolstered by the observation that double membranes surround the plastids of euglenids, likely relicts of the phagocytic event that entrapped them. The green plastids were probably acquired secondhand from partly digested green algae.

In some classifications, euglenids are placed with the green algae (Phylum Pr-28, Chlorophyta) because of their grass-green chloroplasts. However, they differ from chlorophytes, even in their chloroplast pigmentation. The plastids of euglenids, like those of chlorophytes and plants, contain chlorophylls *a* and *b* only, as well as beta-carotene and the carotenoid derivatives alloxanthin, antheraxanthin, astaxanthin, canthoxanthin, echinenone, neoxanthin, and zeaxanthin. In addition, though, euglenid plastids have the carotenoid derivatives diadinoxanthin and diatoxanthin, which are not

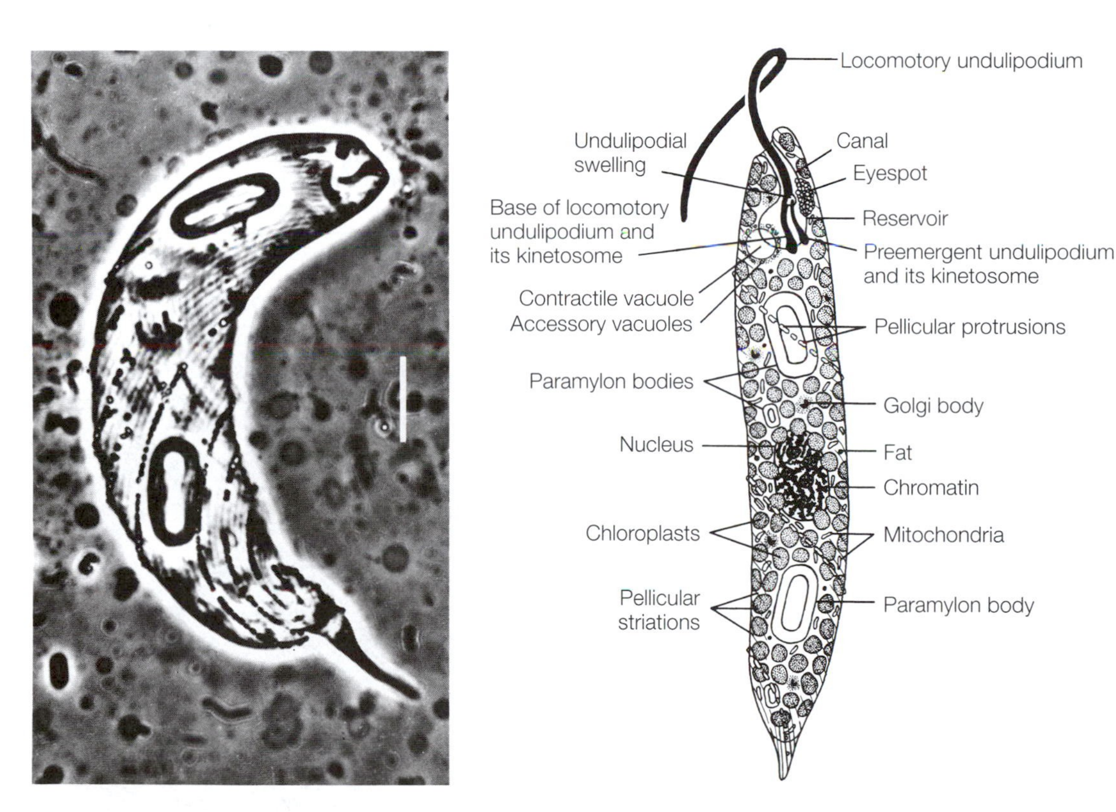

B *Euglena spirogyra,* a freshwater euglenid from England. LM (phase contrast), bar = 10 μm. [Photograph courtesy of G. F. Leedale (1967); drawing by R. Golder; information from G. F. Leedale.]

present in chlorophytes (or in plants). Euglenids lack rigid cellulosic walls; instead, they have pellicles, finely sculpted outer structures made of protein. These pellicles are usually very flexible, so euglenids typically can change shape easily. They do not store starch; instead they store paramylon, a glucose polymer having the β-1–3 linkage of the monosaccharides.

Euglenid reproduction is not sexual; all attempts to find meiosis and gametogenesis in euglenids have failed. The nuclei of different individual organisms of the same species may have different amounts of DNA. The nuclei contain large karyosomes, also called endosomes, which are structures homologous to the nucleoli of other cells. Like nucleoli, the endosomes are composed of RNA and protein combined in bodies that are precursors to the ribosomes of the cytoplasm. Euglenids have a nearly invisible mitotic spindle composed of a few intranuclear microtubules. Many lack distinct, countable chromosomes; in cell division, their chromatin granules do not move in a single mass as in standard anaphases. The chromatin granules do not split at metaphase; rather, no metaphase plate is formed, and each granule autonomously proceeds to one of the nuclear poles, where the newly replicated undulipodia are located. Before nuclear division, the two kinetids move toward the anterior end of the cell, causing the cell to distort in a way characteristic of division as the cell divides lengthwise.

The species *Euglena gracilis* (Figure C) has proved to be a fine tool for analyzing cell organelles. Euglenas can be found both with and without chloroplasts, which may be permanently lost or temporarily "turned off." Thus, the effect of light, chemical inhibitors, temperature, and many other agents on the development of chloroplasts and other organelles can be beautifully observed. For example, if the cells are placed in the dark, the chloroplasts regress; after a little growth, the euglenids turn white and become entirely dependent on external food supplies for growth (Figure D). These "animal" cells can be reconverted into "plants." If dark-grown white euglenas are reilluminated, they turn light green within hours. Their chloroplasts go through a series of developmental changes induced by the light, and, after three days or so, they recover their bright green color.

Euglena gracilis is the only species known that can be genetically "cured" of chloroplasts without killing the organism. Although mutants of other photosynthetic protoctists and plants that lack the capacity for photosynthesis can be produced, *Euglena* is the only chloroplast-containing organism that survives and reproduces in-

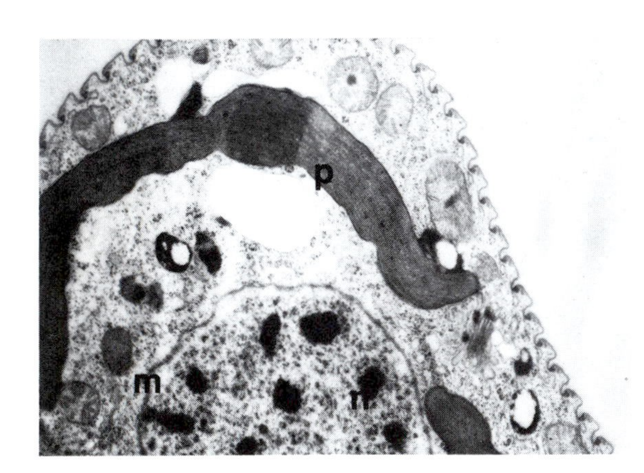

C A thin section of *Euglena gracilis* grown in the light, showing the well-developed chloroplast (p). m = mitochondrion; n = nucleus. TEM, bar = 1 μm. [Courtesy of Y. Ben Shaul.]

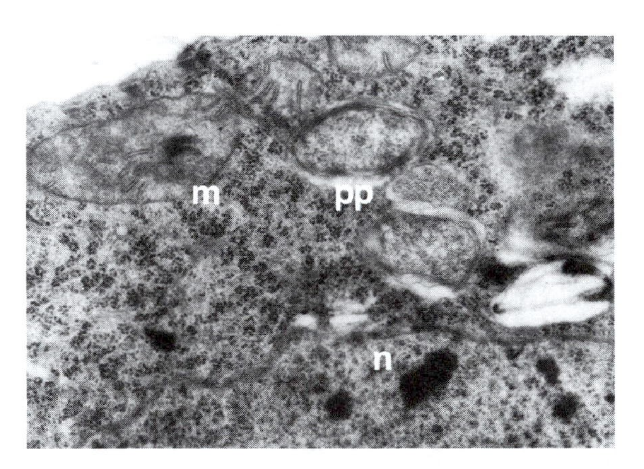

D The same strain of *Euglena gracilis* as that shown in Figure C, grown for about a week in the absence of light. The chloroplasts dedifferentiate into proplastids (pp). This process is reversible: proplastids regenerate and differentiate into mature chloroplasts after about 72 hours of incubation in the light. m = mitochondrion; n = nucleus. TEM, bar = 1 μm. [Courtesy of Y. Ben Shaul.]

dependently of chloroplast DNA. If *E. gracilis* cells are treated with ultraviolet light or with a number of other treatments to which the chloroplast genetic system is more sensitive than is the nucleocytoplasmic system, the genetic entities responsible for chloroplast development can be lost permanently. The euglenas then lose all their plantlike characteristics and become irreversibly dependent on food. By this treatment, the metabolism of the nucleocytoplasm can be studied in detail separately from that of the plastid.

As mentioned earlier, all members of the class Pseudociliata belong to the genus *Stephanopogon.* The lack of nuclear dimorphism in their 16 nuclei, which are homokaryotic, and the presence of kinetids entirely unlike those of ciliates indicate that these undulipodiated marine benthic (seafloor dwelling) beings do not belong with the ciliates in Phylum Pr-8. Feeding on diatoms, smaller protoctists, and bacteria, these organisms have undulipodia that are in rows primarily on the ventral surface of the cell. Reproduction, always in the absence of any sign of sexuality, takes place within a cyst as a form of multiple fission (palintomy). Mitosis, which takes place within an intact (closed) nuclear membrane, is acentric, with an intranuclear spindle but no centrioles or centrosomes.

Pr-13 Chrysomonada
(Chrysophyta)

Greek *chrysos,* golden; *phyton,* plant

EXAMPLES OF GENERA

Chattonella
Chromulina
Chrysobotrys
Chrysocapsa
Chrysosphaerella
Dinobryon
Echinochrysis
Epipyxis
Gonyostomum
Hydrurus
Mallomonas
Merotricha
Monas
Nematochrysis
Ochromonas
Phaeothamnion
Raphidomonas
Reckertia
Rhizochrysis
Sarcinochrysis
Synura
Thallochrysis
Thaumatomastix
Trentonia
Vacuolaria

A new grouping of Phyla Pr-13 through Pr-21, from chrysomonads through hyphochytrids, has been established on the basis of similarity in ribosomal RNA gene sequences, which suggests that they have common ancestry. All organisms in these nine phyla produce cells with mastigonemate (hairy) undulipodia in the heterokont style (anteriorly attached and of unequal length; see Figure C). These phyla are called stramenopiles, "straw bearers," referring to the hollow mastigonemes that decorate their undulipodia.

The chrysomonads form a large and complex group of algae whose plastids contain golden yellow pigments. As heterokonts, at some stage of their life history, their cells have two anteriorly attached undulipodia of unequal sizes. Chrysomonads are ubiquitous in fresh, temperate lakes and ponds. Only one widespread group of marine planktonic chrysomonads, the silicoflagellates (silicomastigotes), are marine. Of the primarily unicellular protoctists (Phyla Pr-1 through Pr-13), chrysomonads have the strongest tendency to multicellularity. Morphologically even more diverse than the green algae (Phylum Pr-28, Chlorophyta), some form large, branching colonies. However complex they become, because their cells retain the characteristic ultrastructure with golden yellow plastids, they unequivocally can be identified as chrysomonads.

Chrysomonads lack germ cells; neither a sexual stage nor meiosis has ever been reliably documented. They form heterokont zoospores (swarmer cells) that swim away to develop into new colonies. As an alternative form of reproduction, an entire colony fragments into two or more young colonies, each containing hundreds or thousands of cells. Offspring colonies float away from each other and establish themselves at new sites. Waves and other mechanical disturbances in the water apparently enhance this reproductive mode.

Chrysomonads interact strongly with minerals, especially with silica and iron. The marine planktonic silicomastigotes form tests, or shells, from silica scavenged from seawater. These coverings have complex and often strikingly beautiful patterns that are used to distinguish genera. The scavenging activity of chrysomonads, diatoms (Phylum Pr-16), actinopods (Phylum Pr-27), and glass sponges (Phylum A-2) depletes soluble silica in the surface waters of the ocean, so this element is not easily detected by chemical means. How silicomastigotes use dissolved silica to elaborate tests is poorly understood because of the difficulty of capturing organisms alive and rearing them in the laboratory. Thanks to the fact that some tests have been preserved as fossils, chrysomonads are known from about 500 million years ago, in the early Paleozoic.

Freshwater chrysomonads that must overwinter or survive desiccation typically make structures called statocysts—the cellulosic membrane of the cell becomes endogenously silicified within an outer coat. Sometimes the statocyst coat becomes heavily encrusted with iron minerals. Within the statocyst are often seen two conspic-

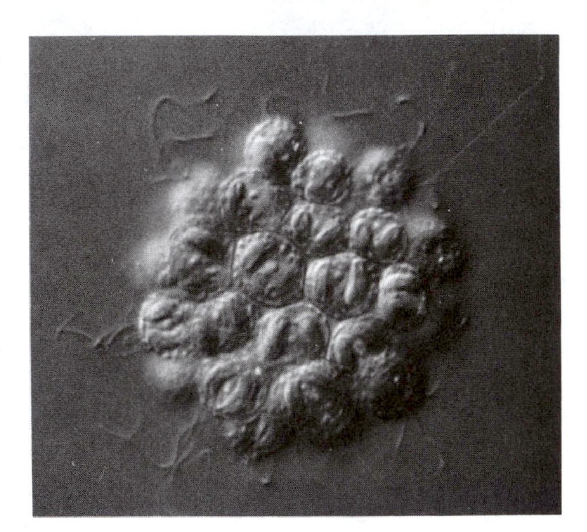

A *Synura* sp., a living freshwater colonial chrysomonad from Massachusetts. LM; each cell is about 18 μm in diameter. [Courtesy of S. Golubic and S. Honjo.]

B A siliceous surface scale from a member of the *Synura* colony shown in Figure A. SEM; greatest diameter is about 1 μm long. [Courtesy of S. Golubic and S. Honjo.]

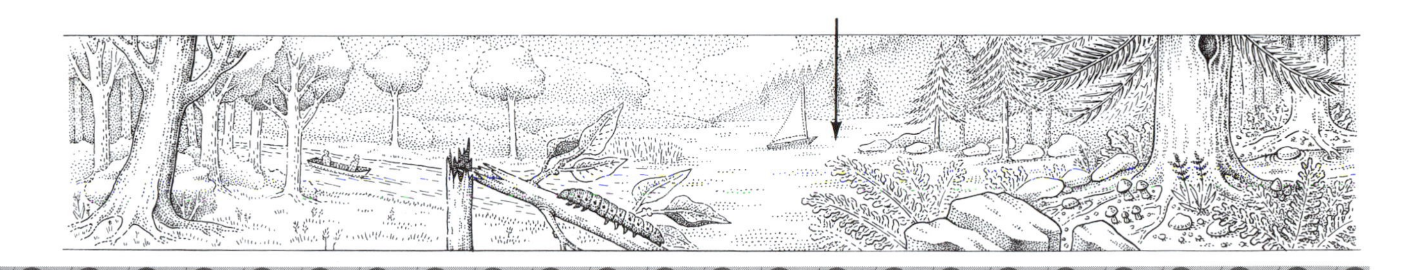

uous granules of a storage oil, called leucosin or chrysolaminarin. At the top of the statocyst is a pore, often surrounded by a tapering, conical collar and plugged with material that dissolves when the germinating ameboid chrysomonad emerges. The cyst walls may be uneven in thickness. Some fossil chrysomonad cysts have been preserved well enough to be identified by micropaleontologists; others may be among the pre-Phanerozoic acritarchs, abundant spheroid microfossils that are difficult to identify.

The chrysomonads fall into three classes: Chrysophyceae, Pedinellophyceae, and Dictyochophyceae. The vast majority of chrysomonads are members of the chrysophyte class. The pedinellophytes have a ribbonlike undulipodium with only one row of mastigonemes, and a cell morphology characterized by radiating cytoplasmic extrusions ("tentacles"). The dictyochophytes (silicomastigotes) have a basketlike siliceous external test and little in common with the other chrysomonads except the chrysoplast itself. Class Chrysophyceae consists of three large groups: Chrysomonadales, Chrysosphaerales, and Chrysotrichales. Although in some literature these groups are orders, we consider them to be subclasses here.

The chrysophyte subclass Chrysomonadales contains two orders: Chrysomonadineae, mainly unicellular forms and their direct multicellular, colonial descendants, and Chrysocapsineae, all multicellular genera whose members form either compact spherical colonies or loosely branched ones. The well-known *Synura* (Figures A and B), *Chromulina, Mallomonas, Ochromonas* (Figure C), *Chrysobotrys, Monas, Rhizochrysis,* and *Dinobryon* are members of the Chrysomonadineae. Each of these genera typifies a large group, considered a family—for example, Synuraceae. In each family, both unicellular genera and colonial genera clearly related to them abide. The evolutionary transition from unicellular forms to multicellularity occurred independently many times in the chrysomonads.

Species of Subclass Chrysosphaerales (for example, *Chrysosphaerella brevispina*) typically form complex, differentiated, spherical colonies that release heterokont zoospores.

The Chrysotrichales include *Nematochrysis, Thallochrysis,* and *Phaeothamnion,* among many others. The members of this subclass show every sort of variation on the theme of the filamentous habit. Some genera are composed of long intertwined threads of algae; close inspection reveals each compartment of a thread to be a typical chrysophyte cell. Others form flaccid, highly branched treelike structures reminiscent of branching cyanobacteria (Phylum B-5), yellow-green algae (Phylum Pr-14), or green algae (Phylum Pr-28). Such branching organization evolved independently many times in different groups of microbes.

An obscure group of large unicellular algae that, like chrysomonads, contain plastids with chlorophylls *a* and *c,* is misnamed Chloromonads. These mud, lake, and marine dwellers are not green algae at all. The accepted current name for the group is Raphidomonads, named after the best-known genus *Raphidomonas.* This small group may soon be raised to phylum status. We arbitrarily include them in the Chrysomonada phylum until more information about them becomes available. The other genera in the group are *Chattonella, Gonyostomum, Merotricha, Thaumatomastix, Reckertia,* and *Vacuolaria.* Each bears two undulipodia, apically inserted, one forward and one trailing. Like chrysomonads, raphidomonads have large nuclei at the anterior ends of their cells and store food as fat. *Vacuolaria virescens* is a naked ovoid or pear-shaped cell that forms spherical cysts with thick envelopes of mucus. *Thaumatomastix* and *Reckertia* are colorless genera that lack plastids.

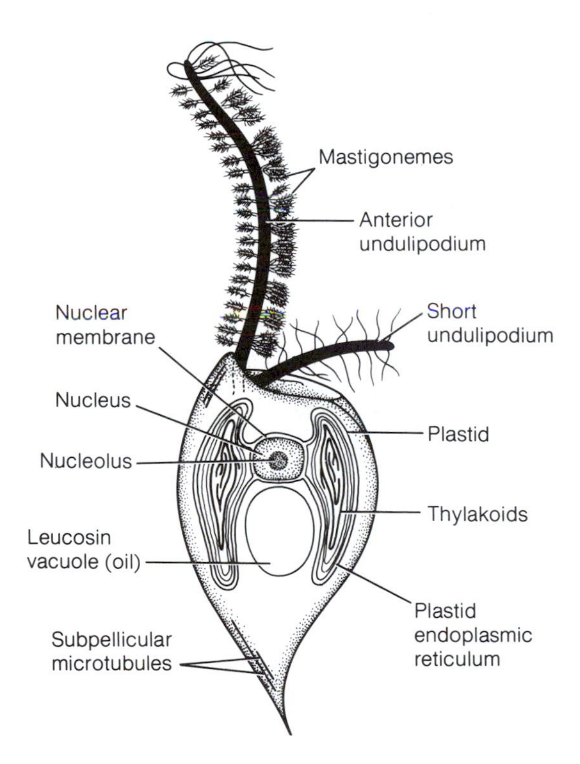

C The freshwater, single-cell chrysomonad *Ochromonas danica;* the ultrastructures of *Ochromonas* cells and of single *Synura* cells are similar. [Drawing by M. Lowe.]

Pr-14 Xanthophyta

Greek *xanthos*, yellow; *phyton*, plant

EXAMPLES OF GENERA

Botrydiopsis
Botrydium
Botryococcus
Centritractus
Characiopsis
Chlamydomyxa
Chloramoeba
Chloridella
Heterodendron
Mischococcus
Myxochloris
Ophiocytium
Rhizochloris
Tribonema
Trypanochloris
Vaucheria

Xanthophytes, like eustigmatophytes (Phylum Pr-15), are yellow green. However, the unique organization of their cells and their tendency to form strange colonies suggest that they are related to eustigs only by pigmentation. Their photosynthetic organelles, xanthoplasts, probably share ancestry with those of the eustigs, but the nonplastid part of the cell is far more like that of the chrysomonads (Phylum Pr-13). In fact, on the basis of a common morphology of the nonplastid part, some phycologists prefer to group xanthophytes with chrysomonads and phaeophytes (Phylum Pr-17)—excluding haptomonads (Phylum Pr-10)—in a single phylum called Heterokonta. In all members of this Heterokonta phylum, there are two anteriorly inserted undulipodia. One is directed forward and is mastigonemate—that is, hairy; the other, trailing, undulipodium is shorter and smooth. Although the nonplastid parts of the cells in these three Heterokonta groups probably do have common ancestry, the differences among these groups of protoctists and the uniformity within each group seem to us marked enough to justify the raising of each group to phylum status. Production of heterokont zoospores by xanthophytes, which resembles the pattern in certain mastigote molds, such as oomycotes (Phylum Pr-20) and plasmodiophorans (Phylum Pr-19), has led us to place Phyla Pr-19 and Pr-20 near xanthophytes on our phylogeny.

The plastid pigmentation of the xanthophytes, like that of the eustigs, consists of chlorophylls a, c_1, c_2, and e. Several xanthins are found in the best-studied members: cryptoxanthin, eoxanthin, diadinoxanthin, and diatoxanthin; heteroxanthin and beta-carotene have been detected by spectroscopic methods. Oils are the storage products of photosynthesis—starch is absent. At least some of the glucose monomers in these oils are linked by β-1–3 linkages.

Xanthophytes typically have pectin-rich cellulosic walls made of overlapping discontinuous parts. Many cells are covered with scales characteristic of the species. In winter and under other adverse conditions, many species form cysts that are encrusted with iron or in which silica is embedded.

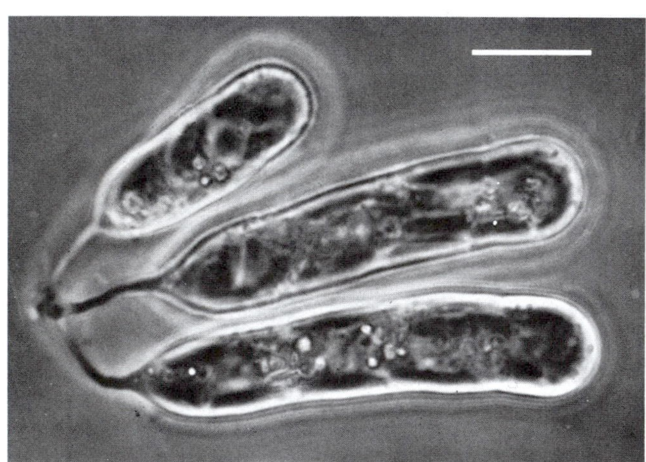

A Vegetative cells of *Ophiocytium arbuscula*, a freshwater xanthophyte from alkaline pools in England. LM (phase contrast), bar = 10 μm. [Courtesy of D. J. Hibberd.]

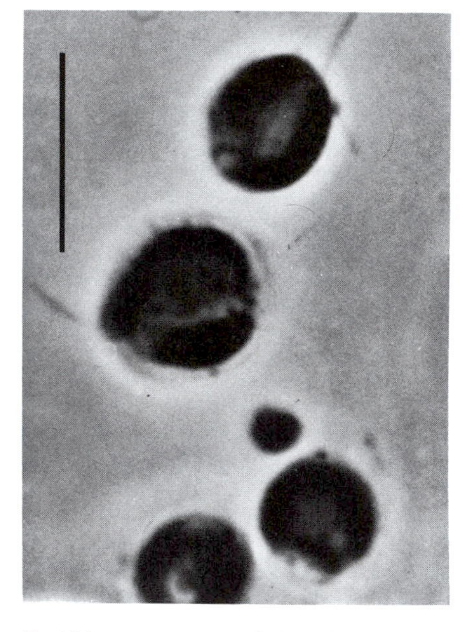

B Living zoospores of *Ophiocytium majus*. LM, bar = 10 μm. [Courtesy of D. J. Hibberd and G. F. Leedale, *British Phycological Journal* 6:1–23 (1971).]

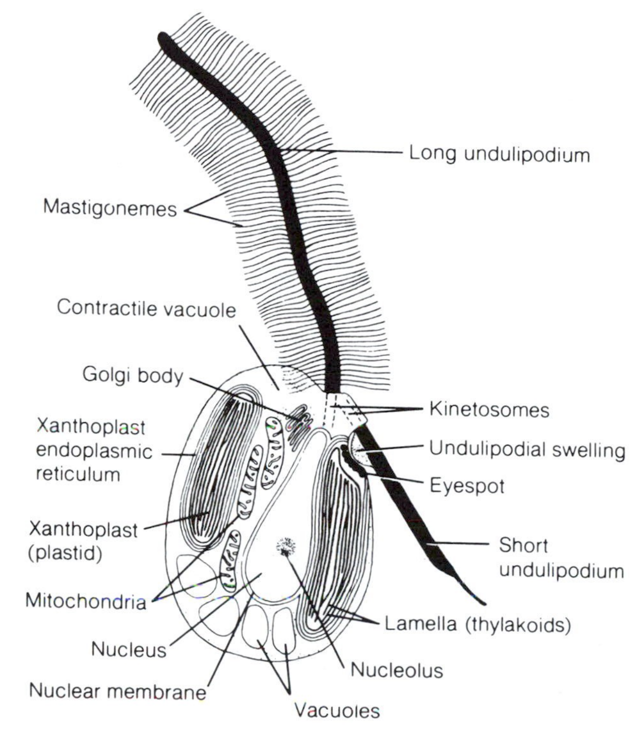

C Zoospore of *Ophiocytium arbuscula,* showing typical heterokont undulipodia. [Drawing by R. Golder.]

Xanthophytes populate fresh waters (Figure A). They are found in a variety of highly structured multicellular and syncytial (multinucleate) forms; some produce amebas or undulipodiated zoospores (Figures B and C). Although complex differentiation patterns are known, meiotic sexual cycles have not been reported.

The phylum contains four major subgroups, which we raise to the status of class. They are the Heterochloridales, Heterococcales, Heterotrichales, and Heterosiphonales. Each of the classes, traditionally reported as orders, has many genera.

Class Heterochloridales contains the morphologically least complex xanthophytes. It comprises two major groups, here called orders: Heterochlorineae (motile unicells) and Heterocapsineae (palm-shaped flattened colonial forms). *Botryococcus,* a very common pond-scum organism, is perhaps the best-known member of the second group.

Class Heterococcales, which includes *Ophiocytium,* consists of coccoid cells. Their genera take various colonial forms—filamentous, branched, or bunched—reminiscent of the colonies formed by coccoid cyanobacteria (Phylum B-5), chrysomonads (Phylum Pr-13), and chlorophytes (Phylum Pr-28). However, the single coccoid unit cell has an internal organization characteristic of the xanthophytes. The well-known genus *Botrydiopsis* looks like a bunch of grapes.

Class Heterotrichales includes many complex multicellular organisms, most of them being variations on the theme of filaments. They include the highly branched, flaccid, tree-shaped alga called *Heterodendron.*

Class Heterosiphonales contains the most morphologically complex xanthophytes. They can be quite formidable in appearance. *Botrydium,* for example, develops a collapsible balloonlike multicellular thallus in drying muds, looking superficially like a chytrid (Phylum Pr-29), may become encrusted with calcium carbonate, and may grow to nearly a meter in length. It has an extensive system of branched rhizoids in which resistant, hard-shelled cysts develop. The cysts, when rehydrated, germinate into unicellular, heterokont zoospores typical of xanthophytes. The zoospores disperse, germinate, and grow into thalli, completing the life cycle. All zoospores are without sex and competent by themselves for further development.

Although there are fewer than 100 well-documented species, this phylum of algae, best known as unsightly messes in muddy water, probably has many other members that have not yet come under scrutiny.

Pr-15 Eustigmatophyta

Greek *eu,* true, original, primitive; *stigma,* brand put on slave (as refers to eyespot), mark, spot; *phyton,* plant

ALL GENERA
Chlorobotrys
Ellipsoidion
Eustigmatos
Monodopsis
Nanochloropsis
Pleurochloris
Polyedriella
Pseudocharaciopsis
Vischeria

Because they are yellowish green, form immotile coccoid vegetative cells, and propagate by motile elongated asexual zoospores, eustigmatophytes were originally lumped together with the xanthophytes (Phylum Pr-14), which they resemble. Electron microscopic studies, however, reveal a distinctive eyespot and organization of the eustigmatophyte cell, distinguishing the morphology that justifies recognition of eustigmatophytes as a unique set of photosynthetic motile protoctists warranting their own phylum. Only the genera listed above are known to be in Phylum Eustigmatophyta for sure, but this is due far more to lack of study at the ultrastructural level than to a paucity of these organisms. At present, *Pleurochloris, Polyedriella, Ellipsoidion,* and *Vischeria* (Figure A) are the major genera of "eustigs," as they are fondly called. Although multicellular eustigs exist—for example, colonies of *Chlorobotrys,* cells surrounded by layered mucilage that form no zoospores—the majority are independent single cells. They live primarily in fresh waters.

In the pigmentation of their plastids, eustigs are indeed very much like the true yellow-green xanthophytes. Their plastids, called xanthoplasts, contain chlorophyll *a,* as all oxygen-eliminating organisms do; in addition, they contain chlorophylls c_1, c_2, and *e*. They lack chlorophyll *b*. They contain beta-carotene and several oxygenated carotenoids—which ones depends on the genus. Violaxanthin is commonly present; epoxanthin, diadinoxanthin, and diatoxanthins also may be present. Eustigs store glucose not as starch but as a solid material (not yet chemically identified) that lies outside the plastid. In some vegetative cells, a conspicuous polyhedral crystalline body constitutes the pyrenoid of the plastid; it is typically attached to the thylakoids by a thin stalk.

Although the pigments of eustigs are like those of xanthophytes (Phylum Pr-14), the cell organization is not. Most eustigs have only a single mastigonemate (hairy) anterior undulipodium, at the base of which is a conspicuous undulipodial swelling, T shaped in trans-

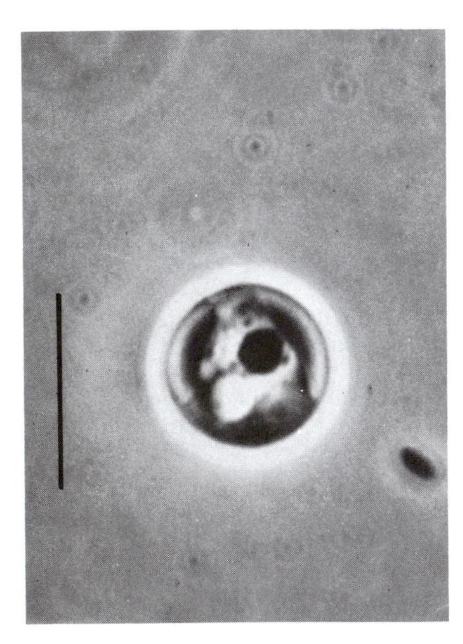

A Growing cell of *Vischeria* (*Polyedriella*) sp. LM, bar = 10 μm. [Courtesy of D. J. Hibberd, from D. J. Hibberd and G. F. Leedale, *Annals of Botany* 36:49–71 (1972).]

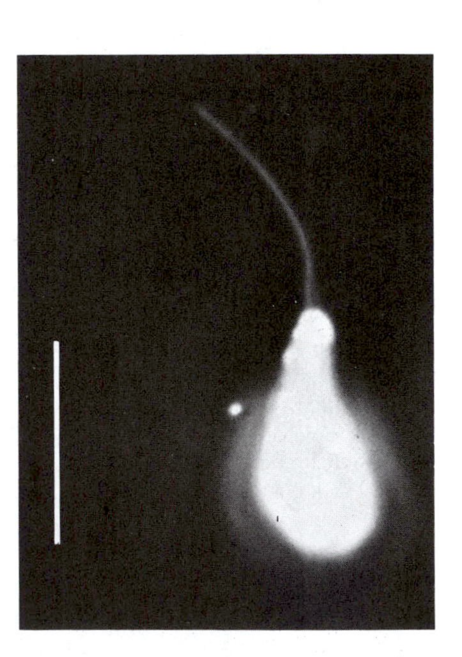

B Zoospore of *Vischeria* sp. LM, bar = 10 μm. [Courtesy of D. J. Hibberd, from D. J. Hibberd and G. F. Leedale, *Annals of Botany* 36:46–71 (1972).]

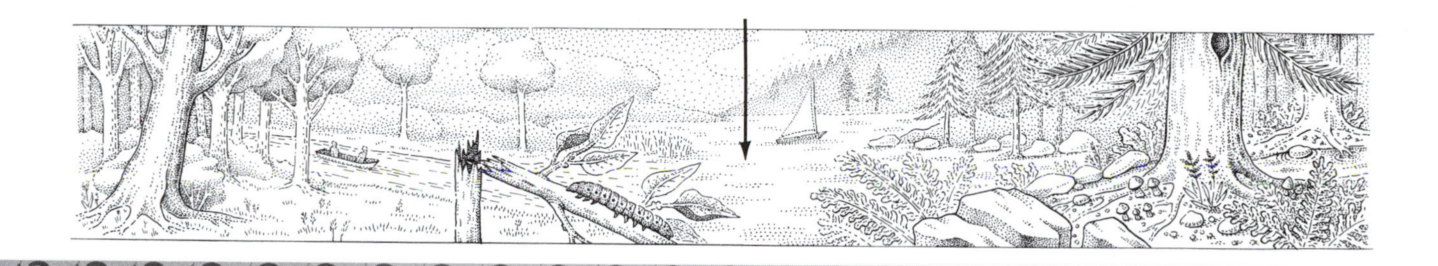

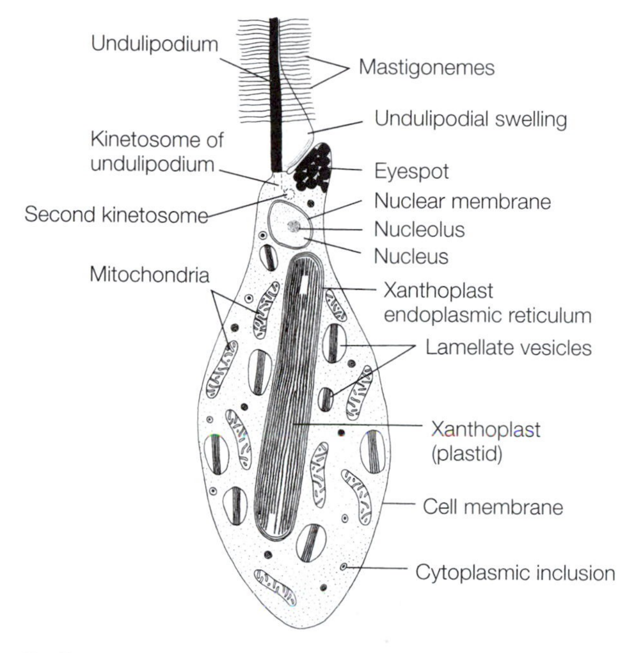

C Zoospore of *Vischeria* sp. [Drawing by R. Golder.]

verse section (Figures B and C). An adjacent swelling filled with drops of carotenoids forms the eyespot; it probably communicates somehow with the undulipodium to direct the cell to optimally lighted environments. The eyespot is not associated with the plastid, nor is it membrane bounded. The xanthophyte lacks such a swelling on its anterior undulipodium but has a swelling on the posteriorly directed second undulipodium, which is apposed to a specialized part of the plastid. Some eustigs (for example, *Ellipsoidion*) have a second, smooth undulipodium.

The yellow-green eustig plastid is single and long; it lies in the center or at the posterior end of the cell and fills some two-thirds of its volume. The thylakoids are stacked inside, rather like the grana of plant plastids. Nearly all the endoplasmic reticulum (ER) of eustigs is associated with the plastid. The xanthoplast ER is not associated with the nuclear membrane, as it is in many other algae, and there is little developed free ER in the cytoplasm. This morphological arrangement suggests an integrated metabolic relation between the products of photosynthesis and the nucleocytoplasm-directed biosyntheses. The cell wall is entire—that is, it completely surrounds the cell—and, in some cases, it contains silica deposits. The cell divides directly into two offspring cells, and sexual processes are unknown in the group.

These planktonic algae, at the base of aquatic food chains, are eaten by other protoctists and animals. Scientists have studied them only very little, however, so not much is known about their natural history.

Pr-16 Diatoms
(Bacillariophyta)

Latin *bacillus,* little stick; Greek *phyton,* plant

EXAMPLES OF GENERA

Achnanthes
Amphipleura
Arachnoidiscus
Asterionella
Biddulphia
Coscinodiscus
Cyclotella
Cymbella
Diatoma
Diploneis
Eunotia
Fragilaria
Hantschia
Melosira
Meridion
Navicula
Nitzschia
Pinnularia
Planktoniella
Rhopalodia
Surirella
Tabellaria
Thalassiosira

Beautiful aquatic protists—perhaps 10,000 living species—diatoms are single cells or form simple filaments or colonies. Some 250 genera of these gorgeous protoctists are commonly described, and specialists recognize as many as 100,000 species, including those in 70 fossil genera. Valves, the two parts of the diatom test (shell), extend into the Mesozoic fossil record; the first ones appeared in the Lower Cretaceous period.

Each valve of the diatom test is composed of pectic organic materials impregnated with silica (hydrated SiO_2), in an opaline state. The valves may be extremely elaborate and beautiful; their elegantly symmetrical patterns are used to test lenses for optical aberrations. Diatoms require dissolved silica for growth; they are so competent at the removal of silica from natural waters that they can reduce the concentration to less than one part per million, below the value detectable by chemical techniques.

Diatoms, important at the base of marine and freshwater food chains, are very widely distributed in the photic (illuminated) zones of the world. Some species are found in hypersaline ponds and lagoons, others in clear fresh water, and others in moist soil. Their cysts, empty tests, and dying cells can be found in unlighted regions of the ocean. Most species under study are obligate photosynthesizers, although some also require organic substances, such as vitamins, for growth. Some strains of *Nitzschia putrida* are saprobes.

Diatoms are generally brownish; for years, they were classified with the golden yellow algae, the chrysomonads (Phylum Pr-13). The xanthoplasts (plastids) of diatoms contain the pigments chlorophyll *a,* chlorophyll *c,* beta-carotene, and xanthophylls, including fucoxanthin, lutein, and diatoxanthin. The photosynthetic food reserve of diatoms, like that of chrysomonads, is the oil chrysolaminarin. Nevertheless, in life history, cell structure, and division, the diatoms differ greatly from the other golden yellow algae. The diatoms make up such an easily distinguished and large natural group that, in the light of modern information, we provide them a phylum separate from the other organisms that have golden brown plastids.

The two great classes of diatoms are the Coscinopiscophyceae (Centrales) and the Bacillariophyceae (Pennales). The Coscinopiscophyceae, or centric diatoms, have radial symmetry, like that of *Thalassiosira* and *Melosira* in Figures A and B. The Bacillariophyceae, or pennate (featherlike) diatoms, have bilateral symmetry; many of them are boat shaped (Figure C) or needle shaped. The pennate diatom has a slit, called a raphe, between the valves. The raphe exudes cytoplasmically produced slime in which the diatom glides. The centric diatoms lack raphes and are never motile. In spite of the correlation of the raphe and movement, the detailed mechanism of gliding in the pennate diatoms is not known. The centric diatoms usually have numerous small plastids, whereas, in the pennate diatoms, the plastids are fewer, as a rule, and larger.

The centric diatom class comprises nine major groups, recognized now as subclasses. Some examples are the discoid subclass,

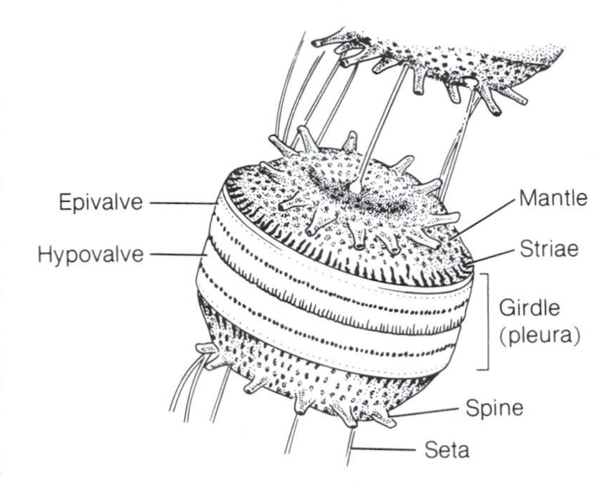

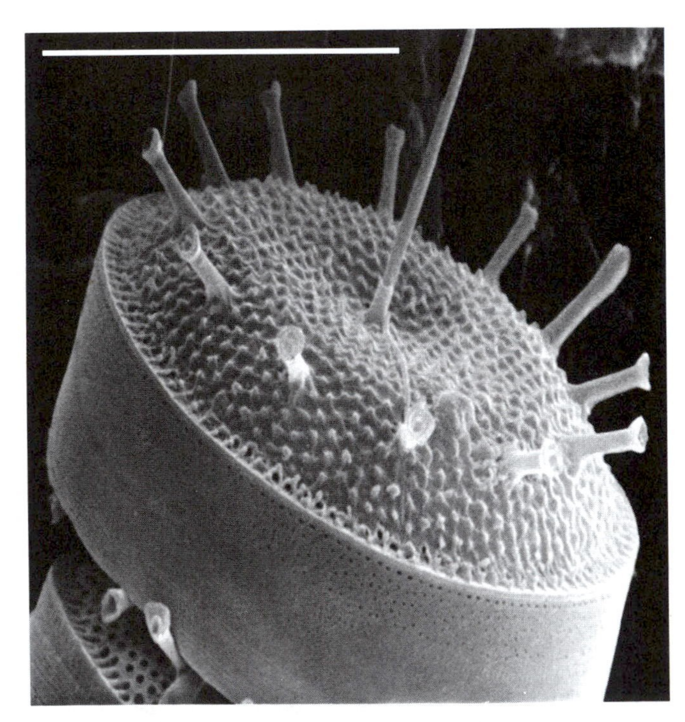

A *Thalassiosira nordenskjøldii,* a marine diatom from the Atlantic Ocean. SEM, bar = 10 μm. [Drawing by E. Hoffman; photograph courtesy of S. Golubic.]

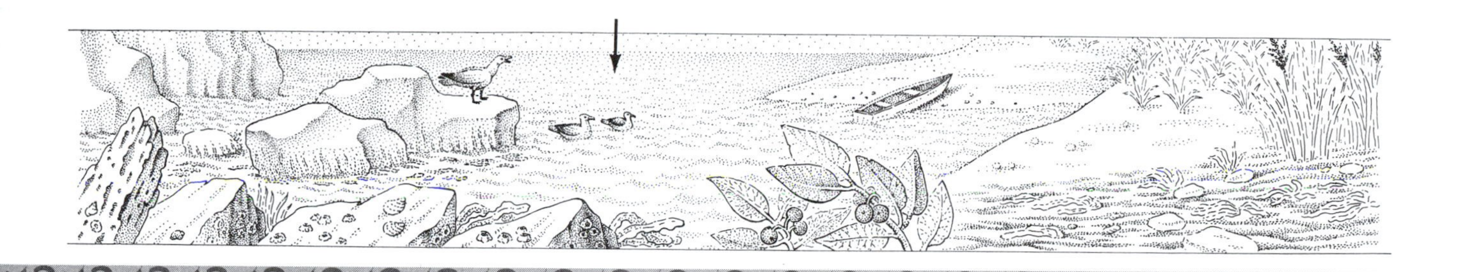

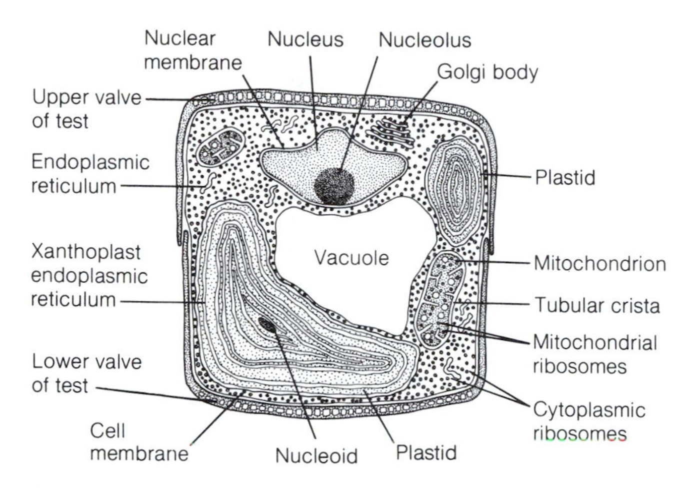

B *Melosira* sp., a centric diatom. [Drawing by L. Meszoly.]

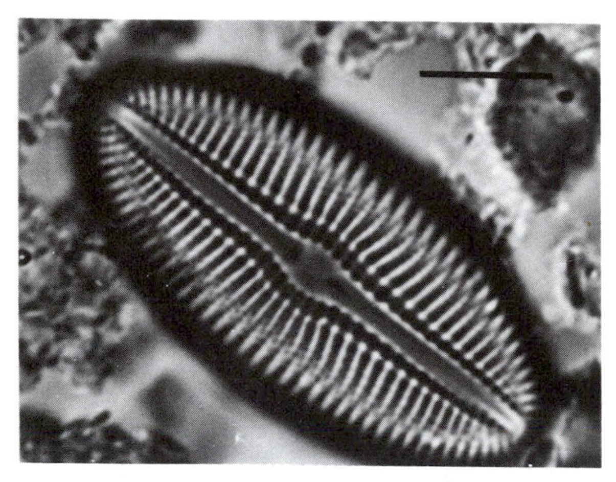

C *Diploneis smithii,* a pennate (naviculate, or boat-shaped) diatom from Baja, California. With the light microscope, only the silica test, which has been cleaned with nitric acid, is seen. LM, bar = 25 μm.

which includes *Thalassiosira,* a colonial diatom that extrudes threads, called setae, of chitin, which hold individual diatoms together in chains; the solenoids, elongate diatoms, cylindrical or subcylindrical in shape; the box-shaped biddulphioids, most with horns or other decorations on their tests; and the rutilarioids, with their naviculoid valves (that is, they depart from strict radial symmetry by having a boatlike shape with pointed ends that have radial or irregular markings).

The pennate diatoms are classified into two subclasses: one small, Eunophycidae; and the other enormous, Bacillariophycidae. In classifying the four orders of Subclass Bacillariophycidae, attention is given to the presence and development of the raphe. Araphids lack a true raphe; raphids show the beginnings of a raphe; monoraphids have a fully developed raphe but only on one of the valves; biraphids have a fully differentiated raphe on each valve.

Diatoms are highly sexual organisms that, like animals, spend most of their life cycle in the diploid state. Meiosis occurs just before the formation of haploid gametes. After fertilization, the diploid zygote develops into the familiar diatom.

Diatoms reproduce by mitotic division into two offspring cells. In many, each offspring cell retains one valve of the parent shell and produces one new valve, which fits into the parental valve. Hence, the two offspring cells are each slightly smaller than the parent. This tendency for diatoms to decrease in size is counteracted by auxospore formation. An auxospore, not really a spore in the sense of being able to resist adverse conditions, is an enlarged, shell-less diatom formed when a protoplast (test contents) is released into the water column. The size of the auxospore is a characteristic of each species. Freed from the inexorable sequence of shrinking, the auxospore secretes two new, large valves.

In the pennate diatoms, auxospore formation is usually triggered by the sexual confrontation: it follows fertilization and zygote formation. Haploid male gametes (sperm), bearing single anterior undulipodia (Figure D), fertilize immotile haploid female protoplasts. The valves of the tests sometimes open for the purpose. The zygote becomes the large auxospore; reproduction by mitotic division follows. In the centric diatoms, auxospores may be formed without the intervention of the sexual processes of meiosis and fertilization.

Diatomaceous earth, which is composed of deposits of diatom cell walls, is used for abrasives, filters, chalk, and talc.

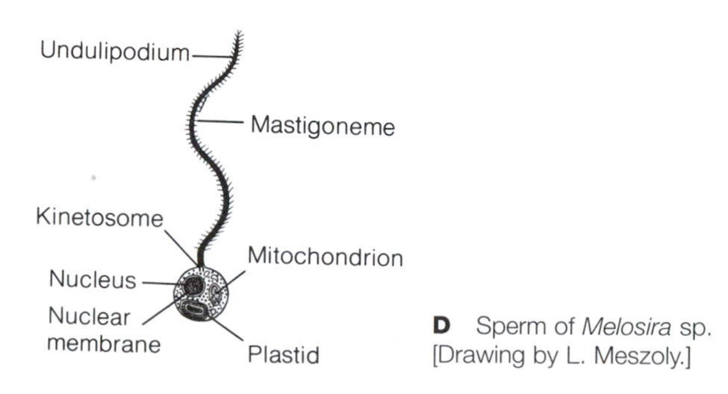

D Sperm of *Melosira* sp. [Drawing by L. Meszoly.]

Pr-17 Phaeophyta

(Brown algae)

Greek *phaios,* dusky, brown; *phyton,* plant

EXAMPLES OF GENERA

Alaria
Ascophyllum
Chorda
Chordaria
Cutleria
Cytoseira
Dictyosiphon
Dictyota
Durvillaea
Ectocarpus
Fucus
Himanthalia
Ishige
Laminaria
Macrocystis
Nereocystis
Notheia
Pelagophycus
Pelvetia
Sargassum
Seirococcus
Tilopteris
Zanardinia

The phaeophytes are the brown seaweeds; nearly all are marine. They are the largest protoctists. The giant kelps, for example, are as much as 10 m (more than 30 feet) long. Some 900 species in 250 genera have been described, all of them photosynthetic. Most phaeophytes live along rocky coastal seashores, especially in temperate regions. They dominate the intertidal zone, where they form great seaweed beds. Their communities are referred to as "forests of the sea." They are crucial primary producers, providing habitat and food for other protoctists, marine animals, and microbes.

Although details of the life cycle vary with genus, the pattern of sexuality in most phaeophytes is similar. The thallus (algal "leaf") forms eggs and biundulipodiated sperm. The sperm has one forward mastigonemate (hairy) undulipodium and one trailing smooth (whiplash) one. Thus, the morphology of phaeophyte sperm resembles that of the heterokont phyla, the chrysomonads (Phylum Pr-13) and the xanthophytes (Phylum Pr-14). The brown algae are thought to have evolved from single-celled chrysomonads.

The fertilized egg of phaeophytes typically germinates in response to light; rhizoids (algal "roots") grow out away from the source of light. A diploid thallus arises from a fertilized egg and differentiates into a blade ("leaf"), a stipe ("stem"), and a holdfast ("root") and may grow to enormous size. These phaeophyte thalli lack the complex tissue organization characteristic of land plants, even though some do develop sieve tubes for conducting water and photosynthate, which in many cases is the sugar alcohol mannitol. The tubes are analogous, not homologous, to the sieve tubes of plants.

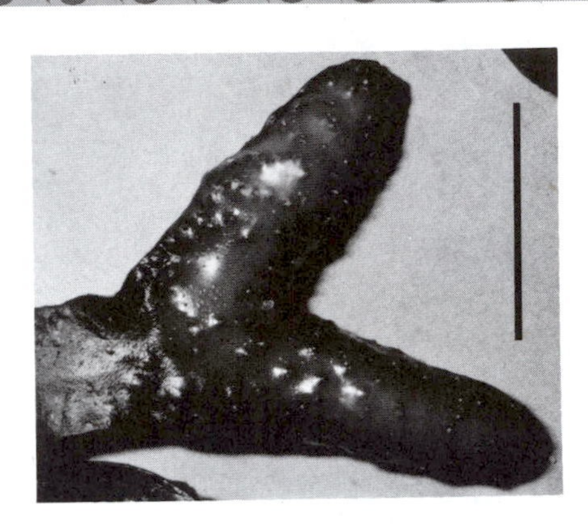

B Receptacles on sexually mature apices of *Fucus vesiculosus* thallus. Bar = 1 cm. [Courtesy of W. Ormerod.]

A Thallus of *Fucus vesiculosus* taken from rocks on the Atlantic seashore. Bar = 10 cm. [Courtesy of W. Ormerod.]

Haploid and diploid generations alternate in phaeophytes. The diploid thallus that develops from a fertilized egg is a sporophyte, the agamont organism. Sporangia (spore-producing structures) form on the sporophyte thallus and release zoospores (swarmer cells), not gametes. If meiosis precedes the release of the zoospores, they are haploid; otherwise, they are diploid. Although they resemble the phaeophyte sperm, zoospores are not gametes and do not require a fertilization step to develop and grow into the multicellular alga.

After withdrawing their undulipodia, some zoospores develop into male thalli that will later produce sperm or into female thalli that will produce eggs. These thalli are thus gamonts, because they are individual organisms capable of gamete formation. If meiosis has preceded zoospore formation, the thalli that grow from the haploid zoospores are haploid and thus are gametophytes. Male and female gametophytes produce sperm and eggs directly from their haploid tissue by mitotic division in gametangia. In thalli that develop from diploid zoospores, haploid gametes (sperm or eggs) are generated by meiosis in the gametangia. In either case, the female gametangium is called an oogonia, and the male gametangium an antheridium. These structures are analogous but not homologous to similar structures in rhodophytes (Phylum Pr-25), chlorophytes (Phylum Pr-28), and plants, illustrating the convergent trends so common in the reproductive morphology of protoctists.

The alternation of haploid and diploid generations in phaeophytes is analogous but not homologous to the alternation of generations in chlorophytes and plants. In some phaeophytes, the gametophytes (male and female haploid gamonts) are indistinguishable from the diploid agamont sporophytes except for their chromosome number. The generations in others are so dissimilar that their gametophytes and sporophytes were first thought to be entirely different species.

Brown algae contain a distinct set of photosynthetic pigments: their plastids, called phaeoplasts, contain chlorophylls *a* and *c,* but never *b.* Fucoxanthin is generally the prominent carotenoid derivative. This xanthophyll is responsible for the brown or olive-drab color of the thallus. The carbohydrate stored as food is called laminarin (named after *Laminaria,* a well-known kelp genus). Phaeophyte cells also store lipids. Many are capable of synthesizing elaborate, ecologically significant, organic compounds, called secondary metabolites because they are not the primary metabolites essential for development through the life history. The usefulness of the rapidly growing seaweeds as sources of drugs, food, and fuel is yet to be determined.

Although most frequently found along seashores, some brown algae, such as the well-known *Sargassum* found in the Sargasso Sea, form immense floating masses far offshore. These algae provide food and shelter for unique communities of marine animals and microbes and lead to rates of primary productivity in the open oceans far higher than would be possible if the phaeophytes were absent. Attempts are being made to increase the ocean's productivity by using these prodigious algae—for example, by planting *Macrocystis* on rope rafts off the coast of California.

Fucus, rockweed, is very common on temperate rocky seacoasts. As in other members of the order Fucales, the thallus is flattened; it branches in one plane (Figure A). The sex organs on the gamonts may be on the same thallus or on separate thalli. The gametangia are contained in dark clotlike bodies, called conceptacles, scattered on the surface of heart-shaped swellings called receptacles (Figures B and C). Within a female gametangium, each oogonium produces one, two, four, or eight eggs. Oogonia containing ripe eggs are released into the water, where they are fertilized by sperm produced by antheridia in male gametangia. No embryo is formed: the thallus, unsupported by maternal tissue, develops directly. For these reasons alone, brown algae are not considered to be plants.

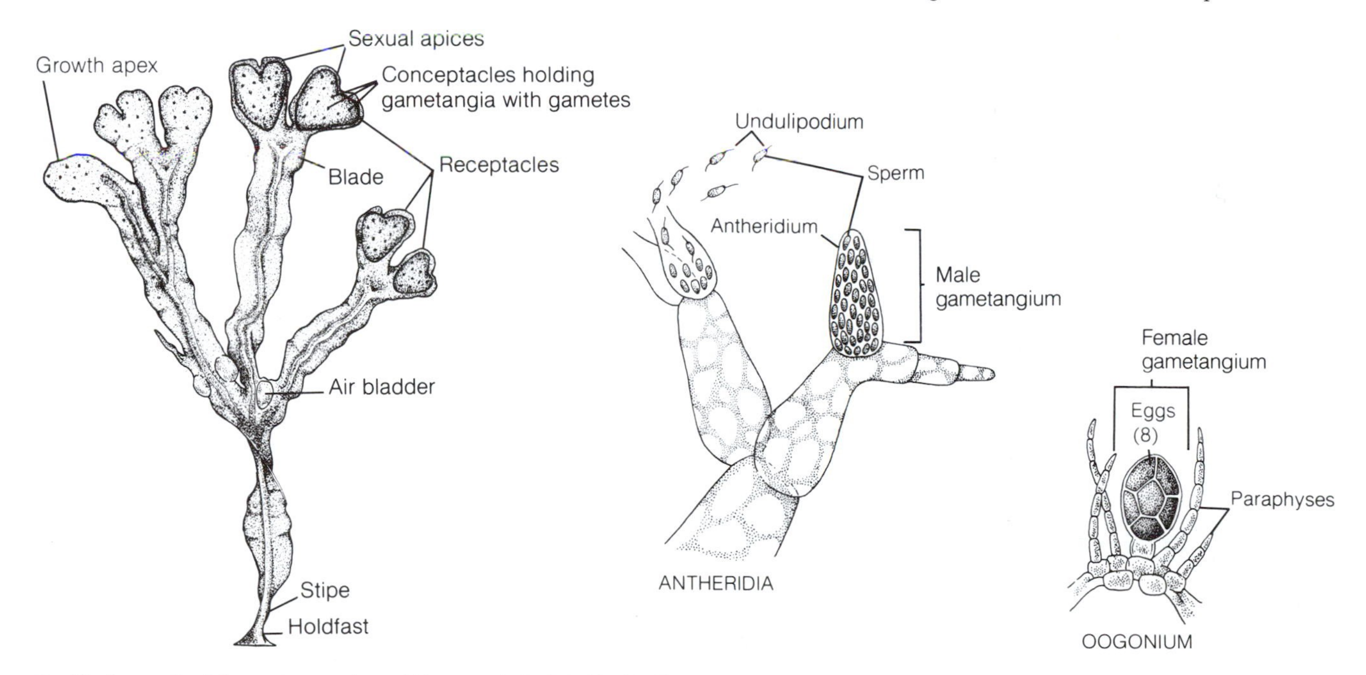

C Thallus, antheridia, and oogonium of *Fucus spiralis* hybrid with *Fucus vesiculosus*. [Drawings by R. Golder.]

Pr-18 Labyrinthulata
(Slime nets and thraustochytrids)

Latin *labyrinthulum*, little labyrinth

KNOWN GENERA
Althornia
Aplanochytrium
Japonochytrium
Labyrinthula
Labyrinthuloides
Schizochytrium
Thraustochytrium
Ulkenia

Molecular studies of ribosomal RNA have confirmed the relation between the two classes of this phylum: the thraustochytrids and the slime nets, or labyrinthulids. The phylum is defined as protoctists that use cellular organelles called bothrosomes (sagenogenosomes) to produce extracellular (ectoplasmic) slime matrix. The ectoplasmic matrix is continuous with the plasma membrane at the bothrosome.

The thraustochytrids, an obscure group of encysting marine protoctists that form swollen zoospore cases (sori) and superficially resemble chytrids (Phylum Pr-29), have traditionally been classified as fungi. Because they produce undulipodiated cells and also have other features that demonstrate an indusputable relation to labyrinthulids, these protoctists—seven genera with nearly three dozen species—belong as a class here. Their slime matrix does not surround each cell, as it does in the labyrinthulids; rather, this rich proteinaceous polysaccharide slime matrix emanates from the base of the developing sorus when it arises from single or clustered bothrosomes (sagenogens). The walls of the thick-walled sorus ("sporangium") contain large quantities of L-galactose; in some thraustochytrid walls, a high-sulfate galactan was detected.

All labyrinthulids have been united into a single genus, *Labyrinthula,* with at least eight distinguishable species. Labyrinthulids form colonies of individual cells that move and grow entirely within the slime matrix or slimeway, which is a slime track (Figures A and B). This track, probably mucopolysaccharides and actomyosin proteins, is laid down in front of the cells (Figures B and C). They have been called slime net amebas in some texts, but there is nothing ameboid about them.

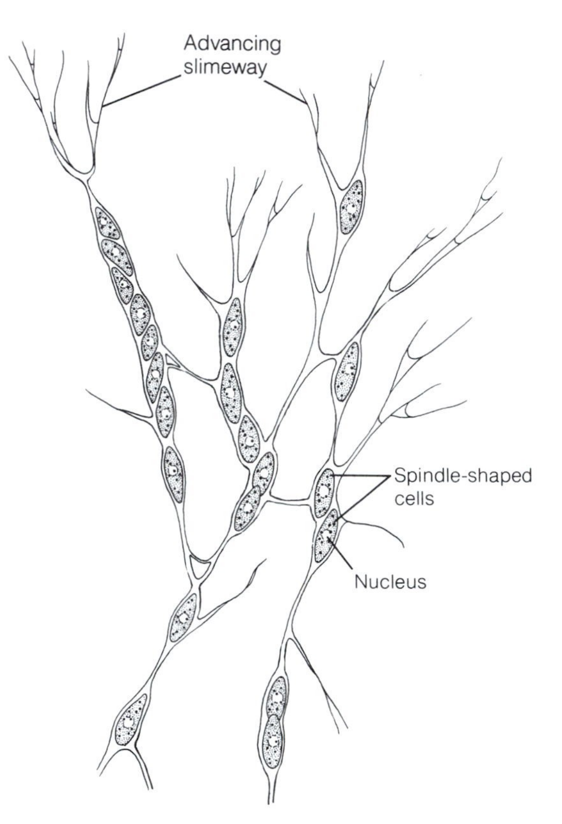

B *Labyrinthula* cells in a slimeway. [Drawing by R. Golder.]

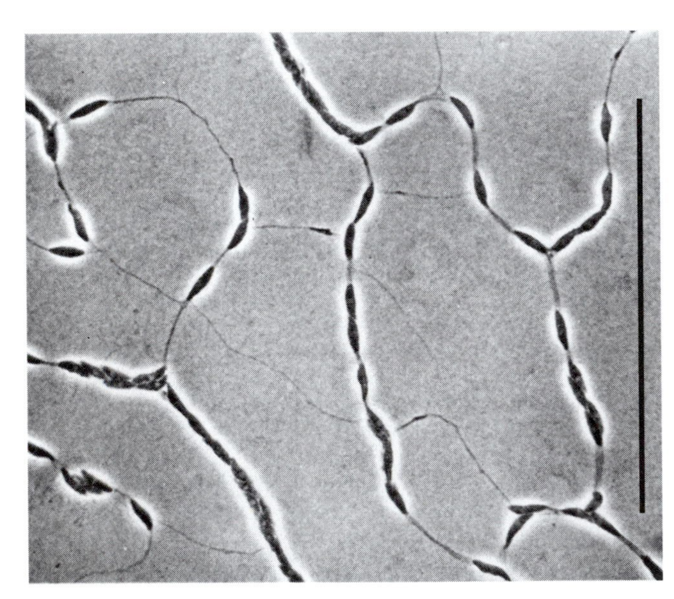

A Live cells of *Labyrinthula* sp. traveling in their slimeway. LM, bar = 100 μm. [Courtesy of D. Porter.]

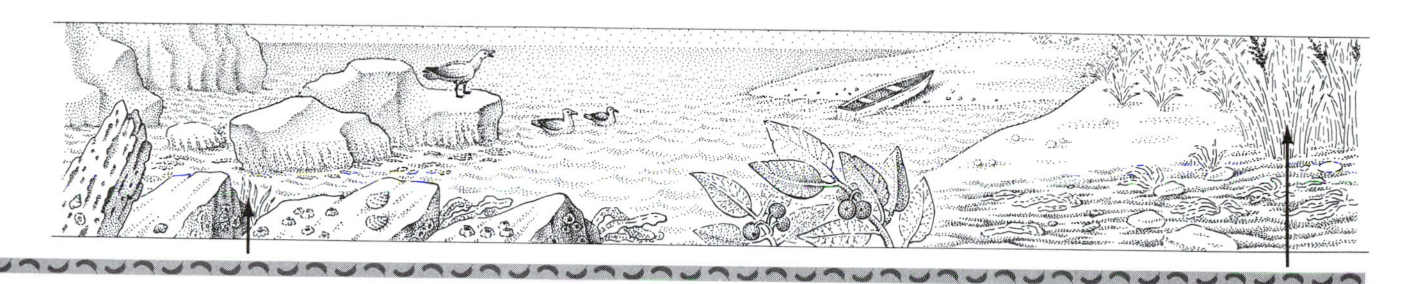

The labyrinthulids form transparent colonies—they may be centimeters long. With the unaided eye they look like a slimy mass on marine grass. Under the microscope, spindle-shaped cells can be seen migrating back and forth in tunnels within the slime, like little cars in tracks (Figures D and E). Labyrinthulid cells cannot move at all unless they are completely enclosed in the slime track. The mechanism of movement is unknown; it certainly does not require undulipodia or pseudopods. Calcium ions regulate contraction of actomyosin filaments in the slimeways. It seems that force is generated by the actomyosin musclelike movement along the surface of the slimeways, external to the cells in the slime layer, causing cells in the slimeway to travel at quite a rate—several micrometers per second.

The movements of trains of cells together back and forth in the track may look random. However, when a potential food source is sensed, such as a bacterial or yeast colony, the cells of the slime net move toward the source. Labyrinthulids are osmotrophic; they do not ingest the bacteria or yeast. Instead, extracellular digestive

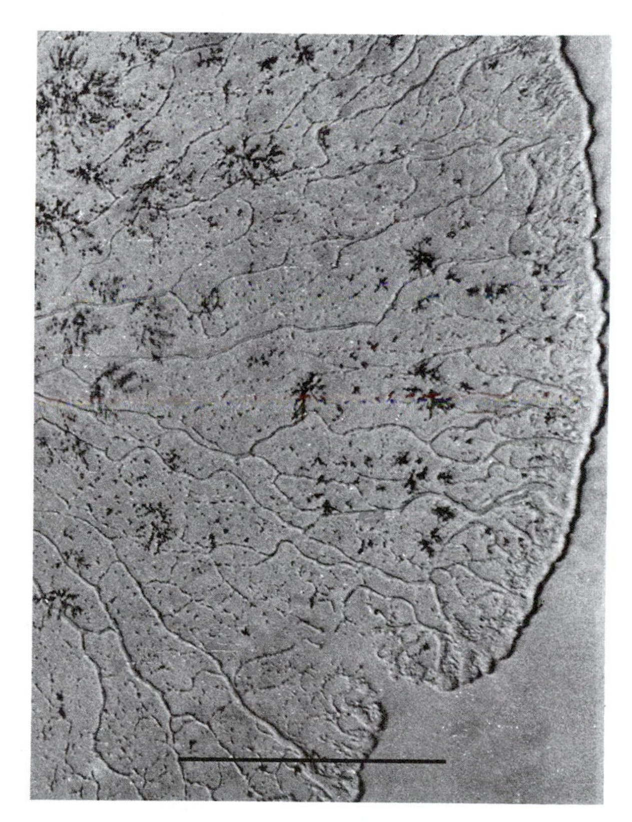

C Edge of a *Labyrinthula* colony on an agar plate. Bar = 1 mm. [Courtesy of D. Porter.]

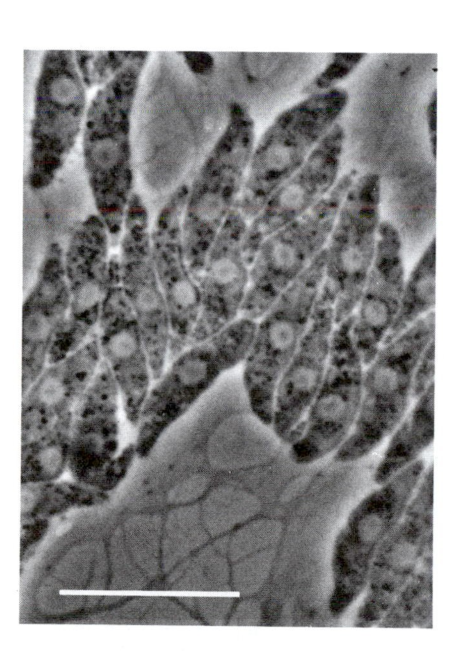

D Live *Labyrinthula* cells in their slimeway. LM, bar = 10 μm. [Courtesy of D. Porter.]

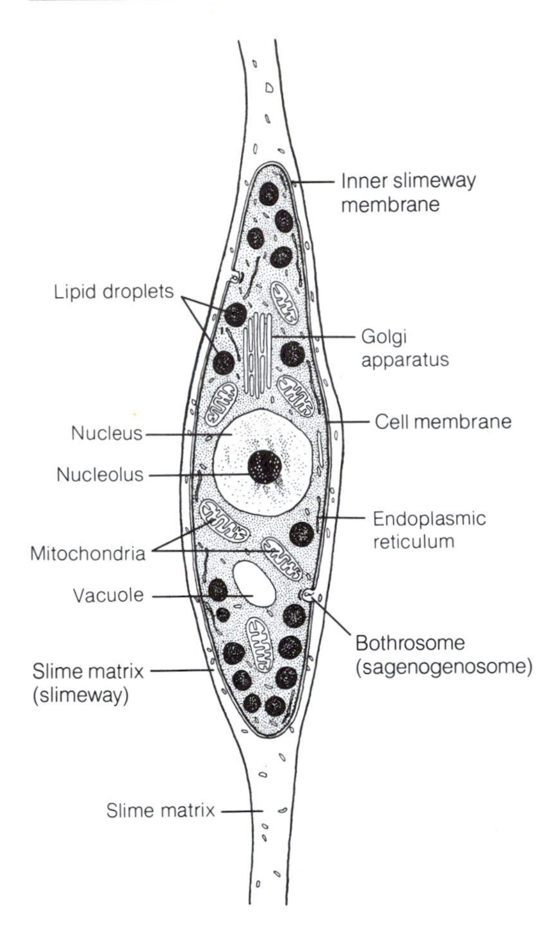

E Structure of a single *Labyrinthula* cell. [Drawing by R. Golder.]

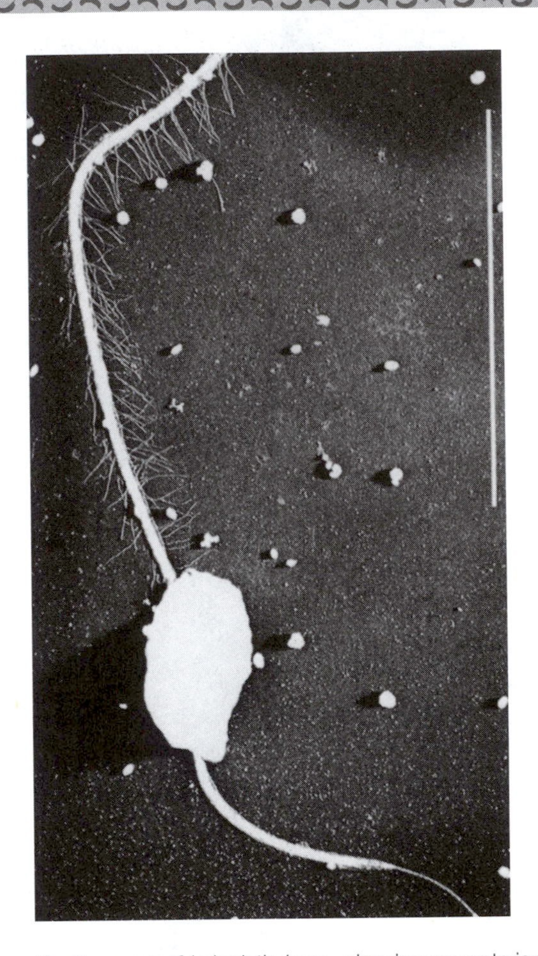

F Zoospore of *Labyrinthula* sp., showing one anterior undulipodium with mastigonemes and one posterior undulipodium lacking them. SEM, bar = 10 μm. [Courtesy of F. O. Perkins, from J. P. Amon and F. O. Perkins, *Journal of Protozoology* 15:543–546 (1968).]

enzymes are released and the resulting small food molecules diffuse into and through the slime to nourish the colony. The labyrinthulid colony changes shape and size so extensively—that is, it is so pleiomorphic—that confusion surrounds its classification on the basis of morphological characteristics. Growth is by the mitotic division of cells inside the slime track, followed by separation of the offspring cells, capable of excreting slime.

When conditions become drier, labyrinthulids adopt a desiccation strategy. In the older parts of the colony, cells aggregate and form hardened, dark cystlike structures, clumps of cells surrounded by tough membrane. These cysts have no precise size or shape. The cysts are resistant to desiccation; they wait until moisture and food become more abundant and then rupture, liberating small spherical cells that grow again into spindle-shaped cells in a slime matrix that they produce. Undulipodiated stages, zoospores, of *Labyrinthula marina, L. vitollina,* and *L. algeriensis* have been found (Figure F). Some seem to be products of meiosis that are isogametes. There is no difference between male and female gametes, which fuse to form a zygote. The zygote undergoes mitosis; its offspring apparently develop into the multicellular net.

Labyrinthula is best known because it grows on the eel grass *Zostera marina* (Phylum Pl-12, Anthophyta) and on many algae, such as *Ulva* (Phylum Pr-28, Chlorophyta). *Zostera* is very important to the clam and oyster (Phylum A-26, Mollusca) industries along Atlantic shores because it is the primary producer of ecosystems that include the clam and oyster beds. Blooms of *Labyrinthula* have been associated with diseased *Zostera.*

Pr-19 Plasmodiophora

New Latin *plasmodium,* multinucleate mass of protoplasm not divided into cells; Greek *pherein,* to bear

ALL GENERA

Ligniera
Membranosorus
Octomyxa
Plasmodiophora
Polymyxa
Sorodiscus
Sorosphaera
Spongospora
Tetramyxa
Woronina

Plasmodiophorans, 10 genera and 29 species, all osmotrophic microbes, are obligate symbiotrophs. Most species live inside plants. Because the feeding, or trophic, stage of plasmodiophorans is a multinucleate plasmodium that lacks cell walls and because zoospores with two anterior undulipodia are formed, these organisms were sometimes aligned with plasmodial slime molds (Phylum Pr-6). However, the two groups are quite different in life history and cell structure.

The plasmodiophoran cell forms a mitotic spindle and from a genetic point of view, it certainly has true mitosis. However, the way in which it divides is peculiar. The nuclear membrane persists during the mitotic stages. It constricts to form the two offspring nuclei, and, during the constriction, the very conspicuous dark nucleolus elongates and divides in half. At metaphase, the chromosomes form a folded ring inside the nucleus, which gives the nucleus in light-microscopic preparations a characteristic "cross-shaped," or cruciform, appearance. When duplicate chromosome sets separate, the two resultant rings of chromosomes pass to opposite ends of the nucleus along with the offspring nucleoli.

Although the details of the life cycle of most species are not entirely understood, the protoplasts (plasmodia) develop into sporangia that form either dispersed swimming zoospores or cystosori. The cystosori are aggregations of thick-walled uninucleate cells (cysts) that withstand desiccation. The life cycle of *Plasmodiophora brassicae* is the best known. A uninucleate cyst of this species persists in soil and then germinates into a primary zoospore bearing two smooth undulipodia of unequal lengths. When the zoospore swims, the short undulipodium is directed forward and the long tapered one is directed backward. This zoospore attaches to and penetrates the cabbage plant. Inside the victim plant cell, each zoospore develops by nuclear divisions into a small primary plasmodium that forms a wall. This plasmodium cleaves into uninucleate segments called secondary sporangia (Figure A), and each nucleus then divides by mitosis to produce at least four secondary zoospores resembling the primary ones. The secondary zoospores exit the plasmodium through an opening in its wall or through an exit tube formed by the plasmodium.

The secondary zoospores escape into the soil. Whether they actually become gametes and fuse by pairs is not confirmed. In any case, after a time in the soil, they reinfect the plant roots (Figure B).

A Uninucleate stage (secondary sporangium) of *Plasmodiophora brassicae* in a cell of a cabbage root hair. Pl = plasmodium; HC = host cytoplasm; G = Golgi apparatus (of host); M = mitochondrion (of host); E = envelope of plasmodiophoran cell; C = cytoplasm of plasmodiophoran cell; N = nucleus of plasmodiophoran cell; CH = chromatin. TEM, bar = 0.5 μm. [Courtesy of P. H. Williams.]

B Cabbage infected with *Plasmodiophora brassicae* (left) and normal cabbage (right). [Courtesy of P. H. Williams.]

Inside the infected plant cells, they develop by nuclear divisions into secondary plasmodia, which feed, move, and grow. As a secondary plasmodium matures, its nuclei become less distinct, probably owing to changes in the function and morphology of the nucleoli. After what have been purported to be two meiotic divisions of its nuclei, the plasmodium cleaves out uninucleate cysts; in each cyst, a spore forms. The cyst walls apparently lack cellulose, but the cysts can remain viable in soil for years before they land on an appropriate plant and begin the cycle again (Figure C).

Plasmodiophorans feed primarily by absorbing dissolved nutrients, but zoospores have also been seen to engulf particles. *Plasmodiophora brassicae* and *Spongospora subterranea* are of economic importance because they cause serious diseases of food plants: *P. brassicae* causes the club root disease of cabbage and other cruciferous plants (Phylum Pl-12); *S. subterranea* causes the powdery scab of potatoes. However, most plasmodiophorans do no obvious harm to the plants in which they reside. The plasmodiophoran forms a cytoplasmic extension, the adhesorium, in which a dense penetrating organelle, the stachel, protrudes. The spent stachel is discarded into the host cytoplasm, which is then transported and taken in by the aggressor, the swelling, growing former zoospore.

The plasmodiophoran genera are distinguished by differences in such characteristics as the arrangement of the cyst in the host cells, the presence and morphology of cystosori, and the development of plasmodia, their exit tubes, and papillae (raised bumps on the exit tubes).

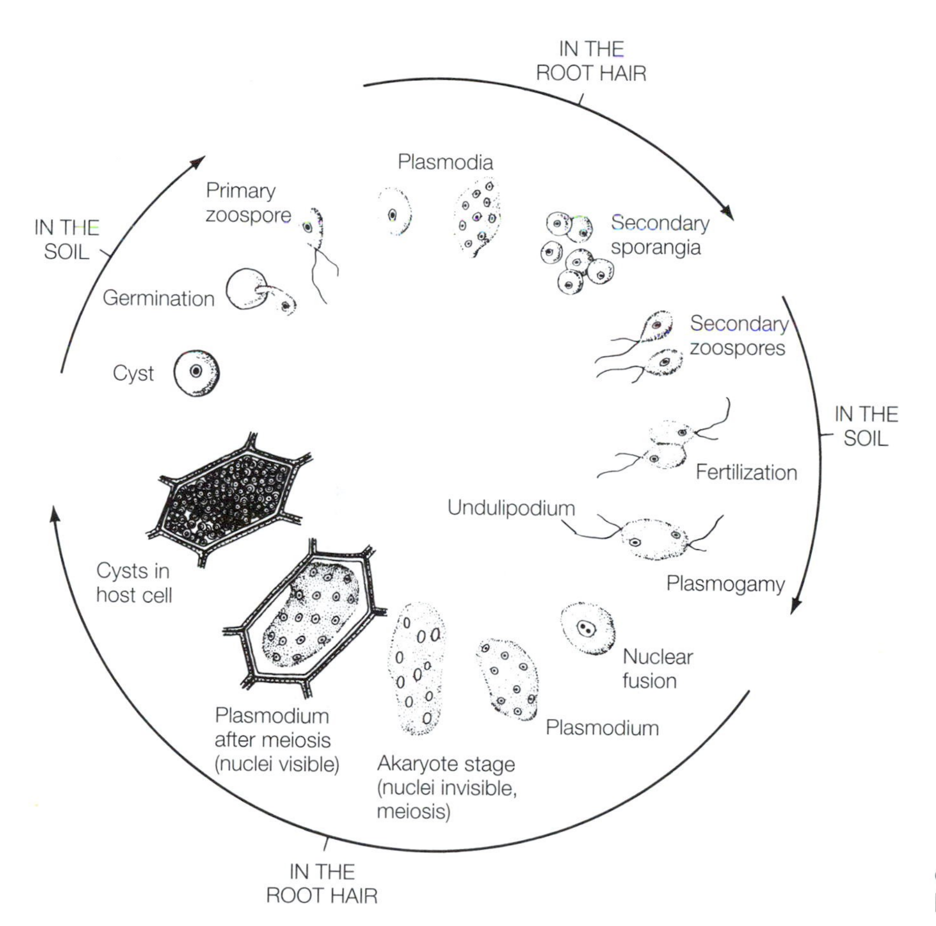

C Life cycle of *Plasmodiophora brassicae*. [Drawing by L. Meszoly.]

Pr-20 Oomycota

(Oomycetes, oomycotes)

Greek *oion*, egg; *mykes*, fungus

EXAMPLES OF GENERA

Achlya
Albugo
Aphanomyces
Apodachlya
Dictyuchus
Isoachlya
Lagenidium
Myzocytium
Peronospora
Phytophthora
Plasmopara
Pythium
Rhipidium
Saprolegnia

The oomycote zoospore morphology and sequence of nucleotides in its ribosomal RNA gene confirm these organisms as stramenopiles ("straw bearers"). Oomycotes are variously called water molds, white rusts, and downy mildews. Some 50 genera in two classes are recognized. The more diverse and pathogen-filled class comprises the peronosporans, which assort into five orders: leptomites, rhipidians, sclerosporans, pythians, and the peronosporans themselves. The free-living class consists of the saprolegnians.

Like the hyphochytrids (Phylum Pr-21) and chytrids (Phylum Pr-29), oomycotes are either saprobes or symbiotrophs. They feed by extending funguslike threads, or hyphae, into victimized tissues, where they release digestive enzymes and absorb the resulting nutrients. All are coenocytic—the growing oomycote mass lacks even the partial septa of the chytrids. Only the reproductive organs are separated from the other hyphae by septa, but septa may form by wall and membrane hypertrophy in response to injury. The cell walls are made of cellulose. Mycelia of oomycotes show less regularity in their branching patterns than do the mycelia of chytrids and fungi. Most live in fresh water or moist soil; a few are symbiotrophs of land plants and rely on air currents to disperse their offspring.

The oomycotes are distinguished from the other funguslike protoctists by kinetids and other details of zoospore structure and by the nature of their sexual life cycle. In the style of all heterokonts, the zoospores bear two undulipodia of unequal length. One of them is directed forward during swimming and is mastigonemate, whereas the other is smooth and trails behind. The zoospores are produced and released by a sporangium, an asexual reproductive organ that differentiates at the tip of a vegetative hypha. Either immediately or after some transformation, a zoospore germinates to grow a new oomycote thallus.

Sexuality is well developed in this phylum, but it never includes undulipodiated gametes. The tips of the growing hyphae produce specialized male and female structures, called by botanical analogy antheridia and oogonia, respectively. An antheridium in contact with an oogonium produces fertilization tubes, which penetrate the oogonial tissue and provide a conduit through which the male nuclei migrate.

Fertilization takes place inside the walled oogonium (Figure A), which usually contains a few large spheres called oospheres: haploid egg cells. Most oospheres are uninucleate; those that are multinucleate are referred to as compound oospheres. By fertilization, the oospheres become zygotes. They may develop thick chitinous walls and become resistant to starvation and desiccation. Such developed zygotes are called oospores or, sometimes, zygospores, after an analogous structure of zygomycotes (Phylum F-1). The walls of the oospores are usually dark and sculpted in patterns peculiar to each species.

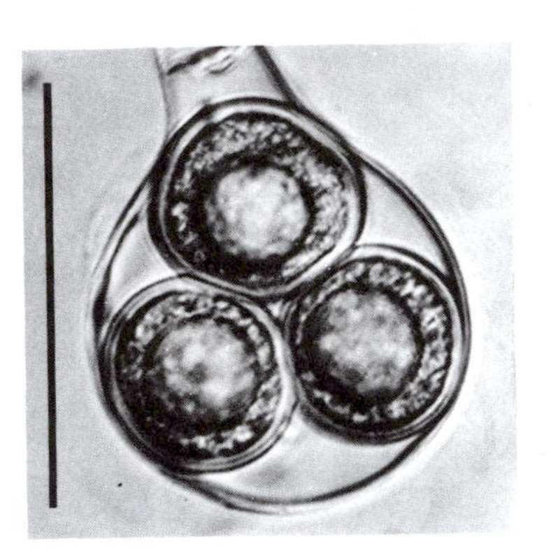

A Oogonium of *Saprolegnia ferax,* an oomycote from a freshwater pond. LM, bar = 50 μm. [Courtesy of I. B. Heath.]

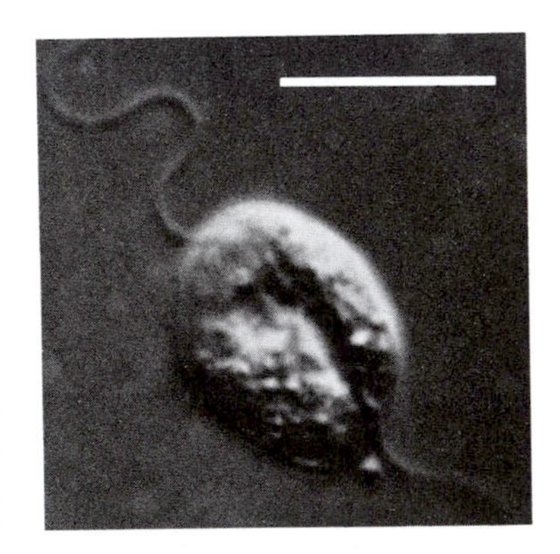

B Zoospore of *Saprolegnia ferax.* LM, bar = 10 μm. [Courtesy of I. B. Heath.]

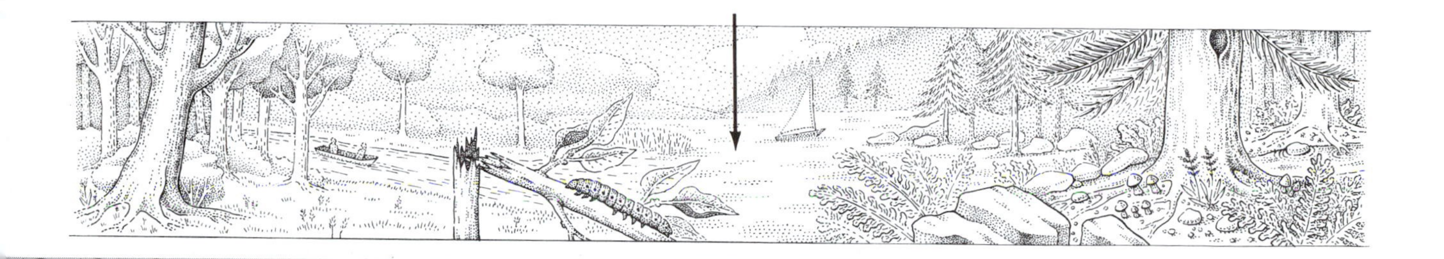

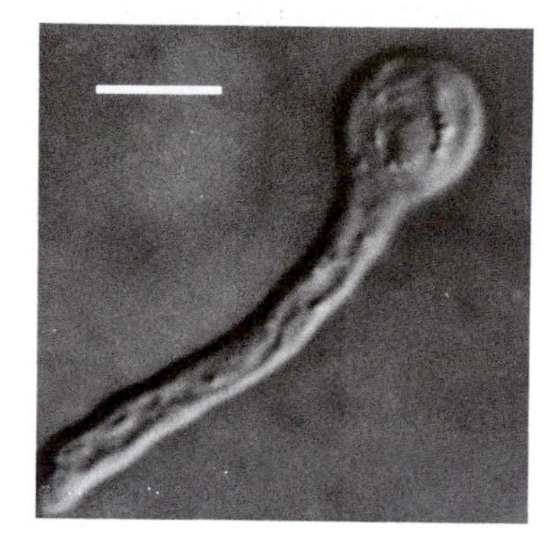

C Germinating secondary cyst of *Saprolegnia ferax*. LM, bar 10 = μm. [Courtesy of I. B. Heath.]

In some oomycotes, such as *Saprolegnia*, an oospore can germinate to form a new thallus directly. Alternatively, meiosis may take place inside the oospore, followed by karyogamy between adjacent haploid offspring nuclei so that eventually, diploid zoospores (Figure B), which then grow into thalli, are released. The diploid phase of the life cycle, by implication, is dominant.

Hundreds of species have been reported. Every crop plant species (Phylum Pl-12, Anthophyta) seems to have its own threatening oomycote species, such as *Pythium capsicum* on peppers, or *Phytophthora infestans* on potatoes. The simplest produce merely a unicellular thallus and the characteristic reproductive structures. Other species produce a highly branched, profuse, and abundantly growing thallus. The most complex of the oomycotes live within specific plants and depend on wind to disseminate the oospores or the sporangia full of zoospores. Even wind-disseminated oomycotes produce zoospores at some stage in their life cycle, indicating their affinities with free-living and less complex genera. Some species have two types of zoospores. In *Saprolegnia ferax*, for example, the ovoid primary zoospore bears two undulipodia at its apex (see Figures B and E), and the pear-shaped secondary zoospore bears oppositely directed undulipodia at its side (Figures C and E). The two types are produced in succession, not simultaneously. In eucarpic oomycotes, only part of the thallus differentiates for the reproductive function, whereas, in holocarpic oomycotes, the entire thallus becomes a reproductive sporangium. In some, the zoospores are formed within the sporangium (Figure D); in others, within an evanescent vesicle produced by the sporangium.

The free-living class of oomycotes, the saprolegnians, are widely distributed in freshwater environments. Representative genera of saprolegnians include *Saprolegnia, Achlya, Isoachlya,* and *Dictyuchus,* most of which live on seeds and other decomposing vegetation. The many species of the genus *Saprolegnia* include *Saprolegnia parasitica,* very common in freshwater aquaria as a white fuzz on fish fins. This species in the extensive genus attacks fish and their eggs. *S. ferax* is cultured in the laboratory and has proved itself useful in the study of developmental and mitotic processes. The hyphae of saprolegnians are profusely branched. At the hyphal tips, sporangia form, inside of which zoospores develop. Some

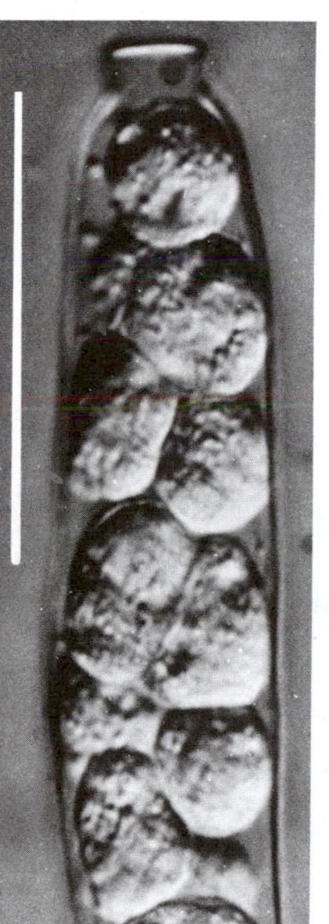

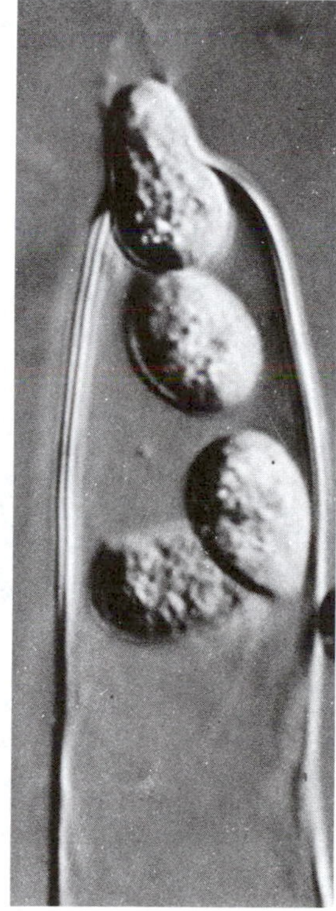

D Zoospores in zoosporangium (left) and their release (right). LM, bar = 50 μm. [Courtesy of I. B. Heath.]

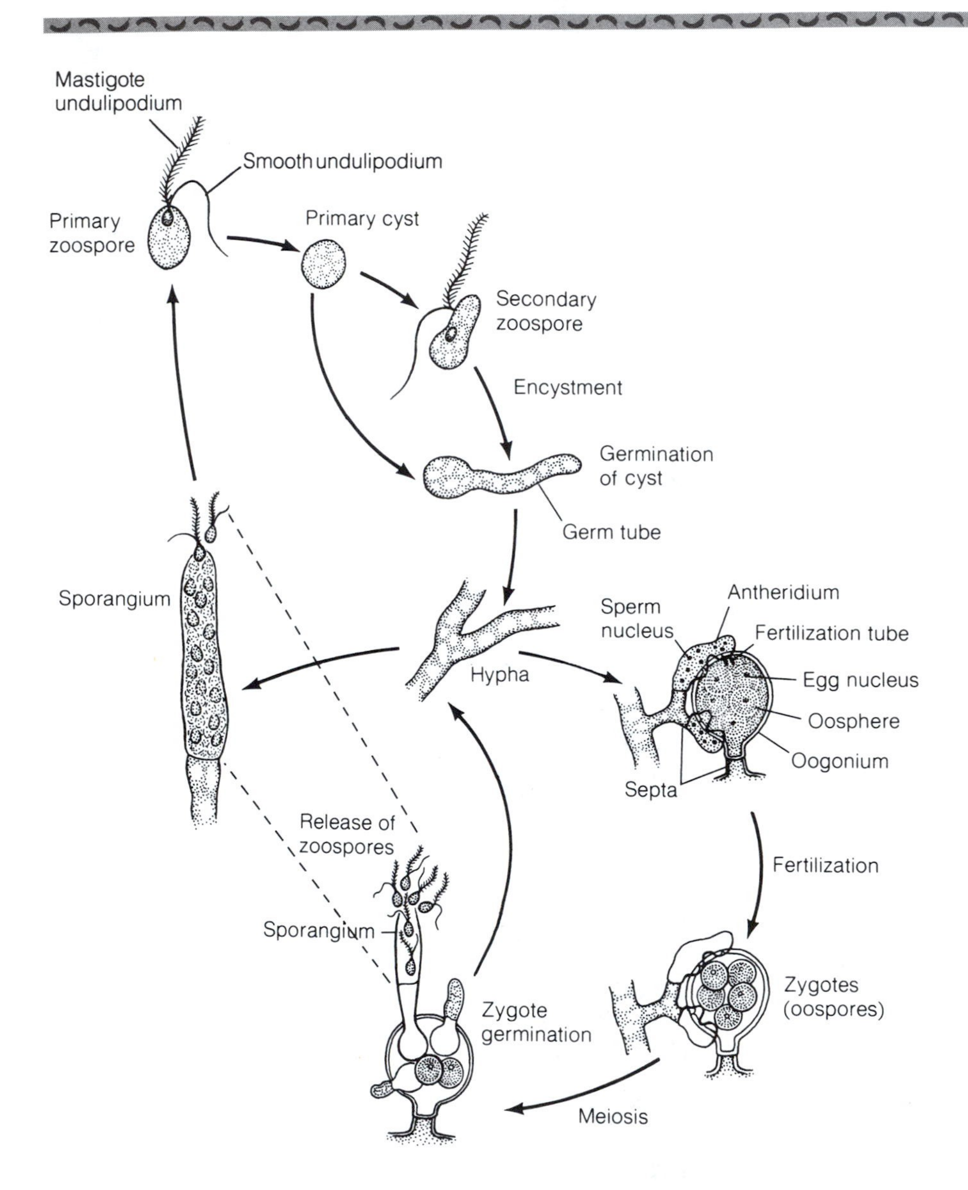

E Life cycle of *Saprolegnia.* [Drawing by L. Meszoly; information from I. B. Heath.]

saprolegnians have an additional type of asexual reproduction: they produce chlamydospores, or gemmae. These irregular bits of protoplasm separate from the mycelium (mass of hyphae) and grow by germ tubes. These germ tubes grow into hyphae bearing typical sporangia (Figure E).

The peronosporan class of oomycotes includes *Apodachlya* and *Rhipidium* in the leptomite order. Members of the peronosporan order in this class have been of utmost importance in history. The infamous *Phytophthora infestans* is associated with the late blight of potatoes. In the nineteenth century, this oomycote destroyed the entire potato harvest of Ireland and Germany and caused mass migrations of people from their homelands. *Plasmopara viticola*, another dangerous peronosporan, causes mildew of grapes (*vitis*). Other genera in this order are *Pythium*, *Albugo*, and *Peronospora*.

Pr-21 Hyphochytriomycota

Greek *hyphos*, web; *chytra*, little earthen cooking pot; *mykes*, fungus

EXAMPLES OF GENERA
Anisolpidium
Canteriomyces
Hyphochytrium
Latrostium
Rhizidiomyces

Hyphochytrids, with chytrids (Phylum Pr-29) and oomycotes (Phylum Pr-20), have traditionally been considered fungi. These aquatic osmotrophs do resemble fungi in their mode of nutrition, which may be either symbiotrophic or saprobic. Thin threads or filaments invade host tissue or dead organic debris where they release digestive enzymes and absorb the resulting nutrients (Figure A). However, they differ from fungi in that they produce undulipodiated cells. The kinetid structure, often inferred from the detailed description of motility, distinguishes these "mastigote molds" or "zoosporic fungi." The single anteriorly directed undulipodium that confers rapid swimming, quick changes in direction, and a wide spiral path is enough to distinguish hyphochytrids.

Hyphochytrids either are parasites on algae and fungi or live on insect carcasses and plant debris. The body, or thallus, of a hyphochytrid may be holocarpic (that is, where the entire body converts into a reproductive structure) or eucarpic (a part of the thallus develops into a reproductive structure while the remaining part continues its somatic function). In the holocarpic species, the thallus is formed inside the tissues of the host. In these species, the thallus consists of only a single reproductive organ bearing the branched rhizoids, rootlike tubes that penetrate the substrate, or hyphal feeding tubes that grow out of the substrate, in which cross walls, or septa, may or may not develop. The thallus in eucarpic species may reside on the surface rather than inside the host tissues.

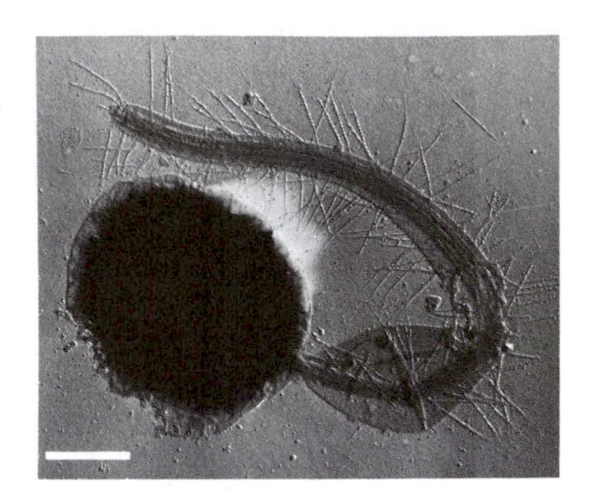

B Zoospore of *Rhizidiomyces apophysatus*, showing mastigonemate undulipodium (right). TEM (negative stain), bar = 1 μm. [Courtesy of M. S. Fuller.]

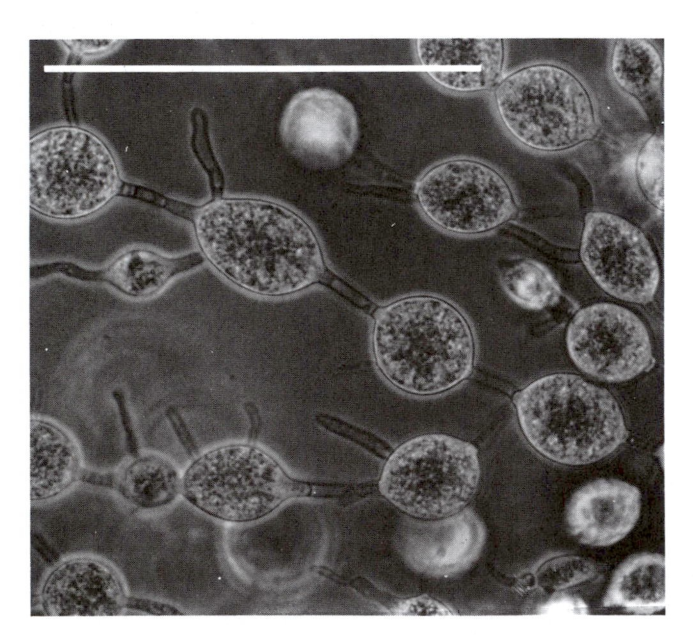

A Filamentous growth of *Hyphochytrium catenoides* on nutrient agar. LM, bar = 0.5 μm. [Courtesy of D. J. S. Barr.]

From the reproductive organ, or zoosporangium, zoospores emerge through discharge tubes. Hyphochytrid zoospores are very active swimmers. True to their classification as stramenopiles, each bears one mastigonemate anterior undulipodium that moves in helical waves as well as in whiplash fashion (Figure B). The zoospores swim to new food sources. Zoospores stop swimming, encyst, withdrawing their undulipodia, and produce cell walls as they develop again into a thallus. Reproduction, asexual by zoospores, is the norm—and explains their need for aquatic habitats. Neither sexuality nor resistant spore formation has been confirmed.

All hyphochytrids live in fresh water and have been isolated primarily from soil and tropical fresh waters. However, they are probably distributed all over the world wherever their hosts are found. The five genera containing 23 species here are grouped into three classes: Anisolpidia, Rhizidiomycetae, and Hyphochytria.

The best-known hyphochytrid, *Rhizidiomyces apophysatus*, parasitizes water molds, such as *Saprolegnia* (Phylum Pr-20, Oomycota). After *Rhizidiomyces* zoospores come to rest on their victims, they transform into spheres, withdrawing their undulipodia. The

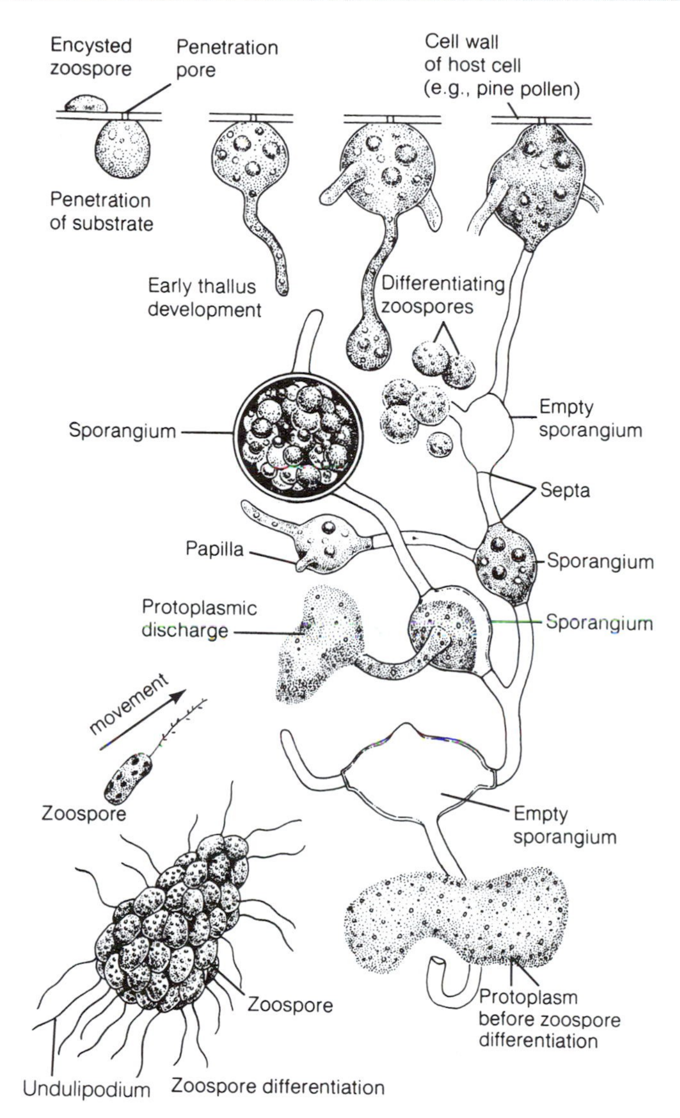

C Life cycle of *Hyphochytrium* sp. [Drawing by R. Golder.]

sphere does not divide but germinates, extending a germ tube into the host. With continued growth, the tube ramifies into a branching system of rhizoids in the host tissue. Between the rhizoids and the surface of the host, a swelling develops into a baglike structure called the apophysis of the sporangium. The sporangium itself grows at the outer end of the apophysis; it enlarges and forms an exit papilla—a raised bump with a hole in it—that becomes a discharge tube. Karyokineses (nuclear divisions) take place in the sporangium to form a plasmodial mass. Mitotic division figures in hyphochytrids are unmistakably distinguished from those in plasmodiophorans (Phylum Pr-19), oomycotes, and chytrids. The mass of multinucleate protoplasm passes through the discharge tube and emerges from it. The protoplasm then cleaves into a mass of individual zoospores that develop outward-directed undulipodia. The zoospores swim away to begin the cycle again (Figure C).

The details of the life cycles of most hyphochytrids are not known, partly because they are difficult to observe. *Hyphochytrium catenoides,* for example, grows mainly on the pollen of conifers; the thallus stages develop inside empty pollen grains (Figure D). The cell walls of the hyphochytrids that have been studied are composed of chitin, but, in some species, they contain cellulose as well.

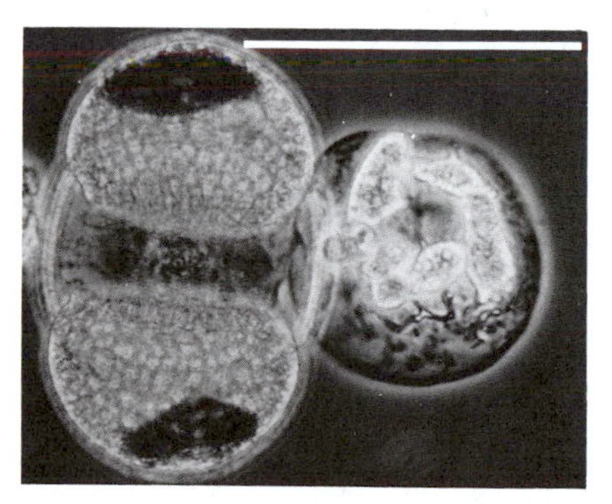

D Sporangium (right) of *Hyphochytrium catenoides* on a ruptured pine pollen grain (left; Phylum Pl-10). LM, bar = 0.5 μm. [Courtesy of D. J. S. Barr.]

Pr-22 Haplospora

Greek *haplo,* single, simple; Latin *spora,* spore

ALL GENERA
Haplosporidium
Minchinia
Urosporidium

Convergent evolution that led to symbiotrophic, small, dark structures in animal tissue unites this phylum, as well as Phyla Pr-23 (Paramyxa) and Pr-24 (Myxospora)—three phyla of propagule-forming (for example, cyst-forming) parasites—with an alveolate phylum, the Apicomplexa (Phylum Pr-9), and other "parasitic protozoa" into the "Sporozoa" category. The haplosporosome-forming parasites (Haplospora) are known mainly from fish and other marine animals; the paramyxans are nesting-cell parasites; and the multicellular myxosporans are the far better known traditional "sporozoa." Sporozoa, some of which are associated with serious diseases, were considered a class of animals in the phylum "Protozoa" when all the small heterotrophic protoctists were classified according to their importance to people. Fine-structure analyses with the electron microscope and complementary molecular studies of sequences of nucleotides, especially in ribosomal RNA, and of amino acids in proteins make it abundantly clear that "sporozoa" have little in common except their habitat. The investigation of these many very different organisms has thus been inappropriately restrained by ignorance of their great differences.

Clearly the "sporozoan habit" evolved convergently in several free-living protist lineages: small protists invaded animal tissues and took up residence. The majority are now coevolved benign inhabitants with life histories tightly coupled to those of their hosts, but some remain necrotrophic (for example, they kill the host tissues in which they reside). Those evolving innovative modes of transfer from the bodies of their benefactors succeeded in leaving more offspring.

Only three genera with a total of 33 species are placed in Haplospora, a phylum of marine animal tissue symbiotrophs. Because the phylum definition requires the propagule—spore—to have an anterior pore (opening or aperture) covered with a hinged cap or piece of wall material folded into the opening, other genera and species are suspected to exist in which this particular defining feature has not been seen. Knowledge of the group comes mainly from observations of live infected marine animals coupled with electron microscopy of their tissues.

Haplosporans constitute a phylum of tissue and body-cavity symbiotrophs in marine invertebrate animals; the haplosporan is unicellular and begins its life history uninucleate. The cell contains tubulovesicular mitochondria with a small number of tubules and a

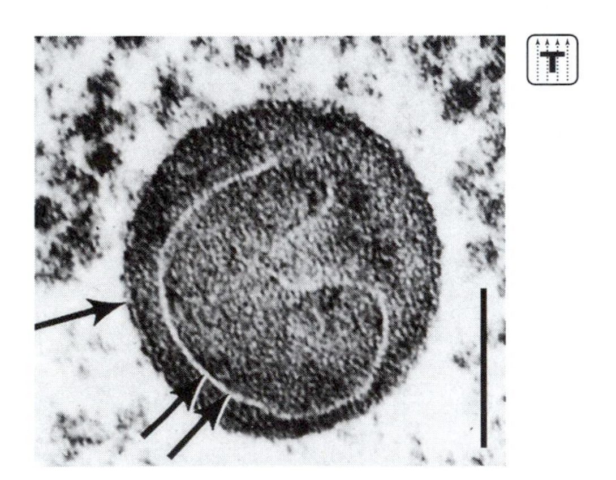

A Haplosporosome of *Haplosporidium nelsoni* in which a limiting membrane (arrow) and internal membrane (double arrow) are visible. TEM, bar = 0.1 μm. [Micrograph by F. O. Perkins. Reprinted from the *Handbook of Protoctista* (Jones and Bartlett, 1990).]

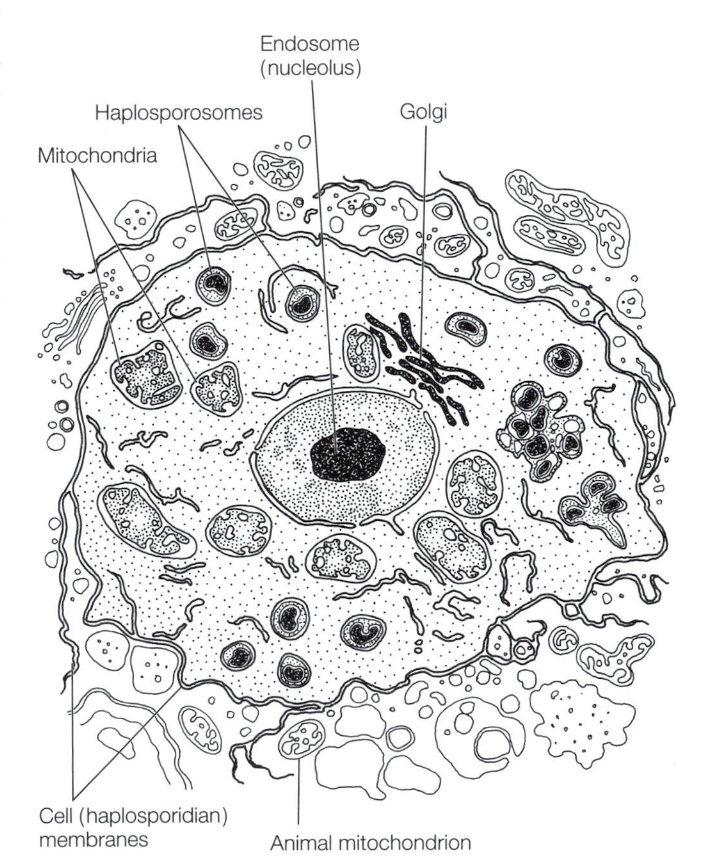

B A generalized haplosporidian. Plasmodium with haplosporosomes in host tissue. [Drawing by K. Delisle.]

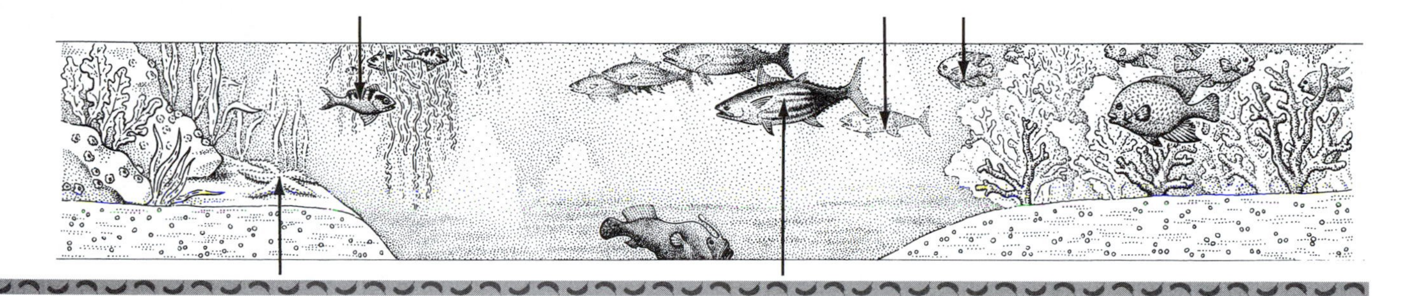

distinctive organelle of unknown function, the haplosporosome, for which the group is named. Haplosporosomes, scattered throughout the cytoplasm, are electron-dense, generally spherical [but sometimes vermiform (wormlike), elliptical, or of other shape] objects that range in size from 0.07 to 0.25 μm in diameter. Both a unit membrane and an inner looser membrane delimit this organelle (Figure A).

Haplosporans lack walls and undulipodia at all stages in their life history. They have a paucity of ribosomes and endoplasmic reticulum but apparently do have Golgi bodies (Figure B). They have never been seen outside tissues of the animals in which they reside. Their spores seem to be shed into the water and, either there or in tissue, the spore lids probably open to release sporoplasm, but the entire life history has never been described for any species.

Before the formation of spores, haplosporans grow in tissue as uninucleate or multinucleate plasmodia (Figure C). The nuclear membrane does not disintegrate during karyokinesis (nuclear division; Figure D). The first sign of sporulation is the deposition of a thin wall around the larger multinucleate cells that then become, by definition, sporonts. The sporont undergoes cytokinesis (whole-cell division) in a manner that subdivides it into uninucleate cells called sporoblasts. A kind of sexual fusion then occurs in the development of the spores: pairs of uninucleate sporoblasts fuse to form binucleate sporoblasts. The nuclei of these dikarya fuse (karyogamy), and the resulting, presumably diploid, cell assumes the shape of an hourglass. A strange happening ensues: the anucleate half of the cell nearly entirely engulfs the nucleate half, forming the epispore cytoplasm (the former enucleate sporoblast) and the sporoplasm (the former nucleate part of the sporoblast). In the epispore cytoplasm, a new cup-shaped spore wall is formed that has an anterior constricted opening (aperture). The aperture is covered by a hinged lid or a tucked-in tongue of wall material. Decorations with distinguishing substructure are formed inside the epispore cytoplasm and transported to the outer surface of the spore wall. This distinctive ornamentation, prominent extensions of the spore, are only fuzz at the level of light microscopy. Spore decorations are well resolved by electron microscopic analysis.

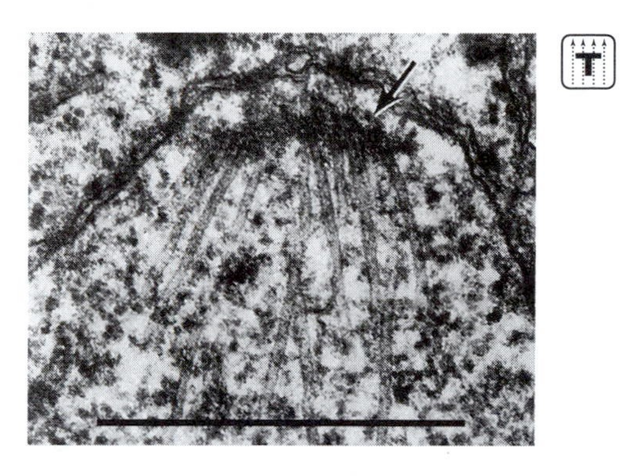

D Spindle pole body (arrow) of *Haplosporidium nelsoni* in mitotic nucleus with attached microtubules. TEM, bar = 1 μm. [Micrograph by F. O. Perkins. Reprinted from the *Handbook of Protoctista* (Jones and Bartlett, 1990).]

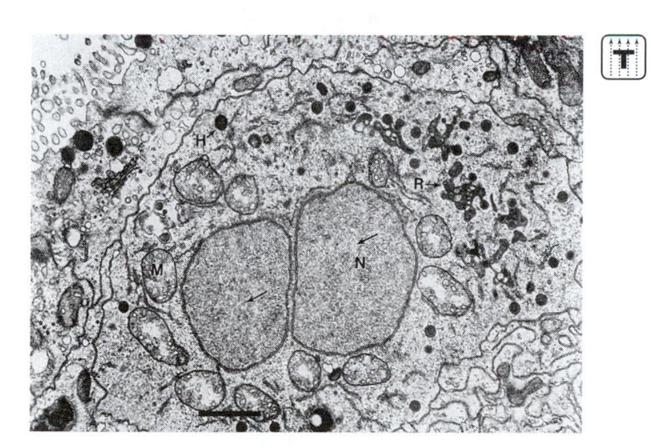

C Plasmodium of *Haplosporidium nelsoni*. Nuclei (N), free haplosporosomes (H), mitochondria (M), microtubules (arrows) of the persistent mitotic apparatus, and membrane-bounded regions in which haplosporosomes are formed (R) are visible. TEM, bar = 1 μm. [Micrograph by F. O. Perkins. Reprinted from the *Handbook of Protoctista* (Jones and Bartlett, 1990).]

Animals in which these symbiotrophs thrive include limpet molluscs (Phylum A-26) and worms: nematodes (Phylum A-10), trematodes (Phylum A-5), and polychaetes (Phylum A-22). Haplosporans are easiest to find as parasites of parasites, which are called hyperparasites—that is, symbiotrophs of symbiotrophs. For example, *Urosporidium* species are detected after they enter, presumably from the digestive tract, into the hepatopancreas of trematode worms. The worms themselves are symbiotrophs in bivalve molluscs such as oysters (Phylum A-26). After spore development begins, the worms change color to brown or black, indicating the presence of haplosporans. The worm-parasitized soft tissue of the mollusc becomes watery; its greater transparency provides the investigator with a clue to the whereabouts of haplosporans. Infected tissue may be replete with haplosporan plasmodia having small, eccentrically placed nuclei with conspicuous endosomes (nucleoli, sites of ribosome synthesis). The Brownian movement of the haplosporosomes in the cytoplasm of the haplosporan plasmodia can be seen in live material, another way by which the existence of these obscure organisms is inferred.

Pr-23 Paramyxa

Greek, *para,* alongside of; *myxa,* mucus

ALL GENERA
Marteilia
Paramarteilia
Paramyxa

Formerly ignored or lumped with "sporozoa," the symbiotrophic paramyxans are immediately distinguished from all other organisms by their "nesting cell" behavior. Their propagules, called "spores" (as are other small spherical compact structures capable of further growth), consist of several cells enclosed inside one another

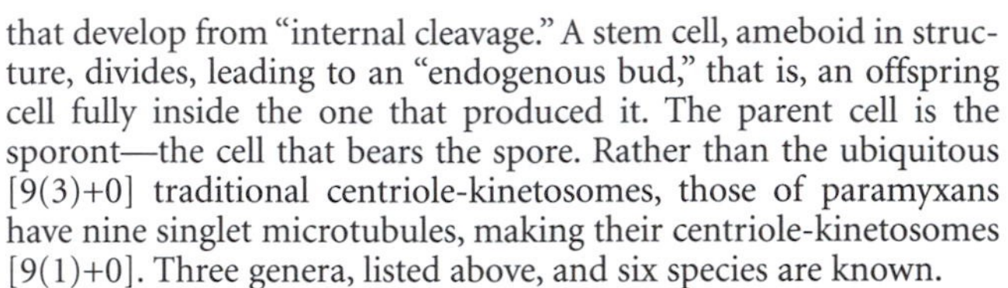
that develop from "internal cleavage." A stem cell, ameboid in structure, divides, leading to an "endogenous bud," that is, an offspring cell fully inside the one that produced it. The parent cell is the sporont—the cell that bears the spore. Rather than the ubiquitous [9(3)+0] traditional centriole-kinetosomes, those of paramyxans have nine singlet microtubules, making their centriole-kinetosomes [9(1)+0]. Three genera, listed above, and six species are known.

The presence of haplosporosomes, organelles as described for other propagule-forming parasites, has led to the placement of paramyxans with *Haplosporidium* and other haplosporans (Phylum Pr-22). However, the unequivocal differences in the reproductive biology and other aspects of cell structure have led to the separation

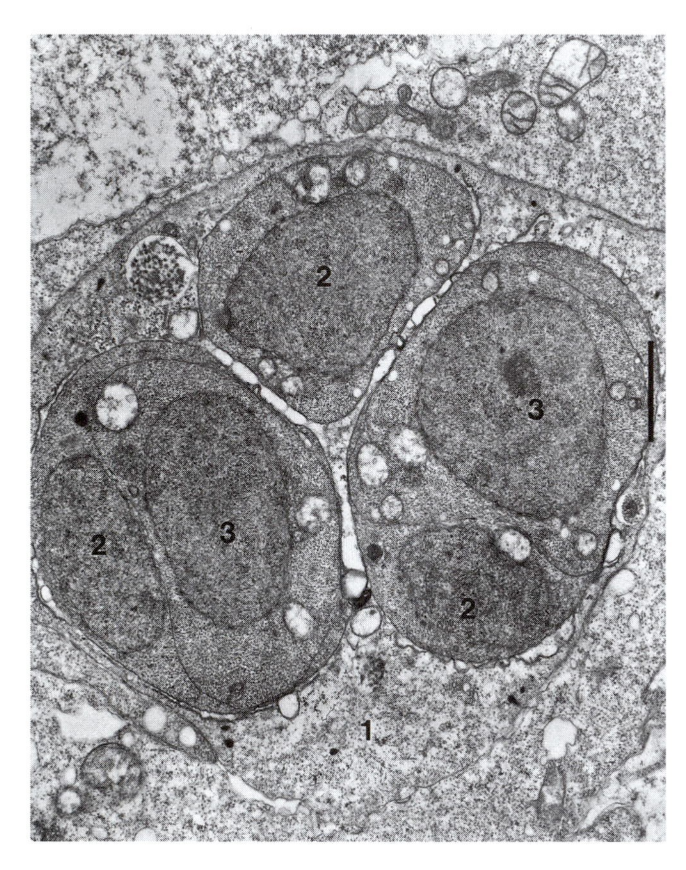

A The stem cell of *Paramarteilia orchestiae* (1) containing three sporonts (2). In two of them, the tertiary cell (3) is already differentiated. This stage can be observed in all paramyxeans. TEM, bar = 1 μm. [From T. Ginsburger-Vogel and I. Desportes, "Étude ultrastructurale de la sporulation de *Paramarteilia orchestiae* gen. n., sp. n., parasite de l'amphipode *Orchestia gammarellus* (Pallus)." *Journal of Protozoology* 26:390–403 (1979).]

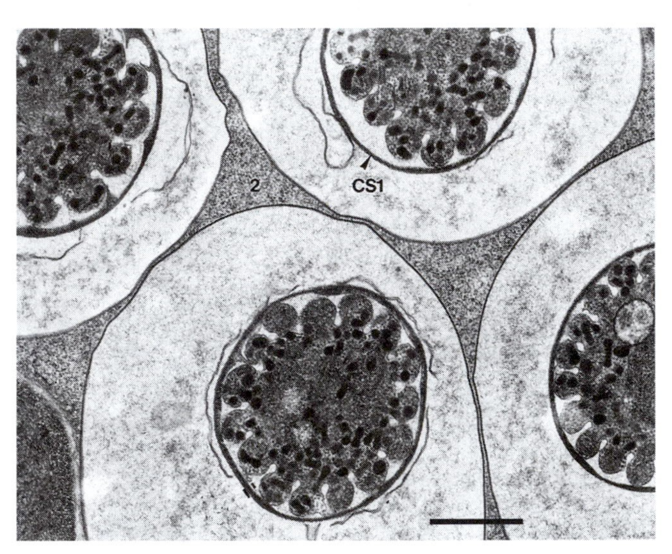

B Transverse sections of four mature spores of *Paramyxa paradoxa*. The outer sporal cell (CS1) is reduced to a thin cytoplasmic layer (arrowhead). Infoldings and dense bodies of the secondary sporal cell can be seen. The light area around each spore results from its retraction in the sporont cytoplasm (2). TEM, bar = 1 μm. [From I. Desportes, "Étude ultrastructurale de la sporulation de *Paramyxa paradoxa* Chatton (Paramyxida) parasite de l'annelide polychete *Poecilochaetus serpens.*" *Protistologica* 17:365–386 (1981).]

of this paramyxan phylum from haplosporans and other phyla of symbiotrophs that form protected propagating stages in animal tissue. Thus, they cannot be placed in other traditional "sporozoa" phyla, because they lack the apical complexes of apicomplexans (Phylum Pr-9), the polar filaments of microsporans (Phylum Pr-2), the capsules of myxosporans (Phylum Pr-24), and all other criteria for placement elsewhere.

The various genera of paramyxans are distinguished by two criteria: the number of spores and cells developing into spores in the sporont (such as the three illustrated in Figure A) and the taxonomic position of the animal on whose tissue the paramyxan depends. Paramyxans live in various invertebrate hosts: *Paramyxa* (Figure B) in the intestinal cells of annelids (Phylum A-22), *Paramarteilia* in the testes of crustaceans (Phylum A-21), and *Marteilia* in the hepatopancreas of bivalve molluscs (Phylum A-26).

The development of *Paramyxa paradoxa* in the cytoplasm of cells of a marine animal is shown in Figure C.

Although paramyxans have mitochondria and therefore probably some oxidative metabolism, their status as obligate intracellular symbiotrophs has precluded their cultivation. No metabolic or genetic studies are available, and no sexual behavior or motile stages of any kind have been reported.

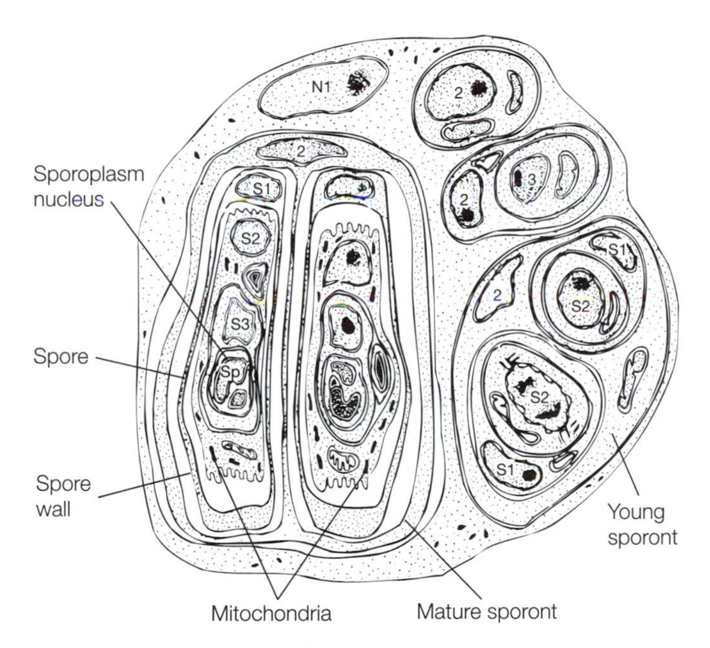

C The development of *Paramyxa paradoxa* is shown here in the cytoplasm of cells of a marine animal. Only two of the four spores are shown in the young sporont and in the mature sporont. 2, nucleus of secondary (stem) cell; 3, tertiary cell nucleus; N1, stem cell nucleus; S1, S2, S3, nuclei of sporal cells 1, 2, 3. [Adapted from I. Desportes, "The Paramyxa Levine 1979: an original example of evolution towards multicellularity," *Origins of Life* 13, 343–352 (1984).]

Pr-24 Myxospora

(Myxozoa, myxosporidians)

Greek *myxa*, mucus; Latin *spora*, spore

EXAMPLES OF GENERA

Alatospora	*Hexacapsula*	*Sinuolinea*
Auerbachia	*Ichthyosporidium*	*Sphaeractinomyxon*
Aurantiactinomyxon	*Kudoa*	*Sphaeromyxa*
Ceratomyxa	*Myxidium*	*Sphaerospora*
Chloromyxum	*Myxobolus*	*Telomyxa*
Coccomyxa	*Myxostoma*	*Triactinomyxon*
Fabespora	*Ormieractinomyxon*	*Trilospora*
Guyenotia	*Ortholinea*	*Unicapsula*
Helicosporidium	*Parvicapsula*	*Unicauda*
Henneguya	*Septemcapsula*	*Wardia*

This phylum consists of the better-known traditional "sporozoa," multicellular organisms that some believe are animals. These aerobic "spore" formers are often confused with the amitochondriate microsporans (Phylum Pr-2) because members of both groups have polar filaments, but their respective filaments differ greatly in structure and function.

Phylum Myxospora comprises two classes: Myxosporea, the larger, more diverse group, and Actinosporea (actinomyxids), limited to intestinal or coelomic tissue of benthic (seafloor dwelling) marine sipunculans (two species, Phylum A-23) or freshwater oligochetes (about 35 species, Phylum A-22). More than 1100 species of myxosporans, a thousand of which live in fish (Phylum A-37) and none of which have been cultured in vivo or in vitro, belong to about 40 genera. These symbiotrophs parasitize animals, and all produce some sort of polar filament or thread.

Most true myxosporans (members of Class Myxosporea) parasitize fish and other aquatic animals and do not cause serious diseases. The coevolution of myxosporans with teleost fish is so complex and complete that the evolution of the symbiotrophs is thought to be closely linked to the evolution of late Cretaceous bony fish called Actinopterygii. The lack of a myxosporan fossil record and the benign nature of the association make speculation of evolutionary history difficult.

The feeding stage, which can be a huge plasmodium, contains many generative cells floating freely in the cytoplasm. Centrioles and undulipodial stages are always lacking. The polar filaments of myxosporan "spores" are formed when the feeding stage—trophozoites—forms spores. The spores, triradially symmetrical, are always packed in groups of eight inside a two- to four-celled envelope called a pansporoblast. The myxosporan spore is multinucleate. It has a wall composed of two or more shell valves connected by sutures; the infective sporoplasm emerges from between the valves. An anterior polar capsule, the cnidocyst, encloses the polar filament. The polar capsule is covered by the shell valves, except for the gap through which the polar filament discharges. The polar filaments are used not for injecting (as they are in microsporans) but for anchoring the spores to the host tissue.

Myxosporans penetrate the integument of nearly any part of the animal body and make their way to the intestine, where they release uninucleate ameboid forms called amebulinas. These penetrate the tissues and presumably are carried by the blood to target organs, where they often form large unsightly growths as much as several centimeters in diameter. They seem to favor hollow organs, such as urinary and gall bladders, but they also infest gills, muscles, intestines, liver, brain, bone, and skin. *Myxobolus pfeifferi* causes boil disease of the European barbel. The twist disease of salmon is caused by *Myxostoma cerebralis,* which lives in cartilage and forms tumors that exert pressure on the central nervous system.

The striking resemblance between the cnidocysts of myxosporans and the nematocysts of coelenterates (Phylum A-3) in morphology, developmental pattern, and function suggests that the myxosporans evolved from coelenterate animals. However, the myxosporan's lack of a blastula embryo, lack of eggs and sperm, and absence of any histogenesis to produce tissues prompt us to retain the myxosporans with the protoctists.

Although members of Class Actinosporea (the actinomyxids) have been placed in this phylum because they are more like members of Class Myxosporea than like microsporans, they are so distinctive that some favor raising them to phylum status. Actinomyxids are the more poorly known of the two classes. They are parasites, in particular of annelids (Phylum A-22). Their spores are cnidocysts that produce the filaments. *Sphaeractinomyxon gigas* is 40 μm long and lives in the coelom of the oligochaete worm *Limnodrilus*. *Triactinomyxon* lives in the gut of *Tubifex,* also an oligochaete annelid. In both actinomyxid genera, the spore valves are elongated into large hooks for attachment to the host.

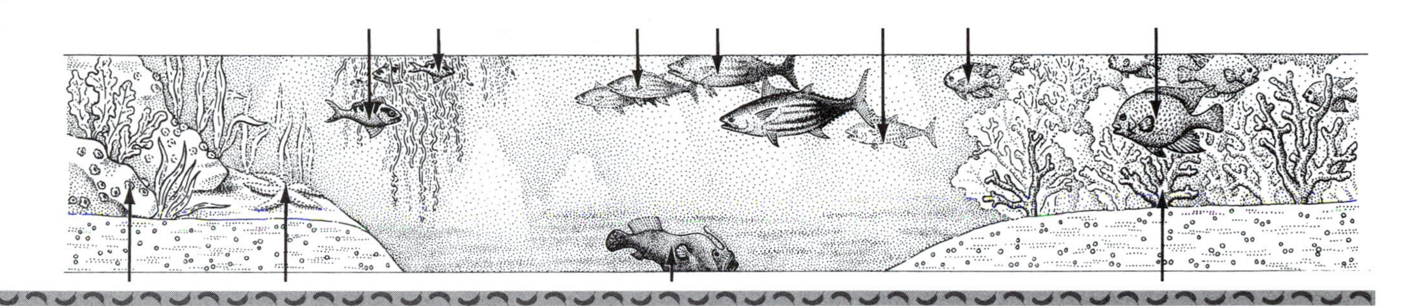

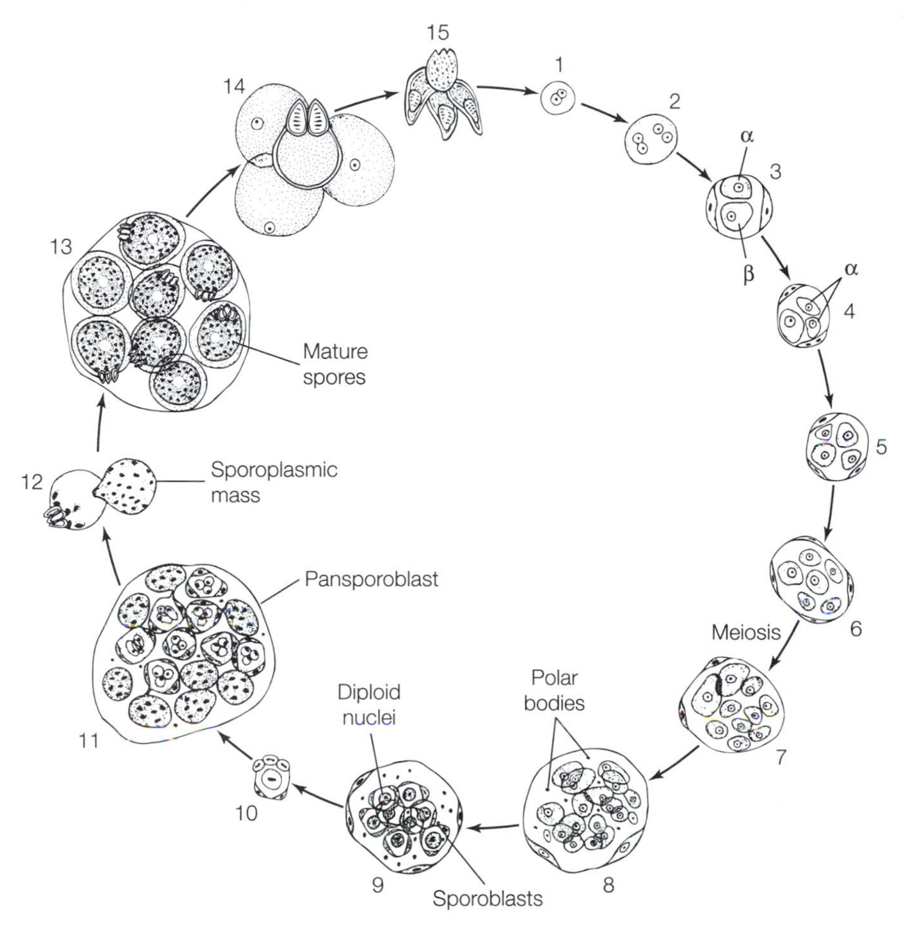

Aurantiactinomyxon eiseniellae developmental cycle, in the oligochaete worm Phylum (A-22) in which it lives. 1. Binucleate stage. 2. Four-cell stage. Karyokinesis, but not cytokinesis, has occurred. 3. Differentiation into two pansporoblast cells and two inner cells, α and β. 4. Pansporoblast cell divides; the α cell has divided. 5. Both α and β cells have divided. 6. Pansporoblast is now four cells; inner cells divide. 7. Eight α and two β cells; meiotic division produces haploid gametes. 8. Eight α and eight β gametes. This is the prezygotic phase of the life cycle. Black dots are expelled polar bodies. 9. Eight sporoblasts, which are produced from the eight diploid zygotes. Fusion of the gametes restored the diploid state. Inside each of the sporoblasts are future sporoplasms. 10. Young sporoblast isolated: the three upper cells divide to give rise to capsulogenic and valvogenic cells, and two endospore cells encircle the future sporoplasm. 11. Pansporoblast with eight primordial spore envelopes, including polar capsules and eight primordial sporoplasms, both connected by thin bridges. 12. Isolated sporoblast; the sporoplasmic mass penetrates the spore membrane. 13. Pansporoblast with eight mature spores. 14. Spore released in water with inflated round projections of the shell valves (not drawn to scale). 15. Mature spore after release.

Pr-25 Rhodophyta

(Red algae)

Greek *rhodos*, red; *phyton*, plant

EXAMPLES OF GENERA

Agardhiella
Amphiroa
Bangia
Batrachospermum
Callophyllis
Chantransia
Chondrus
Corallina
Dasya
Erythrocladia
Erythrotrichia
Gelidium
Goniotrichum
Gracilaria
Hildebrandia
Lemanea
Lithothamnion
Nemalion
Polysiphonia
Porphyra
Porphyridium
Rhodymenia

The sexual organisms of this phylum and of the Gamophyta (Phylum Pr-26) probably evolved their peculiar conjugating mating systems independently of each other and independently of the fungi, most of which show similar behavior. Whereas the red seaweeds in Phylum Rhodophyta are a huge and important group that deserves (and has received) many books of its own, the gamophytes—conjugating green algae, the desmids, and their kin—form a small group that has been removed from the great diverse phylum of other green algae (chlorophytes) because of a peculiar sexual system. Unlike sexual processes in other green algae, no undulipodiated gametes (motile sperm) are ever formed by any members of the phyla Rhodophyta and Gamophyta.

The red algae (rhodophytes), along with the phaeophytes (Phylum Pr-17), are the largest and most complex of the protoctists (Figure A). The largest rhodophytes are somewhat smaller and less complex than the largest phaeophytes, and some 50 genera, comprising about 100 different species of rhodophytes, grow symbiotrophically only on other red algae. Rhodophytes commonly inhabit the edges of the sea and are cosmopolitan in distribution. In the tropics, particularly, they abound on beaches and rocky shores. About 675 genera with 4100 species are known, the vast majority of which are marine. There are two subclasses, the Florideae and the Bangiales. (Only one class, Rhodophyceae, is recognized, and all species are placed in it.) Although rhodophytes are primarily marine, some taxa are restricted to freshwater habitats and others live on land. Marine taxa live in littoral and benthic (seafloor) habitats where suitable substrata such as rocks and jetty pilings exist for attachment.

The red algae form a natural group—all species display several traits that characterize the phylum. They range from microscopic unicells and filaments (of single cells in rows or of multiple aligned rows of cells) to large (as much as 1 m), cell-packed, branched or unbranched, cylindrical, leaflike thalli (Figure B), including crustose (flat) and erect forms, some of which are calcified. Rhodophytes are distinguished by reddish plastids, rhodoplasts, with accessory, water-soluble pigments—allophycocyanin, phycocyanin,

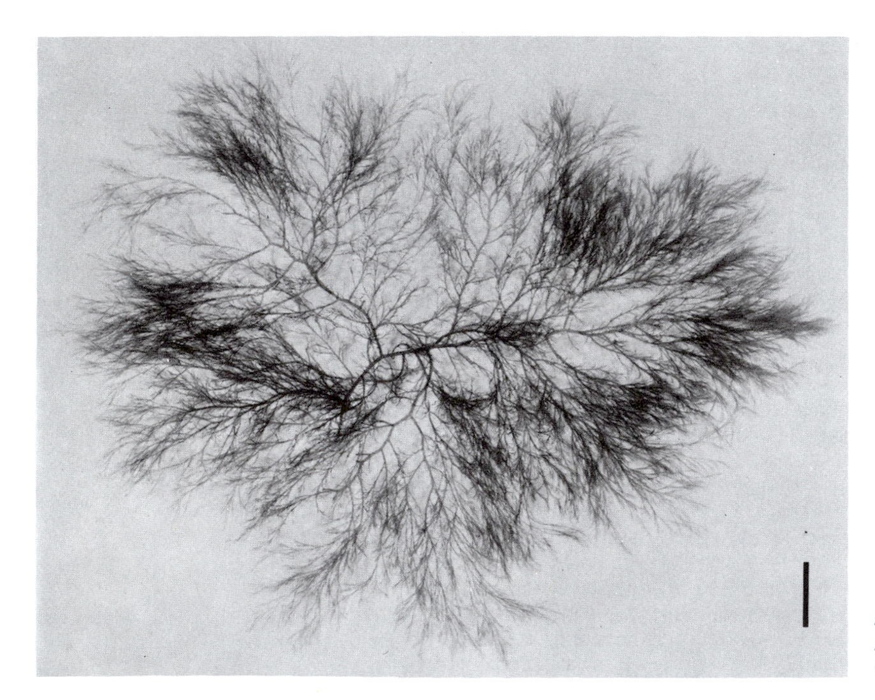

A *Polysiphonia harveyi* from rocky shore, Atlantic Ocean. Bar = 1 cm. [Courtesy of G. Hansen.]

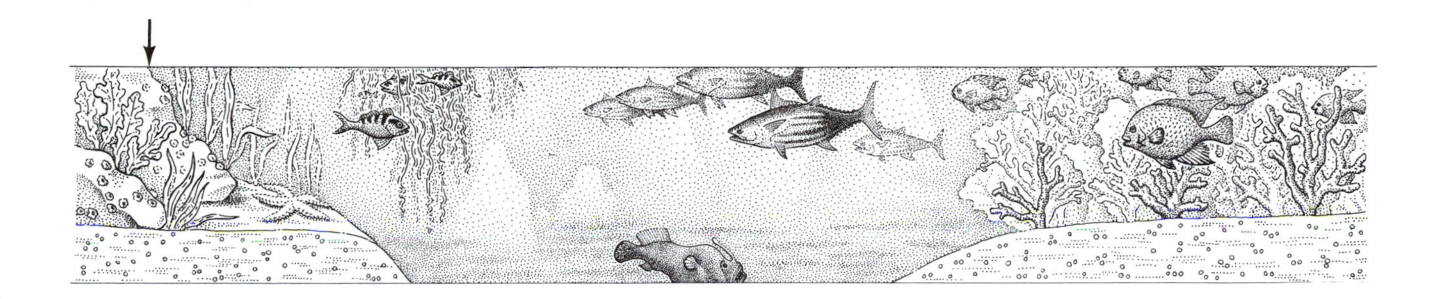

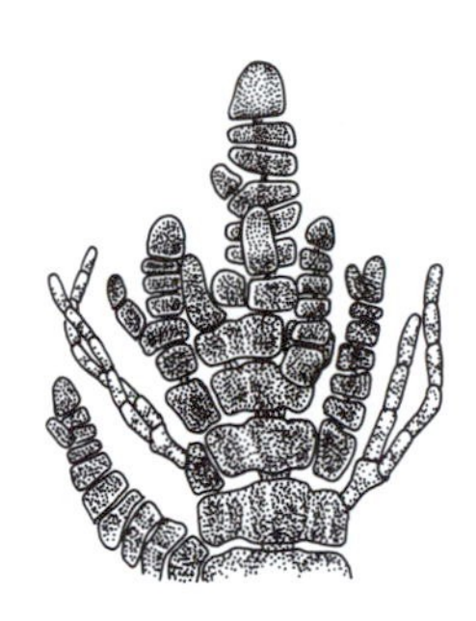

B Apex of male thallus. [Drawing by R. Golder.]

and phycoerythrin—localized in structures termed phycobilisomes found on the outer faces of the plastid photosynthetic lamellae (thylakoids). Other rhodoplast pigments include chlorophyll *a*, alpha- and beta-carotene, lutein, and zeaxanthin. Thylakoids are present as single lamellae (that is, not stacked) in the rhodoplasts. Food reserves are stored as floridean starch (α-1–4-linked glucan) in granules outside the plastid.

Characteristic of some red algae are "pit connections" between cells (a misnomer, because they do not connect cells, but rather are proteinaceous plugs deposited in the pores that result from incomplete centripetal wall formation—Figure C); mitochondria associated with the forming faces of the membranous Golgi bodies; plastids surrounded by one or more encircling thylakoids; and a life history consisting of an alternation of two free-living and independent generations called gametophyte (Figure D) and tetrasporophyte, respectively. A third generation, the carposporophyte, is present on the female gametophyte.

Even though none has undulipodia at any stage in the life history, all reproduce sexually. Reproduction is oogamous: a large egg cell is formed in a special female organ, the oogonium, which bears a long neck that is receptive to the male gamete. The male organ, the antheridium, produces a single male "sperm," which lacks an undulipodium and is incapable of locomotion. After male gametes are released near the female, at least one attaches to the neck of the female structure and moves down the neck to fertilize the egg. The physiology of this process is poorly known.

After fertilization, meiosis may take place immediately to form haploid spores. Alternatively, meiosis does not take place and diploid spores, called carpospores, are formed. They are often formed in bunches of threads that grow out of the oogonium. In some of the more elaborate life cycles, the carpospores are formed in a special organ (cystocarp) that establishes a connection with the oogonium. The carpospores of some species develop into complex little "plants," called carposporophytes (Figure E), that bear organs called tetrasporangia. Meiosis takes place in the tetrasporangia; the four meiotic products, the tetraspores, are released into the sea. They germinate into haploid thalli that eventually produce oogonia or antheridia. Predominantly haploid or alternating haploid and diploid life cycles are common in the red algae; the details of most of the cycles have yet to be worked out.

When we note that many red algae calcify, we mean that they become encrusted with calcium carbonate. *Lithothamnion* looks like reddish circular crust on rocks, and *Corallina* looks like an encrusted tree. A single genus may have both calcifying and noncalcifying species. The propensity for calcification has produced a good fossil record for the phylum; mineralized forms of coralline algae first appear in the Lower Paleozoic era. The first red algae in the fossil record are filaments dating to the late Proterozoic eon.

Rhodophytes show a marked parallelism of forms with other groups of algae—the chrysomonads (Phylum Pr-13), the chlorophytes (Phylum Pr-28), and the phaeophytes (Phylum Pr-17). As in these other phyla, there are heterotrichous filaments (*Chantransia*), prostrate discs (*Erythrocladia*), cushions (*Hildebrandia*), elaborate erect structures (*Batrachospermum*), compact tissuelike types (*Lemanea*), and delicate, many-branched forms (*Polysiphonia* and *Porphyra*).

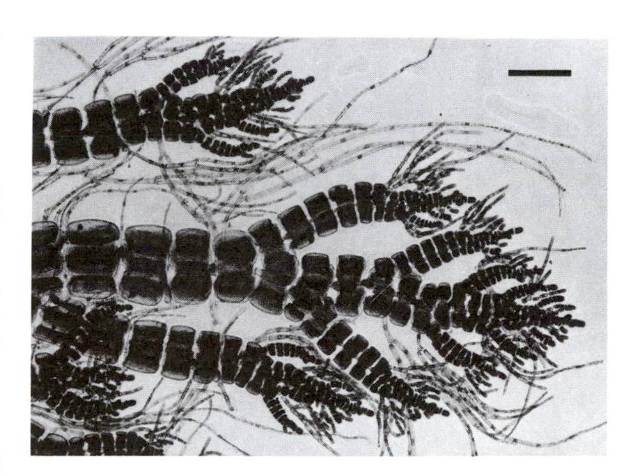

C Apex of thallus, showing cells and pit connections. LM, bar = 0.1 μm. [Courtesy of G. Hansen.]

Agar, the substance used to firm the broth on which colonies of microorganisms are grown so that they can be isolated and studied, is extracted from red algae. Agarose, so important to the gels of molecular biology and biochemistry, is extracted from *Gracilaria,* a member of the Gigantinales, one of the 11 orders in the Florideae subclass. Polysaccharides from these seaweeds are also used in the manufacture of ice cream and other food products, toothpaste, cosmetics, and pharmaceuticals. The leafy dulce of New England's rocky shore is dried and eaten whole.

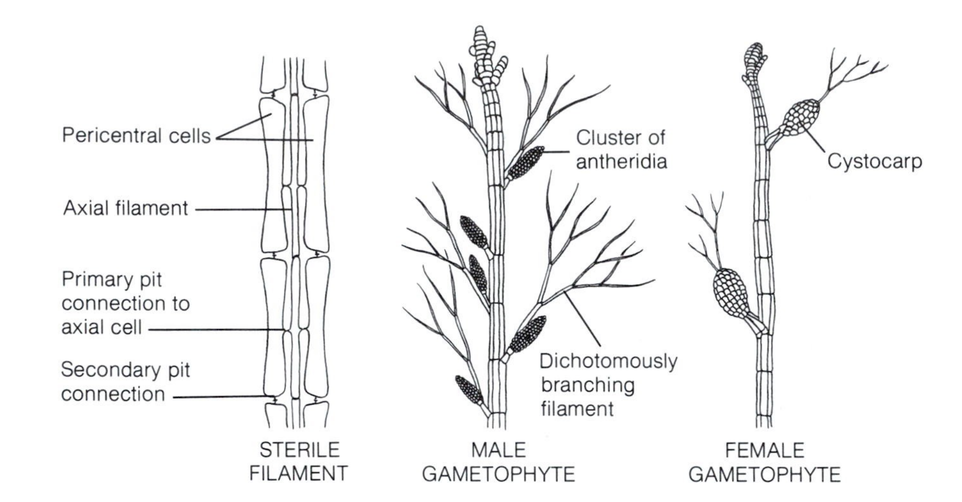

D Sterile and sexually mature apices of thalli. [Drawings by R. Golder.]

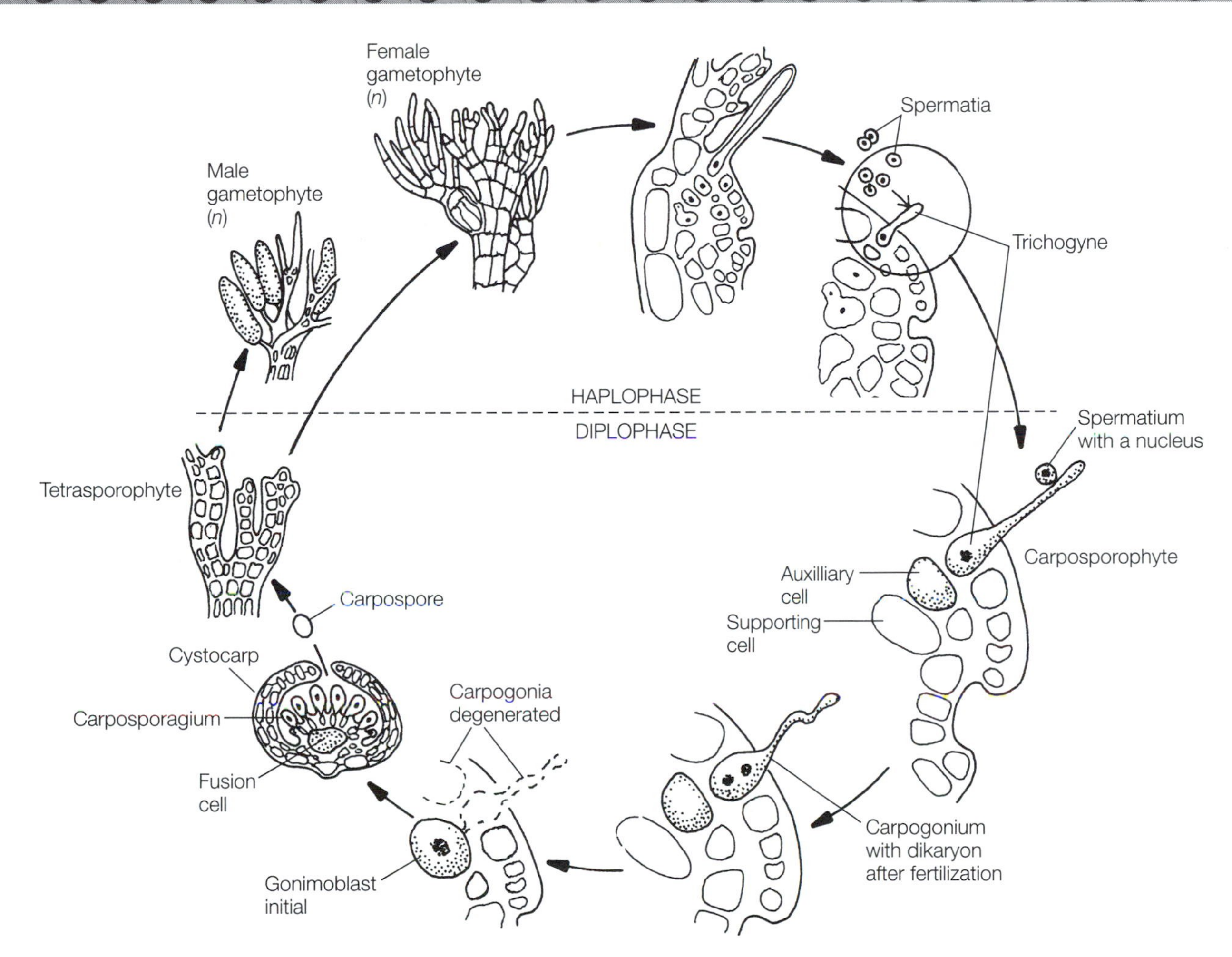

E *Polysiphonia:* fertilization of carpogonia on the female gametophyte. [Drawing by K. Delisle.]

Pr-26 Gamophyta
(Conjugaphyta, conjugating green algae)

Greek *gamos,* marriage; *phyton,* plant

EXAMPLES OF GENERA

Bambusina	*Gonatozygon*	*Penium*
Closterium	*Hyalotheca*	*Spirogyra*
Cosmarium	*Mesotaenium*	*Staurastrum*
Cylindrocystis	*Micrasterias*	*Temnogyra*
Desmidium	*Mougeotia*	*Zygnema*
Genicularia	*Netrium*	*Zygogonium*

Gamophytes are green algae that lack undulipodia at all stages of their life history. Without motile sperm, any other sperm, or other means of locomotion, they engage regularly in sexual processes. They have symmetrical cells containing complex chloroplasts, which are usually aligned down the long axis of the cell. One large and conspicuous nucleus tends to be found in each cell. These conjugating green algae are found in ponds, lakes, and streams; no truly marine forms have been reported. To reproduce, the haploid growing cells either divide directly (asexually) or produce equal ameboid gamete cells that fuse to form a zygote. This usually develops into a resistant and conspicuous structure called a zygospore. Zygotic meiosis takes place within the zygospore, and haploid algal cells eventually emerge.

In their pigmentation, gamophytes are similar to all the other green algae: they have chlorophylls *a* and *b,* and most are grass green in color. They are often classified with the chlorophytes (Phylum Pr-28).

Two classes are in the phylum Gamophyta as presented here, all oxygenic photoautotrophs: Euconjugatae (true conjugating algae) and Desmidioideae (the desmids).

The euconjugates, which are generally filamentous forms, consist of one order (Zygnematales). In this order, three families are recognized: Mesotaeniaceae (for example, *Cylindrocystis, Mesotaenium,* and *Netrium*), Desmidaceae, and Zygnemataceae. The last two orders, the desmids and zygnemids, are the best known. Most zygnemids, such as *Zygnema, Spirogyra,* and *Mougeotia,* form pond scums—stringy masses of long unbranched filaments. In *Zygnema* and *Spirogyra,* the chloroplast is helically wound along the length of the long cylindrical cell; in *Mougeotia,* a single large, flat, plate-shaped chloroplast extends the full length of the cell, as illustrated here. These organisms grow rapidly by mitosis, the filaments breaking off fragments that start new filaments, thus forming a bloom or scum in a few days.

During sexual union, two filaments, which are haploid, come to lie side by side. Protuberances grow and join to form conjugation tubes that link cells in opposite filaments. Through the conjugation tubes the cells of the "male" filament flow, chloroplasts and all, to fuse with the cells of the "female" filament. Each fusion results eventually in a dark, spiny diploid zygote in a chamber of the female filament. Because fertilization is often simultaneous, rows of such zygotes, seen as black zygospores, are common. After a period of dormancy, the zygotes are released into the water; they undergo meiosis and germinate to produce new haploid filaments.

The largest of these three families are the desmids. In our classification, subfamilies of the desmids, each named for its best-known genus are Penieae (*Penium*), Closterieae (*Closterium*), and Cosmarieae (*Cosmarium*). Several thousand species are known. Most are single cells—more precisely, they are pairs of cells whose cytoplasms are joined at an isthmus (Greek *desmos,* bond). The isthmus is the location of the single shared nucleus. Some desmids are colonial. In many desmids, the chloroplasts are lobed or have plates or processes that extend from the center toward the periphery of the cell. The outer layers of the cell wall form a shell, typically decorated with spines, knobs, granules, or other protrusions arranged in lovely designs. These outer layers are composed of cellulose and pectic substances and, in many cells, are impregnated with iron or silica; the inner layer, on the other hand, is composed of cellulose only and is structureless at the light-microscopic level of magnification.

The outermost layer of the desmid cell is a mucilaginous sheath, sometimes thin and sometimes very thick and well developed. It is secreted through pores in the cell wall. The slow gliding movement of desmids is thought to be due to actin protein secretions in this mucilage.

In the typical desmid, the two "half cells" are mirror images of each other, and each has its own chloroplast. In asexual reproduction, after the nucleus divides, the two partners simply separate. Each one grows a new half cell replacement. During sexual conjugation, the two partners also separate; both leave their shells and fuse outside, either with each other or with a liberated protoplast from another desmid, to form a dark, spiny zygote reminiscent of the zygotes of the Euconjugatae and of the zygomycotes (Phylum F-1). In *Desmidium cylindrium,* only one partner, the "male," leaves its shell; conjugation takes place inside the shell of the "female," and that is where the zygote is lodged.

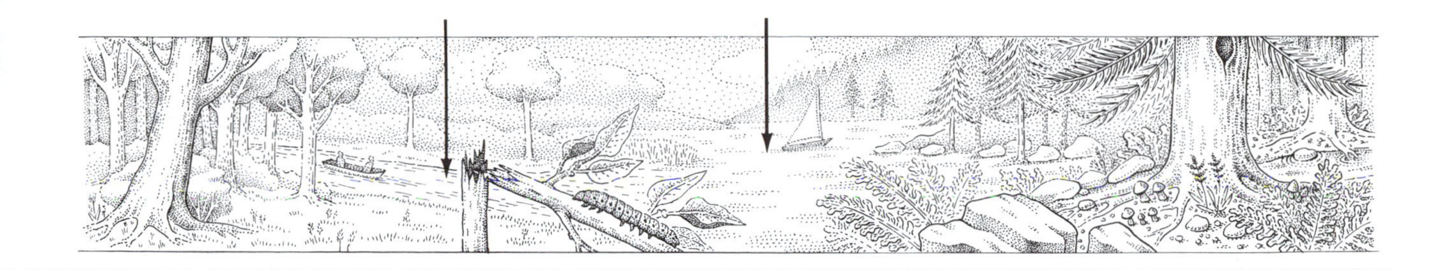

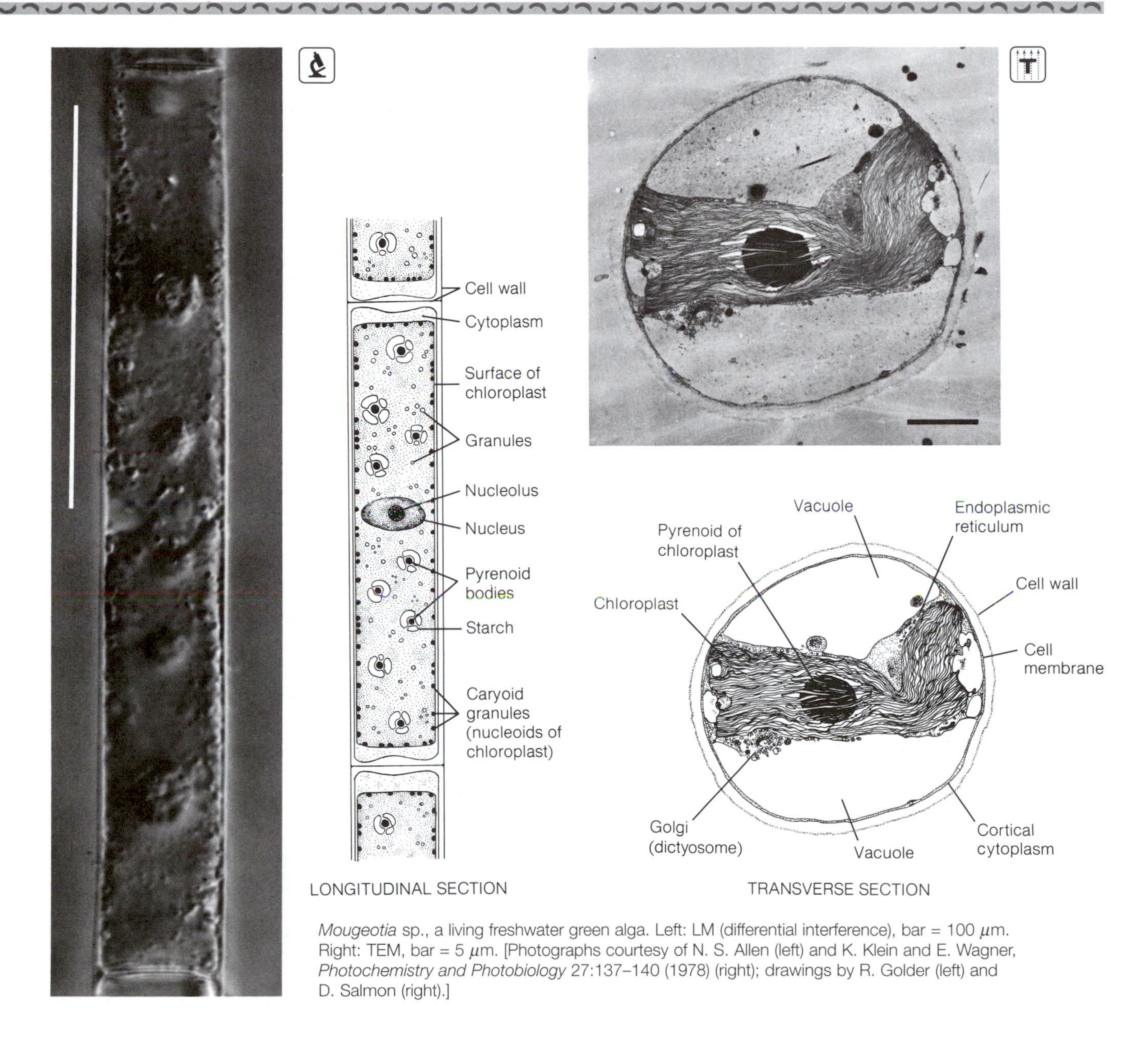

Mougeotia sp., a living freshwater green alga. Left: LM (differential interference), bar = 100 μm. Right: TEM, bar = 5 μm. [Photographs courtesy of N. S. Allen (left) and K. Klein and E. Wagner, *Photochemistry and Photobiology* 27:137–140 (1978) (right); drawings by R. Golder (left) and D. Salmon (right).]

Pr-27 Actinopoda

Greek *actinos,* ray; *pous,* foot

EXAMPLES OF GENERA

Acantharia
Acanthocystis
Acanthometra
Actinophrys
Actinosphaera
Actinosphaerium (Echinosphaerium)
Aulacantha
Challengeron
Ciliophrys
Clathrulina
Collozoum
Heterophrys
Pipetta
Sticholonche
Thalassicola
Zygacanthidium

The marine protists that Ernst Haeckel traditionally called "radiolarians" and other superficially similar plankton, large protists with some radial symmetry, are grouped as classes in the phylum Actinopoda for convenience and pedagogy. That actinopods represent convergently evolved lineages more related to certain zoomastigotes (Phylum Pr-30) than they are to each other is likely but, in the absence of comprehensive information, we retain the traditional actinopod grouping, with its four classes. The first class is Heliozoa, freshwater sun animalcules. The second is the mostly deep dwelling Phaeodaria, and the third is the more open ocean Polycystina, which two classes together constitute the traditional Radiolaria. The fourth class is Acantharia (sometimes also grouped in Radiolaria), with their strontium sulfate skeletons.

Actinopods, heterotrophic protoctists, are distinguished by their long slender, cytoplasmic axopods, also called axopodia (Figure A). These fine projections are stiffened by a bundle of microtubules running down the axis of the structure called an axoneme (Figures B and C). Each axoneme has an often quite elaborate arrangement of microtubules characteristic of that actinopod group, and the microtubules are often cross linked. Electron-microscopic studies have shown that the classes Acantharia, Polycystina, and Phaeodaria, all considered marine "radiolarians," are products of evolutionary convergence and are only remotely related to one another.

Heliozoans are primarily freshwater plankton, although estuarine, marine, and benthic (seafloor dwelling) species are known. Thirty-four genera and nearly 100 species are known. Many use their axopods to catch prey. The axopods radiate out into the water, surrounded along their length by plasma membrane. In some heliozoans, the axonemes grow out directly from the endoplasm; in others, each axoneme grows out from its own structure, the axoplast, located next to the nucleus. In a group called the centrohelidians, all the axonemes arise from a single axoplast, called a centroplast, whose center often contains a clearly defined organelle (see Figure B).

The rowing actinopod illustrated in Figure D, *Sticholonche zanclea* Hertwig, has been an enigma for taxonomists. Its peculiar skeleton, the placement of its axopods on the nuclear membrane, and the hexagonal pattern of the axopods in cross section have justified its placement as the only species in the isolated order Sticholonchidea (Figures E and F). That order was originally thought to be radiolarian (as suggested by A. Hollande, M. Cachon, and J. Valentin in 1967), but it is more likely a marine heliozoan (as suggested by Cachon in 1971). *Sticholonche* is found rowing in the Mediterranean with the splendor of a Roman galley: it has microtubular oars and sets of moveable microfibrillar "oarlocks" (Figures G and H). Unfortunately, it does not grow in the laboratory.

Many heliozoans have siliceous or organic surface scales or spines. In a few species, a spherical organic or siliceous cage encloses the entire cell. The cage has bars arranged in a repeating hexagonal pattern through which the axopods penetrate.

Except in the order Desmothoraca (for example, *Clathrulina elegans*), reproduction in heliozoans by zoospores or swarmer cells is unknown. Sexual reproduction has been rarely seen; cells reproduce by binary or multiple fission or budding. In some multinucleate species, the nuclear and cytoplasmic divisions are not synchronized. In uninucleate forms, the axopods retract so that the organism does not move or feed during cell division. Retraction is caused by disassembly of the microtubules in the axonemes.

A kind of autogamy (self-fertilization) has been reported in some heliozoans. A mature cell forms one or more cysts inside the cell. Meiosis apparently takes place in the cysts, and certain nuclei degenerate. Two of the final meiotic products in each cyst then fuse—their haploid nuclei form a new single diploid nucleus. The only surviving product of the two meiotic divisions and fusion emerges from the cyst as a mature heliozoan. Whether this inbred sort of reproduction is common is not known, because of the paucity of study. Heterogamy (fusion of nuclei from different individuals) may occur. In *Actinophrys,* two cells (but not their nuclei)

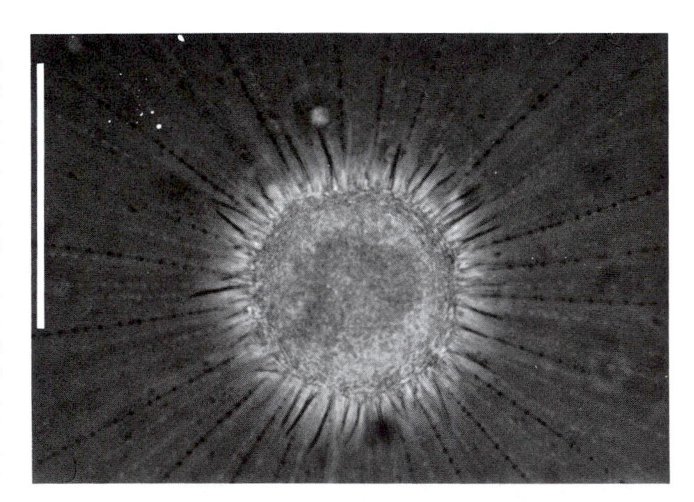

A *Acanthocystis aculeata,* a freshwater centrohelidian heliozoan, showing axopods and spines. LM, bar = 50 μm. [Courtesy of C. Bardele.]

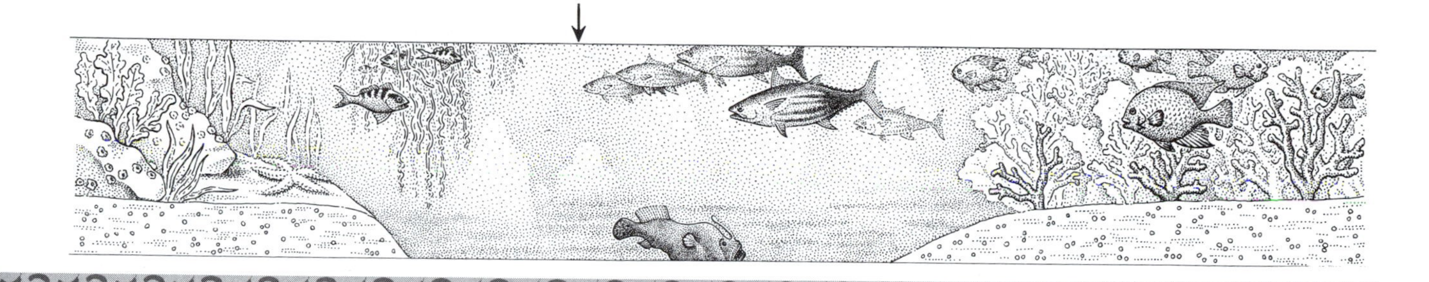

may fuse just before they undergo autogamy. Gametes originating from one of the two cells have been seen to fuse with gametes originating from the other. Cell fusion is common in heliozoans, but whether it constitutes meiosis and fertilization is not known.

Both polycystines and phaeodarians often have strikingly beautiful opaline skeletons made of hydrated amorphous silica; they are extremely common in tropical waters. Of the more than 4000 actinopods described in the literature, some 500 are estimated to be polycystines. Along with diatoms, silicoflagellates, and sponges, they are responsible for the depletion of dissolved silica in surface waters.

Polycystines and phaeodarians differ in many ways. The polycystine skeleton is made of opal (hydrated amorphous silica); the phaeodarian skeleton is made of silica plus often a large quantity of organic substances of unknown nature. The polycystine skeletal elements look solid under the light microscope; however, electron microscopy reveals tiny canals and pores in their skeletons. The skeletal elements of phaeodarians look hollow even under the light microscope: their spines are tubular and the continuous shells of many species have a bubbly "styrofoam" ultrastructure barely visible by light microscopy and conspicuous by electron microscopy.

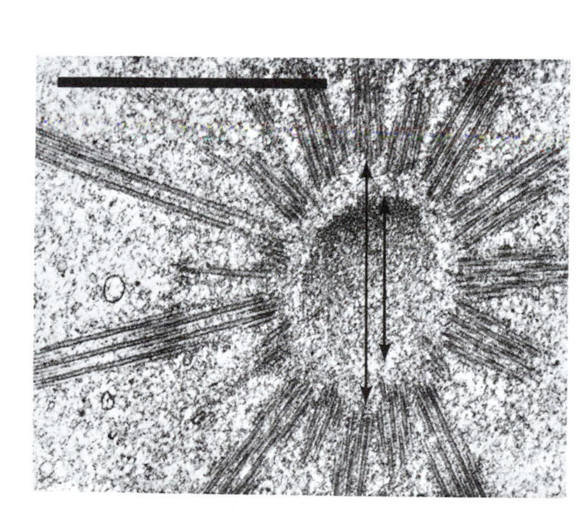

B The cell center of *Acanthocystis penardi,* showing the microtubules of the axonemes and the central organelle (short arrow) of the centroplast (longer arrow). TEM, bar = 5 μm. [Courtesy of C. Bardele.]

Endoplasm
Ectoplasm
Axoplast
Zooxanthellae (photosynthetic symbionts)
Central capsule "membrane" (microfilaments)
Filopodia
Fusule
Cell membrane
N
Axopod
Axoneme (microtubules)
Extracellular space
Skeletal spine
Cortex

C A generalized polycystine actinopod in cross section. Thick black lines, skeleton; stippled areas, cytoplasm; N, nucleus. [Drawing by L. Meszoly; information from Georges Merinfeld.]

Pr-27 Actinopoda
(continued)

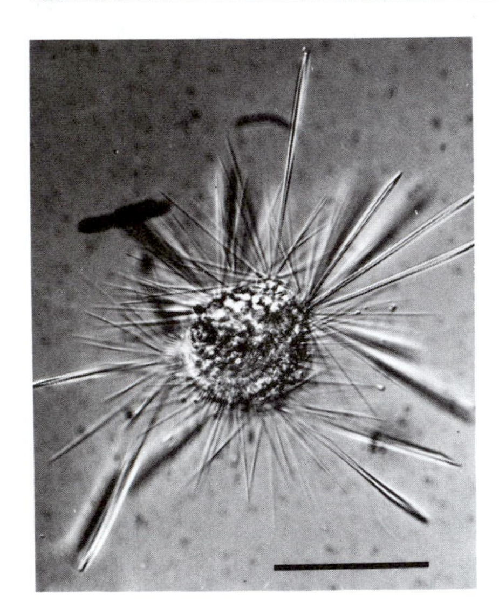

D A living *Sticholonche zanclea* Hertwig, taken from the Mediterranean off Ville Franche sur Mer Marine Station. LM, bar = 100 μm. [Courtesy of M. Cachon.]

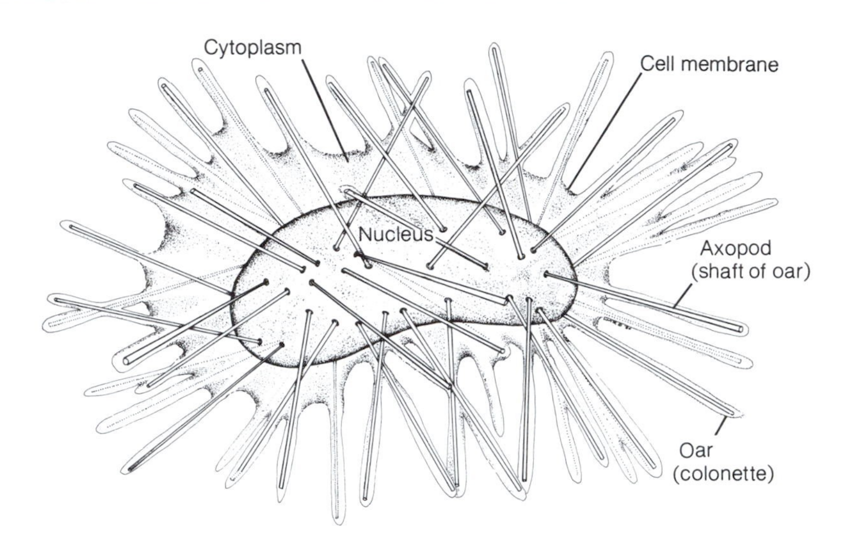

E *Sticholonche zanclea,* showing the placement of the microtubular oars on the nuclear membrane. [Based on electron micrographs of A. Hollande, M. Cachon, and J. Valentin, *Protistologica* 3:155–170 (1967); drawing by L. M. Reeves.]

Crystals, but not skeletal components, of strontium sulfate ($SrSO_4$) are secreted by some adult polycystines in their endoplasm and perhaps by all of them in their undulipodiated swarmers, whereas $SrSO_4$ is unknown in phaeodarians.

The capsule enclosing the central mass of cytoplasm in both polycystines and phaeodarians is not a flimsy microfibrillar open mesh net (as in acantharians) but is made of massive organic material. The polycystine capsule, probably composed of mucoproteins or mucopolysaccharides, is made of numerous juxtaposed plates, like the pieces of a jigsaw puzzle separated by narrow slits, whereas the phaeodarian capsule is a single continuous structure. The polycystine capsule grows in diameter during the life of the organism; the phaeodarian capsule cannot increase in diameter after it has formed—it can only thicken its wall.

The axonemes of the polycystine axopods studied so far are made of parallel microtubules aligned in geometrical arrays, with bridges between microtubules. Most species have many such axopods per cell. Polycystines usually have one axoplast from which all axonemes originate, but some groups have other arrangements—for example, individual axoplasts, one per axoneme, are located near the nucleus. In phaeodarians, only two axonemes penetrate the capsule. They originate from separate axoplasts just inside the capsule. The microtubules in the basal part of these axonemes are not linked by bridges. Light microscopy reveals a cortex of many thin peripheral pseudopods, which are perhaps branches of the two axopods. No polycystine axoneme is known to branch.

Polycystine orifices called fusules are complex mufflike structures each filled with a dense plug that permits the passage of the axonemal microtubules, if they originate inside the endoplasm, but that hampers the circulation of cytoplasm between the endoplasm and the extracapsular pseudopodial network. The phaeodarian capsule normally has only three orifices of two kinds: a wide, complex astropyle, which is an opening that ensures exchange between the endoplasm and whatever cell parts lie outside the capsule; and two,

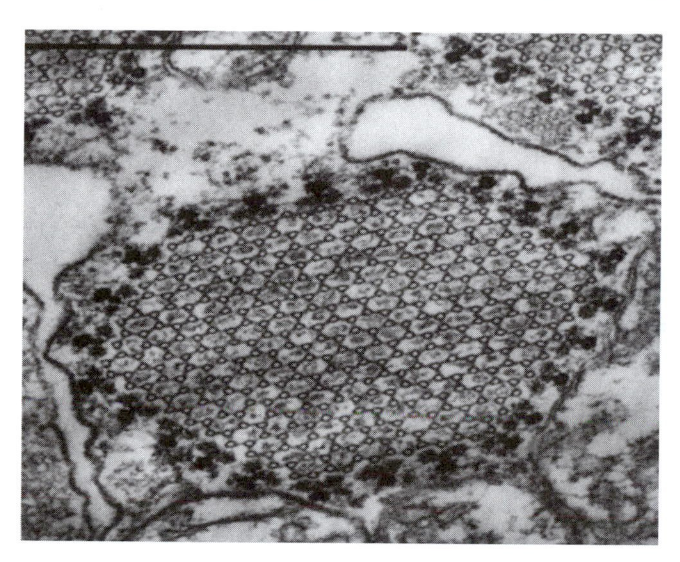

F Transverse section through an oar of *Sticholonche zanclea,* showing the hexagonal pattern of the axopods in cross section. SEM, bar = 1 μm. [Courtesy of J. Cachon and M. Cachon.]

rarely more, parapyles. These openings, simpler than polycystine fusules, allow the passage of the two thick cell axonemes. At each parapyle there is a cup-shaped axoplast from which an axoneme originates. Outside the capsule in front of the astropyle of many phaeodarians is a mass of predigested food called the phaeodium. The polycystines lack the phaeodium.

In the phaeodarian endoplasm are numerous strange tubes, called rodlets, about 200 nm wide, having a complex repeating ultrastructure. Their role is unknown (perhaps they take part in the secretion of the capsule). No such rodlets are known in the polycystines.

Polycystines supplement heterotrophy by photoautotrophy in symbiotic yellow or green algae (zooxanthellae or zoochlorellae; Phyla Pr-14 and Pr-28); phaeodarians lack algal symbionts.

Most polycystines and all phaeodarians have only one nucleus, large and polyploid. Only the phaeodarian nucleus undergoes an extraordinary equational division, superficially resembling classical mitosis, in which two monstrous "equatorial plates" are formed, each with more than 1000 chromosomes.

Class Polycystina is divided into the orders Spumellaria and Nassellaria. The spumellarian has fusules scattered all over its central capsule membrane; thus, its axopods radiate in all directions. The protist is usually spherical, ellipsoidal, or flattened, and so, naturally, is its skeleton. Some spumellarians form large colonies in which hundreds of individual organisms are embedded in a common mass of jelly. The fusules of nassellarians, which never form colonies, are clustered at one pole of the capsule membrane; their axopods are grouped in a conical bunch that leaves the cell at that pole.

Acantharians, generally spherical organisms, have a unique radially symmetrical skeleton composed of rods of crystalline strontium sulfate ($SrSO_4$). The skeleton usually has 10 diametrical (20 radial) spines, called spicules, inserted according to a precise rule, known as Moller's law, which was discovered by Johannes Moller in the nineteenth century. The acantharian cell is a globe from whose center the spicules radiate and pierce the surface at fixed "latitudes" and "longitudes." If there are 20 spicules, then there are five quartets—one "equatorial," two "polar," and two "tropical"—that pierce the globe at the latitudes 0°, 30° N, 30° S, 60° N, and 60° S. For the equatorial and both polar quartets, the longitudes of the piercing points are 0°, 90° W, 90° E, and 180°; for the tropical quartets, 45° W, 45° E, 135° W, and 135° E. Even in acantharians that do not have the general shape of a globe, these orientations are strictly observed, although some spicules are thicker and longer than the others. Some species have many more than 20 spicules, as many as several hundred, but they are always grouped by some elaboration of Moller's law.

Acantharian cells are made of distinct layers. The innermost layer, coarsely granulated with many small nuclei, is the cell's central mass. Immediately surrounding the central mass is a perforated, flimsy network of microfilaments called the central capsule membrane. Through the central capsule membrane, the central mass extends several kinds of cytoplasmic outgrowths: cytoplasmic sheaths surrounding the skeletal spines; reticulopods, which are cross-connected netlike pseudopods lacking axonemes; filopods, which are thin pseudopods stiffened by one or very few microtubules, and a number of axopods (usually 54, but in some acantharians there may be several hundred) arising from axoplasts between the spines.

At the periphery is the cortex, a thin, flexible layer of microfilaments, which may be arranged in intricate designs. The cortex is underlaid by a network of reticulopods. Where the strontium

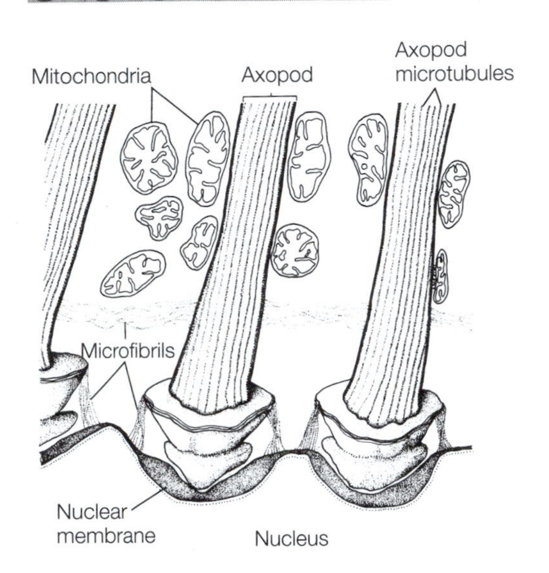

G The axopods of the oars (colonettes) of *Sticholonche,* showing their relation to the nucleus (central capsule) and the mitochondria. [Drawing by L. M. Reeves.]

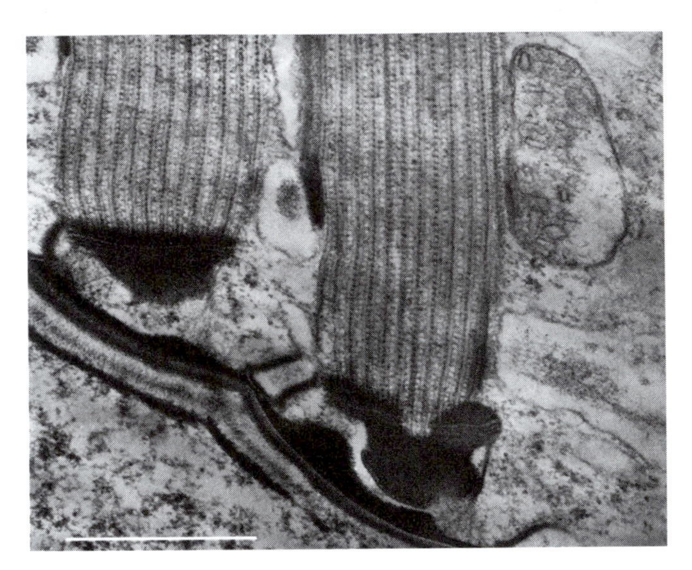

H Longitudinal section through two oarlocks and the attached oars of *Sticholonche zanclea.* SEM, bar = 1 μm. [Courtesy of J. Cachon and M. Cachon.]

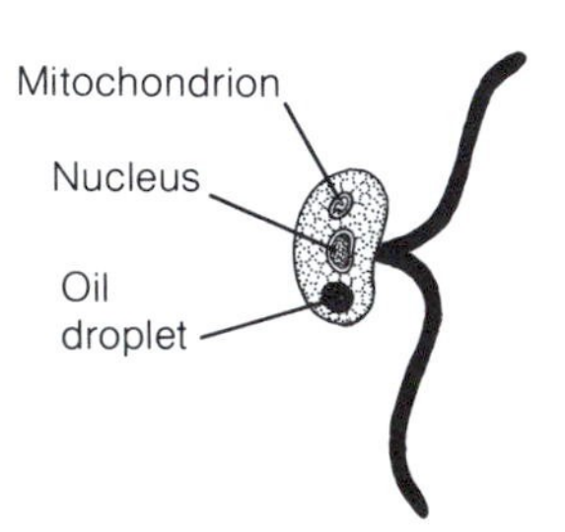

I Generalized swarmer cell, as can be found in some acantharian actinopods. [Drawing by L. Meszoly.]

sulfate skeletal spines pass through, the cortex is pushed out, like a tent stretched out over tent poles. At these points are filaments, the myonemes, that apparently control the tension of the cortex and bind it to the skeletal rods.

The delicate axopods increase the amount of cell surface exposed to the sea. They retard sinking and perhaps allow efficient scavenging of nutrients from the water. Prey, generally protoctists and small animals, adhere to the axopods. Cytoplasm from the axopods then engulfs the prey and cytoplasmic flow transports it down the axopods toward inner parts of the cell, where it is digested.

Acantharians produce many small swarmer cells, each containing a drop of oil reserve and a crystal and bearing two [9(2)+2] undulipodia (Figure I). The undulipodia originate from kinetosomes in the anterior part of the swarmer cell. Some acantharians round up to form cysts in which they undergo mitotic divisions. Swarmers develop and are later released from these cysts. Little about the development process is known because swarmers have been devilishly difficult to culture in the laboratory. Meiosis has not been seen.

Most acantharians are effectively photoplankton because they harbor many haptomonads (Phylum Pr-10) that live and grow in them. The haptomonads are grass green in color, and photosynthetic. The symbiotrophy permits the acantharians to obtain their energy and food by photosynthesis in the nutrient-poor open ocean. The acantharian wastes provide nitrogen and phosphorus for their haptomonad symbionts.

Pr-28 Chlorophyta

(Green algae)

Greek *chloros,* yellow green;
phyton, plant

EXAMPLES OF GENERA

Acetabularia
Bryopsis
Caulerpa
Chaetomorpha
Chara
Chlamydomonas
Chlorella
Chlorococcum
Cladophora
Cylindrocapsa
Dunaliella
Enteromorpha
Fritschiella
Gonium
Halimeda
Klebsormidium
Lamprothamnium
Microspora
Nitella
Oedogonium
Penicillus
Platymonas
Pseudobryopsis
Pyramimonas
Spongomorpha
Stigeoclonium
Tetraspora
Tolypella
Trebouxia
Ulva
Urospora
Volvox

We group together the three phyla likely to have hiding in their midst the ancestors of the other three kingdoms of eukaryotes: Phylum Chlorophyta, mastigote green algae and their relatives, from which green plants may have arisen; Chytridiomycota (Phylum Pr-29), which may have given rise, by loss of undulipodia, to the ancestors of the fungal lineage; and Zoomastigota (Phylum Pr-30), a miscellany of mastigote microbes, one lineage of which is likely to be ancestral to animals.

Chlorophytes are algae that have grass-green chloroplasts surrounded by two membranes and that form zoospores or gametes having undulipodia, usually at least two, of equal length. About 500 genera with as many as 16,000 species have been described. Within the phylum, several evolutionary lines have led from unicellular forms to multicellular organisms. Their chloroplasts contain chlorophylls *a* and *b* as well as the carotenoid derivatives astaxanthin, canthoxanthin, flavoxanthin, loraxanthin, neoxanthin, violaxanthin (which tends to convert into zeaxanthin in dark-grown algae placed in the light), and the xanthophyll echinenone. Starch, the a-1–4-linked glucose polymer, is the carbohydrate reserve stored by most green algae. Botanists agree that somewhere in this extremely diverse group lie the ancestors of the plants.

Phylum Chlorophyta here excludes the gamophytes (Phylum Pr-26), which lack undulipodia, but unites the siphonales, charales, and prasinophytes with the chlorophytes in the strict sense because all are green algae that have undulipodia at some stage in their life history. Although this is a somewhat arbitrary plan, it emphasizes the tendency of the unicellular, biundulipodiated algae to give rise to impressive and cohesive classes of reproductively and morphologically complex water "plants." Some of them are at least periodically resistant to desiccation; that one or several such algae were the progenitors of the land plants seems incontrovertible.

Chlorophytes are a major component of the freshwater photoplankton; it has been estimated that they fix more than a billion tons of carbon in the oceans and freshwater ponds every year.

The cell walls of green algae, like those of land plants, are composed of cellulose and pectins or of polymers of xylose (*Bryopsis* and *Caulerpa*) or mannose (*Acetabularia,* Figure A) linked to protein. The walls in many genera are encrusted with calcium carbonate, silica, and, less frequently, other minerals such as iron oxides.

Sexuality is rampant in this group: there is a trend from isogamy, in which two motile gametes of like size and shape conjugate and fuse, toward oogamy, in which a large immotile egg is fertilized by a small motile sperm. The sperm is very much like the individual adults (*Chlamydomonas,* Figure B, and *Dunaliella*), zoospores, or isogametes of many species in the phylum. In *Acetabularia,* a diploid zygote is the product of fertilization. It immediately undergoes meiosis to regenerate the haploid stage in the life cycle.

Within the Chlorophyta as presented here are four major classes and other groups of uncertain status. They are Class Chlorophyceae, with 11 orders, including Volvocales, Oedogoniales, and Chaetophorales; Class Ulvophyceae, which includes, among others, orders Ulotrichales, Siphonocladales, Ulvales, and Caulerpales; Class Charophyceae; and Class Prasinophyceae. Within each of these classes, except the prasinophytes, which are unicells, trends from unicellular forms to various types of complex colonies can be seen.

Class Chlorophyceae is very diverse and probably polyphyletic. It includes the ubiquitous tree-scum alga *Chlorococcum* and both symbiotic and free-living *Chlorella* species. Chlorellas grow like weeds in the laboratory; they are frequently used to study the physiology of photosynthesis. The "water nets," the Hydrodictyaceae, are another family in this class.

The volvocales include *Chlamydomonas.* Probably more is known about the genetic control of mating, undulipodia, photosynthesis, and mitochondrial metabolism in *Chlamydomonas* than in any other protoctist.

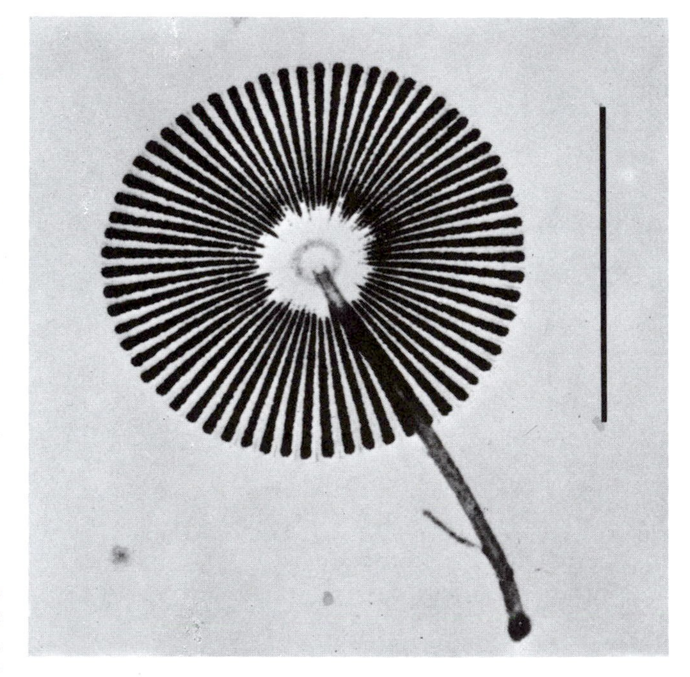

A *Acetabularia mediterranea,* a living alga from the Mediterranean Sea. Bar = 1 cm. [Courtesy of S. Puiseux-Dao.]

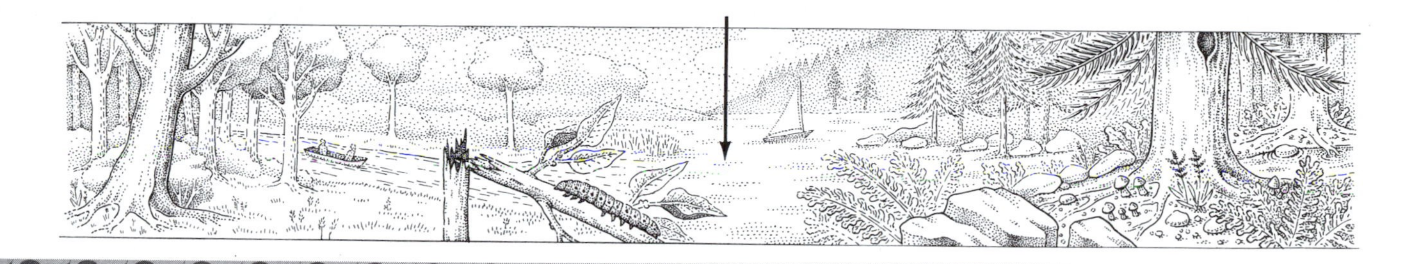

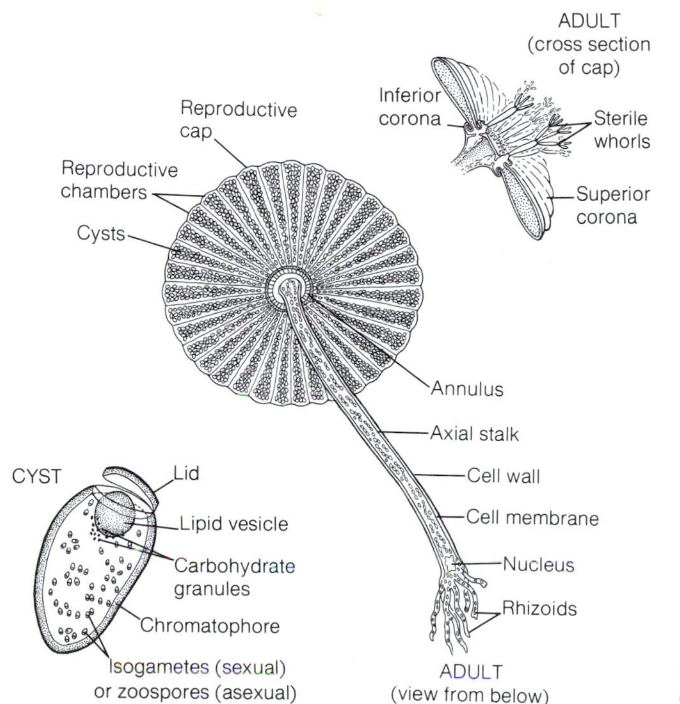

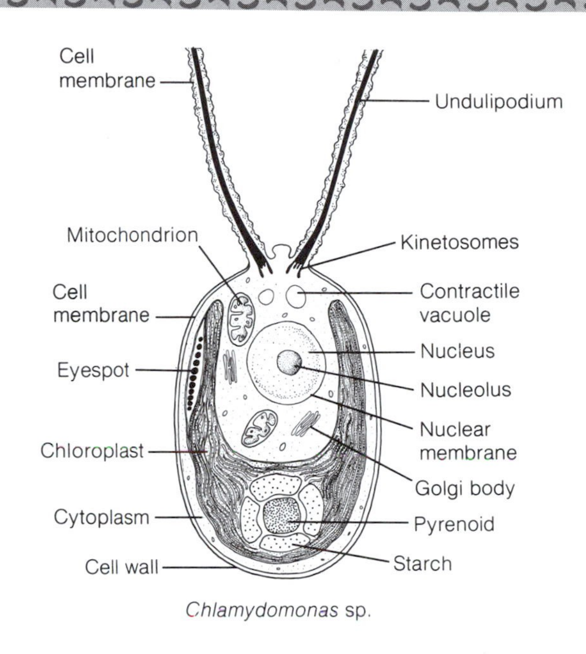

B *Chlamydomonas* is similar in structure to the zoospores of *Acetabularia*. [Drawings by L. Meszoly.]

The oedogoniales produce zoospores having an unusual ring of undulipodia; they have a unique method of cell division and an elaborate style of sexual conjugation. The relationship of *Oedogonium* and other members of the Oedogoniales order to other Chlorophyceae is not well understood.

The chaetophorales (for example, *Stigeoclonium* and *Fritschiella*) are mainly branched multicellular algae; they are differentiated into prostrate and upright thalli.

In the Class Ulvophyceae, the ulotrichales are primarily filamentous (*Microspora* and *Cylindrocapsa*) or thalloid (*Ulva,* called sea lettuce, and the common estuarine form *Enteromorpha*).

Order Siphonocladales includes the family Cladophoraceae, green algae typically composed of multinucleate elongate cells. Algae of the Cladophoraceae may be unbranched (*Urospora* and *Chaetomorpha*) or branched (*Cladophora* and *Spongomorpha*). Many species of common green seaweeds, such as *Codium* and *Acetabularia,* are in this group. Many are quite large, even though all are syncytial: no cell membranes form, and so millions of nuclei and chloroplasts share the same cytoplasm.

Those in Class Charophyceae are multicellular and live in fresh or brackish water. Their cells are long and either uninucleate or multinucleate. *Nitella* and *Chara,* delicate pondweeds, are favorite experimental organisms.

The last class, Prasinophyceae, whose members are unicellular, differs a good deal from all the others. Prasinophytes lack the typical chlorophyte gametes and sexual life cycle. Their cell structure differs from that of the standard chlorophyte. The prasinophyte has an anterior pit or groove, for example, from which emerge four undulipodia, and it bears scales on its cell surface. Only a few genera have been described, including *Platymonas,* a regular tissue symbiont of the green photosynthetic flatworm *Convoluta roscoffensis* (Phylum A-5, Platyhelminthes). Further work may raise these green algal protoctists to phylum status.

Pr-29 Chytridiomycota

Greek *chytra,* little earthen cooking pot; *mykes,* fungus

EXAMPLES OF GENERA
Allomyces
Blastocladiella
Chytria
Cladochytrium
Coelomomyces
Monoblepharella
Monoblepharis
Neocallimastix
Olpidium
Physoderma
Polyphagus
Rhizophydium
Synchytrium

Recent ribosomal RNA gene sequence data have strengthened the hypothesis that the chytrids evolved, by loss of undulipodia, into the ancestors of the fungal lineage. Although the chytrids may be ancestral to the fungi, the details of their cell structure, life history, and other biology make it apparent that chytrids themselves are not fungi. The differences between the chytridiomycotes and the superficially similar protoctist stramenopiles ("straw bearers"), or heterokonts (Phyla Pr-13 through Pr-21), also are clear, definitively establishing that the stramenopiles are not ancestral to fungi. We recognize, too, that it is just as possible that conjugating, plastid-bearing protists (such as the many symbiotrophic white members of the Rhodophyta, Phylum Pr-25) were fungal ancestors. Loss of plastids would have led to organisms greatly resembling fungi such as the ascomycote *Taphrina* (Phylum F-3). A third strong possibility exists: both the fungi and the red algae might have common conjugating heterotrophic ancestors, one lineage of which (the algae group) acquired plastids by ingesting, in a phagotrophic mode, cyanobacteria that were not digested and became plastids. In this scenario, the chytrids are not ancestral to the fungi, and their similarity is only by evolutionary convergence. These evolutionary puzzles ought to be solved by continued study of the biology, including the molecular biology, of these protoctists.

Chytrids are microbes that live in fresh water or soil. Like the hyphochytrids (Phylum Pr-21), the oomycotes (Phylum Pr-20), and the saprobes or symbiotrophic fungi, they grow and feed by extending threadlike hyphae or rhizoids into living hosts, recently dead bodies, or other organic debris. They secrete extracellular digestive enzymes and absorb the resulting nutrients. The simplest chytrids grow and develop entirely within the cells of their hosts. The more complex produce reproductive structures on the host's surface, even though the vegetative and feeding parts of the chytrid body, or thallus, are sunk deep in the host's tissues. Some chytrids are necrotrophs (disease agents) of plants; *Physoderma zea-maydis,* for example, causes the brown-spot disease of corn.

Like members of our Kingdom Fungi and some hyphochytrids (Phylum Pr-21), chytrids synthesize the amino acid lysine by the aminoadipic pathway. Other funguslike protoctists, including other hyphochytrids and the oomycotes (Phylum Pr-20), metabolize by the diaminopimelic acid pathway.

The cell walls of all chytrids are composed of chitin; some contain cellulose as well. The chytrid thallus is coenocytic; that is, its many nuclei are not separated by cell walls. However, a septum, which is a solid plate composed of cell-wall material, separates each reproductive organ from the thallus. In addition, the hyphae of many chytrids have pseudosepta—regular partitions composed of substances different from the composition of the outer chytrid wall. Pseudosepta are rather more pluglike than platelike and, typically, are not complete partitions.

All chytrids have motile stages that develop as zoospores inside a single- or multicelled, thick-walled structure—the thallus or sporangium. The distinctive kinetids and the single posteriorly directed undulipodium of their zoospores distinguish chytrids from look-alike protoctists.

Chytrids, unlike hyphochytrids, are sexual. The gametes that fuse to form a zygote may be identical with each other or may differ; at least one is motile. A motile chytrid gamete has one smooth posterior undulipodium. The zygote, by withdrawing its undulipodium, converts itself into a resting structure. In some species, this structure eventually releases zoospores that germinate into new chytrid thalli; in others, it germinates directly and grows into a new thallus.

We recognize four classes in Phylum Chytridiomycota, only the first of which includes the chytrids in the strict sense: Chytridiales (*Polyphagus*); Spizellomycetales (*Neocallimastix*); Monoblepharidales (*Monoblepharella*); and Blastocladiales (*Allomyces, Blastocladiella*). Because zoospore kinetid structure in each class is distinct, ultrastructural analyses of the kinetosome part of the cell are useful for taxonomy and reconstruction of evolutionary history. About 1000 species in 100 genera grouped into a total of 17 families are currently recognized.

The chytridiales are unicellular, lacking any well-developed mycelia (hyphae). Some species do produce a rhizomycelium, a system of branched hyphae that, like rhizoids, emerge from the posterior of the chytrid body. Other species produce no mycelia at all.

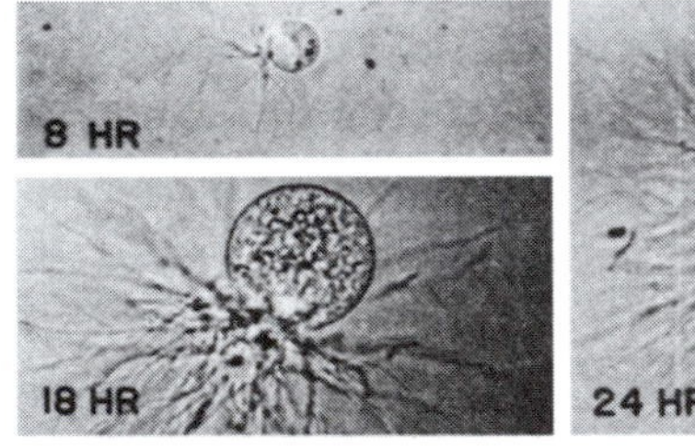

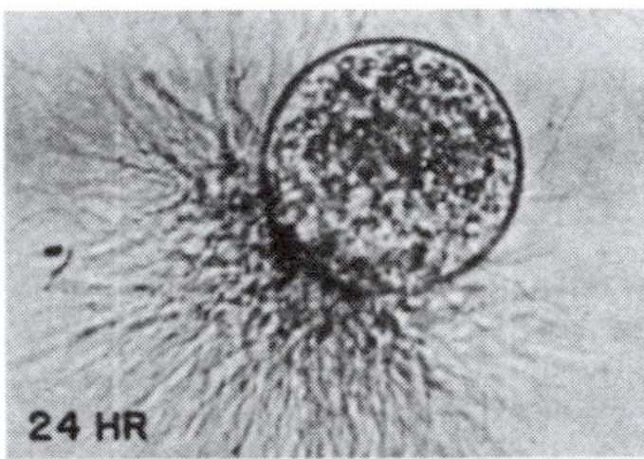

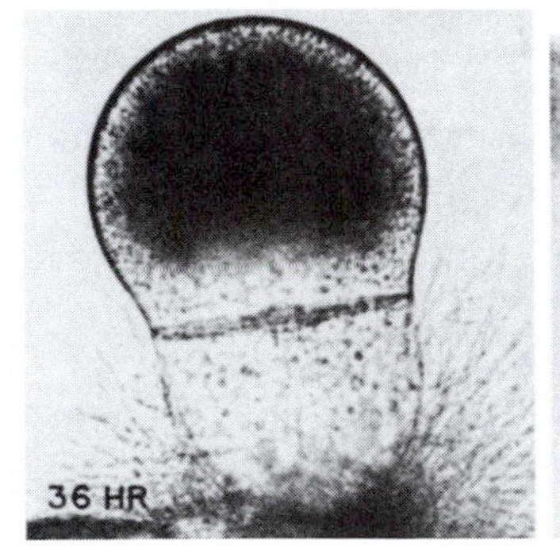

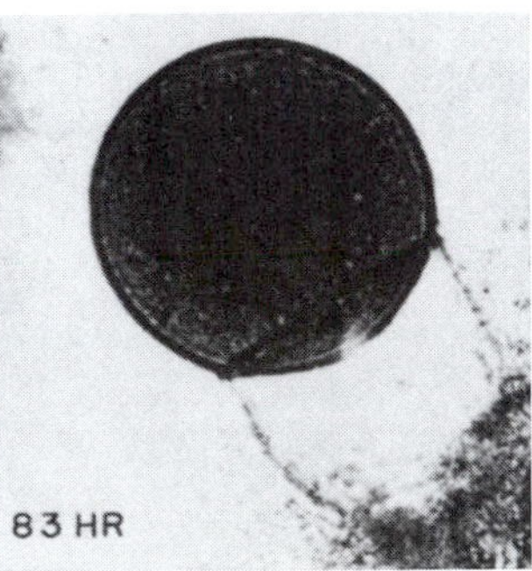

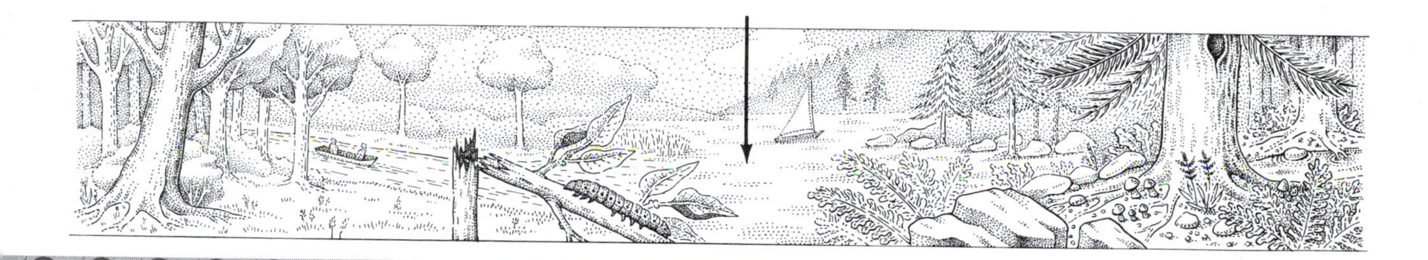

In sexual reproduction, equal zoospore-like gametes fuse to form a zygote.

The blastocladiales have well-developed branching mycelia. Many of them have complex life cycles with several alternative developmental pathways. *Blastocladiella emersonii,* for example, produces zoospores that have three distinct developmental options: a zoospore can form an ordinary colorless thallus, a stiff and resistant dark thallus, or a tiny thallus that releases only a single zoospore. Which option the organism takes depends on the quantity of food, moisture, and carbon dioxide in the medium. These factors, in turn, are related to the degree of crowding. Blastocladiales also reproduce sexually by the fusion of undulipodiated cells called planogametes (Greek *planos,* wandering), which look like the asexual zoospores. The planogametes are formed inside a thick-walled sporangium.

Blastocladiella zoospores have an unusual feature that can be seen by light microscopy. Virtually all the ribosomes of the members of this group are packed near the nucleus in a membrane-bounded structure called the nuclear cap (see zoospore drawing). The significance of this arrangement is not known.

In the monoblepharidales, male planogametes are released from a specialized part of the thallus called the male gametangium or antheridium. Other specialized hyphae of the thallus produce a walled female gametangium, also called the oogonium. Inside the oogonium, the protoplasm differentiates into a uninucleate oosphere, a fancy name for what is really an egg. This large, nonundulipodiated, nearly nonmotile female gamete fuses entirely with a motile male gamete. The zygote hardens over to form a structure traditionally misnamed a zoospore, in which meiosis takes place. The zoospore germinates to begin growing the hyphae of a mycelium; presumably, the nuclei in the mycelium are haploid. The mass grows; eventually, either it produces sporangia that release asexual zoospores or it differentiates male and female gametangia. In *Monoblepharis polymorpha,* the same thallus that produces asexual sporangia at certain temperatures is capable of producing the male and female gametangia and releasing compatible gametes at slightly higher temperatures.

In members of Class Spizellomycetales, the nucleus is connected to the kinetosomes by microtubules or by a rhizoplast (a fibrous "nuclear connector"). Rumposomes (honeycomb-like organelles—whose function is unknown—consisting of regularly fenestrated cisternae) are absent in the spizellomycetales, though common in zoospores of chytridiales and monoblepharidales. Spore germination in members of Class Spizellomycetales is "exogenous"; that is, the nucleus moves into the germ tube as it divides on germination. The thallus that ultimately forms is monocentric: it has a single central structure into which nutrients flow and from which reproductive structures are initiated.

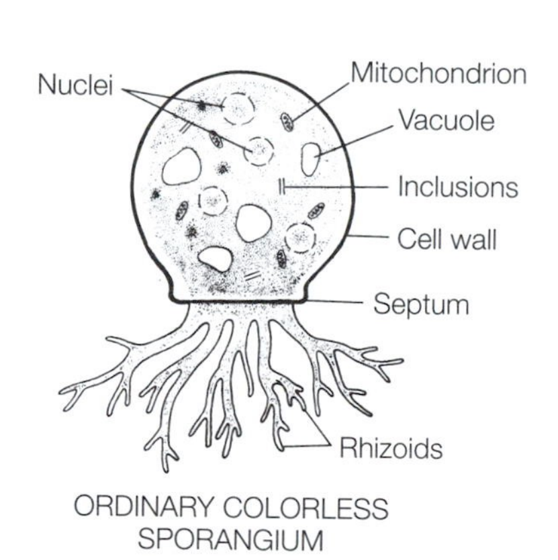

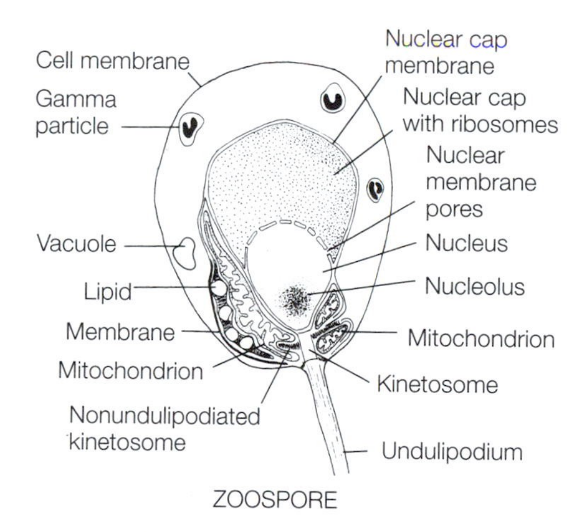

Development of the ordinary colorless sporangium of *Blastocladiella emersonii*. The hours are time elapsed after water was added to an initial small dry sporangium. After 18 hours, there is swelling and a proliferation of rhizoids. After 36 hours, the protoplasm has migrated into the anterior cell that becomes the sporangium. After 83 hours, the sporangium has thickened and zoospores have begun to differentiate from the coenocytic nuclei inside. LM, bar = 1 μm. [Photographs courtesy of E. C. Cantino and J. S. Lovett; drawings by R. Golder.]

Pr-30 Zoomastigota

(Zoomastigotes, zooflagellates)

Greek *zoion*, animal; *mastix*, whip

EXAMPLES OF GENERA

Acronema
Cafeteria
Cepedea
Desmarella
Hexamastix
Histiona
Histomonas
Jakoba
Karotomorpha
Monosiga
Oikomonas
Opalina
Proteronomas
Protoopalina
Reclinomonas
Zelleriella

Zoomastigotes are phagotrophic or osmotrophic aerobic swimming unicells. Former members that are amitochondriates have been removed to Archaeprotista (Phylum Pr-1). Those with flattened, cristate mitochondria are now in Discomitochondriates (Phylum Pr-12), leaving here a miscellany of unicells. Each bears at least one undulipodium; some bear thousands. They may be either free living or symbiotrophic, sexual or asexual, but all are heterotrophs with tubular mitochondrial cristae and without plastids. Some, had they plastids, would be classified among the green or stramenopile (heterokont) algae on the basis of their cell structure.

In two-kingdom classifications, zoomastigotes were placed in the order Protomonadina in the kingdom Animalia. In our classification, Phylum Zoomastigota is placed in Kingdom Protoctista, but it is likely that this phylum contains among its members the ancestors of the animals. The phylum consists of five classes: jakobids, bicosoecids, proteromonads, opalinids, and choanomastigotes. This classification is tentative; too few are known in sufficient detail to ascertain the relationship between these and members of the Archaeprotista and Discomitochondriates phyla.

Because they are small, jakobids are a newly discovered class of zoomastigotes that required electron microscopy to be visualized. All are mononucleate, biundulipodiated bacterivores. The anteriorly inserted undulipodia are equal in length. Although neither undulipodium bears hairs or scales, one undulipodium is directed forward and the other trails in typical heterokont style. The trailing undulipodium is pressed along, but not actually attached to, the ventral surface of the cell. A cytoplasmic flap, called the "lip," forms a groove on the right ventral side of the cell, and the posterior undulipodium lies in this groove. The groove serves as a mouth only in the sense that bacteria are ingested in it; there is no cytostome. Binary fission cell division is longitudinal as in mastigotes generally, and at least one offspring is an active swimmer with a reduced or no lip. Sexuality is unknown in the group. Only three genera have been reported. Cells of the genus *Jakoba* lack a lorica, or shell. An inhabitant of marine sediments, *Jakoba* may attach to the sediment surface by its anterior undulipodium. Freshwater mastigotes in the genus *Reclinomonas* have a hyaline lorica, which is covered with spiny scales. The cell resides dorsal side down inside the lorica—hence the name, *Reclinomonas. Histiona* also dwells in a lorica in freshwater habitats. The hyaline lorica of *Histiona* lacks scales; the cell lies upside down, with its posterior protruding from the lorica. The protruding "lip" on the posterior ventral side of the cell forms a sail-like extension.

In the bicosoecid, a marine organism that bears a shell and two undulipodia, one undulipodium extends forward from the cell and is mastigonemate—it bears numerous tiny appendages called tinsel or flimmers. The other lies along the surface of the cell and is smooth. The bicosoecid kinetid is unmistakable; it has two rows of associated microtubules, one long (ten tubules) and one short (three tubules). Bicosoecids tend to form colonies. Sexuality has not been reported in this group. One newly described species, *Acronema sippewissettensis* (Figures A and B), is a tiny healthy eater that tolerates a few days of desiccating conditions by forming a walled stage. It has been grown in culture on bacterial food. Coming from a sulfurous mud flat, it easily survives anoxia. These colorless mastigotes resemble some of the chrysomonads (see Phylum Pr-13, Fig-

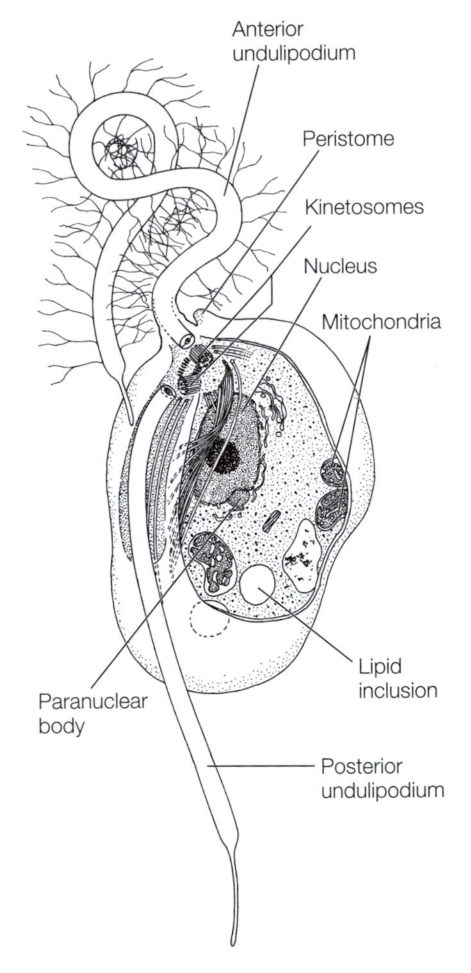

A *Acronema sippewissettensis.* Note the acronemate (tapering) undulipodia. The mastigonemes (Y-shaped hairs) on the anterior undulipodium are the basis for this organism's taxonomic placement with the stramenopiles (Phyla Pr-13 through Pr-21). [Drawing by K. Delisle.]

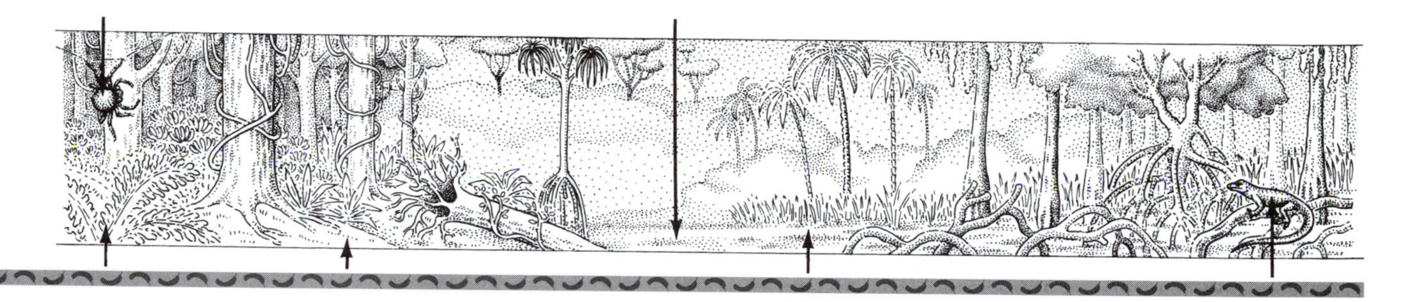

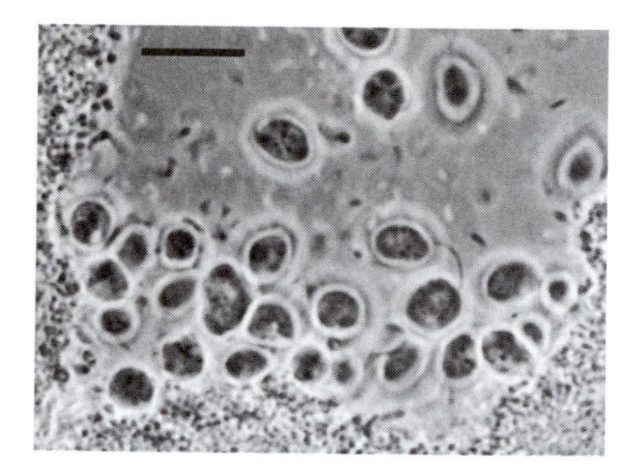

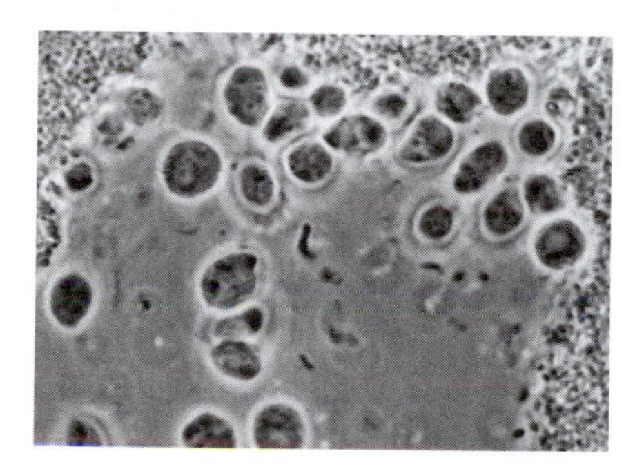

B *Acronema sippewissettensis.* Live mastigotes clustering around a clump of actively feeding bacteria. LM, bar = 10 μm.

ure C), and some chrysomonads may have evolved from bicosoecids by acquiring plastids. Alternatively, bicosoecids may have evolved from chrysomonads by losing plastids.

Proteromonads, with two genera—*Proteromonas* and *Karotomorpha*—and five species, are small mastigotes limited to the posterior intestine of amphibians, reptiles, and some mammals, especially rodents. The proteromonad is a nonmastigonemate, heterokont cell with a rhizoplast that connects the nucleus and a large Golgi body to the anterior kinetosomes. Proteromonads form resistant cysts that pass from the intestines to the soil. At least one species undergoes multiple fission.

The opalinids are characterized by the falx, a unique structure composed of several rows of hundreds of close-set kinetosomes. The falx lies along the anterior part of the long axis of the cell, and undulipodia emerge from it. As in euglenids (Phylum Pr-12), a patterned proteinaceous skin, called a pellicle, lies just beneath the plasma membrane. Like jakobids and bicosoecids, the opalinid lacks a cytostome, so it absorbs dissolved nutrients directly through its pellicle. Opalinids have sexual stages: dissimilar haploid whole-cell gametes fuse to form diploid cells. Nuclear division can take place without cytoplasmic division, giving rise to bi- or multinucleate cells. Opalinids are large symbiotrophs in the large intestine, or rectum, of fish, reptiles, and amphibians, mostly of frogs and toads. Opalinids have flattened bodies and swim with spiral movements. Four genera are recognized. The *Protoopalina* are considered the most primitive; each cell has two large nuclei. *Zelleriella* cells also each have two large nuclei, which lie on opposite sides of the cell's long axis (Figure C). *Cepedea* species are very large multinucleate cells, as long as 2.8 mm. They have small nuclei and a lengthy falx. *Opalina* species, not as large, are also multinucleate and have an extensive falx; the pellicle is pleated.

Some choanomastigotes, such as *Monosiga,* are colorless. Others, such as *Desmarella,* contain green plastids. They are distinguished by having a hard lorica structure from which the single undulipodium emerges. The lorica has ribs, called choastes. The undulipodium is tapered and smooth; it can be retracted into the cell. Many choanomastigotes stand on peduncles, cell stalks that contain longitudinal fibers. The organization of their cells is remarkably similar to that of the body cells of sponges (Phylum A-2, Porifera). In fact, it is generally thought that choanomastigotes, some of which are organized into colonies, are direct ancestors of the sponges, but not of any other metazoan phyla.

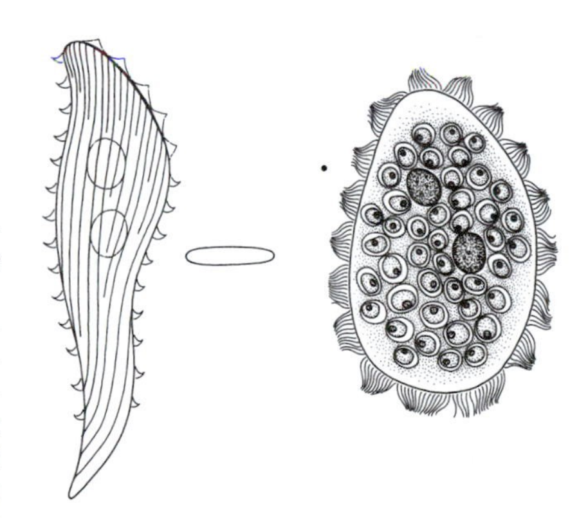

C *Zelleriella* is always binucleate (left) but is flat in cross section (middle). Shown here is the trophont stage from a frog (Phylum A-37) hindgut. Heavy infection with *Entamoeba* in its uninucleate cyst stage is shown at right. [Redrawn and modified from H. Wessenberg, "Opalinata." In *Parasitic protozoa, intestinal flagellates, histomonads, trichonomonads, amoebae, opalinids and ciliates,* J. P. Kreier, ed., Vol. 2, pp. 551–581 (Academic Press, New York, 1978), and R. M. Stabler and T.-T. Chen, "Observations on an endamoeba parasitizing opalinid ciliates." *Biological Bulletin* 70:56–71 (1936).]

Bibliography: Protoctista

General

Abbott, I. A., and E. Y. Dawson, *How to know the seaweeds,* 2d ed. Brown; Dubuque, IA; 1978.

Copeland, H. F., *The classification of lower organisms.* Pacific Books; Palo Alto, CA; 1956.

Dyer, B. D., and R. A. Obar, *Tracing the history of eukaryotic cells.* Columbia University Press; New York; 1994.

Ebringer, L., and J. Krajcovic, *Cell origin and evolution.* Czechoslovak Society of Microbiology; Prague; 1995.

Fenchel, T., and B. J. Finlay, *Ecology and evolution in anoxic worlds.* Oxford University Press; Oxford; 1995.

Fritsch, F. E., *Structure and reproduction of the algae,* 2 vols., 2d ed. Cambridge University Press; Cambridge and New York; 1961.

Grell, K. G., *Protozoology,* 2d ed. Springer-Verlag; New York, Heidelberg, and Berlin; 1973.

Hogg, J., "On the distinctions of a plant and an animal, and on a fourth kingdom of nature." *Edinburgh New Philosophical Journal* 12:216–225; 1861.

Hülsmann, N., "Undulipodium: End of a useless discussion." *European Journal of Protistology* 28:253–257; 1981.

Lee, J. J., S. H. Hunter, and E. C. Bovee, *An illustrated guide to the protozoa.* Allen Press; Lawrence, KS; 1985.

Leedale, G. F., "How many are the kingdoms of organisms?" *Taxon* 23:261–270; 1974.

Lipps, J. H., *Fossil prokaryotes and protists.* Blackwell; New York; 1993.

Margulis, L., "Archaeal-eubacterial mergers in the origin of Eukarya: Phylogenetic classification of life." *Proceedings of the National Academy of Sciences,* USA 93:1071–1076; 1996.

Margulis, L., *Symbiosis in cell evolution,* 2d ed. W. H. Freeman and Company; New York; 1993.

Margulis, L., and D. Sagan, *Origins of sex: Three billion years of genetic recombination,* 2d ed. Yale University Press; New Haven; 1991.

Margulis, L., J. O. Corliss, M. Melkonian, and D. J. Chapman, eds., *Handbook of Protoctista: The structure, cultivation, habitats and life histories of the eukaryotic microorganisms and their descendants exclusive of animals, plants and fungi.* Jones and Bartlett; Boston; 1990.

Margulis, L., H. I. McKhann, and L. Olendzenski, *Illustrated glossary of Protoctista.* Jones and Bartlett; Boston; 1992.

Margulis, L., and K. V. Schwartz, *Five kingdoms: Introduction to the phyla of life on Earth.* CD-ROM (IBM- or Macintosh-compatible). Ward's Natural History Establishment; Rochester, NY; 1994.

Margulis, L., K. V. Schwartz, and M. Dolan, *The illustrated five kingdoms: Guide to the diversity of life on Earth.* Addison-Wesley-Longman; Menlo Park, CA; 1994.

Raikov, I. B., *The protozoan nucleus: Morphology and evolution.* Springer-Verlag; Vienna; 1982.

Round, F. E., *The ecology of algae.* Cambridge University Press; Cambridge; 1981.

Sagan, D., and L. Margulis, *Garden of microbial delights: A practical guide to the subvisible world.* Kendall/Hunt; Dubuque, IA; 1993.

Sapp, J., *Evolution by association: A history of symbiosis.* Oxford University Press; New York; 1994.

Swofford, D. L., and G. O. Olson, "Phylogeny reconstruction." In D. M. Hillis and C. Moritz, eds., *Molecular systematics.* Sinauer; Sunderland, MA; 1990.

Van De Graaff, K. M., S. R. Rushforth, and J. L. Crawley, *A photographic atlas for the botany laboratory.* Morton; Englewood, CO; 1995.

Pr-1 Archaeprotista

Daniels, E. W., and E. P. Breyer, "Ultrastructure of the giant amoeba *Pelomyxa palustris.*" *Journal of Protozoology* 14:167–179; 1967.

Gunderson, J., G. Hinkle, D. Leipe, H. G. Morrison, S. K. Stickel, D. A. Odelson, J. A. Breznak, T. A. Nerad, M. Müller, and M. L. Sogin, "Phylogeny of trichomonads inferred from small-subunit rRNA sequences." *Journal of Eukaryotic Microbiology* 42:411–415; 1995.

Hollande, A., and J. Valentin, "Appareil de Golgi, pinocytose, lysomes, mitochondries, bactéries symbiontiques, attrac-

tophores et pleuromitose chez les hypermastigines du genre *Joenia:* Affinités entre Joeniides et Trichomonadines." *Protistologica* 5:39–86; 1969.
Jeon, K. W., ed., *The biology of Amoeba.* Academic Press; New York and London; 1973.
Lee, J. J., S. H. Hutner, and E. C. Bovee, *Illustrated guide to the protozoa,* 2d ed. Society of Protozoologists, Allen Press; Lawrence, KS; 1997.
Whatley, J. M., and C. Chapman-Andresen, "Phylum Karyoblastea." In L. Margulis, J. O. Corliss, M. Melkonian, and D. J. Chapman, eds., *Handbook of Protoctista.* Jones and Bartlett; Boston; 1990.

Pr-2 Microspora

Canning, E. U., "Phylum Microspora." In L. Margulis, J. O. Corliss, M. Melkonian, and D. J. Chapman, eds., *Handbook of Protoctista.* Jones and Bartlett; Boston; 1990.
Sprague, V., "Annotated list of species of Microsporidia." In L. A. Bulla and T. C. Cheng, eds., *Comparative pathobiology,* Vol. 2: *Systematics of the Micosporidia.* Plenum Press; New York; 1977.

Pr-3 Rhizopoda

Bonner, J. T., *The cellular slime molds,* 2d ed. Princeton University Press; Princeton, NJ; 1967.
Grell, K. G., *Protozoology,* 2d ed. Springer-Verlag; New York, Heidelberg, and Berlin; 1973.
Martinez, J. A., *Free-living amebas: Natural history, prevention, diagnosis, pathology and treatment of disease.* CRC Press; Boca Raton, FL; 1985.
Olive, L. S., *The mycetozoans.* Academic Press; New York, San Francisco, and London; 1975.
Page, F. C., *Marine gymnamoebae.* Institute of Terrestrial Ecology; Cambridge, United Kingdom; 1983.
Page, F. C., *A new key to freshwater and soil gymnamoebae.* Institute of Terrestrial Ecology; Cambridge, United Kingdom; 1988.

Pr-4 Granuloreticulosa

Cushman, J. A., *Foraminifera: Their classification and economic use,* 4th ed. Harvard University Press; Cambridge, MA; 1948.
Lee, J. J., "Living sands." *Bioscience* 45:252–261; 1995.
Lee, J. J., "Phylum Granuloreticulosa (Foraminifera)." In L. Margulis, J. O. Corliss, M. Melkonian, and D. J. Chapman, eds., *Handbook of Protoctista.* Jones and Bartlett; Boston; 1990.

Pr-5 Xenophyophora

Gooday, A. J., B. J. Bett, and D. N. Pratt, "Direct observation of episodic growth in an abyssal xenophyophore (Protista)." *Deep Sea Research* 40:2132–2143; 1993.
Tendal, Ø. S., "Phylum Xenophyophora." In L. Margulis, J. O. Corliss, M. Melkonian, and D. J. Chapman, eds., *Handbook of Protoctista.* Jones and Bartlett; Boston; 1990.

Pr-6 Myxomycota

Alexopoulos, C. J., and C. W. Mims, *Introductory mycology,* 3d ed. Wiley; New York; 1979.
Frederick, L., "Phylum Plasmodial Slime Molds, Class Myxomycota." In L. Margulis, J. O. Corliss, M. Melkonian, and D. J. Chapman, eds., *Handbook of Protoctista.* Jones and Bartlett; Boston; 1990.
Hagelstein, R., *The mycetozoa of North America.* Hagelstein; Mineola, NY; 1944.
Olive, L. S., *The mycetozoans.* Academic Press; New York, San Francisco, and London; 1975.
Stephenson, S. L., and H. Stempen, *Myxomycetes: A handbook of slime molds.* Timber Press; Portland, OR; 1994.

Pr-7 Dinomastigota

Burkholder, J. M., and H. B. Glasgow, Jr., "Interactions of a toxic estuarine dinoflagellate with microbial predators and prey." *Archiv für Protistenkunde* 145:177–188; 1995.
Burkholder, J. M., E. J. Noga, C. H. Hobbs, and H. B. Glasgow, Jr., "New 'phantom' dinoflagellate is the causative

agent of major estuarine fish kills." *Nature* 358:407–410; 1992.

Fensome, R. A., F. J. R. Taylor, G. Norris, W. A. S. Sarjeant, D. I. Wharton, and G. L. Williams, *A classification of living and fossil dinoflagellates. Micropaleontology* Special Publication No. 7. Sheridan Press; Hanover, PA; 1993.

Fritsch, F. E., *Structure and reproduction of the algae,* 2d ed., Vol. 1, Cambridge University Press; Cambridge and New York; 1961.

Hutner, S. H., and J. J. A. McLaughlin, "Poisonous tides." *Scientific American* 199:92–98; August 1958.

Scagel, R. F., R. J. Bandoni, G. E. Rouse, W. B. Schofield, J. R. Stein, and T. M. C. Taylor, *An evolutionary survey of the plant kingdom.* Wadsworth; Belmont, CA; 1966.

Taylor, F. J. R., "On dinoflagellate evolution." *Biosystems* 13: 65–108; 1980.

Pr-8 Ciliophora

Corliss, J. O., *The ciliated protozoa: Characterization, classification, and guide to the literature,* 2d ed. Pergamon Press; London and New York; 1979.

Curds, C. R., *British and other freshwater ciliated protozoa,* Part 1, *Ciliophora: Kinetofragminophora.* Cambridge University Press; Cambridge; 1982.

Curds, C. R., M. A. Gates, and D. M. Roberts, *British and other freshwater ciliated protozoa,* Part 2, *Oligohymenophora and Polyhymenophora.* Cambridge University Press; Cambridge; 1983.

Fenckel, T., *Ecology of protozoa: The biology of free-living phagotrophic protists.* Science Tech; Madison, WI; 1987.

Lynn, D. H., and E. B. Small, "Phylum Ciliophora." In L. Margulis, J. O. Corliss, M. Melkonian, and D. J. Chapman, eds., *Handbook of Protoctista.* Jones and Bartlett; Boston; 1990.

Olive, L. S., and R. L. Blanton, "Aerial sorocarp development by the aggregative ciliate, *Sorogena stoianovitchae.*" *Journal of Protozoology* 27:293–299; 1980.

Small, E. B., and D. H. Lynn, "Phylum Ciliophora Doflein, 1901." In J. J. Lee, S. H. Hutner, and E. C. Bovee, eds., *Illustrated guide to the protozoa,* 2d ed. Society of Protozoologists, Allen Press; Lawrence, KS; 1997.

Pr-9 Apicomplexa

Hammond, D. M., and P. L. Long, *The coccidia:* Eimeria, Isospora, Toxoplasma, *and related genera.* University Park Press; Baltimore; 1973.

Kreier, J., *Parasitic Protozoa,* Vol. 3, *Gregarines, Haemogregarines, Coccidia, Plasmodia and Haemoproteids;* Vol. 4, *Theileria, Myxosporida, Microsporida, Bartonellaceae, Anaplasmataceae, Ehrlichia* and *Pneumocystis.* Academic Press; New York; 1977.

Noble, E. R., and G. A. Noble, *Parasitology,* 5th ed. Lea & Febiger; Philadelphia; 1982.

Pr-10 Haptomonada

Green, J. C., and B. S. C. Leadbeater, eds., *The haptophyte algae.* Systematics Association, Clarendon Press; Oxford; 1994.

Green, J. C., K. Perch-Nielsen, and P. Westbroek, "Phylum Prymnesiophyta." In L. Margulis, J. O. Corliss, M. Melkonian, and D. J. Chapman, eds., *Handbook of Protoctista.* Jones and Bartlett; Boston; 1990.

Pr-11 Cryptomonada

Gillott, M., "Phylum Cryptophyta." In L. Margulis, J. O. Corliss, M. Melkonian, and D. J. Chapman, eds., *Handbook of Protoctista.* Jones and Bartlett; Boston; 1990.

Pr-12 Discomitochondriates

Anderson, O. R., *Comparative protozoology: Ecology, physiology, life history.* Springer-Verlag; New York; 1987.

Leedale, G. F., *Euglenoid flagellates.* Prentice-Hall; Englewood Cliffs, NJ; 1967.

Pr-13 Chrysomonada

Kristiansen, J., "Phylum Chrysophyta." In L. Margulis, J. O. Corliss, M. Melkonian, and D. J. Chapman, eds., *Handbook of Protoctista.* Jones and Bartlett; Boston; 1990.

Round, F. E., "The Chrysophyta: A reassessment." In J. Kristiansen and R. A. Andersen, eds., *Chrysophytes: Aspects and problems.* Cambridge University Press; New York; 1986.

Pr-14 Xanthophyta

Hibberd, D. J., Phylum Xanthophyta." In L. Margulis, J. O. Corliss, M. Melkonian, and D. J. Chapman, eds., *Handbook of Protoctista.* Jones and Bartlett; Boston; 1990.

Pr-15 Eustigmatophyta

Hibberd, D. J., "Phylum Eustigmatophyta." In L. Margulis, J. O. Corliss, M. Melkonian, and D. J. Chapman, eds., *Handbook of Protoctista.* Jones and Bartlett; Boston; 1990.

Hibberd, D. J., and G. F. Leedale, "Observations on the cytology and ultrastructure of the new algal class Eustigmatophyceae." *Annals of Botany* 36:49–71; 1972.

Pr-16 Diatoms

Round, F. E., R. M. Crawford, and D. G. Mann, *The diatoms: Biology and morphology of the genera.* Cambridge University Press; New York; 1989.

Round, F. E., R. M. Crawford, and D. G. Mann, *A scanning electron microscope atlas of diatom structure and diatom genera.* Cambridge University Press; New York; 1993.

Pr-17 Phaeophyta

Clayton, M. N., "Phylum Phaeophyta." In L. Margulis, J. O. Corliss, M. Melkonian, and D. J. Chapman, eds., *Handbook of Protoctista.* Jones and Bartlett; Boston; 1990.

Pr-18 Labyrinthulata

Porter, D., "Phylum Labyrinthulomycota." In L. Margulis, J. O. Corliss, M. Melkonian, and D. J. Chapman, eds., *Handbook of Protoctista.* Jones and Bartlett; Boston; 1990.

Pr-19 Plasmodiophora

Dylewski, D., "Phylum Plasmodiophoromycota." In L. Margulis, J. O. Corliss, M. Melkonian, and D. J. Chapman, eds., *Handbook of Protoctista.* Jones and Bartlett; Boston; 1990.

Karling, J. S., *The Plasmodiophorales,* 2d rev. ed. Macmillan (Hafner Press); New York; 1968.

Sparrow, F. K., Jr., *Aquatic phycomycetes,* 2d ed. University of Michigan Press; Ann Arbor; 1960.

Pr-20 Oomycota

Dick, M. W., "Phylum Oomycota." In L. Margulis, J. O. Corliss, M. Melkonian, and D. J. Chapman, eds., *Handbook of Protoctista.* Jones and Bartlett; Boston; 1990.

Erwin, D. C., S. Bartnicki-Garcia, and P. H. Tsao, eds., Phytophthora: *Its biology, taxonomy, ecology, and pathology.* American Phytopathological Society; St. Paul, MN; 1983.

Pr-21 Hyphochytriomycota

Fuller, M. S., "Phylum Hyphochytriomycota." In L. Margulis, J. O. Corliss, M. Melkonian, and D. J. Chapman, eds., *Handbook of Protoctista.* Jones and Bartlett; Boston; 1990.

Stevens, R. B., *Mycology guidebook.* University of Washington Press; Seattle; 1974.

Pr-22 Haplospora

Perkins, F. O., ed., "Haplosporidian and haplosporidian-like diseases of shellfish." *Marine Fisheries Review* 41:25–37; 1979.

Perkins, F. O., "Phylum Haplosporidia." In L. Margulis, J. O. Corliss, M. Melkonian, and D. J. Chapman, eds., *Handbook of Protoctista.* Jones and Bartlett; Boston; 1990.

Pr-23 Paramyxa

Desportes, I., and F. O. Perkins, "Phylum Paramyxea." In L. Margulis, J. O. Corliss, M. Melkonian, and D. J. Chapman, eds., *Handbook of Protoctista*. Jones and Bartlett; Boston; 1990.

Pr-24 Myxospora

Lom, J., "Phylum Myxozoa." In L. Margulis, J. O. Corliss, M. Melkonian, and D. J. Chapman, eds., *Handbook of Protoctista*. Jones and Bartlett; Boston; 1990.

Sprague, V., "Myxozoa." In S. P. Parker, ed., *Synopsis and classification of living organisms*. McGraw-Hill; New York; 1982.

Pr-25 Rhodophyta

Gabrielson, P. W., D. J. Garbary, M. R. Sommerfeld, R. A. Townsend, and P. L. Tyler, "Phylum Rhodophyta." In L. Margulis, J. O. Corliss, M. Melkonian, and D. J. Chapman, eds., *Handbook of Protoctista*. Jones and Bartlett; Boston; 1990.

Hoshaw, R. W., "Phylum Conjugaphyta." In L. Margulis, J. O. Corliss, M. Melkonian, and D. J. Chapman, eds., *Handbook of Protoctista*. Jones and Bartlett; Boston; 1990.

Pr-26 Gamophyta

Pickett- Heaps, J. D., *Green algae*. Sinauer; Sunderland, MA; 1975.

Pr-27 Actinopoda

Hollande, A., J. Cachon, and M. Cachon-Enjumet, "L'infrastructure des axopods chez les Radiolaires Sphaerellaires periaxoplastidies." *Comptes Rendu Hebdomedaire Seances Academie des Science* (Séries D) 261:1388–1391; 1965.

Pr-28 Chlorophyta

Graham, L., "The origin of the life cycle of land plants." *American Scientist* 73(2): 178–186; March–April 1985.

Puiseux-Dao, S., Acetabularia *and cell biology*. Springer-Verlag; New York, Heidelberg, and Berlin; 1970.

Raven, P. H., R. F. Evert, and S. Eichhorn, *Biology of plants*, 6th ed. Worth; New York; 1998.

Pr-29 Chytridiomycota

Barr, D. J. S., "Phylum Chytridiomycota." In L. Margulis, J. O. Corliss, M. Melkonian, and D. J. Chapman, eds., *Handbook of Protoctista*. Jones and Bartlett; Boston; 1990.

Cantino, E. C., "Morphogenesis in aquatic fungi." In G. C. Ainsworth and A. S. Sussman, eds., *The fungi: An advanced treatise*, Vol. 2. Academic Press; New York and London; 1966.

Noyes Mollicone, M. R., and J. E. Longcore, "Zoospore ultrastructure of *Monoblepharis polymorpha*." *Mycologia* 86: 615–625; 1994.

Sparrow, F. K., Jr., *Aquatic phycomycetes*, 2d ed. University of Michigan Press; Ann Arbor; 1960.

Pr-30 Zoomastigota

Brugerolle, G., and J. P. Mignot, "Phylum Zoomastigina, class Proteromonadida." In L. Margulis, J. O. Corliss, M. Melkonian, and D. J. Chapman, eds., *Handbook of Protoctista*. Jones and Bartlett; Boston; 1990.

Buck, K. R., "Phylum Zoomastigina, class Choanomastigotes (Choanoflagellates)." In L. Margulis, J. O. Corliss, M. Melkonian, and D. J. Chapman, eds., *Handbook of Protoctista*. Jones and Bartlett; Boston; 1990.

Corliss, J. O., "Phylum Zoomastigina, class Opalinata." In L. Margulis, J. O. Corliss, M. Melkonian, and D. J. Chapman, eds., *Handbook of Protoctista*. Jones and Bartlett; Boston; 1990.

Dyer, B. D., "Phylum Zoomastigina, class Bicoecids." In L. Margulis, J. O. Corliss, M. Melkonian, and D. J. Chapman, eds., *Handbook of Protoctista*. Jones and Bartlett; Boston; 1990.

O'Kelley, C., "*Jakoba*." *Archives of Protistology* 1:2–100; 1994.

CHAPTER THREE

ANIMALIA

A-20 *Limenitis archippus.* [Courtesy of P. Kromholz.]

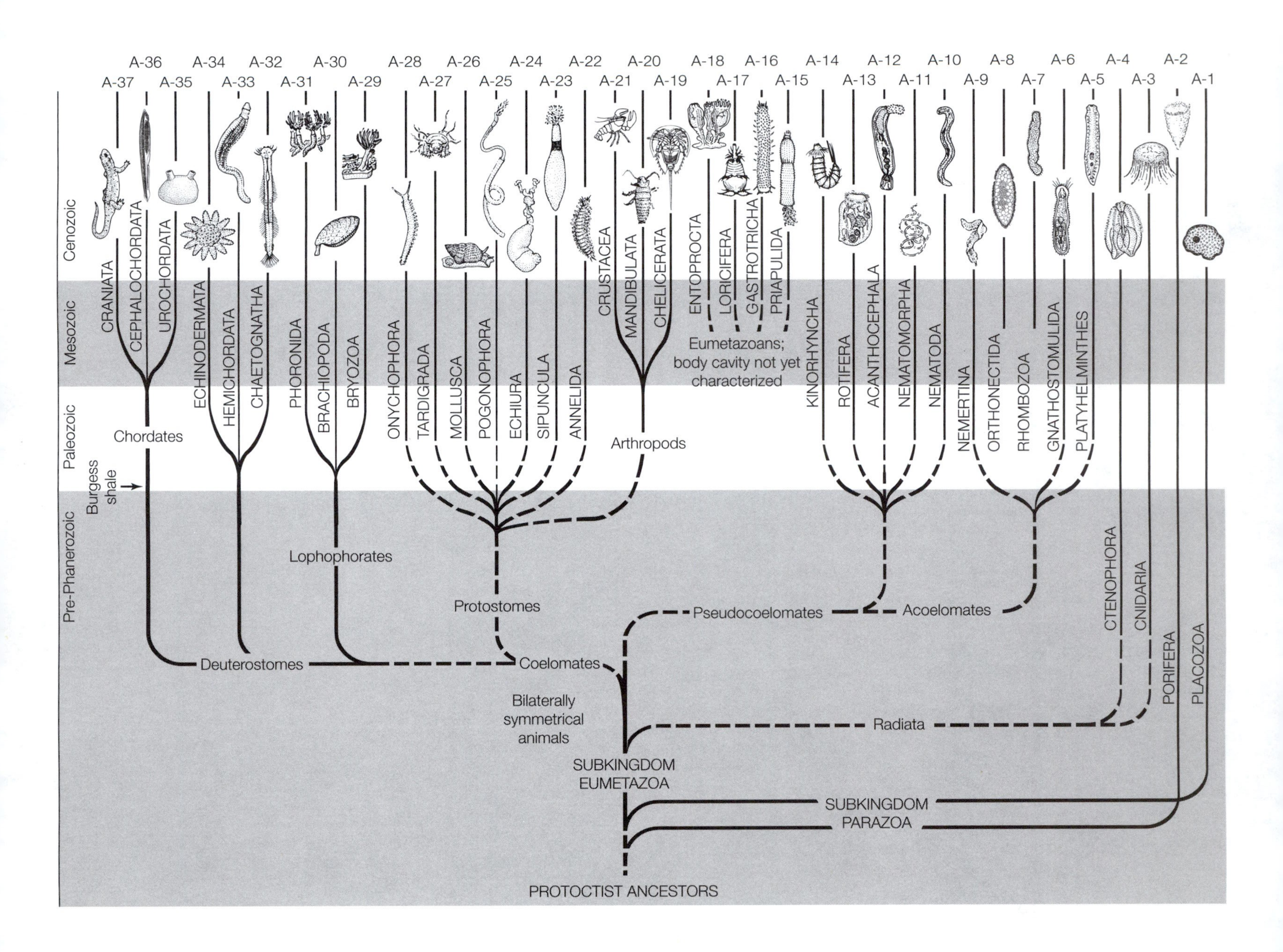

A-1
A-2
A-3
A-4
A-5
A-6
A-7
A-8
A-9
A-10
A-11
A-12
A-13
A-14
A-15
A-16
A-17
A-18
A-19
A-20
A-21
A-22
A-23
A-24
A-25
A-26
A-27
A-28
A-29
A-30
A-31
A-32
A-33
A-34
A-35
A-36
A-37
Cenozoic
Mesozoic
Paleozoic
Pre-Phanerozoic
Burgess shale
CRANIATA
CEPHALOCHORDATA
UROCHORDATA
ECHINODERMATA
HEMICHORDATA
CHAETOGNATHA
PHORONIDA
BRACHIOPODA
BRYOZOA
ONYCHOPHORA
TARDIGRADA
MOLLUSCA
POGONOPHORA
ECHIURA
SIPUNCULA
ANNELIDA
CRUSTACEA
MANDIBULATA
CHELICERATA
ENTOPROCTA
LORICIFERA
GASTROTRICHA
PRIAPULIDA
Eumetazoans; body cavity not yet characterized
KINORHYNCHA
ROTIFERA
ACANTHOCEPHALA
NEMATOMORPHA
NEMATODA
NEMERTINA
ORTHONECTIDA
RHOMBOZOA
GNATHOSTOMULIDA
PLATYHELMINTHES
CTENOPHORA
CNIDARIA
PORIFERA
PLACOZOA
Chordates
Arthropods
Lophophorates
Protostomes
Pseudocoelomates
Acoelomates
Deuterostomes
Coelomates
Bilaterally symmetrical animals
Radiata
SUBKINGDOM EUMETAZOA
SUBKINGDOM PARAZOA
PROTOCTIST ANCESTORS

ANIMALIA

Latin *anima,* breath, soul

Diploid organisms that develop from embryos (blastulas), which form by fusion (fertilization: cytogamy and karyogamy) of haploid egg and sperm (anisogametes). Meiosis of gametes yields anisogametes.

In the two-kingdom (animal and plant) classification—older and not used in this book—animals composed of many cells (multicellular) were referred to as Metazoa to distinguish them from Protozoa (one-celled animals). In our system, there are no one-celled animals; traditional protozoans are placed in the Protoctista kingdom. We define animals as heterotrophic, diploid, multicellular organisms that usually (except sponges) develop from a blastula. The blastula, a multicellular embryo that develops from the diploid zygote produced by fertilization of a large haploid egg by a smaller haploid sperm, is unique to animals.

Because animal gametes—the egg and sperm—differ in size, they are called anisogametes. The diploid zygote produced by fertilization divides by mitotic cell divisions, resulting in a solid ball of cells that usually hollows out to become a blastula (Figure 3-1). In many animals, the blastula develops an opening called the blastopore, which is the opening to the developing digestive tract and will be the site of the mouth in animals belonging to some phyla or the anus in animals belonging to some of the other phyla. Animals in some phyla show neither of these two patterns; rather, some animals with spiral cleavage produce a blastula (stereoblastula) that is a solid ball of cells—their affinities remain unclear until more is known of their biology. Cephalopod molluscs (Phylum A-26), which have much yolk, lack blastocoels (embryonic cavities). Cell differentiation and cell migrations transform the blastula into a gastrula, an embryo with a dead-end indentation that is the embryonic digestive tract in most animals.

The details of further embryonic development differ widely from phylum to phylum. Nevertheless, common developmental patterns provide clues to relationships between the phyla. In many phyla, developmental details are known for very few species so far; in some phyla, for no species. Because development is intricate and complex, we cannot summarize it in a few words. For similar reasons, concise, accurate definitions of the phyla cannot always be given. Our descriptions are more informal.

Multicellularity is not unique to animals; multicellular organisms abound in all the kingdoms. Examples include most Cyanobacteria (Phylum B-5) and Actinobacteria (Phylum B-12) in Kingdom Bacteria; Phaeophyta (Phylum Pr-17), Oomycota (Phylum Pr-20) and Rhodophyta (Phylum

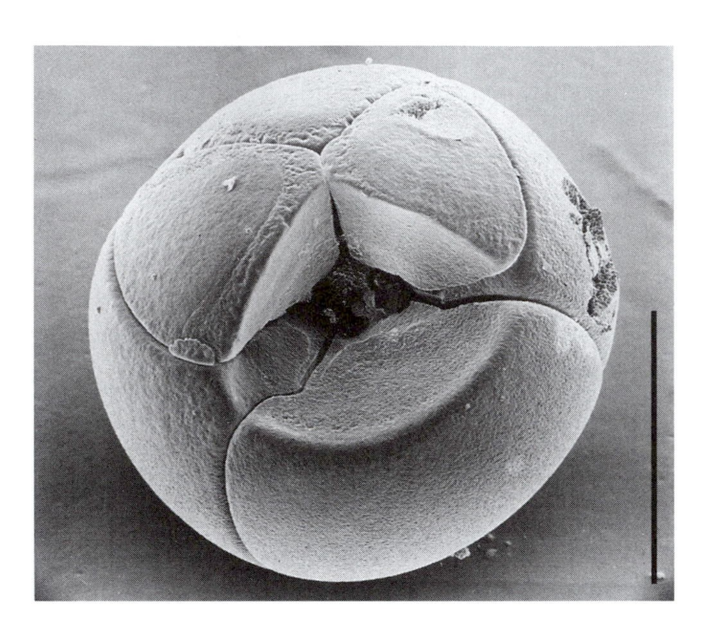

FIGURE 3-1 Blastula, the embryo that results from cleavage of the zygote of *Xenopus,* the clawed frog. In many animal species, a sphere of cells surrounds a liquid-filled cavity, the blastocoel. One cell has been removed from this eight-cell embryo. SEM, bar = 0.5 mm. [Courtesy of E. J. Sanders, *Biophysical Journal* 15:383 (1975).]

Pr-25) in Kingdom Protoctista; most members of Kingdom Fungi; and all members of Kingdom Plantae. However, multicellularity is most diverse in the animals; that is, many cells having highly specialized functions are grouped into tissues and tissues into organs. Complex junctions link cells into tissues in most phyla; two types of junctions unique to animals are desmosomes and gap junctions, which regulate communication and flow of materials between cells. Cell-to-cell connections can be seen with an electron microscope.

Most animals ingest nutrients. Many animals take food into their bodies through an oral opening and then either engulf solid particles into digestive cells by phagocytosis ("cell eating") or, for liquid droplets, pinocytosis ("cell drinking") or absorb food molecules through cell membranes.

Parasites, such as the orthonectids (Phylum A-8) and gordian worms (Phylum A-11), often lack digestive systems. Solar-powered animals, such as *Convoluta paradoxa* (a platyhelminth, Phylum A-5) and *Elysia* (a mollusc, Phylum A-26) acquire photosynthesizing symbionts, just as did the protoctists that became plants (Phylum Pr-28).

Animals that inhabit deep-sea black smokers (hydrothermal vents) and cold seeps (cold water rising through the sea floor) do not directly depend on sunlight for energy. Rather, the energy that powers their symbioses comes from inorganic compounds such as sulfides and methane that seep up through vents in the sea floor. Tube worms, clams, and other vent and cold-seep animals are nourished by symbioses with internal chemolithoautotrophic bacteria. A chemolithoautotroph is a self-feeding bacterium that uses energy released by inorganic chemical oxidations as the source of energy for its life processes, including synthesis of organic molecules from CO_2. Vent and seep animals either digest the bacteria directly or absorb organic molecules synthesized by their symbiotic partners. These vent and seep communities are rare today but were likely typical of Earth's environment 3 billion years ago.

Animals exhibit behavior of various kinds, such as attraction to light, avoidance of noxious chemicals, and sensing of dissolved gases and temperature. Such behavior is found in members of all five kingdoms, but animals have most elaborated this theme. Early in the history of the animal kingdom, more than a half billion years ago, nervous systems including brains evolved in several lineages. Organisms in no other kingdom have nervous systems or brains.

In form, the animals are the most diverse of all organisms. The tiniest animals are termed microbes. Smaller than many protoctists, these animals require a microscope to be seen. Many of these minute animal species make up the heterotrophic fraction of the plankton (Greek *planktos,* wandering); planktonic animals—together with photosynthesizing planktonic species—constitute the base of freshwater and marine food webs.

The largest animals today are whales, sea mammals in our own class (Mammalia) and phylum (Craniata, Phylum A-37). The members of most animal phyla inhabit shallow waters. Truly land dwelling forms are found in only four phyla: chelicerates such as spiders (Phylum A-19), mandibulates (uniramians) such as insects (Phylum A-20), crustaceans such as sowbugs (Phylum A-21), and craniates such as reptiles, birds, and mammals (Phylum A-37). Species that live on land in the soil (for example, earthworms) belong to several phyla, but, requiring constant moisture, they have not freed themselves from an aqueous environment. In fact, animals of most phyla are aquatic worms of one kind or another, except insects and others of Phylum Mandibulata. Probably more than 99.9 percent of all the species of animals that have ever lived are extinct and are studied in paleontology rather than zoology.

Of all organisms, only the animals have succeeded in actively invading the atmosphere. Representatives of all five kingdoms (for example, spores of bacteria, fungi, and plants) spend significant fractions of their life cycles airborne in the atmosphere, but none in any kingdom spends its entire life history in the air. Active flight evolved only in animals. Locomotion of animals through the air independently evolved several times but in only two phyla: Mandibulata, class Insecta, and Craniata, classes Aves (birds), Mammalia (bats), and Reptilia (several extinct flying dinosaurs).

For many years and even now, some biologists assign animals to one of two large groups: the invertebrates—animals without backbones, and the vertebrates—backboned animals. All animals, except members of our own phylum, Craniata, are invertebrates. Today, about 98 percent of all living animals are invertebrates. This invertebrate-vertebrate dichotomy amply accounts for our skewed perspective. Our pets, beasts of burden, and sources of food, leather, and bone—that is, terrestrial animals closest to our size and most familiar—are members of our own phylum. From a less human-centered point of view, traits other than lack of a backbone are better indicators of early evolutionary divergence. We prefer to describe these mostly marine animals by their unique traits rather than to collectively dismiss them as invertebrates.

The animal phyla are described here in approximate order of increasing morphological complexity. Two phyla of animals, Placozoa (Phylum A-1) with its single genus and Porifera, the better-known sponges (Phylum A-2), constitute Subkingdom Parazoa ("alongside animals"); members of these phyla lack tissues organized into organs and most have an indeterminate form. The Rhombozoa and Orthonectida (Phyla A-7 and A-8) are animals whose evolution seems to have been independent of the true metazoa; rhombozoans and orthonectids do not fit the criteria of either Parazoa or Eumetazoa, the other animal subkingdom. The other 33 phyla constitute Subkingdom Eumetazoa (true metazoans); most have tissues organized into organs and organ systems.

Broad overviews of organ systems that carry out circulation, respiration, digestion, support, and reproduction will be discussed in each phylum essay. Certain organisms have open circulation with blood circulating at least partially in body spaces rather than in veins, arteries, and capillaries. In other organisms, blood is confined to arteries, capillaries, and veins in what are referred to as closed circulatory systems. A circulatory system transports dissolved gases—oxygen and carbon dioxide—whereas an excretory system functions to rid an organism of toxic wastes, such as nitrogenous wastes and salts. In regard to reproductive systems, some species are monoecious (one house, hermaphroditic), having both sexes within one individual organism; other species are dioecious (two houses), with separate male and female organisms. Monoecious organisms may be either simultaneous hermaphrodites or sequential hermaphrodites—first male and then female or first female, then male.

The Eumetazoa comprises two branches: radially symmetrical and bilaterally symmetrical animals. The Radiata—radially symmetrical organisms—are Cnidaria (Phylum A-3) and Ctenophora (comb jellies), biradially symmetrical organisms (Phylum A-4). Many species in these phyla are planktonic. Comb jellies and cnidarians encounter a uniform aquatic environment on all sides; their bodies are radially symmetrical both internally and externally. All other eumetazoan phyla have bilateral symmetry, at least at some time in their life cycles. For example, echinoderms often are radially symmetrical as adults, but all echinoderms are bilaterally symmetrical as larvae.

Characteristics of the body cavity including its embryonic origin allow most of the bilaterally symmetrical phyla to be assigned to one of three groups, but about eight phyla cannot yet be assigned because too little is known about the nature of the origin of their body cavities. Those that lack a body cavity between gut and outer body wall musculature are Acoelomata (Phyla A-5 through A-9), although neither Phylum A-7 nor Phylum A-8 has outer body wall musculature; those that have a body cavity called a pseudocoelom—not a coelom—are Pseudocoelomata (Phyla A-10 through A-14); and those that develop a true coelom are Coelomata (Phyla A-19 through A-37). What is the difference between a pseudocoelom and a coelom? Gastrulation leads to the development of two, and eventually three, tissue layers in all animals more complex than the placozoans (Phylum A-1), sponges (Phylum A-2), cnidarians (Phylum A-3), ctenophores (Phylum A-4), rhombozoans (Phylum A-7), and orthonectids (Phylum A-8). The three tissue layers are called the endoderm, mesoderm, and ectoderm (in order from the inside out) and are the masses of cells from which the organ systems of animals develop. In general, the intestine and other digestive organs develop from endoderm; the muscle, skeletal, and all other internal organ systems except the nervous system develop from mesoderm; and the nervous tissue and outer integument develop from the ectoderm. In the coelomates, the embryonic mesoderm opens to eventually form an internal body cavity lying between the digestive tract and outer body wall musculature. This body cavity is called the coelom. A pseudocoelom is also an internal body cavity lying between the outer body wall musculature and gut. Unlike the coelom, though, it is lined by loose cell masses rather than with mesoderm, and the pseudocoelom generally forms by persistence of the blastocoel, the embryonic cavity of the blastula.

Problematic phyla that do not easily fit these categories include nemertines (Phylum A-9). If one accepts that the nemertine proboscis cavity and blood vessels are homologues of the coelom, then Phylum Nemertina may be assigned coelomate status and located beside sipunculans (Phylum A-23) on the phylogenetic tree.

For animals of eight phyla whose members are bilaterally symmetrical—particularly for kinorhynchs (Phylum A-14) and loriciferans (Phylum A-17) and a newly proposed Cycliophora phylum—the nature of the body cavity is uncertain. Embryological studies are needed to determine the origin of the body cavity for members of these phyla before they can be placed in any of these groups. For priapulids (Phylum A-15), gastrotrichs (Phylum A-16), entoprocts (Phylum A-18), ectoprocts, or bryozoans (Phylum A-29), brachiopods (Phylum A-30), and phoronids (Phylum A-31), relationships to other phyla in the animal phylogeny are uncertain; some of the uncertainties are discussed in relevant essays. In establishing classification schemes, when we cannot place a species with any previously established phylum—Loricifera, for example—we create a new place for it. In creating a new phylum, we set that phylum forth as a hypothesis to be tested by studying relationships. Investigation of the origin of the body cavity is one mode of studying relationships between organisms. Affinities of the new Cycliophora (*Symbion*) phylum to existing phyla Entoprocta and Ectoprocta warrant testing.

Two groups of coelomate animals are distinguished according to the fate of the blastopore—the site of invagination of the blastula. In protostome ("first mouth") animals (Phyla A-19 through A-28), the blastopore is the site of the mouth of the adult. In deuterostomes (Phyla A-32 through A-37), the blastopore becomes the anus—the rear end of the intestine; the mouth forms as a secondary opening at the end opposite the anus. The deuterostome phyla are thought to have common ancestors more recent than their protostome ancestors. This protostome-deuterostome divergence occurred at least 520 million years ago, as judged from the presence of both protostomes and deuterostomes in the Lower (early) Cambrian fauna. The relationship of lophophorates (Phyla 29, 30, and 31), for example, to either deuterostome or protostome coelomates is not established.

Most biologists agree that animals evolved from ancestral protoctists. Which protoctists, when, and in what sort of environments are questions that are still debated. Earl Hanson (Wesleyan University) amassed much information on the protoctist–animal connection and suggested that the question remains open. Patricia Wainright of Rutgers University (and colleagues) did the same for the fungus–animal connection. The Porifera (Phylum A-2) evolved from the choanomonads (Class 5 in Phylum Pr-30, Zoomastigota), deduced from both molecular systematics (ribosomal nucleotide sequences) and details of fine structure of the cells. It is likely that the animal phyla other than poriferans, especially the eumetazoans, had different ancestors among the protoctists.

A-1 Placozoa
(Trichoplaxes)

Greek *plakos*, flat; *zoion*, animal

GENUS
Trichoplax

Only one species, *Trichoplax adhaerens*, is known in this phylum. Soft and so small as to be barely visible to the naked eye, trichoplaxes are among the least complex of animals. Members of Subkingdom Parazoa, they lack specialized sensory, digestive, excretory, respiratory, and reproductive tissues and organs and a body cavity. *Trichoplax* looks and behaves like a very large ameba—its grayish body continuously alters shape. Under higher magnification, though, it can be seen that the "ameba" is really an animal composed of a few thousand cells but no distinct tissues. Trichoplaxes are between 0.2 and 1.0 mm in diameter.

The entire surface of this mobile, minute animal is ciliated. Flat cells bear scattered undulipodia and shiny lipid droplets, making up the dorsal epithelium. The ventral surface is composed of evenly ciliated columnar cells. Between the dorsal and the ventral cell layers, ameboid mesenchyme cells containing lumpy bodies interconnect in a layer of fluid. *Trichoplax* has a dorsal and a ventral side, but it lacks both anterior-posterior (head-tail) and right-left symmetry. The animal crawls along belly side down, dorsal side up. Even fragments can right themselves to this orientation.

Trichoplaxes replicate sexually and asexually. In one asexual mode, large trichoplaxes divide by fission into two multicellular organisms. In another asexual mode, round buds—which contain dorsal, ventral, and mesenchyme cells—form and detach. Cut trichoplaxes heal rapidly, and a complete animal regenerates from any cutaway part. *Trichoplax* is interpreted to be a diploid animal that develops from eggs fertilized by sperm, but details of its life cycle have not yet been described.

In their sexual mode, trichoplaxes develop from eggs that probably come from the layer of cells on the ventral surface. Eggs form in profusion if the population of animals becomes dense. Spermlike, small cells lacking tails have been seen in aquaria containing *Trichoplax*. A membrane rises up around the eggs; this raised membrane is assumed to be a manifestation of fertilization, although meiosis and fertilization have not been seen. The fertilized egg cleaves: the two-cell stage becomes the four-cell and then the eight-cell stage, and so on, until a ball of cells—a blastula—is formed. However, because blastulas that have been watched have always stopped developing, the later stages of development are not known.

The smallest trichoplaxes—probably just recently developed from eggs—can swim, though the cilia beat asynchronously. Larger, adult trichoplaxes creep along by using their undulipodia, probably contracting with the aid of fibrous cells in the middle fluid layer. The periphery is the most mobile body region. How these animals feed is a mystery: they have no mouths at all. Sometimes, a broad pocket is formed on the ventral side over organic debris and algae thought to be food. After settling over the area on which it will feed, *Trichoplax* may secrete enzymes into this pocket, thus forming a temporary "stomach."

Trichoplax was discovered in the seawater aquarium of the Graz Zoological Institute in Austria in 1883 but has been meagerly investigated since then, even though it can be cultured in the laboratory. *Trichoplax* has always been found in seawater—in the Red Sea, in Hawaii, at the Plymouth Marine Station on the southern coast of England, at the Rosenstiel marine station at Miami, Florida, on walls of the marine aquarium at Temple University in Philadelphia, and at Woods Hole, Massachusetts. This distribution probably corresponds to the distribution of interested zoologists rather than the distribution of the little trichoplaxes.

A *Trichoplax* cell has little DNA, of the order of 10^{10} daltons (molecular mass units), which is more like that of a bacterial nucleoid or the nucleus of some small protoctists than like that of the nuclei of other animal cells. Compared with those of most animals, its chromosomes are very small—from 0.6 μm to 1.0 μm in length, the size of a bacterium. Nuclei of epithelial (diploid) cells contain 12 chromosomes. Cells of the mesenchyme layer have twice as much DNA as do epithelial cells; thus, either mesenchyme cells may have an extra set of chromosomes or the extra DNA may belong to intracellular bacteria.

Because the organization of the body of *Trichoplax* is superficially quite like that of the planulae (free-swimming larvae) of cnidarians (Phylum A-3), trichoplaxes were misidentified until careful studies showed them to be adults of their own unique group. They may have evolved from planula-like ancestors of the cnidarians; other workers suggest that trichoplaxes and sponges might be related. Comparison of ribosomal RNA sequences suggests that—as shown in our animal phylogeny—the sponges diverged from ancestral protoctists, followed by trichoplaxes, cnidarians, and ctenophores. If this is so, trichoplaxes may have lost specialized cell junctions. There is no fossil record of Placozoa that might shed light on the earliest metazoan relationships and so, until more is known about *Trichoplax*, it remains an intriguing beast of unknown ancestry.

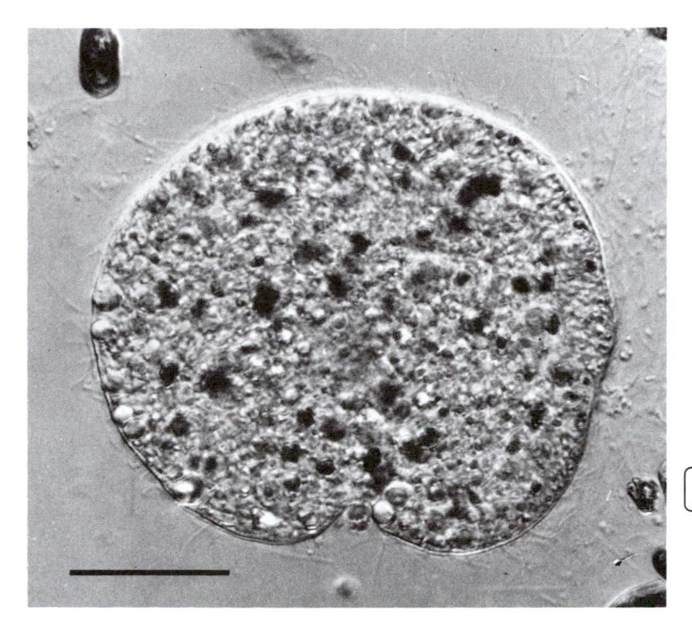

Trichoplax adhaerens, the simplest of all animals, found adhering to and crawling on the walls of marine aquaria. LM, bar = 0.1 mm. [Photograph courtesy of K. Grell; drawing of cutaway view by L. Meszoly; information from R. Miller.]

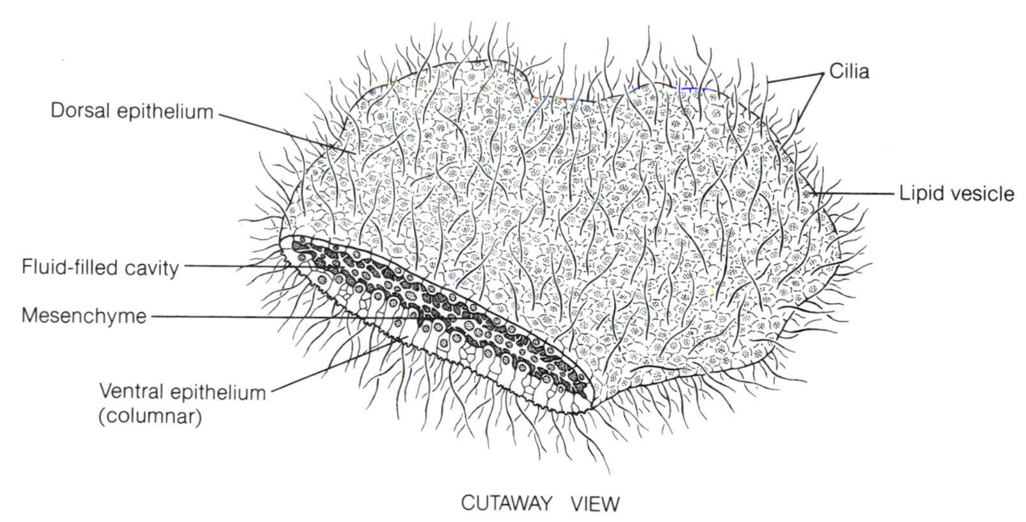

A-2 Porifera

(Sponges, poriferans)

Latin *porus,* pore; *ferre,* to bear

EXAMPLES OF GENERA

Agelas
Asbestopluma
Euplectella
Gelliodes
Grantia
Halisarca
Hippospongia
Leucosolenia
Microciona
Scolymastra
Speciospongia
Spongia
Spongilla
Stromatospongia

Poriferans, the sponges, are named for their pores. Most sponges have thousands of pores into which water flows; thus their body forms contribute in large part to current-induced water flow through their bodies. Because poriferans have no tissues or organs, like the placozoans (Phylum A-1), they belong to the Parazoa subkingdom. Also like placozoans, most sponges lack anterior-posterior and left-right symmetry. Between 5000 and 10,000 species of sponges are known; all are aquatic and only 150 species live in fresh water. Their fantastic forms—fans, cups, crusts, and tubes—range from a few millimeters wide to the 2 m tall *Scolymastra joubini,* a barrel-like glass sponge of the Antarctic. Most classes encompass various body forms.

The classification of sponges is being revised because, like certain extinct sponges, recently described living sponges—chaeteiids, stromatoporids, and sphinctozoans—deposit calcium carbonate in patterns; eventually, the number of classes recognized in this phylum may increase or decrease. Currently, the sponges may be grouped in four classes: Calcarea, Demospongiae, Sclerospongiae, and Hexactinellida. The Calcarea, calcareous sponges such as *Leucosolenia,* comprise about 500 marine species bearing spicules of calcium carbonate. Sponges in other classes have siliceous spicules fashioned from opaline silica (SiO_2). The SiO_2 is extracted from seawater, concentrated, and laid down in delicate patterns inside sclerocytes, cells that secrete spicules. The Demospongiae, about 4000 marine and freshwater species of horny sponges (Figure A), are supported by spongin, fibers of protein related to the keratins of hair and nails. Whether demosponges contain silica spicules but not $CaCO_3$ depends on the species. The commercial bath sponges *Spongia* and *Hippospongia* are demosponges; *Hippospongia* has no spicules. Coralline sponges (Sclerospongiae, 15 marine species, sometimes included in Demospongiae or Calcarea, depending on the species) have a stony calcareous skeleton of $CaCO_3$, siliceous spicules, and spongin. Corallines are tropical reef sponges, such as *Stromatospongia.* The Hexactinellida, 600 species of glass sponges, contain six-rayed siliceous spicules. Most of the elegant glass sponges (for example, *Euplectella*) dwell in the deep sea. Some biologists propose a unique phylum for glass sponges because of their unique features, such as the syncytial outer net and syncytial choanoderm (flagellated collar cells) layers; syncytia are made up of nuclei not separated by cellular membranes.

Sponges are made of two cell layers; between the layers lies an acellular, gel layer called the mesohyl, which contains ameboid cells (amebocytes) and support spicules (skeletal needles) or spongin fibers (Figure B). Amebocytes are also called archaeocytes because they can differentiate into sperm, eggs, nutrient storage cells, spicule-secreting cells, spongin-secreting cells, and waste-eliminating cells. Although both inner and outer layers contain some specialized cells, the sponge body lacks the cell cohesiveness and coordination typical of tissues in true metazoans (Phyla A-3 through A-37 but not Phyla A-7 and A-8). As multicellular organisms, sponges have intercellular junctions (desmosomes and septate junctions), basement-membrane-like structures, and matrix components (collagen and fibronectin). Nearly all sponges are sessile or encrusting, permanently attached. Water turbulence and the amount of available space tend to determine their shape and size. Many coat rocks and logs and are so shapeless that they are not recognizeable as animals.

Sponges lack mouths, intestines, muscle, nerves, and respiratory and circulatory organs. A nonnervous conducting system capable of halting the excurrent water flow in response to stimuli has been demonstrated, but it is not made up of nerves. Oxygen diffuses through the sponge body wall. Food is filtered from the copious quantities of water that flow through the sponge body. Plankton, such as dinomastigotes (Phylum Pr-7), makes up about 20 percent of their food; another 80 percent consists of detrital organic particles. Choanocytes (Figure B), distinctive cells having a collar of microvilli and a flagellum, are partly responsible for moving water (body form passively generates water flow in part) into the porocyte-lined pores and along incoming waterways called inhalent canals. Food particles either stick to the choanocyte collars or are directly engulfed by ameboid phagocytic cells lining the canals. When food is inside a phagocyte, these cells digest nutrients and allow other cells to absorb food. Wastes either diffuse out of the sponge directly through the body wall or flow out of the spongocoel (body cavity) through the excurrent opening.

Most sponges are hermaphrodites; mature organisms bear both eggs and sperm. In a few species, however, the sexes are separate. Choanocytes develop into sperm. Eggs develop either from choanocytes or amebocytes. In most sponges, clouds of sperm released through the excurrent opening of one sponge enter another sponge with incoming water. Choanocytes capture the sperm and then tranform into ameboid cells, which convey captive sperm to eggs in the mesohyl. Thus, fertilization is internal in most species. In synchrony with sperm release, the tube sponge *Agelas* releases a mass of eggs in strands of mucus to the reef to which it is attached. Zygotes develop into ciliated, free-swimming, multicellular, nonfeeding larvae. Some sponges with internal fertilization release their developing larvae into the water; others retain larvae for some time. Sponge larvae are of two main types: parenchymula or amphiblastula. Parenchymula larvae are solid and almost entirely ciliated. Amphiblastula larvae are hollow and ciliated at one end.

The amphiblastula larvae turn inside out, bringing their formerly internal cilia outside. Development of sponge larvae is not homologous to the development of other animals, although sponges form blastulae. Eventually, sponge larvae metamorphose into adults.

Some sponges reproduce asexually. Fragments may break off, be wafted away by water currents, and then grow to be individual sponges. Because of this capacity for regeneration, sponge cuttings are used to restock Florida's depleted sponge beds. Many freshwater and a few marine sponges also reproduce by means of gemmules. These overwintering buds—composed of nutrient-laden ameboid cells surrounded by a tough outer layer of epithelial cells with spicules—disperse and, after dormancy, grow to form new sponges.

Many sponges have symbiotic algae that provide food and probably oxygen, remove waste, and screen the sunlight. Symbionts may be photosynthetic cyanobacteria (Phylum B-5) or red (Phylum Pr-25), green (Phylum Pr-28) or brown (Phylum Pr-17) algae. In some species, symbionts are transmitted from adult sponges to their offspring by adhering to gemmules. Because of their algal symbionts and their own pigments, sponges may be red, orange, yellow (most with carotenoids), green, blue, purple, brown, or white. Some sponges luminesce.

Parasitic sponges have not been described, and most animals prefer other sources of food. Sponge spicules are evidently unsavory—a needle-sharp deterrent to predators, with the exception of snails and nudibranch molluscs (Phylum A-26), a few sea stars (Phylum A-34), and a few fish (Phylum A-37). Feather stars perch on sponges to feed.

A carnivorous demosponge has recently been discovered in a nutrient-poor, deep, cold, sea cave. *Asbestopluma* passively snares crustaceans (Phylum A-21) on filaments covered with hook-shaped spicules. Subsequent to capture, thin new filaments envelop the prey, digesting externally, and transferring nutrients into the body. This Mediterranean carnivore lacks choanocytes and canals and does not filter water.

A *Gelliodes digitalis,* one of the simpler sponges, shown live in its marine habitat. Water enters through its pores and exits through a single excurrent opening, called the osculum, on top. Bar = 10 cm. [Courtesy of W. Sacco.]

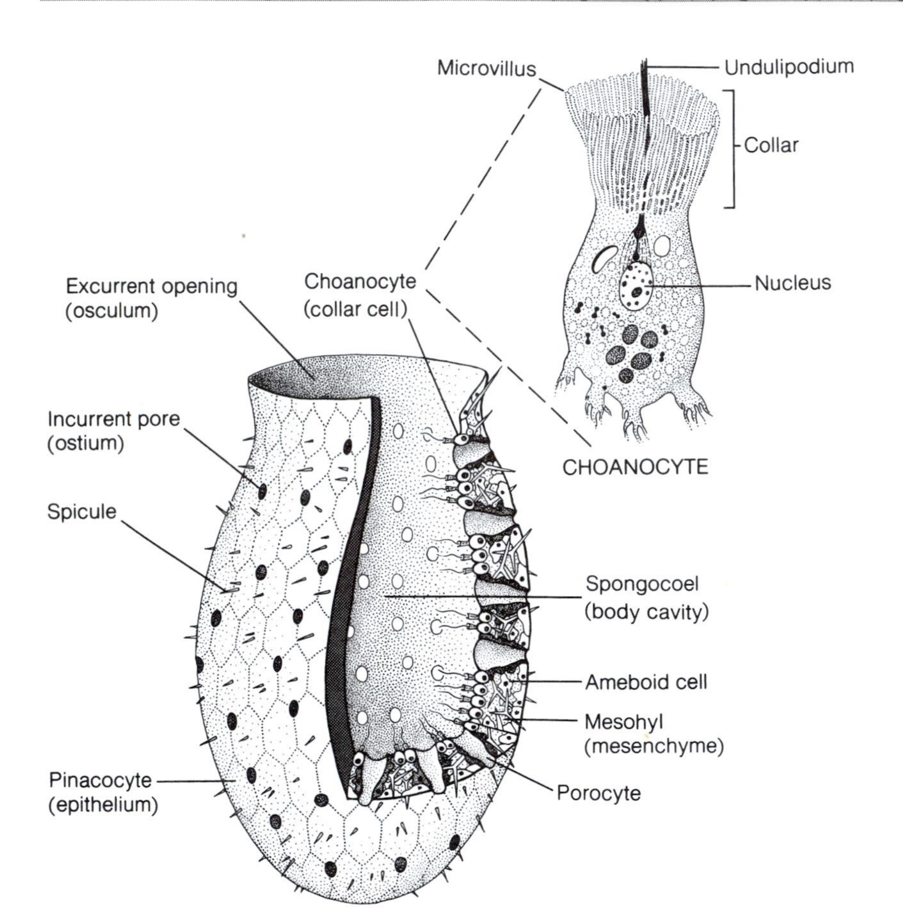

B Cutaway diagram of simple sponge. The water current through the sponge is generated, in part, by two factors: (1) sponge shape (architecture) passively generates flow through incurrent pores and out at the osculum and (2) choanocytes (collar cells) actively generate additional water flow and capture food and sperm. [Drawing by L. Meszoly; information from S. Vogel and W. Hartman.]

The oldest sponges in the fossil record, the well-preserved Burgess shale sponges, date from the early Cambrian period, about 500 million years ago. Sponges seem to form a dead end, not ancestral to other metazoans. Of all animals, poriferans are the easiest to relate directly to their protoctist ancestors—free-living colonial choanomastigotes (Phylum Pr-30) that are remarkably similar to sponge choanocytes. Sponges may have been the earliest animal lineage to diverge from protoctists, as inferred from ribosomal RNA data.

Sponges harvested off the West Indies, Florida, Mexico, and the Philippines are used by lithographers, potters, silversmiths, and bathers. Oysters and other shellfish are sometimes destroyed by sponges boring into their shells as the sponges excavate living space in the shell.

A-3 Cnidaria

(Cnidarians, hydras)

Greek *knide,* nettle; *koilos,* hollow; *enteron,* intestine

EXAMPLES OF GENERA

Acropora
Alcyonium
Antipathes
Atolla
Aurelia
Branchioceranthus
Chironex
Corallium
Craspedacusta
Cryptohydra
Cyanea
Dendronephthya
Ediacara
Haliclystus
Heliopora
Hydra
Metridium
Millepora
Obelia
Physalia

(continued)

Sea anemones, jellyfish, hydras, and corals are among the 9400 species of Cnidaria. These radially symmetrical invertebrates are the least morphologically complex members of Subkingdom Eumetazoa, the true metazoa. The term coelenterate is used only in reference to both the Cnidaria and the Ctenophora phyla; it is no longer used as a synonym for Cnidaria. All cnidarians are aquatic and nearly all are marine. The four classes of cnidarians are the Hydrozoa (hydras), the Scyphozoa (true jellyfish), the Anthozoa (most corals and sea anemones), and the Cubozoa (sea wasps and several other genera named for their cube-shaped bodies). Cnidarian tentacles and oral arms are replete with stinging cells called cnidoblasts, each containing an intracellular nematocyst, unique to this phylum.

Cnidarians have two basic body patterns—polyp and medusa (Figures A and B); because they have more than a single body pattern, cnidarians are said to be polymorphic. Polyps, such as *Hydra,* are cylindrical animals that live mouth upward (Figures C and D). Some polyps are sedentary; others glide or somersault or employ their tentacles as legs. Medusae—*Craspedacusta,* for example—usually swim free, the Frisbee-, umbrella-, or box-shaped bell pulsating, mouth downward, with tentacles trailing (see Figures A and B). The tentacles of medusae resemble the snaky locks of the mythical Medusa. Some cnidarian species are exclusively polyps; others are medusae only; and still others alternate between polyp and medusa forms.

Cnidarian cells are assembled into tissues, in comparison with placozoans, which lack tissues. Between a cnidarian's outer layer of epidermis and its inner layer of gastrodermis lies an intermediate layer called the mesoglea. This mesogleal layer contains translucent secretions and often loose cells but is not organized as a tissue. The gastrodermis lines the gastrovascular cavity, or stomach.

Cnidarians have nerve nets but no central nervous system. The pacemaker of the nerve net maintains the swimming rhythm of medusae. Motion and light-sensitive cells on the edges of many medusae enable them to detect light and orient themselves. The nerve fibers of cnidarians are the only truly naked nerves in the

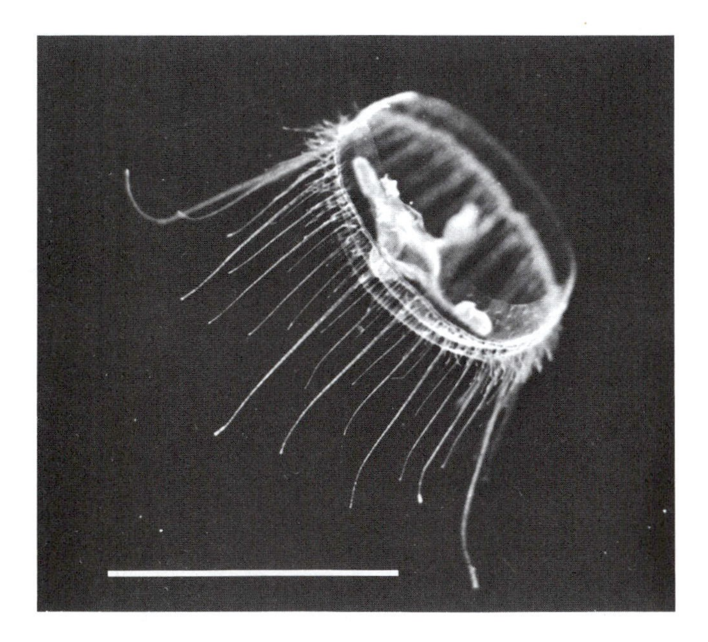

A *Craspedacusta sowerbii,* the living medusa of a freshwater cnidarian. Contraction of the bell expels water, thereby propelling the medusa. Class Hydrozoa. Bar = 10 mm. [Courtesy of C. M. Flaten and C. F. Lytle.]

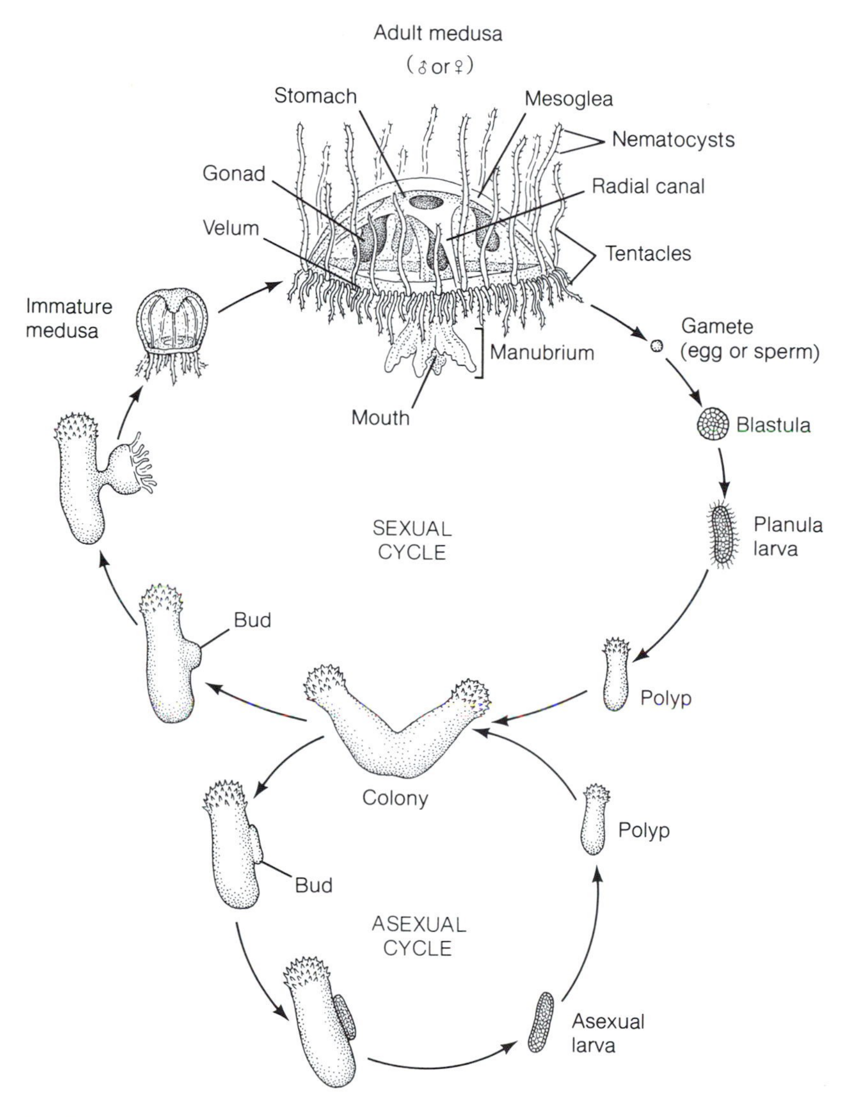

B The life cycle of *Craspedacusta sowerbii,* a freshwater hydrozoan, and the anatomy of the adult medusa. The mouth of the medusa opens at the external end of the manubrium; the stomach is at the internal end. [Drawing by L. Meszoly; information from C. F. Lytle.]

Psammohydra
Renilla
Tripedalia
Tubipora
Tubularia
Velella

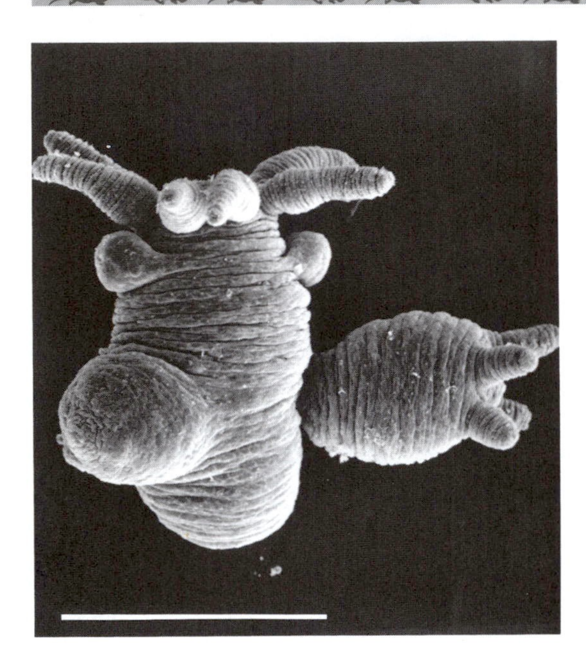

C A sexually mature *Hydra viridis* (Ohio strain). The tentacles are at the top of the upright sessile form, two spermaries are located below the tentacles, a large swollen ovary is shown at the lower left in this picture, and an asexual bud is at the right. *Hydra viridis* is normally about 3 mm long when extended, but this one shrank by about 1 mm when it was prepared for this photograph. SEM, bar = 1 mm. [Courtesy of C. F. Lytle.]

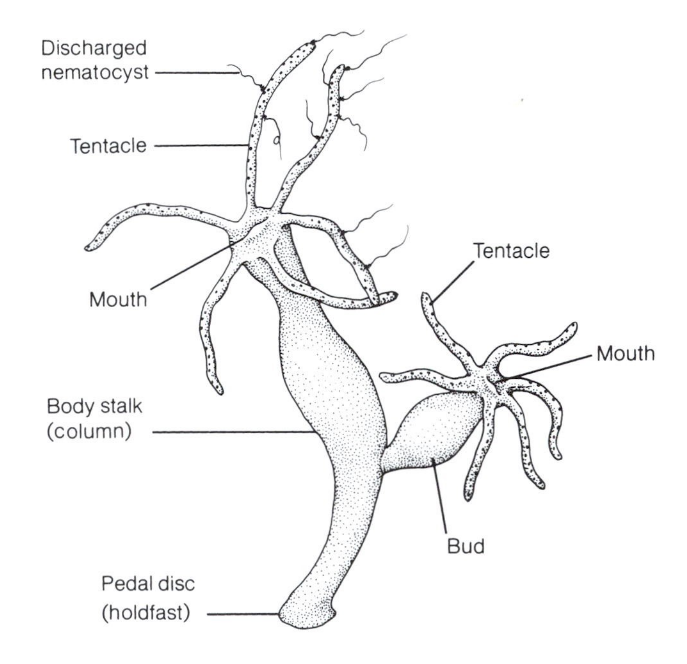

D Overall view of a typical *Hydra,* between 0.5 mm and 10 mm long, depending on species. [Drawing by L. M. Reeves.]

animal kingdom—all others are covered by sheaths of insulating material, such as myelin. Most cnidarian nerve junctions transmit impulses in two directions, whereas nerve junctions of all other animals transmit impulses in only one direction.

The contractile system of polyps consists in part of a layer of epitheliomuscular cells; at the base of the epitheliomuscular cells are contractile fibers that run longitudinally and are anchored in the mesoglea. In addition, cnidarians have nutritive-muscular cells on the inside (below the epithelium) with contractile fibers that run circularly; nutritive-muscular cells contract, taking up and digesting food particles from the gastrovascular cavity. A medusa has muscle fibers in its swimming bell. Although cnidarians have no bones, polyps and medusae are stiffened by fluid pressure in the gut itself and by the mesoglea with its collagen matrix. Hydrocorals (Hydrozoa) and true corals (Anthozoa) secrete calcium carbonate exoskeletons within which their soft polyps shelter. Medusae lack a carbonate skeleton.

Cnidarians are usually carnivores. An herbivorous soft coral (*Dendronephthya*) has recently been described; this reef inhabitant lacks symbiotic algae and has poorly developed nematocysts but feeds almost exclusively on phytoplankton. Plant-digesting enzymes such as laminarinase and amylase are present in the soft coral *Alcyonium*. The cubozoans may pursue fish prey. Cnidarians sting when they contact active prey—worms, crustaceans, fish, comb jellies, diatoms, and other protoctists. Cnidoblasts triggered by touch or by chemical stimulation or by both forcibly discharge their stings—nematocysts—for food getting or defense (Figure E). Unlike sponges (Phylum A-2), which lack a stomach, cnidarians digest food within their single digestive sac, the gastrovascular cavity, or stomach, which opens through the mouth. The mouth squirts out waste and serves as an anus—which cnidarians lack. Radial canals distribute dissolved oxygen and carbon dioxide as well as nutrients from the stomach to the periphery of the medusa.

Within the cells of most shallow-water corals, horny corals, some sea anemones, and a few medusae, algae are symbiotic partners. The algae sustain the cnidarian partner with photosynthate and oxygen. Symbionts are not common in scyphozoans.

Hydrozoans, with about 3100 described species, include the freshwater hydras, marine colonial hydroids such as Portuguese man-of-war, and fire corals. A velum—characteristic of most hydrozoan medusae—forms a rim around the umbrella margin. *Psammohydra* is interstitial in sea sand, completely cilia covered, and less than 1 mm in length. Hydras are named for the nine-headed dragon slain by Hercules in Greek myths. Hydrozoans can be polyps; many have small or abortive medusae or lack medusae altogether. Hydrozoan polyps usually reproduce by budding daughter polyps to form polyp colonies and medusae, whereas hydrozoan medusae reproduce sexually, releasing eggs and sperm from gonads along the radial canals (see Figure B). Hydrozoan polyps make either eggs or sperm in most species; polyps are hermaphrodites in a few species. The zygote develops from a fertilized egg into a microscopic blastula and then into a free-swimming, ciliated, solid mouthless larva called a planula. Planula larvae metamorphose into polyps.

All 200 or so species of Scyphozoans are marine—most are free-swimming medusae. *Haliclystus* and a few other scyphozoans are sessile medusae. Medusae of scyphozoans and hydrozoans are frequently called jellyfish because their mesoglea is thick relative to that of other cnidarians. Scyphozoan medusae do not have vela. Sexual reproduction by scyphozoan medusae produces zygotes that grow into planulae and then usually into sessile polyps that reproduce asexually, giving rise to stacks of tiny, swimming, incipient medusae called ephyra. Ephyra eventually develop into adult medusae. Planulae of some open-ocean scyphozoans bypass the polyp stage, metamorphosing directly into medusae. Many scyphozoans brood their larvae in pouches or in oral arms.

Cubozoa consists of the sea wasps, bearing one or a group of tentacles at each of the four corners of their glassy medusae. *Tripedalia*, *Chironex*, and other sea wasps are active swimmers in tropical and subtropical seas. In human encounters, nematocysts usually cause just nasty stings, although sea wasps have been fatal to people. Lenses, light-sensing pigments, and retinas make sea wasp eyes among the most complex invertebrate eyes. The cubomedusan planula gives rise to a polyp; the polyp subdivides into additional polyps (rather than into multiple medusae, as in scyphozoans). Cubozoan medusae reproduce sexually, but few life cycles of these venomous cnidarians are described.

Anthozoans, as solitary or colonial marine polyps, include about 6500 species—sea anemones, sea pens, sea fans, sea pansies, stony (true or hard) corals, and soft corals. Anthozoans form polyps exclusively, never medusae. Some anthozoan species are hermaphroditic, whereas others have separate sexes. Fertilized anthozoan eggs usually develop into planula larvae that settle, attach, and then metamorphose into polyps, cemented by secretions of their pedal disc. Some anemones are viviparous: offspring of sexually mature polyps are "born." Anemones may also reproduce asexually by splitting in two, by budding, or by pedal laceration—splitting off part of the pedal disc. Without their symbionts, most anthozoans can survive, but they grow faster in sunlight with partners and corals deposit limestone faster.

Coral reefs—underwater limestone ridges in shallow tropical seas—usually form by combined secretions of several species of cnidarians and other carbonate-precipitating organisms, such as chlorophytes (Phylum Pr-28) and rhodophytes (Phylum Pr-25).

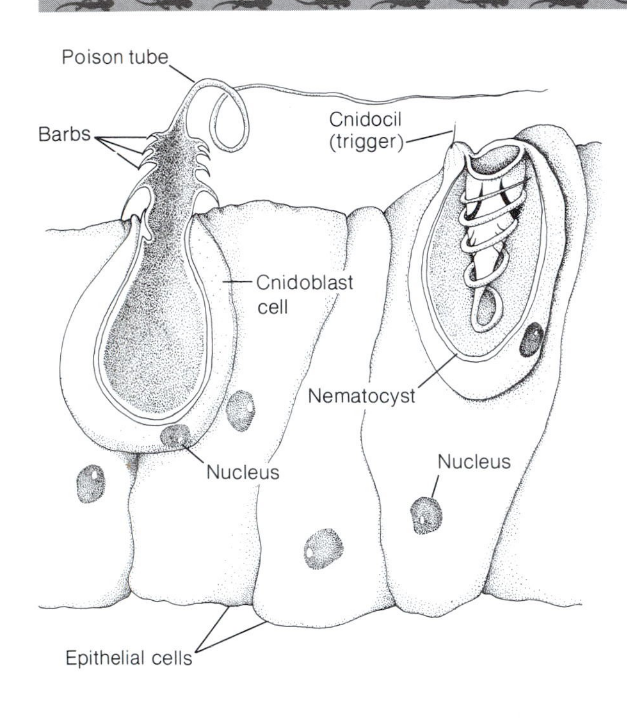

E Discharged and undischarged nematocysts. Toxin is injected through the poison tube. The undischarged nematocyst is about 100 μm long. [Drawing by L. M. Reeves.]

Soft corals predominate in Atlantic reefs; a soft coral does not lay down an exoskeleton, only internal spicules. Hard corals are more important in the Pacific. Below a depth of 60 m, corals do not form reefs because the shortage of light limits photosynthesis by the algal symbionts. Most symbionts of anthozoans are dinomastigotes (Phylum Pr-7); for example, *Gymnodinium adriaticum,* also called zooxanthellae, are yellow in color. Zooxanthellae inhabit coral polyp tissue in densities as high as 5 million/cm^2. Corals that lack algae can live to a depth of about 3000 m. Reefs support life on land by serving as barriers that reduce waves. Reefs provide a sea haven for protoctists, fish, and other marine animals. Coral reefs form islands, such as Bermuda, the Bahamas, and St. Croix.

Cnidarians are eaten in Korea, Japan, and China. Jewelry has been carved from the internal limestone skeletons of the red coral *Corallium rubrum* since pre-Roman times, from black coral *Antipathes,* and from the blue coral *Heliopora.* Overcollecting of corals prompted the United States to forbid the importation of coral. Biocoral, a biomaterial derived from natural coral, is being used for jaw and face bone grafts in Europe and the United States; the porous structure of coral facilitates movement into the substitute bone graft by cells that form bone (osteoblasts). The biocoral graft is partly replaced by normal bone when the graft is resorbed.

Many of the newly discovered deep-sea medusae bioluminesce, sparkling blue green. One species sheds its bioluminescent tentacles on attackers; the bell then pulses off into the wine dark sea and eventually regenerates tentacles. *Branchiocerianthus,* a hydrozoan polyp, may reach 2 m in length. This giant is a seafloor-deposit feeder. The sea blubber, or lion's mane, *Cyanea,* is the largest medusa—its bell is more than 3.6 m wide with tentacles more than 30 m long. The tiniest cnidarians, such as *Cryptohydra,* are polyps and jellyfishes smaller than 2 mm in diameter.

Cnidarians are among the oldest fossil metazoans. *Ediacara,* a well-known South Australian fossil found as a sandstone imprint, has been interpreted to be a hydrozoan—it is nearly 700 million years old. Imprints of scyphozoans, hydrozoans, and medusae are found in Burgess shale, Cambrian rocks 500 million years old. Fossil corals abound from the Ordovician through the Devonian (from 500 million to 430 million years ago.)

Major differences between cnidarians and ctenophores (Phylum A-4) are in tentacle morphology (hollow tentacles in cnidarians, solid tentacles in ctenophores), pattern of development, alternation of generations (cnidarians are polymorphic, whereas ctenophores are monomorphic, lacking polyps), and the presence of colloblasts (adhesive, nonstinging cells) but not nematocysts in ctenophores. Whereas cnidarians are radially symmetrical, ctenophores are biradially symmetrical. The middle layer (mesoglea) differs between cnidarians and ctenophores: cnidarians have loose cells and muscle fibers in mesoglea; ctenophores have muscle fibers. The two phyla are no longer believed to derive from common ancestors; similarities of cnidarians and ctenophores appear to be due to convergence rather than close phylogenetic relationships. Although at least one theory of the origin of the rest of the phyla of metazoan animals has them evolving from Cnidaria, ribosomal RNA data suggest that sponges and ctenophores branched from protoctist ancestors first, followed by trichoplaxes and cnidarians.

A-4 Ctenophora
(Comb jellies)

Greek *kteis,* comb; *pherein,* to bear

EXAMPLES OF GENERA

Bathyctena
Beroë
Bolinopsis
Cestum
Coeloplana
Ctenoplana
Euchlora
Eurhamphea
Lampea (Gastrodes)
Mertensia
Mnemiopsis
Ocyropsis
Pleurobrachia
Thalassocalyce
Tjalfiella
Velamen

Sea walnuts, sea gooseberries, cat's eyes, and all other comb jellies belong to Phylum Ctenophora. Ctenophoran bodies are flexible and mobile, the consistency of soft jelly in a membrane bag. Paddlelike comb plates (ctenes) unique to ctenophorans sweep these translucent, biradially symmetrical invertebrates through the sea. Ctenophores—like nudibranch molluscs and turbellarian flatworms—retain cilia as locomotor organs throughout life. Thousands of individual cilia fused together at their bases make up a comb plate; eight rows of comb plates extend along the length of the ctenophore (Figure A). Each comb plate (ctene) mechanically activates the next to lift in sequence from mouth to the opposite—aboral—end to propel the animal mouth end forward. Many comb jellies swim with a combination of comb action and muscle movements; some also swim like jet-propelled cnidarian medusae. Benthic comb jellies crawl over other organisms on the seafloor. Comb jellies are weak swimmers. Currents, tides, and wind carry planktonic ctenophores through open seas from the Antarctic to the Arctic. Tides and currents sometimes concentrate comb jellies in vast numbers. When beached, the fragile bodies collapse and quickly dry.

Comb jellies range in size from 0.4 cm to more than 1 m in length. These predators compete with commercial fish for copepods (crustaceans) and are themselves food for sea turtles, fish, medusae, and other ctenophores. Multicolored flashes in the night sea often originate from bioluminescent comb jellies. By day, comb jellies are iridescent; their plates refract light. When touched, *Eurhamphaea* releases sparkly blue green ink. Tropical comb jellies are tinted delicate violet, rose, yellow, or brown by symbiotic algae. Deep-sea comb jellies may be purple or red.

About 100 comb jelly species have been described in two classes, Tentaculata and Nuda. Most are grouped in Class Tentaculata and, like the *Bolinopsis* shown in Figure A, have two tentacles that retract into pits in larval stages and generally in adulthood. The twofold symmetry of the tentacles is superimposed on the comb jelly's radial symmetry. The solid tentacles are usually extended into sinuous fishing nets as much as 100 times the length of the body. Most lobate comb jellies—such as *Bolinopsis*—have elaborate large tentacles that are less apparent because their tentacles are hidden by the lobes. Tentacles snare living prey but are not used in swimming. However, creeping ctenophores extend their tentacles and raft with sea currents.

Not all of these ocean floaters bear tentacles; comb jellies lacking tentacles are grouped into the other class of ctenophores, Class Nuda. The North Sea thimble jelly, *Beroë gracilis,* has no tentacles at any stage in its development (Figure B). *Beroë* gulps in comb jellies and salps (soft urochordates, Phylum A-35) by rapidly opening its mobile, muscular lips lined with "macrocilia." Under the electron microscope, macrocilia are actually multicilia—composed of thousands of interconnected cilia arising from a single cell. Macrocilia shaped like sawteeth take bites from or trap prey. After feeding, *Beroë* zips its lips shut with reversibly adhesive strips of cells having highly folded membranes; closed lips probably reduce drag as *Beroë* swims mouth-forward to seek prey.

Comb jellies of both classes have the same general body plan; a comb jelly can be sliced through its long axis—from mouth to aboral end—to produce mirror images. A cut through the long axis rotated 90 degrees results in different identical halves. Ctenophores have a biradial body plan. The central cavity is lined with gastrodermis. Interposed between epidermis and gastrodermis, mesoglea provides buoyancy, stores food reserves, and contains ameboid cells that develop into smooth muscle within the mesoglea. Oxygen diffuses in and carbon dioxide out, across the body wall. Comb jellies lack any special circulatory or respiratory organs.

The comb jelly nervous system is a diffuse epidermal net that coordinates without a central nervous system. An aboral apical sense organ (a statocyst) determines the orientation of the animal (see Figure A). Tilting the comb jelly brings several hundred calcareous stones called statoliths, each 5 to 10 μm in diameter, to bear on one of four fused groups of cilia called balancers, which support the statocyst as on a spring. From each balancer, a ciliated groove branches to two comb rows and controls them. The pair of comb rows corresponding to each balancer beat faster or slower, depending on the tilt, and thus restore the animal to an upright position. The apical sense organ and the nerve net synchronize swimming.

Adhesive cells called colloblasts or lasso cells stud the tentacles and—in some species—oral lobes. Colloblasts (Figure C) entangle live fish, crustaceans, fish eggs, and other zooplankton. The sticky head of the colloblast attaches to two filaments, one of which spirals around the other and acts as a spring. This design prevents struggling prey from tearing free. Lobate comb jellies, such as *Bolinopsis,* gather prey with oral lobes covered with colloblasts and mucus; a pair of ciliated flaps called auricles guides food into the mouth (see Figure A). Whatever the food—live larvae, eggs, tiny fish, arrow worms (Phylum A-32), copepods—cilia and mucus carry it into the digestive cavity. Ciliated gastrodermis lines the digestive cavity of all comb jellies. Comb jellies with tentacles wipe their tentacles across their mouths to unload captive food. Perhaps poisonous or anesthetic mucus such as that secreted by the colloblast head facilitates prey capture. Unlike nematocysts of cnidarians, colloblasts do not penetrate prey.

Enzymes secreted into the digestive cavity break down food. The resulting nutrients are distributed along canals that run through the mesoglea, the thick jelly that lies between the epidermis and the digestive cavity. Some of the dead-end canals run beneath the comb plates. Undigestible food is voided through the

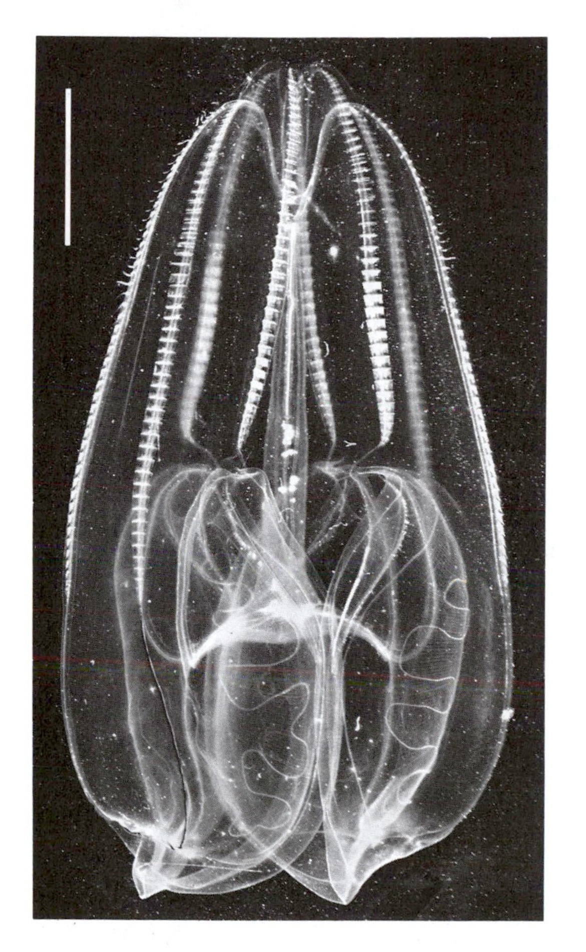

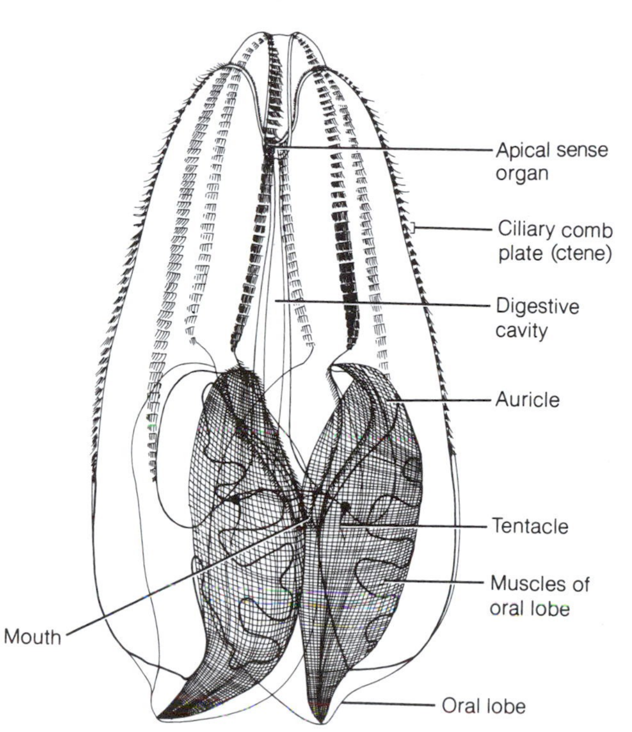

A *Bolinopsis infundibulum* (alive), a common northern comb jelly. A planktonic species of Class Tentaculata, *Bolinopsis* swims vertically with its mouth end forward—at the bottom of this photograph—and enmeshes prey in mucus on its extended ciliated oral lobes. Tentacles are short in this adult but long in the young. Bar = 1 cm. [Photograph courtesy of M. S. Laverack; drawing by I. Atema; information from M. S. Laverack.]

mouth and through two anal pores beside the statocyst. No functional excretory system has been demonstrated.

Ctenophores are sequential or simultaneous hermaphrodites. Ovaries and testes develop under the comb rows along each digestive canal. Gametes, shed into the digestive cavity, are released through the mouth or through gonopores between comb plates. In most, fertilization takes place externally; creeping nonplanktonic comb jellies have internal fertilization. Fertilized eggs develop into free-swimming cydippid larvae. Cydippids of all tentaculate ctenophores are oval with two tentacles; those of Class Nuda lack tentacles. Cydippids develop into sexually mature forms in some species; they metamorphose into adults gradually in other species. *Tjalfiella,* a viviparous comb jelly, is an exception that broods its young in special pouches and then releases free-swimming young. *Tjalfiella* bear comb plates only as juveniles and live as commensals (benefiting from a relation with another species without either

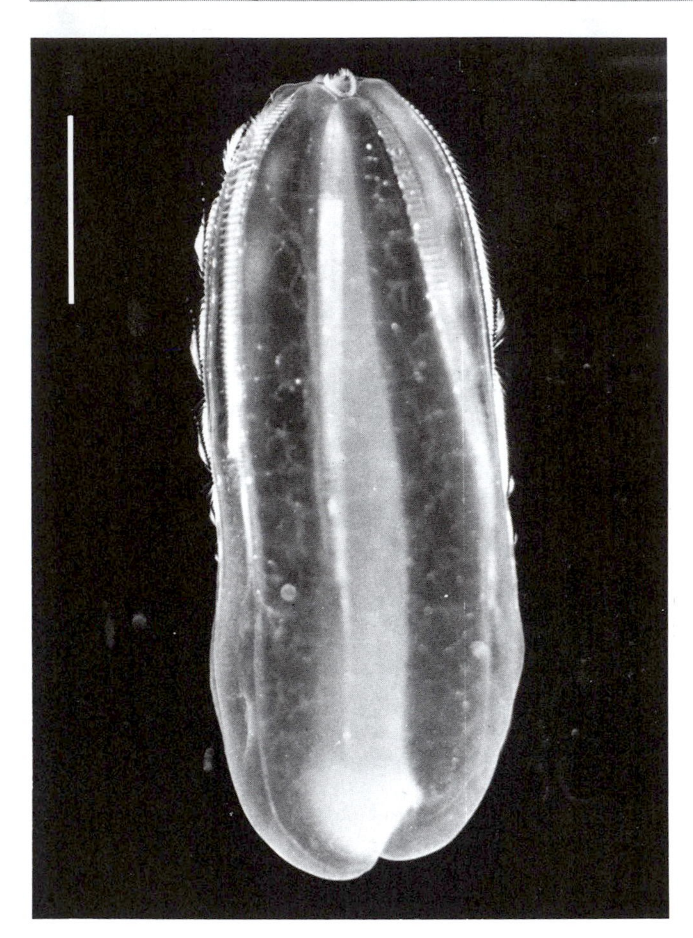

B A living comb jelly, *Beroë cucumis,* which lacks tentacles. A member of Class Nuda, *Beroë* is common in plankton from Arctic to Antarctic seas. It engulfs prey with muscular lips visible here at the bottom of the animal. Bar = 1 cm. [Photograph courtesy of M. S. Laverack.]

benefiting or harming the host) on the seafloor. *Mnemiopsis* can reproduce asexually by regeneration from fragments. Tentacles, combs, and colloblasts regenerate readily in most ctenophores. In a process similar to pedal laceration of sea anemones (Phylum A-3), platyctenes—flat benthic ctenophores of Order Platyctenida, Class Tentaculata—reproduce asexually.

Large—as much as 2 m long—ribbonlike Venus's girdle (*Cestum veneris* and *Velamen*) undulate through the water, rolling up and unrolling by contraction of muscle fibers in the mesoglea. They feed with small tentacles along the ribbon's edge. *Ocyropsis* swims both with comb plates and by flapping its oral lobes. *Ocyropsis* surrounds prey with an oral lobe, drawing food to its mouth. Some comb jellies live at depths of several kilometers. Not all comb jellies are planktonic. Leaflike *Ctenoplana* creeps on the water surface or on the floor of shallow tropical seas and swims in the plankton. *Coeloplana,* a flattened, tentacled creeper found in the Indo-Pacific Ocean, loses its combs as an adult. *Lampea pancerina*—a free-living bottom dweller—is the only parasitic (symbiotrophic) ctenophore described so far. Its juvenile form, which has a different name, *Gastrodes parasiticum,* bores into and feeds on the floating tunicates *Cyclosalpa* and *Salpa* (Phylum A-35).

Because ctenophora have such soft and fragile bodies, their potential for preservation as fossils is very low, and clues to their evolution must be inferred primarily from living forms. X-rays have detected a single fossil comb jelly 400 million years old in Devonian slate, complete with comb rows, statocyst, and tentacles. Some suggest that ctenophores evolved from medusa-shaped cnidarians (Phylum A-3), but the similarity of medusae to ctenophorans seems only superficial. Comb jellies and hydrozoan cnidarians are more generally thought to share a common ancestor. Limited sequence data portray sponges and ctenophores as having branched from protoctists, with cnidarians branching later. Cnidarians and ctenophores have evolved similar features: mesoglea interposed between epidermis and gastrodermis, carnivory, bioluminescence, a netlike nervous system, and the ability to regenerate body parts. However, in fundamental morphology, cnidarians and ctenophores differ: the muscle fibers within the mesoglea of ctenophores are lacking in cnidarians; colloblasts of comb jellies differ from nematocysts of cnidarians; and only ctenophores bear ctenes. Ctenophore muscle fibers develop from ameboid cells in mesoglea rather than from epidermal and gastrodermal cells as in cnidarians. Most ctenophores swim with a combination of comb action and muscle movements; a few comb jellies use the jet propulsion of medusae. Some cnidarians are polymorphic; ctenophores are monomorphic, lacking a polyp form. Cnidarians are radially symmetrical; ctenophores are biradial.

Comb jellies are underrepresented in descriptions of the open-sea ecosystem because their delicate bodies are difficult to net and preserve. With new tools—such as the remotely operated vehicles—comb jellies are revealed to be abundant and diverse. Diving in submersibles, marine biologists gather ctenophores by using a slurp gun, a gentle vacuum cleaner used to collect plankton. A slurp gun consists of a collecting cylinder connected at one end to a flexible hose, all mounted on a submersible vehicle. The free end of the hose is positioned near an animal by the robot arm of the submersible and the comb jelly is sucked into the cylinder. From field observations, fragile comb jellies are considered the most abundant planktonic animals between 700 and 400 m below the surface.

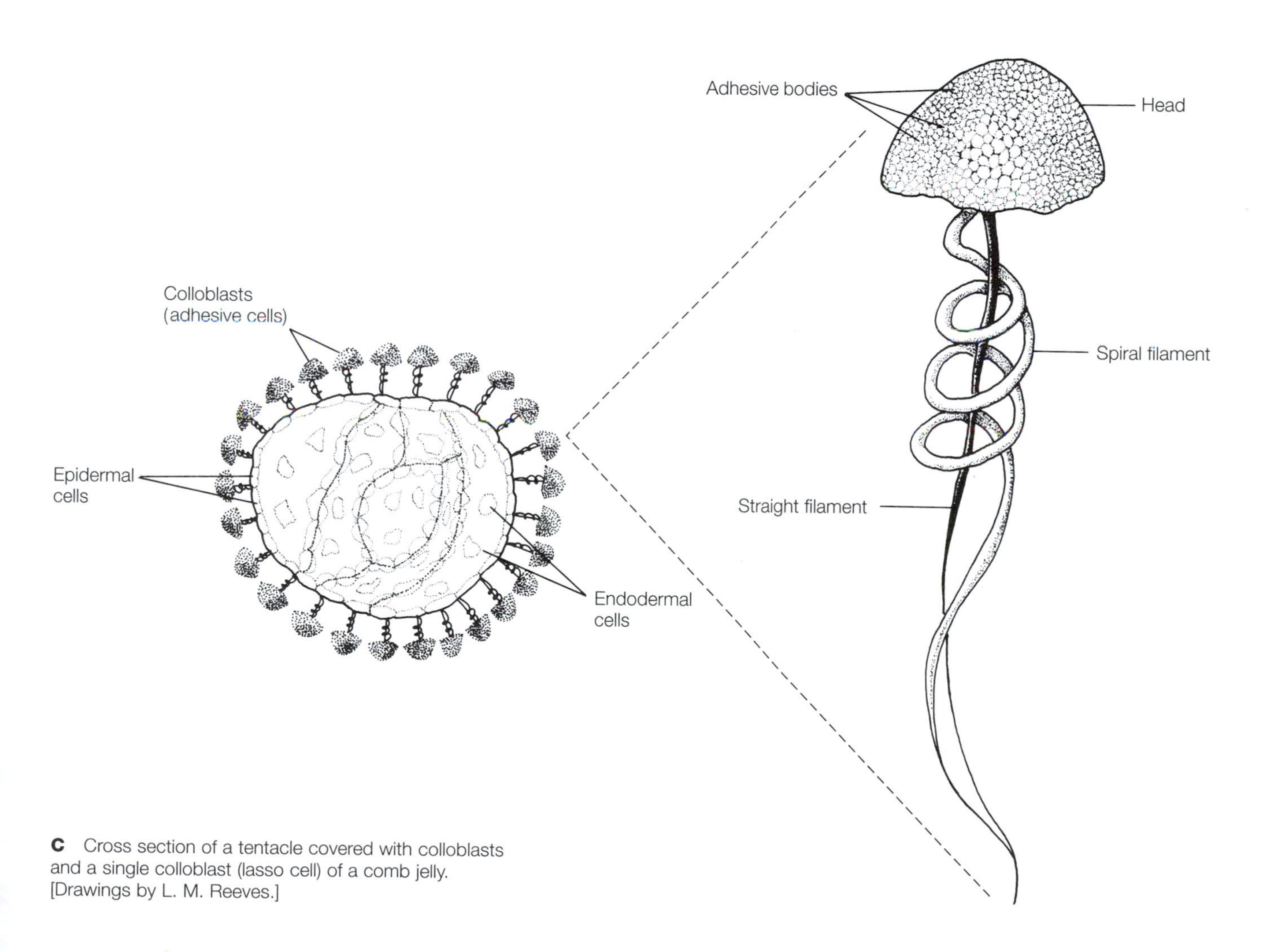

C Cross section of a tentacle covered with colloblasts and a single colloblast (lasso cell) of a comb jelly. [Drawings by L. M. Reeves.]

A-5 Platyhelminthes
(Flatworms)

Greek *platys*, flat; *helmis*, worm

EXAMPLES OF GENERA

Bothrioplana	*Echinococcus*	*Planaria*
Convoluta	*Fasciola*	*Procotyla*
Dipylidium	*Hymenolepis*	*Schistosoma*
Dugesia	*Opisthorchis*	*Taenia*

Platyhelminthes are the ribbon and leaf-shaped flatworms. The soft body of the flatworm is bilaterally symmetrical. Structures for capturing and consuming prey are localized in the anterior end except in turbellarian flatworms, in which the mouth is ventrally located. Flatworm organs are composed of tissues and are organized into systems. Flatworms have distinct mesoderm tissue in comparison with the gelatinous mesoglea middle layer of cnidarians and ctenophores. The platyhelminth, like the cnidarian, lacks an anus. The flatworm middle tissue layer—a loose mesoderm called parenchyma—never splits into a cavity (coelom) in which internal organs are suspended. Flatworms and other animals without a coelom are called acoelomates. Flatworms—having three tissue layers—are triploblastic, have spiral cleavage of their eggs, and are among the least complex of bilaterally symmetrical true metazoans.

There are three classes of flatworms: Turbellaria, including brightly colored free-living forms as well as some commensal and parasitic species; Trematoda (flukes); and Cestoda (tapeworms). Some workers consider monogenean (having a single host) trematodes a fourth class, the Monogenea consisting of gillworms and parasites in coral reef fish. The four-class system considers digenean (having two hosts or more) trematodes Class Digenea. We retain three classes of platyhelminths, with monogeneans and digeneans in Class Trematoda. Flatworms are masters of adaptation, exploiting an enormous variety of habitats. Some live in bat guano, others in the mantle fold of limpets. Members of many animal phyla—certainly an enormous number of vertebrates—play host to ubiquitous flatworm parasites. In sediments low in molecular oxygen, a few flatworms utilize energy by oxidizing hydrogen sulfide. A survey of the phylum reveals that flatworms tolerate an immense temperature range from minus 50° to plus 47°C.

Some free-living flatworms are marine, many are freshwater, and a few dwell in moist soil. Soil flatworms are mainly tropical, whereas aquatic forms are more abundant in temperate than in tropical waters. Flatworms frequent certain submerged plants and are easily collected by baiting with meat. Platyhelminths have several cilia per cell; gnathostomulid worms (Phylum A-6) have a single cilium for each cell. By simultaneously sweeping ventral cilia through secreted mucus and generating muscular waves, free-living flatworms glide over surface films on water, plants, and soil. On the ventral surface of free-living flatworms, duoglands (adhesive and releaser cells) secrete either adhesive that attaches the worm to its substrate or a substance that detaches it. Most free-living aquatic species are benthic, a few swim with undulations or loop along substrate like caterpillars, and some are plankton.

There are about 20,000 species of flatworms altogether. Some species are richly colored. Others are bright green; they harbor symbiotic algae called zoochlorellae. Tapeworms are the largest platyhelminths—some reach a length of more than 30 m. The smallest are less than 1 mm in length.

Turbellarians are detritus feeders, carnivores, and scavengers. They eat insects, crustaceans, tunicates, bivalve molluscs, other worms, bacteria, mastigotes (Phylum Pr-30), ciliates (Phylum Pr-8), and diatoms (Phylum Pr-16). Most turbellarians are marine; some inhabit the digestive tract of sipunculans and echinoderms; a few are terrestrial in damp habitats or are freshwater species. Digestive systems of turbellarians range from a straight or branched gut to absence of a gut; food moves from the pharynx of acoel (lacking a gut) turbellarians into loose digestive cells. Some jab food by using a proboscis separate from the mouth. Others "vacuum out" soft parts of their prey by using a tubular, muscular pharynx, which may project through the mouth on the ventral side. Digestive enzymes secreted into the gut begin digestion; intestinal cells continue digestion by engulfing food in food vacuoles.

All flukes (trematodes) are internal or external parasites, usually of vertebrates. Some but not all have a life cycle that includes several types of larvae and sometimes an intermediate host or hosts. Trematode larvae include miracidium, sporocyst, redia, cercaria, and metacercaria. Schistosomiasis (bilharziasis), caused by several species of the blood fluke *Schistosoma,* is currently the second most prevalent infectious disease worldwide (malaria is first). Cercariae—distinctive swimming larvae with a tail and sucker—are carried by snails that spread schistosomiasis. Snails release cercariae; the cercariae swim, attach to and penetrate human skin between the fingers and toes, and then mature into adult worms and migrate to take up residence in the liver and other organs of the human host. The disease results from the human host's immune response to schistosome eggs deposited in host tissues—by activation of lymphocytes and other immune cells; urinary tract and bowel blockage also can result. Trematodes have one or two suckers; some trematodes feed through their oral suckers. Suckers of tapeworms serve only for attachment to the host.

Tapeworms (cestodes) are exclusively internal parasites that usually attach inside the gut of vertebrates. Tapeworms lack their own gut and so are obligatory parasites. Microvilli—minute tissue projections—absorb nutrients (amino acids and sugars) from the hosts parasitized by tapeworms and trematodes. The epithelium of cestodes and trematodes lacks the cilia found in free-living flatworms. The tapeworm body plan consists of a head usually with hooks and suckers followed by repeated segments, each with reproductive organs; these sexually reproducing segments bud from the tapeworm's neck. Flukes and turbellarians are unsegmented externally. Like flukes, some tapeworms have intricate life cycles with several distinctive larval types.

The large surface area of free-living and parasitic flatworms relative to their volume has physiological implications. Oxygen, carbon dioxide, and ammonia exchange across the body surface. Like cnidarians and ctenophores, flatworms are blood-, lung-, and heartless. In parasitic flatworms, gases and nutrients diffuse into tissues of the flatworm from the host digestive system or from water. Nonparasitic flatworms ingest food. Platyhelminths that have a gut discharge solid waste through their mouths because they lack an anus. Protonephridia are the organs of excretion and osmoregulation in flatworms, except in turbellarians, that lack a gut. Protonephridia are composed of ciliated cells that—like primitive kidney cells—collect dissolved wastes. Protonephridia regulate water and ions by wafting liquid through ducts that exit to the outside through pores.

The simplest flatworm nervous system consists of light sensitive pigment-cup eyespots (either single or in groups) connected to a cluster of nerve cells (brain) in the head and ventral, longitudinal nerve cords. The nervous system of flatworms ranges in complexity from this simple system to the nerve net of acoel turbellarians, resembling that of cnidarians and ctenophores. Free-living flatworms detect chemicals, food, objects, and currents with sensory pits or tentacles on the sides of the head. When flatworms wander away from a scent source, they turn from side to side more frequently and so eventually home in on the source.

Triclad turbellarians and cestodes have prodigious powers of regeneration and reproduce sexually or asexually. Slices of *Dugesia*, a triclad, regenerate to form entire worms. Planarians (freshwater species of triclad turbellarians, an order of turbellarians characterized by a gut having three branches) and taeniid cestodes commonly reproduce asexually. Almost all flatworm species are simultaneous hermaphrodites. Each individual flatworm bears ovaries and testes but self-fertilization is rare in turbellarians and common in cestodes. Mating pairs of hermaphrodites mutually copulate—a copulatory bursa receives sperm or, in some turbellarians, a hypodermic-like penis injects sperm through the body wall into the body of the mate. Some flatworm sperm have no tails; others have a [9(2)+1] arrangement or a [9(2)+0] arrangement (see Figure I-2). Ribbons of fertilized turbellarian eggs are laid in cocoons. Freshwater flatworms "glue" eggs to stones. Eggs of most turbellarians hatch into miniature adults. A few marine turbellarians develop ciliated larvae known as Müller's larvae. Parasitic flatworms frequently have complex reproductive cycles, with a succession of larval stages. *Schistosoma*, a fluke, is dioecious. *Bothrioplana*, a turbellarian, is parthenogenetic—females asexually produce females.

The origin of this phylum is uncertain because flatworms fossilize poorly. Molecular data suggest that the bilaterally symmetrical, triploblastic, acoelomate pattern of early turbellarians was ancestral to the coelomates. Turbellarians probably were the ancestors of other platyhelminths, of nematodes (Phylum A-10), of gastrotrichs (Phylum A-16), and of molluscs (Phylum A-26).

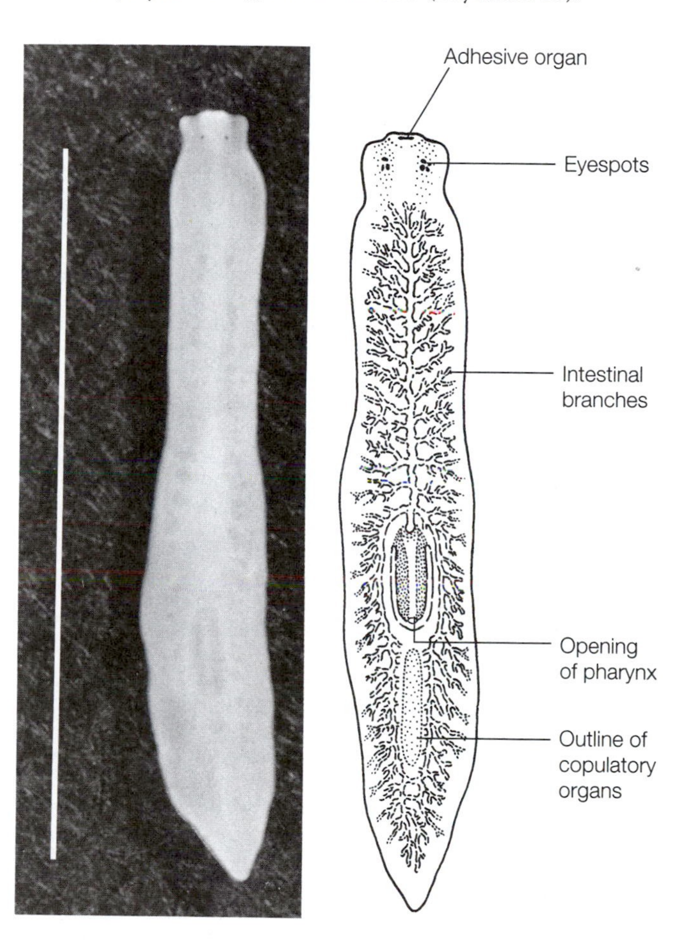

Dorsal view of gliding *Procotyla fluviatilis*, a live freshwater turbellarian flatworm from Great Falls, Virginia. Its protrusible pharynx connects to a branched intestine visible through its translucent body. Bar = 1 cm. [Photograph courtesy of R. Kenk; drawing by L. Meszoly; information from R. Kenk.]

A-6 Gnathostomulida
(Jaw worms)

Greek *gnathos,* jaw; *stoma,* mouth

EXAMPLES OF GENERA
Austrognatharia
Gnathostomaria
Gnathostomula
Haplognathia
Mesognatharia
Nanognathia
Onychognathia
Problognathia
Pterognathia
Semaeognathia

Gnathostomulids are translucent jaw worms, characterized by unique, toothed jaws near their ventral mouths. These free-living worms graze on fungi, bacteria, and protoctists among grains of sea sand. Like platyhelminths, jaw worms are bilaterally symmetrical, acoelomate eumetazoans. As triploblastic animals—having three body layers—the middle layer of mesodermally derived muscle is exterior to the digestive cavity and interior to the epidermis.

About 80 species in 20 genera of gnathostomulids have been described, from Maine, the Florida Keys, the Bahamas, the Caribbean, California, the Indo-Pacific, and the White Sea. Probably more than a thousand species of these worms live today in shallow oceans down to a depth of several hundred meters throughout the world. Jaw worms stick to sand particles or live on leaves of marine plants, such as the eel grass *Zostera,* turtle grass *Thallassia,* and marsh grass *Spartina* (Phylum Pl-12), and on thalli (leafy parts) of algae. Because they are recognizable only when living, the natural history of gnathostomulids remained unknown for a long time, until sophisticated techniques were devised to pull jaw worms off the surfaces on which they live. In California, jaw worms live near the roots of surf grass, *Phyllospadix,* in anaerobic sand. Some inhabit sulfureta, communities of organisms in black, fine, sulfide-rich sediments. Often deep and underneath the white oxidized layer of marine sand bottoms, the sulfuretum smells like rotten eggs. The odor emanates from hydrogen sulfide, a gas produced under anoxic conditions (in absence of molecular oxygen) by marine bacteria. Gnathostomulids in sulfureta tolerate low levels of O_2 and high quantities of sulfide. In certain sediments, gnathostomulids may outnumber even nematodes (Phylum A-10), with population densities of more than 6000 gnathostomulids per liter of sediment.

Gnathostomulids are hermaphrodites. The single ovary produces large eggs that mature one at a time. Posterior to the ovary are paired testes. Gnathostomulid species have a variety of male organs. Sperm may be undulipodiated, mushroom shaped, or dwarf (spherical). During copulation, one worm can inject hundreds of sperm within a mucus ball beneath the skin of a second worm. A female genital pore is present in certain species. The penis of some species, like the *Problognathia* shown here, is stiffened with a stylet, facilitating hypodermic impregnation. Sperm in the prebursa—the mucus ball filled with sperm—seem to migrate to a storage sac called a bursa, part of the female reproductive apparatus, and fertilize the most mature egg. Afterward, the fertilized egg ruptures the body wall and is released. Because development is from egg to adult without a larval stage, we say that development is direct. In at least some species, a nonsexual feeding stage may alternate with a distinct nonfeeding sexual stage, taking a year to complete the cycle to the mature adult.

The gnathostomulid body lacks an external cuticle, its average length is 1.5 mm, the range being from 0.3 to 3.5 mm. A slight constriction separates the bristly head from the trunk in some; in others, an elongate rostrum forms the anterior end. Circulatory and respiratory organs, a coelom, and a skeleton are lacking, as in platyhelmiths (Phylum A-5). Gas exchange takes place across the body wall of the jaw worm's minute body. The modest nervous system is made up of longitudinal nerve fibers and ganglia (frontal, buccal, caudal, and penile). In some species, the head bears well-

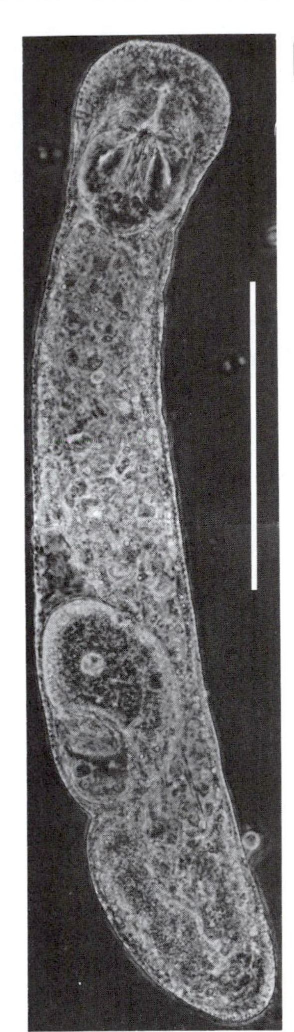

An adult *Problognathia minima.* It glides between sand grains in the intertidal zone and shallow waters off Bermuda. LM (phase contrast), bar = 0.1 mm. [Photograph courtesy of W. Sterrer, *Transactions of the American Microscopical Society* 93:357–367 (1975); drawing by L. Meszoly; information from W. Sterrer.]

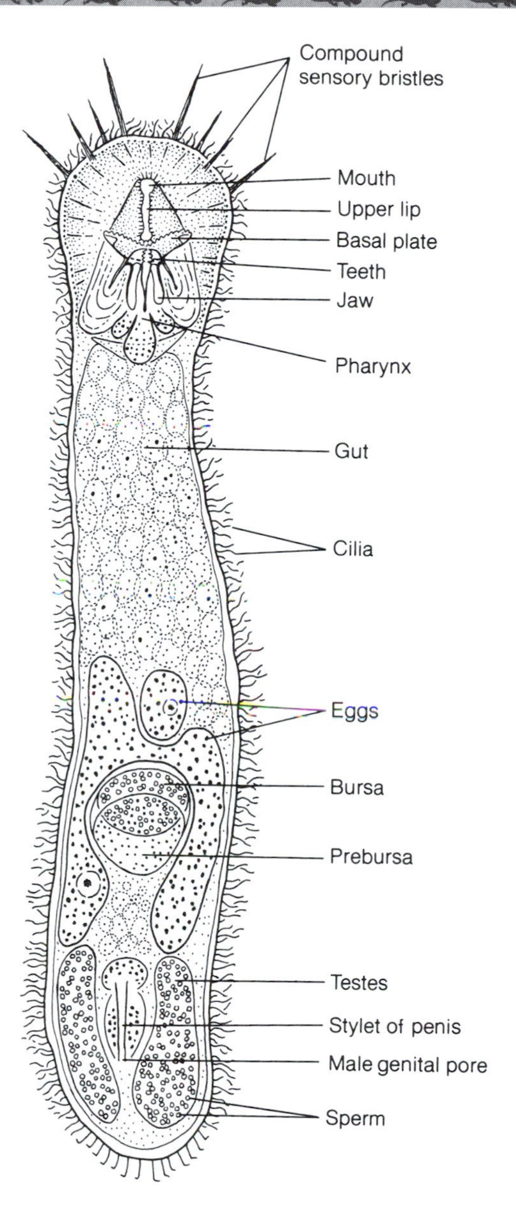

developed tactile organs collectively called the sensorium—stiff bundles of sensory bristles and pits lined with cilia. On the belly side of the head are feeding structures: a cuticular comblike basal plate that can be protruded through the mouth and a pair of toothed lateral jaws. Contraction of the muscular pharynx snaps the jaws open and closed in about a quarter of a second. Food particles are passed into the digestive cavity, which has a single opening; undigested solid waste may leave through a temporary dorsal connection to the epidermis—a temporary anus—or by the mouth. Dissolved waste is disposed of by ciliated excretory cells—protonephridia—that open to the exterior through pores.

Gnathostomulids move by using cilia. They nod their heads from side to side, swim, glide, and twist. Under the epidermis lie weak circular muscle fibers, with three or four paired longitudinal muscles underneath; the body-wall muscles do not take part in locomotion but do shorten the body. Parenchyma is almost absent. Polygonal cells of the external epithelium bear only one cilium each, and the ciliary propulsion can reverse direction. These traits distinguish gnathostomulids from turbellarian flatworms (Phylum A-5), which gnathostomulids otherwise resemble. In addition, gnathostomulids' anterior bristles and less flattened bodies distinguish them from flatworms.

Gnathostomulids have been recognized as an independent phylum since 1969. They are grouped with Platyhelminthes (Phylum A-5) and Nemertina (Phylum A-9) as triploblastic, acoelomate invertebrate animals. Both jaw worms and flatworms are externally ciliated, lack a coelom, have a dead-end gut and lack an anus (at least, a permanent anus), have protonephridia, and are hermaphroditic. Evidence against a close relationship between jaw worms and flatworms is that the sperm tails of gnathostomulids are typical undulipodia with the [9(2) + (2)] cross section (see Figure I-2) or sperm with no tails, whereas sperm tails of flatworms have a different arrangement or there is no tail. Their jaw structure may relate gnathostomulids to rotifers (Phylum A-13); on the other hand, monociliated epithelium has been found only in gastrotrichs (Phylum A-16) and gnathostomulids. Gnathostomulid adhesive cells are not homologous to platyhelminth or gastrotrich dual-adhesive glands.

Fossils called conodonts were once thought to be remains of the tough parts of ancient gnathostomulids found in rocks that extend in age from about 580 million to 200 million years. However, the basal plates of modern gnathostomulids differ from those of conodonts in that gnathostomulid toothlike feeding structures are made of acellular rather than cellular organic material. Conodonts, in contrast, are made of cells that—like bone—precipitate calcium phosphate. Conodonts are now assigned a fossil phylogenetic position in the Craniata (Phylum A-37) and probably are the earliest known vertebrate hard tissue.

A-7 Rhombozoa

(Rhombozoans)

Greek *rhombus*, a spinning top; *zoion*, animal

EXAMPLES OF GENERA
Conocyema
Dicyema
Dicyemennea
Microcyema
Pseudicyema

Two groups of invertebrates—Rhombozoa and Orthonectida—were earlier designated as two classes—Rhombozoa (dicyemids and heterocyemids) and Orthonectida (orthonectids)—within Phylum Mesozoa. In our classification system, we now consider each of the two former classes to be a separate phylum because, until and unless closer affinities between the rhombozoans and the orthonectids are demonstrated, they do not seem closely enough related to each other to be a single phylum. These two phyla are neither in Subkingdom Parazoa nor in Subkingdom Eumetazoa. Except for placozoans (Phylum A-1), rhombozoans (and orthonectids) are the least morphologically complex animals—with a minimum of tissue and organ complexity. The name of the former phylum Mesozoa indicates that they were at one time considered intermediate between the protoctists, which lack tissues, and the more complex metazoans, which contain millions of cells integrated into tissues, organs, and organ systems. Rhombozoans are small wormlike, bilaterally symmetrical animals. Although they are multicellular, their cells are arranged in one or two layers, so they cannot be forced into the category of either two- or three-layered animals. Moreover, the inner cell layer of rhombozoans does not correspond to endoderm. Rhombozoans lack a body cavity and circulatory, respiratory, skeletal, muscular, nervous, excretory, and digestive systems.

The outer layer of rhombozoans (Figure A) consists of from 20 to 30 "jacket" cells, a constant number in each species. Like flatworms (Phylum A-5), rhombozoans are ciliated. These mostly ciliated cells form a jacket, enclosing one or several long, cylindrical axial cells. Within the axial cell or cells are from 1 to about 100 cells called axoblasts, each containing a polyploid nucleus. The function of the large specialized axial cell is solely to surround the axoblast cells (also called agametes). The axoblast cells are reproductive cells that ultimately will produce eggs and sperm.

Phylum Rhombozoa encompasses both dicyemids (*Dicyema, Pseudicyema*) and heterocyemids (*Conocyema, Microcyema*). Stages in the rhombozoan life cycle include rhombogen, nematogen, vermiform larvae, and infusoriform larvae. The nematogen of heterocyemids has a syncytial, unciliated outer cell layer, whereas the nematogen of dicyemids (see Figure A) has a cellular outer layer. Rhombozoans include sexual and asexual generations.

Adult dicyemids (Figure B) range in size from less than 0.5 mm to more than 5.0 mm in length in the nematogen stage (see Figure A). About 65 species of rhombozoans have been described. All rhombozoans are parasites or endosymbionts (symbiotrophs) that live in the bodies of benthic (seafloor dwelling) cephalopods mainly in temperate waters, especially cuttlefish and octopods, and probably evolved from more complex free-living ancestors. Because they frequently inhabit hosts that are widespread in shallow seas, rhombozoans are common. Dicyemids live in the kidneys of octopus and other benthic cephalopod molluscs (Phylum A-26) but not pelagic (open ocean–dwelling) squid. The dicyemid microhabitat is the interface between urine and mucus-covered epithelial kidney tissue of the mollusc host. The dicyemid attaches loosely by its anterior end to the host's kidney with the rest of the dicyemid hanging free in urine. These minute parasites absorb nutrients directly from the urine of their host.

Octopus urine has organic solutes that may sustain the dicyemids. When cultured, dicyemids consume glucose in culture media—thus they are aerobes. Dicyemids contribute to the acidification of urine, facilitating ammonia excretion by their hosts; the dicyemid-cephalopod relation is thus symbiotic.

Because the heterocyemid life history has not been completely described, we will discuss the life cycle of dicyemids here. The dicyemid life cycle is not completely known, but let us begin with the infusoriform larvae. These larvae are the dispersal stage of the dicyemid. The spherical or top-shaped dicyemid infusoriform larva is about 0.04 mm long. It consists of 28 cells—each of the four interior cells contains another cell. Like sets of Chinese boxes, one cell is packed inside the other. The larva grows by differentiation and enlargement of its existing cells rather than by mitotic cell division. Free-swimming dicyemid larvae, called infusoriform larvae (see Figures A and C), hatch from fertilized eggs within the axial cells of adults. Soon after hatching, the larvae are shed into the sea and are weighted to the sea bottom by two cells filled with a high-density substance, magnesium inositol hexaphosphate. The larvae are acquired somehow by young bottom-dwelling cephalopods. (The dotted arrow in Figure A indicates a possible intermediate host.) The fate of larvae in the ocean is unknown; it is possible but not likely that intermediate hosts transfer rhombozoans to other cephalopods. Infusoriform larvae have been found in cephalopods younger than 3 weeks old. Larvae seem not to infect older cephalopods. The entry route of the larva into the cephalopod is obscure, although experimental infection in laboratory aquaria has been achieved. The larvae enter the kidneys of their host and attach lightly by their anterior cells. Larvae develop into the adult form at the interface between kidney epithelial cells and urine in the host kidney. Infusoriform larvae develop into nematogens only, not into rhombogens.

Adult dicyemids have morphologically similar but functionally distinct reproductive phases called nematogen, the nonsexually reproductive stage, and rhombogen, the sexually reproductive stage. Nematogen adults reproduce mitotically—in the absence of sexual reproduction, they produce and release immature vermiform (wormlike) larvae into the urine of the young mollusc hosts. Vermiform larvae attach to the kidney–urine interface and grow into nematogen adults. As long as the cephalopod is immature, new generations of nematogens are asexually produced in this way. In

older cephalopods—at high densities of nematogens—rhombogens develop and there is either a mixture of nematogens and rhombogens or just rhombogens.

Rhombogens develop from nematogens and are sexual adults. All rhombozoans have only one organ—a hermaphroditic gonad; in fact, this gonad is a single cell, rather than a multicellular organ. When a population of adult dicyemid rhombogens becomes dense, the rhombogens develop—nested within their axial cells—nonciliated, sexual infusorigens. An infusorigen is a hermaphroditic gonad, producing both eggs and ameboid sperm, that does not emerge from the enclosing rhombogen. Self-fertilization takes place within the axial cells of the rhombogen parent. Oocytes do not complete the first meiotic nuclear division until after fertilization. The resulting zygotes develop into infusoriform larvae. These ciliated infusoriform larvae escape from their parent into the molluscan host urine and are shed into the sea, thus completing the life cycle.

Some scientists have suggested that rhombozoans evolved by simplification from flatworms (Phylum A-5). Like some flatworms, rhombozoans are parasites (endosymbionts). Evidence that rhombozoans are not degenerate flatworms includes properties that are unique to rhombozoans and orthonectids: the cell-

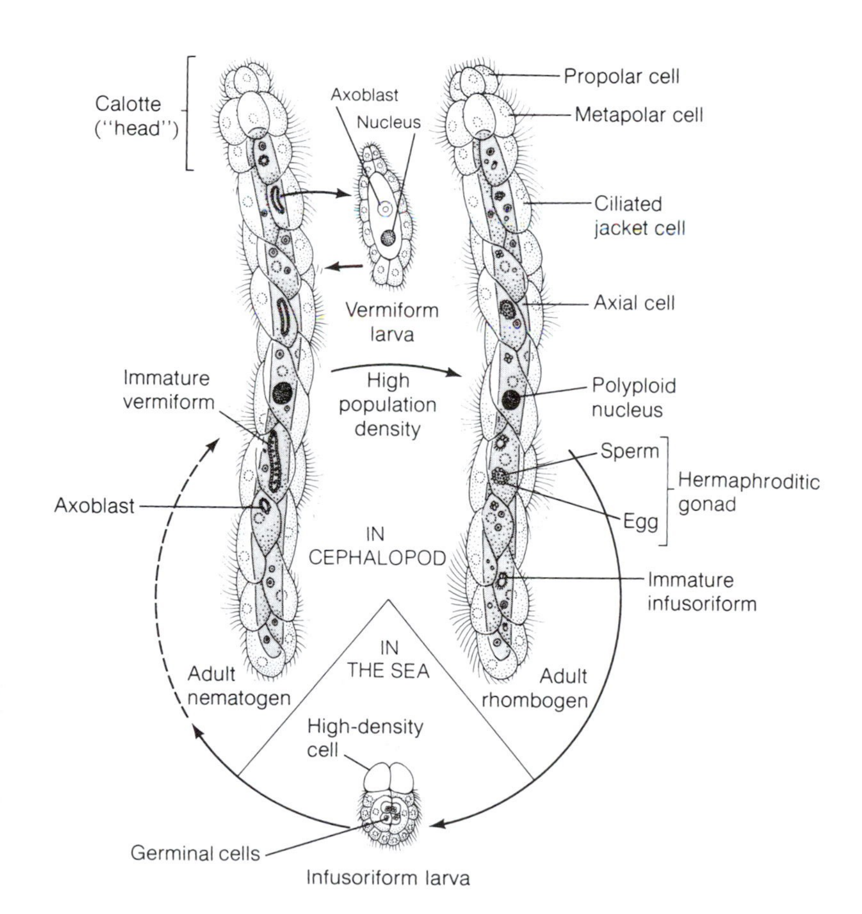

A *Dicyema truncatum* life cycle. The dashed arrow indicates the unknown mode by which infusoriform larvae infect cephalopod hosts. [Drawing by L. Meszoly; information from E. Lapan.]

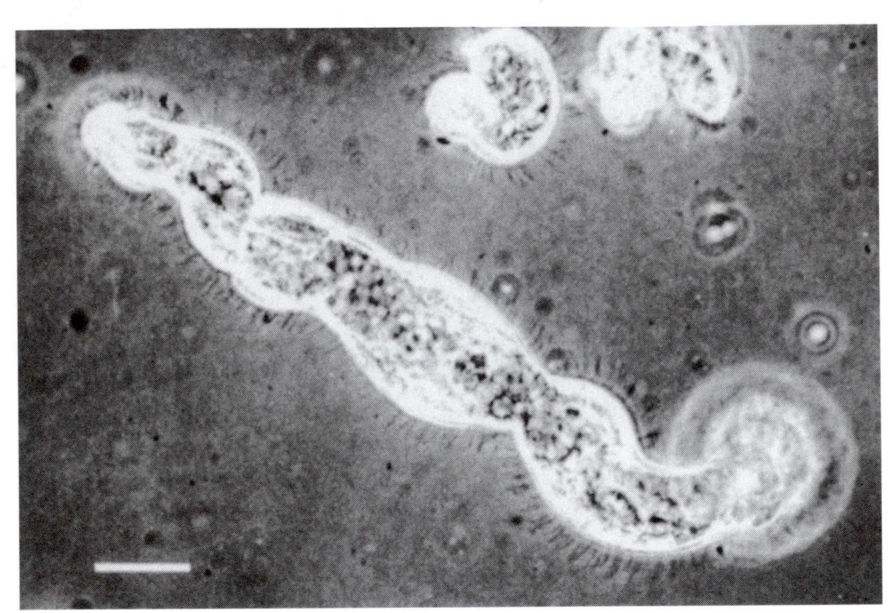

B An extended adult *Dicyema truncatum,* with a small contracted one above. LM, bar = 10 μm. [Courtesy of H. Morowitz.]

C *Dicyema truncatum* larva found in the kidneys of cephalopod molluscs. Free-swimming larvae disperse the dicyemids. LM, bar = 100 μm. [Courtesy of H. Morowitz.]

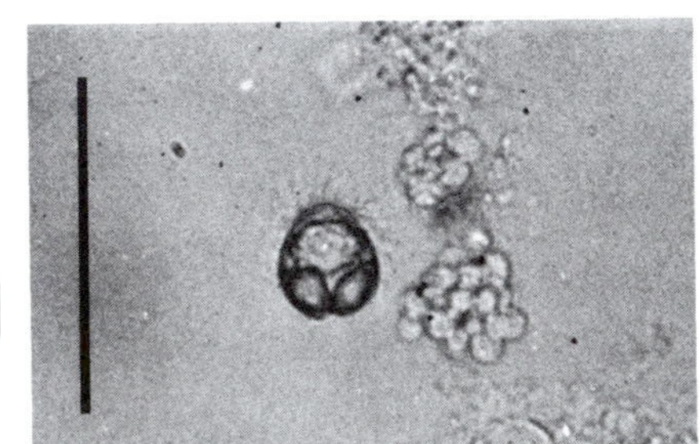

within-a-cell arrangement of rhombozoans and—in both rhombozoans and orthonectids—intracellular development of embryos; polyploid nucleus of the axoblast; and alternation of asexual and sexual generations. In many other respects, though, what is known of rhombozoan biology suggests that rhombozoans and orthonectids may have originated independently of each other. Furthermore, the percentage of combined guanine and cytosine in the DNA of flatworms (35–50 percent) is considerably higher than that measured in dicyemids (*Dicyemennea* has 23 percent).

A-8 Orthonectida
(Orthonectids)

Greek *orthos*, straight; *nektos*, swimming

EXAMPLES OF GENERA
Ciliocincta
Rhopalura
Stoecharthrum

Orthonectids, like members of Phylum Rhombozoa (Phylum A-7), were formerly considered a class in Phylum Mesozoa. A closer look affords orthonectids phylum status based on their unique characteristics. The adults swim with their cilia on a straight path, at least when artificially released from their host, accounting for the derivation of the phylum name. Most orthonectids are dioecious and have one or two cell layers. Externally, the adult wormlike body consists of as many as 40 rings of jacket cells. For a given species, cell arrangement and number is rather constant. Most of the outer jacket cells are ciliated (Figures A and B). The one or two cell layers of orthonectids do not correspond to the two layers of cnidarians and comb jellies. Orthonectids, like rhombozoans, are multicellular, are ciliated larvae and adults, have sexual and asexual generations, and have a marine habitat. Unlike rhombozoans, most orthonectids are dioecious.

From oceans throughout the world, about 20 species of orthonectids have been described. All are endoparasites of invertebrates; that is, all are symbiotrophs that live in echinoderms such as brittle stars (Phylum A-34), in nemertines (Phylum A-9), in turbellarian flatworms (Phylum A-5), in polychaete annelids (Phylum A-22), and in clams (Phylum A-26). Orthonectids reside in the gonads of their hosts. Host tissue fluids are probably the sole source of nutrition for orthonectids. Orthonectids—less benign than rhombozoans—castrate their hosts in certain species.

Orthonectids are microscopic; none is larger than about 300 μm. In most species, females almost completely filled with eggs tend to be several times longer than males of the same species. An intriguing facet of the life history of *Rhopalura ophiocomae* is the existence of two female types within the same species: larger elongate (Figure C) and smaller ovoid (Figure D). Both types exist with the same frequency but not simultaneously in the same individual host. Whatever factor determines whether a female is elongate or ovoid has not been discovered.

Like rhombozoans, orthonectids do not have digestive, respiratory, skeletal, excretory, or nervous organs at any stage of the life cycle. In some species, cells believed to be aggregations of contractile fibers considered to be muscle are located underneath the jacket cells and around the gonads (testes or oocytes); such fibers are not present in rhombozoans.

The sexual generation of orthonectids consists of tiny, ciliated, free-swimming males (see Figure B) and females (see Figure D) swimming in the ocean. The outer jacket cell layer encloses either eggs in the female (see Figure C) or sperm with tails in the male (see Figure A). A few species are hermaphrodites and self-fertilize. Orthonectids mate in the sea by bringing their genital openings—which are located about midway along their bodies—side by side. The male releases sperm into the female's genital pore; fertilization is internal.

The fertilized eggs develop into ciliated larvae with two cell layers. These larvae—the dispersal stage in the life history—exit into the ocean, perhaps through the female's genital pore, and somehow locate the appropriate marine animal host. Orthonectid larvae may shed their outer layer of cells when they have reached the genital duct of their marine victims. Only the orthonectid larva's inner cells enter the host gonad, though there is doubt in this regard. In the host gonad, each orthonectid larval cell divides by mitosis to form a plasmodium, which is an ameboid multinucleate syncytium (a single cell with many nuclei). (This facet of orthonectid life history bears scrutiny because the "plasmodium" of *Rhopalura ophiocomae* is a hypertrophied muscle cell of its host.) Plasmodia that are too immature to sexually reproduce themselves may be considered the juveniles of the orthonectid life history.

Each plasmodium develops ciliated, tiny male and female adults within itself. Because this development takes place inside the host gonad, the gonad loses its fertility. Eventually, the orthonectid adults abandon the plasmodium and their invertebrate host gonad for life in the sea, commencing a new sexual generation.

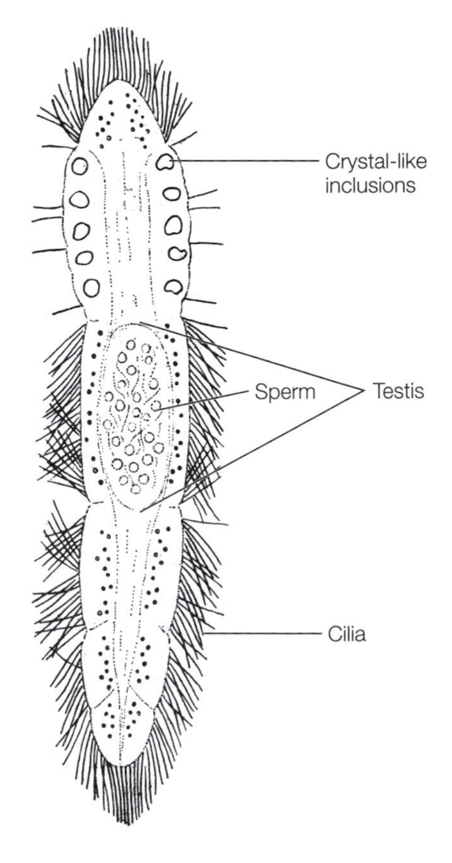

A *Rhopalura ophiocomae.* An optical section brings a shallow slice of a living mature male into crisp focus and shows lipid inclusions, testis, and ciliated cells. In mature males, living sperm with tails moving appear to completely fill the testis. [Drawing by E. Kozloff; courtesy of *Journal of Parasitology* 55:172 (1969).]

Orthonectids, like rhombozoans, have been proposed as distant relatives of the flatworms (Phylum A-5). Because of the many differences between orthonectids and flatworms, such as the unique body plan of orthonectids, this relationship is not likely. The orthonectid features of producing reproductive cells and embryos that develop within other cells probably indicates an evolutionary line independent of all other animals, even of rhombozoans (Phylum A-7). Orthonectids most likely have not given rise to any other animal line. The relationship between orthonectids and other animals is enigmatic; as a result, orthonectids are not grouped as eumetazoans or as parazoans. Additional information regarding the unique orthonectid life history, genetic system, and molecular biology is needed to clear up the puzzle.

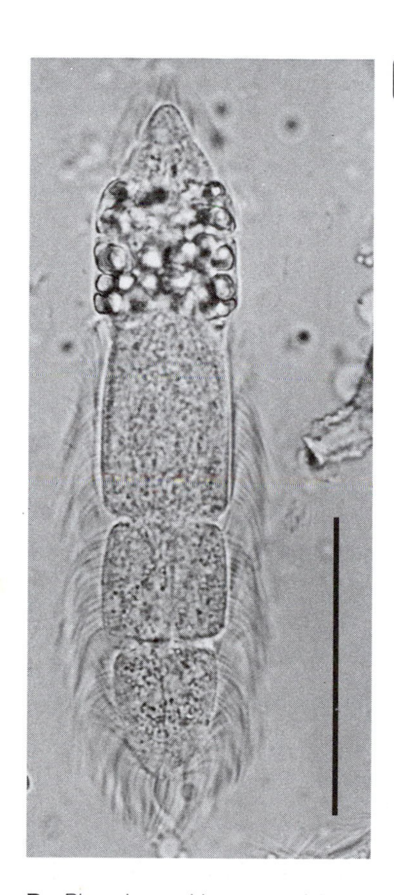

B *Rhopalura ophiocomae* adult male showing ciliated outer jacket. This orthonectid feeds on tissue fluids of its echinoderm host, the brittle star *Amphipholis squamata* (Phylum A-34). LM, bar = 50 μm. [Courtesy of E. Kozloff.]

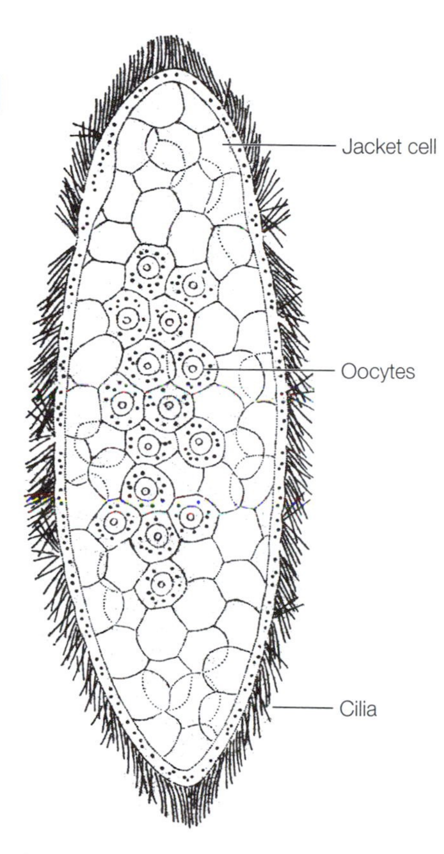

C Living mature female *Rhopalura ophiocomae,* as seen in optical section. The species has two types of females: elongate (shown here) and ovoid. Both types mate and then incubate fertilized eggs until larvae develop. [Drawing by E. Kozloff; courtesy of *Journal of Parasitology* 55:186 (1969).]

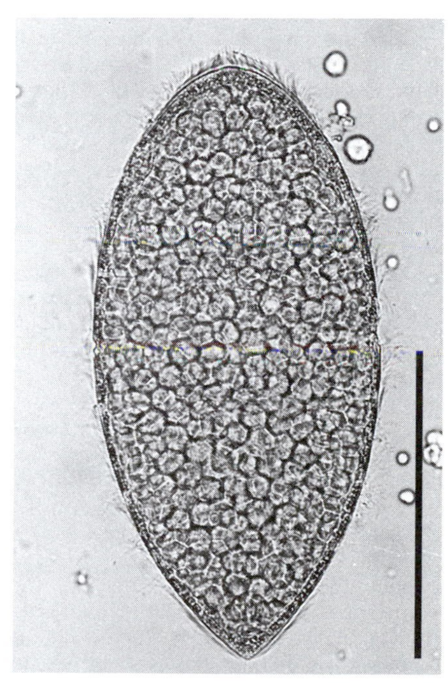

D Ovoid type of mature female *Rhopalura ophiocomae* packed with oocytes. Near the anterior end, underlying the jacket cells, many small cells encircle the oocytes. LM, bar = 120 μm. [Courtesy of E. Kozloff.]

A-9 Nemertina
(Ribbon worms, Nemertines, Nemertea, Rhynchocoela)

Greek *Nemertes,* a sea nymph

EXAMPLES OF GENERA

Amphiporus	*Emplectonema*	*Nectonemertes*
Carcinonemertes	*Geonemertes*	*Paranemertes*
Carinina	*Gononemertes*	*Prostoma*
Cephalothrix	*Lineus*	*Tubulanus*
Cerebratulus	*Malacobdella*	

Nemertina is a phylum consisting mostly of free-living worms found in marine, freshwater, and soil habitats. Their common name ribbon worm refers to their flat bodies and the brilliant color patterns of many species. A long, sensitive anterior proboscis that is separate from the digestive tract characterizes nemertines. This unusual organ is branched in some species. The proboscis resides in a body cavity (rhynchocoel), from which the worm rapidly everts its proboscis as much as three times the length of its body. Muscular pressure on the fluid-filled proboscis chamber forces explosive eversion of the proboscis. So accurate is its aim that another common name for these worms is nemertine, based on the Greek term meaning the unerring one. Nemertines use the proboscis to explore the environment, to capture prey, to defend themselves, and for locomotion. Annoyed nemertines release their proboscises, which they then regenerate.

Most ribbon worms live in the sea, in the intertidal marine sands, and in estuaries and are more abundant in temperate than in tropical oceans. *Carinina* is found in the abyss down to 4000 m. Some species are symbionts; *Gononemertes* lives in the branchial chambers of tunicates (urochordates, Phylum A-35). *Tubulanus* secretes a mucus dwelling tube, whereas *Lineus* takes over empty burrows of *Chaetopterus* and other marine polychaete annelids (Phylum A-22). *Prostoma* (Figure A) lives on aquatic plants in quiet fresh water along the Atlantic, Gulf, and Pacific coasts, in the U.S. Midwest, and in Europe. The terrestrial ribbon worm *Geonemertes* inhabits moist soil of subtropical forests, between pandanus leaf bases, and, when introduced, thrives in greenhouse soils. Parasitic *Carcinonemertes* lives on crustaceans and feeds on the host's developing eggs as well as the host.

About 900 nemertine species are known. They range from less than 0.5 mm to 30 m in length. *Lineus longissimus,* the iridescent bootlace worm, about 30 m long, is one of the longest invertebrates known. *Cerebratulus lacteus* (Figure B) can extend itself from about 1 to 10 m. *Emplectonema* is the only bioluminescent nemertine described. Most nemertines, especially the bottom dwellers, are pale, though a few are striped, speckled, or marbled multicolor ribbons.

Nemertines are abundant in the intertidal zone, although rarely seen; most are active at night, burrowing and feeding. They burrow by everting their proboscises into the mud; then they dilate the proboscis, forming an anchor. Nemertines pull themselves into their burrows by contracting body and proboscis muscles, pulling their bodies through the sediment. Like many other boneless animals, the nemertine's support is the incompressible liquid enclosed within its body wall. On tidal mud flats, nemertines can be found among algae, mussels, and tube-dwelling annelids. Nemertines sometimes creep out of seaweed placed in a dish of seawater.

Pelagic species (open ocean dwellers), such as *Nectonemertes,* tend to be more leaf-shaped and have less well developed muscles than do benthic (seafloor dwelling) nemertines. Pelagic nemertines float passively or swim with lateral undulations. Benthic nemertines crawl by muscular contractions, secreting a slime track. The smallest glide with their cilia against the resistance of their viscid mucus. Rarely, nemertines use their proboscises to attach themselves to an object and pull themselves forward.

Nemertines, unlike flatworms (Phylum A-5), have a blood vascular system through which contractile vessels and body muscle contractions pump blood. Unidirectional valves and heart are lacking. The heme-containing blood cells in a few species may carry oxygen. Nemertine blood may be colorless, red, yellow, purple, or green, depending on the species. Some rhythmically take water into the vascularized foregut, presumably for gas exchange; most respire through the epidermis, like flatworms. The excretory system of these worms consists of protonephridia with flame cells that regulate ions, water, and possibly dissolved waste, which exits through lateral pores. Nemertina is the first phylum with openings at both ends of the digestive tract; solid waste leaves through the anus.

The nemertine nervous system resembles that of flatworms: a bilobed cerebral ganglion (brain) and longitudinal nerve cords with connecting nerves (see Figure B). Their light-sensitive eyespots number from zero to as many as several hundred. Functions suggested for the cephalic slits (those in the head) include chemotactic, auditory, excretory, respiratory, and endocrine. A cerebral organ opens into the cephalic slits. Papillae on the anterior end are sensory.

Three of the four orders of nemertines are predators, feeding on a wide variety of prey: annelids (Phylum A-22), crustaceans (Phylum A-21), flatworms (Phylum A-5), molluscs (Phylum A-26), roundworms (Phylum A-10), and even small fish (Phylum A-37). One of the three orders of predaceous nemertines has stylets. The proboscis of predaceous *Prostoma* (see Figure A) has a venomous stylet with which the worm repeatedly stabs and paralyzes prey. The sticky proboscis wraps around prey and transfers captured prey to the mouth. Mouth and proboscis may share a common opening, depending on the species. Prey are sucked whole into the mouth; however, if the prey is too large to swallow, juices are sucked out of it instead. Cilia move food from the mouth along the foregut. Phagocytosis and extra- and intracellular digestion take place in the intestine, which has numerous pouches (diverticula). *Malacobdella* is the unique filter-feeding nemertine, living commensally within the mantle cavity of clams (Phylum A-26). *Malacobdella* filters bacteria, algae, diatoms, and other protoctists from water within its host's mantle cavity through ciliated papillae in its foregut.

Nemertines are the prey of crustaceans, annelids, and other marine invertebrates. Along the Atlantic coast of North America, ribbon worms serve as fish bait, but people do not eat ribbon worms.

Nemertines' prodigeous regeneration is a potential research model for tissue culture. The worms reproduce asexually by fragmentation—each fragment regenerates a complete worm. *Carcinonemertes* reproduces by parthenogenesis. Nemertines can also reproduce sexually. In most species, the sexes are separate. In sexual reproduction, numerous temporary gonads form during the breeding season in mesenchyme tissue between intestinal pouches. Each gonad opens to the outside through its own surface pore. Eggs are laid in gel strings. Fertilization typically takes place in the water but is internal in some species. The eggs develop either directly into adults or first into pilidium (free-swimming) larvae, which look like ciliated caps with ear flaps and apical tuft. Still other species have Iwata larvae or Desor larvae—named for embryologists. Desor larvae are characteristic of *Lineus* and other heteronemertines. The Desor is a ciliated, oval, postgastrula stage that stays within the egg membrane (unlike pilidium larvae); it lacks oral lobes and the apical tuft of pilidia. A few species are protandric hermaphrodites—each individual is first male and then becomes female. Members of the hermaphroditic terrestrial genus *Geonemertes* bear live young. In *Nectonemertes,* a genus of active swimmers, males clasp females with special attachment organs during mating. Knotted balls of about 30 *Cephalothrix* have been observed—perhaps mating—in breeding season beneath stones along the Yorkshire coast.

The fossil record of nemertines is sparse. The Cambrian *Amiskwia* was regarded as a nemertine fossil or as a chaetognath (Phylum A-32). Its phyletic position is obscure. Structures common to both nemertines and flatworms are parenchyma tissue that encloses organs; lack of body cavity, respiratory organs, and segmentation; ciliated epidermis that moves the animal along a mucus track; similar sensory and excretory organs; and multiple reproductive organs. Some flatworms, annelids, and molluscs also have anterior proboscises but only those of nemertines are in fluid-filled, cell-lined cavities separate from the gut. Common features were thought to point to a close relationship between flatworms

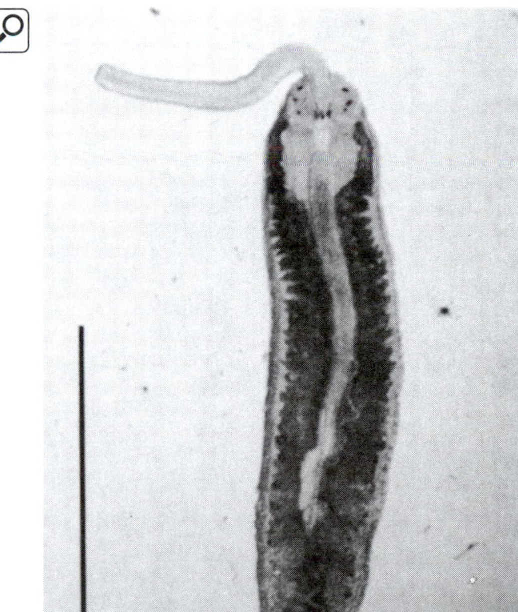

A *Prostoma rubrum,* a live nemertine taken from Peck's Mill Pond in Connecticut. Representative of the only freshwater nemertine genus, *Prostoma rubrum* has a proboscis armed with a stylet, unlike *Cerebratulus,* shown in Figure B. Bar = 1 cm. [Courtesy of J. Poluhowich.]

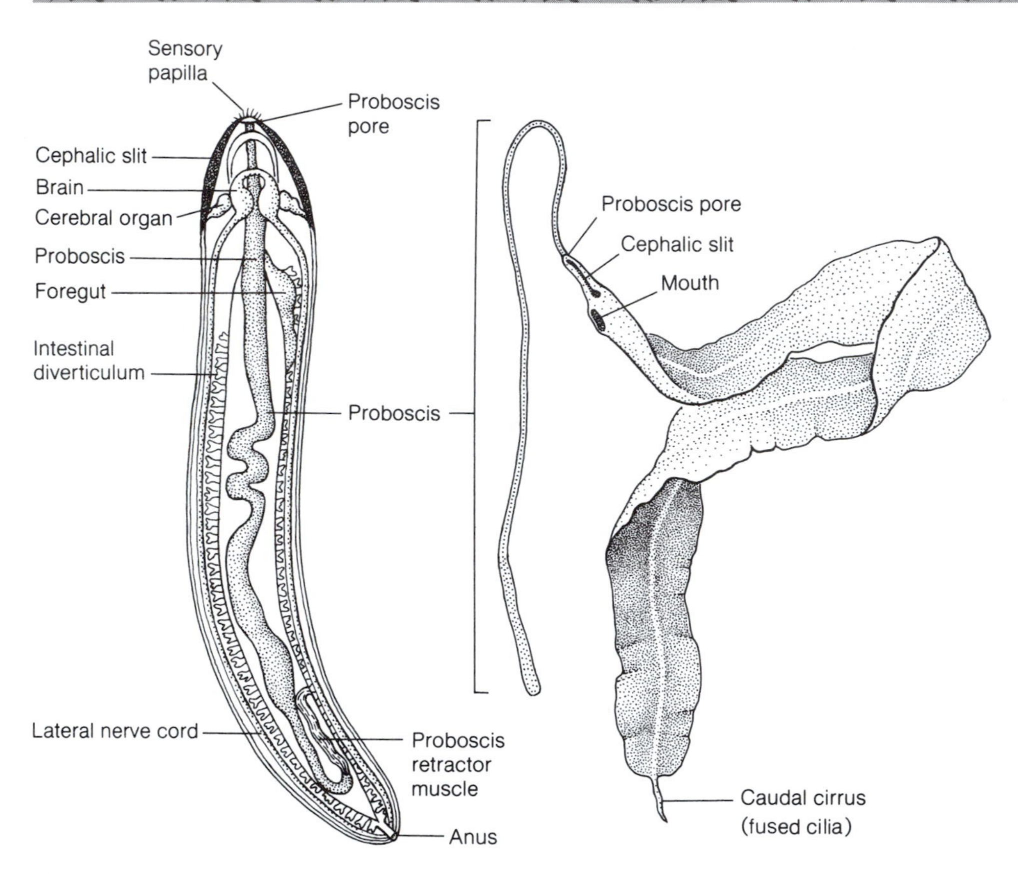

B *Cerebratulus,* a marine nemertine, shown in dorsal view with proboscis retracted (cutaway view, left) and with unarmed proboscis extended (external view, right). This free-living ribbon worm swims small distances by undulating. [Drawing by L. Meszoly; information from P. Roe.]

and nemertines despite their differences in food getting, digestive systems (the one-way nemertine gut is assumed more efficient than the dead-end flatworm gut), and oxygen–carbon dioxide exchange. However, comparison of ribosomal RNA sequences places *Cerebratulus*—the nemertine studied—near sipunculids, annelids, and molluscs (protostomous coelomates—the embryonic blastopore is the site of the adult mouth) and more distant from flatworms (acoelomates). These molecular data are supported by ultrastructural evidence that places nemertines as protostomous coelomates; the blood vascular system, gonadal sacs, and the rhynchocoel cavity are modified coelomic cavities, originating from mesoderm. It appears that the acoelomate attributes of nemertines may have evolved secondarily from a more typical coelomate ancestor. If this affiliation is accepted, nemertines will move to the protostome coelomate position in the phylogenetic tree.

A-10 Nematoda

(Nematodes, thread worms, roundworms)

Greek *nema,* thread

EXAMPLES OF GENERA

Ancyclostoma
Ascaris
Caenorhabditis
Dioctophyme
Diplolaimella
Dirofilaria
Dracunculus
Enterobius
Leptosomatum
Necator
Pelodera
Rhabdias
Trichinella
Trichodorus
Tricoma
Wuchereria

Nematodes are unsegmented pseudocoelomate worms inconspicuous until they capture our attention by infesting us, our plants, or our animals. Because nematodes are round in cross section, they are also called roundworms. Their body cavity is a pseudocoel, defined as a space between embryonic endoderm and ectoderm; the pseudocoel lacks a peritoneum—the mesodermal lining of the coelomic body cavity. Nematodes are probably the most abundant animals living on Earth. About 80,000 species of nematodes have been described in the scientific literature; researchers estimate that nearly 1 million living species exist. These worms range from only 0.1 mm (100 μm) to about 9 m in length. The female giant nematode *Dioctophyme renale* is 1 m in length; the male is only half as long. Free-living nematodes are slender and cylindrical, tapering at both ends, typically about 1 mm in length. Parasitic nematodes (Figure A) have a variety of shapes, many saclike; the longest is a 9 m long parasite from a sperm whale.

Members of this phylum are grouped into two classes: Adenophorea and Secernentea. The Adenophorea lack phasmids and therefore are also called Aphasmida. *Trichinella* is a parasitic member of this class. Phasmids are sense organs, possibly chemoreceptors, found in the tail region particularly of parasitic (symbiotrophic) roundworms. Secernentea, also called Phasmida, do have phasmids. Many members of this class—in fact, entire suborders—live in vertebrates, insects, or plants. Hookworms such as *Necator* and *Ancyclostoma,* some gapeworms, hairworms, stomach worms, lungworms, *Ascaris,* pinworms, and the filarial worms that cause filariasis (tropical diseases—elephantiasis and river blindness) are all phasmidians.

Typically nematodes are transparent, covered with a noncellular, patterned cuticle of collagen—a fibrous protein. Nematodes move with a characteristic flip of their bodies; unique oblique longitudinal muscles encircle their bodies, but nematodes lack circular muscles and so cannot extend and contract as segmented (annelid) worms do. Many parasitic nematode species look like microscopic dragons—they have well-developed teeth and are predaceous. Free-living nematodes devour rotifers and tardigrades. Parasitic species, such as the hookworms, have evolved specialized mouthparts with which they hook onto the intestinal wall or other tissues of their host.

Nematodes form three layers—ectoderm (ecto-, outside), mesoderm (meso-, middle), and endoderm (endo-, inside)—during embryonic development and so are triploblastic. Ventral, dorsal, and lateral nerve cords are present, as is a nerve ring around the pharynx. The nematode's digestive tract forms a tube complete with mouth and anus (Figure B) within the worm. The gut lacks cilia. The muscular pharynx pumps fluids into the gut. The pharyngeal pump counters internal hydrostatic pressure generated by contraction of the nematode's longitudinal muscles and by its non-

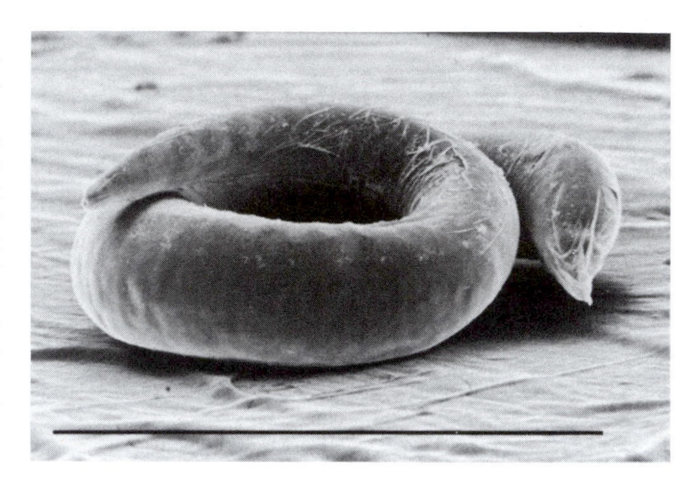

A *Rhabdias bufonis* (female), a nematode belonging to Class Secernentea, parasitic in the lung of the leopard frog, *Rana pipiens.* SEM, bar = 1 mm. [Courtesy of R. W. Weise.]

expandable cuticle. This pumping mechanism differs from the circular muscle contractions that propel food through the human gut. Solid waste is eliminated from the anus (in nematode females) or cloaca (Latin *cloaca,* sewer), from which waste and gametes leave the males. Nematodes have no specialized organs for circulation or excretion. Dissolved oxygen and carbon dioxide diffuse through the permeable body wall.

The sexes are separate in almost all species, the male being smaller than the female. *Caenorhabditis* is a simultaneous hermaphrodite. Reproduction is always sexual with internal fertilization. Males have copulatory spicules near the posterior end; some grip the female during mating. Single or paired ovaries connect to the outside through a midventral gonopore (genital pore) in females (see Figure B). In males, a single testis produces ameboid sperm, lacking undulipodia, that leave by way of the cloaca. The reproductive capability of nematodes is prodigious: some females have been known to contain 27 million eggs and extrude a quarter million fertilized eggs a day. Nematodes lack a free-swimming larval stage—they hatch from the eggs as miniature adults. "Dauer larvae" of nematodes are actually juveniles in developmental arrest. The young nematode molts its cuticle four times as it develops into the adult form. Even as an adult, a nematode grows and continues to shed and resecrete its cuticle.

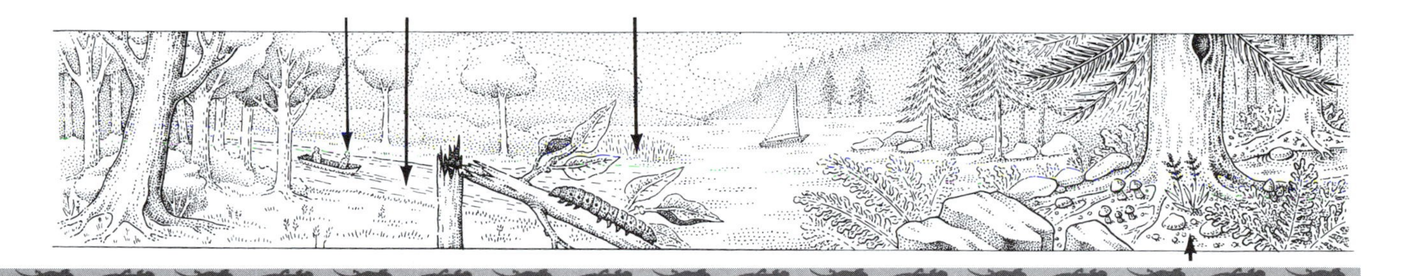

About one of five humans harbor hookworms worldwide. Hookworms produce anticlotting molecules, proteases that dissolve antibodies released by the infected person, and antioxidant enzymes that neutralize oxidizing molecules secreted by the victim; these and at least half a dozen additional mechanisms enable hookworms to thwart the body's immunological and vascular protective mechanisms. Vaccines against hookworm are being tested.

Many nematodes that are plant parasites live in plant tissue; some form root galls—abnormal swellings of tissue—whereas others live in fruits, roots, leaves, and stems or in bark crevices. Most nematodes produce hardy eggs that are well able to pause in their development indefinitely until harsh environmental conditions improve.

Because they parasitize human guts and domesticated plants and animals, some nematodes are being intensively studied; *Trichodorus,* which infects plant roots, is one of them and is a carrier of plant viruses. The heartworm, *Dirofilaria immitis,* infests cats and dogs. Trichinosis is probably the most familiar disease caused by nematodes in the United States; *Trichinella spiralis* is the parasitic nematode responsible. Trichinosis can be acquired by eating infested pork or other meat that has not been cooked sufficiently. The minute juveniles of these worms are harbored as cysts in the striated muscles of pigs, cats, dogs, rats, and bears. If the flesh of an infested animal is eaten by another, the nematode cysts are digested, liberating the juvenile worms into the intestine of a new host. About 2 days after their release, the nematodes mature sexually; male and female mate in the intestine. The male then dies. The females, about 4 mm long, then burrow into the muscles of the intestinal wall. Females are ovoviviparous in this species—they produce eggs that hatch within the mother's body. Their eggs release hundreds of live juvenile worms, which enter the lymph and are carried to the bloodstream. From there, juveniles burrow into skeletal muscles and any other organs, where they coil up and become enclosed in cysts. Juveniles develop into adulthood only if they are in skeletal muscle. Cysts may remain dormant for months, even years, or the host may deposit calcium salts in the infected tissues, calcifying and killing the nematodes. Only when the skeletal muscle is eaten are the cysts passed on to another host; that is, the cyst is the infective stage of *Trichinella.*

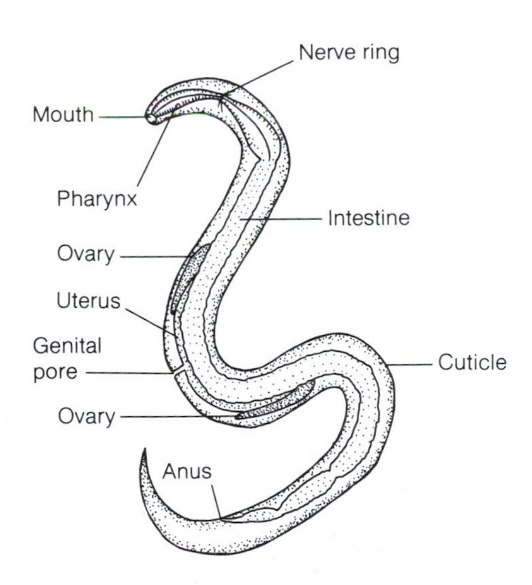

B Diagram of a female nematode showing well-muscled pharynx with which these worms pump liquid food into their digestive tracts. [Drawing by I. Atema; information from R. W. Weise.]

Although nematode-caused disease attracts public attention, most nematodes are beneficial. Free-living roundworms live almost everywhere. They inhabit moist soils, beach sand, salt flats, the ocean, hot springs, and lake water in prodigious numbers. As many as a billion roundworms per acre have been counted in the top 2 cm of rich soil. Free-living nematodes aerate soil, consume detritus, and circulate mineral and organic components of soil and sea sediments. Research on the free-living nematode *Caenorhabditis elegans* has revealed much of what we currently know about the expression of genes during development, genetic manipulation, and inheritance. *Caenorhabditis* is a good research animal; in *Caenorhabditis,* as in all nematodes, cell fates are permanently determined at the first cleavage of the zygote. The final fate of each of the embryonic cells of *Caenorhabditis* has been traced.

Nematodes are distinctive, placed in this phylum apart from other worms, although nematodes, rotifers (Phylum A-13), gastrotrichs (Phylum A-16), kinorhynchs (Phylum A-14), nematomorphs (Phylum A-11), and acanthocephalans (Phylum A-12) were formerly placed with aschelminthes or pseudocoelomates. Nematodes lack the circular body muscles and segmentation characteristic of annelid worms (Phylum A-22) and the eversible proboscis of ribbon worms (Phylum A-9), acanthocephalans, and certain flatworms (Phylum A-5). Lack of locomotory cilia sets nematodes apart from rotifers (Phylum A-13); nematodes do have nonmotile cilia in their sense organs. It seems likely that nematodes gave rise to no other phyla.

A-11 Nematomorpha

(Gordian worms, horsehair worms, nematomorphs)

Greek *nema,* thread; *morphe,* form

EXAMPLES OF GENERA

Chordodes	*Gordius*	*Paragordius*
Chordodiolus	*Nectonema*	
Gordionus	*Parachordodes*	

Nematomorphs are commonly called horsehair worms. Their name stems from the once-held belief that these slender, cylindrical worms, observed in horse-watering troughs, spring from horsehairs. Adult nematomorphs coil and tangle with each other, so they are also known as gordian worms, after Gordius, king of Phrygia. Gordius tied a knot, declaring that whoever untied his intricate knot should rule Asia. Alexander the Great cut the Gordian knot with his sword and added Asia to his Greek Empire.

Although 240 species are known, nematomorphs are grouped into very few genera. Gordian worms live all over the world in shallow oceans and lakes, temperate and tropical rivers, ditches, alpine streams, moist soil, and stock tanks. The only marine species in the continental United States is *Nectonema agile,* distinguished from other horsehair worms by a row of slate-colored bristles on each side of its gray-yellow or pale whitish body. Chances of observing *Nectonema* are highest in late summer, from July to October on moonless nights when the tide is receding. *Nectonema's* geographical distribution is poorly known. In the Gulf of Naples, Vineyard Sound (Massachusetts), Norway, and the East Indies, nematomorphs can be seen coiling and winding in shallow coastal waters. In fresh and marine water, they make up only a small fraction of the plankton.

Nematomorph adults are free living, usually in fresh or salt water. All nematomorph species are endoparasitic (internal symbiotrophic parasites) for a part of their lives. Nematomorphs are rarely found in the human urethra or digestive system, and they do not seem to cause human disease. Hosts of nematomorphs include leeches (Phylum A-22); beetles, crickets, grasshoppers, and cockroaches (Phylum A-20); hermit crabs and spiders (Phylum A-19); and true crabs (Phylum A-21).

Nematomorphs are leathery, unsegmented invertebrates—stiff as wire—and generally brown, black, gray, or yellowish in color (Figure A). The head end is distinguishable by being a lighter color than the rest of the body. A pair of caudal lobes posterior to the anus distinguishes the posterior from the anterior end. Nematomorphs range from 10 to 70 cm in length and from 0.5 to 2.5 mm in diameter, depending on the species. Body length also depends on the sex of these dioecious worms: females are longer than males. Polygonal or round thickenings ornament the hard, noncellular cuticle having fibrous layers—probably collagen—secreted by the epidermis. As they grow, nematomorphs molt the cuticle.

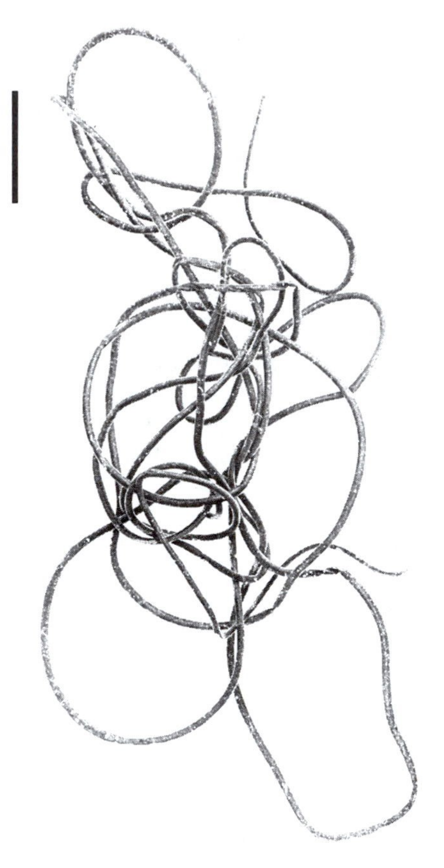

A An adult female *Gordius villoti,* a horsehair worm. Bar = 1 cm. [Courtesy of Trustees of the British Museum (Natural History).]

Neither adults nor larvae ingest food; although the hind part of the digestive tube, the cloaca, is used in reproduction, the anterior part of the gut is degenerate. Because the posterior end of the nematomorph digestive tract receives gametes, it is called a cloaca; in other animals, both gametes and waste usually exit through the cloaca, as in male nematodes. During its parasitic, larval phase, instead of ingesting food, a nematomorph absorbs nutrients across its body wall from its host animal. Respiratory, circulatory, and excretory organs are absent, as in many parasites. Digestive, reproductive, and nervous systems are embedded in a matrix of collagen fibers packed with parenchyma. The nematomorph nervous system resembles that of kinorhynchs (Phylum A-14): a nerve ring encircles the pharynx, a single nerve cord runs down the ventral side, and some adults have eyespots composed of innervated sacs

lying beneath transparent cuticle and backed by a pigment ring. Larvae lack eyes (Figure B) and have protrusible, spiny proboscises that resemble the proboscis of acanthocephalans (Phylum A-12). Like nematodes (Phylum A-10) and kinorhynchs (Phylum A-14), nematomorphs have only longitudinal muscles that permit whip-like swimming and serpentine coiling. Fluid in the pseudocoelom (body cavity) serves as a hydraulic skeleton.

Nematomorphs are dioecious. Eggs produced by ovaries or sperm produced in spermaries pass to the cloaca and then out through the anus. Adult male worms crawl or swim, especially in winter. In contrast, females are less active. The male wraps around the female, deposits sperm near her cloacal opening, and soon dies. The eggs are fertilized internally. The female drapes millions of fertilized eggs in gelatinous strings around aquatic plants. From 15 to 80 days later, the eggs hatch as tiny motile larvae. Nematomorphs lack an asexual reproductive mode.

The larvae of nematomorphs enter the body cavities of arthropods or leeches in a way that has not yet been observed. Larvae may be accidentally eaten or drunk by any of nematomorph's host animals and may bore through a host's gut by using piercing mouthparts borne on their proboscises. Larvae of marine, soil, and freshwater nematomorphs metamorphose within their hosts; then the sexually immature worms burst out of their hosts near or in water or during rain. How nematomorphs that spend part of their life cycle in water induce terrestrial hosts to seek water is a mystery. The exit of the larvae kills the host. The same larvae may pass through one or more hosts, the number of hosts depending on the species of nematomorph. If worms mature in autumn, they form cysts on waterside grasses and reenter water in spring. As a result, development from egg to adult worm may take a short time (2 months) or as long as 15 months.

The nematomorph body cavity is a pseudocoel, a body cavity that lacks a mesodermal lining. Nematomorphs probably did not evolve directly from any other pseudocoelomates—rotifers (Phylum A-13), kinorhynchs (Phylum A-14), acanthocephalans (Phylum A-12), or nematodes (Phylum A-10). Rather, each phylum of pseudocoelomates is thought to have evolved from acoelomates (lacking a body cavity between the gut and the outer body wall musculature) at different times in several different ways. The pseudocoel is not a stage in the development of the true coelom; it developed independently.

Nematomorphs seem to be of no veterinary or medical importance.

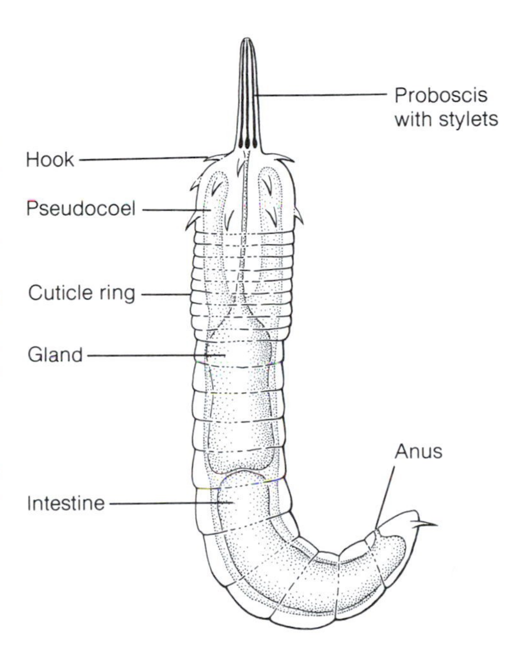

B Parasitic larva of gordian worm with its proboscis extended. The larva is about 250 μm long. [Drawing by L. Meszoly; information from L. Bush.]

A-12 Acanthocephala

(Thorny-headed worms)

Greek *akantha*, thorn; *kephale*, head

EXAMPLES OF GENERA

Acanthocephalus
Acanthogyrus
Echinorhynchus
Gigantorhynchus
Leptorhynchoides
Macracanthorhynchus
Moniliformis
Neoechinorhynchus
Plagiorhynchus
Polymorphus

Thorny-headed worms, or acanthocephalans, are parasites that live in the gut of vertebrates (Phylum A-37) and—earlier in their life cycle—within invertebrates. The thorny protrusible proboscis is globular or cylindrical (Figures A and B); the body also may bear spiny thorns. These parasitic worms anchor themselves with their proboscises and body spines to the gut wall of a host.

More than a thousand species of these bilaterally symmetrical, pseudocoelomate worms have been described from intermediate invertebrate hosts and final vertebrate hosts. Adult females are usually about 2 cm in length; the longest is more than 1 m long and some are as short as 1 mm, depending on the species. Females tend to be larger than males. Thorny-headed worms may be wrinkled or smooth and whitish, yellow, orange, or red.

The habitat of their intermediate and final hosts—therefore of these parasites—is soil, fresh water, or salt seas worldwide. Although hundreds of adult worms may be present in a single animal, thorny-headed worms are seldom seen, because they lack a free-living stage. Vertebrate hosts of acanthocephalans include dogs, hyenas, seals, squirrels, shrews, moles, pigs, rats, dolphins, insectivores, birds, fish, amphibians, snakes, and turtles. Vertebrates become infected when they consume an intermediate invertebrate host that harbors the acanthocephalan. Intermediate hosts are often the prey of other invertebrates. Thus, acanthocephalan larvae often are transferred to two or three different invertebrate hosts.

The ability to alter host behavior may characterize all thorny-headed worms. The acanthocephalan *Polymorphus paradoxus* changes the behavior of its amphipod intermediate host (Phylum A-21); infected amphipods skim the surface of the pond or move toward the light, whereas uninfected amphipods burrow into the pond bottom. This behavioral change increases the acanthocephalan-infected amphipod's chance of being ingested by mallard ducks along with the acanthocephalan parasite. In this way, the acanthocephalan enters its second and final host, a bird such as the mallard duck.

Circulatory, digestive, and respiratory organs, as well as mouth and anus, are absent from adult and larval thorny-headed worms. The body cavity lacks a peritoneum, the mesodermal layer typically present in coelomate animals. Thus, the acanthocephalan's body cavity is a pseudocoel. The acanthocephalan likely absorbs nutrients from its host(s) and circulates nutrients within its thick body wall through canals that open neither to the exterior nor to its pseudocoel. In some species, excretory organs consist of ciliated flame bulbs (protonephridia, like those in flatworms) with collecting tubules that drain urine into the sperm duct, then into the penis in males or the uterus in females, and from these locations to the outside. Wastes may also diffuse through the body wall of the worm.

The nervous system consists of a sensory pit at the proboscis tip, a touch receptor on the reproductive organs, and a ventral cerebral ganglion in the proboscis sheath (see Figure A), from which nerves extend to tissues.

The innermost body wall contains longitudinal and circular muscles. When these muscles contract the body, hydrostatic pressure in fluid reservoirs called lemnisci everts the proboscis; muscles

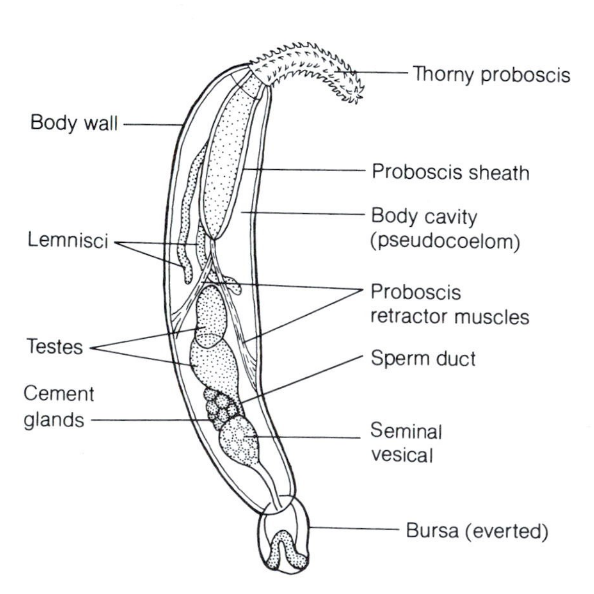

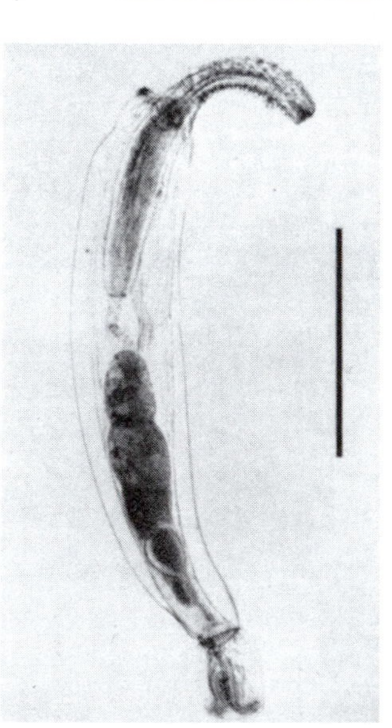

A *Leptorhynchoides thecatus,* a young male parasitic acanthocephalan from the intestine of a large-mouth black bass *Micropterus salmoides*. LM (worm fixed and stained), bar = 1 mm. [Photograph courtesy of S. C. Buckner; drawing by L. Meszoly; information from B. Nickol.]

retract the proboscis directly. Larvae move by contracting proboscis muscles.

Thorny-headed worms reproduce only by egg-and-sperm fertilization and are dioecious. Male and female worms mature sexually in the gut of their host. The female produces eggs in an ovary. The original ovaries then break up; the resulting ovarian balls either attach posterior to the proboscis within the female's pseudocoelom or lie free in her pseudocoelom. A funnel called the uterine bell opens to the female's pseudocoel at one end and to her uterus at the other end. Males have a pair of testes from which sperm ducts (see Figure A) lead to a penis. During copulation, the male injects seminal fluid from his seminal vesicle into the wall of his bursa. The bursa everts and holds the female—which lacks a bursa—during copulation. Sperm are released through the penis, which extends into the male's everted bursa, into the female's gonopore. Cement-gland secretions of the male, released through the penis, cap the female's posterior end, preventing loss of sperm. Sperm travel from the female's gonopore into the vagina, into the uterus, and through the uterine bell into the female's pseudocoel, where fertilization and embryonic development take place.

After the embryonic stages, the female discharges larvae enclosed in a shell—called acanthor larvae—through her gonopore into the vertebrate host's gut. The acanthor has spines on its anterior end and body. Larvae leave the host with the host's feces. Acanthor larvae are picked up from water or soil by an intermediate invertebrate host, which may be an insect such as a roach or grub (Phylum A-20), an aquatic crustacean (Phylum A-21), or a snail (Phylum A-26). Within this intermediate host, acanthors metamorphose into a second larval stage—an acanthella larva—and then hatch from their shells. The acanthella has all of the organs of an adult acanthocephalan except reproductive organs. The "naked" acanthella in the intermediate host eventually bores through the gut wall of the intermediate host, lodges in the host's body cavity, and withdraws its proboscis, before encysting and entering a resting stage called a cystacanth. Several intermediate invertebrate hosts may intervene; the acanthocephalan does not develop but remains infective in these intermediate "transport host(s)." A vertebrate acquires an encysted cystacanth in its gut as part of its dinner of intermediate transport host(s). After reaching its final, vertebrate, host, the cystacanth larva emerges from its cyst and the acanthocephalan reaches adulthood.

Acanthocephalans and other pseudocoelomates have in common an external cuticle and absence of respiratory and circulatory systems. Highly specialized parasites tend to lose features that their ancestors might have had in common with other phyla. As specialized parasites, acanthocephalans have only an elusive link to other phyla. The ultrastructures of acanthocephalan sperm, body wall, and muscles resemble those of rotifers (Phylum A-13), nematodes (Phylum A-10), kinorhynchs (Phylum A-14), nematomorphs (Phylum A-11), and two phyla in which the origin of the body cavity is not clear—gastrotrichs (Phylum A-16) and priapulids (Phylum A-15). Fossils of priapulids discovered in the Burgess shale (about 500 million years before the present) provide evidence that thorny-headed worms may be related to priapulids, as reported by Sidney Conway Morris and D. W. T. Crompton, both of the University of Cambridge. The suggestion has been made that thorny-headed worms may have common ancestors with the marine, free-living priapulids. Thorny-headed worm adults also resemble nematodes in having a species-specific constant number of cells or nuclei. However, unlike nematodes, the thorny-headed worm lacks a gut and has a proboscis, circular body wall muscles, a ciliated excretory organ, nonnematode developmental features, and a body divided into three regions—proboscis, neck, and trunk.

Acanthocephalans seldom parasitize humans; in the United States, *Moniliformis dubious* infests the house rat and cockroaches and, accidentally, humans, probably when humans ingest roach- or rat-contaminated food. Heavy infestations of these thorny-headed worms in domestic animals and wildlife interfere with the vertebrate host's digestion.

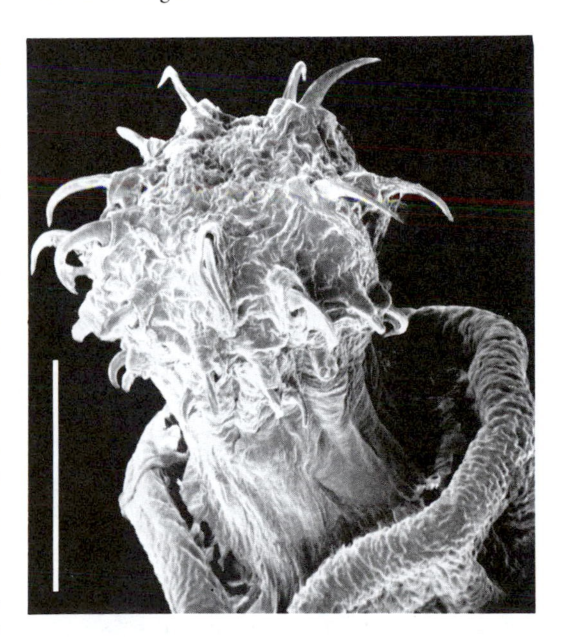

B The larva of *Macracanthorhynchus hirudinaceus*, showing its everted proboscis. The hooks are used to penetrate tissues of the thorny-headed worm's beetle hosts. SEM, bar = 0.4 mm. [Courtesy of T. T. Dunagan, *Transactions of the American Microscopical Society* 90:331 (1969).]

A-13 Rotifera

(Rotifers)

Latin *rota,* wheel; *ferre,* to bear

EXAMPLES OF GENERA

Albertia
Asplanchna
Brachionus
Chromogaster
Conochilus
Cupelopagis
Embata
Euchlanis
Floscularia
Notommata
Philodina
Proales
Seison
Stephanoceros
Synchaeta

The rotifer is a minute aquatic animal named for an optical illusion—waves of beating cilia on its head appear to be a rotating wheel. The cilia are dual purpose: they propel the swimming rotifer and direct food currents to the mouth. Behind the anterior mouth lies a feeding apparatus unique to rotifers: rigid jaws called trophi are manipulated by and embedded in the mastax—a muscular pharyngeal region between the mouth and esophagus. Trophi suck, grab, or grind food.

About 2000 rotifer species have been reported; only about 50 species are marine, being benthic or pelagic. Some rotifers are epizoic (living on other animal species) or parasitic (symbiotrophic). Many marine rotifers live interstitially (between sand grains) with loriciferans, tardigrades, kinorhynchs, and gastrotrichs. Most are free swimming, although *Conochilus* swims as a revolving colony of about 100 individual organisms anchored in a jelly sphere. *Stephanoceros* is sessile, permanently attached in a secreted jelly case. *Floscularia* molds pellets of mucus and feces to form an exquisite dwelling tube. Rotifers are the most abundant and cosmopolitan of the freshwater zooplankton. They inhabit bogs, sandy beaches, lakes, rivers, glacial muds, bird baths, gutters, ditches, moss and lichen pads, rocks, and tree bark. Rotifers resist desiccation by secreting a protective envelope of gel around their bodies.

Rotifer animal hosts are exclusively invertebrate. *Seison* lives exclusively on the leptostracan crustacean *Nebalia. Seison* is toeless and attaches to its host with a posterior adhesive disc. It moves leechlike over its host's gills, where it is nourished both by eating its host's eggs and by food in the ocean current. *Proales* lives on *Daphnia,* the common freshwater water flea (Phylum A-20), and in snail eggs (Phylum A-26), the heliozoan *Acanthocystis* (Phylum Pr-27), *Vaucheria* filament tips (Phylum Pr-14), cnidarians (Phylum A-3), and *Volvox* (Phylum Pr-28). Other rotifers are endoparasites of annelids and shell-less molluscs. The rotifer *Albertia* is a wormlike obligatory parasite living in the coelom of annelids (Phylum A-22).

Free-living rotifers typically have eclectic diets, feeding on bacteria, suspended organic matter, protoctists, and other small ani-

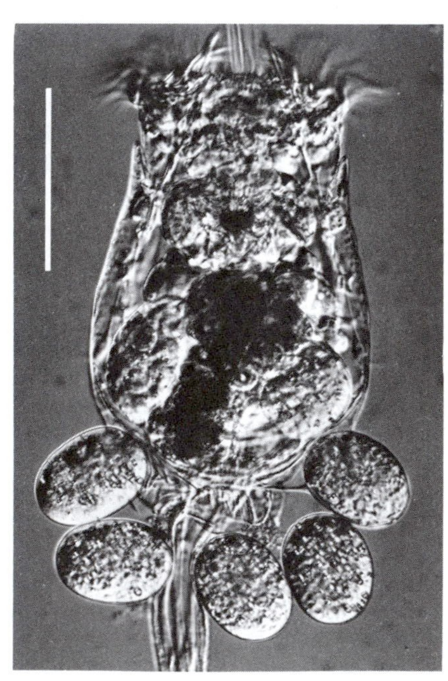

Living *Brachionus calyciflorus,* a freshwater female rotifer. Thin filaments attach the eggs to the female until they hatch. LM (interference phase contrast), bar = 0.1 mm. [Photograph courtesy of J. J. Gilbert; drawing by L. Meszoly; information from J. J. Gilbert.]

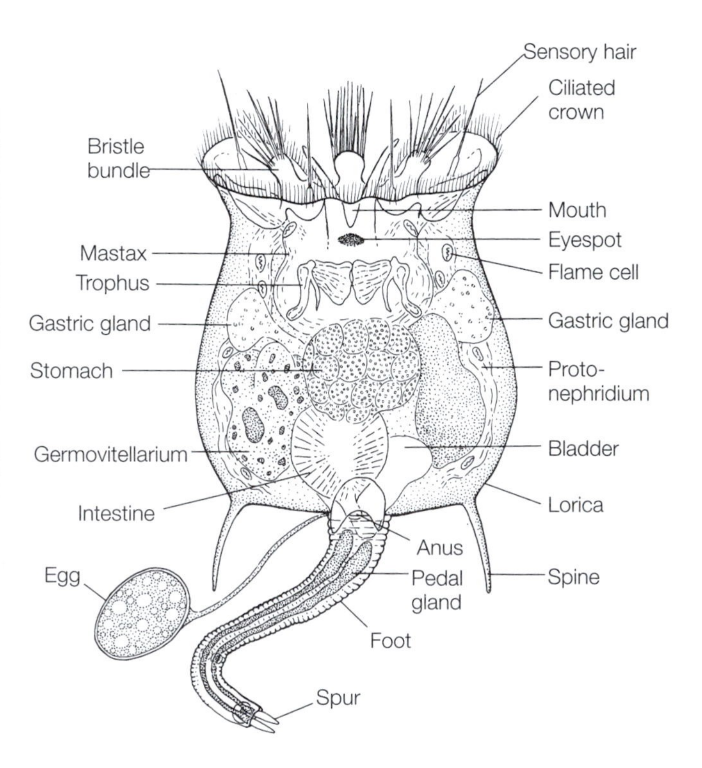

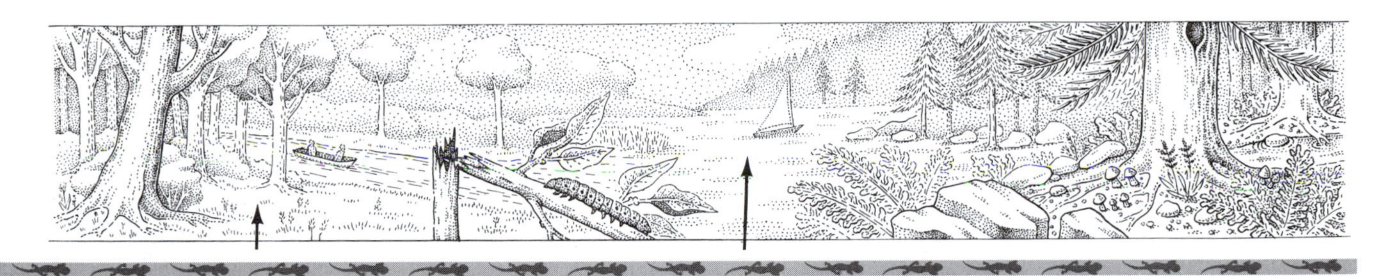

mals including rotifers. A few sessile (permanently attached) rotifers are trappers; *Cupelopagis* traps protoctists by means of a retractable funnel. *Chromogaster* specializes in grabbing dinomastigotes (Phylum Pr-7) with its pincerlike mastax, drilling through the test (the firm covering) of the prey and sucking it dry.

The flexible body of some rotifers is covered with a nonchitinous layer that is not molted. In species of Class Monogononta, the thick cuticle is called a lorica. In many species, the lorica can collapse and expand like a portable telescope. Rotifer shapes range from trumpetlike to spherical. Free-swimming rotifers, such as *Brachionus,* have spines that enhance flotation and swim in spirals propelled by their ciliated crowns. The head cilia may be fused into tentacles or platelike structures or they may be lacking altogether in some species. Other free-living species stick temporarily and sessile species attach permanently to substrates with secretions from the pedal cement glands on their contractile toes; moving like inchworms, free-living rotifers attach and reattach. On the posterior end, rotifers may have nonadhesive (lacking cement glands) spurs and as many as four pairs of adhesive toes. Rotifers are usually translucent. Some are colored green, orange, red, or brown from food in the gut, which can be seen through the translucent body wall. The largest rotifers reach 2.0 mm in length; the smallest are about 0.04 mm long.

The crown of rotifers that feed on suspended food directs water to a food groove; the ciliated groove conveys particulate food to the mouth. Prey is ground by the trophi as it passes through the mastax into the ciliated stomach, into which gastric glands secrete digestive enzymes. Digestion is mostly extracellular. After passing through the ciliated intestine, solid waste leaves the anus. Dissolved wastes are collected within a pair of protonephridia, convoluted tubules that drain to a bladder, then to the cloaca or to nephridiopores, and out to the environment. Beating cilia of protonephridia maintain water flow and salt balance through the excretory system in freshwater rotifers; marine species lack protonephridia. Striated muscles move rotifer appendages and enable looping over the substrate. Rotifers have smooth muscle also, for more sustained contraction. Dorsal to the mastax is the small brain, from which nerve pairs extend to muscles and organs. Hair cells respond to mechanical and chemical sensation. *Asplanchna brightwelli* has photoreceptors containing membranes layered like cabbage leaves, but, unlike vertebrate photoreceptors, rotifer photoreceptors are not formed from cilia. The *Euchlanis* photoreceptor contains a lens. Rotifers respire through the body surface and lack a blood circulatory system.

Reproduction in rotifers differs among the three classes—Seisonidea, Bdelloidea, and Monogononta—in this phylum. Most free-living rotifers are females, which reproduce parthenogenetically. Eggs mature in a germinarium (ovary). Embryos develop in a vitellarium from which they are released from their mothers. The combined site for the germinarium and vitellarium is a germovitellarium. In most, young are born live through the female's cloaca. All species in Class Bdelloidea are free living. Adult bdelloid rotifers produce diploid eggs that develop into females; males are unknown and so, consequently, is sexual reproduction. Some, such as *Brachionus* (Class Monogononta), carry their eggs on an external filament. Monogonont rotifers reproduce either asexually by means of parthenogenesis or sexually, depending on environmental stimuli. Monogonont rotifers produce two kinds of eggs: diploid (containing two sets of chromosomes) and haploid (containing one set). Unfertilized diploid eggs hatch into female adults, 40 or more generations may form per year. However, if ponds dry up, haploid eggs form by meiosis. These haploid eggs may be fertilized. If not fertilized, haploid eggs hatch into small, degenerate males, incapable of feeding but able to produce sperm that may fertilize haploid eggs. The males produce two distinct types of sperm: (1) ordinary sperm, which penetrate the female body wall and then fertilize the egg in the ovary, and (2) rod-shaped bodies thought to assist the regular sperm. Heavy-shelled resting, or winter, eggs, the products of monogonont rotifer fertilization, enter developmental arrest—cryptobiosis—until spring. Winter eggs hatch into females that produce female young asexually. A very few monogonont females produce both diploid eggs by means of mitosis and haploid eggs by meiosis. In Class Seisonidea, the sexes are separate. *Seison*—a dioecious ectoparasite—and others in this class reproduce exclusively sexually by either hypodermic copulation (males inject sperm into the female's pseudocoelom) or cloacal copulation; from the cloaca, sperm move up to the ovary. All rotifers, including sessile species, bear free-swimming young.

Membranes between somatic cells disappear in the epidermis of adult rotifers; such tissues are said to be syncytial. Like those of gastrotrichs, organs of rotifers of each species have a constant number of nuclei. Nuclear constancy and failure of cells to divide in the mature adult may account for the lack of healing and regeneration in rotifers.

For other animals in freshwater communities, rotifers are a major food source. Rotifers also aid in soil decomposition. Rotifer fossils are unknown. Bilateral, ciliated flatworms (Phylum A-5) were likely ancestral to rotifers because pharynx, cilia, and protonephridia are similar in these two phyla. Rotifers are informally grouped with nematodes (Phylum A-10), kinorhynchs (Phylum A-14), acanthocephalans (Phylum A-12), nematomorpha (Phylum A-11), and perhaps gastrotrichs (Phylum A-16) as pseudocoelomates. Although all pseudocoelomates lack a peritoneum-lined body cavity and circulatory and respiratory organs and have specialized pharyngeal regions, a one-way gut (except acanthocephalans, which lack a gut), and an external cuticle, it now appears that, at various times in the past, the pseudocoelomate phyla may have evolved independently from ancestors that were acoelomate.

A-14 Kinorhyncha
(Kinorhynchs)

Greek *kinein,* to move; *rynchos,* snout

EXAMPLES OF GENERA
Campyloderes
Cateria
Centroderes
Echinoderes
Kinorhynchus
Pycnophyes
Semnoderes

Kinorhynchs are free-living marine animals generally 1 mm or smaller, somewhat larger than rotifers and gastrotrichs. A kinorhynch moves ahead by forcing body fluid into its head and thereby everting, anchoring with scalids—spines—on its head in sand grains or mud and then hauling itself forward as it retracts its head. Kinorhynchs do not swim. The name kinorhynch, meaning moveable snout, refers to the kinorhynch head. Microscopic neck plates cover the head when the kinorhynch inverts its protrusible snout. Some kinorhynch species retract both head and neck.

Kinorhynchs have been collected all over the world on seafloors, in estuaries, and on muddy marine beaches between the intertidal zone and to a depth of 5000 m as far north as Greenland, as far south as Antarctica, and in the Black Sea. About 150 species have been described. Most are colorless or yellow brown, perhaps colored by food (diatoms) in their gut. Kinorhynchs feed on bacteria, minute algae, diatoms (Phylum Pr-16), and organic debris. No kinorhynchs are known to be parasites (symbiotrophs), although some seem to be commensal with hydrozoans (Phylum A-3), bryozoans (Phylum A-29), and sponges (Phylum A-2). Kinorhynchs are food for shrimp, snails, and other bottom-feeding marine animals.

A cross section of the body is triangular in one of the two classes (Homalorhagida), oval in the other (Cyclorhagida). Segmented plates of tough cuticle armor the kinorhynch with one curved dor-

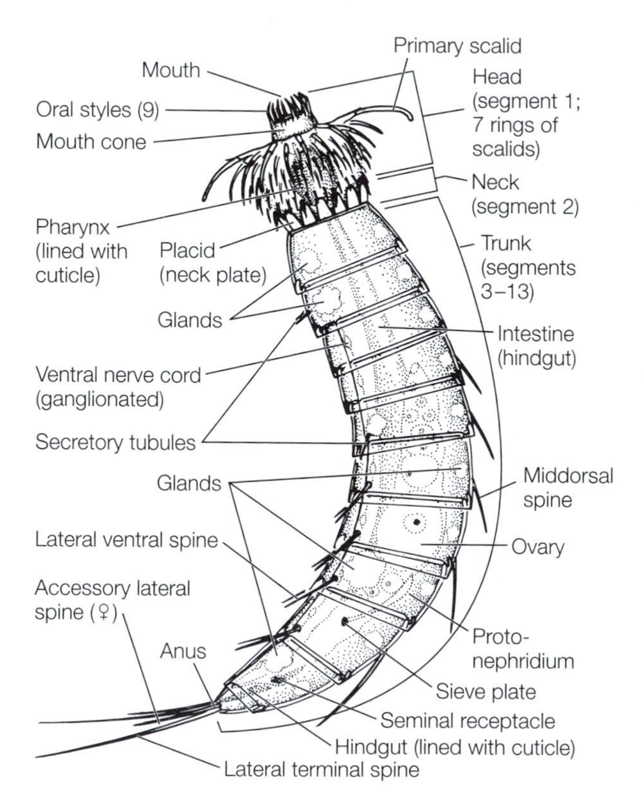

An adult kinorhynch, *Echinoderes kozloffi,* with its head extended. LM, bar = 0.1 mm. [Photograph courtesy of E. Kozloff; drawing of female *Echinoderes* courtesy of R. P. Higgins.]

sal plate and two flat ventral plates on each trunk segment. Flexible cuticle lies between the plates. The eversible head is a single segment, as is the neck; there are 13 segments altogether. The dorsal plates and the forked hind end bear moveable, hollow spines and secretory tubules (adhesive tubes). Smaller spines cover the body.

The kinorhynch body cavity—a pseudocoelom (false body cavity)—is not really a cavity; rather, it is filled with cellular material, as in rotifers (Phylum A-13), nematodes (Phylum A-10), acanthocephalans (Phylum A-12), nematomorphs (Phylum A-11), and loriciferans (Phylum A-17). Like nematodes, kinorhynchs lack circular body muscles. Fluid within the pseudocoelom circulates dissolved oxygen, carbon dioxide, and nutrients. This fluid also serves as a hydraulic skeleton, enabling kinorhynchs to burrow slowly through sediment. The segmented cuticle lacks free cilia. Kinorhynch muscles and nervous system also are segmented.

The oral styles are probably used to grasp microscopic algae. Then the muscular, cuticle-lined pharynx sucks food into the terminal mouth, from which the food moves to the foregut where extracellular digestion is believed to take place. At the posterior end, undigested matter passes out through the anus. Protonephridia, a pair of rudimentary blind-ended tubules with a single undulipodium, collect dissolved waste and discharge it through an excretory pore (sieve plate) in the eleventh segment. This constitutes the kinorhynch's water-balance and excretory system.

The kinorhynch nervous system is composed of a nerve ring circling the pharynx, a ventral double nerve cord, scattered clusters of ganglia ("brain"), and sensory bristles on the trunk. In some species, red pigmented light sensors called ocelli (singular: ocellum) lie behind the mouth cone.

Individual kinorhynchs are either male or female. No obvious external features distinguish the sexes from each other, though there is minor sexual dimorphism in copulatory spines at the rear of males. Females, such as that illustrated in the adjoining drawing, have paired ovaries; males have paired testes. From these gonads, a gonoduct opens on the terminal segment. The ovary contains two types of nuclei: (1) germinal nuclei that form eggs and (2) nuclei that nourish the eggs. Males and females have not been observed copulating, but females have a seminal receptacle and fertilization is believed to be internal. Males of some species deposit a sperm packet called a spermatophore in females. Juveniles molt their cuticles at least six times as they develop into adults. Kinorhynchs lack a larval stage. A complete life cycle has not yet been observed for any kinorhynch species.

Because ancient kinorhynchs left no fossils, kinorhynch evolution is inferred by comparing living organisms. Kinorhynchs resemble priapulids (especially *Tubiluchus* larvae), rotifers, gastrotrichs, nematodes, and loriciferans. The cuticle of kinorhynchs and priapulids is chemically unique. The kinorhynch spiny cuticle, adhesive tubes, and (in some species) forked posterior end resemble those of gastrotrichs; however, the kinorhynch cuticle lacks external cilia in comparison with the gastrotrich ciliated epithelium. Kinorhynchs are likely descendants of free-living flatworms (Phylum A-5) and have been aligned with priapulids (Phylum A-15).

A-15 Priapulida
(Priapulids)

Latin *priapulus*, little penis

EXAMPLES OF GENERA
Halicryptus *Meiopriapulus* *Priapulus*
Maccabeus *Priapulopsis* *Tubiluchus*

Priapulids are short, plump, exclusively marine worms. The priapulid proboscis terminates in a mouth and inverts as it retracts, so the proboscis (presoma) is called an introvert. Spiny papillae called scalids stud the introvert. The trunk consists of 30 to 100 superficial rings (bands of circular body wall muscle) covered with spines and warts. *Tubiluchus corallicola* has a long, contractile, postanal appendage—called a tail, in some priapulids—that may anchor or retract the trunk (Figure A). Species of the genus *Priapulus*, after which the phylum is named, have one or two retractable caudal appendages into which the body cavity extends. *Meiopriapulus, Halicryptus*, and *Maccabeus* lack tails. Seventeen species of priapulids have been described in this phylum. All are free living. About half are meiobenthic (seafloor dwellers less than 0.5 mm in length), and half—all cold-water forms—are macrobenthic (longer than 0.5 mm). The smallest priapulid is *Tubiluchus corallicola*, 0.05 cm long, and the largest is a new species of *Halicryptus*, 32 cm long. Body and tail length very much depend on the individual worm's state of contraction.

Priapulid worms have been collected from a variety of depths in the sea—from waters as shallow as intertidal pools and as deep as the ocean abyss. *Tubiluchus corallicola* lives in coral sand, silt, and mud in warm shallow waters of the Caribbean, Bermuda, Cyprus, and Fiji. Cold water priapulids live in the Arctic Ocean, off North America north of Massachusetts and California, in the Baltic and North Seas south to Belgium, around Patagonia, in the Antarctic, and in cold deep waters off Costa Rica. In an Alaskan bay, *Priapulus caudatus* larvae are abundant members of the macrofauna (organisms greater than 0.5 mm), as many as 85 adults per square meter; larvae have been found with densities of 58,000 larvae per square meter. Priapulids' spotty distribution in the seas may be an artifact of collecting.

Priapulids burrow by alternately anchoring their anterior and posterior ends; longitudinal muscles that line the body wall push the body through the sediment. *Maccabeus cirratus* and *M. tentaculatus* are filter feeders, utilizing hollow tentacles; they lack an eversible proboscis. *Maccabeus* builds a permanent tube that is open at both ends and buried in the seafloor. The tube of *Maccabeus* is flimsy, built of secreted material in which plant fragments are encased in a longitudinal pattern. Unlike some other burrowing worms, priapulids do not maintain water currents through their temporary burrows. Most priapulids lie with their mouths flush with the sea bottom, their bodies in sediment.

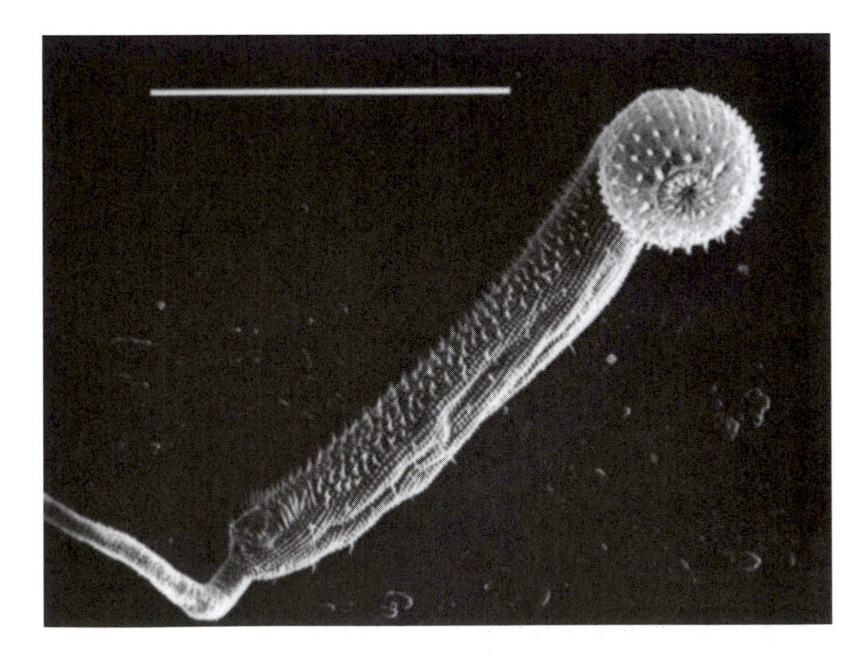

A *Tubiluchus corallicola,* an adult priapulid taken from the surface layer of subtidal algal mats at Castle Harbor, Bermuda. SEM, bar = 0.5 mm. [Courtesy of C. B. Calloway; from *Marine Biology* 31:161–174 (1974).]

Smaller priapulid species—the meiofaunal *Meiopriapulus*, for example—feed on bacteria that coats sand grains. Larger priapulids are carnivorous; using the spines that line their mouths, they seize polychaete annelids (Phylum A-22) and other priapulids. Both the spine-studded proboscis (Figure B) and the terminal mouth roll inside and out again, passing prey whole into the muscular, toothed pharynx. Food is digested as it passes down the straight intestine, which is surrounded by longitudinal and circular muscle and leads to the rectum. Nutrients may be distributed by body cavity fluid. Solid waste is evacuated from the anus.

The cuticle-covered priapulid body lacks internal and external segmentation. The body wall contains longitudinal and circular muscles. *Priapulus caudatus* has red cells containing the pigment hemerythrin—as does nemertine blood—and amebocytes that circulate in the body fluid. Hemerythrin is an iron-containing protein that stores or carries oxygen; hemerythrin is also present in brachiopods and polychaete annelids. Priapulid respiration is not well understood. The tail of *Priapulus* may function in gas exchange or chemoreception, but its function has not yet been demonstrated. However, removal of its tail does not kill a priapulid (which regenerates the appendage), so other modes of gas exchange must exist.

A circumpharyngeal nerve ring connects to a single, ventral nerve cord, which runs down the body and from which peripheral nerves extend. Raised bumps (papillae or tubercules) on the body surface seem to be sensory. Tiny flower-shaped flosculi (composed of microvilli) of unknown function are scattered on the trunk. The gonads in both sexes are tubular (Figure C). The excretory system consists of a pair of protonephridia—waste-collecting, ciliated tubules. The protonephridia and gonads share ducts that open into nephridiopores, one on each side at the posterior end of the trunk. The body cavity may be a pseudocoel or a coelom.

Because any individual worm is either male or female, priapulids are dioecious ("two houses"). Whether females are similar to males or differ from them in external appearance depends on the species. Some priapulids, such as tube-dwelling *Maccabeus*, probably reproduce parthenogenetically, because only females have been found. Along the Scandinavian coast, spawning takes place in winter; eggs and sperm are shed into the sea. Macrobenthic priapulids such as *Priapulus* and *Halicryptus* are assumed to have external fertilization; meiobenthic priapulids such as *Meiopriapulus*,

B The presoma of *Tubiluchus corallicola*, showing the retractile proboscis everted. SEM, bar = 0.1 mm. [Courtesy of C. B. Calloway; from *Marine Biology* 31:161–174 (1974).]

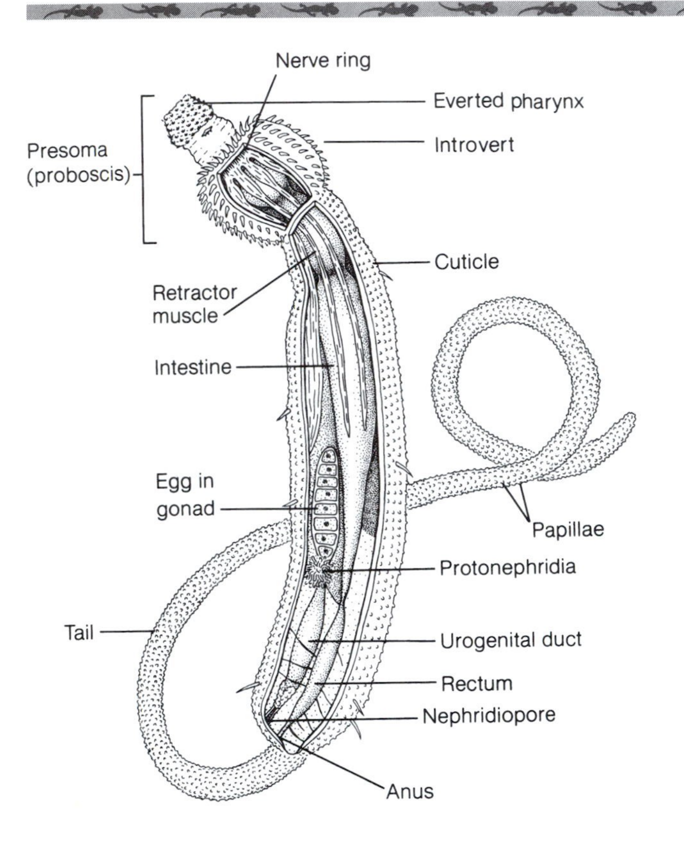

C Morphology of an adult female *Tubiluchus corallicola*, a minute priapulid of the meiobenthos. [Drawing by L. Meszoly; information from C. B. Calloway.]

Tubiluchus, and *Maccabeus* are assumed to have internal fertilization. After external fertilization, eggs of most species develop into larvae that are smaller versions of the adults. *Meiopriapulus* is the only genus that lacks a larval stage. Of all the other species, *Tubiluchus* larvae have a longitudinally ridged cuticle, but other larvae have a lorica—a firm cuticle—of plates. Larvae adhere to mud and other sediment with adhesive from their hollow toes and are never free swimming. A series of larvae (morphologically indistinct stages) molt as they attain adult size. Adults also shed their partly chitinous cuticle. *Meiopriapulus fijiensis* females brood their young embryos and then release postembryonic young from the urogenital pore; *Meiopriapulus* lacks larvae—these juveniles develop directly into adulthood.

Middle Cambrian fossil priapulids abound in the Burgess shale, deposited about 500 million years ago. When polychaete annelids (Phylum A-22) with jaws evolved during the Ordovician, about 440 million to 500 million years ago, polychaetes displaced priapulids from their role as abundant marine carnivores.

The relationship of priapulids to other marine animal phyla continues to be uncertain. Until the early 1900s, priapulids were placed with echiurans (Phylum A-24) and sipunculid worms (Phylum A-23) in Phylum Gephyrea. In 1961, when the spacious priapulid body cavity was discovered to be seemingly lined with mesentery and thus considered a coelom, some researchers moved priapulids from pseudocoelomate to protostome coelomate status. Investigation of priapulid embryonic development is needed to clarify whether the priapulid body cavity is indeed a coelom, because other zoologists group priapulids with pseudocoelomate animals—that is, animals having a body cavity not lined with mesoderm. Priapulids resemble nematomorphs (Phylum A-11), nematodes (Phylum A-10), and juveniles of kinorhynchs (Phylum A-14) and loriciferans (Phylum A-17) in that they molt their cuticles from time to time. Also like nematomorphs, priapulids are dioecious and have a cylindrical unsegmented body without internal septa (cross walls). In 1980, newly revealed morphological features of priapulids, nematomorphs, kinorhynchs, and loriciferans led V. V. Malakhov, a Russian invertebrate zoologist, to propose that these four animals be included in a newly created phylum named Cephalorhyncha. Priapulids may also be related to rotifers (Phylum A-13) or to acanthocephalans (Phylum A-12).

A-16 Gastrotricha
(Gastrotrichs)

Greek *gaster,* stomach; *thrix,* hair

EXAMPLES OF GENERA

Acanthodasys
Chaetonotus
Dactylopodola
Lepidodermella
Macrodasys
Tetranchyroderma
Turbanella
Urodasys

Bristles, spines, or scales ornament the backs and sides of gastrotrichs but not their bellies (Figure A). This phylum derives its name from the cilia that cover the underbellies—gastrotrich is compounded from the Greek words for "hairy stomach." Cuticle pattern, body shape, and tracks of ventral cilia are distinctive enough to be useful in taxonomic classification. Gastrotrichs temporarily glue themselves belly side down to grains of sea sand or freshwater plants or sediment with adhesive from tubes (2–250) that project from their sides, concentrated near their posterior ends. Thus secured belly side down to sand, the worm is protected from abrasion by the rough exterior of its sides and top. Soft scales, ventral cilia, and adhesive tubes characterize most gastrotrichs. The minuscule, flat, transparent body with lobed head ranges from less than 0.1 to 3.5 mm in length, usually about 0.5 mm.

All 400 or so species of gastrotrichs are free-living members of the meiofaunal community—animals that pass through collecting screens having 1 mm mesh but remain behind on screens having 0.042 mm mesh. Sandy beach crevices and aquatic environments are gastrotrich habitats, from the frigid Arctic to Chile. Most marine species live in interstices of intertidal and subtidal sands. A few glide over coral. Freshwater gastrotrichs prefer the quiet water of bogs, mossy pools, and plant-choked ponds, crawling over water lily leaves or the underside of duckweed. Like free-living flatworms (Phylum A- 5), a gastrotrich can move by fixing its anterior end in place, pulling up its posterior, and then detaching its anterior tubes. These bristly invertebrates have a dual-gland arrangement that allows such movement: glands that cement the animal to substrate are paired with releaser glands that secrete solvent to dissolve the "glue" as needed.

Gastrotrichs exchange oxygen and carbon dioxide by diffusion with their aquatic habitat. Circulatory and respiratory organs are absent. A thin unsegmented cuticle of lipoprotein and nonchitinous polysaccharide covers the entire body. Like rotifers, gastrotrichs do not shed their cuticles. Their cilia direct water currents that bear organic debris, bacteria, algae, foraminiferans (Phylum Pr-4), and diatoms (Phylum Pr-16) to the mouth, from which the muscular pharynx pumps food into the stomach-intestine. Food is digested intracellularly. Undigested material leaves through the anus. In some species, gametes also exit through the posterior opening, which is then—by definition—a cloaca.

Most gastrotrichs, especially freshwater species, have a pair of protonephridia with a midventral excretory pore. Marine species lack protonephridia. Gastrotrichs swim short distances, steering with longitudinal muscles. Muscles that encircle the body move the bristles and adhesive tubes. A large brain surrounds the pharynx, with a pair of longitudinal, lateral nerve cords. On the head and trunk are sensory bristles and tufts of sensory cilia. Red spots on both sides of the brain may be photoreceptors.

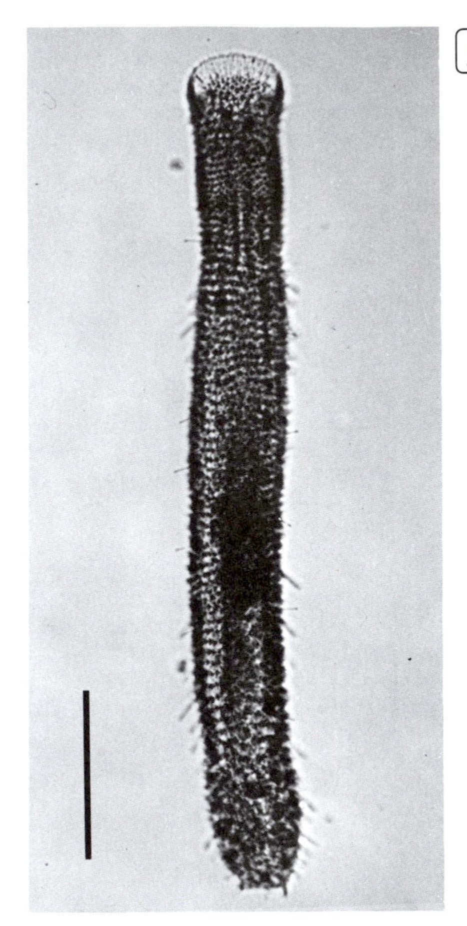

A Living adult *Tetranchyroderma* from a New England beach. Adhesive tubes secrete glue that temporarily anchors it to sand in the intertidal zone. LM, bar = 0.1 mm. [Photograph courtesy of W. Hummon; drawing by I. Atema; information from W. Hummon.]

Many individual gastrotrichs are simultaneous hermaphrodites, producing both eggs and sperm, particularly marine gastrotrichs. Some species are sequential hermaphrodites. In *Dactylopodola* and *Urodasys,* sperm packets called spermatophores are transferred from individuals that behave as males to individuals that behave as females. The end of the sperm duct serves as a penis for sperm transfer. Fertilization is internal. Freshwater gastrotrich populations including *Lepidodermella* are entirely female; they reproduce parthenogenetically. The diploid eggs develop into females of the next generation without fertilization. However, rodlike sperm are found in some *L. squammata* individuals that also

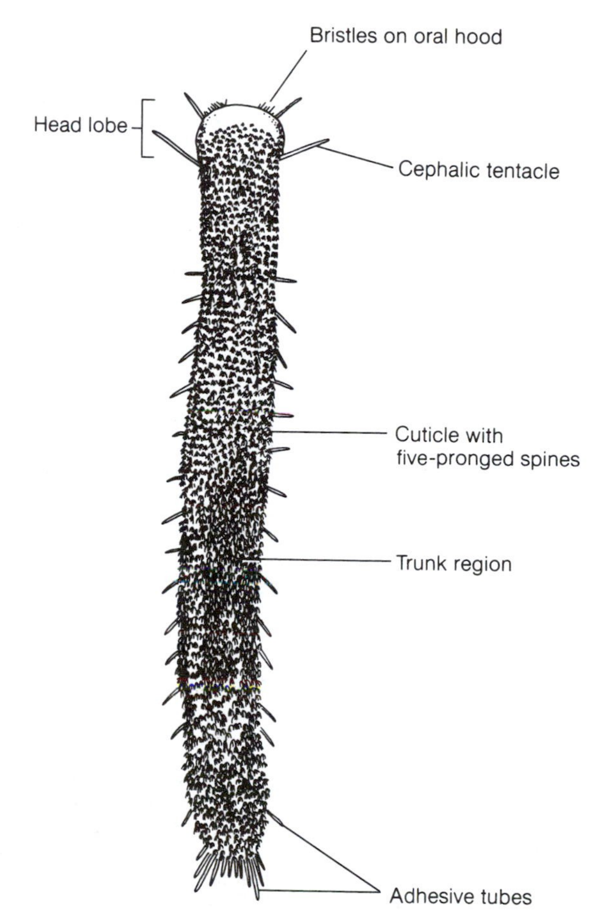

contain eggs, suggesting that this genus evolved from hermaphroditic ancestors. Female gastrotrichs lay from one to five large eggs in a lifetime, depositing eggs on algae, discarded skeletons, or pebbles (Figure B). When water touches the egg, an eggshell forms. Freshwater gastrotrichs produce two egg types. One egg type, usually laid at the end of the more favorable season (summer compared with winter, for example), may remain dormant for as long as 2 years and must be exposed to drying, freezing, or other harsh conditions before it will develop. The second egg type is thin walled and develops immediately after laying. Gastrotrichs develop directly and hatch at almost adult size; they lack larvae.

Because they scavenge dead bacteria and plankton from beaches and estuaries, gastrotrichs are indirectly important to humans. The healthiest marine sands are those with the greatest biodiversity of meiofauna, such as gastrotrichs. In aquatic food webs, gastrotrichs are themselves food for amebas (Phylum Pr-3), hydras (Phylum A-3), turbellarian flatworms (Phylum A-5), insects (Phylum A-20), crustaceans (Phylum A-21), and annelids (Phylum A-22).

Adult gastrotrichs lack body cavities; various biologists have claimed that gastrotrichs are acoelomate or pseudocoelomate or coelomate. Currently, many biologists align gastrotrichs with acoelomates. Turbellarian flatworms—acoelomates—may be ancestral to gastrotrichs. The gastrotrich body form, musculature, and protonephridia are like those of rotifers (Phylum A-13), which are pseudocoelomates. However, unlike rotifers, gastrotrichs have neither mastax nor ciliary crown. Both gastrotrichs and kinorhynchs (Phylum A-14), also pseudocoelomates, have similar structures: ornamented cuticles; a muscular pharynx triangular in cross section; complete gut with mouth and anus; protonephridia; and adhesive tubes. An adult member of any single gastrotrich species has a predetermined number of cells in the body resulting from the fixed number of cell divisions; a similar condition is present in nematodes (Phylum A-10), acanthocephalans (Phylum A-12), and rotifers (Phylum A-13). Monociliated epithelium is present in gnathostomulids (Phylum A-6) and some gastrotrich species. Gastrotrichs also resemble turbellarian flatworms (Phylum A-5); at least some members of both phyla glide on cilia, are hermaphroditic, digest food intracellularly, have protonephridia, have sense organs that are ciliated pits on the sides of the head, and have dual-gland adhesive tubes.

B *Acanthodasys,* a marine gastrotrich and simultaneous hermaphrodite. After fertilized eggs are laid through a temporary opening in the body wall, the wall heals. LM, bar = 0.25 mm. [Photograph copyright © David Scharf/Peter Arnold, Inc.; information from W. Hummon.]

A-17 Loricifera
(Loriciferans)

Latin *loricus,* corset, girdle; *fero,* carry, bear

EXAMPLES OF GENERA
Nanaloricus
Pliciloricus
Rugiloricus

A loriciferan is a microscopic marine animal with a mouth cone—a flexible anterior mouth tube—a head carrying club-shaped and clawlike spines, a neck (thorax), and a girdle of plates called a lorica that covers the abdomen. The head and neck can be inverted into the abdomen. Because the mouth is terminal, the spiny invertible head is, by definition, an introvert. The neck telescopes down to half its length when a loriciferan is disturbed. Heavily sculptured ventral lorica plates or longitudinal folds close over the retracted animal. Longitudinal lorica plates are sculptured and patterned with pores. The head and anterior neck are armed with spines called scalids, which fold together like the ribs of an umbrella. Adults are from about 100 to 400 μm in length, ranking in size with small rotifers.

The phylum was first described in 1983, and about 25 species are currently known. Preliminary studies suggest at least 100 species, including many as yet undescribed. *Nanaloricus* (Latin *nana,* dwarf), illustrated in Figures A and B, has been found in shelly gravel dredged near the Atlantic shore in North Carolina and Florida, the Gulf of Mexico in the United States, and the Azores Islands. Other loriciferans have been collected from the Arctic Ocean, off Denmark, in the Mediterranean, and in the Coral Sea from depths of 10 to 480 m. The animals are thought to be a cosmopolitan part of the interstitial fauna—that living between particles of coarse sand.

What loriciferans eat is still unknown. Adult loriciferans are sedentary and may be either ectoparasites (ectobionts)—living attached to other animals—or attached to gravel grains. Larvae are probably free living and travel over the ocean floor by using two or three ventral locomotory spines (see Figure A). The leaflike toes of *Nanaloricus* bear spines and rotate in ball-and-socket joints. These toes serve as paddles for swimming and enable larvae to change direction. Other species have straight toes. Glands that open into each hollow toe probably secrete adhesive. The mouth cone of the larva retracts into the introvert and, in turn, can be inverted into the neck (thorax). Unlike the adult mouth cone, the larval mouth cone lacks oral stylets in some species. A disturbed larva retracts its many-plated neck (thorax), like accordian bellows.

Researchers are undecided whether the body cavity of loriciferans is a pseudocoelom or a coelom—an understanding of the loriciferan body cavity during embryonic development is needed to resolve this question. The fluid in the body cavity serves both circulatory and respiratory functions.

The flexible head, called an introvert, everts, and has a terminal mouth with extrusible stylets (see Figures B and C). The mouth opens into a buccal canal into which salivary glands open in turn. The buccal canal is somewhat folded within the mouth cone. The mouth tube surrounds the extruded buccal canal; both can be extruded (see Figure C) through the mouth. Mouth, buccal canal,

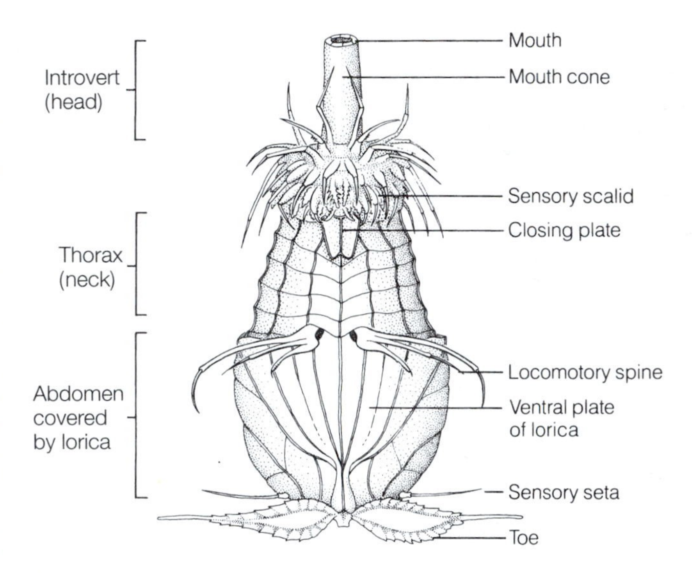

A Ventral view of larva of *Nanaloricus mysticus,* a loriciferan. The toes are swimming appendages. [Drawing by L. Meszoly; information from R. P. Higgins.]

and muscular pharyngeal bulb retract in some species. A short esophagus passes into a midgut, followed by a short rectum. Both anus and saclike gonads open at the posterior end. One pair of protonephridia constitutes the excretory system.

The loriciferan nervous system consists of a large brain within the introvert. Nerves run to each scalid. A large ventral ganglion innervates the thorax and probably continues into the abdomen. Flosculi, each a rosette of five microvilli or a single papilla, are found on the posterior dorsal side of the abdomen of some species; a sensory function has been proposed for flosculi.

Like the muscles of kinorhynchs (Phylum A-14), at least some of the large muscles that retract the introvert are cross-striated. The excretory system comprises a pair of protonephridia that may open through nephridiopores.

Differences in the most anterior rows of spines distinguish male from female loriciferans. In males, these spines are presumed to be olfactory. Males have two large dorsal testes, containing sperm, that fill the lorica. Although females have paired ovaries, only a single enormous egg develops (see Figure B). Fertilization is likely to be

internal, judging by *Nanaloricus.* The female may have a seminal receptacle near the lorica hinge (see Figure B).

Each in a series of larval stages molts, then secretes a fresh cuticle. Young loriciferans are named "Higgins' larvae," after Dr. Robert P. Higgins, expert zoologist, formerly at the National Museum of Natural History, Smithsonian Institution. From his knowledge of animal communities, Dr. Higgins had predicted the existence of organisms similar to loriciferans before their discovery.

Loriciferans are unique in their combination of characters. The mouth cone with stylets and cross-striated introvert muscles resembles those of kinorhynchs (Phylum A-14). Similarities between the flexible mouth cone with stylets of adult loriciferans and that of the tardigrade *Diphascon* (Phylum A-27) are examples of convergence. The unciliated cuticle of loriciferans resembles that of kinorhynchs and priapulids (Phylum A-15); the rotifer cuticle is intracellular (Phylum A-13). Larvae of loriciferans and of nematomorphs (Phylum A-11) bear hook-shaped ventral spines. If loriciferan glands prove to be adhesive, they may resemble the adhesive glands of kinorhynchs. Sensory organs (flosculi) are similar in loriciferans, priapulids (especially the larvae of *Tubiluchus*), and kinorhynchs. No fossil record for loriciferans has been reported. Loriciferans have so many characteristics in common with kinorhynchs, nematomorphs, and priapulids that some zoologists favor placing all four in one phylum. Until more details of loriciferan biology are known, their classification status will remain ambiguous.

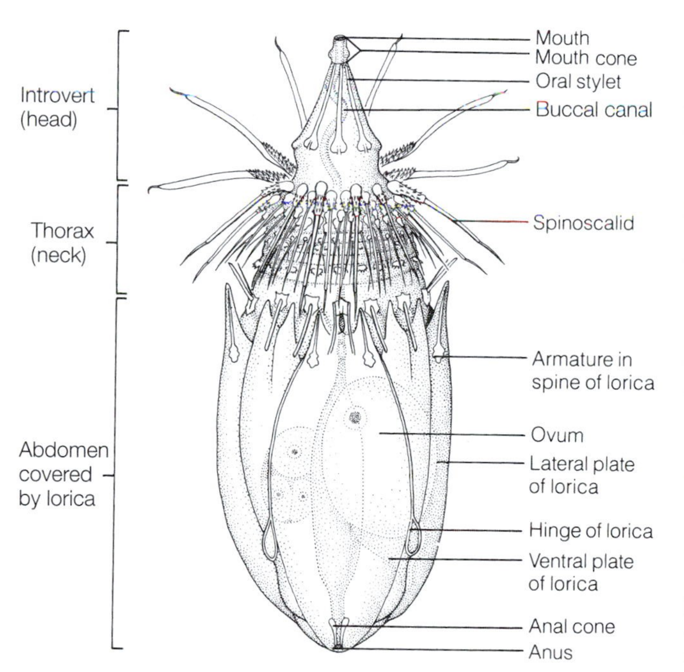

B *Nanaloricus mysticus,* adult female loriciferan. The head and neck can be inverted into the abdomen. [Drawing by L. Meszoly; information from R. P. Higgins.]

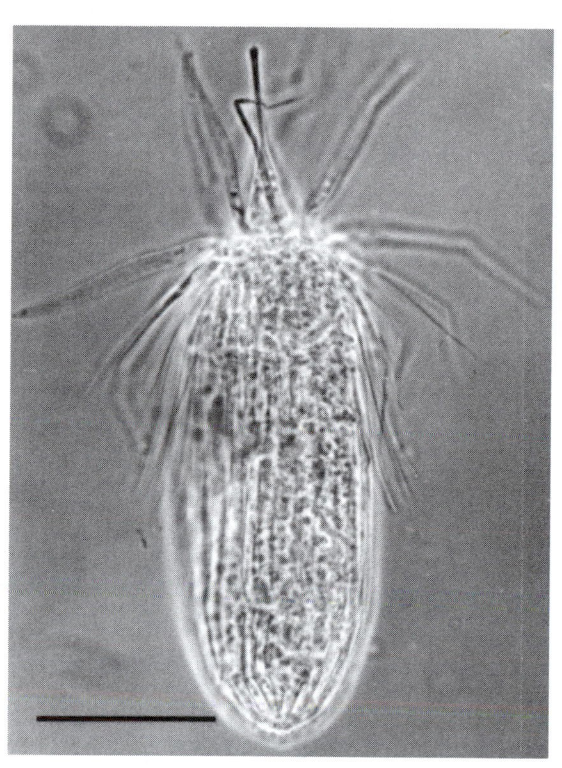

C *Pliciloricus enigmaticus.* The mouth cone with its long mouth tube extended is visible centered on the head. LM, bar = 100 μm. [Courtesy R. P. Higgins, Smithsonian Institution.]

A-18 Entoprocta
(Entoprocts)

Greek *entos,* inside; *proktos,* anus

EXAMPLES OF GENERA

Barentsia
Loxosoma
Loxosomella
Pedicellina
Pedicellinopsis
Urnatella

Entoprocts are tiny, transparent animals, primarily marine. Most are sessile, living in colonies permanently and firmly attached by stalks, horizontal stolons, and basal discs to rocks, pilings, shells, algae, or other animals. A crescent of ciliated tentacles that contract and fold over the mouth and anus is located at the free end of the entoproct (Figures A and B). An individual entoproct has from 6 to 36 tentacles, depending on its age and species. A conspicuous muscle bulb at the stalk base permits the flicking movements of individual entoprocts (Figure B). The lightly tinted, cup-shaped body is called a calyx. Some entoprocts look and creep like hydrozoan cnidarians (Phylum A-3) or like those bryozoans (Phylum A-29) that can creep; but, unlike either bryozoans or cnidarians, entoprocts fold the tentacles of their cup-shaped bodies when disturbed. Entoprocts do not withdraw their tentacles into their bodies. The whole entoproct animal—tentacles, calyx, and stalk—may be as long as 10 mm.

About 150 entoproct species are known. Many form large entoproct colonies that form "animal mats" on seaweeds, rocks, shells, and other animals in shallow marine waters. Crabs, shrimp, sipunculans (Phylum A-23), sponges (Phylum A-2), bryozoans, polychaetes, and hydrozoan cnidarians are among the animals to which entoprocts attach. A few entoproct species are solitary, such as *Loxosomella davenporth,* one of the few that is mobile as an adult. *Loxosomella* somersaults with its tentacles over its basal disc. Entoprocts are widely distributed along seacoasts of Africa, South and North America, Asia, Europe, and the Arctic. *Urnatella,* the only freshwater form, is found in the United States, France, Germany, Hungary, the former Soviet Union, the Congo, South America, India, and ponds near Tokyo, Japan.

Using their tentacles and uncoordinated beats of the lateral cilia on the tentacles, entoprocts filter food suspended in water (see Figure B). Entoproct tentacles set up a feeding current that enters be-

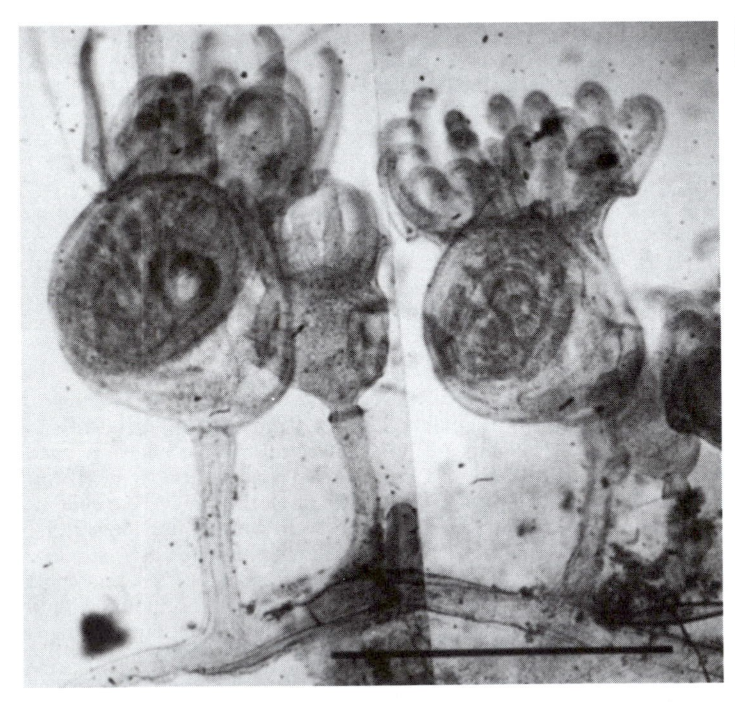

A A living laboratory culture of *Pedicellina australis,* with tentacles folded. A marine colonial entoproct, part of a fixed colony; from Falkland Islands. LM, bar = 1 mm. [Courtesy of P. H. Emschermann.]

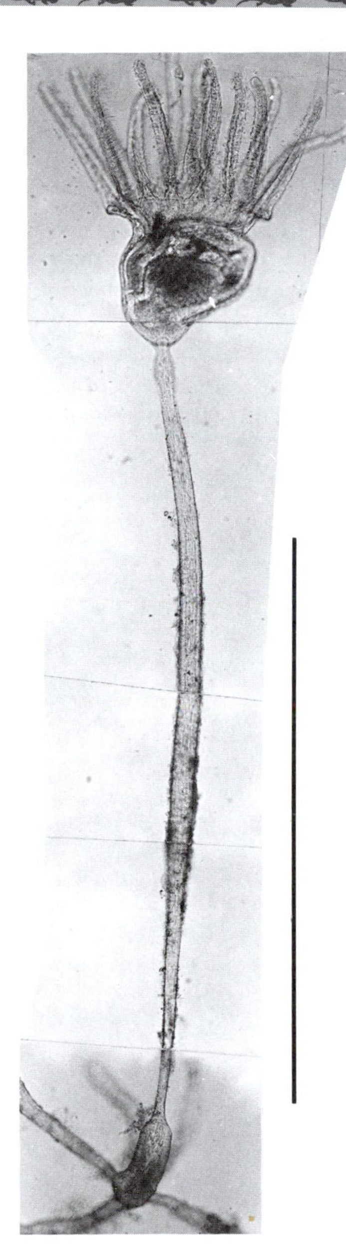

B An individual entoproct, *Barentsia matsushimana*. Rows of cilia are visible on the extended tentacles. LM, bar = 1 mm. [Courtesy of P. H. Emschermann.]

tween the tentacles and leaves at the top of the body. Diatoms (Phylum Pr-16), desmids (Phylum Pr-26), other plankton, and detritus—food for entoprocts—are trapped in mucus on extended tentacles, swept along ciliated tracts into their mouths, and then proceed through a ciliated, U-shaped digestive tract complete with anus (Figure C).

The body cavity of entoprocts may be a pseudocoelom. Gelatinous material containing cells fills the entoproct body cavity and extends into tentacles. The entoproct lacks a heart and blood vascular system. The star cell apparatus (heart) consists of muscle cells that may—by their contractions—move nutrients from the calyx to the stolon growing tips. Uric acid and guanine—usually part of the soluble waste excreted by protonephridia in other phyla—are discharged by exocytosis from the stomach surface into the stomach cavity (see Figure C). Additional dissolved waste is collected in a ciliated flame cell, called a protonephridium, and discharged through nephridiopores, located in most species on either side of the inner part of the calyx—the atrium. (The vestibule within the crescent of tentacles is called the atrium). *Urnatella* has a unique nephridiopore distribution: one in the calyx and many in the stalk.

Entoproct muscles include both longitudinal cross-striated muscles and smooth muscles in the tentacles and a muscular ring (atrial sphincter) in the tentacle membrane around the calyx. Muscle bulges on the stalk permit the characteristic "nodding" activity of entoprocts.

The nervous system of entoprocts is simple, consisting of a ganglion lying above the loop of the digestive tract with nerves extending to tentacles, stalk, and body. The horizontal stolon (see Figure A) lacks muscles and nerves. Specialized cells about the body are thought to be sensitive to chemicals, light, and vibrations.

Entoprocts reproduce both sexually and asexually. In the absence of sexual reproduction, colonies commonly arise by budding. Many of the sexual entoprocts are dioecious, whereas only a few are monoecious ("one house"), having both male and female in a single entoproct. Both simultaneous and sequential hermaphroditism have been observed. In sequential hermaphroditism, an individual is first one sex and then the other in either one of two patterns: protogyny (Greek *protos,* first; *gyne,* female) or protandry (*andros,* man). Reproductive cells exit the body through a gonopore (not shown in Figure C) located in the vestibule, the region surrounded by the tentacles. The entoproct yolk-rich eggs extruded into the brood pouch are probably internally fertilized in the ovaries by undulipodiated sperm. How sperm reach eggs is unknown. The zygotes are incubated in a brood pouch between the gonopore and the anus. Embryos develop, without parental nourishment, into ciliated, free-swimming larvae that resemble the larvae of some bryozoans (ectoprocts; Phylum A-29). Entoproct larvae leave the brood pouch, settle on substrate, attach with secretions, and then metamorphose into adults. During metamorphosis, the organs change position so that the larval ventral side ends up in the adult atrium. Larvae of some entoproct species form trochophore-like larvae, not considered true trochophores. Some entoprocts can shed their calyxes under unfavorable conditions and then regenerate new calyxes when conditions return to normal; the shed calyx is not a viable offspring.

Invertebrates in four phyla—entoprocts, phoronids (Phylum A-31), bryozoans (Phylum A-29), and brachiopods (Phylum A-30)—have crowns of ciliated tentacles. Researchers debate the relationship of these four phyla to one another. In phoronids, bryozoans, and brachiopods, the anus lies outside the tentacle crown—a defining feature of lophophorates; but, in entoprocts, the anus lies within the tentacle circle (see Figure C). With a hand lens, a curious naturalist can discern both anus and mouth within the entoproct's tentacle circle and thus distinguish the entoproct from bryozoans (Phylum A-29), hydrozoan cnidarians, and other invertebrates with which entoprocts might be confused. Some biologists believe that bryozoans evolved from entoprocts because their larvae and patterns of settling on substrate are similar. Entoproct larvae, however, do not undergo the cataclysmic metamorphosis of bryozoan larvae. A very good case has been made for reuniting phyla Entoprocta and Bryozoa; because of the radical metamorphosis that causes features of the bryozoan larva to disappear, it cannot be ascertained with certainty that the body cavity of the adult bryozoan is a coelom and, in any case, the bryozoan larva does not have a coelom. The body cavity of the entoproct may be a pseudocoelom. However, Entoprocta and Bryozoa phyla differ in many ways. The entoproct mouth and anus both open within the

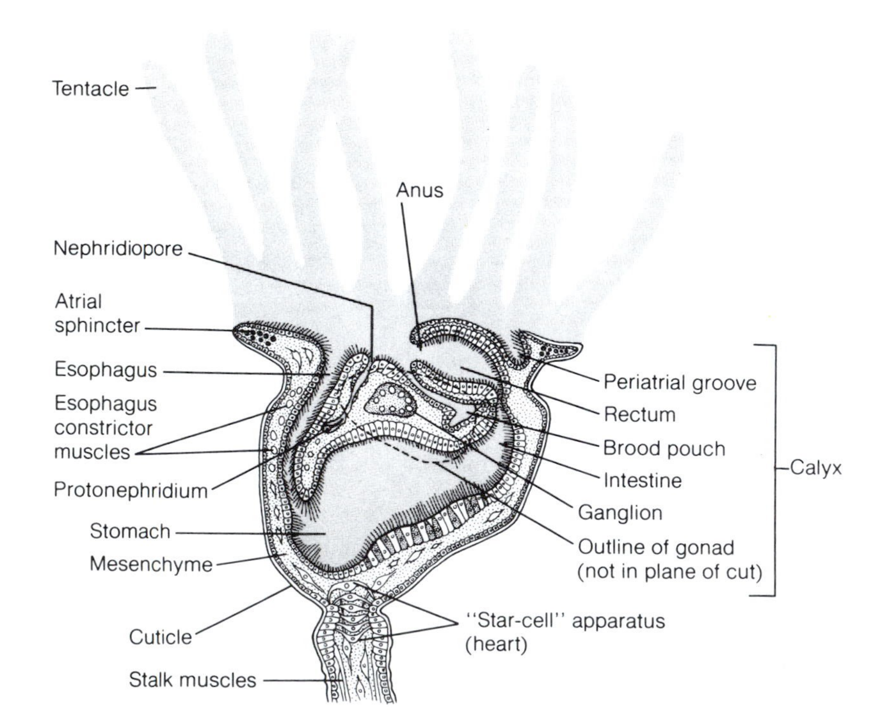

C *Barentsia matsushimana.* A vertical section shows digestive, nervous, excretory, and muscle systems within the cup-shaped calyx. [Drawing by L. Meszoly; information from P. H. Emschermann.]

circle of tentacles; water flows in between the tentacles and then carries waste upward and away from the atrium. In contrast, the bryozoan anus opens outside the circle of tentacles, and water flows outward between the tentacles. Bryozoan cilia beat in regular waves; entoproct cilia beat irregularly. Entoprocts have protonephridia; bryozoans lack excretory systems. The characteristic nodding action of entoprocts is absent from bryozoans. We favor separate phyla for these two groups at this time.

The soft-bodied entoprocts have left no recognizable fossils. Some zoologists suggest that, because of resemblances in body shape and the U-shaped intestine, rotifers (Phylum A-13) and entoprocts have a fairly recent common ancestor.

A-19 Chelicerata

(Chelicerates)

Greek *cheli*, claw

EXAMPLES OF GENERA

Acanthoscurria
Argiope
Carcinoscorpinus
Chelifer
Dermacentor
Ixodes
Latrodectus
Leiobunum
Limulus
Loxosceles
Lycosa
Nymphon
Psoroptes
Scorpio
Trachypheus

The chelicerates, which number about 75,000 species, include three classes: the horseshoe crabs (Class Merostomata), sea spiders (Class Pycnogonida), and spiders, scorpions, mites, ticks, chiggers, and harvestmen, also called daddy-longlegs (Class Arachnida). Members of phylum Chelicerata have segmented bodies, jointed appendages, and a chitinous exoskeleton in common with other arthropods (crustaceans, Phylum A-20; and insects, Phylum A-21). Chelicerates lack the antennae of crustaceans and insects; chelicerates also lack the mandibles—biting tips of the jaws for grinding and chewing food formed from the distal part of appendages that are beside the mouth—of crustaceans and insects. Chelicerae are the clawed, most anterior pair of jointed appendages characteristic of all chelicerates. Although some place chelicerates with insects, centipedes, crustaceans, and other joint-footed animals in one vast Phylum Arthropoda, other zoologists, as we do, prefer to recognize chelicerates as a phylum, a unique and cohesive lineage. Chelicerates', mandibulates', and crustaceans' consistent distinguishing features imply that they are three independent evolutionary lineages.

The chelicerates are coelomate protostomes. The head and thorax of chelicerates are fused into a single unit called the cephalothorax or prosoma (Latin *pro,* forward; Greek *soma,* body). The prosoma bears a pair of feeding appendages called chelicerae—the most anterior appendages; posterior to the chelicerae is a pair of pedipalps—the first pair of walking legs—and four or more additional pairs of walking legs. The abdomen, or opisthosoma (Greek *opisthen,* behind) is distinct from the cephalothorax. Abdominal appendages are variously used for gas exchange (for example, the book gills—named for their resemblance to pages of an open book—of the horseshoe crab and book lungs of some arachnids), for reproduction, or for silk extrusion, depending on the species.

Members of Class Pycnogonida, sea spiders, include about a thousand species of marine chelicerates that resemble long-legged, slow, slender spiders. Because *Nymphon* and other sea spiders lack excretory and respiratory organs, the sea spiders are not true spiders. Sea spiders have been collected in habitats ranging from oceans 6800 m deep to shallow seas, from the Arctic to the Antarctic. Many sea spider young are commensals or parasites of medusae (Phylum A-3) and echinoderms (Phylum A-34). Most sea spider adults are free living and dull in color. Some deep-sea species are red. The bodies of sea spiders range from less than 1 to more than 10 cm in length, the largest with legs spanning almost 80 cm. The largest are deep-sea and polar species.

The sea spider cephalothorax has a muscular proboscis with a terminal mouth flanked by paired appendages: one pair of chelicerae; one pair of pedipalps; and a pair of ovigers—appendages often absent in females. Sea spiders clean their bodies with the ovigers and hold eggs on them. The nervous system consists of four eyes on a short projection above a brain in the cephalothorax, ventral nerve cords with ganglia, and sensory hairs. Four pairs of walking legs are usually clawed. Carnivory is the rule; through the terminal mouth on the proboscis, the pharynx pumps soft sponges (Phylum A-2), sea anemones and other cnidarians (Phylum A-3), and ectoprocts (Phylum A-29) into the extensive digestive tract.

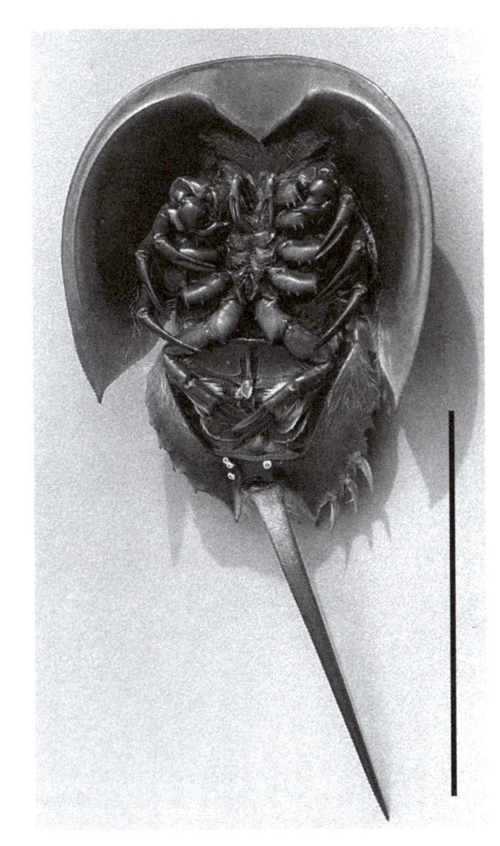

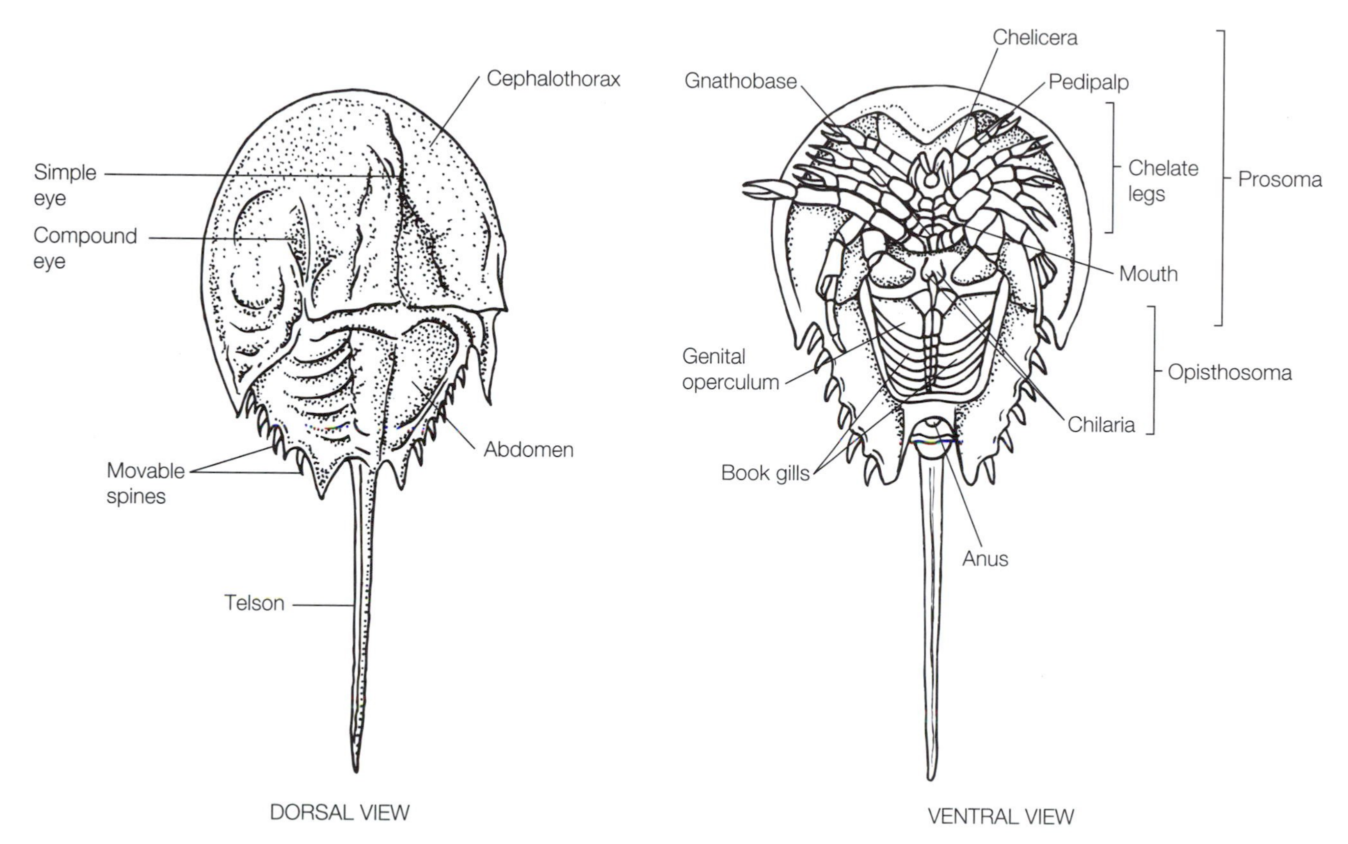

Beachcombers often encounter *Limulus polyphemus* along beaches from Nova Scotia to the Yucatan Peninsula in Mexico. In spring mating season, scores of the harmless horseshoe crabs become stranded as mature females come out of the shallows to lay eggs. The smaller male—like the one in this photograph—hitchhikes clasped on the female's abdomen and deposits sperm as the female drags him over the sandy nest. This adult male from the Florida Keys bears the clawed appendages that characterize chelicerates. Bar = 120 mm. [Photograph courtesy of C. N. Shuster, Jr.].

Compared with that of the horseshoe crab, the abdomen is very small relative to the cephalothorax; the digestive and reproductive systems fill the stumpy abdomen as well as much of the four to seven pairs of walking legs that attach to the cephalothorax. The sea spider's large surface area may account for the lack of organs of excretion and gas exchange.

Reproduction in sea spiders is modified by the body form; because of the small cephalothorax, reproductive organs extend into the legs, where eggs are produced in the ovaries of females and sperm are produced in the testes of males. Gonads open through pores on the legs. Males gather fertilized eggs, produce adhesive that holds the amassed eggs on the ovigers, and shelter eggs until larvae hatch from them. Larvae become parasitic on colonial hydrozoans (Phylum A-3), stay on the ovigers of the male, or leave. After a series of molts, the larvae attain adult form.

Class Merostomata (merostomates) comprises three marine genera: the genus *Limulus* consisting of a single species, *L. polyphemus*—the horseshoe crab (illustrated here); and two additional genera from the shallow seas of the Indo-Pacific Ocean—*Trachypheus (Tachypleus)* and *Carcinoscorpinus (Carcinoscorpius)*. The horseshoe crab carapace is shaped like the iron shoe of a horse. *Limulus* lives in shallow waters of the Gulf of Mexico and the eastern seaboard from Nova Scotia to the Yucatan. *Limulus,* with its flexible, hornlike shell of chitin and protein, is not a true crab; true crabs are crustaceans and have a brittle, hard shell. *Limulus,* a familiar member of the shore community, reaches a length of 60 cm. The remaining three merostomate species are found in the Southeast Asian sea from India to Korea. Horseshoe crabs consume clams, other small animals, and plant material encountered in mud. The last of the horseshoe crab's five pairs of walking legs is specialized for paddling, for cleaning its gills, and for shoving mud as it burrows; these legs lack claws. The other four pairs of legs are clawed. *Limulus* uses not only its walking legs, but also its pedipalps for walking. The anteriormost appendages are a pair of small chelicerae that gather food, pass it to spines on the leg bases, which grind the food, and pass it to the mouth. The mouth is located between the pedipalps. *Limulus* regulates its internal ion concentration by means of excretory glands and tubules; a bladder passes urine through an excretory pore at the base of the last pair of legs.

The horseshoe crab brings its formidable tail spine into play when it rights its flattened, leathery body or shoves forward. Vast numbers of horseshoe crabs migrate to the shallows for nocturnal spawning. Females excavate a depression into which they lay about 300 eggs. Gonads open through genital pores that are under a flap on the abdomen. Fertilization is external; the smaller males clasp egg-laying females with modified pedipalps. The fertilized eggs hatch as swimming trilobite larvae, so-called because they resemble the extinct trilobites, once common in Cambrian seas and now found as beautiful fossils in Utah, Nevada, Ohio, and worldwide. The name trilobite is applied to these larvae because the dorsal side of the cephalothorax has three longitudinal lobes. Horseshoe crab larvae molt more than a dozen times and attain maturity in about 3 to 10 years. Small horseshoe crabs can swim upside down. Larger crabs are bottom dwellers. Intertidal-dwelling horseshoe crabs tend to have a green tail because *Microcoleus* and *Spirulina* (Phylum B-5), *Thiocapsa* (Phylum B-3), and other photosynthetic bacteria adhere to it. If the crab happens to drag its tail over a suitable habitat, these bacteria can be seeded onto the substrate, leading to the formation of a microbial mat.

The *Limulus* body cavity is a hemocoel (blood-filled sinus in tissue); the coelom is very small. The *Limulus* circulatory pattern follows the general circulatory system of arthropods with variations in mode of gas exchange. A long tubular heart with ostia (perforations) pumps blood through an extensive arterial system of closed vessels into ventral sinuses from which blood flows into the book gills. The book gills contact seawater, underneath five pairs of gill flaps—abdominal appendages modified for gas exchange. Gill activity pumps blood back to the sinus around the heart. As the heart expands, blood is drawn in through ostia. *Limulus*' blue blood contains hemocyanin, a respiratory pigment, and amebocytes that facilitate clotting. A coagulating agent in *Limulus* blood is used to diagnose human bacterial infections. Bacteria produce endotoxins; when the horseshoe crab coagulating agent (a lysate) reacts with endotoxins, clotting of the horseshoe crab blood is triggered. The presence of fever-producing substances called pyrogens in intravenously administered medicines also is detected with the use of this same clotting reaction of the *Limulus* lysate. Another substance extracted from horseshoe crabs, lobster, shrimp, and crab shells is chitin, a polysaccharide that is useful in surgical sutures and implantable drug containers because chitin does not elicit allergic reactions in humans.

The horseshoe crab nervous system includes a brain that encircles the esophagus, a ganglionated ventral nerve cord, a pair of compound eyes and a median simple eye, and a chemosensory organ (frontal organ). Merostomates are the only chelicerates with compound eyes. *Limulus* eyes probably sense movement but do not form images. Research studies of *Limulus* compound eyes have contributed to our knowledge of the physiology of vision.

Limulus has been virtually unchanged since it dwelt in shallow Silurian seas, about 425 million years ago. In the Silurian, sea scorpions—chelicerates—appeared as well as trilobites of earlier lineages. Fossil merostomates, diverse and abundant in Paleozoic oceans, may be ancestors of Class Arachnida.

By far the majority—more than 60,000 species—of chelicerates belong to the Class Arachnida, which is the only air-breathing class of Chelicerata. Arachnids take their name from Arachne of Greek mythology, a girl turned into a spider for challenging the goddess Athena to a weaving competition. Spiders (*Lycosa,* the wolf spider; *Argiope,* the orb weaver spider; *Loxosceles,* the brown recluse spider; *Latrodectus,* the black widow spider) make up half of this class, which also contains mites (many, such as *Psoroptes,* or mange mite, are parasites), ticks (also parasites, such as *Ixodes,* which transmits Lyme disease), harvestmen, and scorpions. Many spiders engage in elaborate courtship. Like *Limulus,* the arachnid body consists of prosoma (cephalothorax) and abdomen. In spiders, the abdomen is usually distinct from the prosoma, linked by a narrow pedicle;

whereas, in mites and ticks, the abdomen is fused to the prosoma. Scorpion bodies have 18 segments: 6 segments in the prosoma, 7 segments of anterior abdomen, and 5 segments of posterior abdomen. Chelicerae—the most anterior appendage pair—tear food and function as fangs. The second appendages—pedipalps—kill prey, handle food, or have sensory or reproductive functions. The final four appendage pairs on the prosoma are walking legs. Some arachnid species have an abdomen with internal book lungs (similar to horseshoe crab book gills) that open to the environment by closeable orifices called spiracles. Instead of book lungs, tissues of some tiny arachnids are linked to the outside by a network of branched tubing called tracheae; body movements pump air through the tracheae. Most spiders are terrestrial, and all are air breathers; a few freshwater spiders capture bubbles from which they breathe air. Arachnids lack a blood circulatory system.

Spiders secrete silk proteins from glands within their abdomen; silk extruded through tiny orifices in abdominal appendages—spinnerets—at the posterior end of the spider abdomen solidifies in air. The spider uses silk to build nests, to snare insects, to weave sacs for spiderlings, to wrap food gift packages during courtship, and to spin threads for sailing on the winds. Spider silk is very strong and has been used as uniform, tough threads in bomb sights of airplanes. Arachnids are carnivorous predators that dine on liquids. A spider bites prey and paralyzes it with poison from its chelicerae, predigests the prey's insides by extruding enzymes from the arachnid digestive tract into the prey, and pumps in the liquid meal with the muscular action of its stomach.

A few arachnids such as scorpions have abdominal stingers, producing toxic venom. The bite of a black widow spider injects toxic, but seldom deadly, venom. Arachnids feed mostly on insects; many arachnids are predators of agricultural insect pests.

Ticks and mites number more than 30,000 species. Certain mite species inhabit the dust in our houses and the follicles of our eyelashes. These chelicerates live in myriad terrestrial, marine, and freshwater habitats. Some species are vectors of Lyme disease and encephalitis. Other mites and ticks are pests of domestic animals. Mites and ticks feed on vertebrate blood, invertebrates, fungi, and plants.

The chelicerate exoskeleton, the small body size of many species, and the extraordinary diversity of form and habitat contribute to the success of chelicerates as a phylum. Scorpions, present in the Carboniferous (about 345 million to 290 million years ago), are considered the most primitive arachnids. The affiliations between chelicerates, crustaceans (Phylum A-21), and insects (Phylum A-20) are unclear. All arthropods may have derived from annelid ancestry, but this theory is currently contested.

A-20 Mandibulata (Uniramia)

(Mandibulates, mandibulate arthropods)

Latin *mandere,* to chew; *mandibulum,* a jaw

EXAMPLES OF GENERA

Anopheles
Apis
Blatta
Bombyx
Chironomus
Cryptocercus
Drosophila
Incisitermes
Kalotermes
Limenitis
Locusta
Melanopus
Musca
Pauropus
Pediculus
Periplaneta
Polyxenus
Pterotermes
Rothschildia
Scolopendra
Spirobolus
Tenebrio

Of the more than 30 million species of animals estimated to exist now (only about 10 million of which have been formally described and named), Phylum Mandibulata claims the largest number. There are at least 750,000 species of insects alone—not to mention centipedes, millipedes, and less familiar species—in this phylum. Uniramia (one branch) indicates unbranched appendages (Figure A). Both unbranched (uniramous), as in insects, and branched (biramous) appendages, as in crustaceans, consist of a linear series of joints, as do all arthropod appendages. The mandibulate body has three distinct parts: head (cephalum), thorax, and abdomen (Figure B). As the mandibulate changes size or form, it secretes a new exoskeleton and sheds the outgrown armor. Although many mandibulates are dioecious, parthenogenesis is common in insects. Many species of mandibulates metamorphose: the fertilized egg hatches a larva that develops into a series of juvenile forms that differ considerably from the adult.

Features common to mandibulates and all other arthropods (Phyla A-19 through A-21) include a cuticle containing chitin and protein. The adult arthropod coelom is very small. The principle body cavity is a hemocoel—a cavity in which blood pumped by the tubular heart circulates; from the hemocoel, blood enters the heart through slit-shaped ostia characteristic of arthropod hearts. Food is moved through the digestive tract by muscle action, and nutrients are distributed by the blood. Most mandibulates are terrestrial arthropods with internal gas exchange; through thin-walled tracheae—minute tubes that penetrate their tissues—body activity moves the gases. Air enters tracheae through pores in the cuticle (spiracles), passing through hairs that filter out small particles. Many species can close their spiracles, retarding moisture loss. The mandibulate nervous system is typically arthropod—a dorsal brain, ventral nerve cord ganglionated in each segment, and compound or single-unit (simple; that is, nonimage forming) eyes or both. Jointed appendages and segmented bodies characterize all mandibulates, chelicerates, and crustaceans as arthropods. A mandibulate has a single pair of antennae (see Figures A and B); a crustacean has two pairs of antennae, and a chelicerate lacks antennae. A comparison of body parts reveals that mandibulates and most crustaceans have head, thorax, and abdomen; the chelicerate has a cephalothorax and abdomen. Mandibulates and crustaceans have mandibles (Figure C)—mouthparts that crush food; chelicerates lack mandibles. Here Phylum Mandibulata includes insects and myriapods; crustaceans are placed in Phylum Crustacea.

The two largest classes of mandibulate arthropods are Hexapoda (Class Insecta is an alternative name)—the largest class of arthropods by far—and Myriapoda, centipedes and millipedes (Figure D). The other two classes in this phylum are Symphyla and Pauropoda. Symphyla, garden centipedes and their relations, are soft-bodied (covered with soft chitinous exoskeleton) mandibulates with one pair of antennae, twelve pairs of jointed legs with claws, and a single pair of unjointed posterior appendages. Less familiar are members of Class Pauropoda, such as *Pauropus,* tiny soft mandibulates distinguished by branched antennae. They are similar to millipedes *(Spirobolus)* but with only eleven or twelve segments and nine or ten leg pairs.

All member of Class Hexapoda are insects, with three pairs of legs, three body sections, generally one or two wing pairs, and one pair of antennae. Insects are the only terrestrial animals living, be-

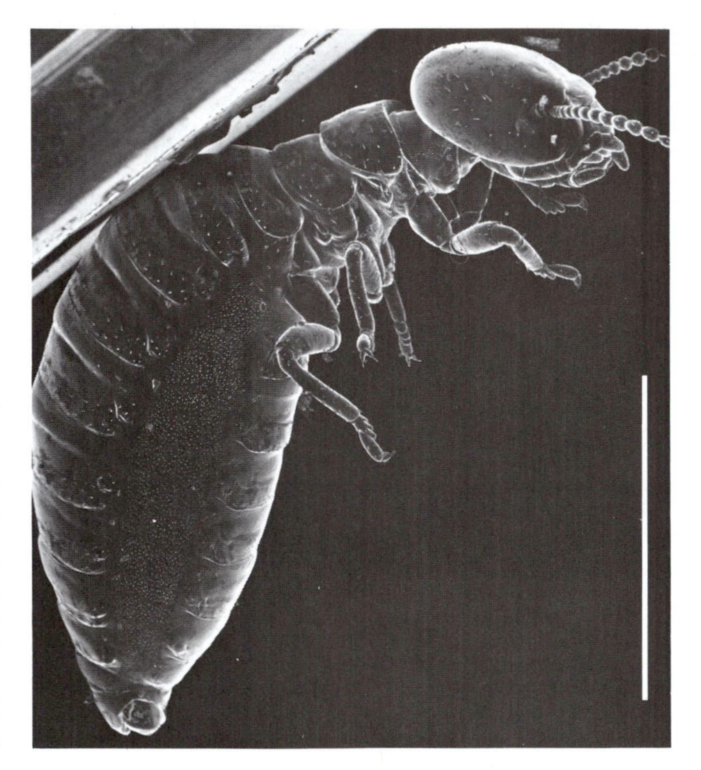

A *Pterotermes occidentis,* the largest and most primitive dry-wood termite in North America. Its colonies are limited to the Sonoran Desert of southern Arizona, southeastern California, and Sonora, Mexico. The swollen abdomen of this pseudergate (worker) covers the large hind gut, which harbors millions of microorganisms responsible for the digestion of wood. SEM, bar = 0.5 mm. [Courtesy of C. Spada.]

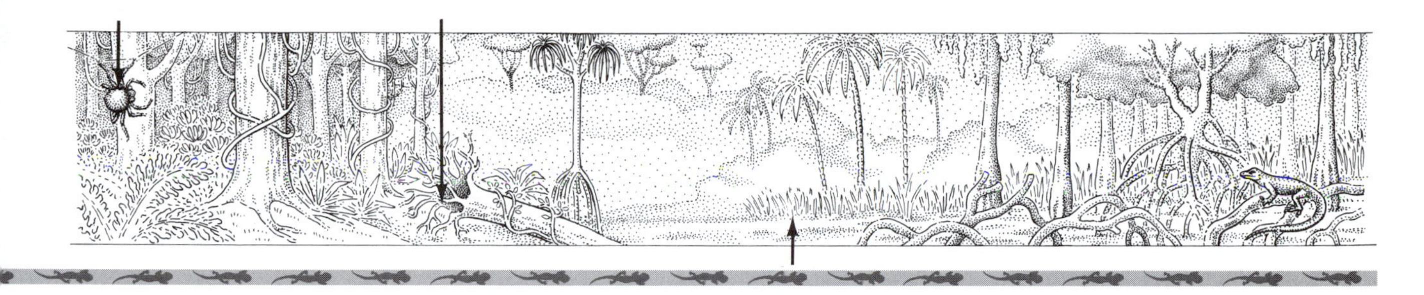

sides birds and bats in our own phylum Craniata, that evolved flight capability (Figure E), for which the earliest evidence comes from winged insects preserved as fossils. Flight by insects is possible because the combination of small body size and highly specialized striated muscles for rapid, strong contraction enabled the predator avoidance afforded by flight (insect flight evolved before the evolution of insect-hunting bats and humans wielding butterfly nets). Water loss increases during flight because of increased movement of air past the insect body; dehydration is slowed by the waxy cuticle, by closeable spiracles, and by excretion of solid waste as almost dry uric acid by the Malpighian tubules—the excretory organs located in the insect hemocoel. The vast array of muscle and exoskeletal designs in many species of flying insects coupled with a well-developed respiratory system and adequate storage of energy supply meet flight requirements of as many as 1000 wing beats per second. Finally, well-tuned visual and nervous controls for flight navigation make insect migration, mating on the wing, hovering, and food getting possible.

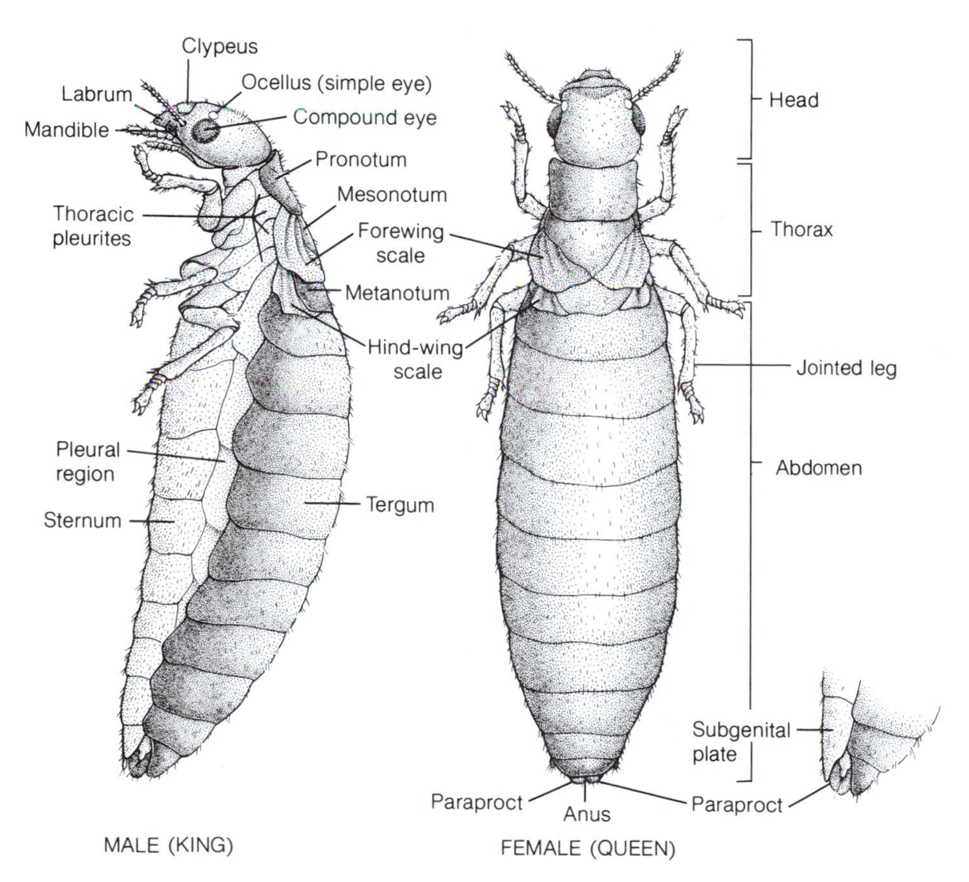

B Adult reproductive forms of the termite *Pterotermes occidentis*. [Drawing by L. Meszoly; information from W. Nutting.]

A-20 Mandibulata
(continued)

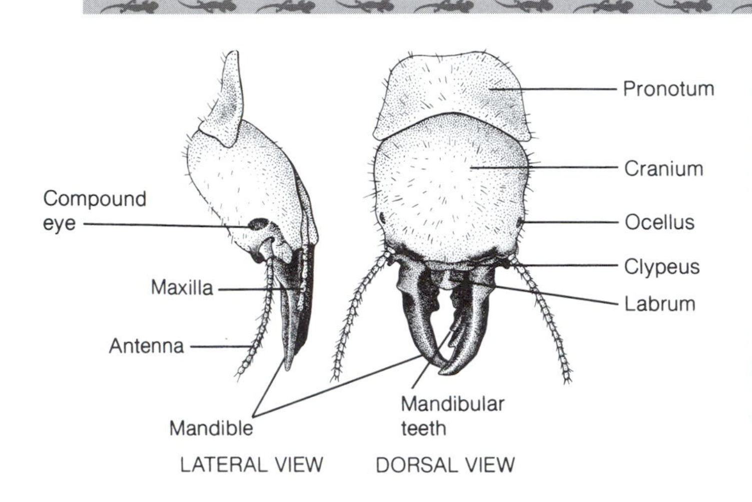

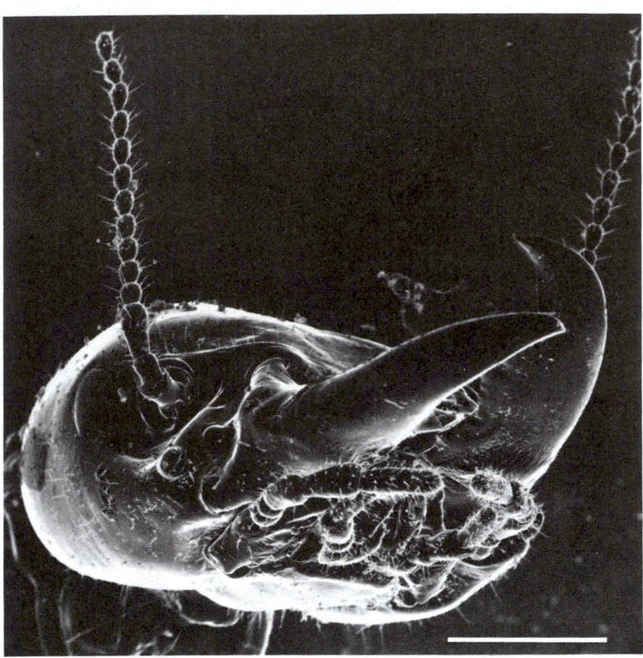

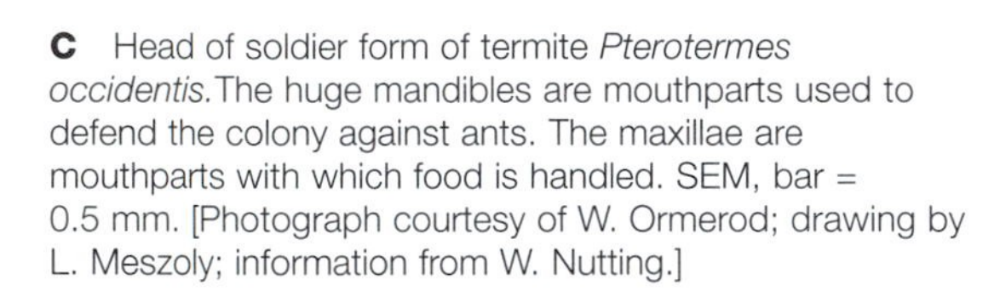

C Head of soldier form of termite *Pterotermes occidentis.* The huge mandibles are mouthparts used to defend the colony against ants. The maxillae are mouthparts with which food is handled. SEM, bar = 0.5 mm. [Photograph courtesy of W. Ormerod; drawing by L. Meszoly; information from W. Nutting.]

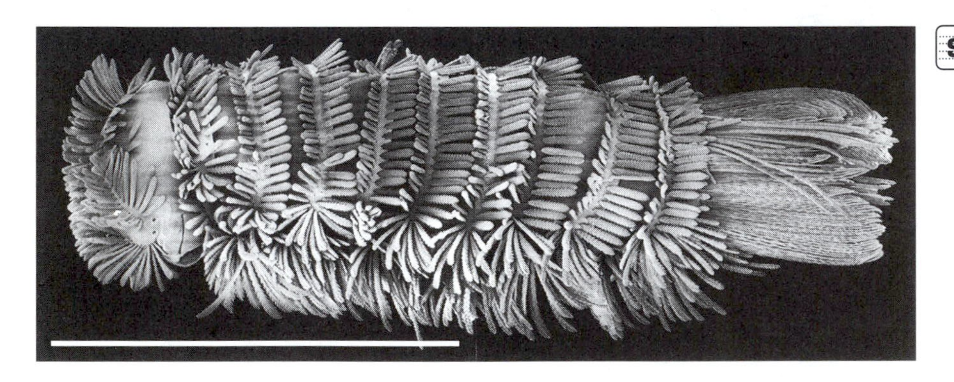

D The millipede *Polyxenus fasciculatus* repels attacks by sweeping its bristly tail tuft into ants and other predators. The bristles detach, tangling the ant's body hairs and thus incapacitating the ant. This millipede was discovered under the bark of a slash pine tree in Florida. SEM, bar = 1 mm. [Courtesy of The *New York Times;* photograph by Maria Eisner.]

Insects, the dominant invertebrates in innumerable terrestrial and freshwater habitats, have evolved few marine species. Only about 200 fully aquatic insects have been documented. Some insects are predators of plant pests, and many pollinate crop vegetables and fruit trees, which evolved simultaneously with their insect pollinators. Many insect species transmit disease associated with bacteria, fungi, or viruses. One such species is the mosquito *Anopheles,* which transmits the microbe that causes malaria. For many carnivorous animals—from ducklings to insect-eating bats—insects are crucial sources of protein. Carnivorous plants often live in nitrogen-poor soils and derive nitrogen from the protein of their insect prey bodies. Silkworms (larvae of moths) are the only entirely domesticated invertebrate animal.

Termites such as *Pterotermes occidentis* are social insects, like some bees, all ants, and some wasps. These termites live in colonies composed of a queen, king, and workers (see Figures A and B), which may be female or male but always sterile, and soldiers (see Figure C). The division of labor among colony members is associated with marked differences in behavior and form. Workers protect and care for the queen and her offspring. The king mates with the queen. Certain workers become soldiers, which develop enlarged mandibles for defense of the colony and aggressive behavior patterns. The reproducing queen is much larger than other members of her colony, with the exception of the king; her role in the colony is reproduction.

All members of Class Myriapoda (many feet) are centipedes or millipedes. Centipedes (Order Chilopoda) comprise about 2500 species of predaceous, nocturnal carnivores; the centipede has a head and from 15 to 177 trunk segments, each with a pair of tiny jointed legs. The largest is *Scolopendra gigantea,* the giant African centipede, about 0.3 m in length. A few are harmless temperate species and most are tropical; they look like armored caterpillars. Centipedes secrete a neurotoxin that they inject from fangs (modified trunk appendages) into prey. The millipede (Order Diplopoda), such as *Spirobolus* and *Polyxenus,* has a head and from 20 to 200 trunk segments, each with two pairs of appendages and one pair of poison glands (see Figure D). Calcium carbonate—like a calcified crustacean exoskeleton—hardens the cuticle of the trunk. Upward of 10,000 millipede species have been described; those that are carnivores sometimes include hydrogen cyanide in their poison armamentarium and—because most millipedes are scavengers—their toxins serve to deter predators. Onychophorans (Phylum A-28), millipedes, and centipedes all employ adhesive defense, feed on fluids dissolved by the hunter's (centipede, millipede, or onychophoran) enzymes, have antennae, secrete cuticles of chitin, have hemocoels, and have tubular hearts. However, these traits may be due to convergent evolution, although analysis of ribosomal RNA suggests that onychophorans are specialized relatives of insects and other arthropods, evolved from a common ancestor.

Chitin is present in all arthropods (Phyla A-19 through A-21), as well as in pogonophoran (Phylum A-25), annelid (Phylum A-22), brachiopod (Phylum A-30), and some echiuran (Phylum A-24) setae (bristles); in onychophoran cuticle (Phylum A-28), the radula (rasping organs) of molluscs (Phylum A-26), and cell walls of fungi and red algae (Phylum Pr-25); and in chytrid cell walls (Phylum Pr-29). Thus, chitin by itself is not a reliable indicator of evolutionary relationships.

Fossils of joint-footed animals are found in Ediacaran (Upper Proterozoic) rocks. Trilobites—common Cambrian (early Paleozoic) fossils having jointed legs and frequently placed with the arthropods, but sometimes granted their own phylum—are extinct. Trilobites were the earliest animals to have eyes. The trilobite also had jointed legs, one pair of antennae, and a carapace (hard covering) with head, thorax, and rear segment. (Trilobites are named for their three body lobes). Insects arose in the late Paleozoic era. At 200 million years ago, cockroaches appear in the fossil record. Arthropods—terrestrial mandibulates, crustaceans (almost all aquatic), and chelicerates—evolved from annelid ancestors (Phylum A-22) but may be polyphyletic (have arisen independently), on the basis of their life histories. Affinities of insects, centipedes, and millipedes to crustaceans and chelicerates are far from certain at present.

E *Limenitis archippus* mating. These orange and brown viceroy butterflies, Class Insecta, are found in central Canada and in the United States east of the Rockies. Insect wings are outfoldings of the body wall, supported by trachaea. A network of veins links the two wing pairs to the circulatory system. Bar = 7 cm. [Courtesy of P. Krombholz.]

A-21 Crustacea

(Crustaceans)

Latin *crustaceus,* having a shell or crust

EXAMPLES OF GENERA

Armadillium
Armillifera
Artemia
Balanus
Calanus
Cambarus
Cancer
Cephalobaena
Clibanarius
Coronula
Daphnia
Euphausia
Gigantocypris
Homarus
Isaacsicalanus
Linguatula
Pagurus
Porocephalus
Raillietiella
Reighardia
Uca
Waddycephalus

The defining feature of Phylum Crustacea is the possession of two pairs of antennae on the head, distinguishing crustaceans from all other arthropods (Phyla A-19 and A-20). Many familiar animals belong to the Crustacea phylum—among them are crabs, lobsters, brine shrimp, and krill. The two pairs of jointed antennae of the crayfish *Cambarus* are typical crustacean features; the presence of antennae distinguish crustaceans from chelicerates (Phylum A-19). Biramous (branched) appendages—such as the five pairs of swimmerets beneath the abdomen of the crayfish in Figure A—distinguish crustaceans from insects and other members of Phylum Mandibulata with their unbranched appendages. Biramous appendages have two branches to each appendage (limb); the two branches connect to a single base, and each of the two branches consists of several jointed segments. The crustacean exoskeleton may be as thick as a lobster's or as thin and flexible as the outer covering of a soft-shelled crab larva. Crustaceans—along with chelicerates (Phylum A-19) and mandibulates (Phylum A-20)—are still considered by some to be part of Phylum Arthropoda. But consistent distinguishing features of these three groups imply that they are three independent evolutionary lineages, and we agree with zoologists who raise each group to phylum status. We agree to place pentastomes—formerly Phylum Pentastoma—as Class Pentastomida within Phylum Crustacea. Pentastomes—highly modified parasites—are regarded by specialists to possibly be crustaceans, similar to branchiurans (marine crustaceans that parasitize fish skin). Crustacea number about 45,000 species.

A Freshwater crayfish *Cambarus* sp., a component of the food web in a New Hampshire lake. The branched appendages are a distinguishing feature of this phylum. Crayfish are noctural lake-bed scavengers, feeding on aquatic worms and plant growth. In turn, loons, herons, black bass, and people prey on these crustaceans. Bar = 5 cm. [Photograph by K. V. Schwartz.]

As protostome coelomates, the crustacean body cavity is a coelom and, during embryological development of protostomes, the mouth originates at or near the blastopore, the opening from the outside of the blastula to its interior. The crustacean body plan resembles that of other arthropods in some respects (nervous system, for example), but most crustaceans have three body sections—head, thorax, and abdomen—as do insects; in comparison, chelicerates have two body sections (fused cephalothorax and abdomen). In some crustaceans, the head is fused with various thoracic segments to form a cephalothorax. The crustacean coelom is small and the main body cavity—a hemocoel, or cavity in which blood circulates—is part of the circulatory system, as in other arthropods. The crustacean mode of gas exchange, however, differs from that in other arthropods; gills—evident in crabs and lobsters—exchange gases with the water or air. A tube-shaped heart with ostia is the circulating pump for blood. In lobsters, one of the mouthparts is modified as a gill bailer, which the animal uses to generate a current of water through its gill chamber, enhancing gas exchange through the gills. Crustaceans that are semiterrestrial (for example, the hermit crab *Clibanarius*) inhabit the intertidal zone of the sea beach, a habitat that provides at least occasional moisture to their respiratory surfaces. Some crustaceans have blood vessels; others have no vessels and pump blood only to the hemocoel. A dorsal brain, ventral nerve cords with ganglia in each body segment, and sensory organs such as the compound eyes of *Daphnia* and statocysts constitute the nervous system. Krill chemically produce light in luminescent organs on their legs and eyestalks; their light displays may function in mating or in protecting them from predators, such as penguins, that are below them in the sea. All crustacean muscle is striated, typical of arthropods, and quicker to contract and relax than the smooth muscle that predominates in other invertebrates. The one-way digestive tract includes a mouth and stomach in which food is ground extensively; food then passes to a midgut sometimes associated with organs that facilitate digestion, such as the "liver" of a lobster. Crustaceans may be dioecious, often with sexual dimorphism that ranges from a difference in size between females and males to sexual dimorphism of appendages (males may have appendages used in mating) to the extreme sexual dimorphism of some isopods. Barnacles are hermaphrodites; the long penis of one barnacle fertilizes another nearby but permanently attached barnacle. A three-segmented, free-living larva called a nauplius is a characteristic crustacean larva. Some barnacles brood embryos to the nauplius larval stage.

Some highly specialized crustacea are so bizarre in form that they can be classified only upon observation of their nauplius larvae. Diverse larvae having forms as exotic as their names—zoea, megalops, phyllosoma, mysis—are present in decapod custaceans, an order of Class Malacostraca that includes crabs, some shrimp, hermit crabs, lobsters, and crayfish. Crustacean larvae are abundant in the ocean's plankton. Other species in Class Malacostraca lack a larval stage; young hatch as juveniles. Some freshwater copepods and members of Class Branchiopoda are parthenogenetically reproducing females. Dissolved nitrogenous wastes diffuse from the crustacean body through the gills. Freshwater crustaceans' green glands or antennal glands excrete water and resorb salts by way of nephridial canals in the anterior of the animal, collecting the resulting liquid in a bladder that discharges through an excretory pore near the antenna base. Chelicerate and crustacean excretory organs are similar; from the hemocoel, fluid is collected and eventually excreted as urine.

Crustaceans have evolved an enormous array of appendage diversity: gills, food collection, food shredding, defense, courtship display, walking, swimming, generating respiratory currents, brooding eggs, and clinging to seaweed. The crustacean head [two pairs of antennae, one pair of mandibles (food crushers), two pairs of maxillae (food handlers and current generators)], thorax (eight pairs), and abdomen (six pairs) all bear appendages. Females of one isopod species carry the male of the species on their antennae, the anteriormost appendages.

The covering of crustaceans provides a rigid skeleton that functions in locomotion, support, and protection; in general, the chitin-protein covering is similar in all arthropods, though less rigid in many species. Calcium carbonate is deposited in crustacean cuticle, strengthening the cuticle; insect cuticle lacks this calcification. Joints between sections of this armor are thinner cuticle, affording flexibility. The outermost layer of the cuticle is protein and often water-impermeable wax, rendering the soft body within somewhat resistant to water loss. At the same time, gas exchange across the body surface is severely reduced; oxygen and carbon dioxide diffuse from gills, which have great surface area and blood supply. Beneath the waxy layer lies a blue, red, yellow, or black pigmented chitin and protein layer underlaid by a calcified layer that hardens the cuticle. The epidermis, which secretes the cuticle, is the innermost layer. The crustacean must shed its rigid cuticle to accommodate growth. Enzymes partly break down the old exoskeleton, and new cuticle is secreted before the old cuticle splits. During ecdysis (escape) the crustacean not only slips out of its exoskeleton, but also jettisons the chitinous, internal linings of its fore- and hindgut.

Phylum Crustacea includes the major classes (largest and most diverse of the classes in number of species) Branchiopoda, Ostracoda, Copepoda, Cirripedia, Malacostraca, and Pentastomida (Linguatulida), as well as other classes including rare species or those in restricted habitats such as Remipedia, which inhabits underground saltwater caves in the tropics. In freshwater, including vernal pools, and saline habitats, crustaceans are the most prominant arthropods. Brine shrimp (*Artemia*), water fleas (*Daphnia*), fairy shrimp, tadpole shrimp, and others in Class Branchiopoda mainly inhabit freshwater, although the brine shrimp withstand highly saline water. Class Branchiopoda also includes commensals (epibionts) that live on other crustaceans, herbivores, detritus feeders, and filter feeders. Ostracods (*Gigantocypris*) live mostly in marine and estuary habitats; a few species are abyssal or freshwater. Copepods (*Calanus*) are enormously abundant in diverse habitats: marine (from ocean surface to 5000 m deep), freshwater, estuarine, interstitial (between sand grains, belonging to the meiofauna); some species are parasitic or commensal (symbiotrophic). Many commensal copepods are color matched to their hosts—crinoid echinoderms. As the base of worldwide food webs that include commercially valuable fishes, copepods affect our own species. *Isaacsicalanus* is a newly described copepod from a hydrothermal vent. Class Cirripedia comprises all barnacles, such as *Balanus.* As adults, cirripeds cement themselves permanently to firm marine substrates. Research on this biomaterial reveals a strong new adhesive protein, promising for cementing artificial spare parts, such as artificial hip replacements, into humans. *Coronula diadema,* a huge planktonic barnacle, settles as an epibiont on *Megoptera,* the humpback whale. Many barnacles are sessile as adults; their calcium carbonate shelter houses the chitin-covered crustacean within, complete with biramous appendages.

Class Malacostraca comprises more than half of the 45,000 crustacean species: amphipods (sand fleas), isopods (sow bugs, pill bugs, woodlice), and decapods [lobsters (*Homarus*), crayfish, crabs (*Cancer,*) some shrimp, hermit crabs (*Pagurus*)]. Isopods dwell in caves, in all damp soil habitats, under logs and rocks, in the depths of the ocean, as parasites of marine animals including whales, molluscs, fish, and other invertebrates, and as commensals with echinoderms, polychaetes, and bivalve molluscs. The krill (*Euphausia*) is one of the most ecologically critical malacostracan species; krill is the primary food for many marine fish species, sea birds, seals, filter-feeding whales, and other crustaceans worldwide. Exploitation of *Euphausia superba,* the krill of Antarctic seas, by humans for food may directly reduce the base of marine food webs that are critical to life on this planet.

Members of the Class Pentastomida (Greek *pente,* five; *stoma,* mouth), called tongue worms because of their shape, have a flat, soft body covered by a nonchitinous cuticle. The relationship of tongue worms to other invertebrates is difficult to infer because the worms have been modified to live in damp passages of vertebrates.

A-21 Crustacea
(continued)

Although the body is not segmented internally, the exterior appears to be composed of about 90 rings, depending on the species. All 110 species are parasites (symbiotrophs) of vertebrates. Two anterior pairs of leglike hooks with hollow retractile chitinized claws hold the worms in place on their hosts. They live embedded in the lungs, nostrils, or nasal sinuses of mammals such as dogs, foxes, wolves, goats, and horses, as well as of reptiles such as snakes, crocodiles, lizards, turtles, and birds. Pentastomes are particularly prevalent in tropical and subtropical hosts.

In the tropics and elsewhere, larval tongue worms are occasionally found in humans, who acquire them from other vertebrates, generally from domestic animals. Pentastome growth in humans is checked, however, because the nasal tissue responds by forming calcareous capsules that surround and isolate the parasites.

Tongue worms range in length from a few millimeters to more than 150 mm, depending on the species. They are colorless, red, or yellow and transparent, except in a pigmented area around their hooks. A projection between the two pairs of hooks bears the mouth. A pentastome sucks mucus, epithelial cells, and lymph from the host animal through its mouth into a swelling foregut. The digestive tract is a straight tube. Excretory, respiratory, and circulatory organs are absent—common adaptations to parasitic life. The nervous system of tongue worms consists of surface sensory papillae, a ventral nerve cord with ganglia, and head ganglia.

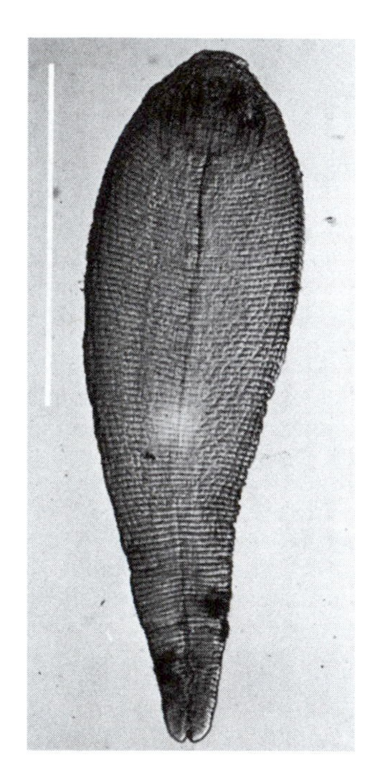

B A living female tongue worm, *Linguatula serrata,* a pentastome that clings to tissues in the nostrils and forehead sinuses of dogs. Bar = 5 cm. [Courtesy of J. T. Self.]

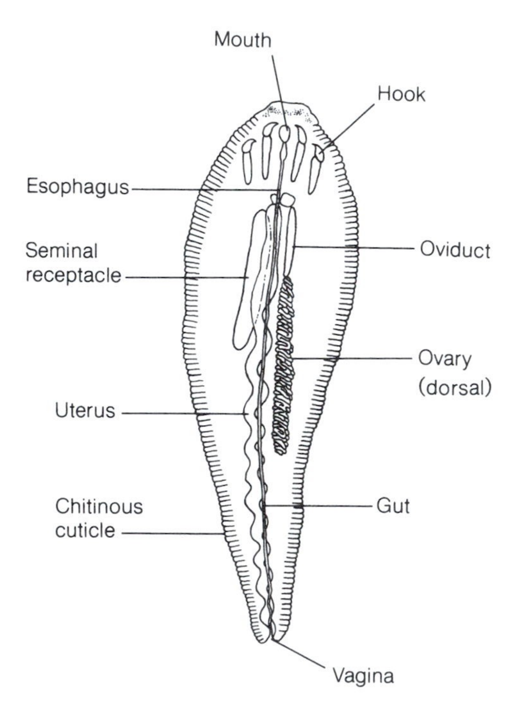

C Ventral view of female tongue worm showing one of each pair of oviducts and seminal receptacles. [Drawing by R. Golder; information from J. T. Self and R. E. Haugerud.]

The sexes are separate in pentastomes, and the male is smaller than the female. In *Linguatula serrata* (Figures B and C), the male and female are 2 and 13 cm long, respectively. Females have a genital pore near the anus, whereas the genital pore of the male is near his mouth. The eggs are minute, thick shelled, and without yolks. Fertilization is internal. The vagina of the female tongue worm is able to hold some half million fertilized eggs. As it fills with eggs, the vagina may stretch until it is 100 times its former size. The fertilized eggs pass through the genital pore onto plants with the nasal secretions and saliva of the host animal. The eggs must be eaten by herbivorous vertebrates to develop.

Tongue worms have three larval stages. The first stage develops within the eggs in a herbivorous intermediate host. Rabbits are the main intermediate hosts for *Linguatula;* fish are intermediate hosts for crocodile pentastomes. In the second larval stage, pentastomes resemble tardigrades (Phylum A-27). These larvae hatch from the eggs in the stomach of the herbivore; they swim and bore with their mouths into the lungs or liver or into the linings of these host organs. In the tissues of the intermediate host, the larvae enter a third stage: they encapsulate to form cysts. It is possible that the cysts are digested when the herbivore host is eaten by a carnivore, and that the pentastome then moves with its hooks from the carnivore's stomach into its nasal or lung passages. However, there is evidence that the pentastome leaves its cyst while it is still in the carnivore's mouth. In either case, the adults embed themselves in the nasal sinuses or lung tissue, they mate, and the cycle begins again.

Crustaceans may be derived from annelid-like ancestors; the annelid–arthropod connection is currently in a state of flux. All appendages of most primitive arthropods are assumed to be similar with serial repetition of similar segments, resembling the bodies of extinct trilobites and of recently discovered remiped crustaceans. Some believe that pentastomes are descended from early crustacean arthropods (Phylum A-21), because both pentastomes and arthropods undergo extensive larval development and molt their cuticles at certain stages in their life cycles; molecular sequence data support this view. Affinities of crustaceans with the two other arthropod phyla Mandibulata (Phylum A-20) and Chelicerata (Phylum A-19) is uncertain. Crustaceans may have an evolutionary origin independent from insects, centipedes, millipedes, and other terrestrial arthropods.

A-22 Annelida

(Annelid worms)

Latin *annellus,* little ring

EXAMPLES OF GENERA

Aphrodite
Arenicola
Chaetopterus
Eunice
Hirudo
Limnodrilus
Lumbricus
Macrobdella
Megascolides
Myzostoma
Nephthys
Nereis
Podarke
Sabella
Tubifex

Annelid worms—polychaetes, earthworms (oligochaetes), and leeches (hirudineans)—are distinguished by linear series of external ringlike segments; the grooves between segments coincide with internal compartments, often separated by transverse sheets of tissue (septa), containing serially repeated nervous, muscle, and excretory systems. Anterior segments bear jaws, eyes, and cirri (singular: cirrus, a slender appendage) in some species; the terminal segment may bear a cirrus (Figure A). Annelids have spacious, mesoderm-lined coeloms—except for leeches, in which tissue packs the coelom—and their coeloms are important in excretion, circulation, and reproduction. Chitinous lateral bristles called setae on each segment are used for locomotion or to anchor the annelid in substrate or burrow; leeches lack setae. Parapodia are unique to polychaetes; these thin, fleshy flaps protrude laterally from each body segment (Figure B). Chitinous cuticle covers the entire body.

Annelids live in soil, fresh water, and oceans—including Antarctic seas. They may be striped or spotted, and pink, brown, or purple. Others are iridescent or luminescent. Some have colorful gills and cirri, modified parapodia. The endangered Australian earthworm *Megascolides,* which is 3 m in length, is the largest species; the smallest annelid is only 0.5 mm.

Active predation or scavenging is the feeding mode for most annelids. Many annelids burrow incessantly, turning over and exposing detritus and soil and aerating anaerobic muds and sands; these activities are known as bioturbation. Swimming annelids catch fish eggs or larvae. Filter-feeding marine annelids capture bacteria and feed selectively on sediment particles within tubes (which they build of mucus-cemented sand grains, calcium carbonate, protein and polysaccharide compounds, and other materials) buried in sand or mud. Some trap plankton on a mucus-covered, ciliated eversible proboscis. Others pop out of their tubes to

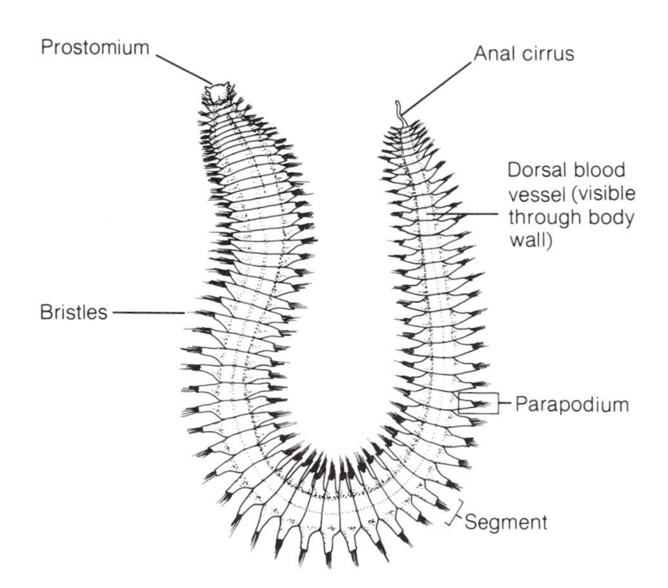

A (Right) An adult *Nephthys incisa,* a polychaete (13 cm long) taken from mud under 100 feet of water off Gay Head, Vineyard Sound, Massachusetts. (Left) External dorsal view of adult *N. incisa,* showing thin parapodia that serve in locomotion and gas exchange in this polychaete. [Photograph courtesy of G. Moore; drawing by I. Atema; information from M. H. Pettibone.]

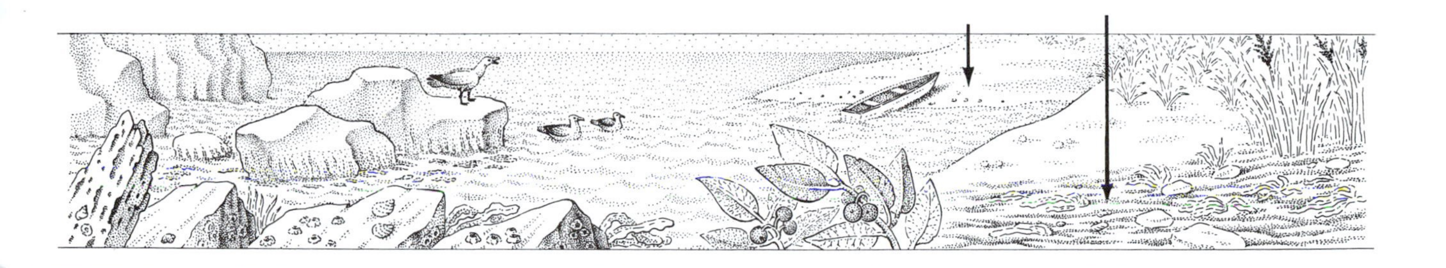

seize prey. Certain species harvest algae growing on their tubes. The sea star *Luidia* (Phylum A-34), hosts the polychaete *Podarke* among its tube feet. Some carnivorous polychaetes have fangs with which they inject toxins into prey.

About 15,000 species of annelids are grouped into three classes: the Polychaeta, mostly marine and a few soil and freshwater bristle worms (see Figure A); the Oligochaeta, terrestrial and freshwater bristle worms; and the Hirudinea, or leeches. There are about 9000 species of polychaetes, including myzostomarians, a group of about 100 species of small polychaetes that live on or in echinoderms (Phylum A-34), 6,000 species of oligochaetes, and 500 species of hirudinids.

In most polychaetes (paddle-footed worms), a fleshy lobe (prostomium) projects over the ventral mouth and bears tentacles. Parapodia of all polychaetes with few exceptions are stiffened by a bundle of chitinous bristles, enabling the parapodia to function as oars and levers. In a few polychaete species, chitinous rods called acicula support parapodia. Some tube-dwelling polychaetes leave their tubes, others do not; all of these tube builders are grouped as sedentary polychaetes. Free-living polychaete species are grouped as errant polychaetes. Marine polychaetes include *Aphrodite,* the hairy sea mice; lugworms (*Arenicola*), which burrow in sand and mudflats; sabellid and serpulid worms, whose tubes encrust shells, rocks, and algae, including peacock worms (*Sabella*), which construct mosaic tubes of sand or shell; economically important bait worms such as *Nereis;* and a few pelagic (ocean dwelling) species. The concretelike tubes of some polychaetes foul ships. There are also some soil and freshwater (in the Great Lakes and Lake Baikal) polychaete species. Oligochaetes include the earthworms and a few small freshwater, estuarine, and recently described deep-sea forms. The hirudinids, with anterior and posterior suckers, are popularly called leeches. Most leeches are free-living predators of frogs, turtles, fish, birds, and invertebrates of soil, foliage, algal thalli, fresh and salt water; a few parasitize vertebrates and invertebrates. Oligochaetes and leeches usually lack distinct eyes, tentacles, and parapodia including gills. Light-sensitive cells and sensory hairs in the earthworm epidermis that connect to the central nervous system alert earthworms to light and other environmental stimuli. Leeches lack setae. Annelids, except for the leeches, regenerate lost body parts from bristles to terminal segments. Some polychaetes reproduce by budding.

Bloodletting by the freshwater leech *Hirudo medicinalis* is used to control swelling subsequent to the reattachment of severed fingers or transplanted tissue. Saliva of bloodsucking leeches contains the anticoagulant hirudin as well as an anesthetic. This leech harbors a symbiotic bacterium *Aeromonas hydrophila;* not only does the bacterium digest blood, it also produces an antibiotic that kills other bacteria.

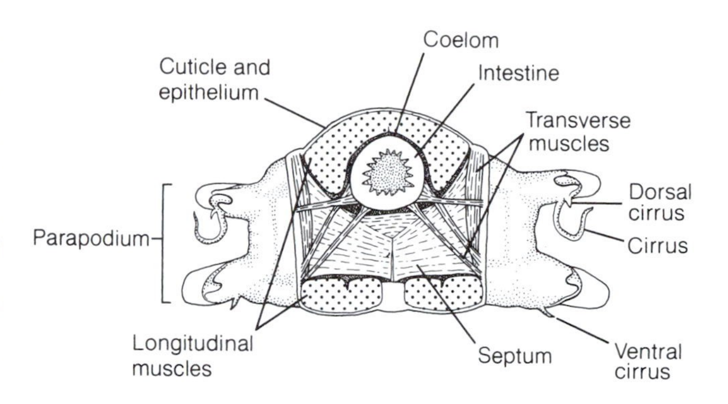

B Cross section of one body segment of the polychaete *Nephthys incisa.* Contraction of the longitudinal muscles shortens the annelid, increasing the diameter of its body. [Drawing by L. Meszoly; information from M. Pettibone.]

Burrowing polychaetes turn over 1900 tons of seafloor sand per acre each year. Charles Darwin calculated that earthworms bring 18 tons of soil to the surface per acre each year. Suction by a muscular pharynx draws soil into the earthworm mouth. Food is passed through the esophagus by the peristaltic movements of digestive tract muscles to a crop, where the food is temporarily stored. The muscular gizzard of the earthworm grinds seeds, eggs, larvae, small animals, and plants ingested with soil. Annelids' longitudinal and circular body-wall muscles work against the coelomic fluid, which is—like all fluids—relatively incompressible (see Figure B); this system functions as a hydraulic skeleton. Thin peritoneal sheets called septa separate adjacent segments. As body-wall muscles contract, colorless coelomic fluid flows from segment to segment through openings in each septum. Food is pushed through the gut from mouth to anus by cilia or by peristaltic contractions of muscles that encircle the digestive tract. The aquatic annelid pumps water through its burrow with peristaltic body waves, cilia, and parapodia. The water current brings in food and dissolved oxygen and removes waste.

Three iron-containing pigments—hemerythrin, hemoglobin, and the green pigment chlorocruorin—transport oxygen in blood vessels and coelomic fluid in most annelids, in corpuscles or in solution in blood. The annelid dorsal blood vessel is contractile with unidirectional valves, forcing blood through five aortic arches ("hearts") that act as pressure regulators and then into the ventral blood vessel. From the ventral vessel, blood moves to the digestive tube wall, the body wall, and the nephridia. Blood carried to the body wall exchanges oxygen and carbon dioxide through highly vascularized parapodia (in polychaetes)—sometimes modified into gills—and through the moist body wall, even through the protective cuticle (see Figure B). The continuous coelom of the leech lacks septa; most circulatory functions in leeches are carried out by coelomic fluid that is transported within the contractile channels and sinuses of the coelom itself. The closed annelid circulatory system, with contractile heart, blood vessels, and capillaries, is quite unlike the circulation pattern in arthropods.

The annelid excretory system consists of nephridia; within most body segments, ciliated paired tubules—nephridia—draw in wastes from coelomic fluid and then discharge dissolved waste through external pores called nephridiopores. Additional waste moves into nephridia from the nephridial blood vessels. Gametes also exit through nephridiopores in many annelids and through the mouth in a few species. Nephridia also regulate the water content of the coelomic fluid. Castings of intertidal polychaetes are sand, cleaned of organic matter as it passed through the gut and out of the anus.

Segmental ganglia, bilobed cerebral ganglia (brain) and single or paired ventral nerve cords are the main components of the nervous system. Most polychaete annelids have eyes, some with retinas and lenses. Chemoreceptors, touch receptors, vibration receptors, and statocysts (balance organs) concentrated at the head end link to the ventral nerve.

Breeding polychaetes swarm by the millions, their hormones triggered by phases of the moon, the tides, or changes in temperature. Polychaetes are usually dioecious, oligochaetes are usually monoecious but cross-fertilize, and leeches are monoecious. Polychaete gametes arise from the coelom walls in a number of body segments; polychaetes lack permanent gonads. Nephridia usually discharge gametes through nephridiopores in addition to urine, and polychaete fertilization is external. Many polychaete adults brood their young; in some species, the male protects and aerates the eggs. Development of most polychaete annelids includes a free-swimming, ciliated, planktonic larva, the trochophore, which is also often a feeding larva (Figure C). In some sedentary polychaete species, individuals are budded off or are transformed into epitokes, which are sexually mature, swimming, gamete-bearing individuals. Epitokes of *Eunice viridis*—the palolo—swarm on the sea surface in the South Pacific, in coincidence with lunar cycles. Male epitokes—stimulated by a pheromone from female epitokes—shed sperm. Responding to sperm, females release eggs. Samoans and other South Pacific people gather palolo epitokes to eat.

In contrast with polychaetes, leeches and oligochaetes are usually hermaphroditic (monoecious); each copulates with another individual. Sperm and eggs are produced in ovaries and testes, rather than in the peritoneal coelom lining. Oligochetes transfer sperm from one worm to its partner, which stores sperm in seminal receptacles until egg laying. The oligochaete lays eggs and releases stored sperm into a secreted mucus band; embryos develop in a secreted cocoon until a juvenile worm escapes. Some leeches attach to their partners with suckers and forcibly drive spermatophores—packets of sperm—into their mates' bodies; fertilization is internal. Leeches and oligochaetes incubate eggs in a cocoon—an adaptation to terrestrial life. They hatch as miniature adults, without a free-living larval stage.

Those free-living polychaetes that evolved in the Cambrian seas are the most ancient annelid group. From polychaetes, oligochaetes evolved. Leeches are of oligochaete ancestry. Fossil polychaetes have been found in rocks of the late Proterozoic eon that contain the Ediacaran fossil assemblages (about 700 million years old) and are well preserved in the middle Cambrian rocks of the Burgess shale (500 million years old) of western Canada. Terrestrial oligochaetes probably did not evolve before the Cretaceous, when angiosperms, which contributed humus in which earthworms live, arose. Annelids are possible ancestors of sipunculans (Phylum

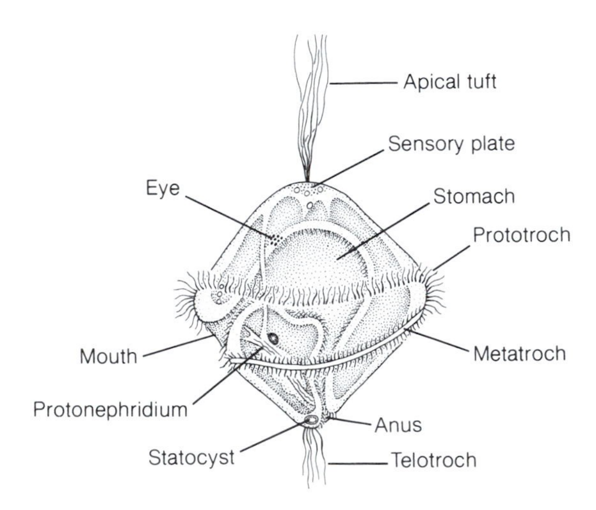

C Annelid trochophore, free-swimming larva that is the dispersal form of marine polychaete annelids. Ciliary bands—telotroch, metatroch, and prototroch—are distinctive features of trochophore larvae. [Drawing by L. Meszoly; information from M. Pettibone.]

A-23) and echiurans (Phylum A-24)—at least some species of these phyla have trochophore larvae, produce gametes from peritoneal tissue rather than ovaries and testes, and have similar body-wall anatomy, nervous systems, excretory systems, and patterns of gamete production. Immunological data support annelids as closer relatives of sipunculans than molluscs are; however, the evidence from paleontology, biochemistry, and embryology supports molluscs as closer relatives of sipunculans than are annelids. Data from ribosomal RNA indicate that pogonophorans (Phylum A-25) may have evolved from annelids. Molluscs (Phylum A-26), which also have trochophore larvae, may have evolved from annelid ancestors or from sipunculans; alternatively, the trochophore larvae may have arisen independently, an example of convergent evolution. Unlike wormlike nemertines (Phylum A-9), nematodes (Phylum A-10), nematomorphs (Phylum A-11), and acanthocephalans (Phylum A-12), annelids have linearly segmented bodies, coeloms, a ventral nerve cord, and distinct eyes (in polychaetes). Onychophorans and annelids may have a common ancestor (see Phylum A-28). The phylogenetic relationship between annelids and arthropods is currently controversial.

A-23 Sipuncula
(Sipunculans, sipunculids, peanut worms)

Latin *siphunculus,* little pipe

EXAMPLES OF GENERA
Aspidosiphon
Golfingia
Lithacrosiphon
Nephasoma
Onchnesoma
Phascolion
Phascolopsis
Siphonosoma
Sipunculus
Themiste

Sipunculans are crevice-dwelling and burrowing, unsegmented sea animals. The contractile introvert ranges in length from less than half of the trunk length to several times the trunk length in some species. Minute spines or papillae stud the distal part of the introvert in some species. Ciliated bushy or fingerlike mucus-covered tentacles encircle the terminal mouth (Figure A) in Sipunculidea, one of the two classes. Small tentacles are arrayed in a crescent dorsal to the mouth in the other class, Phascolosomatidea. Sipunculans can slip closed like telescopes. With the introvert drawn into the plump trunk, some species resemble very firm peanuts, and so they are also called peanut worms. *Sipunculus* swims briefly by contracting its body. Sipunculans form temporary burrows in sea sediments by thrusting the introvert forward and dilating its tip; by contracting longitudinal retractor muscles, the body pulls forward. Hydraulic pressure exerted by body-wall muscles on coelomic fluid thrusts the introvert out; retractor muscles retract it. Sipunculans are protostome coelomates; their mouth forms near the site of the embryonic blastopore.

There are about 150 species of sipunculans. Commonly, the trunk is from 15 to 30 mm long. *Phascolion psammophyllum* is only a few millimeters long; others are 0.5 m long. Their rubbery, cuticle-covered body walls are iridescent pearl gray, yellow, or dark brown. The cuticle is mostly collagen (a protein), contains no chitin, and lacks cilia and setae. Sipunculans line their dead-end burrows with mucus. Some live inside large *Spheciospongia* sponges or among mangrove roots, eel grass, or reef-forming coral. Others live beneath rocks or in polychaete annelid tubes. *Aspidosiphon muelleri* inhabits empty gastropod mollusc shells. Larvae of the Indo-Pacific coral *Heteropsammia* and *Heterocyathus* settle on the sipunculan-occupied shell. Many sipunculans live at or just below the low-tide mark of warm seas. *Themiste lageniformis* (see Figure A) is subtropical and circumtropical. Certain species are found in polar seas and others in the abyss as far as 7000 m down. Recent collections for which fine mesh equipment was used suggest that species diversity of sipunculans in deep oceans is not far behind that in tropical and subtropical shallow seas.

Sipunculan distribution is patchy, tending to greatest abundance on rocky shores. The highest density recorded is about 4000 sipunculans per square meter in Indian River lagoon, Florida. Sipunculans are active in bioturbation—like some polychaetes, they disrupt and aerate the sediment in shallow bay bottoms in much the same way that earthworms aerate soil. Because sipunculans burrow, they are the most important eroders of coral, along with polychaete annelids.

Gas exchange takes place with seawater through the tentacle–contractile vessel complex and through the body wall. The body wall has fluid-filled epidermal canals—somewhat like a radiator—so that coelomic fluid in the canals is close to seawater for gas exchange. The fluid-filled compartments in the hollow tentacles are separate from the spacious body cavity, the coelom. Fluid in the tentacles is stored in a reservoir called the contractile vessel attached to the esophagus; the tentacle cavity and contractile vessel communicate. In some sipunculans, such as *Themiste,* the oxygen-carrying respiratory pigment hemerythrin is dissolved in coelomic fluid; together with myohemerythrin in retractor muscles, this can be considered a closed vascular system. Heart and blood vessels are not present, but oxygen and nutrients are distributed by coelomic fluid.

Some sipunculans feed on detritus in mud; others, on diatoms, other protoctists, and larvae. Sipunculans inhabiting coral scrape algal films with hooks on their introverts. A few filter feeders trap food from seawater on mucus-covered tentacles; cilia on the tentacles waft the food to the mouth. Dissolved organic matter in seawater may supply as much as 10 percent of sipunculan nutrition. Detritus feeders are selective deposit feeders, selecting nutrients from ingested mud. A ciliated groove conveys food through part of the length of the U-shaped, spirally coiled intestine, which loops forward on itself to open in an anus at the introvert base (Figure B). With this arrangement, sipunculans need not turn in their burrows to defecate.

Metanephridia collect dissolved wastes from the coelom. Ammonia is sipunculans' main nitrogenous waste. In some species, urn cells form on the peritoneum. These ciliated groups of cells are of two types—fixed or free; urn cells gather waste from coelomic fluid, then leave the worm through the nephridia or leave the waste on the body wall.

The sipunculan nervous system consists of a bilobed brain (cerebral ganglion) located above the esophagus and connected to an unsegmented ventral nerve cord. Body and tentacle surfaces bear protrusible ciliated cells, presumably sensory. Sipunculans also have pigmented, ciliated photoreceptor cells and likely chemoreceptors near the mouth.

Sexes are generally separate in sipunculans, but males and females cannot be distinguished externally. Seasonal gonads form temporarily and inconspicuously at the base of the retractor muscles; permanent gonads are absent. *Nephasoma minutum* is the only monoecious species; it fertilizes itself. Of the dioecious sipunculans, females tend to be more abundant than males. Mature females shed gametes into the coelom, where the oocytes accumulate yolk and become ova. Males release sperm into the coelom from which sperm are discharged from nephridiopores into the sea; these sperm induce females to discharge ova through nephridiopores. Fertilization takes place in seawater. Sipunculans have three developmental patterns, depending on species. Some develop directly into adults without passing through a larval stage; most form free-swimming, nonfeeding trochophore larvae (see trochophore larvae of annelids, Phylum A-22) that eventually settle to the bottom, where they metamorphose to adulthood. Still others form

trochophore larvae succeeded by pelagosphera larvae, which may feed several months on plankton and which may survive transport across oceans. *Aspidosiphon* can reproduce asexually; the animal constricts and separates the posterior end of its trunk to form a new individual, including a brain.

The only direct value of sipunculans to humans is limited to their food value. *Sipunculus* and *Siphonosoma* are reportedly eaten in the Indo-Pacific, and *Phascolosoma* is eaten in China. However, observations that the sipunculan introvert and ventral nerve cord regenerate if cut or damaged may enhance our understanding of nerve regeneration.

Sipunculans, like annelids (Phylum A-22) and echiurans (Phylum A-24) have similar trochophore larvae, body-wall muscles, nervous systems (the annelid nerve cord has ganglia in each segment, whereas that of sipunculans does not), excretory systems, and gamete production in seasonal gonads. Undoubted fossil sipunculan tubes appear in fossil coral. Sipunculans may have evolved from annelid ancestors before annelids developed segmentation; this theory is supported by immunological evidence. Morphological, paleontological, biochemical (nucleic acid), and embryological data suggest that molluscs are closely related to sipunculans and that annelids are more distantly related.

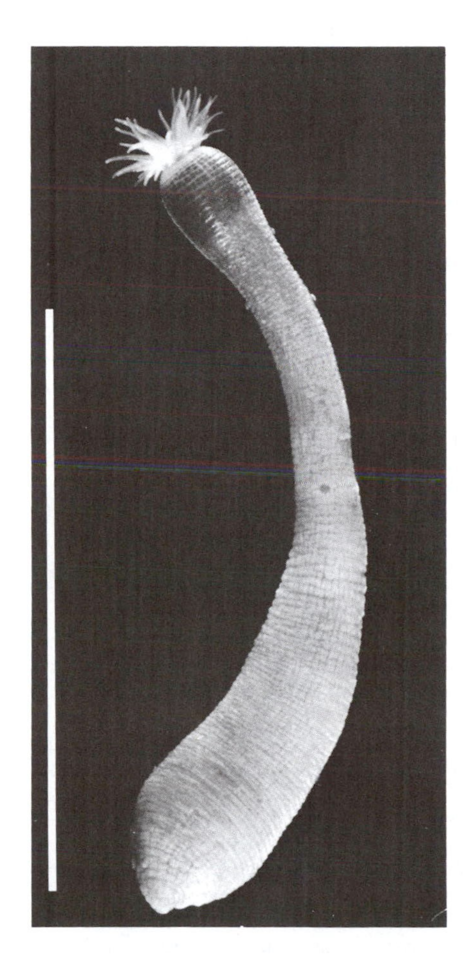

A *Themiste lageniformis,* a peanut worm from the Indian River, Fort Pierce, Florida, with introvert and tentacles extended. Bar = 0.5 cm. [Courtesy of W. Davenport.]

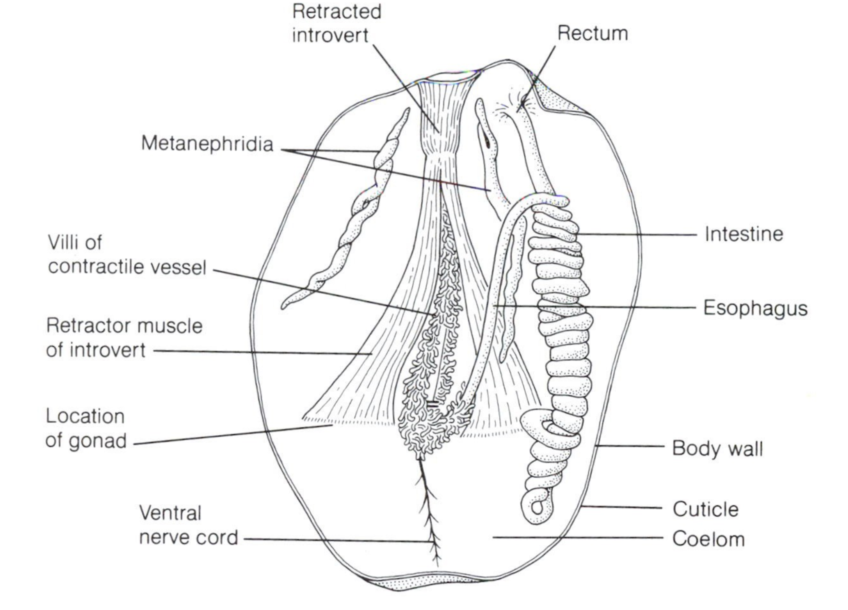

B A cutaway view of a *T. lageniformis* with its introvert retracted. Contractile vessels push fluid into the hollow tentacles, causing them to extend. Contraction of the introvert retractor muscles forces fluid back into the contractile sacs. [Drawing by L. Meszoly; information from M. E. Rice.]

A-24 Echiura

(Spoon-worms, echiurans, echiurids)

Greek *echis*, snake; Latin *-ura*, tailed

EXAMPLES OF GENERA

Bonellia	*Lissomyema*	*Prometor*
Echiurus	*Listriolobus*	*Tatjanellia*
Ikeda	*Metabonellia*	*Thalassema*
Ikedosoma	*Ochetostoma*	*Urechis*

About 140 species of known echiurans are named for their flexible proboscises, which may extend as far as 1.5 m from their plump, soft trunks (Figure A). The contracted proboscis, shaped like a spoon in *Urechis* and some other genera, gives them the common name spoon-worm. The mouth is at the base of the proboscis, not at the tip. A disturbed echiuran can cast off its proboscis and later regenerate it. The proboscis does not withdraw into the trunk and is therefore not an introvert. Echiuran bodies, which can be rough or smooth, measure from a few millimeters to about 40 cm long. Their cylindrical or ovoid bodies may be dull brown, gray, red, rose, yellow, or transparent. Green *Bonellia tasmanica* (*Metabonellia tasmanica*) derives its green color from its food algae; "bonellin" is a green porphyrin pigment derived from chlorophyll *a* of algae.

Echiurans dwell in U-shaped tunnels, mud, or rock crevices in estuaries, among mangrove roots, and in all oceans, some at depths of 10,000 m in the abyss. None live in fresh water. Some occupy abandoned shells or tests—calcareous external skeletons. For example, brick-red *Thalassema* lives in discarded sand dollar tests. One of the few known enemies of echiurans, the bat ray (Phylum A-37), uses its flat body like a toilet plunger to pop echiurans out of their burrows. On the shores of the North Sea and in Belgium, fishermen use echiurans for cod bait.

Echiurans have no bones. The skeleton is hydraulic—progressive contractions of the body wall are countered by the resistance of coelomic fluid in the coelom. Peristaltic contractions of body-wall muscles continuously alter the shape of the spoon-worm, pushing food along the gut or powering its burrowing along the seafloor. The body-wall structure is sipunculan- and annelid-like: a chitinous cuticle on the outside protects the epidermis, which overlays the circular muscles that surround longitudinal and oblique muscles. The peritoneum lining the body cavity separates the muscles from the digestive tract within the coelom. *Urechis* and *Echiurus* have setae—chitinous hooks encircling the posterior trunk and near the mouth. To tunnel, an echiuran wedges the anterior of its trunk (not the proboscis) into sand or mud. Then it draws its hind end forward and, after anchoring it again, wedges the anterior forward. *Urechis* and *Echiurus* use their anal setae for anchoring. *Echiurus pallassii* takes some 40 minutes to burrow out of sight; other species may work rather more quickly. Echiurans do not secrete tubes.

Cilia sweep microbes and detritus along a mucus-coated ventral groove of the proboscis into the mouth at the proboscis base (Figure B). Food moves through the pharynx, esophagus, crop, and gizzard to the intestine. When the tide is in, most spoon-worms thrust their proboscises out in search of food—a pencil-size burrow opening remains at the anterior end even though mud may fill the unoccupied part (Figure C). Food is digested mainly in the intestine, and nutrients presumably diffuse into coelomic fluid. Two anal sacs collect coelomic fluid and dump it—perhaps unchanged in composition—into the expanded part of the lower intestine, from which it leaves the anus, along with solid waste. Echiuran nephridia excrete urine, containing soluble waste, through nephridiopores; gametes exit from nephridiopores also. Blood is oxygenated primarily in the gill-like processes of the proboscis. The coelom extends into both proboscis and trunk. Muscles and nucleated coelomic cells carry oxygen bound to hemoglobin, so the coelomic fluid functions as part of the circulatory system. Most echiurans have a closed vascular system—in which blood flows through vessels—that lacks hemoglobin. *Urechis* has hemoglobin in its coelomic fluid and lacks a vascular system.

Urechis caupo, a Pacific Coast echiuran called the fat innkeeper because a fish, a polychaete annelid, several crabs, and a clam are often observed in its burrow, is an unusual species in a number of ways; for example, it anchors by means of a circumanal ring of setae as it burrows. It is the only echiuran that pumps water through its distinctive burrow. The tunnel descends as far as 50 cm on the diagonal, then runs horizontally for 15 to 100 cm, and then rises vertically to the substrate surface. *U. caupo's* feeding strategy is unique among echiurans: it spins a thimble-shaped mucus net, from 5 to 21 cm long, through which it strains particles as small as 0.04 μm wide from seawater in its burrow. When the net is covered with food particles, *U. caupo* swallows it and secretes a fresh net. *U. caupo* pumps seawater in and out of its thin-walled intestine through its anus; oxygen diffuses across the intestinal wall into the hemoglobin-containing coelomic fluid. Oxygen uptake by the worm continues even if the anus is plugged, indicating that respiratory exchange can be maintained across the body wall. *U. caupo* produces eggs and sperm that, in the laboratory, develop year round, an unusual feature, and are thus useful for studies of invertebrate development. The echiuran nervous system includes a nerve loop around the proboscis margin and a ventral, unsegmented nerve cord.

Echiurans only reproduce sexually and are dioecious. Eggs and sperm are formed seasonally by the peritoneum at the base of retractor muscles; sipunculans (Phylum A-23) and polychaetes (Phylum A-22) also have temporary seasonal gonads rather than permanent gonads. Echiurans, like sipunculans, lack permanent gonads. Gametes are shed into the coelom. Although males and females look alike in some species, *Bonellia* may hold the record for the most extreme animal sexual dimorphism. With proboscises extended, *Bonellia* females may be as much as 1 m long. Males, though, are only 1 mm long; and one or more lives inside the

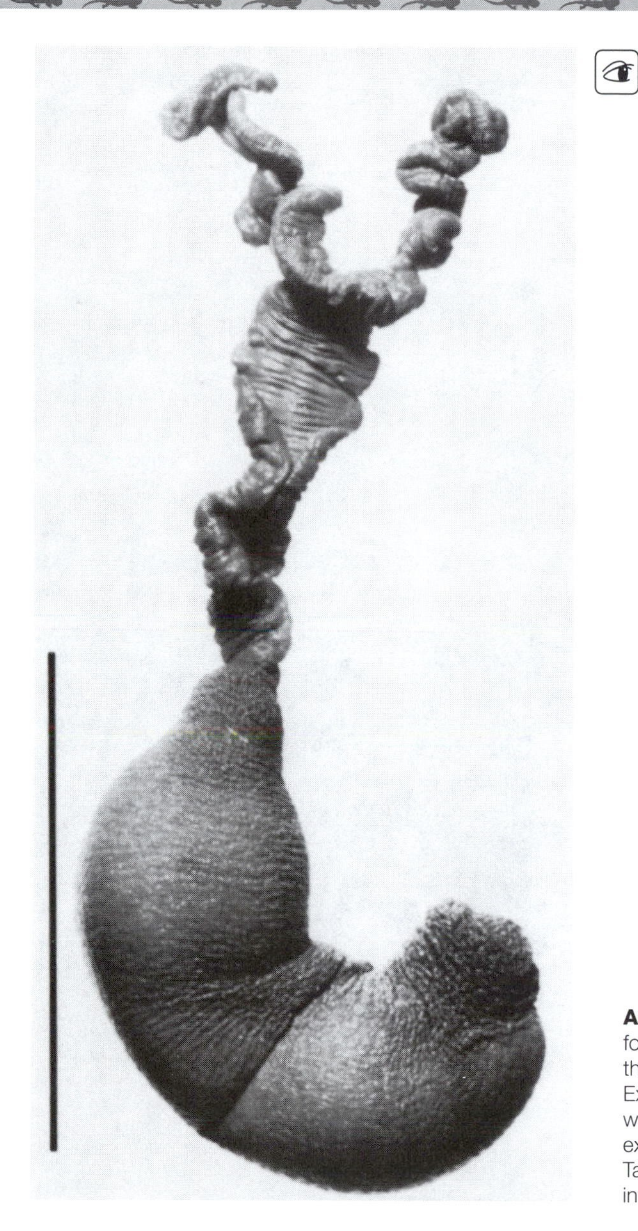

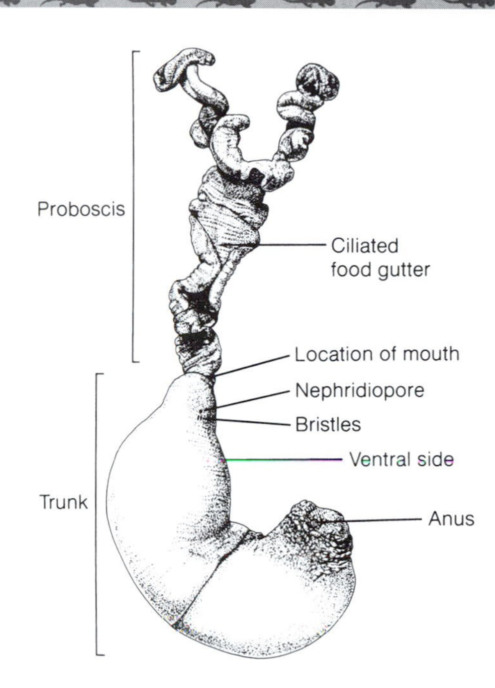

A (Left) A female echiuran, *Metabonellia tasmanica,* found at low-tide mark in sand under rocks in gullies on the coast of southeastern Australia. Bar = 5 cm. (Right) External view showing the proboscis—here contracted—which functions as the spoon-worm's organ of gas exchange. [Photograph courtesy of A. Dartnall and the Tasmanian Museum and Art Gallery; drawing by I. Atema; information from A. Dartnall.]

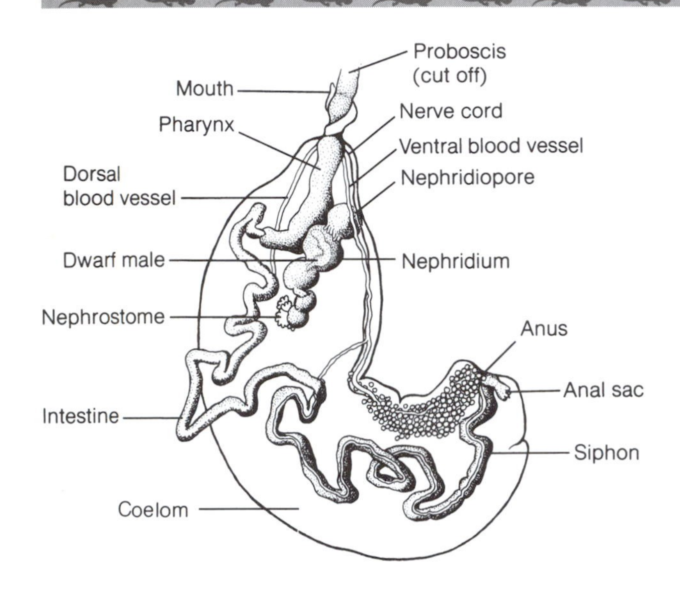

B Internal diagram of a female spoon-worm—*Bonellia*—with dwarf male dwelling inside her nephridium. [Drawing by P. Brady; information from A. Dartnall.]

C (Left) *Listriolobus pellodes.* Bar = 3 cm. (Right) *L. pellodes* proboscis emerging to feed. Bar = 6 cm. [Photographs courtesy of Michael J. Risk.]

D Trace fossil in Silurian sandstone from southern Ontario, Canada (left) and (right) recent feeding trace made by proboscis of *Listriolobus pellodes.* Bar = 6 cm for both. [Photographs courtesy of Michael J. Risk.]

female's nephridium (see Figure B). The nephridium of the female *Bonellia* collects the eggs from her coelom through the nephrostome. The *Bonellia* male—inside the female—fertilizes the eggs, which then exit through the nephridiopore to the sea (see Figure B). Fertilization in other echiuran species usually takes place in the sea after eggs and sperm leave the coelom—producing trochophore larvae (Phylum A-22). The ciliated, free-swimming larvae are pelagic for several weeks and then progressively metamorphose into the sedentary adult form. In the genus *Bonellia,* larvae that mature in the sea without coming into contact with adult females of the same species become females. *Bonellia* larvae that do contact adult females are induced to develop into tiny parasitic (symbiotrophic) males, which lodge in the nephridia of the females.

In Silurian sandstone from 408 million to 438 million years old, imprints of echiuran burrows have been found. These fossil burrows remarkably resemble those of contemporary echiurans, suggesting that the phylum was well established by this time. (Figure D). Echiurans may have evolved from ancestral polychaete annelids (Phylum A-22). Annelids and many echiurans have similarities: chitinous setae, closed circulatory systems with blood vessels, body-wall construction, nervous system layout, a complete digestive system that is open at both ends, metanephridia, gamete formation (gametes arise from the temporary gonads on the peritoneum, mature in the coelom, and leave through nephridiopores), and trochophore larvae. Echiurans lack the segmentation that is evident in the annelid body. Until evidence resolves their phylogeny, the position of echiurans as protostome (the mouth forming near the site of the blastopore) or deuterostome (the adult mouth forming as a secondary opening opposite the blastopore) coelomate animals is enigmatic.

A-25 Pogonophora

(Beard worms, pogonophorans, tube worms)

Greek *pogon,* beard; *pherein,* to bear

EXAMPLES OF GENERA

*Escarpia**
Heptabrachia
*Lamellibrachia**
Lamellisabella
*Oasisia**
Oligobrachia
Polybrachia
*Ridgeia**
*Riftia**
Siboglinum
Spirobrachia
*Tevnia**
Zenkevitchiana

Pogonophorans, the beard worms, are sessile benthic marine worms that live in fixed upright chitin tubes that they secrete on sediments, shell, or decaying wood on the ocean floor. This is the only phylum of free-living multicellular animals in which all lack digestive tracts as adults. The phylum contains about 120 species in two classes, Class Perviata (perviates) and Class Obturata (vestimentiferans). Meredith Jones of the Smithsonian Institution assigns the six vestimentiferan genera, marked with asterisks in the list of examples above, to a proposed separate Vestimentifera phylum that includes all 10 or so vestimentiferan genera, leaving the remaining pogonophorans in Phylum Pogonophora.

Pogonophorans are probably distributed throughout the world in cold (2°–4°C), deep ocean waters, in shallow Arctic and Antarctic seas, as well as in hot (10°–15°C) submarine vents (hydrothermal vents), where the water contains hydrogen sulfide and methane. Pogonophorans of these deep, dark communities do not feed—they derive nutrients and energy from symbiotic chemoautotrophic bacteria (Phylum B-2). Although hydrogen sulfide is usually toxic to animals, pogonophorans in this hot vent habitat use their red extracellular hemoglobin to carry both hydrogen sulfide—bound so that it cannot absorb the hemoglobin site that binds oxygen—and oxygen from their tentacles to the site of bacterial oxidation. Hydrogen sulfide thus reversibly bound is not toxic to the worm.

Beard worms are collected along continental slopes, usually in seas deeper than 100 m and, more rarely, as shallow as 20 m. The first beard worms, in Class Perviata, were dredged up off the coast of Indonesia in 1900; members of Class Obturata were not described until 1969. The greatest diversity of pogonophorans has been found in the western Pacific, as far south as New Zealand and Indonesia. Along the Atlantic coast, beard worms have been brought up from near Nova Scotia down to Florida, the Gulf of Mexico, the Caribbean, and Brazil. In the eastern Atlantic, pogonophorans are reported from Norway to the Bay of Biscay.

Long and skinny, most pogonophorans in Class Perviata are from 10 to 40 cm long and less than 1 mm wide. These beard worms move up and down inside their tough tubes, which are usually open at both ends (Figure A). Tubes are often banded yellow or brown. Each tube is anchored upright in soft sediment. Perviate bodies have three sections: a short forepart, a long trunk, and a short rear part (Figure B). The forepart includes a cephalic lobe containing the brain and bearing the beardlike tentacles that give the phylum its name. A given species may have a single ciliated spiral tentacle or thousands of them (see Figure A). Also in the forepart and posterior to the cephalic lobe is a glandular region that secretes the chitinous, sometimes flexible, sometimes stiff, tube. The long trunk bears papillae (bumps), cilia, and belts of toothed, chitinous bristles called setae (see Figure B). Circular and longitudinal striated muscles make up the body wall. The hind region, called the opisthosoma (Greek *opisthen,* hind; *soma,* body), is composed of 5 to 25 short segments with setae. Because the perviate body is longer than its tube, the opisthosoma facilitates burrowing; the extended narrow opisthosoma protrudes from the bottom end of the tube and is pushed into the sediment. As blood flows in, the opisthosoma expands and the setae protrude, anchoring the posterior end, which allows the upper body to be pulled further into the substrate, anchoring the perviate. Because most perviate pogonophorans are collected by dredging the ocean floor, it took a generation to discover that all have an opisthosoma—this missing region is easily lost during dredging.

Vestimentiferans (Figure C), which include about ten species, are heavy-bodied pogonophorans from 2.5 to 4.0 cm wide and as

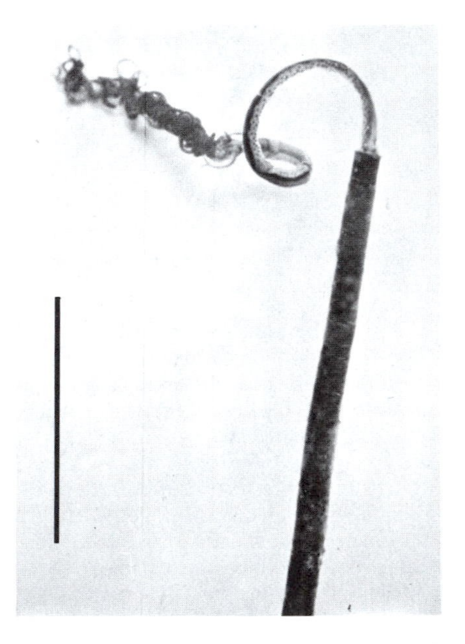

A Front end of body and tentacle crown of *Oligobrachia ivanovi*, a perviate, partly dissected out of its tube. Bar = 1 cm. [Courtesy of A. J. Southward.]

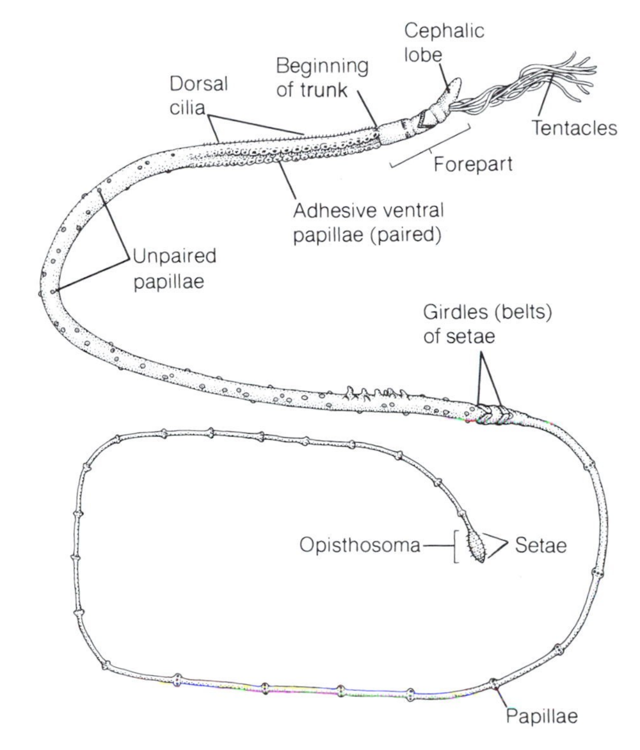

B Diagrammatic and shortened view of pogonophoran removed from tube. This thin beard worm, belonging to Class Perviata, has a segmented hind region—the opisthosoma—that bears chitinous setae. [Drawing by L. Meszoly.]

much as 2 m long; they are much more plump than most other pogonophorans. The vestimentiferan body has four regions. At the top of the first region is the branchial plume—a crown of tentacles fused into sheets that is supported by a structure called the obturaculum. The obturaculum plugs the tube when the worm draws in. The second region bears body-wall folds that form an external collar called the vestimentum; inside are a heart, a brain, genital apertures, and glands that secrete tube material. Excretory pores open on the vestimentum. The third and longest region is a trunk that contains the gonad and a heavily vascularized trophosome, brown spongy tissue packed full of symbiotic chemolithoautotrophic bacteria from which these tube worms obtain nutrients. Sulfide and methane utilized by these bacteria diffuse into the animal across the vestimentiferan's tentacles and are delivered by dissolved hemoglobin (not in cells) in both blood vessels and in coelomic fluid to the trophosome. There the reversibly bound sulfide and methane are available to the bacteria. The fourth and posterior region, called the opisthosoma, has setae that correspond to internal segments. Vestimentiferans have two compartments in each segment of the opisthosoma (in comparison with one compartment per segment in perviate pogonophorans). Unlike the perviate opisthosoma, the vestimentiferan opisthosoma secretes partitions that wall off the tube base, so the bottom end of the tube is closed. The vestimentum and obturaculum of vestimentiferans are lacking in perviate pogonophorans.

The coelom of pogonophorans extends into all three body segments in perviates and into all four body segments in vestimentiferans, including the tentacles. The trunk coelom of pogonophorans contains gonads and the trophosome; the trunk coelom is unseptate, and the trunk is unsegmented. The opisthosoma coelom of all pogonophorans is segmented by muscular septa (walls) that divide it into separate compartments that correspond to the posterior region's external segmentation. The nervous system includes ventral ganglionated nerve cords including giant axons that may innervate muscles by which pogonophorans retract into their tubes.

The pogonophoran adult lacks a mouth, digestive tract, and anus. Gathering evidence for its mode of nutrition is a challenge, considering that many live in the abyss. Thin, small perviate pogonophorans are thought to feed by extending their tentacles from the tube to gather organic detritus and plankton. Suspended food particles may be trapped on the tentacles—cilia may drive seawater through a funnel formed by the tentacles. The worms probably take up nutrients—amino acids, glucose, and fatty acids—through tentacles including their cilia by phagocytosis and pinocytosis. Absorptive feeding by active uptake of dissolved organic matter is the rule—at least for small pogonophorans—even when the pogonophoran does not extend its tentacles out from the tube.

Studies of perviates have shown that absorption of dissolved organic matter does not suffice to meet the metabolic requirements of all species; so an additional mechanism is probably used. But, for the vestimentiferans, which are considerably larger than the perviates, the trophosome—loaded with as many as 1 billion chemosynthetic bacteria per gram of trophosome tissue—is the source of nutrition. The bacteria oxidize methane or sulfide to eventually synthesize organic molecules from carbon dioxide.

C *Riftia pachyptila,* vestimentiferans in their flexible tubes. Taken at a 2500 m depth off the Galapagos Islands, this is the first photograph of a live colony in situ. Bar = 25 cm. [Courtesy of J. Edmond; information from M. Jones.]

Whether the vestimentiferan derives nutrients from this process by directly digesting its symbiotic bacterial partner or by absorbing soluble organic molecules synthesized by the bacteria is not known. One method or the other method may be employed in various cases. Juveniles may take in free-living marine bacteria and transport them through a transient duct to the trophosome; a duct that could function in this manner has been described in *Lamellibrachia, Riftia, Ridgeia,* and *Oasisia* juveniles. Once the sulfide-oxidizing bacteria are in the larval gut, epithelial cells may phagocytose the bacteria to establish the symbiotic association.

A closed blood vascular system with dorsal and ventral vessels extends into each tentacle. A distinct heart and nearby paired excretory organs are located in the anterior region. The respiratory surface of the microvillus-covered tentacles is multiplied by pinnules (protrusions of the surface), major sites of gas exchange for beard worms. Their respiratory pigment is hemoglobin dissolved in blood and in coelomic fluid.

The sexes are usually separate and usually externally indistinguishable. One hermaphroditic species is known. In the trunk coelom are two cylindrical gonads. Perviate pogonophore males package sperm into packets (spermatophores). Sperm are found in bundles of several hundred—not in spermatophores—in *Riftia pachyptila* and other vestimentiferans. Sperm morphology may be related to the mode of sperm transfer from male to female, suggesting the possibility of direct transfer of sperm or internal fertilization or both. Fertilization has not yet been observed. Ciliated embryos have been taken from inside the tubes of females of at least some species, indicating that they brood their embryos. Larvae leave the tube, disperse, settle, and metamorphose into the adult form.

Studies of gastrulation in *Siboglinum* provide evidence that pogonophorans are protostome coelomates related to annelids (Phylum A-22): the mouth forms near the site of the blastopore and the body cavity is a coelom. Early embryos of two species of hydrothermal-vent pogonophorans—*Riftia* and *Ridgeia,* both vestimentiferans—are ciliated trochophore larvae. These annelid-like larvae have mouth-to-anus complete digestive tracts; later in development, the gut is lost.

A fossil *Hyolithellus* from the Lower Cambrian rocks in North America, Greenland, and northern Europe has been assigned to

Pogonophora. It has been suggested that pogonophorans derived from a "protoannelid" during the early Cambrian.

Some researchers include pogonophorans in Phylum Annelida (Phylum A-22). Similarities that suggest that pogonophorans descended from annelid ancestors are as follows: muscular septa compartmentalizing the coelom; chitinous setae; hemoglobin dissolved in body fluid rather than in blood cells; trochophore larvae; and ribosomal RNA. Other specialists consider pogonophorans to be deuterostome coelomates. Therefore, they group Phylum Pogonophora—including vestimentiferans and perviates—with other deuterostome (the anus forming at the blastopore) coelomates: hemichordates (Phylum A-33), echinoderms (Phylum A-34), and chordates (Phyla A-35 through A-37). We reserve judgment regarding classification of vestimentiferans as a separate phylum from perviate pogonophorans until more of their biology is known. Whether vestimentiferans and perviates diverged independently or have a common ancestor is not clear.

A-26 Mollusca

(Molluscs)

Latin *molluscus*, soft

EXAMPLES OF GENERA

Aplysia	*Conus*	*Haliotis*
Arca	*Crepidula*	*Helix*
Architeuthis	*Cryptochiton*	*Littorina*
Argonauta	*Dentalium*	*Loligo*
Busycon	*Doris*	*Mercenaria*
Chaetoderma	*Dreissena*	*Murex*
Charonia	*Elysia*	(continued)

From the jet-propelled (but very slow moving) chambered nautilus to the poison-producing cone shell, diversity within the molluscs is remarkable. Most molluscs have an internal or an external shell, a muscular foot, and an unsegmented, soft body. A mantle, which is a fold of the body wall, lines the external shell and secretes the calcium carbonate and protein of which the shell is made. A tubular extension of the mantle called the siphon, when present, directs cilia-generated water currents into the mantle cavity. The current carries food as well as dissolved gases to the gills of the mollusc. Some molluscs gather food by boring or scraping with a hard, chitinous ribbon called the radula, a unique molluscan structure. Other molluscan species are predators; gastropod *Charonia* consumes sea star *Oreaster.* Still other molluscs live as parasites or commensals with sea squirts (ascidians, Phylum A-35), annelids (Phylum A-22), echinoderms (Phylum A-34), or crustaceans (Phylum A-21). Molluscs fill an extraordinary array of ecological niches: mud and sand flats, forests, soil, fresh water, deserts, driftwood, and the abyss of the sea.

About 50,000 species of molluscs have been described, making Mollusca the third largest phylum after Mandibulata (Phylum A-20) and Chelicerata (Phylum A-19). Mollusc shells come in a beautiful array of colors and patterns. Certain shell-less gastropods are given a green color by algae, from the chloroplasts of the marine algae on which they feed. One slug bioluminesces as a consequence of feeding on bioluminescent sea pansies (anthozoans, Phylum A-3). Some molluscs produce ink, which in some cases is luminescent. Molluscs range in size from no bigger than a sand grain to the giant clam *Tridacna,* which can be 1.3 m wide. The giant squid *Architeuthis,* the largest invertebrate ever known, measured lengths approach 20 m, including 16 m long tentacles and a 4 m long body.

Molluscs are grouped into seven classes. Members of Class Monoplacophora have one cap-shaped shell over a flattened foot. They live in deep water off the west coast of the Americas, in the Gulf of Aden, in the South Atlantic, and off Antarctica. Monoplacophorans are better known from the fossil record—only about 10 living species have been described. *Vema* (Figures A and B), the first of this class to be observed live, was taken from a submarine ridge off San Diego. Aplacophora, the second class—all wormlike and shell-less with external spicules of calcium carbonate—includes the deep-sea solenogasters (*Chaetoderma* and *Neomenia*). Class Polyplacophora includes chitons (*Cryptochiton*). Eight shell plates at least partly buried in the mantle overlie the large flat foot with which a chiton clings to its rock, often in heavy surf. Members of class Bivalvia (Pelecypoda) are the clams, mussels, scallops (*Pecten*), oysters, and wood-boring molluscs. A bivalve is usually sedentary and has two lateral shells called valves with a dorsal hinge. The foot in bivalves can be flattened like a snail's foot, fingerlike, wedge shaped, or absent. They lack tentacles and heads. The largest class, Gastropoda, comprises abundant marine snails

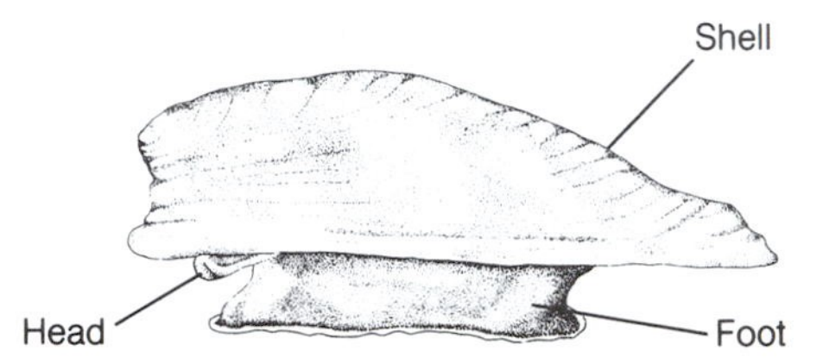

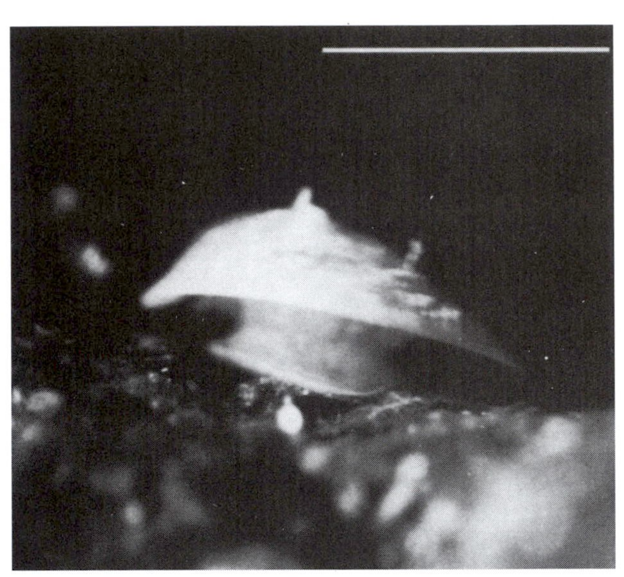

A *Vema hyalina,* probably the first live monoplacophoran mollusc ever to have been photographed. This mollusc was photographed at about 400 m down, off Catalina Island, California. Although Paleozoic *Vema* and *Neopilina* shells are known, these species, the most primitive living molluscs, were first seen live in 1977. Bar = 1 mm. [Photograph courtesy of H. A. Lowenstam and C. Spada; drawing by L. Meszoly.]

such as *Busycon* (Figures C and D), sea slugs and their relations, and all terrestrial molluscs. If present, the gastropod shell is usually spirally coiled. Shell-less gastropods include the nudibranchs. The class Scaphopoda comprises the tooth shells (*Dentalium*), which have tusk-shaped shells open at both ends. They burrow into mud or sea sand foot first, leaving the narrow end of the shell exposed above the sand or mud. Class Cephalopoda includes octopods, which lack an internal shell; squids, which have an internal

shell (pen) secreted by the mantle; and *Nautilus,* which has an external shell. A pen is chitinous and lacks calcium carbonate. The head and prehensile arms (tentacles) of a cephalopod encircles beaklike jaws.

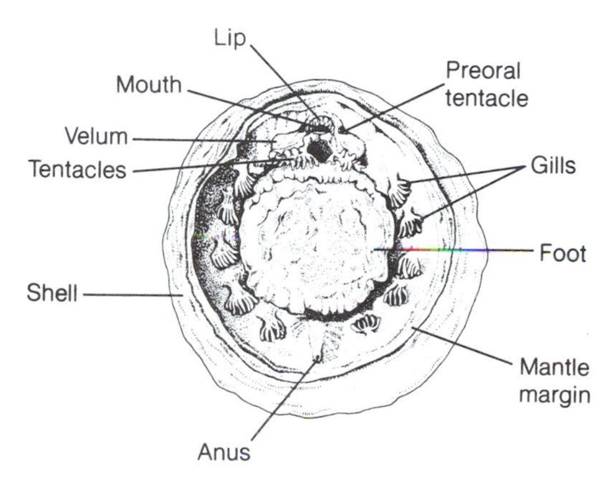

B View of the underside of a monoplacophoran showing multiple gills. In *Neopilina sp.,* the paired gills manifest segmentation. [Drawing by L. Meszoly; information from J. H. McLean.]

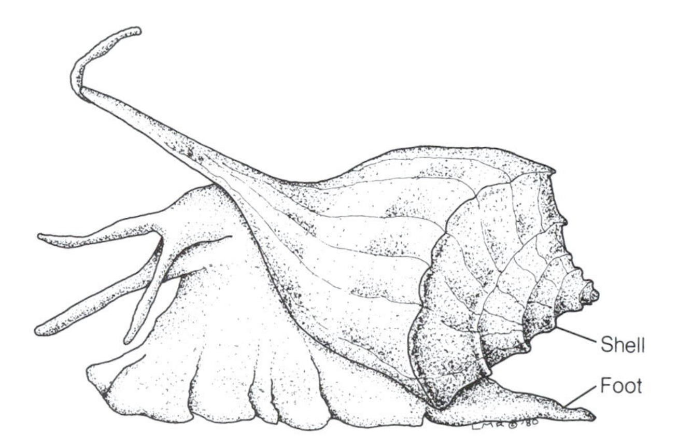

C *Busycon perversens* in motion. The gastropod moves forward by contracting vertical and transverse muscles in the foot, thereby generating waves running from front to rear (at right). [Drawing by L. M. Reeves.]

Separate sexes that shed gametes into the water are common in most molluscs. But many terrestrial and freshwater species fertilize internally. Land snails, nudibranchs, and some bivalves usually cross-fertilize; they are monoecious. Certain species—all nudibranchs—are simultaneous hermaphrodites, a single individual simultaneously produces sperm and eggs. Others are protandric hermaphrodites; an individual male develops into a female as it ages. Oysters reverse their sex, from male to female and back again, several times in a season. After courtship, a male cephalopod releases packets of sperm (spermatophores) from his penis and transfers the spermatophore by using a specialized tentacle to the female's mantle cavity.

Molluscs generally deposit eggs in gel clusters, bubble rafts, horny capsules, or sand collars. Female octopuses clean, aerate, and defend their eggs. The cephalopod *Argonauta* female produces a delicate paper case in which she lives while brooding her embryos. Free-swimming planktonic trochophore larvae hatch from fertilized mollusc eggs. In gastropods, bivalves, and scaphopods, the trochophore larvae develop into veliger larvae (Figure E) before metamorphosis to the adult form. Cephalopod species often have a juvenile stage called a paralarva between the trochophore larval stage and adulthood. Some freshwater clams and chitons brood young. Some periwinkles even "give birth" to live young. The eggs

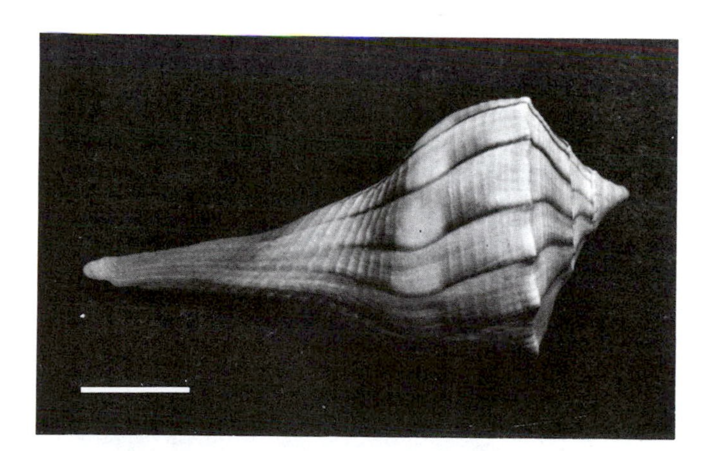

D The shell of *Busycon perversens* Linnaeus. Called the lightning whelk or left-handed whelk, the shell was used as a ceremonial cup for black drink, an emetic brewed from *Ilex vomitoria* (yaupon, a holly relative) by native Americans of the Carolinas, Florida, and Georgia. Bar = 1 cm. [Courtesy of T. Reeves.]

Mya *Mytilus* *Nautilus* *Neomenia* *Neopilina* *Pecten* *Sepia* *Teredo* *Tridacna* *Vema*

of terrestrial molluscs develop directly into adults without a free-living larval stage.

Bivalves and some snails are sedentary filter feeders; their mucus-covered gills entangle food particles and convey them to ciliated flaps, called palps, where particles are sorted, and into the nearby mouth. The hypobranchial gland produces mucus that binds particles transported by the gills. On its head, *Vema* has a tiny mouth with a U-shaped anterior lip, having a radula within (see Figure B). Many large molluscs feed on crustaceans, fish, echinoderms, annelids, and other molluscs; for example, *Busycon* levers its foot between the shells of clams and scrapes away its prey's tissues with its toothed radula (Figure F). A firm flap, the operculum, closes the shell of some retracted molluscs. The ciliated tentacles of scaphopods reach into sand for foraminiferans (Phylum Pr-4). Chitons, snails, and slugs project their toothed radulae to rasp algae and then retract the radulae into their mouths (see Figure F). These teeth are capped with magnetite, an iron oxide, in some chitons. In many bivalves and gastropods, a crystalline style releases digestive enzymes and is rotated by cilia to pull food-laden mucus strings into the stomach. The style is inside the digestive tract and is a habitat for bacteria such as the giant spirochete (Phylum B-4). Food is moved through the mollusc gut mainly by the action of cilia, except in cephalopods, and digestion is intracellular. The hepatopancreas secretes enzymes. Solid waste is discharged from the anus into the mantle cavity, from which waste leaves with the excurrent flow. The digestive tract of certain bivalves from sulfide-rich habitats is reduced, suggesting that the chemolithoautotrophic bacteria in their gills are contributing to the nutrition of the host mollusc.

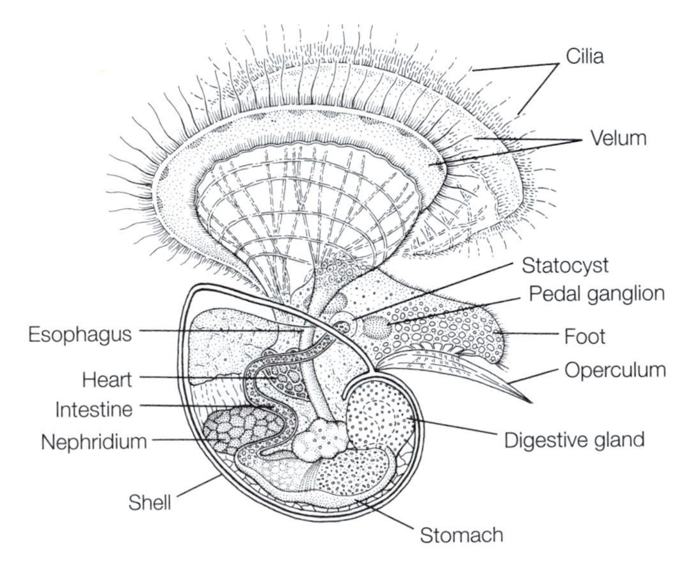

E Veliger larva of a gastropod mollusc; lateral view in swimming position. Generalized but based on *Crepidula*. [Drawing by L. Meszoly; information from K. E. Hoagland.]

Aquatic molluscs exchange gases through the body surface. Gills (ctenidia) usually lie within the mantle cavity between mantle and viscera (see Figure F), function in gas exchange, and may also sort food, depending on the species. Monoplacophorans have multiples of gills, multiple auricles and ventricles (heart chambers) and multiple nephridia—features that relate these segmented molluscs to annelids. The spectacular dorsal tufts of nudibranchs exchange respiratory gases and contain extensions of the digestive tract. The moist mantle cavity of some garden and freshwater snails functions as an air-breathing, highly vascularized lung. In most molluscs, a dorsal heart pumps blood through vessels and an extensive hemocoel. Cephalopods have a completely closed circulation with capillaries, arteries, and veins; squids have accessory hearts as well, and these accessory hearts pump blood to the gills in series with the main heart. Scaphopods lack hearts, gills, and blood vessels; the scaphopod mantle cavity is the respiratory surface with foot contractions moving blood through sinuses—spaces in the tissues. The mollusc kidneys (metanephridia) collect soluble waste from the coelom and discharge it into the mantle cavity through the renal pore.

The mollusc nervous system generally—except for that of cephalopods—consists of three to five pairs of ganglia with connections between them. Balance organs called statocysts are found in pelagic (ocean-dwelling) molluscs. Incoming water passes a chemoreceptor called the osphradium (Greek, *osphra,* a smell) before it reaches gills. The cephalopod brain is more complex than that of any other invertebrate; cephalopods have a complex memory and learning behavior, as well as image-forming, focusable eyes with cornea, lens, iris, and retina. Giant nerve fibers of squid synchronize jet-propelled escape responses.

Molluscs are preyed on by many animals—toothed whales, haddock, walrus, cod, and the sea star *Asterias* among them—and hundreds of mollusc species are sources of food for humans. The first recorded aquaculturist, Sergius Orata, a Roman, cultured oysters in the first century B.C. Tools, trumpets, sacred and decorative objects, cameos, and buttons are carved from mollusc shells. Freshwater mussels and pearl oysters secrete pearl over irritating particles. The internal calcareous shell (pen) of a cephalopod, the cuttlefish *Sepia,* is used as the cuttlebone on which caged birds condition their beaks; the cuttlefish is benthic (bottom dwelling) and uses its cuttlebone to regulate buoyancy. Sepia, a brown pigment used by artists, is prepared from cuttlefish ink. A yellow secretion of *Murex,* a marine gastropod, was the source of the dye royal, or

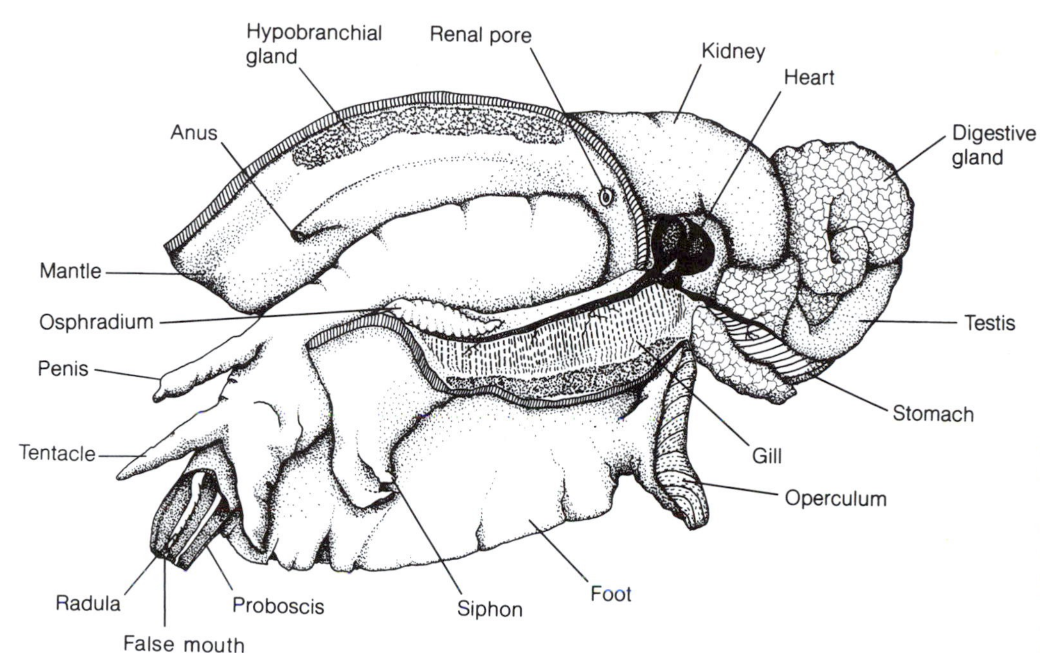

F Internal morphology of *Busycon perversens,* male. The osphradium (Greek *osphra,* a smell) is a chemical and touch receptor that hangs next to the gills in in-flowing water. The true mouth is at the base of the proboscis. [Drawing by L. M. Reeves.]

Tyrrhenian, purple used by Phoenicians. Tusk shells served as wampum for native Americans near the West Coast. East Coast native Americans drilled "roanoke," a hard substance used as money, from cockles. *Mercenaria,* the familiar edible clam, or quahog, was the source of purple wampum. Abalone shells were so widely used for ornaments, including bracelets, that trade routes have been traced by their distribution from California into the interior.

Some freshwater snails are intermediate hosts of *Schistosoma* (Phylum A-5)—a trematode parasite—giving rise to the disease schistosomiasis (bilharzia). Oyster drills and rock- and wood-boring bivalves, such as shipworms, riddle pilings and wooden ships. *Dreissena,* zebra mussel, carried to the United States from Europe in ship ballast water, clogs water pipes. Land slugs and snails can be garden pests.

Because shells preserve well, molluscs are abundantly documented in the fossil record. Monoplacophoran shells are abundant in the middle Cambrian period. Bivalves appear by the middle Cambrian; chitons, cephalopods, and gastropods by the Upper Cambrian. The oldest tooth-shell fossils appear in the middle Ordovician, about 450 million years before the present. Octopuses appear in the fossil record rather recently—about 65 million years ago, in the Cretaceous. Squids, the most modern cephalopods, appear later, in the Tertiary.

The relationship of molluscs to other phyla is debated. Molluscs have a true coelom, reduced to the small pericardial cavity around the heart and gonad. Although adult annelids, flatworms, and molluscs appear very different from one another, most marine molluscs and annelids pass through a swimming, ciliated trochophore larval stage; for this reason, molluscs are believed by some zoologists to share ancestors with annelids (Phylum A-22) and flatworms (Phylum A-5). Digestion within cells rather than in a stomach cavity, the plan of the nervous system, external cilia, and gliding and creeping with ventral waves relate molluscs more strongly to flatworms than to annelids. The monoplacophoran *Vema* and the closely related *Neopilina* are thought by researchers to closely resemble ancestral middle Cambrian fossil molluscs. Others believe that the segmentation of living monoplacophorans derived from secondary replication of body parts and that ancestral molluscs may have been soft-bodied organisms like their trochophore larvae. The intriguing theory that molluscs may be closely related to sipunculans (Phylum A-23) is discussed in the sipunculan essay.

A-27 Tardigrada
(Water bears, tardigrades)

Latin *tardus,* slow; *gradus,* step

EXAMPLES OF GENERA

Archechiniscus
Batillipes
Coronarctus
Echiniscoides
Echiniscus
Halobiotus
Hypsibius
Macrobiotus
Milnesium
Pseudechiniscus
Tetrakentron
Thermozodium

Because of tardigrades' pawing locomotion, the nineteenth-century English naturalist Thomas Huxley called them water bears, a name that stuck. Tardigrades range in length from 50 to 1700 μm and are typically from 100 to 500 μm long. Their color—red, purple, blue, olive, yellow, brown, or translucent—depends in part on their food; an example is a species colored bright orange by carotenoid pigments in the lichen on which it feeds. Tardigrades walk on four pairs of nonjointed, stumpy legs, each having moveable claws, pegs, or adhesive disks. The head and four body segments of the tardigrade are generally covered with chitinous cuticle, highly sculptured in some species or smooth or forming plates but lacking cilia. The *Echiniscus* in the adjoining photograph is an armored form, having cuticular dorsal plates typical of terrestrial tardigrades belonging to Class Heterotardigrada (Figure A). Marine heterotardigrades lack plates. Most other tardigrades belong to Class Eutardigrada and are naked, lacking plates, such as *Macrobiotus* (Figure B). Eutardigrades can be distinguished by their complexly surfaced egg shells. Most eutardigrades are freshwater or terrestrial; a few are marine. The third class—Mesotardigrada—consists of *Thermozodium,* from hot springs. Retractible sharp feeding stylets protrude from the mouth; these rigid, slender protrusions are newly secreted at each molt of the cuticle. Although tardigrades are probably coelomates, the coelom is limited to a cavity around the gonad only. As early as 1776, the Italian biologist Lazzaro Spallanzani observed that tardigrades possess remarkable powers of surviving extremes of heat and drying by entering dormancy, a state called cryptobiosis ("hidden life").

As many as 750 species in 92 genera—depending on who is counting—of tardigrades have been described from the Arctic, the Antarctic, the Tropics, and hot springs. *Coronarctus* is a deep-sea water bear; its reduced legs are an adaptation to its interstitial environment. All water bears are aquatic. Most live in water films on moss, liverworts, lichens, algae, roof shingles, and forest litter—habitats that take up water. Miniature ponds captured in the spiral of the bromeliad leaf base also are a tardigrade habitat. Shore and underwater fauna of lake and marine sands—interstitial fauna—commonly include tardigrades. Although tardigrades cannot swim, they require moisture to lumber about and to exchange gases. Many species are cosmopolitan in unpolluted habitats. Counts reveal as many as 300 tardigrades per square centimeter of alpine meadow dug down to 10 cm. *Tetrakentron synaptae* is commensal on sea cucumber tentacles (Phylum A-34); another—*Echiniscoides sigismundi*—lives in the mantle cavity of *Mytilus edulus,* edible mussel (Phylum A-26). Moss (Phylum Pl-1) is not a benign habitat for water bears; predatory fungi attack and consume water bears, rotifers, and nematodes in moss cushions.

Gases and food diffuse in water bear body fluids. Oxygen and carbon dioxide diffuse directly through their permeable cuticles; tardigrades have no respiratory organs. A long tubular dorsal heart is located in the main body cavity; that cavity may be either a hemocoel—cavity containing blood—or a pseudocoelom and is separate from the coelom. Tardigrades have both striated and smooth muscles. Legs and body are flexed by thin muscles attached to the underside of the cuticle in opposition to the pressure of

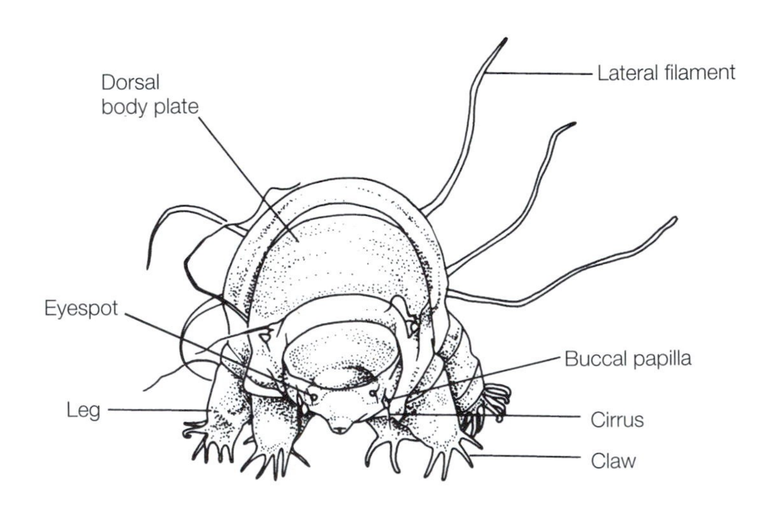

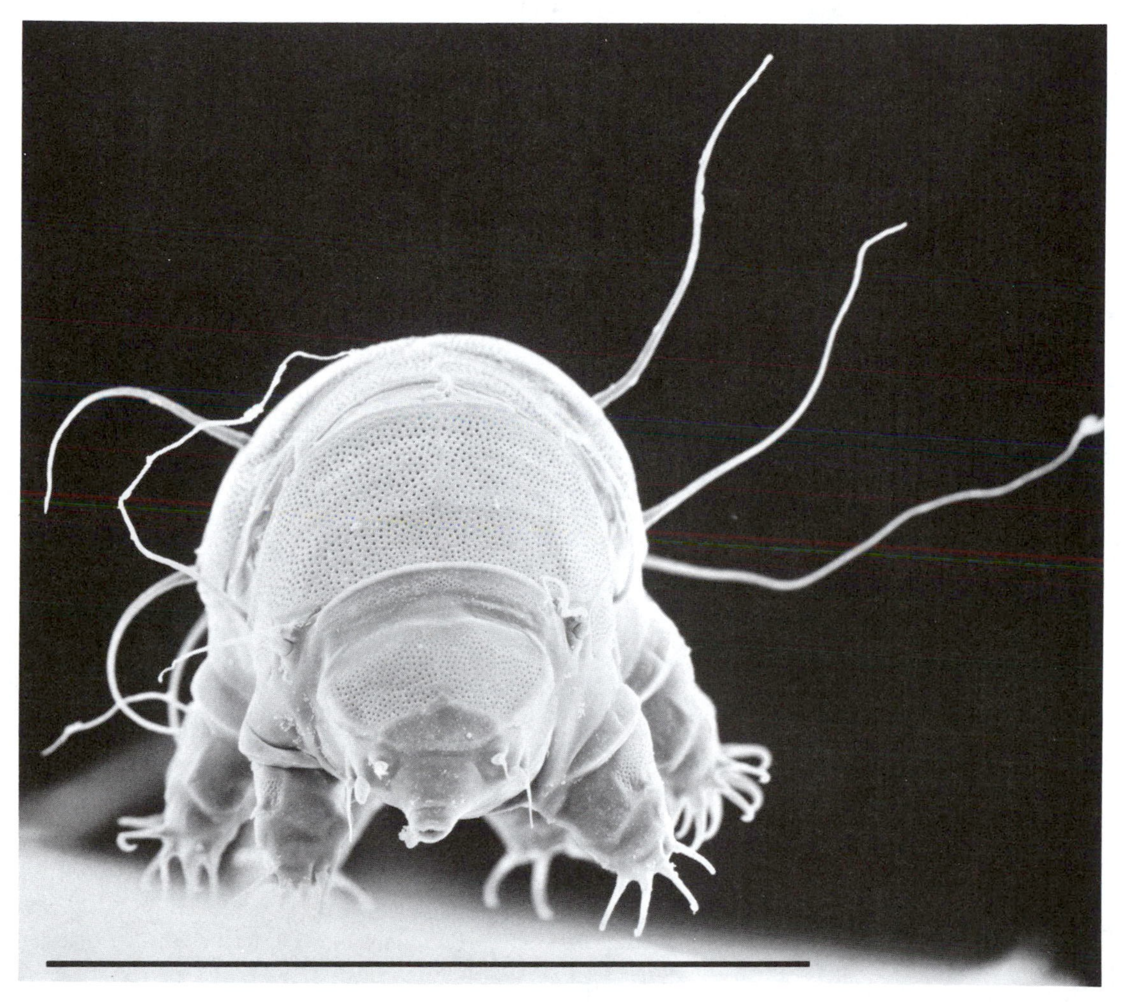

A *Echiniscus blumi,* a land-dwelling water bear from Auburn, Placer County, California. Its moveable claws enable it to cling to moss and lichens. SEM, bar = 0.5 mm. [Photograph courtesy of R. O. Schuster; drawing by I. Atema; information from R. O. Schuster.]

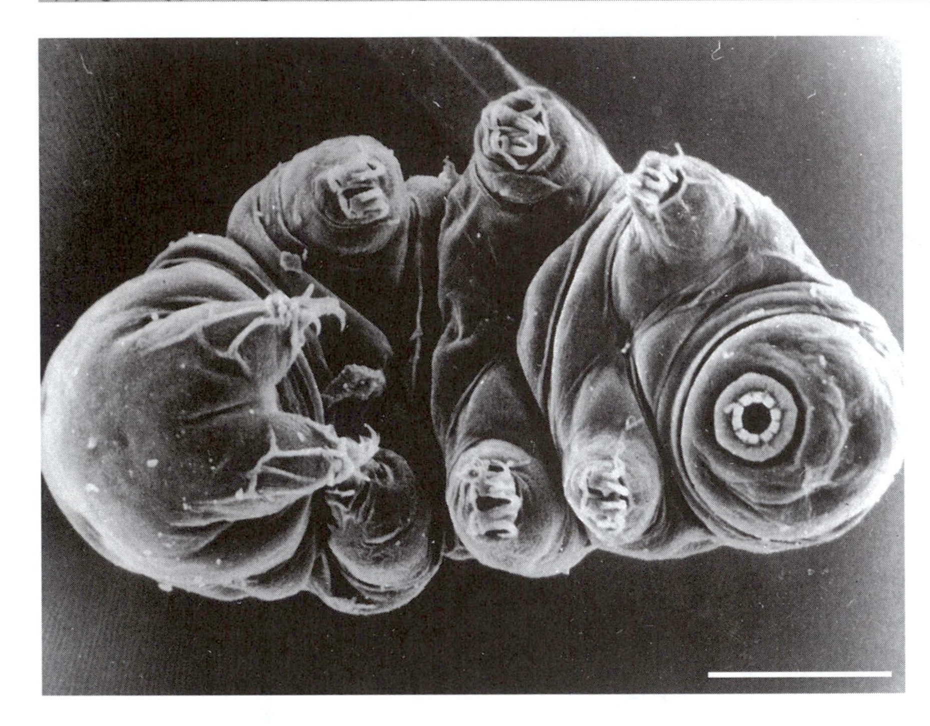

B Ventral view of *Macrobiotus sp.*, showing its stumpy nonjointed legs and round mouth. This tardigrade was collected from a pond in Athens, Georgia. SEM, bar = 0.1 mm. [Scanning electron micrograph courtesy of Jerome J. Paulin.]

body-cavity fluid; water bears lack circular body-wall muscles. Tardigrades ingest liquid food or detritus. Some pierce plant cells with their protrusible stylets, ingesting the liquid with suction generated by the muscular pharynx, visible through the transparent cuticle of many naked eutardigrades. Tiny predaceous tardigrades pierce and suck the body fluids of rotifers, nematodes, and even other tardigrades. The foregut consists of a buccal tube, a pharynx, and an esophagus. A straight midgut (eutardigrades) or midgut with diverticuli (heterotardigrades) absorbs nutrients and opens into a hindgut that collects solid waste. The Malpighian tubules of freshwater eutardigrades and the marine eutardigrade *Halobiotus* are absent in heterotardigrades; these dorsal glands, supposedly excretory, open into the rectum. The Malpighian tubules of tardigrades may not be homologous to those of arthropods. The heterotardigrade gut terminates in an anus. The eutardigrade gut terminates in a cloaca, which receives both undigested waste and gametes. In heterotardigrades, ventral glands may excrete materials from body-cavity fluid to the cuticle. The nervous system consists of a dorsal brain linked to a subesophageal ganglion. Five ventral ganglia connect by a pair of longitudinal ventral nerves. Most tardigrades have a pair of red or black eyespots. Heterotardigrades have filaments, hairs, spines and other sensory receptors adorning their cuticles; these adornments are not found in eutardigrades.

Tardigrades can survive desiccation. Water, which makes up about 85 percent of their weight, drops to 3 percent in a desiccated tardigrade—one with a lowered metabolic rate. These soil and moss-dwelling water bears develop into dry, barrel-shaped, motionless forms called tuns because of their resemblance to wine casks. They can survive temperatures as high as 151°C and as low as –272°C, nearly absolute zero. In this cryptobiotic state, they survive for as long as 100 years. Tardigrades must synthesize the membrane protectant trehalose—a sugar—and glycerol to preserve the structure of cells and revive from desiccation, the most common cause of cryptobiosis. Tardigrade tuns are radiation resistant—for humans, the dose of X-rays that kills 50 percent of those exposed is 500 roentgens; for tardigrades, the dose is 570,000. When surroundings gradually become unfavorable, freshwater and some moss-loving tardigrades form thick-walled cysts, different from tuns. Inside the cyst, the legs contract, the water bear shrinks away from its cuticle, its internal organs may degenerate, and metabolism drops, though not to the degree that it drops in tuns. Before the cyst forms, the tardigrade accumulates food reserves. When the environs become favorable, the tardigrade can regenerate. Cysts (but not tuns) may form in aquatic environments.

A single ovary or testis lies dorsal to the intestine. Although the sexes are usually separate, the males and females are easy to distinguish in only a few species. Most tardigrades are female. The reproductive strategies of few freshwater species are known. In species that reproduce sexually, fertilization may be internal or external. In some freshwater tardigrades, the male injects sperm into the space between the new and old cuticles; the female then molts the cuticle and lays from 3 to 35 eggs in the shed cuticle, where external fertilization takes place. External seminal receptacles receive the sperm of marine tardigrades. The terrestrial tardigrade male injects sperm into the female's body cavity or cloaca for internal fertilization. A few species of *Pseudoechiniscus* and *Echiniscus* reproduce parthenogenetically: females lay eggs that grow into females without male intervention.

Some species lay eggs of two types: with thin or with thick shells. When conditions for growth are unfavorable, thick-shelled eggs are produced and development is deferred until conditions favor it. Sticky or sculptured eggs can be found attached to moss, algae, and bark. Fertilized eggs of eutardigrades develop directly into miniature adults; heterotardigrades have a brief larval stage. After about 2 weeks of development under favorable conditions, miniature water bears of some species rupture their egg shells with their stylets. After hatching, tardigrades grow by a series of molts and enlargement of existing cells as well as by mitotic division to increase the number of cells.

The middle Cambrian *Aysheaia,* a fossil from the Burgess shale, now considered an onychophoran, may be an ancestor of tardigrades. The first known fossil tardigrades also were discovered in Cambrian rocks formed 530 million years before the present. Tardigrades have left one species—*Beørn*—fossilized in Cretaceous amber. Tardigrades are considered to be related to the arachnids of Phylum A-19, particularly the mites; tardigrades and mites have in common a terminal mouth with stylets, four pairs of legs, no cilia, ventral nerve cords, and segmented bodies. Like mites and other arthropods, tardigrades probably evolved from soft-bodied, segmented annelid-like ancestors (Phylum A-22). Tardigrades have been grouped with pentastomids (Phylum A-21) and onychophorans (Phylum A-28), both of uncertain relationship to arthropods. Evidence suggests that, after the evolution of arthropods from a segmented worm ancestral to both annelids and arthropods, tardigrades and onychophorans diverged from the main arthropod lines. On this basis, some propose the grouping of tardigrades and onychophorans in a subphylum Lobopodia of arthropods.

Some researchers considered tardigrades to be related to nematodes (Phylum A-10) and rotifers (Phylum A-13); members of these phyla greatly resist dehydration and cold. The phylogenic relationships of water bears are puzzling because similarities such as dessication tolerance may indicate phylogenetic affinities or convergent evolution in independent lineages.

Tardigrade biologists continue to debate, but there is considerable structural evidence that major events in the evolution of arthropods independently produced Uniramia, Crustacea, Chelicerata, and other groups, perhaps Tardigrada. A possible polyphyletic scheme for arthropods shows tardigrades within Uniramia (with myriapods, hexapods, and onychophorans). Ultrastructural evidence supports placing onychophorans as intermediate forms in a line from annelids and arthropods, with tardigrades branching off independently at some point along the progression.

A-28 Onychophora

(Velvet worms, onychophorans, peripatuses)

Greek *onyx*, claw; *pherein*, to bear

EXAMPLES OF GENERA
Cephalofovea
Epiperipatus
Macroperipatus
Mesoperipatus
Ooperipatus
Opisthopatus
Peripatoides
Peripatopsis
Peripatus
Speleoperipatus
Symperipatus

Onychophorans are commonly known as velvet worms; their cuticles are studded with minute bumps that feel like velvet. Velvet worms crawl like caterpillars, raising each pair of legs in waves, along with a wave of body contraction (Figure A). Paired claws at the tip of each little foot are the basis of their phylum name, Onychophora. Velvet worms usually walk on walking pads; on rough, hard substrates, they extend their claws. Onychophorans are just 14 to 200 mm long (*Macroperipatus*) and may be mistaken for arthropods (Phyla A-19 through A-21) or annelid worms (Phylum A-22). Females in one species may be 50 percent longer and weigh twice as much as males. The onychophoran bodies may be iridescent green, blue black, orange, red, or whitish, although most are brown. The onychophoran walks at a rate of less than 1 cm/sec with its 14 to 43 pairs of unjointed, hollow legs. Like the muscles of annelids, its circular, longitudinal, and diagonal body-wall muscles are smooth (nonstriated). These muscles work against the hydraulic skeleton to move the velvet worm. Vascular (hemal) channels that encircle the velvet worm body like wire in a vacuum cleaner hose are unique to velvet worms. Hydrostatic pressure maintains the firmness of the legs as leg muscles bend and shorten the limbs; a valve at each leg base enables each leg to be extended independently by altering the pressure in the hemocoel, the main body cavity. Velvet worms are coelomates, but their coeloms are vestigial, having been reduced to gonoducts and tiny sacs surrounding the nephridia.

Velvet worms are terrestrial. The thin chitinous cuticle of the onychophorans permits water loss—to resist desiccation they require high-humidity habitats such as forest litter, the underside of logs and rocks, bromeliads, and the tunnels of termites. Velvet worms are believed to be unable to make tunnels themselves even in soft substrate but can reduce the diameter and increase the length of their soft bodies to fit into small spaces. During rain or at night, onychophorans venture forth to hunt or mate, avoiding drying by sunlight. Some hunt partly exposed from their burrow entrances. Certain species hunt in tree foliage. *Speleoperipatus* in Jamaica and *Peripatopsis alba* in South Africa inhabit caves.

About 10 genera and 100 species have been described. Onychophoran species fall into two natural groups: a northern group in tropical India, the Himalayas, West Equatorial Africa (*Mesoperipatus*), tropical America (*Epiperipatus*) as far north as Mexico, the West Indies (but not Cuba), and Malaya; and a southern group in New Guinea, temperate Australia (*Symperipatus*), New Zealand (*Peripatoides* and *Ooperipatus*), Tasmania, Madagascar, South Africa (*Peripatopsis*), and Chile. Recently, this dichotomy has been

A *Speleoperipatus speleus,* a blind and unpigmented onychophoran, or velvet worm, taken from a cave in Jamaica. This troglodyte (cave-dwelling) species lacks eyes; other nontroglodytic onychophoran species have eyes. Bar = 5 cm. [Courtesy of R. Norton.]

explained as being linked to Jurassic-Pliocene biogeography. These two onychophoran groups diverged some 200 million years ago, when a vast desert separated the northern and southern groups on the ancient southern continental mass called Gondwana. Contemporary distribution of onychophorans corresponds to the modern separated continental remnants of Gondwana. In fact, the fossil record of onychophorans, which traces from the middle Cambrian period, has been used to reassemble the historical pattern of drifting continents.

The onychophoran ventral nervous system consists of two eyes—one at the base of each antenna, a brain, and longitudinal ventral cords having transverse connections but lacking the ganglia (bundles of nerve cell bodies) found in arthropods and annelids. Onychophoran eyes have a chitinous lens and retina and are used to direct viscous adhesive directed at prey and predators. Velvet worms avoid light between 470 and 600 nm; this photonegative behavior may protect them from desiccation. Cave-adapted *Speleoperipatus* and *Peripatopsis alba* are eyeless and lack body pigment. Sensory bristles on body papillae and sensory antennae orient the velvet worm to touch and perhaps to water vapor.

All onychophoran species for which the diet is known are carnivores, attacking and eating isopods and spiders (Phylum A-19), crickets and termites (Phylum A-20), and molluscs (Phylum A-26). When hunting or disturbed, velvet worms squirt secretions from adhesive glands (Figure B), modified nephridia that open through perforations in oral papillae beside the mouth. This spray congeals into bitter, elastic, sticky white threads, entangling and immobilizing their prey. An onychophoran holds prey to its mouth by sucking, then slices off bits with bladelike jaws, and liquifies tissues inside the prey with saliva secreted from salivary glands behind the jaws. While waiting for prey to be liquified, the hunter consumes much of the proteinaceous threads and then sucks in its liquid diet. Food passes through the mouth, pharynx, and esophagus; the internal organs are suspended by mesenteries within the body cavity. The midgut secretes a tubular peritrophic membrane—which encloses the food—and deposits uric acid on the inner side of the peritrophic membrane. Food is digested (whatever was not predigested in the body of the prey) and absorbed across the peritrophic membrane in the intestine, whereas waste and uric acid remain within the membrane and both are expelled—still in the membrane—from the gut through the terminal anus by the onychophoran's swallowing air. Nephridia collect dissolved waste of unknown composition, which is discharged by contractile bladders at the leg bases. Onychophorans decrease water loss by nocturnal

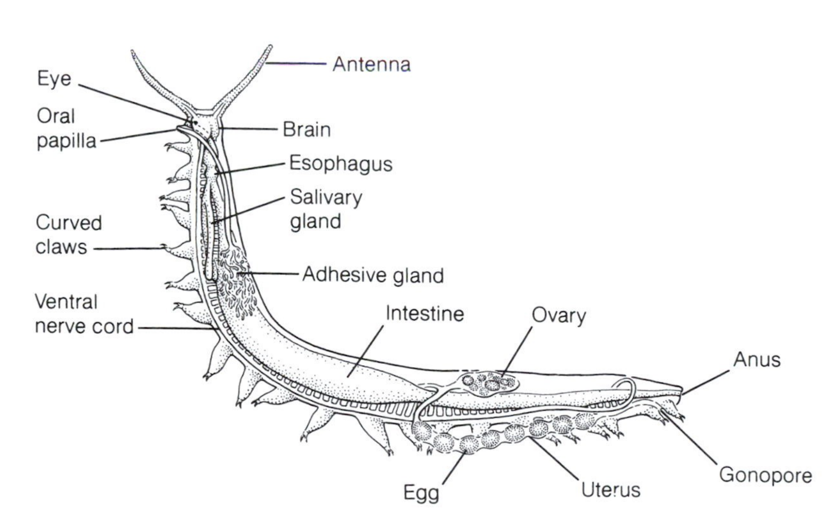

B Cutaway drawing of a female velvet worm. The paired claws are extended when the velvet worm grips or climbs. [Drawing by L. Meszoly; information from R. H. Arnett.]

hunting, by occupying humid habitats, and by resting in pairs in body contact. Young velvet worms have been observed riding on their mother's back.

Like the ostia in an arthropod heart, the muscular tubular dorsal heart has slits in each body segment. Pumping of the heart and body movements circulate colorless blood in the body cavity—which is a hemocoel—and in subcutaneous hemal channels; circulation is open. The hemal channels afford body firmness, which is essential for movement. Oxygen enters over the whole body through the thin cuticle and is transported from hundreds of spiracles (surface pores) to internal organs by tracheae—fine canals that carry oxygen directly to tissues. The spiracles of onychophorans cannot be closed to control water loss, unlike the spiracles of insects. Some species may also respire with thin-walled, fluid-filled, contractile, eversible sacs on their legs. These sacs function roughly as minilungs.

The sexes are separate, and each velvet worm has a pair of testes or ovaries. Females (see Figure B) are usually larger than males. Larger females bear more offspring. In some species, fertilization is external, in others, internal. The *Peripatopsis capensis* male deposits spermatophores—packets of sperm—on the sides or back of the female. Beneath each spermatophore, the cuticle of the female erodes; eventually sperm probably travel through her hemocoel to her ovaries. In other species, the male impregnates the female by injecting sperm through her body wall. The male of *Cephalofovea tomahmontis* moves spermatophores into his cephalic pit, perhaps bending to transfer spermatophores from his genital pore to the pit on his head; the female of that species assists in the mating by pressing the male head to her cephalic pit with her legs. After mating, the blood amebocytes of the female may break down the skin of the female's cephalic pit, providing a path for sperm to travel through her hemocoel fluid to the eggs in her reproductive tract. In any case, she stores sperm in her uterus until they are needed to fertilize eggs. The oviduct is the site of fertilization in species for which it is known. A given female may be impregnated once in her life and may reproduce more than once. Development time is from 6 to 17 months for eggs and from 11 to 13 months for viviparous velvet worms; development is direct. Many tropical species are viviparous; the female nourishes her tiny embryos internally by means of a true placenta attached to the uterine wall. In viviparous velvet worms, material taken from the maternal hemocoel is stored in the placenta and released into the embryo cavity. The viviparous female bears half-inch long live young through the gonopore located between the last or penultimate leg pair. Placental development is an example of convergent evolution of placental mammals and onychophorans. In oviparous species such as *Ooperipatus*, shelled fertile eggs are laid that develop outside the female's body. In species that are ovoviviparous, shelled eggs are retained in the female until the embryos hatch. Ovoviviparous onychophorans are found in temperate areas where food supply and climate are less stable. Females in a population of *Epiperipatus imthurni* in Trinidad are parthenogenetic.

Onychophorans may link mandibulate arthropods (Phylum A-20) and annelids (Phylum A-22). Another view is that onychophorans are a sister group of annelids and that the onychophoran lineage originated from segmented ancestors. Burgess shales from British Columbia contain a middle Cambrian fossil now considered to be an onychophoran—*Aysheaia pedunculata* Walcott. Newly discovered fossilized mandibles of onychophoran-like animals in Yunnan, China, strengthen the evidence that these fossils are onychophorans.

All three groups—arthropods, annelids, and onychophorans—have chitinous cuticles secreted by the epidermis, although the cuticles of many arthropods are nearly impervious. The velvety skin of onychophorans covered with minute papillae is unique to velvet worms. Developmental patterns are similar in arthropods and onychophorans. Both have tracheae—tubes that facilitate gas exchange between tissues and the environment by openings through the body wall (spiracles). Jaws of arthropods and of onychophorans are derived from the differentiation of appendages. The onychophoran has only a single pair of jaws; among the arthropods, the crustacean has more than a single pair of jaws (mandibles and maxillae), chelicerates lack mandibles, and insects have mandibles and maxillae. An open (lacking capillaries) circulatory system with a hemocoel and a tubular, dorsal heart with ostia is similar in arthropods and onychophorans. Velvet worm adhesive secretions—used in defense and to capture prey—resemble those of millipedes and centipedes (Phylum A-20). Jaw muscles are striated in onychophorans and arthropods; onychophoran muscle other than jaw muscle is smooth.

Onychophorans also resemble annelids. Members of both these phyla molt their flexible, thin cuticles in patches; however, the onychophoran cuticle is unsegmented except for the antennae. Both onychophorans and annelids have segmentally arranged, paired nephridia that open at the base of each leg (in each annelid segment); onychophorans lack the Malpighian tubules—the principle excretory organs—of insects, centipedes, and millipedes. Eyes, when present in annelids, are ocelli, like those of onychophorans; thus onychophoran eyes resemble those of annelids more than compound insect eyes. Both annelids and onychophorans have ciliated reproductive tubules. Their chief internal organs, suspended in mesentery, are arranged similarly; onychophorans lack the internal septa of annelids. The fine details of sperm structure of annelids and onychophorans are more similar to each other than to those of arthropods. The skeleton of annelids and onychophorans (and centipedes) is hydraulic. Appendages are unjointed in onychophorans and annelids. A comparison of the mechanics of locomotion in polychaete annelids and onychophorans suggests that polychaetes are probably not direct ancestors of onychophorans, although onychophorans probably arose from annelids. This view is supported by the evidence that onychophoran legs lack joints and internal stiffening bars and are quite different from the appendages of polychaete annelids.

Analyses of mitochondrial ribosomal RNA sequences suggest that velvet worms are highly modified chelicerates or are related to chelicerates, crustaceans, and insects. This interpretation contradicts the other current major hypotheses of onychophoran evolutionary affiliations based on morphological, behavioral, and physiological evidence. On the basis of the present data of reproductive features, biogeography, and phylogeny, it appears that onychophorans may be placed between the polychaete annelids and the arthropods.

A-29 Bryozoa

(Ectoprocta, ectoprocts, moss animals)

Greek *bryon,* moss; *zoion,* animal
Greek *ektos,* outside; *proktos,* anus

EXAMPLES OF GENERA
Alcyonidium
Bugula
Cristatella
Fredericella
Lichenopora
Membranipora
Pectinatella
Plumatella
Selenaria
Stomatopora
Tubulipora

Bryozoans, also called ectoprocts or moss animals, are primarily sessile filter feeders. Like brachiopods (Phylum A-30) and phoronids (Phylum A-31), bryozoans have tentacle-bearing organs called lophophores filled with fluid (Figure A). The lophophore is protruded for food-capturing and gas exchange. Each individual bryozoan is called a zooid. Most ectoprocts are colonial—several million individual zooids can form extensive gelatinous, bushy colonies or hard mats on algae, rocks, ship hulls, driftwood, and rafts. Because bryozoan colonies look much like seaweeds or moss, many people are not aware that ectoprocts are animals.

A A single living zooid of *Plumatella casmiana,* showing the retractile horseshoe-shaped collar, the lophophore, from which ciliated tentacles originate. LM, bar = 0.5 mm. [Courtesy of T. S. Wood.]

All but 50 of the 4000 species are marine, and all are aquatic. Marine bryozoans generally live in the intertidal zone of the beach and below the tide line, as well as on the seafloor to considerable depths but not in the abyss. *Membranipora* colonies grow on kelp fronds below low tidemark. Freshwater bryozoans live, with the appearance of fuzzy jelly, on surfaces such as water-lily leaves and submerged lakeside roots in slow-moving streams. In an ectoproct colony, each zooid is usually less than 1 mm long; colonies may be as large as 0.5 m in diameter. A retractor muscle anchors the soft parts—the individual zooid—to the conspicuous nonliving outer covering of the colony. This covering consists either of (1) a chitinous membrane alone or (2) a chitinous layer with an underlying thick, rigid skeleton made of calcium carbonate or (3) a gelatinous layer, depending on the species.

Members of this phylum are grouped in three classes. Class Phylactolaemata comprises most freshwater ectoprocts, such as *Plumatella* in our illustrations. In this class, the body cavity is incompletely divided into three parts (not visible in Figure B). Most phylactolaemates are sessile, but a *Cristatella* colony can creep with its muscular shared foot. The lophophore is U-shaped in this class. Almost all freshwater ectoprocts produce asexual buds called statoblasts. Statoblasts develop in summer on the funiculus (a tissue cord connecting zooids) as armored balls of cells; the statoblast may have a float and hooked spines (see Figure B). In the fall, statoblasts are dispersed by animals (bird feet), wind, and water as the colony disintegrates. Statoblasts survive desiccation and winter cold, each enclosed by a pair of chitinous convex valves; in spring, the valves open and a zooid grows from the dispersed statoblast. Statoblast formation is unique to phylactolaemates; formation of this resting overwintering stage is ecologically comparable to tardigrade (Phylum A-27) resting eggs, rotifer (Phylum A-13) resting eggs, and freshwater sponge gemmules (Phylum A-2).

Class Stenolaemata comprises marine species, such as *Tubulipora,* having tubular, calcified zooids and circular lophophores. No other class has tubular, calcified zooids. The calcified tube lacks an operculum—the cover to the opening through which the lophophore sticks out. Class Gymnolaemata has circular lophophores, and some species have opercula. Most extant marine ectoprocts, such as *Bugula,* and a few freshwater species are grouped in this class. The walls of the zooids in this class may be gelatinous, membranous (chitinous), or calcified. The body cavities of bryozoans are of three forms: (1) gymnolaemate zooids exchange body fluid through a porous plate between neighboring zooids; (2) in phylactolaemates, a common body cavity is shared; and (3) in stenolaemates, a membrane divides the body cavity into two spaces. Gymnolaemate colonies contain polymorphic zooids, individuals of various shapes and sizes specialized for feed-

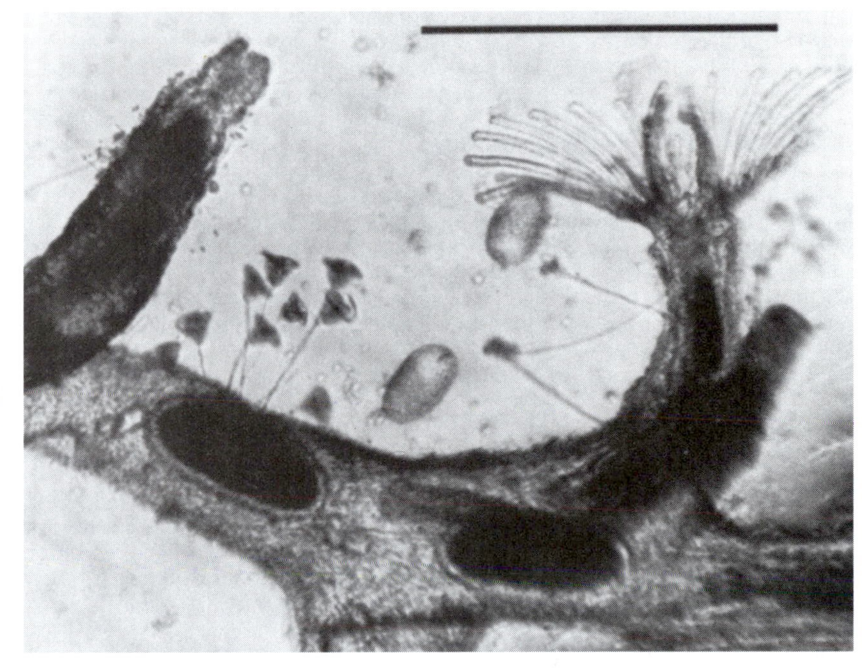

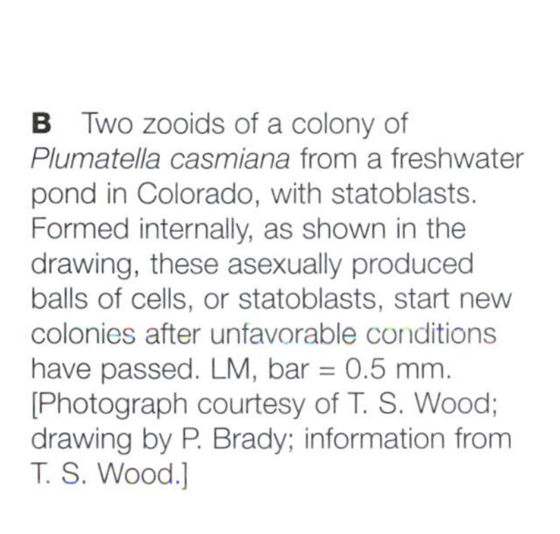

B Two zooids of a colony of *Plumatella casmiana* from a freshwater pond in Colorado, with statoblasts. Formed internally, as shown in the drawing, these asexually produced balls of cells, or statoblasts, start new colonies after unfavorable conditions have passed. LM, bar = 0.5 mm. [Photograph courtesy of T. S. Wood; drawing by P. Brady; information from T. S. Wood.]

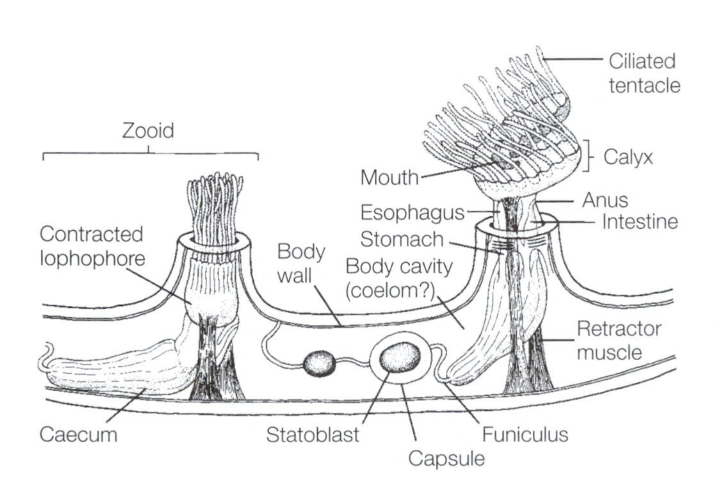

ing, reproduction, cleaning, or active defense. Nonfeeding zooids shaped like tiny beaks are thought to defend their colony. Polymorphism is absent in phylactolaemates. Not all bryozoans are sessile; free-living *Selenaria* with specialized setae can zoom along at 1 m/hour.

Within a colony, each larva is attached to substrate by secreted adhesive and remains attached through adulthood. Each mature zooid is interconnected physically to the other zooids in the colony by mesenchymal tissue called the funiculus (see Figure B). The upright branching *Bugula* colonies attach to substrate with a rhizoid—a rootlike structure. Each zooid has a U-shaped digestive system within the body cavity. Both the mouth and the anus face upward toward the surface; the mouth opens within the lophophore and the anus opens outside the lophophore and just beneath it. A water hydraulic system extends from the body cavity into the ring of hollow tentacles; body-wall muscles contract against the incompressible fluid in the body cavity, forcing the tentacles to protrude. Cilia (and possibly touch receptors) stud the tentacles of the lophophore. With a coordinated beat, cilia waft currents bearing phytoplankton and bacteria into the mouth. Food passes through an esophagus and a stomach, which leads downward to a large cavity called the caecum. Food is digested by intracellar and extracellular modes in the large stomach. The intestine leads back up to the anus, which discharges solid waste. Gas exchange takes place directly from all soft tissues of the zooid to water; adults lack respiratory, circulatory, and excretory organs.

Adults of marine and freshwater bryozoans normally produce a colony by the asexual budding of zooids. A single zooid can form an entire colony—several million contiguous zooids budding from a single ancestor may make up one colony, each genetically identical to the ancestrula (ancestor to the colony by budding). Freshwater ectoprocts bud and reproduce sexually as well. Their zooids typically are protandric hermaphrodites, which means that a given zooid can form sperm and then eggs in sequence but not at the same time. Some zooids are male and supply sperm to females in the colony. Sperm discharged into the body cavity leave through pores in the lophophore. Nearby female zooids somehow gather sperm. Zooids in the colony are usually differentiated such that the lophophore and digestive tract degenerate to provide space for the developing eggs in the actively reproducing members. Some ectoprocts brood large yolky embryos in their tentacles or body cavity. Others brood developing embryos externally in a modified, nonfeeding zooid called an ovicell; the ovicell differs in shape and size from a zooid. Eggs escape from the body cavity either through a pore called the coelomopore or through a pore that is borne on the end of a special organ within the ring of tentacles.

Marine ectoprocts, like freshwater forms, can reproduce both sexually and asexually. Some marine species shed small eggs and sperm directly into the sea. *Bugula* broods its eggs in ovicells on the outside of the calcareous case of the zooid. Fertilized eggs cleave radially and form larvae. For example, *Bugula* produces beautiful coronate (crown-shaped) larvae; these larvae do not feed, though those of a few species do feed. All larvae have a girdle or crown of cilia used for swimming, an anterior tuft of long cilia, and a posterior adhesive sac. At first, larvae are positively phototactic and swim toward light; thus larvae escape the brood chamber and disperse. Older larvae become negatively phototactic; they tend to settle in shadows. Many are specifically attracted to certain substrata, such as *Sargassum* seaweed (Phylum Pr-17). During settling, the adhesive sac everts and secretions fasten the animal to the substratum. After attachment, all larval structures retract, larval tissue is completely resorbed, and a single adult, the ancestrula, develops. Because of this thorough remodeling of larval tissue, we cannot know if the adult body cavity is a true coelom.

Ectoprocts play a minor role in encrusting ships hulls—molluscs and tube-building annelids are the primary villains. Similarly, corals, other animals, and calcareous algae are the primary depositions of reef formation, although ectoprocts make a small contribution.

Phylum Bryozoa has a scanty record in the upper Cambrian rocks. However, thousands of fossil species from the beginning of the Ordovician period have been described. *Bugula*—a genus commonly found clinging to ship hulls and dock pilings—belongs to the order Cheilostomata in the class Gymnolaemata, which first appeared in Cretaceous sediments some 100 million years ago.

Ectoprocts and entoprocts (Phylum A-18) were formerly classified together as Phylum Bryozoa; a strong case has been made for reuniting the two groups. However, because ecto- and entoprocts differ in fundamental ways, each is now assigned a separate phylum by some workers. Although ectoprocts superficially resemble the entoprocts—each has a basal stalk surmounted by a cuplike body bearing tentacles—the tentacular crown of the entoprocts is not considered to be a lophophore, because it surrounds both the anus and the mouth. Furthermore, entoprocts may be pseudocoelomate or acoelomate; the nature of the ectoproct body cavity cannot be determined with certainty, because of the disappearance of all larval structures during metamorphosis.

The phylogenetic relationships of the three lophophore-bearing phyla—Phoronida, Brachiopoda, and Ectoprocta—to one another is uncertain. Neither phoronids nor brachiopods have a resting stage such as the ectoproct statoblast. DNA evidence suggests that brachiopods at least are quite different from ectoprocts and phoronids. In contrast with brachiopods, the bryozoan (and phoronid) lophophore can be retracted. Researchers consider the lophophores in the phoronids, brachiopods, and ectoprocts to be homologous; the lophophore forms a ridge or ring around the mouth and has a coelomic (perhaps) cavity. In adult ectoprocts of class Phylactolaemata, the body cavity (coelom, if that is what it is) is divided into three parts. Such a divided coelom is also characteristic of the deuterostomes (Phyla A-32 through A-37)—animals in which the mouth forms opposite the blastopore—deuterostomes are thus suspected of having an ancestor in common with the ectoprocts.

A-30 Brachiopoda
(Lampshells, brachiopods)

Latin *brachium,* arm; Greek *pous,* foot

EXAMPLES OF GENERA

Argyrotheca
Crania
Dallinella
Glottidia
Hemithyris
Lacazella
Lingula
Megathyris
Notosaria
Terebratula
Terebratulina

Brachiopods (Figure A) are called lampshells because they resemble ancient oil lamps. Like clams and other bivalve molluscs (Phylum A-26), the brachiopod has two apposed hard shells (valves). The soft body of a brachiopod is enclosed by dorsal and ventral tissue folds (mantle). The mantle secretes the pair of valves, which are covered by epithelial tissue called the periostracum (Figure B). The bilaterally symmetrical valves that make up the shell are typically dissimilar. They range in size from 2 mm to nearly 100 mm in shell length and are usually cemented to the substrate by a stalk called the pedicle. *Lacazella* and *Crania* attach directly to the substrate by the ventral shell. Brachiopod symmetry differs from that of clams; the pedicle valve of brachiopods is ventral and the other, smaller shell is dorsal, whereas the shells of bivalve molluscs are arranged on the left and right sides. The anterior part of the space inside the brachiopod shell is occupied by the tentacle-bearing lophophore, which functions as a surface for gas exchange and food getting. The name Brachiopoda stems from the "arms" of the lophophore. Unlike molluscs, the brachiopod has a lophophore; as a consequence, brachiopods are grouped with other lophophorates—bryozoans (ectoprocts; Phylum A-29) and phoronids (Phylum A-31). The brachiopod body occupies the posterior part of the space inside the shell.

Brachiopods are coelomate animals. The large fluid-filled coelom harbors viscera that are supported on thin membranes called mesenteries. The coelom branches into the lophophore.

There are two classes of brachiopods, Articulata (most of the fewer than 350 living species) and Inarticulata, which differ in form as well as development. The valves of Inarticulata such as *Glottidia* and *Lingula* lack hinges (articulations); muscles alone hold the valves together. Inarticulate valves are quite similar to each other and usually consist of of calcium phosphate (or calcium carbonate) and chitin. The lophophore of inarticulates lacks internal support. The inarticulate has a U-shaped gut with an anus. In contrast, members of Class Articulata (*Terebratulina,* for example; see Figure A) support their lophophores with internal calcified structures. Their digestive tracts are dead end. The two valves differ in shape and size and articulate (hence their name) by an interlocking sockets-and-teeth hinge. These valves are made of calcium carbonate (calcite) plus either chitin or scleroprotein. Some articulates lack pedicles.

A Three living articulate brachiopods, *Terebratulina retusa,* dredged from a depth of about 20 m in Crinan Loch, Scotland. Bar = 1 cm. [Courtesy of A. Williams.]

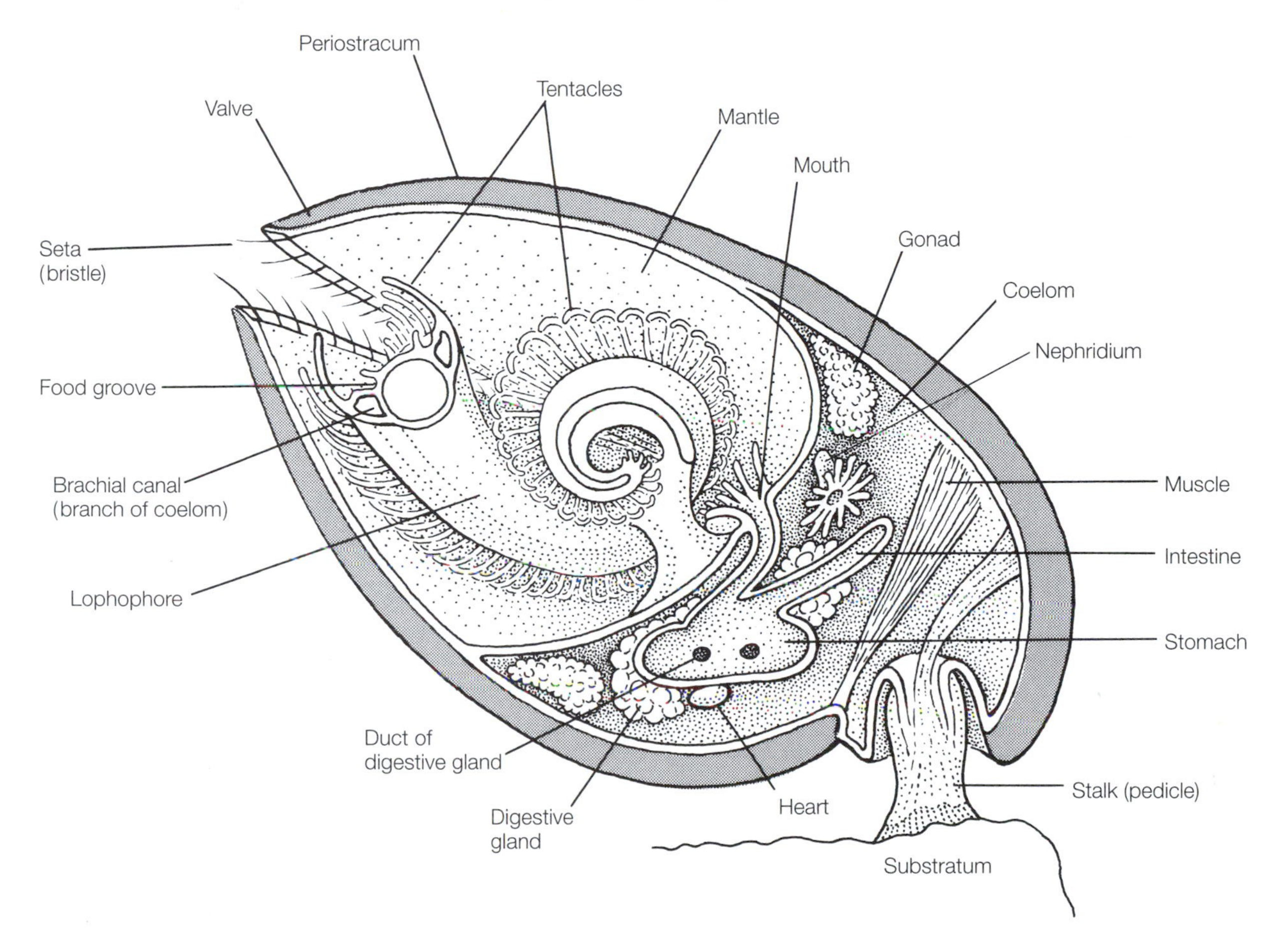

B Cutaway diagram of a generalized brachiopod of Class Articulata depicting the lophophore and internal organs. The lophophore's ciliated tentacles provide a surface for diffusing dissolved oxygen into the brachiopod and dissolved carbon dioxide out, for brooding young, and for generating currents of food-bearing seawater. [Drawing by L. Meszoly; information from A. Williams.]

Brachiopods are cosmopolitan. Most occupy marine habitats between the intertidal zone of the beach and 4000 m deep. Many attach to hard substrates. *Lingula* and a few others burrow; others lie unattached on mud or sand. To shake off sediment, the brachiopod pivots on its flexible stalk.

The mantle of brachiopods is not homologous to the molluscan mantle. Fine papillae of the mantle tissue penetrate the shell. In one class of brachiopods, a thin loop of calcium carbonate supports the coiled lophophore that attaches to the anterior surface of the body and lies in the space between the mantle lobes. Hundreds of tentacles fringe the spiral arms of the lophophore, which can take various shapes of which spirally coiled is common. Beating cilia on the tentacles circulate the water in the mantle cavity, facilitating respiration and excretion. Inhalant currents are drawn in both sides of the lophophore, and a single exhalant current leaves in the center. The cilia also reject nonfood items and whip small food organisms into the ciliary food groove leading to the mouth. Near the base of the lophophore, where it is attached to the body, the mouth opens into a short esophagus leading into a larger stomach; in articulate species, the intestine is dead end and, in inarticulate species, it terminates in an anus through which solid waste is eliminated. One or more pairs of digestive glands open into the stomach. Brachiopods have complex muscles: in one class of brachiopods, one muscle pair opens and closes the valves of the shell; two other pairs turn the brachiopod on its pedicle. The shells of inarticulates open hydraulically rather than by direct muscle action.

The circulatory system is partly open—the blood travels from the small contractile heart through a main dorsal blood vessel and at least one contractile vessel and returns through tissue spaces rather than through veins. The oxygen-binding pigment hemerythrin carries oxygen in the coelomic fluid; the blood lacks pigment. The excretory system consists of one or two nephridia, at each side of the intestine. Dissolved waste from the coelom enters the nephridium through a fringed opening; dissolved waste then drains into the mantle cavity. A ring of nerves encircles the esophagus; the brachiopod lacks a brain. A balance organ in one species and touch and chemical receptors have been reported. Chitinous setae—bristles that edge the mantle of most species—may be sensory.

Reproduction is always sexual. The brachiopod sexes are usually separate: each female has paired gonads that produce eggs, whereas the paired gonads of males discharge sperm into the ocean where fertilization takes place. Gametes are gathered by nephridia and released from the nephridia into the mantle cavity surrounding the lophophore. A few species have internal fertilization; sea currents waft sperm to females. The fertilized egg is initially brooded in the lophophore and then develops into a ciliated larva. The free-swimming, feeding larva generally develops into a solitary, sessile, bilaterally symmetrical adult attached by a pedicle to another animal or to rock.

Brachiopods are relics of a more diverse past; they thrived during the Paleozoic era. Because their calcium phosphate shells preserve well, some 30,000 extinct species have been described. In contrast, only about 335 brachiopod extant species are known. About 250 million years ago, at the boundary between Permian and Triassic periods (also the end of the Paleozoic era and the beginning of the Mesozoic era), the most massive extinction in the history of life on Earth decimated articulate brachiopods. At this time, an estimated 80 to 95 percent of late Permian invertebrate species became extinct. In comparison, in the Cretaceous-Tertiary extinction, which killed off dinosaurs 65 million years ago, only about 47 percent of genera were extinguished. Causes of the mass end–Permian extinction included drop in sea level, increased oxidation of organic matter that reduced atmospheric oxygen and resulted in ocean anoxia, and increased atmospheric carbon dioxide owing to volcanic eruptions. Filter-feeding organisms and attached seafloor animals—reef and shallow-water species including corals, bryozoans, brachiopods, and sea lilies (echinoderms)—were especially hard hit. Members of more recently evolved phyla now occupy most of the niches originally filled by brachiopods. *Lingula,* whose fossils are known from 400 million years ago, may be the oldest genus of animals extant.

Although they are lophophorates, brachiopods differ from the other lophophorates—phoronids and ectoprocts—in a number of ways. A feature common to lophophorates is that their gonads arise from the mesoderm that lines the principal coelom. The partly open circulatory system of brachiopods contrasts with the closed circulation of phoronids and the lack of vascular system and heart in ectoprocts. Whereas the brachiopod respiratory pigment is hemerythrin, phoronid blood corpuscles contain hemoglobin, and ectoprocts lack respiratory pigment and blood corpuscles altogether. Phoronids and brachiopods have metanephridia, whereas ectoprocts rely on diffusion alone. Reproduction is exclusively sexual in brachiopods; all phoronids reproduce sexually, whereas ectoprocts produce colonies asexually but initiate the colony with a sexually produced larva. Ectoprocts lack nephridia, whereas brachiopods and phoronids release gametes through nephridia. A nerve ring that encircles the lophophore is common to all three phyla. All lophophorates produce ciliated swimming larvae, but lophophorate developmental patterns include deuterostome—mouth originating opposite the blastopore site—as well as protostome—mouth originating from the blastopore site—characteristics (in brachiopods, the entire developmental pattern is deuterostome). Thus, the relationship of these three phyla to one another and to other phyla remains unclear.

A-31 Phoronida
(Phoronids)

Greek *pherein*, to bear; Latin *nidus*, nest

GENERA
Phoronis
Phoronopsis

All phoronids are sedentary marine worms, rare in most locales but found throughout the world. The phoronid, a lophophorate, bears a spirally coiled or horseshoe-shaped ridge with ciliated tentacles around its mouth. The lophophores of phoronids, brachiopods, and bryozoans (ectoprocts) are thought to be homologous. This food-gathering and gas-exchange organ, the lophophore, has as many as 1500 tentacles in some species (Figure A). All phoronid worms live permanently and move freely within leathery, chitinous blind tubes, formed from secretions impregnated with calcareous matter or encrusted with sand or shell fragments. Some burrow into mollusc shells and rocks.

With only two genera, phoronids are among the smallest phyla in regard to number of genera in the animal kingdom. Phoronids are about 5 mm wide and from 1 to 500 mm in length, typically 10 cm long. Their bodies may be pink, orange, or yellow. An enlarged end bulb anchors the phoronid in its dead-end tube. Their tubes are generally longer than the worms inside; thus phoronids can withdraw their pink tentacles completely. Some of the 14 species in the two phoronid genera are solitary. Only *Phoronopsis viridis* lives in dense populations (Figure B). The Pacific Coast of North America has the distinction of having the greatest densities of one species; half of the known phoronid species live on that coast. Phoronid tubes can occasionally be observed on rocks or pier pilings, interlaced like miniature hard spaghetti in crevices or in empty shells. The sandy or sticklike tubes are also found in tidal flats or between blades of sea grass on shallow bottoms from low water mark to 400 m down.

Phoronids filter feed on plankton and detritus and absorb dissolved organic compounds. A current generated by cilia on the tentacles draws particles tangled in mucus toward the mouth, which opens within the double row of tentacles (Figure C). Food particles enter the mouth; rejected bits are conveyed by cilia to the tips of the tentacles, where they are returned to the water. The mouth opens into a U-shaped, ciliated gut supported in the coelomic body cavity by thin membranes called mesenteries. Digestion probably takes place within the stomach wall rather than in the stomach lumen (cavity). At the other end of the cilia-lined gut, an anus discharges solid waste from an opening just outside the spiral of tentacles. A fold on the anal side of the lophophore diverts wastes away from the mouth. A pair of metanephridia, tubes lined with cilia, carries dissolved wastes and gametes from the coelom and out through nephridiopores located beside the anus.

Gases diffuse through the phoronid body surface; special respiratory organs are lacking. However, phoronids do have a closed circulatory system with a single blood vessel leading to and from each tentacle of the reddish, hollow lophophore. The two main vessels contract, moving the blood through the body, but a heart is lacking. Nucleated red blood cells contain the oxygen-carrying protein hemoglobin.

A A single *Phoronopsis harmeri* taken from a Pacific Coast tidal flat. This phoronid extends ciliated tentacles from its sand-encrusted tube. Bar = 5 mm. [Courtesy of R. Zimmer.]

The phoronid nervous system lies just under the body wall; a nerve ring circles the mouth, and nerves extend to muscles and tentacles. A giant nerve fiber coordinates the longitudinal muscles; when these muscles contract, the phoronid swiftly withdraws into its tube. The body surface has sensory cells.

Phoronids do not copulate. Some species are dioecious, others monoecious. Phoronid gonads form only during the breeding season. Ovary and testis arise temporarily from the mesodermal lining of the coelom, hanging beside the stomach in the posterior part of the coelom. Gametes are released into the coelom and pass from the coelom out through nephridiopores and then, perhaps by a ciliated exterior furrow, to a space enclosed by the tentacles. In this space, external fertilization takes place in most species. Fertil-

ization is internal in at least one species—*Phoronis architecta.* Eggs are generally fertilized by sperm that come from another phoronid. Adults of some species brood the embryos among their tentacles, probably held by means of secretions from the lophophore organ, or in the parental tube. In most phoronid species, brooded or not, eggs develop into actinotrochs—ciliated larvae that feed and have larval tentacles. The actinotroch's distinctive form with its large anterior lobe cannot be mistaken for a larva of any other phylum. Later, larvae become free-swimming marine plankton. Eventually, actinotrochs test the substrate and then settle to the type of bottom preferred by the species, where they metamorphose into adults. Members of a few phoronid species also reproduce asexually by budding and by transverse fission. All phoronids regenerate lost lophophores.

All phoronids, ectoprocts (Phylum A-29), and brachiopods (Phylum A-30) are lophophorates; as such, they may have ancestors in common. However, some of the adult characteristics that they have in common, such as reduced head, secretion of protective coverings (brachiopods' valves, ectoprocts' chitinous or calcified tube, phoronids' cuticle), and a U-shaped gut (of phoronids and ectoprocts), are adaptations to sessile life and may be at least partly the result of convergent evolution. Unlike phoronids and brachiopods, ectoprocts lack red blood cells, hemoglobin, heart, and closed circulation. Brachiopods have partly open–partly closed circulation with hemerythrin in coelomic fluid; the brachiopod heart is small, and blood corpuscles are lacking. Recent DNA evidence suggests that brachiopods are quite different from ectoprocts and phoronids. Annelids (Phylum A-22) and molluscs (Phylum A-26) seem to have ancestors in common with phoronids; the heavily ciliated phoronid larvae have excretory organs (protonephridia) like those in annelid and molluscan trochophore larvae. Because the ectoproct is so thoroughly remodeled during metamorphosis, we can perhaps never learn whether ectoprocts are coelomates, whereas phoronid development indicates that phoronids may be deuterostome coelomates—the adult mouth forms opposite the blastopore of the embryo—rather than close relatives of ectoprocts.

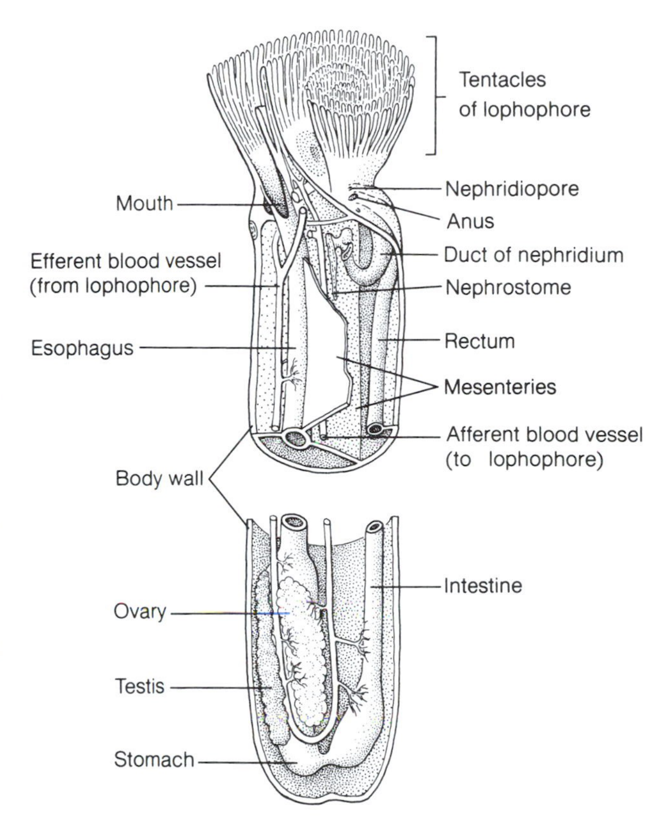

C Cutaway view of a *Phoronis* sp. [Drawing by L. Meszoly; information from R. Zimmer.]

B Several live phoronids, *Phoronis ijima* (=*vancouverensis*), from the Pacific Coast of the United States. Bar = 10 cm. [Courtesy of R. Zimmer.]

A-32 Chaetognatha
(Arrow worms)

Greek *chaite,* hair; *gnathos,* jaw

EXAMPLES OF GENERA
Bathybelos
Bathyspadella
Eukrohnia
Heterokrohnia
Krohnitta
Krohnittella
Pterosagitta
Sagitta
Spadella

The vernacular name arrow worm is frequently given chaetognaths because of their arrow-shaped bodies. The "chaeto" part of their name refers to their moveable hooks, with which chaetognaths grasp living prey. All arrow worms are marine predators that detect prey with vibration sensors. Copepods (Crustacea, Phylum A-21) are their principal food. Chaetognaths also consume other planktonic crustaceans and fish larvae (Phylum A-37). Each hook or adjacent tiny tooth may pierce prey exoskeletons; paralytic neurotoxins released by the arrow worm prevent prey from escaping. These neurotoxins, recently isolated from the heads of arrow worms, paralyze by blocking sodium channels in cell membranes, suggesting how these carnivores can capture prey as large

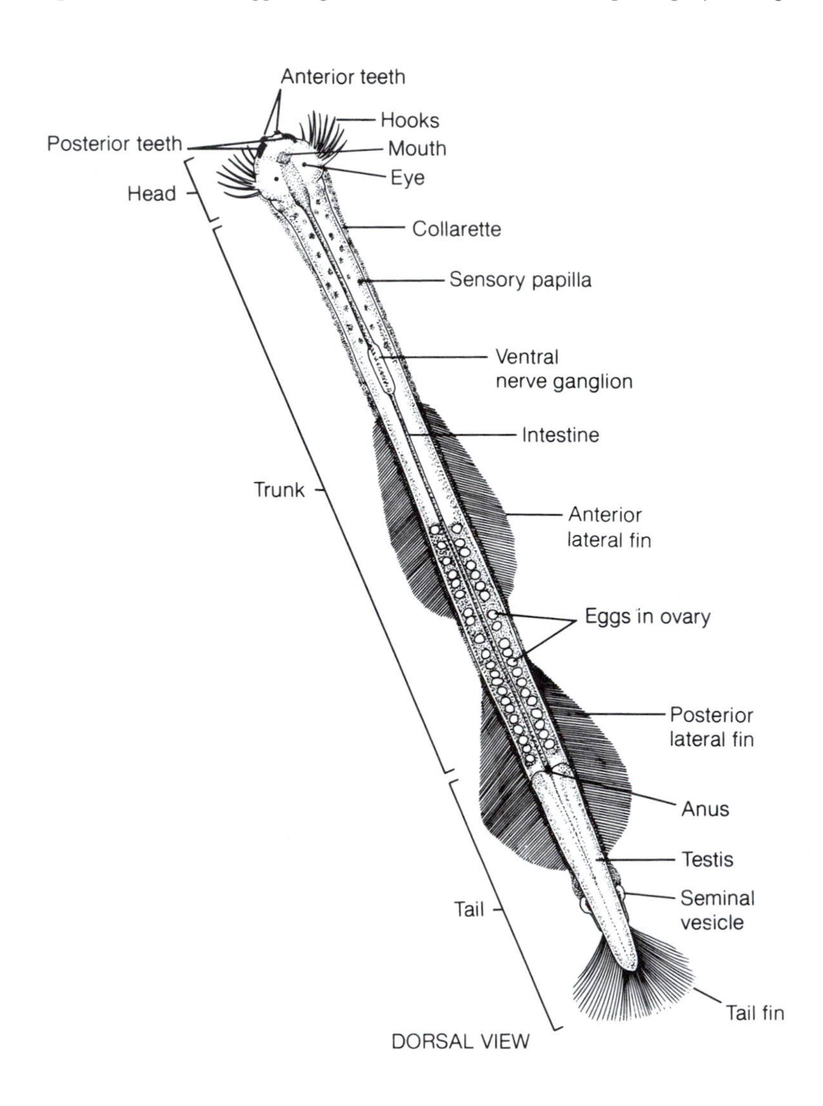

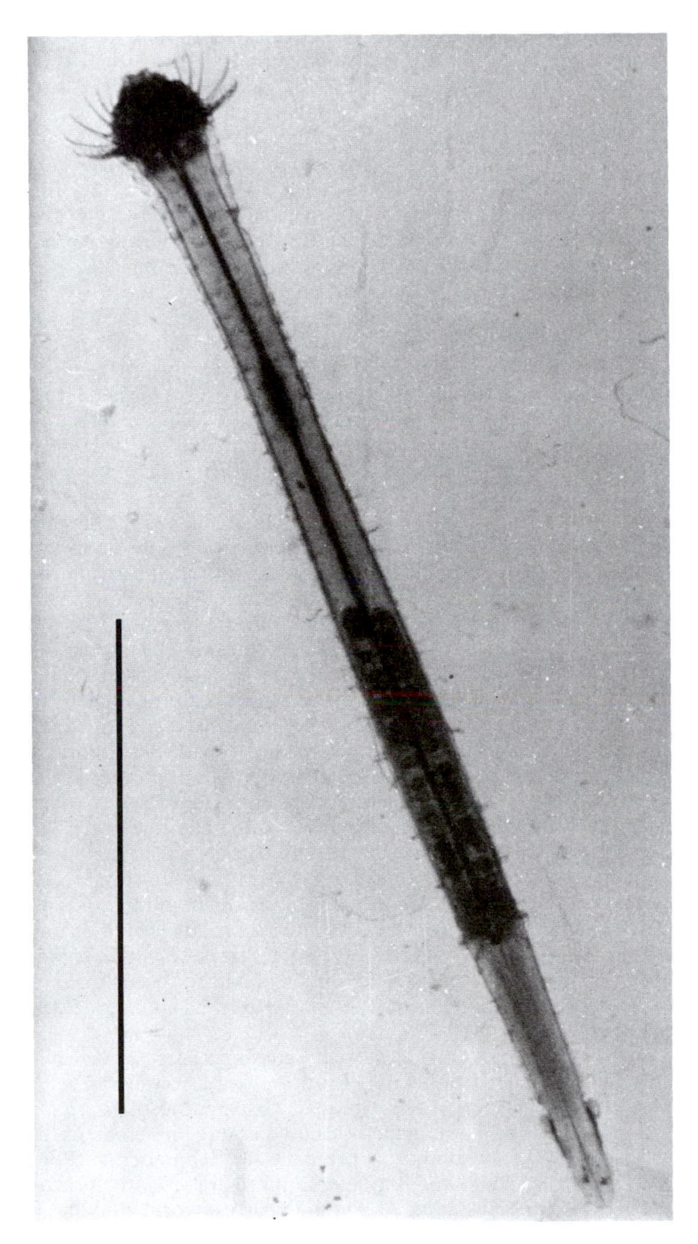

Sagitta bipunctata, a living arrow worm. *S. bipunctata* uses the transparent lateral fins as rigid stabilizers and to maintain buoyancy. The arrow worm shoots forward or backward by flapping its tail fin. Bar = 5 mm. [Photograph courtesy of G. C. Grant; drawing by I. Atema; information from G. C. Grant.]

as themselves. The toxins are likely secreted from pores adjacent to the arrow worm mouth, confirming researcher Robert Bieri's (Antioch College) epithet for arrow worms "cobras of the sea."

Arrow worms range in length from 0.5 to 15.0 cm. The 70 or so arrow worm species are common plankton worldwide in open seas and near shore from Spitsbergen, Norway, to the Indo-Pacific Ocean, especially abundant in warm seas to a depth of about 200 m. In 1844, Charles Darwin observed that "between latitudes 37 degrees and 40 degrees S [off the Atlantic coast of South America], the sea, especially during the night, swarmed with them." In the transparent chaetognaths from upper-ocean levels, food is visible in the intestine through the body wall; several blue chaetognaths—perhaps colored by their blue copepod prey—have been observed. Orange bands sometimes seen on *Eukrohnia* may be derived from copepod prey or microrganisms on the arrow worm's surface. Some deep-water chaetognaths are colored brightly: red, orange, and pink; other deep-water chaetognaths are transparent. *Spadella* is the only benthic (bottom-dwelling) genus; with caudal papillae, it temporarily adheres to substrate and then waits for prey. Although chaetognaths cannot swim against sea currents, many migrate daily, swimming up to surface waters at night and sinking downward by day, perhaps to escape predators; these diurnal migrations vary with vertical distribution of prey and with temperature.

Fine rays of unknown composition internally support one or two pairs of lateral translucent fins and a tail fin. These free-swimming predators dart forward or backward by flapping their tails and contracting striated longitudinal muscles. Lateral fins stabilize arrow worms and do not flap. The arrow worm draws part of its cuticle-covered body wall (cephalic hood) over its spiny head between feeding and sinks gently through the ocean. This hood is believed to decrease friction while the arrow worm is swimming.

The chaetognath body cavity is divided into three compartments by a septum between head and trunk and by a septum between trunk and tail. The chaetognath outer cuticle lacks chitin, and the hooks and teeth are no longer considered chitinous. Prey is swallowed whole. Chaetognath digestion is poorly known; only starch- and glycogen-splitting enzymes have been demonstrated. No proteolytic enzymes have been demonstrated in the digestive tract. Posterior to the teeth is a vestibule leading to a ventral mouth, followed by a pharynx at the beginning of the straight intestine and ending in an anus at the trunk–tail septum. Solid waste is eliminated from the anus. Chaetognaths lack circulatory, respiratory, and excretory organs. Cilia circulate the fluid within each of the three body cavities, distributing nutrients and dissolved wastes. Oxygen enters directly through the body wall, and carbon dioxide leaves the same way.

The nervous system consists of a large aggregation of nerve-cell bodies called the ventral ganglion linked to a large cerebral ganglion. Nerves lead to trunk muscles, tail, gut, hooks, and eyes. Sensitive external papillae (cilia) do not sense touch but rather enable the arrow worm to detect vibrations, chemicals, or water flow or all three. The V-shaped wings of *Pterosagitta draco* may be receptors that can distinguish different frequencies. The hooks, tail fin, and body surface sense touch. Each of two eyes consists of five inverted pigment cup ocelli located dorsally in the head. Arrow worms sense motion and differences in light intensity, although the eyes are probably unable to form visual images.

All reproduction in arrow worms is sexual, and each worm is hermaphroditic. Ovaries along each side of the intestine in the trunk coelom produce eggs; testes in the tail coelom produce sperm. The sperm mature before the eggs; sperm are formed into a single spermatophore (sperm packet) in the seminal vesicles and then released to the outside by rupture of these vesicles. Spermatophores attach themselves to the fins of the worm that produced them or to the fins of a partner. Both self- and cross-fertilization may take place in some species. Others, such as the benthic arrow worm *Spadella,* cross-fertilize. Two arrow worms approach and lie side by side, facing in opposite directions; then each attaches a spermatophore to the neck of the other. The sperm stream along each arrow worm's back and through the opening into the seminal receptacle (tube along ovaries). Fertilization takes place inside the chaetognath's body. The arrow worm oviduct is a tube inside a tube. At the posterior end, the inner oviduct tube expands to form the seminal receptacle; here sperm are received and stored. As an egg matures, a pair of cells from the inner oviduct wall form a hollow attachment stalk to the egg. Through this hollow stalk, the sperm move from the seminal receptacle to the egg. Fertilized eggs lie between the outer and inner oviduct tubes and may reach the ocean through a temporary exit (not through the female gonopore or seminal receptacle). Zygotes are brooded in some species, deposited on the seafloor, or released into the ocean. Embryos in all chaetognaths develop in the sea into diminutive adult arrow worms. Development is direct; there are no larvae. *Spadella* regenerates lateral and tail fins; it is the sole arrow worm species that has been successfully raised in the laboratory from hatching to sexual maturity.

Chaetognaths are phylogenetically puzzling. Fossil evidence is currently considered unhelpful in unravelling arrow worm relationships. The pattern of embryonic development justifies grouping arrow worms with deuterostomes because, in embryos of chaetognaths, hemichordates (Phylum A-33), echinoderms (Phylum A-34), and chordates (Phyla A-35 through A-37), the blastopore site becomes the anus. Although the arrow worm adult has a peritoneum-lined coelom, the formation of the arrow worm embryonic coelom differs from that of other deuterostome coelomates; therefore the phylogenetic affinities of arrow worms are obscure. In another facet of embryonic development, however, arrow worms differ from echinoderms and hemichordates: arrow worms lack ciliated larvae. Like photoreceptors of other deuterostomes, those of chaetognaths develop from cells with undulipodia that contain peripheral double tubules but lack central tubules. An intriguing link between chaetognaths and pseudocoelomates is that nematodes (pseudocoelomates, Phylum A-10), tardigrades (protostomous coelomates, Phylum A-27), and chaetognaths have longitudinal but no circular muscles. As Charles Darwin—the first to accurately describe the action of chaetognath spines—observed, chaetognaths are remarkable for the obscurity of their affinities.

Chaetognaths are important to marine fisheries: arrow worms are food for adult herring. However, larvae of economically valuable fish such as herring occasionally fall prey to arrow worms. Chaetognath species are distributed according to temperature; the species distributions have been used in tracing the course of ocean currents. The distribution of *Sagitta bipunctata,* for example, found in waters of the continental shelf off North Carolina, indicates the location of lateral extensions of the Florida Current.

A-33 Hemichordata

(Acorn worms, pterobranchs, enteroptneusts, tongue worms)

Greek *hemi-*, half; Latin *chorda*, cord

EXAMPLES OF GENERA

Atubaria	*Glandiceps*	*Rhabdopleura*
Balanoglossus	*Glossobalanus*	*Saccoglossus*
Cephalodiscus	*Ptychodera*	*Spengelia*

Hemichordates are small, soft-bodied coelomates characterized by a proboscis, collar, and trunk with gill slits. The most familiar hemichordates—enteropneusts—excavate spiral or U-shaped mucus-lined burrows in shallow sandy or muddy sea bottoms. Other hemichordates—pterobranchs—live in secreted tubes and do not burrow. Adult hemichordates are sedentary, using their proboscises in locomotion mostly in their burrows. Some live sedentary lives in the deep ocean; none are planktonic. Most feed on sediment; others on suspended plankton. Of about 90 species, the best known is *Saccoglossus.* Hemichordates are bilaterally symmetrical, unsegmented invertebrates that range in color from purplish black to white. Because one of their common names, tongue worm, is the same as that for pentastomes (Phylum A-21), we do not use it for hemichordates to prevent confusion.

The two classes of hemichordates, Enteropneusta and Pterobranchia, correspond to two basic hemichordate body plans. Class Enteropneusta comprises about 65 species, including *Ptychodera, Balanoglossus,* and *Saccoglossus.* Enteropneusts have fleshy, cylindrical bodies, muscular contractile proboscises, collars, and nonmuscular trunks having from 10 to more than 100 pairs of gill slits that pierce the pharynx. These solitary animals are also called acorn worms because of the acorn-shaped proboscis and are the only common hemichordates. Most are between 2.5 and 250 cm long, but *Balanoglossus gigas* is 1.5 m long. Most live in shallow seas on soft sediments, such as the Gulf of Mexico and the Carolina coast. Their ciliated epidermis secretes mucus, important in feeding and in lining their burrows. Some species are deposit feeders. Other species are suspension feeders that catch particles in mucus on the proboscis; cilia carry the food-laden mucus to the mouth. From the mouth, located in the collar at the base of the proboscis, a ciliated straight digestive tract runs into an esophagus, where food particles stick to a mucus cord. This cord extends from the esophagus through the pharynx and intestine in the abdominal (hepatic) region to a terminal anus. The enteropneust deposits solid waste from the anus in a coil on the ocean floor outside the exit of its burrow. Enteropneusts lack tentacles. Lateral folds of body wall often shelter ciliated gill slits, anterior to the gonads. Acorn worms are dioecious, have many gonads posterior to the gill slits, and reproduce sexually. Fertile eggs of some enteropneust species develop into ciliated, planktonic larvae called tornaria; other species develop directly into adults.

Class Pterobranchia has a vase-shaped body; a collar, which surrounds the mouth and has an extension that forms pairs of hollow arms bearing ciliated tentacles; a cephalic shield; and a lengthy, contractile stalk. The pterobranch crawls with its muscular cephalic shield—located at the base of its tentacles—within its tube or nearby. The longest are only about 7 mm long, excluding the stalk. Most live in secreted stiff tubes and in colonies no more than 20 mm across. Many pterobranchs live in Antarctic seas, though a few shallow-water forms live in warmer water. Pterobranchs have been collected from depths of 5 to 5000 m in the sea. Because they seem to be rare, little is known about pterobranch biology. The pterobranch *Rhabdopleura* is a suspension feeder, capturing food with the tentacles on its arms. The pterobranch has a U-shaped digestive tract, with mouth near the anus. The pharynx has two gill slits or none, depending on the species. Species lacking gill slits probably exchange oxygen and carbon dioxide across the body surface. Pterobranchs reproduce either sexually or asexually by budding; individual pterobranchs in a colony are produced by budding. The pterobranch larvae differ in form from those of enteropneusts; the pterobranch larva has a U-shaped digestive tract in comparison with the L-shaped gut of the enteropneust tornaria larva.

Class Pterobranchia is grouped into three orders. In the Rhabdopleurida (*Rhabdopleura*), each individual has a single gonad and is enclosed in a secreted tube—the individuals interconnect by a common stalk, or stolon. Members of the second order, Cephalodiscida (for example, *Cephalodiscus*), each have a pair of gonads. Individuals live alone or grouped in a colony covered by a single secreted tube. The third order, Atubarida, has been found only in Sagami Bay, Japan, by dredging. Its single species, *Atubaria,* was twined about a hydroid (Phylum A-3) and is stalked, with feathery tentacles. Only young *Atubaria* and female *Atubaria* are known.

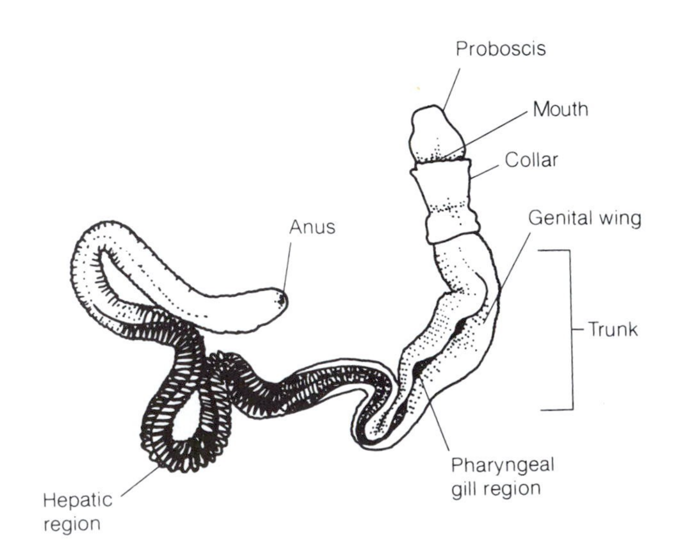

The circulatory system of both classes of hemichordates is mostly open—that is, without capillaries. The coelom is divided into three chambers—anterior, middle, and posterior—called protocoel, mesocoel, and metacoel, respectively. The organ that propels the blood is called a heart vesicle rather than a heart because it never contains blood but rather moves blood indirectly. The pulsating heart vesicle, which resides in the protocoel (the anterior section of the coelom), helps to circulate the blood indirectly by pushing against the dorsal blood vessel that lies alongside it. Blood, lacking pigment, is supplied to all three body regions. Blood flows anteriorly in the dorsal blood vessel, through a blood sinus, is oxygenated in the gills (when gills are present), and returns in the ventral blood vessel. Water is pulled into the mouth and is expelled through the well-vascularized gill slits, where gas exchange presumably takes place. Gases may be exchanged through the body surface as well as through the gill slits. Hemichordates are thought to excrete through a pore near the mouth, but this method of excretion has not yet been demonstrated experimentally. Nerves spread through the epidermis and thicken to form a dorsal and a ventral nerve cord. The two cords are connected by a nerve ring around the gut region; another ring surrounds the proboscis and is connected to the dorsal nerve cord only. The dorsal nerve cord is hollow in some species and is not a homologue of the vertebrate dorsal hollow nerve cord.

Hemichordates are grouped with chordates (Phyla A-35 through A-37), echinoderms (Phylum A-34), and chaetognaths (Phylum A-32) as deuterostomes; the blastopore of the deuterostome embryo develops into the adult anus rather than into the mouth. The hemichordate phylum is ancient—hemichordates may have been the earliest of the deuterostome phyla because fossils similar to an enteropneust and a pterobranch are found in the Burgess shale, middle Cambrian rocks in British Columbia, Canada, especially rich in fossils of soft-bodied animals from 500 million years ago. The tornaria larva of the enteropneust hemichordate resembles the bipinnaria larvae of sea stars (asteroid echinoderms, Phylum A-34).

Hemichordates resemble the chordates in that they have ciliated gill slits in the throat or pharynx. However, hemichordates are not chordates: although the hemichordate nerve cord, called the collar cord, that develops from dorsal epidermis of the embryo is sometimes hollow like that of chordates, the hemichordate nerve cord is not homologous to the vertebrate nerve cord. Hemichordates were thought to have a longitudinal rodlike support formerly called a notochord and so were once classified as chordates. Further study, however, revealed that the so-called notochord of hemichordates is really a buccal pouch—a short anterior diverticulum into the proboscis from the mouth cavity—and is not related to the notochord of chordates. Molecular evidence suggests that enteropneusts and pterobranchs are closely related to each other. Zoologists agree that hemichordates are distinctive enough to deserve a phylum of their own.

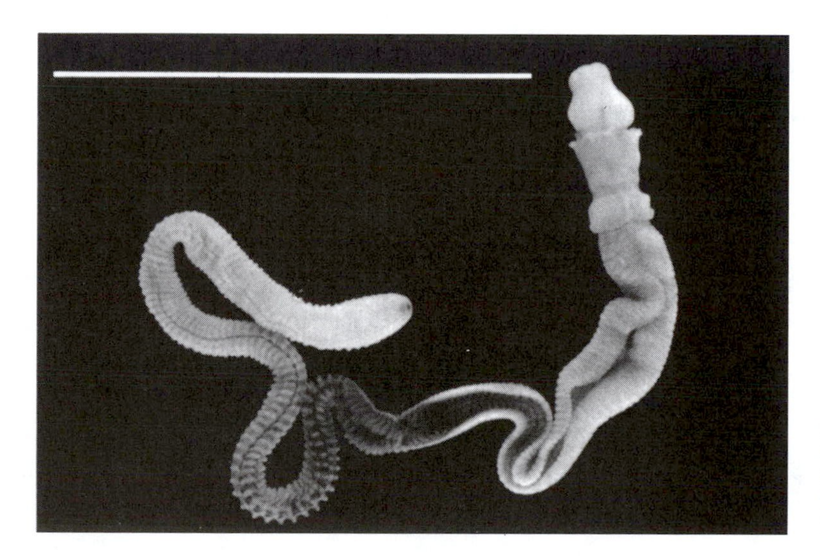

Ptychodera flava, a living acorn worm from subtidal sands near Waikiki beach, Oahu, Hawaii. Bar = 5 cm. [Photograph courtesy of M. G. Hadfield; drawing by I. Atema; information from M. G. Hadfield.]

A-34 Echinodermata

(Echinoderms)

Greek *echinos*, sea urchin; *derma*, skin

EXAMPLES OF GENERA

Arbacia
Asterias
Bathycrinus
Cucumaria
Echinarachnius
Holothuria
Metacrinus
Ophiura
Pisaster
Solaster
Strongylocentrotus
Thyone
Xyloplax

Echinoderm adults are radially symmetrical invertebrates with three unique features: five-part symmetry; an internal calcium carbonate skeleton; and a distinctively divided coelom, including a water vascular system (Figure A). Many have spines, shields, hooks, and scales on both oral and aboral (away from the mouth) surfaces and arms (Figure B). The echinoderm coelom includes a body coelom as well as coelom derivatives—a water vascular system and one or more closed, fluid- or tissue-filled duct systems called the hemal and perihemal systems, which have an array of functions.

Part of the coelom is modified into the water vascular system unique to echinoderms, consisting of seawater-filled canals that project through the body wall as tube feet (Figure C). Seawater enters this system through a central opening called the madreporite (Figure D). Contraction of a flexible bulb called an ampulla—associated with each tube foot—forces seawater through a one-way valve into each tube foot (Figure A) thus hydraulically extending the foot for locomotion, food collection, chemoreception, mucus secretion, and respiration. With suction and ionic interaction, tube feet adhere to molluscs and other prey. Tube feet may terminate in a suction cup, especially in echinoderms on firm substrates; echinoderms on soft substrates tend to lack suction cups. Retractile feeding tentacles that encircle the mouths of sea cucumbers (*Cucumaria, Thyone*) are modified tube feet.

Although about 16 classes of echinoderms containing thousands of species once existed, all of these classes became extinct. Only 6 classes are now extant, with 7000 species organized into two subphyla, Crinozoa and Asterozoa. The first—Subphylum Crinozoa (Pelmatozoa)—consists only of the class Crinoidea, sea lilies (*Bathycrinus*) and feather stars, abundant on coral reefs. The crinoid mouth and anus are both on the upper surface of the cup-shaped body, an adaptation for sessile life. The crinoid has tiny pores that open to the body cavity rather than a madreporite. Most living crinoids are free-living, unstalked feather stars; the rest—sea lilies—attach permanently to a substrate by a long stalk on the surface away from the mouth.

All other living echinoderm classes belong to the subphylum Asterozoa (Eleutherozoa): the sausagelike, burrowing Holothuroidea (sea cucumbers), which has an internal madreporite; the Echinoidea (sea urchins such as *Strongylocentrotus*, heart urchins, and sand dollars), having moveable spines, a madreporite that opens on the dorsal surface, and no arms; the Asteroidea (sea stars, *Asterias*), with five or more hollow arms; the Ophiuroidea (brittle stars

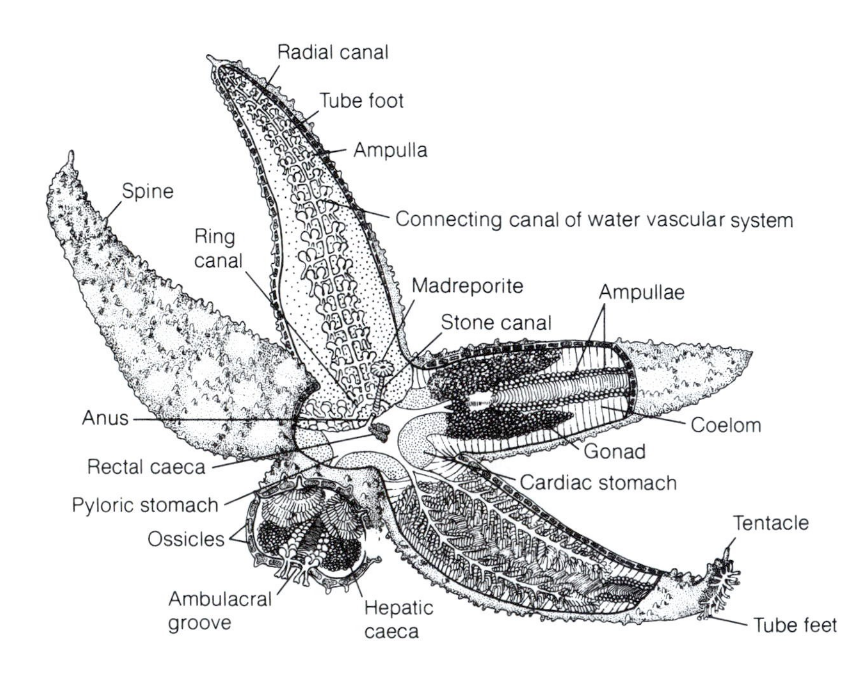

A Morphology of the sea star (starfish) *Asterias forbesi.* [Drawing by E. Hoffman; information from H. B. Fell.]

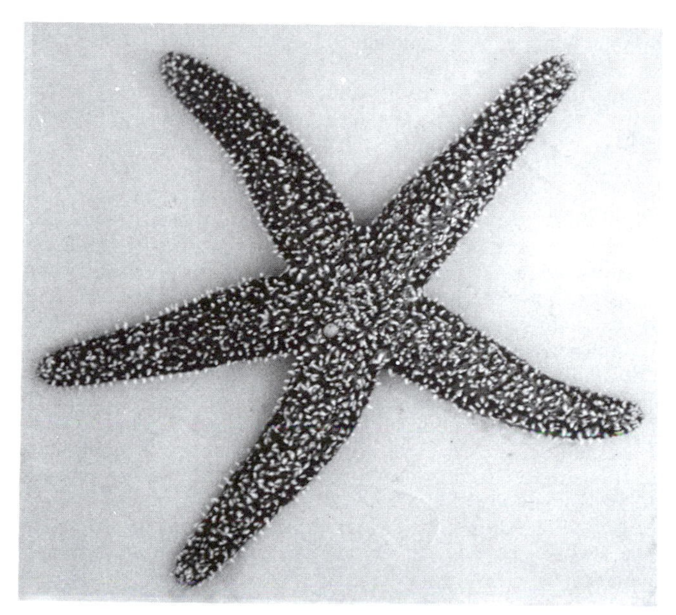

B *Asterias forbesi.* Arm radius of adult = 130 mm. [Courtesy of W. Ormerod.]

and basket stars, with 2000 species, the largest class in this phylum), with flexible, solid arms and a madreporite that opens near the mouth on the ventral surface; and the Concentricycloidea, named for their concentric water vascular rings. Because they were discovered only recently, in 1983, little is known of the biology of the deep-water concentricycloids, *Xyloplax,* collected from submerged wood. Their flattened, disc-shaped, armless bodies edged with marginal spines that look like daisy petals lead to their common name—sea daisies. Asteroids and ophiuroids are sometimes grouped together in Class Stelleroidea. Most species of Subphylum Asterozoa lack a stalk.

Most of the 7000 living echinoderm species are intertidal or subtidal. All are marine, mostly members of shallow-water biota. *Bathycrinus,* a sea lily, lives to depths of 10,000 m. In some abyssal habitats, deposit-feeding sea cucumbers make up 95 percent of the abyssal biomass. Of adult echinoderms, only sea cucumbers (by flexing their bodies with five longitudinal muscles), some brittle stars (by pulling along with their arms), and feather stars (by whipping their arms) swim. Some echinoids trundle along on their tube feet; sand dollars and many urchins move with waves of their spines. The giant of the echinoderms is *Holothuria thomasi,* the tiger's tail sea cucumber more than 2 m in length. Many echinoderms are bioturbators—they rework sediments as they burrow and scavenge. Some sea stars prey on commercially valuable scallops, clams, and oysters. Humans eat sea urchin gonads and make jewelery and ornaments from the rigid skeletons of sea urchins and sand dollars and from sea urchin spines. Sea cucumbers are reared in aquaculture; more than 47,000 metric tons are fished worldwide annually for human consumption, fresh or dried. Dried sea cucumbers are known as trepang or beche-de-mer. *Asterias*—predator of scallops, mussels, and oysters in the North Atlantic—is gathered by Denmark's fisheries for fishmeal, used as poultry food.

The five-part (or multiples of five, depending on the species), symmetrical, multirayed echinoderm body corresponds to its internal organization—coelom, gonads, digestive system, nerves, hemal and perihemal systems, and water vascular canals radiate inside the body. In the rather cylindrical sea cucumbers, the five-part symmetry is evident in muscle bands and some other internal

C Tube feet surrounding the mouth of *Asterias forbesi.* [Courtesy of W. Ormerod.]

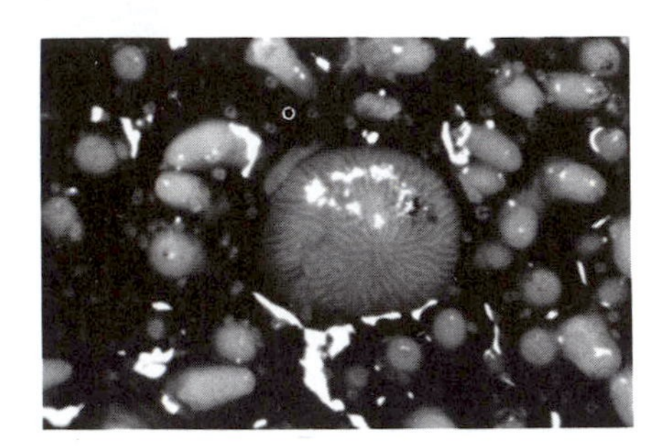

D A madreporite, the opening through which seawater enters the sea star water vascular system. [Courtesy of W. Ormerod.]

structures. The body surface covering is a delicate epidermis stretched over an endoskeleton of firm calcium carbonate plates called ossicles. These moveable or fixed plates are arranged in a genetically determined pattern and bear spines in most species. Beachcombers may come upon the urchin "test"—the skeleton that remains after soft tissues and spines have decomposed after the death of an echinoid. Microscopic ossicles embedded in the leathery or soft sea cucumber body wall are button, basket, granule, rod, sieve, table, rosette, wheel, tack, or anchor shaped. Sea cucumbers, sand dollars, and heart urchins show varying degrees of bilateral symmetry superimposed on the basic echinoderm fivefold symmetry typical of sea stars.

Holothuroids sequester toxins in their body walls, deterring predatory crustaceans, sea turtles, and fish. Other self-defense mechanisms include luminescence, color pigment shift, noctural spawning and feeding, poison pincers (of urchins), clinging with prehensile arms and tube feet to reefs and rock, sharp spines, and self-amputating arms and disc. Echinoderms of many species leave their daytime shelters to venture forth at night and are suspension feeders (sea lilies, feather stars, some sea cucumbers), predators (some sea and brittle stars), grazers (some sea stars and urchins), scavengers (some sea and brittle stars), and deposit feeders (selectively ingesting substrate; some sea cucumbers, some sea stars, and some sea urchins). Some sea cucumbers and brittle stars absorb dissolved organic nutrients from the sea through their skin. The digestive tract of most echinoderms begins with a mouth facing the seafloor and ends with an anus facing the opposite direction. Nutrients move from the digestive tract to the hemal system by an unknown mechanism and finally appear in the gonads. The digestive tract lacks an anus in ophiuroids and one concentricycloid species. In sea star hollow arms, pyloric (hepatic) caeca—part of the digestive tract—absorb nutrients and secrete digestive enzymes that pyloric ducts carry to the stomach, located in the central disc.

The hemal and perihemal coelomic systems—an array of canals and rings—are derived from coelomic compartments; they may or may not be homologous to circulatory systems in other phyla. The hemal system consists of two (oral and aboral) hemal rings and channels situated inside perihemal canals. These hemal and perihemal systems—tubes within tubes—run alongside the water vascular system. Pulsations of contractile channels in the central disc (central body region) of a sea star circulate fluid to gonads and other tissues in the arms. Nitrogenous waste diffuses from coelomic fluid to myriad thin-walled tube feet that function as respiratory surfaces—some echinoderms have more than 2000. In many species, the coelom itself performs the circulatory and respiratory functions. A ciliated peritoneal membrane lines the coelom and circulates coelomic fluid. Coelomocytes, cells sometimes containing hemoglobin, engulf waste particles and bacteria and synthesize collagen, which may repair wounds. Hemoglobin-containing coelomocytes circulate through the coelom and the hemal and water vascular systems of some species. Some sea stars and echinoids respire by minute gills that protrude from the coelom into the sea. Certain sea cucumbers aerate branched cloacal outgrowths called respiratory trees by pumping seawater in and out. The nervous system consists of a nerve ring around the esophagus with nerve nets radiating out into the body, statocysts (organs of balance), eyespots, and surface receptor cells for touch and chemical reception. The bodies of echinoderms are unsegmented. The adult lacks a head, brain, eyes, specialized excretory organs, and—except for a pulsating vessel in some cucumbers—heart.

Most asteroids, crinoids, and ophiuroids regenerate lost arms. Some brittle stars can regenerate broken arms and disc (the pentagonal or round central body region) including stomach and gonads. Sand dollars and urchins repair their calcium carbonate spines. By expelling their digestive tracts, gonads, and respiratory trees, many sea cucumbers satisfy the appetites of predators; the cucumber can later regenerate the eviscerated body parts. Some sea star, sea cucumber, and brittle star adults divide into two or more parts, each of which regenerates missing parts. A few brittle stars are hermaphrodites; all of them self-fertilize and brood young—sometimes at a variety of developmental stages—in bursae (surface slits). Dioecious sexuality is the rule; the sexes are separate, although male and female echinoderms often look alike. Fertilization is external except in sea daisies. Most echinoderms are oviparous (egg laying); some are ovoviviparous (egg yolk provides nutrients to the embryo, which develops within the female reproductive tract and is released from the egg at birth); and some are viviparous (live bearing, with the mother providing nutrients to the embryo directly, rather than from egg yolk). The fertilized egg develops into a bilaterally symmetrical, ciliated, planktonic larva, which passes through several distinct stages before metamorphosing into an adult. Some species develop directly, bypassing metamorphosis. Some brittle stars release gametes into bursal slits in which they brood their embryos; cilia circulate ocean water in these surface slits, facilitating excretion and respiration. Numerous sea cucumbers brood young among tentacles, in special pockets in their body walls, or in the coelom or gonad, where the fertilized eggs have been retained.

Like chordates (Phylum A-35 through A-37), echinoderms are deuterostomes; in embryonic development, the anus forms at or near the blastopore and the mouth forms opposite. The patterns of cleavage of the fertilized egg (blastulation and gastrulation), the formation of the three embryonic germ layers (ectoderm, mesoderm, and endoderm), and the presence of a divided coelom suggest that echinoderms and chordates as deuterostome coelomates have common ancestors. The earliest known echinoderm was preserved 600 million to 500 million years ago in the early Paleozoic. Of all living echinoderms, crinoids appear to have changed from their ancestral echinoderm condition the least. The mass end-Permian extinction decimated echinoderms that lay on the seafloor or were attached to it; sea lilies declined, living today in depths greater than 100 m; the more mobile sea stars flourished after the mass end-Permian extinction. Echinoderms probably evolved from bilaterally symmetric ancestors to radially symmetric sessile animals. Later, certain echinoderms such as sea stars secondarily became mobile.

Sea urchins are collected from wild populations and their roe (eggs) exported, mainly to Japan. Because green sea urchins may be exterminated if their overfishing is not reduced, marine biologists look to aquaculture for alternatives to harvesting wild urchins from the sea. Cultured sea urchins can be induced to increase their egg production, attracting the interest of aquafarmers.

A-35 Urochordata

(Tunicates, sea squirts, ascidians, larvaceans, salps)

Greek *oura,* tail; Latin, *chorda,* cord

EXAMPLES OF GENERA

Ascidia
Bathycordius
Botryllus
Ciona
Cyclosalpa
Dicarpa
Didemnum
Diplosoma
Doliolum
Halocynthia
Herdmania
Lissoclinum
Molgula
Oikopleura
Pyrosoma
Salpa
Trididemnum

Tunicates are considered acraniate chordates because tunicates and chordates have the following features in common: a notochord; a dorsal, hollow nerve cord; and pharyngeal gill slits at some time in their lives. The notochord is a stiff cylinder of cells, each cell containing a fluid-filled vacuole. The notochord of urochordates extends the length of the body but does not persist throughout life except in Class Larvacea. The embryonic notochord of vertebrates is replaced in adult vertebrates by cartilage or bony vertebrae as support. The hollow, dorsal nerve cord is resorbed by tunicates at metamorphosis. Adult tunicates have a small cerebral ganglion but no brain. In vertebrates, the nerve cord becomes the spinal cord and brain. Prominent gill slits are present in both larval and adult tunicates; whereas, in many vertebrates (reptiles, birds, mammals, and most amphibians), gill slits evident in the embryo are transformed and closed in the adult. Fish and a few amphibians retain their gill slits into adulthood. Acraniate chordates—tunicates and cephalochordates (Phylum A-36)—are considered by some to be subphyla within Phylum Chordata.

All urochordates are marine and comprise about 90 percent of the invertebrate (acraniate) chordates. Adult urochordates secrete an external tunic that usually (except in larvaceans) contains tunicin, a polysaccharide related to cellulose. Within the tunic is the body, which contains a pharynx that filters phytoplankton from incoming seawater and exchanges oxygen and carbon dioxide with the seawater. Urochordates use three different mechanisms for generating a seawater current through their bodies, according to their class: beating of cilia on the pharynx (ascidians), beating of the tail to drive water through the "house" (larvaceans), and contraction of circular muscle bands and beating of cilia (thaliaceans). The urochordate continually pumps water into the body, through its pharynx, and out of the body. Fertilization usually takes place in the sea for solitary ascidians and larvaceans and is internal for colonial ascidians and thaliacians. The 1400 species of tunicates grouped in three classes—Ascidiaceae, Larvacea, and Thaliacea—are significant members of oceanic food webs. Some sea stars, for example, feed on tunicates.

Members of class Ascidiaceae (Figures A and B)—commonly called sea squirts, tunicates, or ascidians—have free-swimming, nonfeeding larvae and benthic (seafloor dwelling), sessile adults from 1 to 120 mm long. A few species of ascidians live in soft sediments in beds of sea grass, attached by stalks or filaments to shells or underwater cliffs, and some (*Dicarpa*) are members of the community of the abyss. The sea squirt larva (tadpole) has adhesive suckers (papilli) on its head and a muscular tail, like the frog tadpole that it superfically resembles (Figure C). The larval nervous system includes a dorsal pigment-containing light receptor called an ocellus and a dorsal, hollow nerve cord (neural tube) that runs the length of its tail. The ascidian larva has a nonfunctional digestive tube. After a day of swimming, ascidian larvae adhere with their anterior ends to the bottom—ocean floor, dock pilings, or ship hulls—and undergo extensive metamorphosis into the sessile adult. Tail, nerve cord, and notochord are resorbed as the body rotates about 180 degrees; the oral siphon migrates so as to open opposite the attachment to the substrate; other internal organs also rotate to their adult locations. In the adult ascidian, seawater enters the oral siphon, moves through slits in the pharynx, and leaves through the excurrent siphon; this feeding current is generated by cilia on the pharynx.

Epidermal cells secrete the tunic, which is transparent or purple, red, pink, orange, yellow, or green; the ascidian tunic may be rough, smooth, or hairy or it may bear stiff spicules. The spicule-stiffened tunic and pharyngeal basket (feeding and respiratory structure) are mechanically analogous to pliable synthetic material reinforced with stiff inclusions. The tunic spicules of *Herdmania*

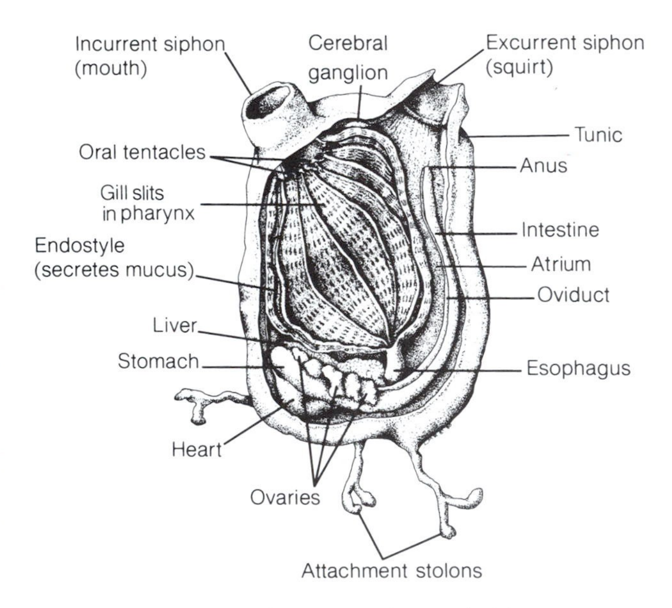

A Cutaway view of a generalized adult noncolonial ascidian tunicate. These tunicates are hermaphroditic, but the testes are not shown here. [Drawing by L. M. Reeves.]

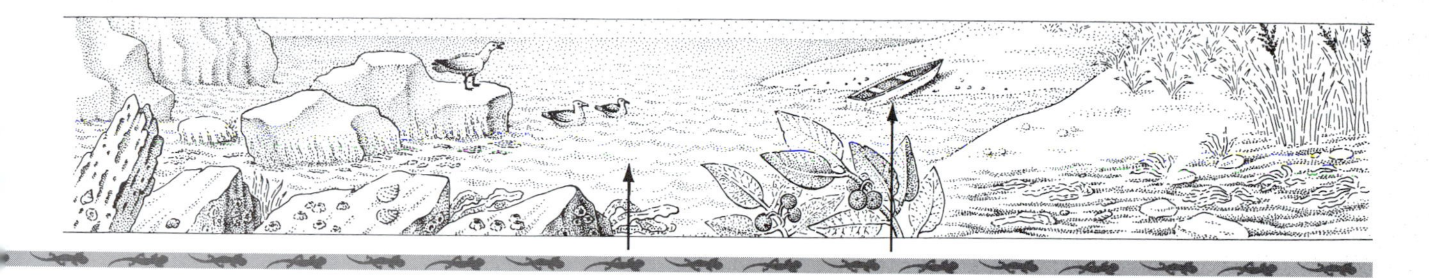

B *Halocynthia pyriformis* (the sea peach), a tunicate belonging to the class Ascidiacea. Sessile adults are found in shallow water along the Atlantic coast of North America from Maine northward. *Halocynthia* is one of the largest ascidians. The sandpapery surface of the tunic is yellow or orange with a tinge of red. Bar = 1 cm. [Courtesy of N. J. Berrill.]

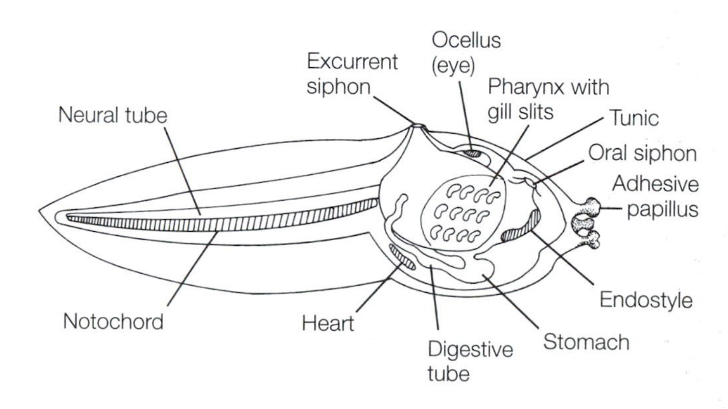

C A tadpolelike tunicate larva has chordate characteristics: a notochord, a hollow dorsal nerve cord, and gill slits. [Drawing by L. M. Reeves.]

form in tunic blood vessels (separated from blood by an envelope that may be a cell) and migrate through the blood vessel and through the tunic to project from the surface of the tunic. Blood cells and, in some species, blood vessels are located in the vase-shaped tunic itself. The blood carries nutrients and oxygen. The circulatory fluid contains no oxygen-binding pigment but does have morula cells (clusters of cells that resemble a mulberry or raspberry) that accumulate vanadium (perhaps a defense mechanism) and deposit it in the tunic. The tubular heart of the ascidian adult is located in the atrium, a cavity between tunic and pharynx. The urochordate heart has a pacemaker at each end. Every few minutes, the heart reverses the direction of its pumping. Cells in the tunic conduct impulses and contract, but the tunic lacks nerves and muscles. Just inside the tunic is a body wall with muscle bands; muscle bands are more extensively developed in thaliaceans. When an ascidian tunicate is disturbed, it contracts the muscle bands in its body wall and closes both siphons. This contraction around the trapped internal water causes the tunic to become turgid, which forces the tunic spicules to project outward, possibly deterring potential predators.

The ciliary feeding and respiratory structure—pharyngeal basket—takes up the greater part of the body of the ascidian tunicate adult; water is pulled in through the oral siphon by cilia (in ascidians). Dissolved oxygen diffuses into the tunicate as the feeding current passes through the pharyngeal gill slits. As the water passes out of the pharynx through these pharyngeal perforations, food particles collect on a mucus sheet secreted by the endostyle—a ciliated groove just below the incurrent (oral) siphon. Then, tiny tentacles (in some species) roll the food-coated sheet into a rope that cilia transport to the esophagus opening. Food is digested extracellularly and absorbed as it moves through esophagus, stomach, and intestine. The filtered water moves through pharyngeal slits into the atrium and out the excurrent siphon. Contraction of body-wall muscles squirts out uncollected food bits and water, giving rise to the common name sea squirt. Waste particles leaving through the anus and gametes leaving through the genital pores are expelled with the outgoing current from the excurrent siphon. Colonial ascidians sometimes share a common excurrent siphon. In most urochordates, most dissolved waste—ammonia is removed by diffusion; the ascidian *Molgula* accumulates uric acid in renal sacs. The tunicate *Lissoclinum,* a coral reef tunicate, hosts a photosynthetic bacterium *Prochloron* (see chloroxybacteria, Phylum B-5) in the wall of its combined excretory and reproductive canal. (In many tunicates, anus and gonopores are separate.)

Ascidian tunicates are simultaneous hermaphrodites with male and female genital pores opening near the anus. Most species release gametes, and fertilization usually takes place in the sea (*Ciona*) but is internal in colonial ascidians. Some ascidians (*Botryllus*) form sessile colonies in a common gel tunic by budding. In other ascidians, buds separate from the parent.

All members of class Larvacea (Appendicularia) are pelagic and solitary, floating free in the open ocean worldwide. Some species of *Oikopleura* bioluminesce. *Bathycordius* can reach 8 cm or more; most larvaceans are smaller than 10 mm in length. The larvacean tunic is not of tunicin. The larvacean adult looks like an ascidian larva; the larval form is retained in adulthood, but the larvacean is a mature, tailed adult with a notochord (an example of retention of larval structure in the adult) and gonads. Like ascidian tunicates, adult larvaceans bear both ovary and testis, heart, and digestive tract; but, unlike ascidians, they form neither buds nor colonies. The adult secretes a soft "house" (case) around itself. Through this house, the larvacean drives ocean water by lashing its tail. Water enters the porous house through mucus-covered external filters that retain food particles; the food is passed to a second, pharyngeal filter and to the mouth and digestive tract. Eventually, water exits through the excurrent siphon. Feces exit through the anus and collect in the soft case. When its external filters clog, the larvacean first secretes a new house, then exits from the old house through an escape hatch, and inflates a new house around itself. The larvacean changes houses several times a day. Although the most abundant of the free-living tunicates, the larvaceans are seldom seen except by an observer in a remote-operating vehicle or by divers, because the fragile larvaceans are destroyed by tow nets.

Members of class Thaliacea, such as *Salpa,* commonly known as salps or chain tunicates, have complex life cycles that alternate between solitary sexual forms and colonial asexual forms. For example, an adult *Doliolum* may bud hundreds of individuals in a chain; these individuals eventually separate as sexually reproducing adults that produce larvae. Large concentrations of *Pyrosoma* and other thaliaceans are a frequent source of ocean bioluminescence. Thaliaceans are free-living translucent urochordates most common in warm ocean surface water. *Pyrosoma* colonies may extend several meters through the sea; individuals range from a few millimeters to 24 cm in length. Their barrel-shaped bodies are banded by muscle, much as hoops band a wooden cask; incurrent and excurrent siphons open at opposite ends of the body, and the muscle bands generate locomotion. The doliolid thaliacean is jet propelled; by shutting one siphon and contracting its circular muscles, it shoots water out of its rear siphon and zips along at a rate of as much as 50 body lengths per second. The anatomy with respect to muscles of doliolid thaliaceans such as *Doliolum* is like that of salps such as

Salpa but with complete muscle bands instead of incomplete bands; neither ascidians nor larvaceans have such muscles. Like ascidians and larvaceans, members of Class Thaliacea are ciliary filter feeders; they secrete mucus with an endostyle and capture food on a slitted, mucus-coated pharyngeal bag or a pharynx reduced to a ciliated bar.

Ascidian spicules are fossilized but the urochordate soft bodies have left no recognizable fossils. Urochordate larvae have left evolutionary footprints—the notochord, pharyngeal gill slits, and hollow, fluid-filled, dorsal nerve cord common to all chordates.

A-36 Cephalochordata
(*Amphioxus*, lancelets, Acrania)

Greek *cephalo*, head; Latin *chorda*, cord

GENERA
Branchiostoma (Amphioxus)
Epigonichthys

Cephalochordates, like tunicates (Phylum A-35), are acraniate chordates; that is, chordates that lack a skull. Cephalochordates range from about 5 to 15 cm in length and live on shallow sandy seafloors. A few of the 23 species in the two genera that make up this invertebrate phylum inhabit estuaries. *Branchiostoma* has a double row of gonads; *Epigonichthys* has gonads on its right side only. Both are fishlike but scaleless and without bones and cartilage. Because cephalochordates are lance shaped, they are also called lancelets. All three defining features of chordates—notochord, hollow dorsal nerve cord dorsal to the notochord, and pharyngeal gill slits—persist in adult cephalochordates. The lancelet notochord persists throughout its life, like that of the larvacean urochordates (Phylum A-35). The gill slits that open in the sides of the pharynx also persist in adult cephalochordates, like those in urochordates. (In terrestrial vertebrates—such as our own species—gill slits appear as transitory embryonic structures.) Like other coelomates (Phyla A-19 through A-37), including urochordates (Phylum A-35) and craniate chordates (Phylum A-37), the cephalochordate has a coelom.

Lancelets swim to escape predators or to move to a new feeding locale. The stiff notochord flexes when the lancelet swims by contracting the longitudinal muscles in its tail; the notochord itself cannot shorten or lengthen. These clearly segmented muscles can be seen through the translucent skin of the tail (omitted in the adjoining photograph). The notochord stiffens the lancelet's body, just as the vertebral column stiffens the body of a swimming fish. Fin rays in the lancelet's dorsal fin may provide additional stiffening. Lancelets lack bony vertebrae. Feeding lancelets shove down into sand, turn, and emerge with their heads protruding above the sand. Twelve tiny tentacles called oral cirri (buccal tentacles) at the top of the lancelet's head screen out large particles from seawater and pass small plankton and organic particles through the pharynx to the mouth and the digestive system. Cilia in the pharynx generate water flow through the pharyngeal gill slits. This mode—ciliary filter feeding—is similar to the way in which some ascidian and thaliacean urochordates (Phylum A-35) waft water through the pharynx. Seawater that has passed through the gill slits continues to flow into the atrium—a chamber around the pharynx—and out to sea through the atriopore, which is an outlet midway along the lancelet's body. As in urochordates, an endostyle secretes a sheet of mucus that coats the gill slits. Food caught on the mucus is wrapped into a mucus-food string, which cilia pass to the intestine. Extracellular digestion and phagocytosis facilitate digestion and absorption of nutrients. A liver (hepatic caecum) extends from the intestine. Excretion is by paired nephridia, like those in annelid worms (Phylum A-22) and craniate chordates (Phylum A-37). The pharyngeal gill slits are vascularized and serve as gas exchangers as well as food gatherers. After being oxygenated in the gill slits, colorless blood is pumped by contractile blood vessels to the rest of

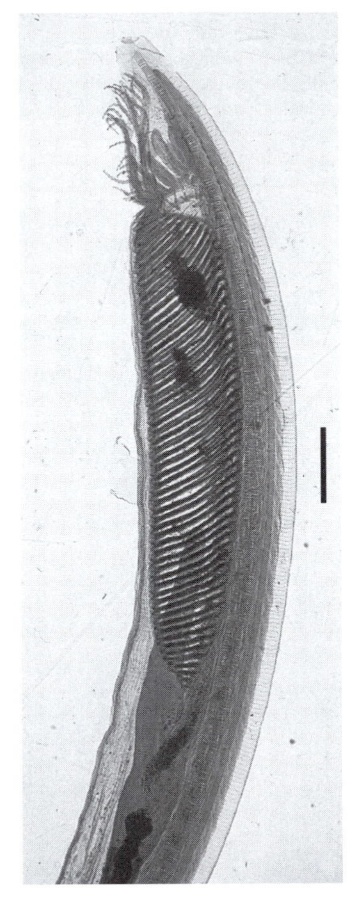

Branchiostoma. This best-known cephalochordate lives with its head projecting out of the sandy bottom of a warm, shallow sea. This lancelet (amphioxus) resembles the larvae of ascidian tunicates (Phylum A-35) and has segmented swimming muscles with nerves in addition to notochord, dorsal hollow nerve cord, and gill slits. Oral cirri on the head (top) sweep in phytoplankton on the water current entering by ciliary action, strain the water through gill slits—visible on the pharynx—into the atrium that leads to the atriopore posterior to the gills. Rays in the dorsal fin (right) are visible, as is the finger-shaped hepatic caecum behind the gills. (Tail not included in this image.) Bar = 0.5 cm. [Courtesy of Ward's Natural Science Establishment, Rochester, NY.]

the body in a pattern similar to the closed circulatory pattern of fish.

The dorsal nervous system of amphioxus contrasts with the solid ventral nerve cords found in nonchordate animals with well-developed nervous systems, such as arthropods (Phyla A-19 through A-21). Nerves from muscles and body wall connect to the hollow dorsal nerve cord of lancelets. Chemosensory cells and touch receptors are located precisely where food-bearing water enters the little body at the anterior end. Pigmented light receptors called ocelli are found along the nerve cord. Cephalochordates lack a cerebral ganglion (brain), unlike ascidian, larvacean, and thaliacean urochordates.

In breeding season, lancelets emerge from the sand to breed. Lancelets are dioecious, unlike the urochordates, which are usually monoecious. Each sex forms gonads beneath the forepart of the intestine. Sperm from males and eggs from females exit through the atriopore into the ocean. Fertilization takes place externally. The parents return to the bottom at dawn, wriggle headfirst into the sea sand, and turn to lie head upward, partly buried. The embryos develop into free-swimming larvae that strongly resemble boneless, fishlike larvae of ascidian urochordates (Phylum A-35) and eventually metamorphose into adults.

Ribosomal RNA comparisons suggest that cephalochordates are the closest relatives of vertebrates, confirming the notochord, pharyngeal gill slits, and hollow dorsal nerve cord that we have in common.

Although most animals consumed by humans are vertebrates, molluscs, and crustacea, the cephalochordates are eaten in China.

A-37 Craniata

Greek *kranion,* brain

EXAMPLES OF GENERA

Alligator
Ambystoma
Amia
Anas
Anser
Aptenodytes
Apteryx
Archilochus
Ardea
Balaenoptera
Bos
Bradypus
Bubalus
Bubo
Bufo
Buteo
Camelus
Canis
Casuarius
Cavia
(continued)

Members of this phylum, our own, are the most familiar of all the animals. Craniates include about 45,000 species, including most animals of direct economic importance, except molluscs and arthropods. As indicated in the table, the eight classes of craniate chordates include Cyclostomata (hagfish and lampreys), Chondrichthyes (sharks and other cartilaginous jawed fish), Osteichthyes (bony fish), Choanichthyes (lungfish), Amphibia (toads, frogs, and salamanders; Figure A), Reptilia (turtles and lizards), Aves (birds), and Mammalia (mammals).

All craniates have a brain that lies within a cranium (skull; Figure B), which distinguishes members of this phylum from the acraniate chordates—urochordates (Phylum A-35) and cephalochordates (Phylum A-36). Craniates are grouped by some as Subphylum Vertebrata in Phylum Chordata, together with Subphyla Urochordata and Cephalochordata. The presence of three defining chordate characteristics in craniates suggests that this phylum and the other chordates have common ancestry. In craniates, the first chordate characteristic—the dorsal, single, hollow (fluid-filled) nerve cord—becomes the brain and spinal cord during embryogenesis. The second characteristic defining chordates—a cartilaginous rod called the notochord—forms dorsally to the gut in the early craniate embryo. This slender flexible cylinder of cells containing a gelatinous matrix and sheathed in fibrous tissue extends the length of the body and persists throughout the life of all members of Classes Cyclostomata and Chondrichthyes. In all other classes of craniates, the embryonic notochord is largely replaced by the bony vertebral column—the backbone—during later development. The third chordate feature is the presence of gill slits in the pharynx at some stage of the life history, as in urochordates and cephalochordates (Phyla A-35 and A-36). Gill slits reveal the aquatic ancestry of the phylum. These slits are present only in the embryo and larva of land-dwelling craniates with few exceptions, such as the axolotl, a Mexican amphibian. Gill slits persist in adults of aquatic craniates in classes Cyclostomata, Chondrichthyes, and Os-

Classes of Phylum Craniata

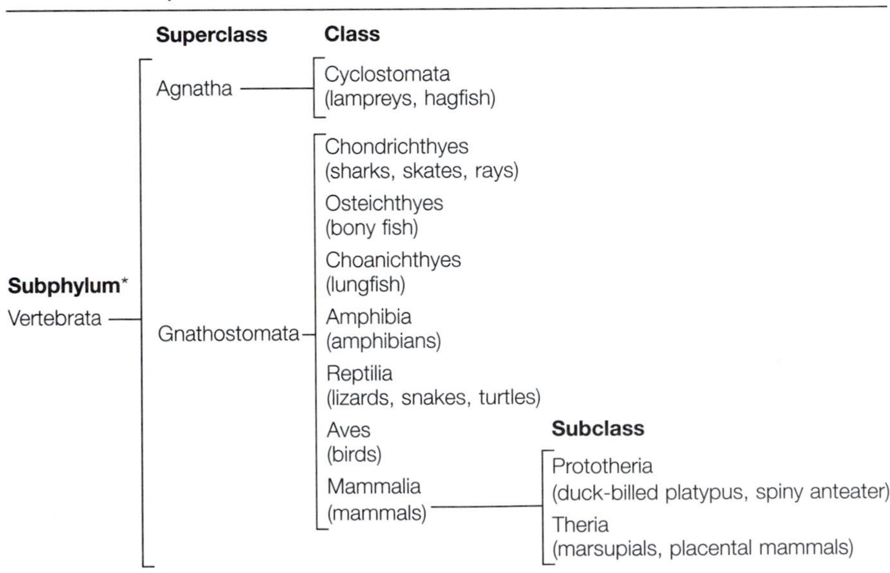

*Subphylum structure of the former Phylum Chordata: Urochordata (now Phylum A-35); Cephalochordata (now Phylum A-36); Vertebrata (now the only subphylum of Phylum A-37).

A *Ambystoma tigrinum,* the tiger salamander, a member of the class Amphibia and the family Ambystomatidae, photographed in Nebraska. *A. tigrinum* is one of the most widespread salamander species in North America and may grow to be more than 20 cm long. The adults are black or dark brown with yellow spots. Bar = 10 cm. [Courtesy of S. J. Echternacht.]

teichthyes. In terrestrial craniates, gill slits grow shut or are transformed into other structures such as the Eustachian tube and jaws, so gill slits are absent as such in the adult.

Craniates, like all chordates, are bilaterally symmetrical animals that develop from three embryonic germ layers: ectoderm, mesoderm, and endoderm. A well-developed coelom lined by a tissue layer called the peritoneum arises from the embryonic mesoderm. Thin membranes called mesentery suspend the internal organs in this coelom. The bodies of craniates are segmented. The backbone is a series of vertebrae (see Figure B) associated with nerves and muscles that are replicated in a series mirroring the segmented muscles present in cephalochordates. All craniates have a digestive tract complete with mouth and anus (Figure C). Although craniates reproduce sexually, a few species also reproduce parthenogenetically. These species consist of uniparental females in which no fertilization of the egg is required for development of the offspring. In the vast majority of species, male and female are separate individuals. Sexual reproduction requires the fertilization of a comparatively large egg by a much smaller, undulipodiated sperm.

Craniates grouped as agnatha lack jaws and paired appendages. All the other craniates are the gnathostomata, which have jaws—facilitating food getting and defense—and paired appendages. The only class of extant agnatha is Cyclostomata, the lampreys (*Petromyzon*), hagfish, and slime eels—aquatic animals that have gill slits and a round mouth like a suction cup with horny teeth on a protrusible tongue. The cyclostome axial skeleton, cartilaginous as in sharks, is essentially a reinforced notochord. The skull is a rigid box of cartilage that protects the brain. Cyclostomates have smooth, scaleless skin and lack the scales and fins of cartilaginous fish (such as *Squalus*) and bony fish. The cyclostomate does not have the ability to regulate its body temperature. The oldest remains of undoubtedly vertebrate animals is that of an ancient armored (jawless) fish with large scales called an ostracoderm, a member of a class of agnathids that is extinct. This class was a transitional form between cyclostomes (jawless fish) and jawed fish. Extant agnathids lack bones.

Of the 25,000 fish species, most—such as *Fundulus* and *Gadus*—are bony. Many species of sharks and bony fish are currently endangered by commercial exploitation and destruction of their near-shore nursery areas.

The classes Chondrichthyes (cartilaginous fish) and Osteichthyes (bony fish) are both gnathostome fish; they retain gills as adults, have paired fins, have jaws, and lack four limbs. Members of

Chordeiles
Chrysemys
Colaptes
Colius
Columba
Corvus
Crotalus
Cygnus
Cynocephalus
Dendrohyrax
Elephas
Equus
Felis
Fregata
Fulica
Fundulus
Gadus
Gallus
Gavia
Geococcyx
Gorilla
Homo
Latimeria
Lepus
Llama
Manus
Megaceryle
Meleagris
Mesocricetus
Micropterus
Mus
Myotis

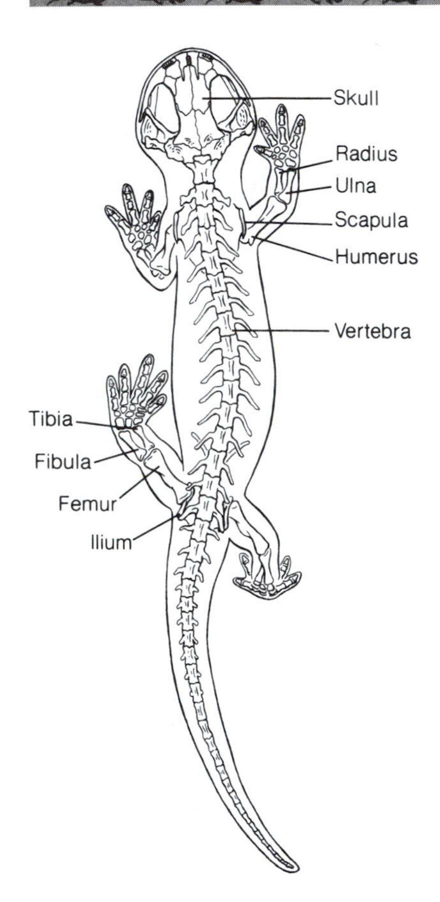

B Skeleton of a generalized salamander, dorsal view. Bony vertebrae enclose the dorsal hollow nerve cord. The cranium (skull) encloses the brain. [Drawing by L. Meszoly; information from R. Estes.]

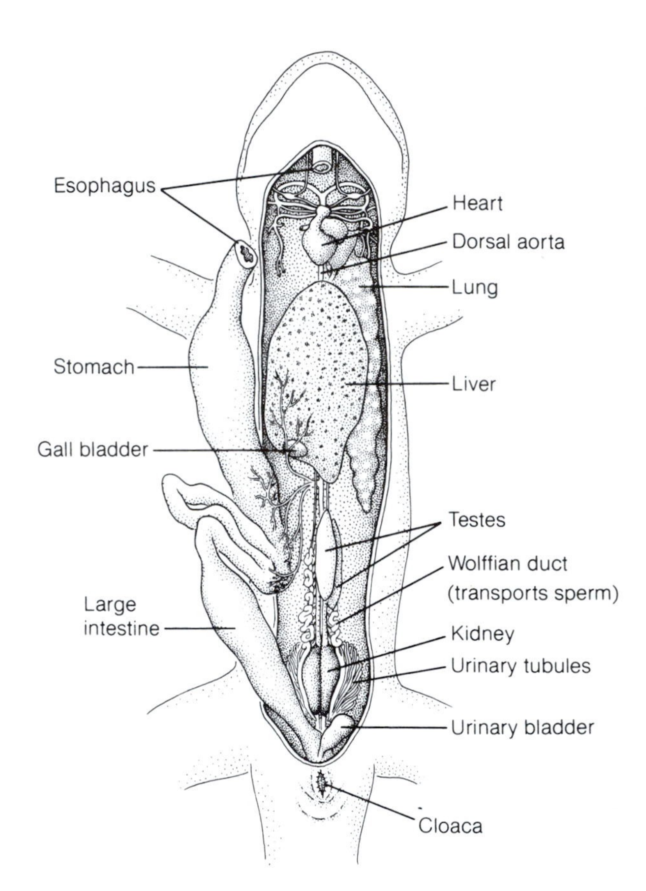

C Internal, ventral view of a generalized male salamander. [Drawing by L. Meszoly; information from R. Estes.]

Chondrichthys lack bones; their skeletons and skull are made instead of a softer but more flexible cell product called cartilage. Marine cartilaginous fish include sharks, skates, and sting rays. Their scales, called placoid scales and composed of a dentine plate covered by enamel, differ in origin from the scales of reptiles and bird legs. The extinct jawed, armored fish—placoderm—is the fossil predecessor of modern cartilagenous fish (Chondrichthyes) and bony fish (Osteichthyes).

Lungfish—Class Choanichthyes—live in African freshwater lakes with low dissolved oxygen. When a lake dries up, lungfish secrete a cocoon, dig into the mud, and gulp air into their exposed mouths.

Numida
Opisthocomus
Ornithorhynchus
Orycteropus
Oryctolagus
Ovis
Pan
Pavo
Petromyzon
Phascolarctus
Phasianus
Phoca
Podiceps
Pongo
Puffinus
Raja
Rangifer
Rhea
Salamandra
Sorex
Sphenodon
Squalus
Sterna
Struthio
Sus
Tachyglossus
Tinamus
Trichechus
Vulpes
Xenopus

All other fish besides cartilaginous fish and lungfish belong to Class Osteichthyes, the bony fish—salmon, tuna, trout, bass—and most saltwater and freshwater fish. Their scales are thin in comparison with placoid scales and are called cycloid or ctenoid, according to whether the outer edge of the scale is rounded (Greek *kyklos*, a circle) or toothed (Greek *ktenos*, comb). Moreover, cycloid and ctenoid scales form internally in the skin and do not emerge from the skin as do the placoid scales of the shark. The heart of a fish pumps blood from the body through a single atrium and a single ventricle, and then to the gills, from which oxygenated blood returns to the body; the circulation is closed.

The remaining classes of gnathostomes, Amphibia, Reptilia, Aves, and Mammalia, are all tetrapods (Greek *tetra*, four; *podos*, foot), having four limbs (except for some reptiles). Modern tetrapods are the most visible animals in terrestrial habitats today. We can trace the evolution of tetrapods by means of fossils and the only extant lobe-finned fish—the coelocanth—to the primitive ancestors that link extinct lobe-finned fishes to their descendants—to amphibians, to reptiles, and subsequently to birds and mammals. The bony fishes and amphibians have given rise to no other class of animals.

Members of Class Amphibia, about 200 described species, including salamanders (*Salamandra*), frogs (*Xenopus*), and toads (*Bufo*), lack scales and respire through their moist, flexible, scaleless skin and across the moist mouth lining, through gills—externally visible in some species—or through relatively small lungs. Some salamander species lack lungs and thus respire entirely across their moist skins. The amphibian heart has two atria but a single ventricle. Many but not all amphibians lay eggs and release sperm in water, where fertilization takes place. Most amphibians spend at least the earlier part of their lives in larval or juvenile form as aquatic beings; a few tropical frogs lack larvae—their embryos develop directly into miniature adults carried on the parent's body or in the minipond formed by a bromeliad. Some amphibians, such as the leopard frog, have aquatic larvae with gill slits (tadpoles). Bullfrogs and leopard frogs live much of their adult lives in fresh water. The Mexican axolotl retains its gilled larval form even as a sexually mature, aquatic adult.

Members of Class Reptilia have dry skin covered with scales—affording protection from desiccation and predation. The tissue layer from which the scales of reptiles develop differs from the tissue layer from which scales of fish develop. Reptiles develop from an egg more adapted to life on land—they are fertilized internally, have a leathery shell, and are retained in the mother until they hatch, in some species—than are the eggs of amphibians. Like embryos of birds, the gilled reptile embryo develops in its own small sea—a fluid-filled compartment enclosed by its eggshell. Reptiles lack aquatic larvae. Turtles (*Chrysemys*), lizards, snakes (*Crotalus*), crocodiles, and alligators together total about 5000 species. The popular Mesozoic era dinosaurs belonged to this class. Living reptiles are poikilothermic (unable to internally regulate body temprature): some species are adapted to a wide range of internal body temperatures; others regulate their temperature with behavioral adaptations such as basking in the sun or taking shelter from direct sun by moving to cooler, moister microhabitats. The leatherback turtle regulates its internal temperature somewhat. Each reptile tooth is generally rather similar to the others. Reptiles breathe by using lungs, although their scaly skin is slightly permeable to gas. The reptilian heart has two atria, and the ventricles are not completely separated (permitting partial mixing of oxygenated and deoxygenated blood), except in crocodiles, which have a complete septum between the left and the right ventricles. Most reptiles are well adapted to terrestrial life. Several sea turtle and snake species are endangered—overharvested for meat, shell, eggs, and skins. Baby sea turtles hatch from eggs laid on land but are disoriented by

D *Cygnus olor,* the mute swan, the swan common in parks and occasionally established in the wild. Swans are members of the class Aves (Latin *avis,* bird). All birds are aviators, except flightless species such as ostriches and the extinct dodo. Bar = 100 cm. [Courtesy of W. Ormerod.]

artificial night lights along the ocean, which they must reach to feed and breed. Floridians and other coastal residents are urged to turn down outdoor night lights to allow turtle hatchlings to reach the ocean.

Feathered reptiles—that is, birds (Figure D)—are placed in their own class, Aves. Nearly 9000 living bird species—among them, *Phasianus, Podiceps, Puffinus,* and *Rhea*—are recognized. To understand avian structure and behavior, it is best to think of birds as feathered reptiles derived from a branch of dinosaur reptiles, the ornithischians; ornithischian hip bones resemble those of contemporary birds. Aves have land-adapted eggs with porous calcium carbonate shells and internal membranes that facilitate gas exchange. Bird embryos deposit insoluble nitrogenous waste during their life within the shell; bird eggs develop externally. The forelimbs of many but not all (*Struthio*) bird species are modified for flight as wings, and their bones are hollow. Their scaly skin is studded with feathers; bird scales (look at the leg of a bird) are modified reptilian scales. The saying "scarce as hen's teeth" is based on biological fact: birds lack teeth. Birds have a four-chambered heart with two atria and two ventricles, and—like mammals—they regulate their internal temperature within a few degrees. Animals with this ability are called homeotherms. *Gavia,* the loon, has a lower body temperature than that of most other bird species.

Class Mammalia contains about 4500 living species, including *Homo sapiens, Bos, Felis, Phoca, Rangifer,* and *Sus.* As mammals, we nourish our young with milk, the nutritious and immunoprotective secretions of the mammary glands of the mother. All mammals are homeotherms—regulators of body temperature. Some species, such as our own, allow only a small range of variation. Others—opossums and hibernators such as woodchucks and ground squirrels—evolved a much broader range of body temperatures. The hair that covers the skin of many mammals at some stage of life is one of several physical and behavioral temperature-control features. The mammalian heart has four chambers. Mammals have complete double circulation: the oxygen-rich blood in the arteries does not mix with the oxygen-depleted blood in the veins. In most mammalian species, an elaborate organ, the placenta, nourishes the developing embryo; fertilized eggs develop inside the female and young are born live. Parental care, although not absent in other classes of vertebrates and even some invertebrates, is well developed in many mammals. Mammals have complex and differentiated teeth.

There are nearly twenty orders of mammals in two great subclasses, Prototheria and Theria. Subclass Prototheria includes all the egg-laying mammals of Australia, New Guinea, and Tasmania. Subclass Theria—all mammals that do not lay eggs—includes all other mammals. The spiny anteater, *Tachyglossus aculeatus,* and the duck-billed platypus, *Ornithorhynchus anatinus*—both prototherians—have lower body temperatures than do most other mammals (28.3°C compared with 38°C). The cloaca of prototherians, like that of birds and reptiles, is a common chamber for digestive waste, excretory products, and eggs or sperm. Both prototherian genera also have horny beaks or bills but lack true teeth. They lay shelled and yolk-rich cherry-sized eggs, have an external pouch, and reptilelike bones. All egg-laying mammals nourish their young with milk from primitive mammary glands after hatching.

Subclass Theria comprises two infraclasses—Metatheria (marsupials such as opposums and *Phascolarctus,* the koala) and Eutheria (placental mammals—*Balaenoptera, Cavia, Pan, Vulpes*). The young of kangaroos and other metatherians are extremely immature at birth after a brief sojourn in the mother's uterus. Their relatively well developed forepaws permit metatherian newborns to crawl into an exterior pouch in which young suckle milk while attached to the mammary glands as they continue development. The metatherian female has a cloaca, two vaginas, and a double (Y-shaped) uterus. The eutherian female has a single uterus and single vagina. Eutherian young undergo considerable development inside the mother's uterus, where they are nourished inside her body by the transfer of nutrients and by gas exchange through the placenta before she gives birth. Eutherian orders include Insectivora (hedgehogs, shrews, moles), Primates (lemurs, tarsiers, monkeys, apes, humans, chimpanzees, gorillas), Hyracoidea (hyraxes), Chiroptera (bats), Dermoptera (flying lemurs), Rodentia (porcupines, mice, squirrels, chipmunks, capybaras), Carnivora (dogs, cats, bears, otters), Pinnipedia (seals, sea lions), Tupaioidea (tree shrews), Edentata (sloths, armadillos, anteaters), Pholidota (pangolins), Lagomorpha (hares, rabbits), Cetacea (whales, dolphins), Tubulidentata (aardvark), Proboscidea (elephants), Sirenia (sea cows, manatees) Perissodactyla (zebras, horses, tapirs), and Artiodactyla (pigs, camels, llamas, deer, cattle, bison).

Bibliography: Animalia

General

Adrianov, A. V., and V. V. Malakhov, "The phylogeny and classification of the phylum Cephalorhyncha." *Zoosystematica Rossica* 3(2): 181–201; 1995.

Barnes, R. D., *The invertebrates.* Blackwell; Oxford, UK; 1995.

Bayer, F. M., and H. B. Owre, *The free-living lower invertebrates.* Macmillan; New York; 1968.

Boldouf, S. L., and J. D. Palmer, "Animals and fungi are each other's closest relatives: Congruent evidence from multiple proteins." *Proceedings of the National Academy of Sciences of the United States of America* 90:11,558–11,562; 1993.

Brusca, R. C., and G. J. Brusca, *Invertebrates.* Sinauer; Sunderland, MA; 1990.

Burton, M., ed., *New Larousse encyclopedia of animal life,* rev. ed. of *Larousse encyclopedia of animal life.* Bonanza Books; New York; 1980.

Cloud, P. E., "Pre-metazoan evolution and the origins of the metazoa." In E. T. Drake, ed., *Evolution and environment.* Yale University Press; New Haven, CT; 1968.

Conway Morris, S., J. D. George, R. Gibson, and H. M. Platt, eds., *The origins and relationships of lower invertebrates.* Systematics Association, Special Volume 28; Clarendon Press; Oxford; 1985.

Dorit, R. L., W. F. Walker, and R. D. Barnes, *Zoology,* Saunders; Philadelphia; 1991.

Erwin, D. H., *The great Paleozoic crisis: Life and death in the Permian.* Columbia University Press; New York; 1993.

Erwin, D. H., "The mother of mass extinctions." *Scientific American* 275(1):72–78. July 1996. Two hundred fifty million years before the present, mass end-Permian extinction led to expansion of mobile marine animals and decrease in immobile forms.

Gilbert, L. E., and P. H. Raven, *Co-evolution of animals and plants.* University of Texas Press; Austin, TX; 1980.

Glaessner, M. F., *The dawn of animal life.* Cambridge University Press; Cambridge, UK; 1984.

Gosner, K. L., *Guide to identification of marine and estuarine invertebrates: From Cape Hatteras to the Bay of Fundy.* Wiley (Interscience); New York; 1974. Books on Demand, Ann Arbor, MI.

Gould, S., *Wonderful life.* Norton; New York; 1989.

Grassé, P.-P., ed., *Larousse encyclopedia of the animal world.* Larousse; New York; 1975.

Grassé, P.-P., ed., *Traité de zoologie: anatomie, systématique, biologie,* 17 vols. Masson; Paris; 1948–continuing.

Hanson, E. D., *Origin and early evolution of animals.* Wesleyan University Press; Middletown, CT; 1977. Books on Demand; Ann Arbor, MI.

Higgins, R. P., and H. Thiel, eds., *Introduction to the study of meiofauna.* Smithsonian Institution Press; Washington, DC; 1988.

House, M. R., ed., *Origin of major invertebrate groups.* Academic Press; London; 1979.

Hyman, L. H., *The invertebrates,* 6 vols. McGraw-Hill; New York; 1940–1967. Vol. 1, *Protozoa through Ctenophora;* 1940. Vol. 2, *Platyhelminthes and Rhynchocoela;* 1951. Vol. 3, *Acanthocephala, Aschelminthes, and Entoprocta;* 1951. Vol. 4, *Echinodermata;* 1955. Vol. 5, *Smaller coelomate groups;* 1959. Vol. 6, *Mollusca,* Part 1; 1967.

Kinchin-Simonetta, A. M., and S. Conway Morris, *The early evolution of metazoa and the significance of problematic taxa.* Cambridge University Press; Cambridge, UK; 1991.

Kozloff, E. N., *Invertebrates.* Saunders; Philadelphia; 1989.

McMenamin, M. A. S., "The emergence of animals." *Scientific American* 256(4):94–102; April 1987.

McMenamin, M. A. S., and D. L. S. McMenamin, *The emergence of animals: The Cambrian breakthrough.* Columbia University Press; New York; 1990.

Meglitsch, P. A., and F. R. Schram, *Invertebrate zoology,* 3d ed. Oxford University Press; New York; 1991.

Nichols, D., J. A. L. Cooke, and D. Whitely, *Oxford book of invertebrates.* Oxford University Press; New York and London; 1971.

Nybakken, J., *Diversity of the invertebrates: A lab manual.* W. C. Brown; Dubuque, IA; 1996.

Nybakken, J., *Marine biology: An ecological approach,* 4th ed. Benjamin Cummings; Redwood City, CA; 1997.

Nybakken, J., and J. L. McClintock, *Diversity of the invertebrates: A laboratory manual,* Gulf of Mexico version. W. C. Brown, Dubuque, IA; 1997.

Parker, S. P., ed., *Synopsis and classification of living organisms,* 2 vols. McGraw-Hill; New York; 1982.

Pechenik, J. A., *Biology of the invertebrates,* 3d ed. W. C. Brown; Dubuque, IA; 1996.

Pennak, R. W., *Collegiate dictionary of zoology.* Krieger; Melbourne, FL; 1987.

Pennak, R. W., *Freshwater invertebrates of the United States,* 3d ed. Wiley (Interscience); New York; 1989.

Ruppert, E. E., and R. D. Barnes, *Invertebrate zoology,* 6th ed. Saunders; Philadelphia; 1994.

Ruppert, E. E., and R. S. Fox, *Seashore animals of the southeast United States.* University of South Carolina Press; Columbia, SC; 1988.

Sterrer, W., *Marine fauna and flora of Bermuda: A systematic guide to the identification of marine organisms.* Wiley (Interscience); New York; 1986.

Thorpe, J. H., and A. P. Covich, eds., *Ecology and classification of North American freshwater invertebrates.* Academic Press; New York; 1991.

Vogel, S., *Life in moving fluids: The physical biology of flow,* 2d ed. Princeton University Press; Princeton, NJ; 1994.

Whittington, H. B., *The Burgess shale.* Yale University Press; New Haven, CT, and London; 1985.

Williamson, D., *Larvae and evolution: Toward a new zoology.* Chapman and Hall; London; 1994.

Willmer, P., *Invertebrate relationships.* Cambridge University Press; New York; 1990.

A-1 Placozoa

Grell, K. G., and G. Benwitz, "Die Ultrastruktur von *Trichoplax adhaerens* F. E. Schulze." *Cytobiologie* 4:216–240; 1971 (English abstract).

Miller, R. L., "Observations on *Trichoplax adhaerens* Schulze 1883." *American Zoologist* 11:513; 1971.

Miller, R. L., "*Trichoplax adhaerens* Schulze 1883: Return of an enigma." *Biological Bulletin* 141:374; 1971.

Ruthmann, A., "Cell differentiation, DNA content, and chromosomes of *Trichoplax adhaerens.*" *Cytobiologie* 15: 58–64; 1977.

A-2 Porifera

Bayer, F. M., and H. B. Owre, *The free-living lower invertebrates.* Macmillan; New York; 1968.

Bergquist, P. R., *Sponges.* University of California Press; Berkeley, CA; 1978.

de Laubenfels, M. W., *A guide to the sponges of eastern North America.* University of Miami Press; Coral Gables, FL; 1953.

Gowell, E. T., *Sea jellies: Rainbows in the sea.* Franklin Watts; New York; 1993.

Newell, N. D., "The evolution of reefs." *Scientific American* 226:54–65; June 1972.

Rasmont, R., "Sponges and their world." *Natural History* 71(3):62–70; March 1962.

Vogel, S., "Current-induced flow through the sponge, *Halochondria.*" *Biological Bulletin* 147:443–456; 1974.

Vogel, S., "Organisms that capture currents." *Scientific American* 239(2):128–139; August 1978.

A-3 Cnidaria (Coelenterata, in part)

Argo, V. N., "Mechanics of a turnover: Bell contractions propel jellyfish." *Natural History* 74(7):26–29; August–September 1965.

Brown, B. E., and J. C. Ogden, "Coral bleaching." *Scientific American* 268(1):64–70; January 1993.

Gao, T. J., "Morphological and biomechanical differences in healing in segmental tibia defects implanted with Biocoral [bone substitute] or tricalcium phosphate cylinders." *Biomaterials* 18:219–223; 1997.

Gould, S. J., "A most ingenious paradox." *Natural History* 93(12):20–29; December 1984.

Gowell, E. T., *Sea jellies: Rainbows in the sea.* Franklin Watts; New York; 1993.

Grange, K., and W. Goldberg, "Fjords down under." *Natural History* 102(3):60–69; March 1993.

Jacobs, W., "Floaters of the sea." *Natural History* 71(7): 22–27; August–September 1962.

Kramp, P. L., "Synopsis of the medusae of the world." *Journal of the Marine Biological Association of the United Kingdom* 40:1–469; 1961.

Lane, C. E., "The Portuguese man-of-war." *Scientific American* 202(3):158–168; March 1960. *Physalia.*

Muscatine, L., and H. M. Lenhoff, eds., *Coelenterate biology.* Academic Press; New York; 1974.

Rees, W. J., ed., *The Cnidaria and their evolution.* Academic Press; New York; 1966.

Roux, F. X., D. Brasnu, B. Loty, B. George, and G. Guillemin, "Madreporic coral: A new bone graft substitute for cranial surgery." *Journal of Neurosurgery* 69(4):510–513; 1988.

Shick, J. M., *A functional biology of sea anemones.* Chapman and Hall; New York; 1991.

A-4 Ctenophora (Coelenterata, in part)

Hardy, A., *Great waters.* Harper & Row; New York; 1967.

Matsumoto, G. I., and G. R. Harbison, "*In situ* observations of foraging, feeding, and escape behavior in three orders of oceanic ctenophores: Lobata, Cestida, and Beroida." *Marine Biology* 117:279–287; 1993.

Purcell, J. E., and J. H. Cowan, Jr., "Predation by the scyphomedusan *Chrysaora quinquecirrha* on *Mnemiopsis leidyi* ctenophores." *Marine Ecology Progress Series* 129: 63–70; 1995.

Robison, B. H., "Light in the ocean's midwaters." *Scientific American* 273(1):60–64; July 1995. Exploring the lives of bioluminescent sea animals by using ROVs (remotely operated vehicles) and submersibles.

Russell-Hunter, W. D., *A biology of lower invertebrates.* Macmillan; New York; 1968.

Tamm, S. L., and S. Tamm, "Reversible epithelial adhesion closes the mouth of *Beroë,* a carnivorous marine jelly." *Biological Bulletin* 181:463–473; 1991. How a pink, blimp-shaped ctenophore feeds.

Vogel, S., "Nature's pumps." *American Scientist* 82(5): 464–471; September–October 1991.

A-5 Platyhelminthes

Cheng, T. C., *Parasitology,* 2d ed. Academic Press; New York; 1986.

Dawes, B., *The Trematoda.* Cambridge University Press; Cambridge, UK; 1946.

Erasmus, D. A., *The biology of trematodes.* Crane, Russak; New York; 1974. Arnold; London; 1972.

Smyth, J. D., *The physiology of cestodes.* W. H. Freeman and Company; New York; 1969.

Smyth, J. D., and D. W. Halton, *The physiology of trematodes,* 2d ed. Cambridge University Press; New York; 1983.

A-6 Gnathostomulida

Briggs, D. E. G., "Conodonts: A major extinct group added to the vertebrates." *Science* 256:1285–1286; 1992.

Fenchel, T. M., and R. Riedl, "The sulfide system: A new biotic community underneath the oxidized layer of marine sand bottoms." *Marine Biology* 7:255–268; 1969.

Gould, S. J., "Nature's great era of experiments." *Natural History* 92(7):12–21; July 1983.

Lammert, V., "Gnathostomulida," pp. 19–39. In F. W. Harrison and E. E. Ruppert, eds., *Microscopic anatomy of invertebrates,* Vol. 4: *Aschelminthes.* Wiley-Liss; New York; 1991.

Sterrer, W., M. Mainitz, and R. M. Rieger, "Gnathostomulida: Enigmatic as ever," pp. 181–199. In S. Conway Morris, D. George, R. Gibson, and H. M. Platt, eds., *The origins and relationships in lower invertebrates.* Oxford University Press; Oxford; 1986.

A-7 Rhombozoa

Lapan, E. A., and H. Morowitz, "The Mesozoa." *Scientific American* 227(6):94–101; December 1972.

McConnaughey, B. H., "The Mesozoa," pp. 557–570. In M. Florkin and B. T. Scheer, eds., *Chemical zoology,* Vol. 2. Academic Press; New York; 1968.

Noble, E. R., and G. A. Noble, *Parasitology,* 5th ed. Lea & Febiger; Philadelphia; 1982.

A-8 Orthonectida

Kozloff, E. N., "Morphology of the orthonectid *Ciliocincta sabellariae.*" *Journal of Parasitology* 57:585–597; 1971.

Kozloff, E. N., "Morphology of the orthonectid *Rhopalura ophiocomae.*" *Journal of Parasitology* 55:171–195; 1969.

Rader, D. N., "Orthonectid parasitism: Effects on the ophiuroid, *Amphipholis squamata,*" pp. 395–401. In J. M. Lawrence, ed., *Echinoderms* [Proceedings of the International Conference, Tampa Bay, 14–17 September 1981]. Balkema; Rotterdam; 1981.

A-9 Nemertina

Gibson, R., *Nemerteans.* Hutchinson University Library; London; 1972.

Kozloff, E. N., *Seashore life of the northern Pacific Coast: An illustrated guide to the common organisms of northern California, Oregon, Washington, and British Columbia.* University of Washington Press; Seattle; 1983.

Roe, P., and J. L. Norenburg, eds., "Comparative biology of nemertines," from symposium of American Society of Zoologists 27–30 December 1983. *American Zoologist* 25:3–151; 1985.

Turbeville, J. M., and E. E. Ruppert, "Comparative ultrastructure and the evolution of nemertines." *American Zoologist* 25:53–71; 1985.

Turbeville, J. M., et al., "Phylogenetic position of phylum Nemertini, inferred from 18s RNA sequences: Molecular data as a test of morphological character homology." *Molecular Biology and Evolution* 9:235–249. 1992.

A-10 Nematoda

Bird, A. F., and J. Bird, *Structure of nematodes,* 2d ed. Academic Press; New York; 1991.

Cheng, T. C., *Parasitology,* 2d ed. Academic Press; New York; 1986.

Goodey, J. B., *Soil and freshwater nematodes.* Wiley; New York; 1963.

Hope, W. D., ed., *Nematodes: Structure, development, classification, and phylogeny.* Smithonian Institution Press; Washington, DC; 1994.

Hotez, P. J., and D. I. Pritchard, "Hookworm infection." *Scientific American* 272(6):68–74. June 1995. Biology of hookworm infection and vaccine development.

Lee, D. L., and H. J. Atkinson, *The physiology of the nematodes,* 2d ed. Columbia University Press; New York; 1977.

Maio, J. J., "Predatory fungi." *Scientific American* 199(1):67–72; July 1958. Nematodes trapped by predatory fungi.

A-11 Nematomorpha

Cheng, T. C., *Parasitology,* 2d edition. Academic Press; New York; 1986.

Croll, N. A., *Ecology of parasites.* Harvard University Press; Cambridge, MA; 1966.

Noble, E. R., and G. A. Noble, *Parasitology,* 5th ed. Lea & Febiger; Philadelphia; 1982.

Poinar, G. O., Jr., "Nematoda and Nematomorpha," ch. 9, pp. 273–282. In J. H. Thorpe and A. P. Covich, eds., *Ecology and classification of North American freshwater invertebrates.* Academic Press; New York; 1991.

A-12 Acanthocephala

Baer, J. G., *Animal parasites.* World University Library; London; 1971. McGraw-Hill; New York; 1971.

Conway Morris, S., and D. W. T. Crompton, "The origins and evolution of Acanthocephala." *Biological Reviews* 57:85–113; 1982.

Crompton, D. W. T., *Parasitic worms.* Taylor and Francis; Bristol, PA; 1980.

Moore, J., "Parasites that change the behavior of their host." *Scientific American* 250(5):108–115; May 1984.

Nicholas, W. L., "The biology of Acanthocephala," pp. 205–206. In B. Dawes, ed., *Advances in parasitology 5.* Academic Press; New York; 1967.

Noble, E. R., and G. A. Noble, *Parasitology,* 5th ed. Lea & Febiger; Philadelphia; 1982.

Olsen, O. W., *Animal parasites: Their life cycles and ecology.* Dover; New York; 1986.

A-13 Rotifera

Barron, G., "Jekyll-Hyde mushrooms." *Natural History* 101(3):47–52; March 1992.

Donner, J., *Rotifers.* Stuttgart; 1956. Reprint, Frederick Warne; London; 1965.

Eddy, S., and A. C. Hodson, *Taxonomic keys to the common animals of the north central states,* 4th ed. Burgess; Minneapolis; 1982.

Edmondson, W. T., "Rotifera," pp. 420–494. In W. T. Edmondson, H. B. Ward, and G. C. Whipple, eds., *Freshwater biology,* 2d ed. Wiley; New York; 1959.

Nogrady, T., R. L. Wallace, and T. W. Snell, *Biology, ecology, and systematics,* Vol. 1. SPB Academic Publishing; the Hague; 1993.

Pennak, R. W., "Ecological affinities and origins of freeliving acoelomate freshwater invertebrates," pp. 435–451. In E. C. Dougherty, ed., *The lower metazoa: Comparative biology and phylogeny.* University of California Press; Berkeley, CA; 1963.

A-14 Kinorhyncha

Dougherty, E. C., ed., *The lower metazoa: Comparative biology and phylogeny.* University of California Press; Berkeley, CA; 1963.

Higgins, R. P., "A historical overview of kinorhynch research." In N. C. Hullings, ed., "Proceedings of the first international conference on meiofauna." *Smithsonian Contributions to Zoology* 76:25–31; 1971.

Higgins, R. P., and H. Thiel, eds., *Introduction to the study of meiofauna.* Smithsonian Institution Press; Washington, DC; 1988.

Morell, V., "Life on a grain of sand." *Discover* 16:78–86; April 1995.

Russell-Hunter, W. D., *A biology of lower invertebrates.* Macmillan; New York; 1968.

A-15 Priapulida

Adrianov, A. V., and V. V. Malakhov, "The phylogeny and classification of the phylum Cephalorhyncha." *Zoosystematica Rossica* 3(2): 181–201; 1995.

Hammond, R. A., "The burrowing of *Priapulis caudatus.*" *Journal of Zoology* 162:469–480; 1970.

Higgins, R. P., V. Storch, and T. C. Shirley, "Scanning and transmission electron microscopical observations on the larvae of *Priapulus caudatus* (Priapulida)." *Acta Zoologica (Stockholm)* 74(4):301–319; 1993.

Morse, P., "*Meiopriapulus fijiensis,* n.gen., n. sp.: An interstitial priapulid from coarse sand in Fiji." *Transactions of the American Microscopical Society* 100:239–252; 1981.

Por, F. D., and H. J. Bromley, "Morphology and anatomy of *Maccabeus tentaculatus.*" *Journal of Zoology* 173:173–197; 1974.

Shirley, T. C., "Ecology of *Priapulus caudatus* Lamarck, 1816 (Priapulida) in an Alaskan subarctic ecosystem." *Bulletin of Marine Science* 47:149–158; 1990.

A-16 Gastrotricha

Brunson, R. B., "Aspects of the natural history and ecology of the gastrotrichs," pp. 473–478. In E. C. Dougherty, ed., *The lower metazoa: Comparative biology and phylogeny.* University of California Press; Berkeley, CA; 1963.

Brunson, R. B., "Gastrotricha," pp. 406–419. In W. T. Edmondson, H. B. Ward, and G. C. Whipple, eds., *Freshwater biology,* 2d ed. Wiley; New York; 1959.

D'Hondt, J.-L., "Gastrotricha." *Oceanography and Marine Biology: An Annual Review* 9:141–192; 1971.

Hummon, W. D., "Biogeography of sand beach Gastrotricha from the northeastern United States." *Biological Bulletin* 141:390; 1971.

Morell, V., "Life on a grain of sand." *Discover* 16:78–86; April 1995.

Thorpe, J. H., and A. P. Covich, eds., *Ecology and classification of North American freshwater invertebrates.* Academic Press; New York; 1991.

A-17 Loricifera

Anonymous, "New phylum found." *Bioscience* 34:321; 1984.

Higgins, R. P., and R. M. Kristensen, "New loricifera from southeastern United States coastal waters." *Smithsonian Contributions to Zoology* 438:1–70; 1986.

Higgins, R. P., and H. Thiel, eds., *Introduction to the study of meiofauna.* Smithsonian Institution Press; Washington, DC; 1988.

Kristensen, R. M., "Loricifera, a new phylum with Aschelminthes characters from the meiobenthos." *Zeitschrift für zoologische Systematik und Evolutionsforschung* 21:163–180; 1983.

Kristensen, R. M., "Loricifera: A general biological and phylogenetic overview." *Verhandlungen der Deutschen Zoologischen Gesellschaft* 84:231–246; 1991.

A-18 Entoprocta

Emschermann, P., "Factors inducing sexual maturation and influencing the sex determination of *Barentsia discreta* Busk (Entoprocta, Barentsiidae)," pp. 101–108. In C. Neilsen and G. P. Larwood, eds., *Bryozoa: Ordovician to recent.* Olsen and Olsen; Fredensborg, Denmark; 1985.

Funch, P., and R. M. Kristensen, "Cycliophora is a new phylum with affinities to Entoprocta and Ectoprocta." *Nature* 378:711–714; 1995.

Nielsen, C. *Animal evolution: Interrelationships of the living phyla.* Oxford University Press; New York; 1995.

Nielsen, C., "The relationships of Entoprocta, Ectoprocta and Phoronida." *American Zoologist* 17:149–150; 1977. Of historical interest. See *Animal evolution* (1995) for his current views.

A-19 Chelicerata

Gupta, A. P., *Arthropod phylogeny.* Van Nostrand Reinhold; New York; 1979.

Henschel, J., "Spider Revolutions." *Natural History* 104(3):36–39; March 1995.

Jack, R. R., "Arachnomania." *Natural History* 104(3):28–31; March 1995.

Kaston, B. J., *How to know the spiders,* 3d ed. W. C. Brown; Dubuque, IA; 1978.

King, P. E., *Pycnogonids.* Hutchinson University Library; London; 1973.

Polis, G. A., ed., *The biology of scorpions.* Stanford University Press; Stanford, CA; 1990.

Polis, G. A., "The unkindest sting of all." *Natural History* 98(7):34–39; July 1989.

Snodgrass, R. E., *A textbook of arthropod anatomy.* Cornell University Press; Ithaca, NY; 1952.

Watson, P. J., "Dancing in the dome." *Natural History* 104(3):40–47; March 1995.

A-20 Mandibulata (Uniramia)

Batra, L. R., and S. W. T. Batra, "The fungus gardens of insects." *Scientific American* 217(5):112–120; November 1967.

Borrer, D. J., and R. E. White, *A field guide to the insects of America north of Mexico.* Houghton Mifflin; Boston; 1970.

Cottam, C. A., and H. S. Zim, *Insects: A guide to familiar American species.* Golden; New York; 1987.

Dillon, E. S., and L. S. Dillon, *A manual of common beetles of eastern North America,* 2 vols. Dover; New York; 1972.

Gupta, A. P., *Arthropod phylogeny.* Van Nostrand Reinhold; New York; 1979.

A-21 Crustacea

Bliss, D. E., ed., *The Biology of Crustacea,* Vols. 1–9. Academic Press; New York. 1982–1987.

Emerson, M. J., and F. R. Schram, "The origin of crustacean biramous appendages and the evolution of Arthropoda." *Science* 250:667–669; 1990.

Fleminger, A., "Description and phylogeny of *Isaacsicalanus paucisetus,* n. gen., n. sp. (Copepoda: Calanoida: Spinocalanidae) from an east Pacific hydrothermal vent site (21 degrees N)." *Proceedings of the Biological Society of Washington* 96(4):605–622; 1983.

Gould, S. J., "Of tongue worms, velvet worms, and water bears." *Natural History* 104(1):6–15; January 1995.

Gupta, A. P., *Arthropod phylogeny.* Van Nostrand Reinhold; New York; 1979.

Haugerud, R. E., "Evolution in the Pentastomidas." *Parasitology Today* 5(4):126–132; 1989.

Nichols, D., J. Cooke, and D. Whiteley, *Oxford book of invertebrates.* Oxford University Press; New York; 1971.

Noble, E. R., and G. A. Noble, *Parasitology,* 5th ed. Lea & Febiger; Philadelphia; 1982.

Schram, F. R., *Crustacea.* Oxford University Press; New York; 1986.

Self, J. T., "Biological relationships of the Pentastomida." *Experimental Parasitology* 24:63–119; 1969.

A-22 Annelida

Brinkhurst, R. O., "Evolution in the Annelida." *Canadian Journal of Zoology* 60:1043–1059; 1982.

Dales, R. O., *Annelids,* 2d ed. Hutchinson University Library; London; 1967.

Darwin, C. R., *The formation of vegetable mould through the action of worms with observations on their habits.* 1881. Reprinted as *Darwin on earthworms: The formation of vegetable mould through the action of worms.* Bookworm Publications; Russelville, AR; 1976.

Edwards, C. A., and J. R. Lofty, *Biology of earthworms,* 2d ed. Chapman and Hall; London; 1972.

Eernisse, D. J., J. S. Albert, and F. E. Anderson, "Annelida and Arthropoda are not sister taxa: A phylogenetic analysis of spiralian metazoan morphology." *Systematic Biology* 41:305–330; 1992.

Laverack, M. S., *The physiology of earthworms.* Macmillan; New York; 1963.

Wells, G. P., "Worm autobiographies." *Scientific American* 200(6):132–142; June 1959. Annelid behavior patterns.

A-23 Sipuncula

Cutler, E. B., *The Sipuncula: Their systematics, biology and evolution.* Cornell University Press; Ithaca, NY; 1994.

MacGinitie, G. E., and N. MacGinitie, *Natural history of marine animals,* 2d ed. McGraw-Hill; New York; 1968.

Rice, M. E., J. Piraino, and H. F. Reichardt, "Observations on the ecology and reproduction of the sipunculan *Phascolion cryptus* in the Indian River lagoon." *Florida Scientist* 46:382–396; 1983.

Scheltema, A. H., "Aplacophora as progenetic aculiferans and the coelomate origin of mollusks as the sister taxon of Sipuncula." *Biological Bulletin* 184:57–78; 1993.

Stephen, A. C., and S. J. Edmonds, *The phyla Sipuncula and Echiura.* British Museum (Natural History); London; 1972.

A-24 Echiura

Kohn, A., and M. Rice, "Biology of Sipuncula and Echiura." *Bioscience* 21:583–584; 1971.

MacGinitie, G. E., and N. MacGinitie, *Natural history of marine animals,* 2d ed. McGraw-Hill; New York; 1968.

Risk, M. J., "Silurian echiuroids: Possible feeding traces in the Thorold sandstone." *Science* 180:1285–1287; 1973.

Stephen, A. C., and S. J. Edmonds, *The phyla Sipuncula and Echiura.* British Museum (Natural History); London; 1972.

Wolcott, T. G., "Inhaling without ribs: The problem of suction in soft-bodied invertebrates." *Biological Bulletin* 160:189–197; 1981.

A-25 Pogonophora

George, J. D., and E. C. Southward, "A comparative study of the setae of Pogonophora and polychaetous Annelida." *Journal of the Marine Biological Association of the United Kingdom* 53:403–424; 1973.

Gould, S. J., "Microcosmos." *Natural History* 105(3):21, 23, 66, 68; March 1996.

Ivanov, A. V., *Pogonophora.* Consultants Bureau; New York; 1963.

Jones, M. L., "The Vestimentifera: Their biology, systematics and evolutionary patterns" (Biology and Ecology Symposium, Paris, 4–7 November, 1985, Proceedings), L. Laubier, ed. *Oceanologica Acta.* Special Volume 8:69–82; 1988.

Nørevang, A., ed., *The phylogeny and systematic position of Pogonophora* (special issue *of Zeitschrift für Zoologische Systematik und Evolutionsforschung*). Verlag Paul Parey; Hamburg and Berlin; 1975.

Southward, A. J., and E. C. Southward, "Pogonophora," pp. 201–228. In T. J. Pandian and F. J. Vernberg, eds., *Animal energetics,* Vol. 2. Academic Press; New York; 1987.

Webb, M., "*Lamellibrachia barhami,* gen. nov., sp. nov. (Pogonophora), from the northeast Pacific." *Bulletin of Marine Science* 19:18–47; 1969.

A-26 Mollusca

Abbott, R. T., *American seashells: The marine Mollusca of the Atlantic and Pacific coasts of North America,* 2d ed. Van Nostrand Reinhold; New York; 1974.

Brooks, W. K., *The oyster.* Johns Hopkins University Press; Baltimore; 1996.

Lane, F. W., *Kingdom of the octopus: The life history of the Cephalopoda.* Jarrolds; London; 1960. Sheridan House; New York; 1960.

Morris, P. A., *A field guide to the shells of the Atlantic and Gulf coasts and the West Indies,* 3d ed. Houghton Mifflin; Boston; 1973.

Morton, J. E., *Molluscs,* 4th ed. Hutchinson University Library; London; 1967.

Page, H. M., C. R. Fisher, and J. J. Childress, "Role of filter-feeding in the nutritional biology of a deep-sea mussel

with methanotrophic symbionts." *Marine Biology* 104:251–257; 1990.
Scheltema, A. H., "Aplacophora as progenetic aculiferans and the coelomate origin of mollusks as the sister taxon of sipuncula." *Biological Bulletin* 184:57–78; 1993.
Solem, A., *The shell makers.* Wiley (Interscience); New York; 1974.
Vogel, S., "Flow-assisted mantle cavity refilling in jetting squid." *Biological Bulletin* 172:61–68; 1987.
Wilbur, K. M., ed., *The Mollusca,* 12 vols. Academic Press; New York; 1983–1988.
Yonge, C. M., *Oysters,* 2d ed. Collins; London; 1966.

A-27 Tardigrada

Clark, K. U., "Visceral anatomy and arthropod phylogeny," pp. 467–547. In A. P. Gupta, ed., *Arthropod phylogeny.* Van Nostrand Reinhold; New York; 1979.
Cooper, K. W., "The first fossil tardigrade *Beørn leggi* Cooper from Cretaceous amber." *Psyche* 71:41–48; 1964.
Crowe, J. H., and A. F. Cooper, Jr., "Cryptobiosis." *Scientific American* 225(6):30–36; December 1971.
Gould, S. J., "Of tongue worms, velvet worms, and water bears." *Natural History* 104(1):6–15; January 1995. Pentastomes, onychophorans, tardigrades.
Kinchin, I. M., *The biology of tardigrades.* Portland Press; London and Chapel Hill, NC; 1994.
Nelson, D. R., "The hundred-year hibernation of the water bear." *Natural History* 84(7):62–65; July 1975.
Pennak, R. W., "Ecology of the microscopic Metazoa inhabiting the sandy beaches of some Wisconsin lakes." *Ecological Monographs* 10:537–615; 1940.
Pennak, R. W., *Freshwater invertebrates of the United States,* 3d ed. Wiley (Interscience); New York; 1989.
Wright, J. C., "Desiccation tolerance and water-retentive mechanisms in tardigrades." *Journal of Experimental Biology* 142:267–292; 1989.

A-28 Onychophora

Boudreaux, H. B., *Arthropod phylogeny with special reference to insects.* Wiley (Interscience); New York; 1979.
Ghiselin, M. T., "A movable feaster." *Natural History* 94(9):54–61; September 1985.
Gould, S. J., "Of tongue worms, velvet worms, and water bears." *Natural History* 104(1):6–15; January 1995.
Manton, S. M., *The Arthropoda: Habitats, functional morphology, and evolution.* Oxford University Press (Clarendon Press); Oxford, UK; 1977.
Monge-Najera, J., "Jurassic-Pleiocene biogeography: Testing a model with velvetworm (Onychophora) vicariance." *Revista de Biologia Tropical* 44(1):159–175; 1996.
Monge-Najera, J., "Phylogeny, biogeography and reproductive trends in the Onychophora." *Zoological Journal of the Linnean Society* 114:21–60; 1995.
Peck, S. T., "A review of the New World Onychophora with the description of a new cavernicolous genus and species from Jamaica." *Psyche* 82:341–358; 1975.
Ross, H. H., *Textbook of entomology,* 3d ed. Wiley; New York; 1965. [Reprinted Krieger; Melbourne, FL; 1991.]
Snodgrass, R. E., "Evolution of the Annelida, Onychophora, and Arthropoda." *Smithsonian Miscellaneous Collection* 97:1–159; 1938.

A-29 Bryozoa (Ectoprocta)

Boardman, R. S., A. H. Cheetham, and W. A. Oliver, Jr., eds., *Animal colonies: Development and function through time.* Dowden, Hutchinson Ross; Stroudsburg, PA; 1973.
Larwood, G. P., *Living and fossil Bryozoa.* Academic Press; New York; 1973.
Larwood, G. P., and B. R. Rosen, eds., *Biology and systematics of colonial organisms.* Academic Press; New York; 1979.
Nielsen, C., *Animal evolution: Interrelationships of the living phyla.* Oxford University Press; New York; 1995.
Pennak, R. W., *Freshwater invertebrates of the United States,* 3d ed. Wiley (Interscience); New York; 1989.

Ryland, J. S., *Bryozoans.* Hutchinson University Library; London; 1970.

Woollacott, R. M., and R. L. Zimmer, eds., *Biology of bryozoans.* Academic Press; New York; 1977.

A-30 Brachiopoda

Erwin, D. H., "The mother of mass extinctions." *Scientific American* 275(1):72–78; July 1996.

Gould, S. J., and C. B. Calloway, "Clams and brachiopods: Ships that pass in the night." *Paleobiology* 6:383–396; 1980.

Jørgensen, C. B., *The biology of suspension feeding.* Pergamon Press; New York; 1966.

LaBarbara, M., "Water flow patterns in and around three species of articulate brachiopods." *Journal of Experimental Marine Biology and Ecology* 55:185–206; 1981.

Richardson, J. R., "Brachiopods." *Scientific American* 255(3):100–106; September 1986.

Rudwick, M. J. S., *Living and fossil brachiopods.* Hutchinson University Library; London; 1970.

Russell-Hunter, W. D., *Biology of higher invertebrates.* Macmillan; New York; 1969.

Williams, A., "The calcareous shell of the Brachiopoda and its importance to their classification." *Biological Reviews of the Cambridge Philosophical Society* 31:243–287; 1956.

Williams, A., et al., *Brachiopoda,* Part H (2 vols.). In R. C. Moore, ed., *Treatise on invertebrate paleontology.* Geological Society of America; Boulder, CO; 1965. University of Kansas Press; Lawrence, KS; 1965.

A-31 Phoronida

Kozloff, E. N., *Seashore life of the northern Pacific Coast: An illustrated guide to the common organisms of northern California, Oregon, Washington, and British Columbia.* University of Washington Press; Seattle; 1983.

MacGinitie, G. E., and N. MacGinitie, *Natural history of marine animals,* 2d ed. McGraw-Hill; New York; 1968.

Zimmer, R. L., "Morphological and developmental affinities of the lophophorates," pp. 593–599. In G. P. Larwood, ed., *Living and Fossil Bryozoa.* Academic Press; New York; 1973.

A-32 Chaetognatha

Alvariño, A., "Chaetognaths," pp. 115–194. In H. Barnes, ed., *Oceanography and marine biology: Annual review 3.* Allen and Unwin; London; 1965.

Bieri, R., D. Bonilla, and F. Arcos, "Function of the teeth and vestibular organ in the Chaetognatha as indicated by scanning electron microscope and other observation." *Proceedings of the Biological Society of Washington* 96:110–114; 1983.

Darwin, C., "Observations on the structure and propagation of the genus *Sagitta.*" *Annals and Magazine of Natural History* 13 (Series 1, No. 81):1–6 and Plate 1; January 1844.

Eakin, R. M., and J. A. Westfall, "Fine structure of the eye of a chaetognath." *Journal of Cell Biology* 21:115–132; 1964.

Ghirardelli, E., "Some aspects of the biology of the chaetognaths." *Advances in Marine Biology* 6:271–375; 1968.

Grant, G. C., "Investigations of inner continental shelf waters off lower Chesapeake Bay 4: Descriptions of the Chaetognatha and a key to their identification." *Science* 4:107–119; 1963.

A-33 Hemichordata

Barrington, E. J. W., *The Biology of Hemichordata and Protochordata.* W. H. Freeman and Company; New York; 1965.

Berrill, N. J., *The origin of vertebrates.* Oxford University Press; New York; 1955.

Harrison, F. W., and E. E. Ruppert, eds., *Microscopic anatomy of invertebrates,* vol. 15: *Hemichordata, Chaetognatha, and the invertebrate chordates.* Wiley-Liss; New York; 1996.

Wada, H., and N. Sato, "Details of the evolutionary history from invertebrates to vertebrates, as deduced from the sequences of 18S rDNA." *Proceedings of the National Academy of Science* USA 91:1801–1804; 1994.

A-34 Echinodermata

Binyon, G., *Physiology of Echinoderms.* Pergamon Press; New York; 1972.

Clark, A. M., and M. E. Downey, *Starfishes of the Atlantic.* Chapman and Hall; London; 1992.

Hendler, G., J. E. Miller, D. L. Pawson, and P. M. Kier. *Sea stars, sea urchins, and allies: Echinoderms of Florida and the Caribbean.* Smithsonian Institution Press; Washington, DC, and London; 1995. Natural history and identification of echinoderms of the Bahamas, Florida, and Caribbean.

MacGinitie, G. E., and N. MacGinitie, *Natural history of marine animals,* 2d ed. McGraw-Hill; New York; 1968.

Nichols, D., *Echinoderms,* 4th ed. Hutchinson University Library; London; 1969.

Nichols, D., *The uniqueness of echinoderms* (Oxford/Carolina Biology Reader). Oxford University Press; New York and London; 1975.

A-35 Urochordata

Alldredge, A., "Appendicularians." *Scientific American* 235(1):94–102; July 1976.

Berrill, N. J., *The Tunicata.* Ray Society; London; 1950. Reprinted 1968.

Deibel, D., and G. Paffenhofer, "Cinematographic analysis of the feeding mechanism of the pelagic tunicate *Doliolum nationalis.*" *Bulletin of Marine Science* 43:404–412; 1988.

Lambert, G., "Ultrastructural aspects of spicule formation in the solitary ascidian *Herdmania momus* (Urochordata, Ascidiacea)." *Acta Zoologica (Stockholm)* 73(4):237–245; 1992.

Lambert, G., and C. C. Lambert, "Spicule formation in the solitary ascidian *Herdmania momus,*" *Journal of Morphology* 192:145–159; 1987.

Madin, L. P., "Field observations on the feeding behavior of salps (Tunicata: Thaliacea)." *Marine Biology* 25:143–147; 1974.

Millar, R. H., "The biology of ascidians." *Advances in Marine Biology* 9:1–100; 1971.

Pearse, V., J. Pearse, M. Buchsbaum, and R. Buchsbaum, "Invertebrate chordates: Tunicates and lancelets," pp. 737–752. In *Living invertebrates.* Blackwell Scientific; Pacific Grove, CA; 1987.

Plough, H. H., *Sea squirts of the Atlantic continental shelf from Maine to Texas.* Johns Hopkins University Press; Baltimore, 1978. Books on Demand; Ann Arbor, MI.

A-36 Cephalochordata

Lytle, C., and J. E. Wodsedalek, *General Zoology Laboratory Guide,* 11th ed. W.C. Brown; Dubuque, IA; 1991.

Pearse, V., J. Pearse, M. Buchsbaum, and R. Buchsbaum, "Invertebrate chordates: Tunicates and lancelets," pp. 737–752. In *Living Invertebrates.* Blackwell Scientific; Pacific Grove, CA; 1987.

A-37 Craniata

Alexander, R. M., *The chordates.* Cambridge University Press; London; 1975.

Barrington, E. J. W., *The biology of Hemichordata and Protochordata.* W. H. Freeman and Company; New York; 1965.

Benton, M. J., ed., *The phylogeny and classification of the tetrapods.* Clarendon; Oxford, UK; 1988.

Berrill, N. J., *The origin of the vertebrates.* Oxford University Press; New York; 1955.

Duellman, W. E., and L. Trueb, *Biology of amphibians.* Johns Hopkins University Press; Baltimore; 1994.

Feduccia, A., "Explosive evolution in Tertiary birds and mammals." *Science* 267:637–638; 1995.

Gee, H., "A backbone for the vertebrates." *Nature* 340:596–597; 1989.

Romer, A. S., and T. S. Parsons, *The vertebrate body,* 5th ed. Saunders; Philadelphia; 1977.

Romer, A. S., *The vertebrate story,* rev. ed. University of Chicago Press; Chicago; 1971.

Ward, D. P., *On Methuselah's trail: Living fossils and the great extinctions.* W. H. Freeman and Company; New York; 1992.

Young, J. Z., *The life of vertebrates,* 3d ed. Oxford University Press; New York; 1981.

CHAPTER FOUR

KINGDOM FUNGI

F-2 *Schizophyllum commune* [Courtesy of W. Ormerod.]

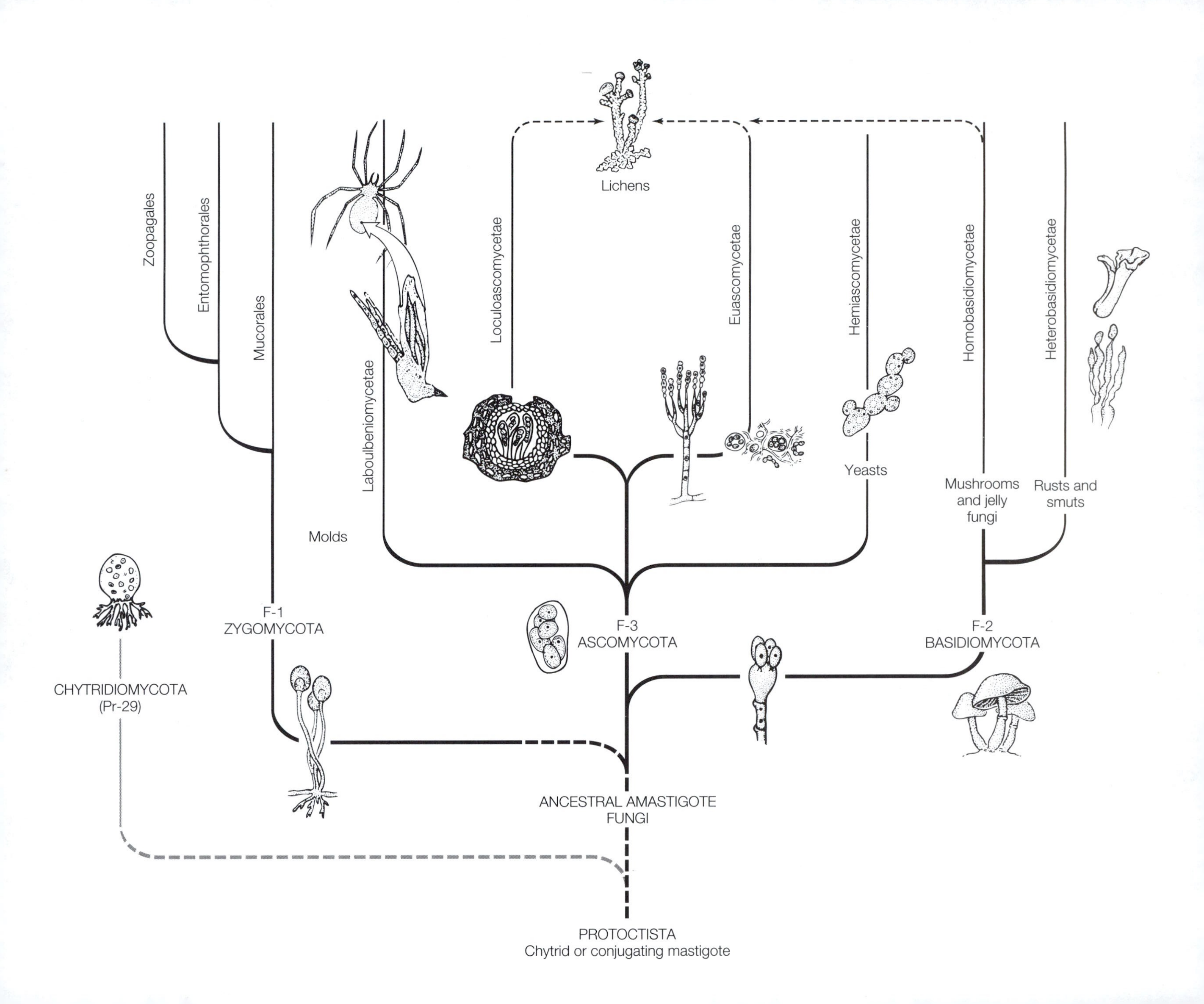
Lichens
Zoopagales
Entomophthorales
Mucorales
Laboulbeniomycetae
Loculoascomycetae
Euascomycetae
Hemiascomycetae
Homobasidiomycetae
Heterobasidiomycetae
Molds
Yeasts
Mushrooms and jelly fungi
Rusts and smuts
F-1 ZYGOMYCOTA
F-3 ASCOMYCOTA
F-2 BASIDIOMYCOTA
CHYTRIDIOMYCOTA (Pr-29)
ANCESTRAL AMASTIGOTE FUNGI
PROTOCTISTA
Chytrid or conjugating mastigote

FUNGI

Latin *fungus,* probably from Greek *sp(h)ongos,* sponge

Conjugating (hypha- or cell-fusing) haploid (monokaryotic) or dikaryotic osmotrophs that develop from resistant nonmotile fungal spores. Cells, including spores, have chitinous walls. Spores are produced by mitosis or by zygotic meiosis. Undulipodia (kinetosomes and axonemes) lacking at all stages. Equivalent to Kingdom Mychota or Eufungi. Fossil record extends from the lower Paleozoic era (450 million years ago) to the present.

Kingdom Fungi, as defined in this book, is limited to eukaryotes that form chitinous, resistant propagules (fungal spores) and chitinous cell walls and that lack undulipodia (that is, are amastigote, or immotile) at all stages of their life cycle. Of the 1,500,000 species of fungi estimated to exist, about 60,000 have been described; most are terrestrial, although a few truly marine species are known. Because fungi often differ only in subtle characteristics, such as the details of structure, pigments, and complex organic compounds, it is likely that many have not yet been recognized as distinct species.

Fossil fungi date from the Ordovician period, 450 million to 500 million years ago (see Figure I-4). The ancestry of fungi is not well understood. Absorptive heterotrophy (derivation of nutrients from digestion of living or dead tissue) associated with multicellularity, syncytia (many nuclei per cell), and propagules (the fungal way of life) has evolved many times and in many protoctist groups: slime molds (Phyla Pr-3 and Pr-6) and slime nets (Phylum Pr-18), oomycotes (Phylum Pr-20), hyphochytrids (Phylum Pr-21), and even such ciliates as *Sorogena* (Phylum Pr-8). True fungi may have descended from conjugating protoctists and thus share an ancestor with the rhodophytes (Phylum Pr-25) or the gamophytes (Phylum Pr-26) through the zygomycotes (Phylum F-1); or, chytrids (Phylum Pr-29) may be the protoctists from which fungi evolved. Molecular systematics supports our classification: only chytrids show a close relation to fungi as the kingdom is presented here. Some classification systems incorrectly place funguslike microbes that include a motile stage (such as plasmodiophorans, oomycotes, and hyphochytrids) in the fungi kingdom. Both molecular evolutionary and morphological studies show that funguslike microbes with motile cells—that is, cells that have undulipodia—are protoctists (Phyla Pr-19 through Pr-21, and Pr-29).

The ascomycotes and basidiomycotes are more closely related to each other than to zygomycotes and probably descended from a common ancestor. Organisms that lack sexual stages were traditionally grouped together as Deuteromycota. Because deuteromycotes clearly descended from either the ascomycotes or the basidiomycotes by loss of sexual stages, we have simply returned them to their relatives and have now abandoned the artificial taxon "Deuteromycota." Lichens no doubt evolved by association, mostly of Ascomycota with either Cyanobacteria (Phylum B-5) or Chlorophyta (Phylum Pr-28) or both. We place lichens in Ascomycota.

Fungi were traditionally aligned with plants, and some classification schemes formerly considered the fungi to be a subkingdom of Kingdom Plantae. However, fungi are clearly more closely related to animals than to plants, considering that chitin is the main component both of fungal cell walls and of the arthropod exoskeleton (Phyla A-19 through A-21). In comparison, plant cell walls instead contain cellulose, a polysaccharide similar to chitin. At any rate, fungi differ from animals and plants in life cycle, in mode of nutrition, in pattern of development, and in many other ways. We thus support here the many mycologists who feel that fungi constitute their own kingdom.

Fungi lack an embryonic stage and develop directly from spores. Often borne in sporangia (singular: sporangium), spores may be of mitotic or meiotic origin. Spores germinate into hyphae or, in the case of yeasts, into single growing cells. The hyphae (singular: hypha) that grow from fungal spores are slender tubes divided into cells by cross walls called septa (singular: septum). The nuclei increase by mitosis as hyphae grow. Each such cell may contain more than one nucleus. The septa seldom separate the cells completely. Thus, cytoplasm, including nuclei, mitochondria, and other inclusions, can flow more or less freely through the hyphae. The hyphae of some fungi have no septa at all. The hyphae of an individual fungus collectively are called a mycelium (plural: mycelia), which is the feeding, growing form of most fungi. As part of an absorptive organism, the fungal mycelium has a morphology that is beautifully fitted to its ecological role: great surface area and growth at the periphery of the tips of hyphae. The yeasts, in comparison, remain as single cells and do not form mycelia.

Spores produced in the absence of any sort of sexual fusion are called conidia (singular: conidium). These propagules can form on the hyphae of zygo- or ascomycotes. The hyphal structures bearing the conidia are called conidiophores (Figure 4-1). Most conidia and many sporangia are dis-

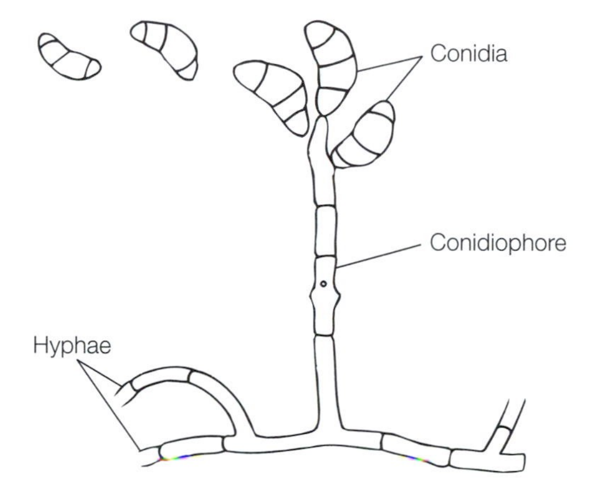

Figure 4-1 Conidiophore and multicellular conidia of a typical ascomycote *Curvularia lunata.* [Adapted from W. B. Kendrick and J. W. Carmichael. "Hyphomycetes," in G. C. Ainsworth, F. K. Sparrow, and A. S. Sussman (eds.), *The Fungi: An Advanced Treatise, vol. IVA: A Taxonomic Review with Keys: Asomycetes and Fungi Imperfecti.* New York: Academic Press (1973).]

persed by the wind and can endure conditions of heat, cold, and desiccation unfavorable to the growth of fungi. Under favorable conditions, the conidia and sporangiospores (spores carried in the sporangia) grow into hyphae and form mycelia. This is the general reproductive pattern, although there are many variations. New individual mycelia also result from fragmentation of mycelia. Yeasts reproduce asexually by budding. Nearly all fungi, even those that have sexual stages, form spores directly—that is, asexually. In fact, most reproduce by mitotically produced (asexual) spores more often than they do sexually.

From time to time, most fungi form a sexual stage, including reproductive structures called gametangia. All basidiomycotes form spores on basidia, zygomycotes form spores in sporangia, and ascomycotes form spores in asci. Fungi bear spores that are the products of meiosis—basidiospores, zygosporangia, or ascospores. The reproductive structures of basidiomycotes are masses of tightly packed hyphae called basidiomata. The reproductive structures of ascomycotes are called ascomata; they sometimes originate from a single cell and are derived from a three-dimensional proliferation of cells. Such spore-bearing structures are commonly noticed as molds, jelly fungi, and mushrooms. The largest and most complex fungi are the large mushrooms and shelf fungi, all of which are products of sexual mergers. Some of these recognizable structures arise from underground mycelia that are kilometers in length and meters in diameter. Many other fungi are microscopic—for example, the unicellular yeasts.

Sexual reproduction in fungi is by conjugation, in which hyphae of complementary mating types come together and fuse (see Phylum F-1, Figure D). One parent fungus can have hyphae of two different mating types. Conjugation always consists of at least two processes: hyphal (cytoplasmic fusion, or cytogamy) and nuclear fusion, or karyogamy. In fungi, after cell or hyphal fusion, the nuclei, which are always haploid, do not immediately fuse. This situation is analogous to the delayed fusion of animal egg and sperm nuclei. In fungi, instead of immediate nuclear fusion, each parental nucleus grows and divides within a common cytoplasm, often for long periods of time. The offspring nuclei remain in pairs, one nucleus descended from each mating type. A hypha containing paired haploid nuclei, whether or not they have been shown to come from separate mating types, is called dikaryotic. A mycelium of such hyphae is called a dikaryon. If the nuclei of each pair differ in genotype, the mycelium is called a heterokaryon. If the paired nuclei are from the hyphae of a single mating type (that is, have the same genotype), the mycelium is called a homokaryon. If the hyphae contain only single, unpaired nuclei, the mycelium is called a monokaryon. The paired nuclei eventually fuse to form diploid zygotes. Diploidy is transient; the zygotes immediately undergo meiosis, which results in the formation of haploid spores and thus reestablishes the haploid state.

All fungi form some sort of spores. In all cases, fungal spores are haploid (although some are dikaryotic) and are capable of germinating into haploid hyphae or yeasts. In most fungi, hyphae of complementary mating types fuse later in the life cycle, and the dikaryotic or sexual stage follows, in which nuclei are transiently diploid. Organisms that lack sexual stages produce spores only by mitosis. The nuclei of hyphae in species that never develop sexual stages seem to be permanently haploid.

Nearly all fungi are aerobes, and all of them are heterotrophs that characteristically absorb their food. They excrete powerful enzymes that break food down into smaller molecules outside the fungus; dissolved nutrients are then transported into the fungus through the fungal membrane.

Fungi are tenacious, resisting severe desiccation and other environmental challenges. Their cell walls, composed of the nitrogenous polysaccharide chitin, are hard and stiff and resist water loss. Some grow in acid; others survive in nitrogen-poor environments such as bogs. Fungal strategies for survival include the production of such complex organic compounds as penicillin derivatives that block the formation of bacterial cell walls and amanita alkaloids that induce hallucinations or even death in mammals. The penicillin derivatives are antibiotics inhibiting bacterial growth; the amanita alkaloids deter feeding. Fungi are the most resilient of the eukaryotes, though by no means invulnerable. For example, red squirrels of the New England forest nip off mushrooms and then carry them up into trees, where the mushrooms dry in the sun, to be eaten later or stored.

Many fungi cause diseases, especially in animals, including humans, and plants. However, many more form constructive, intimate associations with plants. The roots of nearly all healthy tracheophytes (vascular plants; Phyla Pl-4 through Pl-12) have symbiotic relations with fungi. Fungi inhabiting (endomycorrhizal) and coating (ectomycorrhizal) the roots of grasses and trees are especially critical to plants growing in nutrient-poor soils; the fungi function as root extensions. Such mycorrhizal associations enhance the transport of soil nutrients, such as phosphates, nitrates, copper, zinc, and manganese. Fungi inhabiting the roots of trees are responsible for transporting nutrients from the soil to the roots. Most orchid seeds, for example, require specific fungal partners to germinate. (The first plant organ to emerge from the seed is the rootlet, or radicle.) The fungi that transport nutrients through soil may have helped prevent the earliest plants from succumbing to desiccation and direct sunlight as plants made the transition from water onto land. Plant-fungus relations became truly terrestrial. In any case, a strong association between most plants and some fungi has persisted for at least 400 million years.

Many fungi yield products useful to humans. Some fungi are sources of citric acid and pharmaceuticals. The fungal molds *Rhizopus nigricans* and *Curvularia lunata* carry out fermentations in the manufacture of cortisone, hydrocortisone, and prednisone—steroids used to control inflammation. Molds and yeasts are used in the production of cheese, beer, wine, and soy sauce.

Penicillin, which is an antibiotic produced naturally by the mold *Penicillium chrysogenum,* has been joined by several thousand additional antibiotics. An antibiotic is any low-molecular-weight compound, generally excreted by a live organism, that specifically inhibits the growth of other microbes when it is present in low concentration. Microbiological production of pharmaceuticals from molds also includes vitamins, interferons (which prevent rather than cure viral infections), and steroid hormones. Microorganisms carry out syntheses that produce penicillin, cephalosporins from the marine mold *Cephalosporium,* erythromycin, and streptomycin (see Phylum F-3). Antibiotics have been found that interfere with nearly every phase of the bacterial life cycle. Penicillins (several have been discovered) and cephalosporins interfere with construction of the bacterial cell wall. Bleomycins and anthracyclines, isolated from *Streptomyces verticillus* (Actinobacteria; Phylum B-12), interfere with DNA replication; rifamycins block the transcription of DNA into mRNA; erythromycin interrupts the ribosomal synthesis of protein. In the fungal cell that normally contains the antibiotic, it is thought that the antibiotic may inhibit growth of competing organisms. That ability was amplified many thousandfold by treating *Penicillium* with radiation and chemicals, which caused mutations that resulted in increased production of penicillin.

The genetic information to make penicillin is present in the unaltered genome of the mold. To improve antibiotics such as penicillins and cephalosporins, after the starting molecule has been produced by fermentation, a second step—substitution of a part of the original antibiotic molecule—is accomplished synthetically. Such substitutions can broaden the spectrum of organisms against which the antibiotic acts and reduce its toxicity.

Although many substances have been discovered that counter bacterial infections, fewer have been discovered that destroy invading fungi in, for example, the human body. One such substance is the antibiotic amphotericin B, produced by *Streptomyces* and *Streptoverticillium* (Phylum B-12), which interrupts the function of the cell membrane in fungi.

Fungi add flavor, color, protein content, and preservative qualities when used in the production of beverages and food. In the conversion of cabbage, green olives, and cucumbers into sauerkraut, cured olives, and pickles, the microbial actions of lactic acid bacteria are followed by fermentation by the yeasts *Saccharomyces* and *Torulopsis*. The mold *Rhizopus*, acting on soybeans, produces tempeh, or Japanese soybean cakes, and *Aspergillus* produces the Japanese product natto. Soy sauce results from the action of *Aspergillus oryzae* on wheat and soybeans, in addition to fermentation by the yeast *Saccharomyces rouxii* and the bacterium *Pediococcus* (Phylum B-13). Strains of microbes are selected that either increase or decrease flavoring compounds and that change the ability of the microbe to ferment particular carbohydrates; for example, a strain of yeast that ferments dextrin contributes to one facet of beer production.

Yeasts that grow on the surface of Limburger cheese provide aroma; Camembert and Brie owe their ripening to *Penicillium* as well as to yeasts. *Penicillium roquefortii* mold develops flavor in Stilton, Danish blue, Gorgonzola, and Roquefort cheeses. Swiss cheese holes originate with bubbles of carbon dioxide generated by yeasts. In baking, brewing, and wine making, yeast fermentation has been used since long before this one-celled fungus was discovered.

F-1 Zygomycota

(Zygomycotes, zygomycetes)

Greek *zygon,* pair; *mykes,* fungus

EXAMPLES OF GENERA

Basidiobolus	*Endocochlus*	*Mucor*
Blakeslea	*Endogone*	*Phycomyces*
Chaetocladium	*Glomus*	*Pilobolus*
Cochlonema	*Kickxella*	*Rhizopus*
Conidiobolus	*Mortierella*	*Stylopage*

Members of this phylum lack cross walls (septa), except for separations between reproductive structures and the rest of the mycelium. About 1100 species of zygomycotous fungi live on land throughout the world. Many are saprobic, feeding osmotrophically—absorbing nutrients—on decaying vegetation. Some are highly specialized, living on animals, plants, protoctists, or even each other. Zygomycotes may feed on large mushrooms, for example, as can ascomycotes (Phylum F-3).

Two modes of reproduction characterize zygomycotes: (1) direct development, by mitosis, of spores (conidia) and (2) formation of sexual zygosporangia (misleadingly sometimes called "zygospores") in which meiosis occurs to form haploid propagules. In reproduction by direct sporulation, haploid sporangiospores develop inside sporangia, which are often borne aloft on special hyphae, sporangiophores (Figures A, B, and C). The hyphae that develop from these spores are haploid, and each can form a mycelium. In zygomycote sexuality, zygosporangia form by conjugation of hyphae (Figures D and E). Specialized hyphae of complementary mating types (for example, "plus" and "minus") called gametangia are attracted to each other by hormones and grow until they touch. The ends of the hyphae swell, the walls dissolve, and the two sources of cytoplasm fuse (cytogamy). Multiple haploid nuclei from each mating type enter the joined swellings, which develop into a thick-walled zygosporangium. Zygomycote matings more resemble orgies than simple couplings, because many nuclei fuse pairwise (karyogamy), giving multiple diploid nuclei simultaneously. Eventually each diploid nucleus undergoes meiosis during germination. Zygosporangia behave like spores and germinate into haploid mycelia or directly into a sporangiophore bearing sporangia. Clearly, in the fungi, the mammal-centered terms "mating" (in this case, conjugation) and "sexual reproduction" (in reference to number of offspring following conjugation) conceal, rather than reveal, knowledge of sexual practices.

Different taxonomists recognize from three to seven groups (classes) of zygomycotes, which in some classification schemes are considered orders: Mucorales, Entomophthorales, Zoopagales, Glomales, Dimargantales, Kickxellales, and Endogonales. We describe only the four major classes here.

The Mucorales are saprobic organisms; they secrete extracellular digestive enzymes and absorb dead organic matter. They form sporangia mitotically as well as conjugate. Many sporangia, each containing one or many spores, have conspicuous walls that break when desiccated, releasing their contents. Members of this class include the common black bread mold *Rhizopus stolonifer,* and the genera *Mucor, Phycomyces,* and *Pilobolus.* The sensitivity of some members of the genus *Phycomyces* to light is truly astounding—only a few photons are required to initiate the development of the sporangium. The photoreceptor, which absorbs blue light, may be the common B vitamin riboflavin. *Pilobolus* (the "hat thrower")

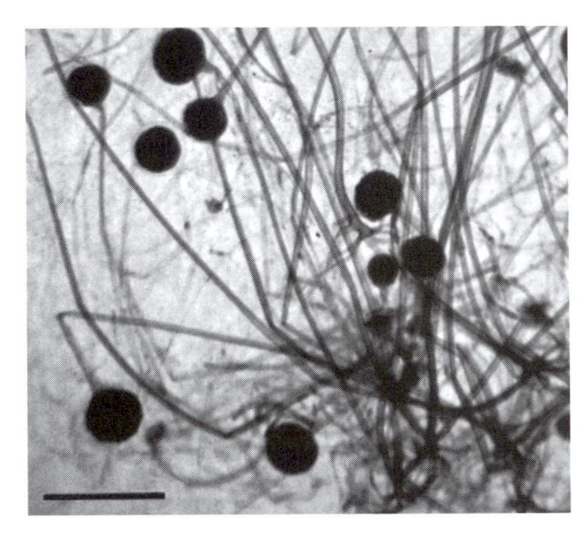

A *Rhizopus stolonifer* hyphae, sporangiophores, and sporangia. LM, bar = 100 μm. [Courtesy of W. Ormerod.]

produces mycelia on horse dung. When mature, it shoots its sporangia as conidiumlike units into the air to a height of 2 m.

Most of the Entomophthorales parasitize animals, mainly insects. They reproduce directly by forming sporangia that, like those of *Pilobolus,* are forcibly discharged as units. *Basidiobolus* is perhaps the best-known member of the Entomophthorales. Found on frog dung, it can easily be grown after it is collected from water that harbors a frog.

The Zoopagales comprise about 65 species in 10 genera, all of which are symbiotrophic on amebas and other protoctists, as well as on nematodes (Phylum A-10) and other small animals. Most of them produce hyphae that enter the plant tissue. They penetrate the surface of the animal, coil inside it, and violently debilitate their victim. From the remains of the host emerge club-shaped gametangia that fuse and form zygosporangia. Among the genera are *Cochlonema* (*C. verrucosum,* a parasite of amebas), *Endocochlus,* and *Stylopage.*

The Glomales include many important symbiotrophs associated with plants. *Glomus* is one of about 100 zygomycote species found in endomycorrhizal symbioses in about 80 percent of vascular plants. The prevalence of mycorrhizal associations in modern terrestrial ecosystems and in plant root fossils (Devonian) indicates that plants already had such fungal assistance when their ancestors successfully colonized the land more than 400 million years ago.

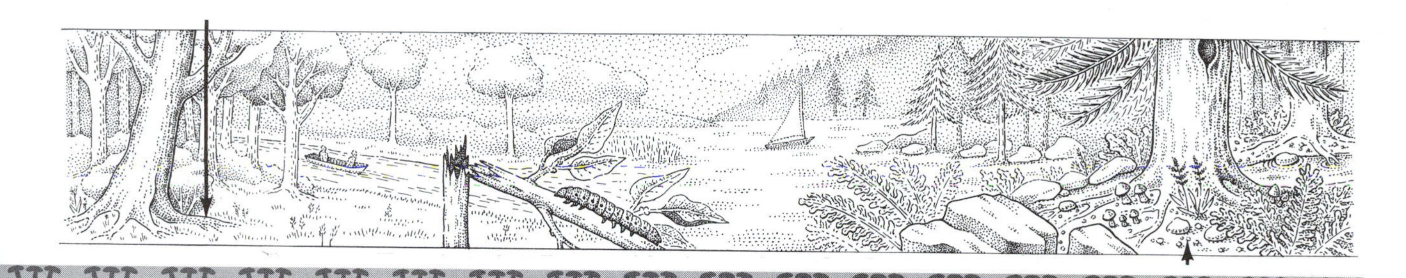

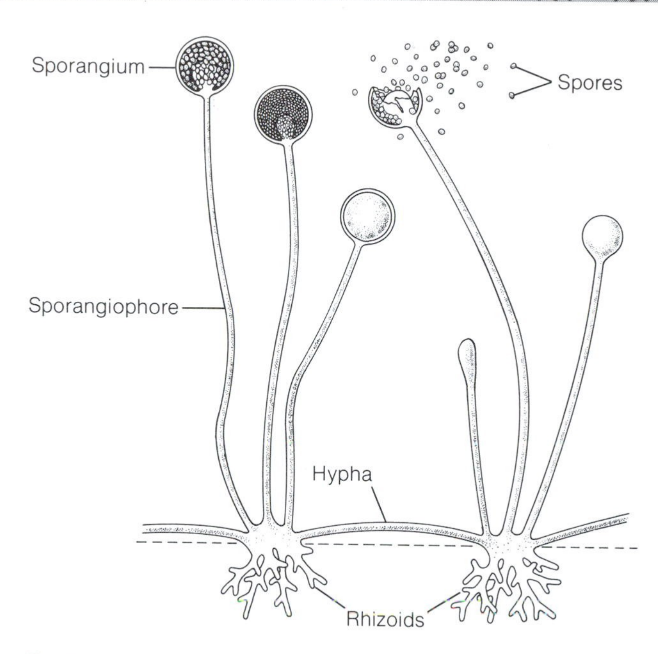

B *Rhizopus* sp., black bread mold. Rhizoids anchor the fungus to the substrate. [Drawing by R. Golder.]

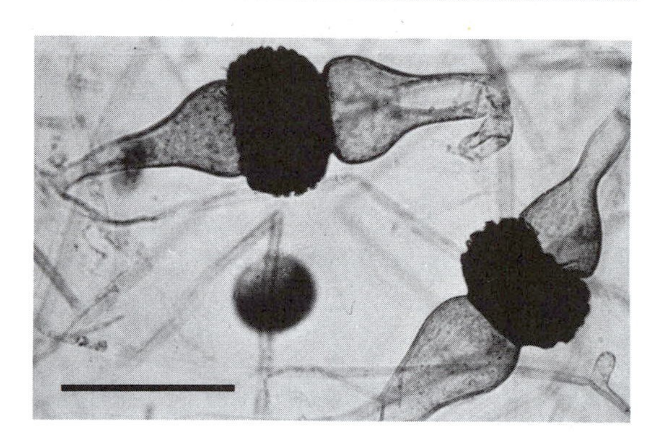

D Zygomycote sexuality: black *Rhizopus* zygosporangia formed by conjugating hyphae of complementary mating types. LM, bar = 100 μm. [Courtesy of W. Ormerod.]

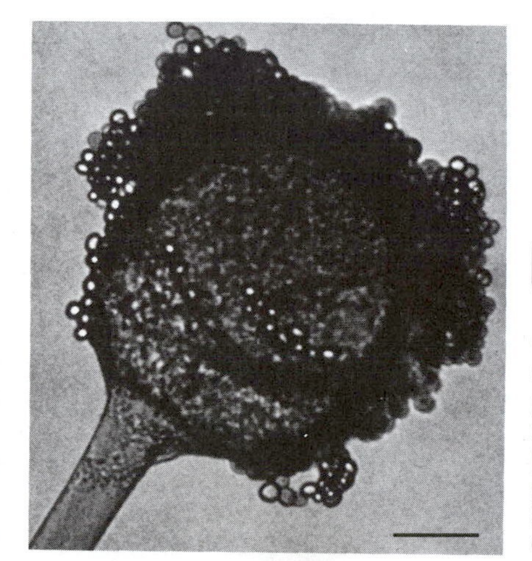

C Broken *Rhizopus* sporangium showing sporangiophore bearing its propagules, spores produced by mitotic cell division. LM, bar = 10 μm. [Courtesy of G. Cope. Reproduced by permission of Elementary Science Study of Education Development Corporation, Inc., Newton, MA.]

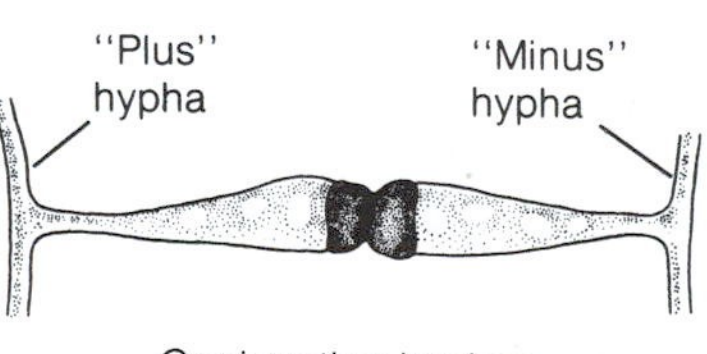

Conjugating hyphae

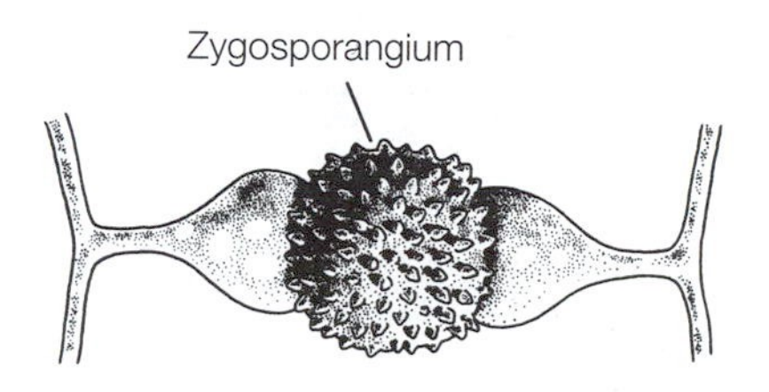

E Conjugation in *Rhizopus.* [Drawing by R. Golder.]

F-2 Basidiomycota

(Basidiomycotes, basidiomycetes)

Greek *basidion,* small base; *mykes,* fungus

EXAMPLES OF GENERA

Agaricus
Amanita
Boletus
Calvatia
Cantharellus
Clavaria
Cyathus
Fomes
Geastrum
Lepiota
Pellicularia
Phallus
Polyporus
Psilocybe
Puccinia
Rhizoctonia
Schizophyllum
Tremella
Ustilago

Basidiomycotes include the smuts, rusts, jelly fungi, mushrooms, puffballs, and stinkhorns. The basidium (plural: basidia), a microscopic club-shaped reproductive structure from which their name is derived, distinguishes basidiomycotes from fungi in other phyla. The basidioma (plural: basidiomata; formerly "basidiocarp"), or mushroom, is the spore-producing body (Figure A). Each basidium usually bears four haploid spores, called basidiospores (Figure B), produced by meiosis. There are about 22,250 species of basidiomycotes. All are heterotrophs, many on agricultural crops and forest trees; some are symbiotrophs. Many basidiomycotes form important symbioses, called ectomycorrhizae ("outside fungus-roots"), with most forest trees and shrubs. Fungi that form ectomycorrhizae create a sheath of hyphae—a fungal mantle—around the root. Fungus-secreted hormones stimulate the root—typically of pine, beech, and willow trees—to branch. Fungal hyphae replace often absent plant root hairs. Mycorrhizal fungi transfer phos-

A Basidiomata of *Boletus chrysenteron,* a yellow bolete mushroom of New England deciduous forest. Bar = 5 cm. [Courtesy of W. Ormerod.]

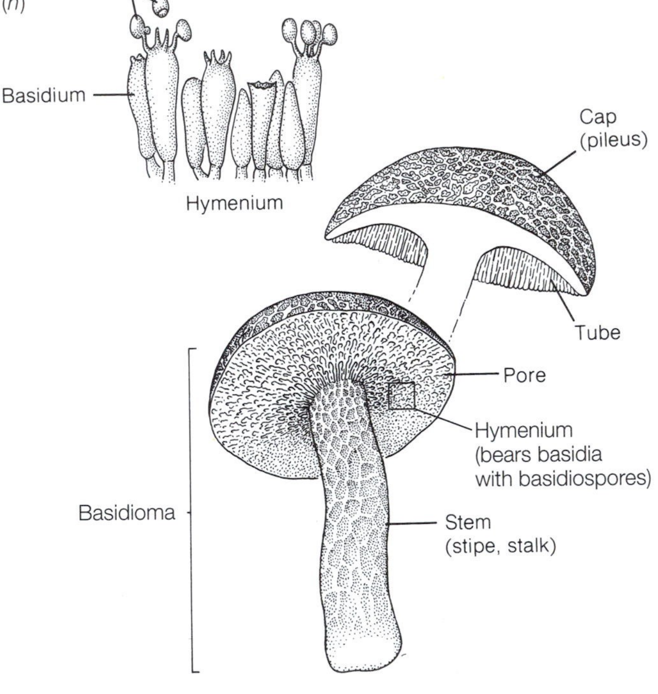

B Basidioma, reproductive structure, of *Boletus chrysenteron.* The basidioma is composed of tightly packed hyphae. Sexual reproduction takes place in basidia; spores form on the basidia, which open in the tubes. [Drawing by L. Meszoly.]

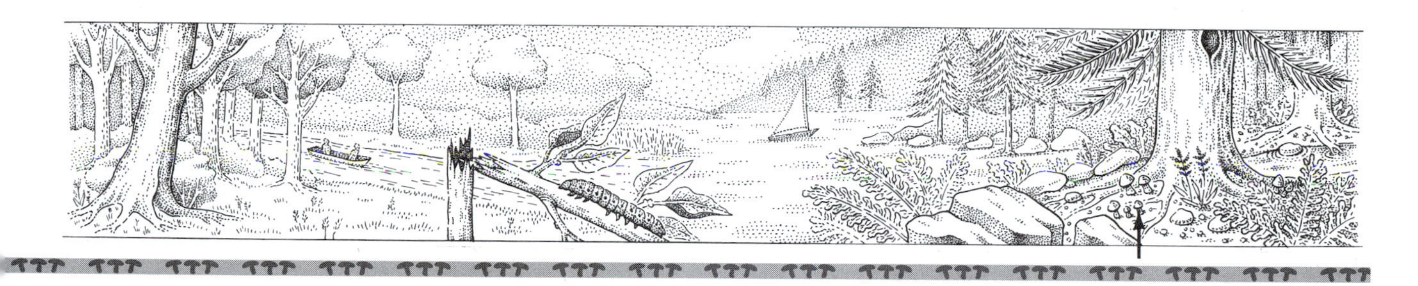

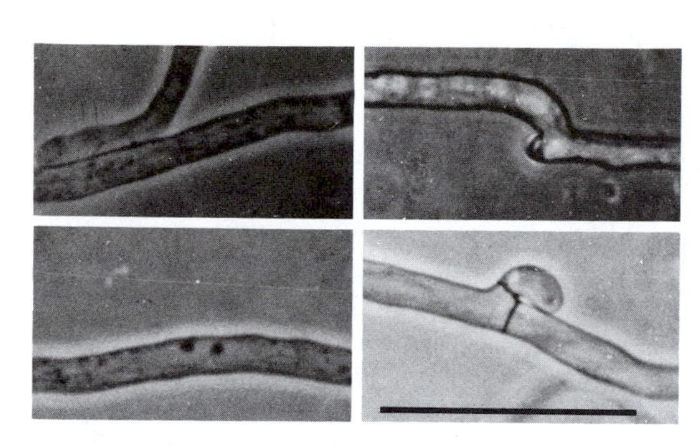

C Conjugation of basidiomycote hyphae of complementary mating types. Sexual reproduction: (left top) approach of hyphae; (right top) incipient fusion; (left bottom) fused hyphae, two nuclei apparent; (right bottom) clamp connection. The species is *Schizophyllum commune,* shown in Figure E. LM, bar = 40 μm. [Courtesy of W. Ormerod.]

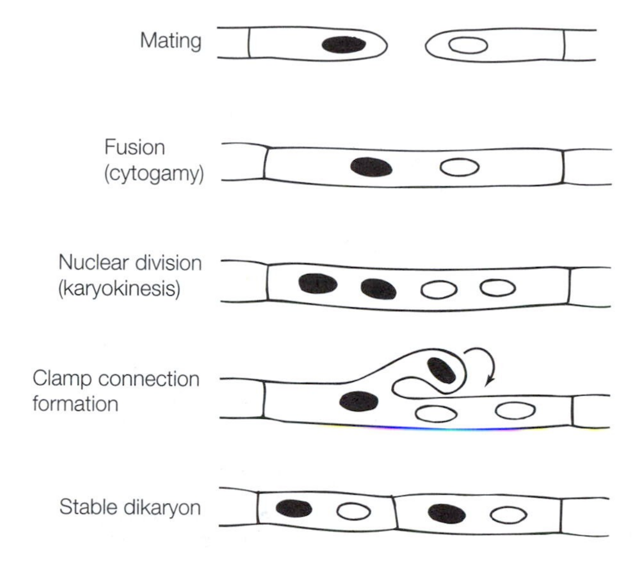

D Conjugation. In the basidiomycete clamp connection, two forming cells of a dikaryotic hypha join laterally, ensuring that each cell contains two dissimilar nuclei.

phorus, nitrogen, and other elements from fungi to plants; mycorrhizal fungi transfer carbohydrates in the opposite direction. Others of this phylum, such as the cultivated button mushroom *Agaricus* and the Mexican delicacy *Ustilago,* are popular in the human diet.

Most basidiomycotes display structures from three stages of development: monokaryotic mycelia, dikaryotic mycelia, and basidiomata, which are also dikaryotic. A basidiospore germinates and grows into a multinucleate mycelium, which lacks septa. Eventually, perforated septa form and divide the hyphae into uninucleate segments. In this monokaryotic mycelium, two segments of complementary hyphae (of complementary mating types) conjugate (Figures C and D). When the mating hyphae have grown from the same basidiospore, this self-fertilizing ability is called homothallism.

From the fused hyphae, a dikaryotic mycelium develops. Dikaryosis, the state of having a pair of haploid nuclei in each hyphal segment, is common in this phylum. A dikaryotic (homo- or heterokaryotic) mycelium grows by the simultaneous division of the two nuclei in a hyphal segment and the formation of a new septum. To ensure that each new segment contains a nucleus descended from each parent, a clamp connection forms (see Figure D). This connection is a short hyphal loop through which one of the four or more nuclei moves to another place in the segment—the result shuffles the order of the nuclei before the new septum forms. When the new septum forms, there are nuclei of two different parental origins on each side.

By differentiation and growth, the dikaryotic, heterokaryotic mycelium forms the reproductive structure, the basidioma. In old literature, these mushrooms are called "fruiting bodies" or "fruits" because they were first well described by botanists. The basidioma reproduces by release of its meiotically produced propagules, the

E Underside of *Schizophyllum commune,* showing the gills. The white double lines of the gills bear the basidia in rows. Bar = 100 μm. [Courtesy of W. Ormerod.]

basidiospores. In basidiomata, such as mushrooms and jelly fungi, or in simpler pustules that are seen in plants infected with smut or rust disease, basidia form on the gills (Figure E) or tube lining (see Figure B), depending on the species. Externally on basidia, meiosis produces haploid basidiospores in a process called basidiogenesis. After spore release, the basidioma disintegrates. Thus, the life cycle of basidiomycotes comes full circle: monokaryotic mycelia, dikaryotic mycelia, spore-producing basidioma.

Two classes of basidiomycotes are recognized. The class Heterobasidiomycetae includes rusts and smuts (Figure F), which typically look like dark protrusions on leaves. All smuts and rusts produce a thick-walled propagule called a teliospore, which is a diploid resting spore in which meiosis takes place. From the teliospore, a septate basidium emerges and develops basidiospores. Some—for example, rust fungi that cause cereal crop diseases—have complex life cycles that are linked to seasonal conditions and the develop-

F *Ustilago maydis,* corn smut. This fungus can be injurious to corn plants when it infests a field heavily. Although infestation is detrimental to the production of field corn, the smut itself has economic importance. The spore masses produced by this fungus are harvested, sold in Mexican marketplaces as cuitlacoche, and fried as a delicacy. In U.S. food shops, corn smut is sold under the name "corn mushrooms." Having been bred for low susceptibility to the smut, corn is now being bred for high susceptibility to produce corn smut as a gourmet food crop. The ear is approximately life size [Courtesy of R. F. Evert.]

mental biology of their host plants. Heterobasidiomycetae are the only basidiomycotes that form septate basidia.

All members of the class Homobasidiomycetae (except for jelly fungi) produce nonseptate basidia, usually containing four basidiospores. One subclass, the Hymenomycetes, includes the common poisonous and edible mushrooms, shelf fungi, jelly fungi, and coral fungi. They bear their basidia and spores on an open structure. In the pore mushrooms, such as the bolete (see Figures A and B), the basidia line the inside of tubes, and spores are released through exterior pores (Figure G). In the gilled mushrooms, basidia line the gills (see Figure E). The other subclass, the Gasteromycetes, includes the puffballs, earthstars, stinkhorns, and bird's-nest fungi. Their spores mature on basidia enclosed within the basidioma, from which they are liberated when it decays or ruptures. For example, puffball spores are produced within the puffball.

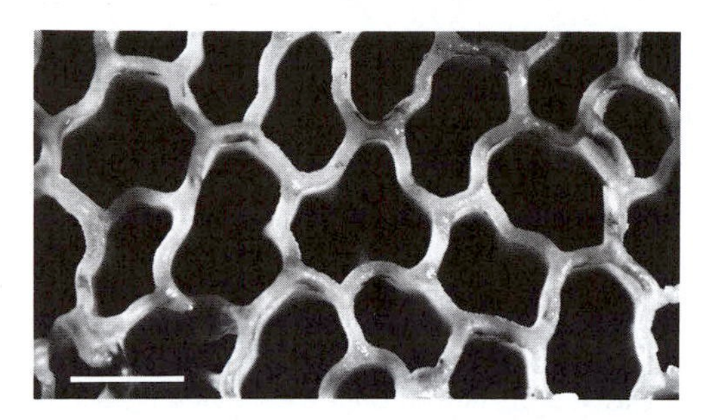

G Pores of *Boletus chrysenteron.* Haploid basidiospores are produced within tubes (into which the pores lead) by meiosis. Spores exit through pores. Bar = 1 mm. [Courtesy of W. Ormerod.]

F-3 Ascomycota

(Ascomycotes, ascomycetes)

Greek *askos*, bladder; *mykes*, fungus

EXAMPLES OF GENERA (d, deuteromycote; l, lichen)

***Alternaria* (d)**
Amorphomyces
***Aspergillus* (d)**
***Calicium* (l)**
***Candida (Monilia)* (d)**
Cephalosporium
Ceratocystis
Chaetomium
***Cladonia* (l)**
Claviceps
***Clypeoseptoria* (d)**
***Collema* (l)**
***Coniocybe* (l)**
***Cryptococcus* (d)**
***Cryptosporium* (d)**
Curvularia
Elsinoe
***Fusarium* (d)**
***Geotrichum* (d)**
***Herpothallon* (l)**

(continued)

Familiar as yeasts, blue-green molds, morels, truffles, lichens, and the former "deuteromycotes," the ascomycotes are a large, diverse, and economically important group of fungi. About 30,000 species are known. Probably more than 10,000 are heterotrophic components of lichens. Symbionts with photosynthesizers [green algae (Phylum Pr-28) or cyanobacteria (Phylum B-5)], ascomycotes make up most of the weight of the lichen. Ascomycotes are distinguished from other fungi by possession of the ascus (plural: asci), a microscopic reproductive structure. Some form hyphae; others (yeasts) are one-celled ascomycotes and lack hyphae and mycelia. Two basic types of asci exist: unitunicate, with a homogeneous wall, and bitunicate, with a double wall and a thick inner wall layer that absorbs water, expanding upward and carrying spores with it. Ascomycotous hyphae form from conjugated—fused of complementary mating types—hyphae or cells. Ascomycotous hyphae have perforated septa; their reproductive structures (spores and gametangia) are cut off by complete, unperforated septa from the rest of the mycelium. In some species, the packed asci and their associated hyphae are so numerous that their organized mass forms a visible body, the ascoma (plural: ascomata; formerly "ascocarp"). Ascomata are multicellular structures that act as platforms from which spores are launched. In some cases, multiple asci form within a reproductive structure called an apothecium, one kind of ascoma (Figure A). The hyphae of ascomycotes are long, slender, branched tubes, and the mycelium that forms is a cottony mass.

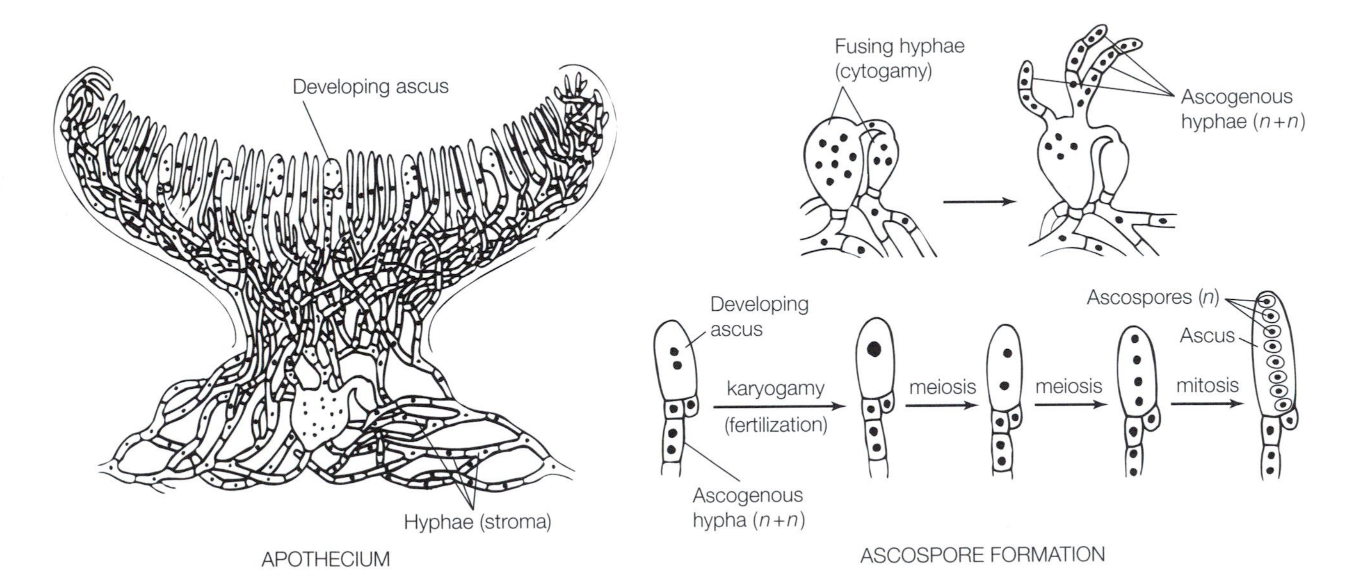

A The ascus, produced as a result of cytogamy, is a specialized ascomycote cell in which meiosis takes place, giving rise at maturity to eight cells, ascospores, in a linear arrangement. In most ascomycotes, the mature ascus bursts to release a cloud of spores; spores may travel as far as 30 cm. Multiple asci form within an apothecium (shown here) or within larger reproductive structures in the morels and cup fungi.

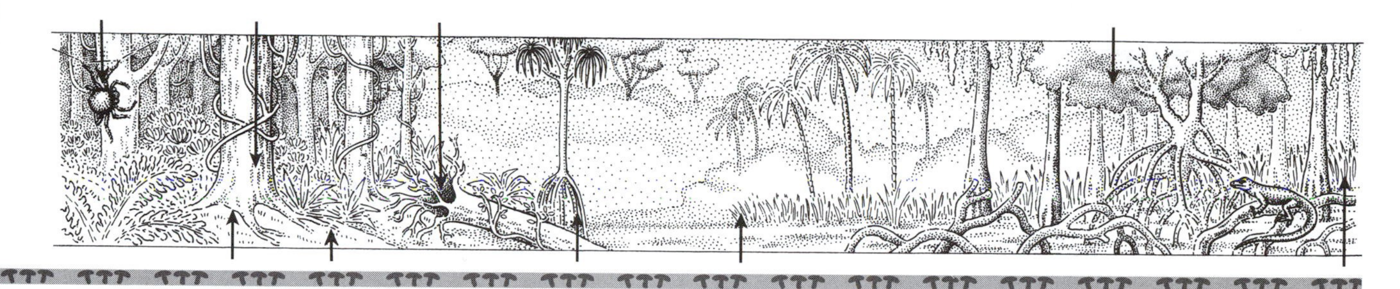

An ascus is produced when two hyphae of complementary mating types conjugate (cytogamy), as illustrated in Figure A. In the ascus, the partners' nuclei fuse (karyogamy), mingling the parental sets of chromosomes temporarily in one zygote nucleus. This transient diploid nucleus undergoes division by meiosis, producing four new haploid nuclei with the same number, but a different combination, of chromosomes as in the nuclei of the parent fungi. These newly formed nuclei then undergo one or two mitotic divisions, and a protective spore wall and some cytoplasm form around each new nucleus. Each walled cell in the ascus formed by meiosis directly or by a mitotic division that follows meiosis is called an ascospore. Ascospores are the kind of propagules that distinguish members of this phylum. Nestled in the ascus like peas in a pod, ascospores are released when mature and may be borne long distances by wind, water, or animals. If ascospores land in an appropriate nutrient-rich place, they germinate and send out hyphae of their own.

The formation of ascospores by the fusion of sexually different hyphae is not the sole means of ascomycote propagation. Reproduction by propagules without sexual fusion is almost universal in ascomycotes. An ascomycote's hyphae, growing by mitosis, simply bud off conidia in succession from specialized cells or they segment into huge numbers of conidia, which, dispersed by wind, water, or animals (often insects), germinate elsewhere. In fact, large numbers of ascomycotes have lost all sexual processes and reproduce only by mitotic production of spores; these ascomycotes are the "deuteromycotes" described later.

Dead or living plant and animal material nourishes ascomycotes; they secrete digestive enzymes into their immediate environment and take in the dissolved nutrients thus formed. Ascomycotes play an essential ecological role by attacking and digesting resistant plant and animal molecules such as cellulose, lignin, and collagen. Valuable biological building blocks—compounds of carbon, nitrogen, phosphorus, among others—locked in such macromolecules are thus recycled. Lichens are photosynthesizers, rather than absorptive heterotrophs.

Ergot fungi, *Claviceps* (Figure B), cause a disease of rye flowers and are poisonous to humans and domesticated animals. Ergot drugs, used to treat migraine and to staunch uterine hemorrhages, are so valued that efforts are being made to cultivate *Claviceps*. Like most symbiotrophs for which the nutrient or genetic contribution of the plant is unknown, *Claviceps* is impossible to grow in culture. Other ascomycotous pathogens have nearly eradicated such trees as the American chestnut and American elm. However, like the basidiomycotes (Phylum F-2), still other ascomycotes form healthy mycorrhizal (fungus-root) associations with trees, shrubs, and other vascular plants. The transfer of inorganic nutrients and water from soil to 80 percent of vascular plants is facilitated greatly by such connections.

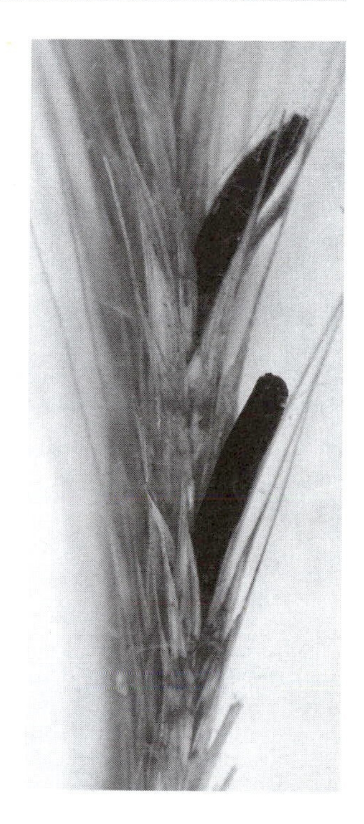

B Grain infected by *Claviceps purpurea.* Airborne spores of the fungus *C. purpurea* infect flowers of cereal grains, replacing normal seeds with a purple-black mass. The plant disease that we call ergot results. Diseased grain or flour, if consumed, produces ergotism in humans and livestock. Temporary insanity, painful involuntary muscle contractions, gangrene, and death result. Bar = 2 cm. [Courtesy of G. Bean.]

There are four classes of ascomycotes: Hemiascomycetae, Euascomycetae, Loculoascomycetae, and Laboulbeniomycetae. All except the last may be symbionts with photobionts [algae (mostly of Phylum Pr-28) and cyanobacteria (Phylum B-5)] to form lichens. Of all fungi, only some laboulbeniomycete yeasts cannot form reproductive propagules called conidia.

F-3 Ascomycota
(continued)

Histoplasma (d)
Lepraria (l)
Lichenothrix (l)
Lichina
Lobaria (l)
Morchella
Mycosphaerella
Neurospora
Ochrolechia (l)
Parmelia (l)
Penicillium (d)
Pneumocystis (d)
Rhizoctonia
Rhizomyces
Saccharomyces
Sarcoscypha
Sordaria
Talaromyces (d)
Torulopsis
Trichophyton (d)
Tuber

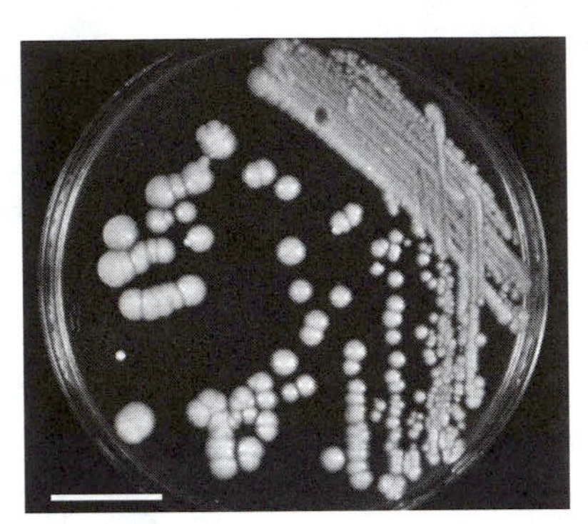

C *Saccharomyces cerevisiae.* Yeast colonies on nutrient agar in petri dish. Bar = 1 cm. [Courtesy of P. B. Moens.]

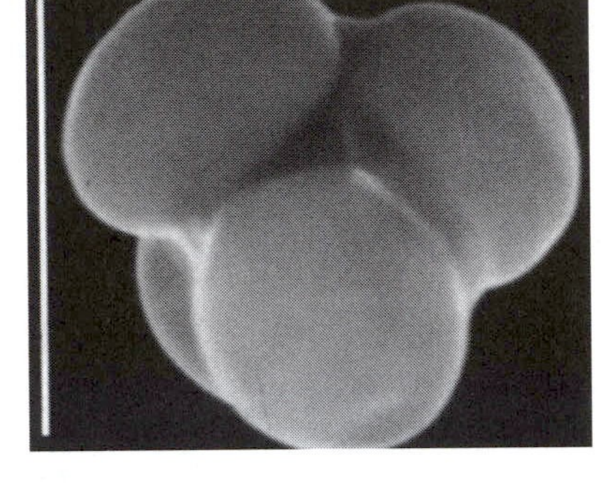

D Tetrad of yeast ascospores formed after fertilization. Sexual reproduction in *Saccharomyces.* Cells of complementary mating types have fused and undergone meiosis. SEM, bar = 10 μm. [Courtesy of L. Bulla.]

The hemiascomycetes are structurally the most simple. They have short hyphae and form small mycelia; the unitunicate ascus is formed directly, rather than on ascogenous (ascus-forming) hyphae that grow from the conjugated cells; and they lack ascomata.

Probably the best known of the hemiascomycetes are the yeasts, represented here by *Saccharomyces cerevisiae* (Figure C). Yeasts, which generally do not grow hyphae or mycelia, have reverted to a single-celled way of life that might seem to place them with the protoctists. Yeasts grow by mitosis; after karyokinesis, the new offspring nucleus is injected, by microtubule spindle elongation, into the bud. The bud enlarges to the size of the parent, and cytokinesis produces two approximately equal offspring cells. Because yeasts form asci, however, their resemblance to protoctists is superficial. Conjugation is by direct fusion of haploid yeast cells to form a diploid zygote, which undergoes meiosis and forms a meiosporangium (ascus). The yeast ascospore arrangement is tetrahedral (Figure D). Typical of yeast cells, the ascospores germinate, on release from the ascus, by budding (Figure E).

Yeasts ferment sugars such as glucose and sucrose to ethyl alcohol; this ability is utilized in the making of wine and beer. In the presence of gaseous oxygen, yeasts oxidize sugars to carbon dioxide, seen as gas bubbles in bread making. Brewer's and baker's yeasts have been cultivated for thousands of years. Now yeasts are being modified by genetic engineering; they are used especially in the construction of artificial chromosomes. Their chromosomal centromeres can be placed on foreign DNA and used to propagate it.

The euascomycetes are the largest and best-known class. Morels, truffles, and most fungal partners in lichens are euascomycetes. The asci generally develop from hyphae, which, in most cases, are part of an ascoma. Ascogenous hyphae are supported by sterile hyphae, tissue called stroma. The asci are unitunicate—the inner and outer layers of the ascus wall are more or less rigid and do not separate when spores are ejected. The genus *Neurospora* is widely used in genetic research. In each ascus, the four products of meiosis divide once by mitosis to form eight cells that remain fixed in a row in the order in which they were formed (see Figure A). Each ascospore in an ascus can be picked up in that order and grown to determine its genetic constitution. The information thus obtained reveals the behavior of chromosomes during a single meiosis and the position of genes on the chromosomes.

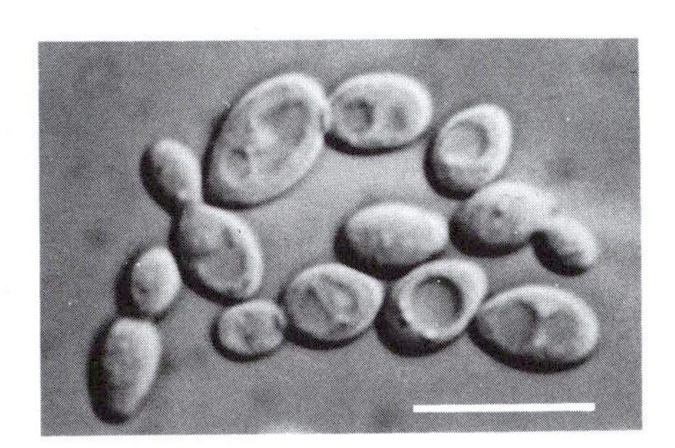

E Budding yeast cells after a day's growth. Cells reproduce by asymmetric mitotic cell division. LM, bar = 10 μm. [Courtesy of P. B. Moens.]

Umbilicaria (l)
Usnea (l)
Verrucaria (l)
Verticillium (d)

The loculoascomycetes have bitunicate asci—the inner wall is elastic and expands beyond the outer wall when spores are released. Ascomata form in a mass of supporting tissue, the stroma. The genus *Mycosphaerella* has about 500 species, including one that forms part of the tissue of a *Fucus*-like seaweed (Phylum Pr-17) called *Ascophyllum*. Others feed on decaying soil organics, including rotting corn. The genus *Elsinoe* includes many pathogenic species, obligate symbiotrophs, that cause diseases of citrus, raspberry, and avocado, among others.

Laboulbeniomycetes are all minute parasites of insects. They are highly host specific—some will parasitize only one sex of the host species or only one body part, such as the legs or the wings. Their ascospores in bitunicate asci germinate directly into reproductive structures between 0.1 and 1.0 mm in diameter. The ascospore forms a septum between two cells. The upper cell differentiates into the male reproductive organ with several vial-shaped cells that produce male gametes called spermatia; the lower cell becomes the female reproductive organ, which is fertilized by the spermatia. The number of cells is fixed for each species. Genera include *Rhizomyces* and *Amorphomyces*. Although some current classifications recognize Laboulbeniales as an order of Euascomycetae, their unique symbiotrophic relation with insects and sexual-only propagule formation justify class status in the minds of some mycologists.

The deuteromycotes (Greek *deuteros*, second; *mykes*, fungus), denoted by "d" in the list of genera, are fungi that lack organs for sexual reproduction. Like the ascomycotes and basidiomycotes, they develop from spores, or conidia, into mycelia whose hyphae are divided by septa.

Although deuteromycotes lack meiotic sexuality, some of these fungi exhibit a parasexual cycle. As documented in the fungal genetics laboratory, they form recombinant mycelia having different inherited traits by fusion of hyphae from two genetically marked, distinct organisms. From these recombinant mycelia, by processes not understood, new true-breeding haploid offspring appear and persist. The parasexual process does not require specialized mycelia.

The term "deuteromycotes" in the literature refers to those ascomycotes or basidiomycotes that have lost their potential to differentiate asci or basidia. A scientific problem is to relate each to its sexual relative—but this is often impossible, because they have so little distinctive morphology. Because they cannot be mated, results of genetic studies are not available. Most of them are thought to have evolved from ascomycotes because of the resemblance of their mycelium and the conidia to those of known ascomycote genera. Sexual structures have been discovered in some species originally classified as Fungi Imperfecti (deuteromycotes), enabling their reclassification with the ascomycotes or basidiomycotes. (The term "perfect" is an old botanical expression that denotes a complete set of sexual reproductive structures.) Indeed, once these species have been reclassified, we are burdened with organisms having two valid names (for example, *Penicillium*, the well-known deuteromycote, and *Talaromyces*, its poorly known sexual ascomycotous stage.)

There are about 15,000 species of deuteromycotes, including some of great economic and medical importance (including, for example, athlete's foot fungus). The deuteromycote organisms are divided into four major groups distinguishable morphologically and functionally analogous to ascomycote classes: sphaeropsids, melanconias, monilias, and Mycelia Sterilia.

The sphaeropsids reproduce by means of structures called pycnidia, hollow structures lined on the inside with conidia-bearing hyphae, or conidiophores. *Clypeoseptoria aparothospermi* is a sphaeropsid that grows inside leaf tissue and forms needle-shaped spores.

The melanconias reproduce by conidia that are borne on short, closely packed conidiophores that form a mat. These mats, called acervuli, can often be seen as flat, disc-shaped cushions on host plants. They are also characteristic of an artificial "form-taxon" called the Melanconiales, from which these particular deuteromycotes are thought to have descended. An example is the melanconia *Cryptosporium lunasporum*, which produces crescent-shaped conidiospores. Form-taxa are convenient groupings of structurally similar, though probably unrelated, organisms. As evolutionary information becomes available, form-taxa are replaced with standard or "systematic" taxa.

The monilias comprise more than 10,000 species. In this group are pathogenic and many other yeasts that form neither asci nor basidia. Monilas reproduce by means of thin-walled conidia that break off from the tips of ordinary hyphae. *Penicillium* (Figures F through

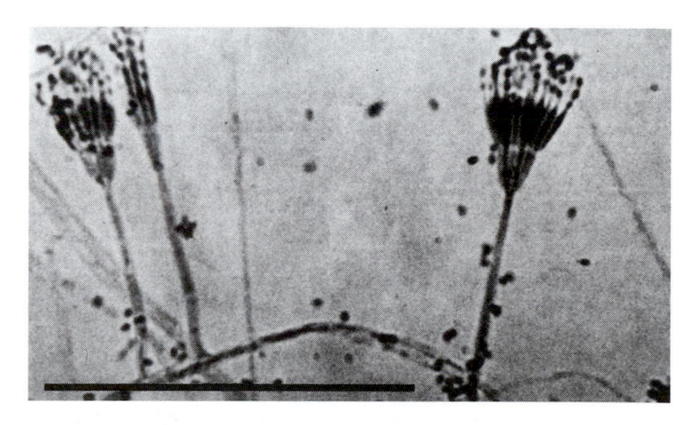

F Hyphae of a *Penicillium* species with several conidiophores bearing conidia (spores) at their tips. LM, bar = 0.1 mm.

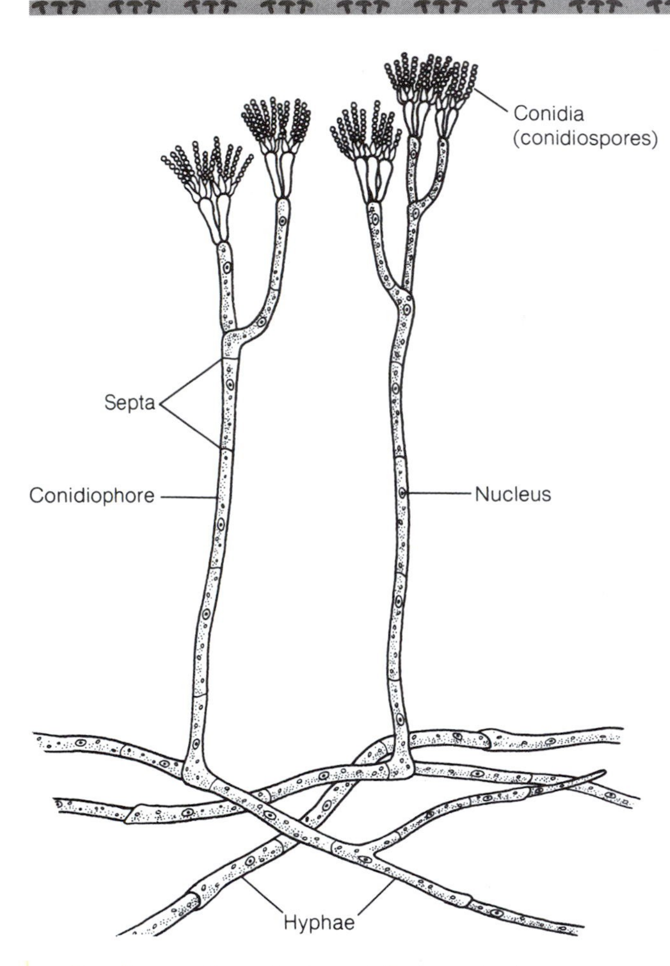

G *Penicillium* sp. The antibiotic penicillin is a natural metabolic product of this mold. [Drawing by R. Golder.]

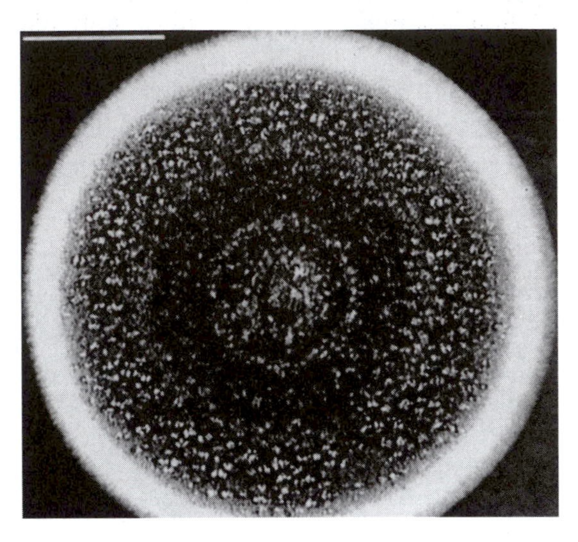

H Colony of *Penicillium* derived from a single conidium growing on nutrient agar in a petri dish. Pigmented conidia form from the center, in the older parts of the colony; only unpigmented newer hyphae, as yet lacking conidia, are at the outer edge. Bar = 1 cm. [Courtesy of W. Ormerod.]

H) belongs to this group. Some monilias reproduce by budding of an offspring cell from its parent. Moniliasis, a common vaginal infection, and thrush, a mucous membrane infection, are caused by the proliferation of the monilia *Candida albicans*. Molecular systematics studies indicate the placement of *Pneumocystis carini* with these monilias. Correlated with fatal pneumonia—the pneumonia that attacks immunocompromised persons, such as AIDS sufferers and patients who have undergone immunosuppressive treatments for cancer or organ transplants—the "cysts" of this organism apparently are fungal spores, because the sequence of nucleotides in the ribosomal RNA gene classifies *Pneumocystis* with the Ascomycota.

The Mycelia Sterilia include those genera that lack any reproductive structures; the mycelia simply grow without visible differentiation. Of the two dozen or so genera belonging to this conglomerate group, the best known is *Rhizoctonia*, a common soil fungus that causes damping off and root rot of plants, including cultivated ones of economic importance. The basidiomycote *Pellicularia filamentosa* has *Rhizoctonia solani* as its imperfect stage.

At least one-quarter of all described fungi, an estimated 12,000 to 20,000 species, can enter symbiotic associations with photosynthetic microbes such as algae to form lichens. Some 13,500 lichen species are formally described. At least 40 genera of photosynthetic partners have been found. To the unaided eye, the lichen fungus partner and the isolated alga are entirely different from their associated counterparts (Figure I). The fungus dominates and varies in the lichen so conspicuously that, invariably, less attention is paid to the algal partner. Some lichens superficially resemble a plant such as a moss. For many years, lichens were called "pioneer plants" because they often grow on bare rock or sterile soil and are among the first organisms to cover burnt-out or newly exposed volcanic regions.

I *Cladonia cristatella,* the British soldier lichen of New England woodlands. Bar = 1 cm. [Courtesy of J. G. Schaadt.]

Although many lichens are pioneers, they are not plants. All lichens, upon analysis, are symbiotic partnerships between a fungus, usually an ascomycote (the mycobiont), and a photosynthetic organism, most often a chlorophyte (Phylum Pr-28, usually *Trebouxia* or *Pseudotrebouxia*) or a cyanobacterium (Phylum B-5) such as *Nostoc* (the photobiont or phycobiont); but one xanthophyte (Phylum Pr-14) has been reported. We informally maintain lichens as a group only because of their common features and easy recognition in nature.

Although lichen fungi are found either unassociated or with algal partners (that is, lichenized) in nature, lichen algae are rarely, if ever, found on their own. In the laboratory, few lichen partners have been separated and grown by themselves. For more than a century, biologists have realized that lichens evolved by many independent symbiotic associations. Such associations probably began in response to attack by the fungus on a photosynthesizing protoctist. The attack became a permanent truce, and the two partners persisted in a unique form. Lichens synthesize compounds, such as the lichenic acids and pigments, that are absent in the individual algae and fungi when grown alone.

Because so many different fungi have evolved symbiotic relations with members of the same algal and cyanobacterial genera (for example, with *Trebouxia* or *Nostoc*), most lichenologists have insisted that they be classified with the ascomycote or basidiomycote groups to which the fungus belongs. (In practice, however, such classification efforts are hindered by the distinct structural differences between the lichen-associated fungus and its free-living relatives.) Like the fungi that become lichens, various heterotrophic protoctists and animals, including at least half a dozen flatworms (Phylum A-5) and molluscs (Phylum A-26), independently acquired some sort of photosynthetic partner. Far less diversity exists among the photobionts in relation to the heterotrophs. By analogy with the taxonomic convention for photosynthetic animals and protoctists therefore, when we refer to 13,500 species of lichen-forming fungi, we are actually citing the number of distinct fungal species, the majority of which are ascomycotes.

The lichens can thus be usefully classified, depending on whether the fungal partner is an ascomycote (as most are), a basidiomycote, or a deuteromycote (grouped with ascomycotes in our classification). Lichens composed of ascomycotes include *Lichina, Collema,* and *Cladonia.* Common in the northeastern part of North America, *Cladonia* species are eaten by many animals (and, in emergencies, by people). They are at the base of the Arctic food webs that include caribou (reindeer) and Arctic people. Most lichenized basidiomycotes belong to the homobasidiomycete families Clavariaceae and Coraceae. *Lepraria* and *Lichenothrix,* woodland lichens, are lichen-forming fungi in which no sexual structures are known; they are lichenized deuteromycotes, which we group here with ascomycotes.

Naturalists group lichens according to their external appearance, which is crustose (low and crusty), foliose (leafy), or fruticose (bushy; see Figure I). The thallus—crusty, bushy, or leafy—is the growing part of the organism (Figure J).

Lichen reproduction is fungal. The sexual structure in which meiosis takes place and propagules form usually has a distinct appearance; lichens composed of ascomycotes typically produce open spore-forming asci (Figure K). Sexual conjugation is followed by the formation of ascospores by meiosis. These ascospores germinate, but only in the presence of the proper algae, and produce fungal hyphae associated from the beginning with the algae, usually *Trebouxia* (Phylum Pr-28). Perhaps the most common form of lichen propagation is by propagule in the absence of any sexual process. Lichens release soredia, small fragments consisting of at least one algal cell surrounded by fungal hyphae (Figure L). Soredia are easily dispersed by air currents; in a suitable environment, they develop into new lichens.

Lichens are most abundant in the Antarctic, the Arctic tundra, high mountains, the tropics, and northern old-growth forests; they commonly grow on tree bark or rocks. In the supratidal zone of rocky coasts, lichens become highly diversified in the sea-spray zone. *Verrucaria serpuloides* is a permanently submerged marine

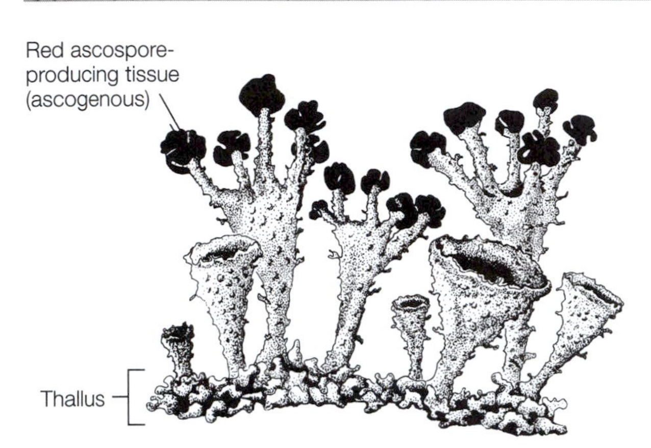

J *Cladonia cristatella* thallus with ascospore tissue. [Drawing by E. Hoffman.]

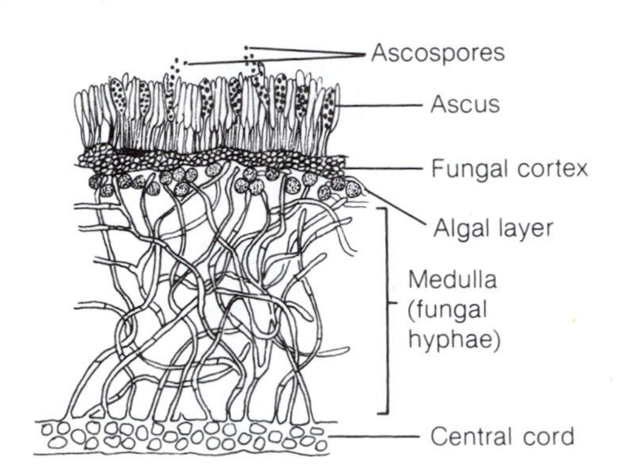

K Enlarged view of a cross section of ascospore tissue. Some lichen fungi form asci. These sexual structures can generate spores that germinate to produce fungal hyphae. [Drawing by E. Hoffman.]

lichen. Lichens are well known for their resistance to desiccation; less well known is that they *must* have alternating dry and wet periods, because continuous drought or continuous dampness kills them, except for marine lichens. Desiccation is accompanied by cessation of photosynthesis; this cessation may enable the lichen to survive otherwise harmful intense sunlight or extremes of heat or cold. The slow rate of growth of lichens is legendary. Many studies of lichens on gravestones and other dated monuments indicate that lichens grow from just 0.1 to 10.0 mm a year.

Lichens increase in total size slowly, but they metabolize quickly. Within a few minutes, carbon dioxide is fixed into organic matter by the photobiont. The photosynthetic partner (photobiont) transfers this photosynthate rapidly as sugar, sugar alcohol, or nitrogenous compounds (such as amino acids) to the fungal partner. Especially in lichens having nitrogen-fixing cyanobacteria, rapid metabolism—chemical production and transfer of nitrogen compounds—can be detected.

Lichens symbiotic with nitrogen-fixing cyanobacteria and perhaps other bacterial nitrogen fixers (Phylum B-5) provide northern old-growth forests with much of their nitrogen. *Lobaria*, one of about 130 lichens recently found as an epiphyte in northern tree canopies, fixes as much as 75 percent of the nitrogen required by this Douglas fir environment. Rain and fog wash soluble nitrogenous compounds from the lichens to the forest floor, where the mycorrhizal fungi in the tree roots absorb them as nutrients. Some tree species even send out roots from their branches into canopy lichens, thereby taking in fixed nitrogen directly.

Lichens are very important initiators of biological succession. By slowly wearing away and dissolving the minerals that compose the rocks on which they establish, lichens prepare the surfaces for the germination of seeds and the formation of rooted plant communities. Lichens thus accelerate weathering and initiate the formation of soils. Despite their hardiness, lichens are very sensitive to certain airborne materials—for example, the sulfur dioxide and volatile metal compounds that are released when coal is burned. Thus, the presence of lichens and the state of their health are used as pollutant indicators.

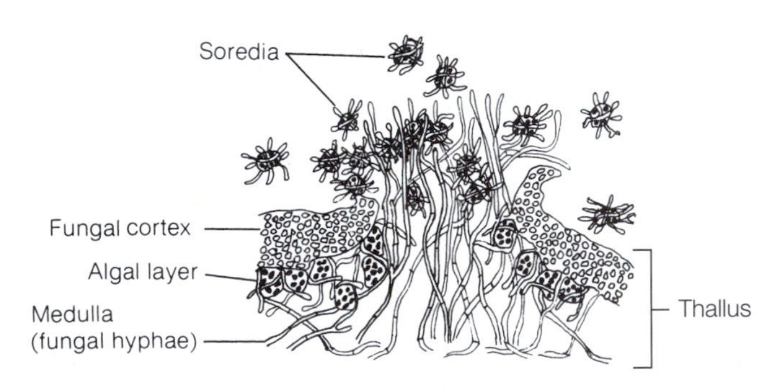

L Enlarged view of a cross section of *Cladonia* thallus. Lichens can also reproduce by means of soredia, which are made up of algae (or cyanobacteria) and fungal hyphae. Dispersal of these soredia establishes new populations of lichens. [Drawing by E. Hoffman.]

Bibliography: Fungi

General

Aharonowitz, Y., and G. Cohen, "The microbiological production of pharmaceuticals." *Scientific American* 245(3): 140–152; September 1981.

Ahmadjian, V., and M. E. Hale, eds., *The lichens.* Academic Press; New York; 1974.

Ahmadjian, V., and S. Paracer, *Symbiosis: An introduction to biological associations.* University Press of New England; Hanover, NH, and London; 1986.

Ainsworth, G. C., *Ainsworth and Bisby's dictionary of the fungi, including the lichens,* 8th ed. Oxford University Press; New York; 1996.

Ainsworth, G. C., *Introduction to the history of medical and veterinary mycology.* Cambridge University Press; New York; 1987.

Ainsworth, G. C., and A. S. Sussman, eds., *The fungi,* 4 vols. Academic Press; New York; 1965–1973.

Bonner, J. T., "The growth of mushrooms." *Scientific American* 194(5):97, 98, 100, 102, 104, 106; May 1956.

Brightman, F. H., *Oxford book of flowerless plants: Ferns, fungi, mosses and liverworts, lichens and seaweeds.* Oxford University Press; New York; 1966.

Emerson, R., "Molds and men." *Scientific American* 186(1): 28–32; January 1952. Potato blight, ergot, wheat rust, athlete's foot, and citric acid from mold fermentation.

Gould, S. J., "A humongous fungus among us." *Natural History* 101(7):10–14, 16, 18; July 1992. *Armillaria bulbosa* covers 2.5 miles square—the largest organism?

Kendrick, W. B., *The fifth kingdom,* 2d ed. Focus Information Group, Inc.; Newburyport, MA; 1992, and Mycologue Publications; Waterloo, Ontario; 1992.

Kosikowski, F. V., "Cheese." *Scientific American* 252(5): 88–99; May 1985.

Large, E. C., *The advance of the fungi.* Dover; New York; 1962.

"Microbes for hire," *Science '85* 6(6):30–46; July/August 1985. A series of articles including: T. Monmaney, "Yeast at work," 30–36; D. Morgan and T. Monmaney, "The bug catalog," 37–41; and P. Preuss, "Industry in ferment," 42–46.

Moore-Landecker, E., *Fundamentals of the fungi,* 3d ed. Prentice-Hall; Englewood Cliffs, NJ; 1990.

Phillips, R., *Mushrooms of North America.* Little, Brown; Boston; 1991.

Pirozynski, K. A., and D. L. Hawksworth, eds., *Coevolution of fungi with plants and animals.* Academic Press; San Diego, CA; 1988.

Rose, A. H., "The microbiological production of food and drink." *Scientific American* 245(3):126–134, 136, 138; September 1981.

Wainwright, P. O., G. Hinkle, M. L. Sogin, and S. K. Stickel, "The monophyletic origins of the Metazoa: An unexpected evolutionary link with Fungi." *Science* 260:340–343; 1993.

F-1 Zygomycota

Alexopoulos, C. J., and C. W. Mims, *Introductory mycology,* 3d ed. Wiley; New York; 1979.

Moore-Landecker, E., *Fundamentals of fungi,* 3d ed. Prentice-Hall; Englewood Cliffs, NJ; 1990.

F-2 Basidiomycota

Alexopoulos, C. J., and C. W. Mims, *Introductory mycology,* 3d ed. Wiley; New York; 1979.

Barron, G., "Jekyll-Hyde mushrooms." *Natural History* 101(3):47–52; March 1991.

Litten, W., "The most poisonous mushrooms." *Scientific American* 232(3):90–101; March 1975.

Smith, A. H., and N. Weber, *The mushroom hunter's field guide.* University of Michigan Press; Ann Arbor; 1980.

F-3 Ascomycota

Amerine, M. A., "Wine." *Scientific American* 211(2):46–56; August 1964.

Bold, H. C., C. J. Alexopoulos, and T. Delevoryas, *Morphology of plants and fungi,* 5th ed. Harper College; New York; 1990.

Bussey, H., "Chain of being." *The Sciences* 36(2):28–33; March/April 1996.

Gumpert, M., "Histoplasmosis: The unknown infection." *Scientific American* 178(6):12–15; June 1948. A fungal disease of lungs caused by *Histoplasma capsulatum.*

Hale, M. E., *The biology of lichens,* 3d ed. Edward Arnold; Baltimore; 1983.

Khakhina, L. N., *Concepts of symbiogenesis: A historical and critical study of the research of Russian botanists.* Yale University Press; New Haven, CT; 1992.

Lawrey, J. D., and M. E. Hale, Jr., *Biology of lichenized fungi.* Greenwood (Praeger); Cornelius, NC; 1984.

Matossian, M. K., "Ergot and the Salem witchcraft affair." *American Scientist* 70(4):355–357; 1982.

Matossian, M. K., *Poisons of the past: Molds, epidemics, and history.* Yale University Press; New Haven, CT; 1991.

Newhouse, J. R., "Chestnut blight." *Scientific American* 263(1):106–111; July 1990.

Phaff, H. J., M. W. Miller, and E. M. Mrak, *The life of yeasts,* rev. ed. Harvard University Press; Cambridge, MA; 1978.

Redfield, R. R., and D. S. Burke, "HIV infection: The clinical picture." *Scientific American* 259(4):90–98; October 1988.

Richardson, D. H. S., *The vanishing lichens: Their history, biology and importance.* Macmillan (Hafner Press); New York; 1975.

Rose, A. H., "Beer." *Scientific American* 200(6):90–100; June 1959.

Rose, A. H., "Yeasts." *Scientific American* 202(2):136–146; February 1960. Yeast, fermentation industries and cell physiology.

CHAPTER FIVE

KINGDOM PLANTAE

Pl-8 *Ceratozamia purpussi* cone
[Photograph by K. V. Schwartz]

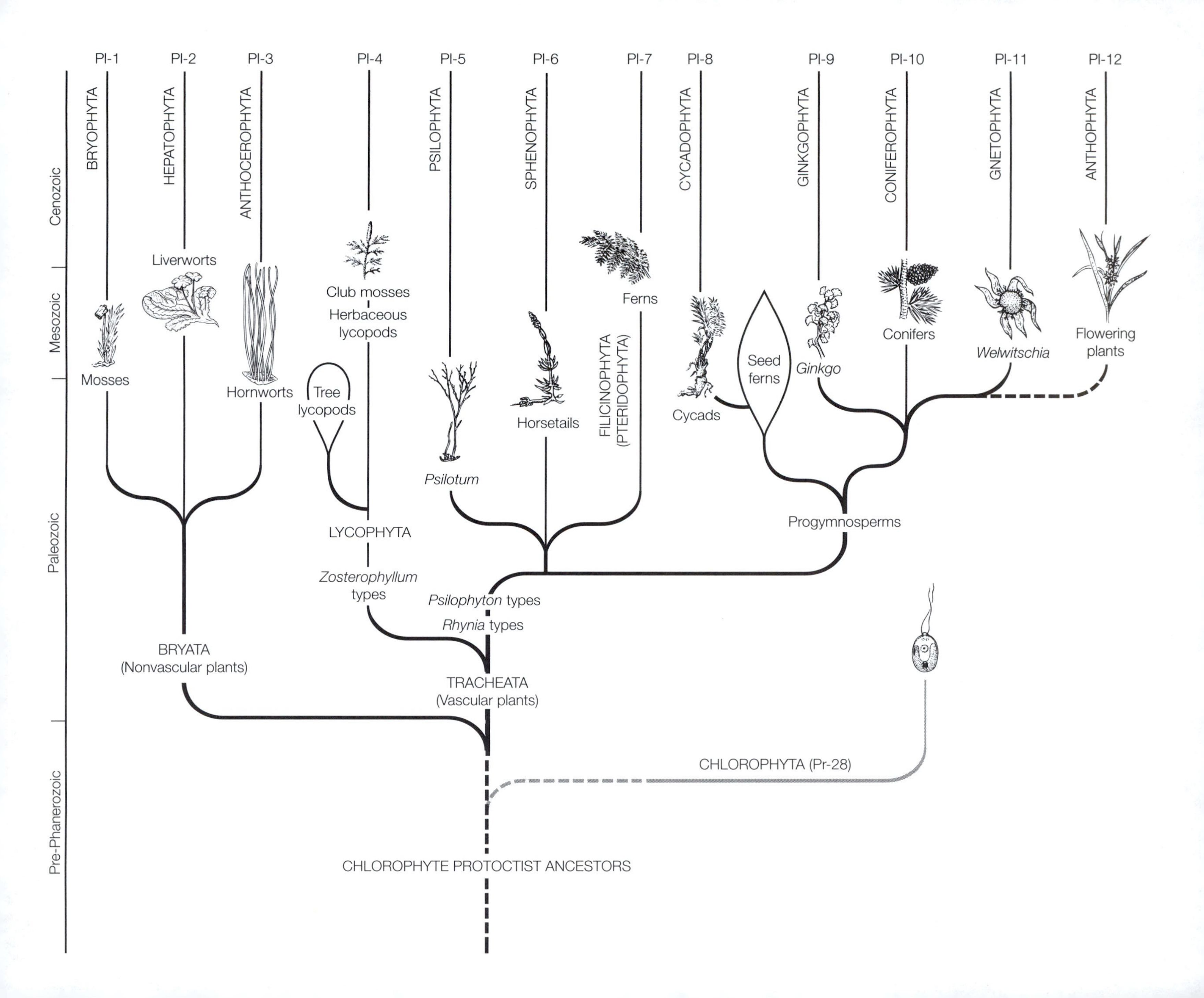

PI-1
PI-2
PI-3
PI-4
PI-5
PI-6
PI-7
PI-8
PI-9
PI-10
PI-11
PI-12
BRYOPHYTA
HEPATOPHYTA
ANTHOCEROPHYTA
PSILOPHYTA
SPHENOPHYTA
CYCADOPHYTA
GINKGOPHYTA
CONIFEROPHYTA
GNETOPHYTA
ANTHOPHYTA
Cenozoic
Mesozoic
Paleozoic
Pre-Phanerozoic
Liverworts
Mosses
Hornworts
Club mosses
Herbaceous lycopods
Tree lycopods
Ferns
FILICINOPHYTA (PTERIDOPHYTA)
Horsetails
Psilotum
Seed ferns
Cycads
Ginkgo
Conifers
Welwitschia
Flowering plants
LYCOPHYTA
Progymnosperms
Zosterophyllum types
Psilophyton types
Rhynia types
BRYATA (Nonvascular plants)
TRACHEATA (Vascular plants)
CHLOROPHYTA (Pr-28)
CHLOROPHYTE PROTOCTIST ANCESTORS

PLANTAE

Latin *planta*, plant

Haploid organisms of complementary sexes grow from spores produced by meiosis (sporogenic meiosis) that takes place in the adult diploid. These haploids produce gametes by mitosis. Fertilization by sperm (cytogamy and karyogamy) or pollen nucleus (karyogamy) leads to diploid embryo retained by the female haploid organism during early development. Fossil record extends from the lower Paleozoic era (450 million years ago) to the present.

Members of the plant kingdom develop from embryos—multicellular structures enclosed in maternal tissue (Figure 5-1). Because all plants form embryos, they are all multicellular. Furthermore, because embryos are the products of the sexual fusion of cells, all plants potentially (although not always in reality) have a sexual stage in their life cycle. In the sexual stage, the male cell (sperm nucleus, haploid) fertilizes the female (egg or embryo sac nucleus, haploid). Many plants grow and reproduce in ways that bypass the two-parent sexual fusion—all must have evolved from ancestors that formed embryos by sexual cell fusion. One example of asexual reproduction is the strawberry plant; plantlets form on extensions called runners extending from the parent plant. A second example is the asexual reproduction of little green balls of cells called gemmae (Latin, "gems" or "buds") by a parent moss or liverwort plant. Evolution of the embryo, protected by maternal tissue from drying and other environmental hazards, was a major factor in the spread of plants from oceans to dry land. Development in green algal (chlorophyte) ancestors of intimate symbioses with fungi may have been another factor in transitions from aquatic to terrestrial life, facilitating uptake of minerals and water by the plant. All plants are composed of eukaryotic cells, many having green plastids (see Figure I-1). We distinguish plants from all other organisms by their life cycles rather than by their capacity for photosynthesis because some plants (beech drops, *Epifagus*, for example) are entirely without photosynthesis throughout their lives. Photosynthesis by plants requires enzymes within membrane-bounded plastids. All plants that photosynthesize produce oxygen. (In comparison, in photosynthetic prokaryote species, enzymes are bound as chromatophores to cell membranes, not packaged separately. Prokaryote patterns of anaerobic and aerobic photosynthesis include formation of end products such as sulfur, sulfate, and oxygen.)

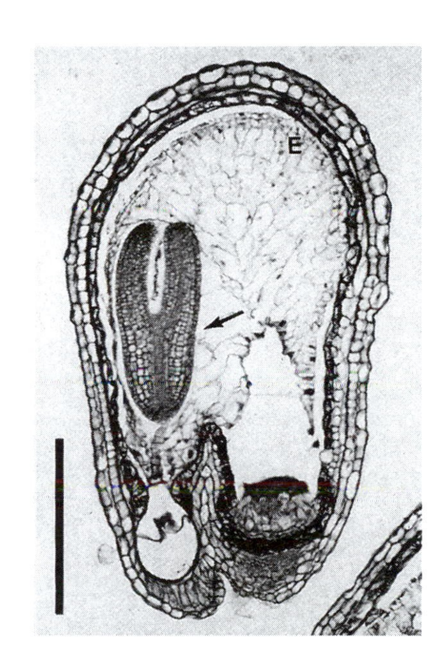

Figure 5-1 Embryo (arrow) of shepherd's purse, *Capsella bursa-pastoris.* The two horseshoe-shaped cotyledons of the embryo have developed within the seed. Stored food (E, endosperm) surrounds and nourishes the body of the young sporophyte plant. LM, bar = 300 mm. [Courtesy of W. Ormerod.]

Plants are adapted primarily for life on land, although many dwell in water during part of their life history. Plants are the organisms most responsible on land and in shallow marine environments for transforming solar energy, water, and carbon dioxide into photosynthate: food, fiber, coal, oil, wood, and other forms of stored energy. (In the open ocean, the plankton protoctists are the primary producers.) Houseplants, trees, and crop plants are members of the plant kingdom. Although most plants are multicellular, green, photosynthesizing organisms, a few genera such as dodder (*Cuscuta*) and Indian pipe (*Monotropa*) lost green pigment in the course of evolution and became parasitic. Many photosynthetic organisms that were once classified as members of the plant kingdom on the basis of color and sedentary habit are no longer considered plants, because they lack embryos and other minimal criteria for plant classification. Cyanobacteria (blue-green bacteria, Phylum B-5), green algae (Phylum Pr-28), all other algae (for example, Phyla Pr-13 through Pr-17, Pr-25, and Pr-26), and lichens (Phylum F-3; fungi with bacteria or protoctist symbionts) are now placed with their relatives in the bacteria, protoctista, or fungi kingdom. Photosynthesis by plants sustains the rest of the biota not only by converting solar energy into food, but also by absorbing carbon dioxide and producing oxygen.

Some half million species of plants have been described. Because new species are found each year, especially in the tropics, probably another half million plants await discovery. Furthermore, this estimate is probably low; many plants resemble each other in form and will be distinguishable as separate species only by chemical analysis.

Two great groups—the nonvascular plants (informally called bryophytes, also called Bryata, Phyla Pl-1 through Pl-3) and the vascular plants (Tracheata, Phyla Pl-4 through Pl-12)—constitute the plant kingdom. We refer to the 12 "phyla" of the plant kingdom, but "division" is the term used by some botanists instead of "phylum." Tracheata, the familiar woody and herbaceous plants, are distinguished by vascular systems—lignified conducting tissues called xylem and phloem. Primary vascular tissues consist of cells derived from apical meristems (undifferentiated cells that give rise to new cells) and their derivatives. An example is a primary vascular bundle of xylem and phloem. Lignin is a complex macromolecule that stiffens the plant, impregnates xylem, and strengthens the wood of woody trees and shrubs. Under the bark of woody plants is a layer of cells (cambium) that generate new xylem and phloem throughout the life of a plant. This so-called secondary growth increases the diameters of tree trunks and shrub stems. Within the ring of cambium lies relatively undifferentiated tissue called pith. Herbaceous plants are nonwoody plants, such as dandelions, ferns, and moss. Xylem cells transport water and ions from the roots through the plant. Phloem cells transport photosynthate—products of photosynthesis—throughout the plant body. Aboveground structures consisting of a shoot (the central upright axis) comprising a stem with branches and leaves and underground structures consisting of roots are unique to tracheophytes. The parts of the plant below ground that are anatomically similar to the stem are called a rhizome (see Phylum Pl-5, Figure A), whereas those that differ from stems anatomically (in pattern of vascular tissue, for example) are roots, anchoring the plant and taking up water with dissolved minerals. A taproot is a single, large root (in carrot or cycad, for example) that may store nutrients (sugar beet) and water. True leaves and roots contain vascular tissues. A true leaf consists of photosynthetic tissue covered by a cuticle (a waxy, water-resistant layer on the external surface) pierced by stomata—openings through which gases pass in and out of the leaf blade (Figure 5-2). Vascular tissue is the plumbing system of the leaf, continuous with stem (and, eventually, root) through the leaf stalk, called the petiole. In comparison, mosses and many other nonvascular plants have leafy structures; a leafy structure lacks a vein of vascular tissue, is only one or a few cells thick, and may even lack cuticle and stomata. Mosses, liverworts, and some vascular plants also lack true roots and instead may have rhizoids, root-hairlike structures that lack vascular tissue veins. The site at which leaves and branches join the stem is called a node, evident in horsetails (Phylum Pl-6); internodes are stem regions between nodes. Branches that subdivide into two smaller branches are said to be dichotomous, as in psilophytes (Phylum Pl-5). Vascular plants may be grouped into seed-bearing (Phyla Pl-8 through Pl-12) and nonseed-bearing vascular plants (Phyla Pl-4 through Pl-7). The vast majority of plants living today are tracheophytes belonging to phylum Anthophyta, the flowering plants (Phylum Pl-12). The current number of gymnosperms (Greek *gymnos,* naked; *sperma,* seed) is about 720 species in 65 genera compared with approximately 240,000 species of flowering plants.

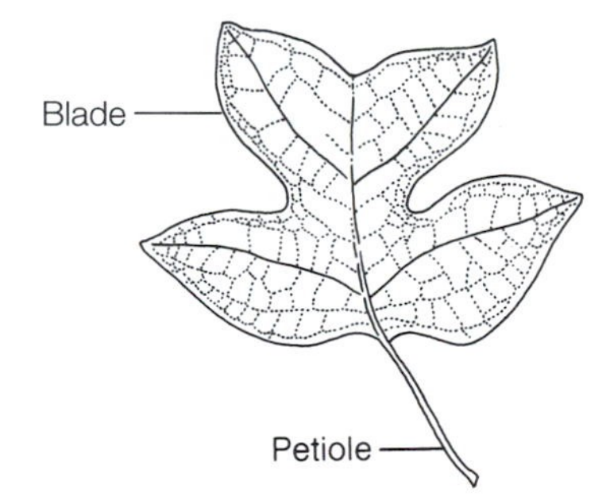

Figure 5-2 Leaf of tulip tree, *Liriodendron tulipifera.* The net of veins is typical of a dicot megaphyll. [Drawing by L. Meszoly.]

Seed-bearing plants develop with exposed (naked) or enclosed seeds. A seed is formed by maturation of the ovule after fertilization; the ovule ("little egg") contains the female gametophyte with its egg cell, both surrounded by integuments. Minimally, the integuments surrounding the ovule form a seed coat. Gymnosperms (Phyla Pl-8 through Pl-11) develop seeds in cones, in comparison with protective, seed-enclosing fruits produced by flowering plants, or angiosperms (Phylum Pl-12). A cone is a reproductive structure that consists of a number of modified leaves clustered at the end of a stem; club mosses, horsetails, and gymnosperms bear cones. Cones are simple or compound. Scales (modified leaves) of a simple cone—a male pine cone, for example—bear sporangia and attach directly to the cone's central axis. In a compound cone—a female pine cone, for example—sporangia attach indirectly by a sterile bract to the cone's axis. Gymnosperms—cycads, ginkgos, conifers, and gnetophytes—produce pollen cones (male) and seed cones (ovule-bearing, female), whereas flowering plants produce pollen and ovules in the flower and, eventually, seed within fruit (mature, ripe ovules).

Four phyla of vascular plants do not produce seeds: club mosses (Phylum Pl-4), whisk ferns (Phylum Pl-5), horsetails (Phylum Pl-6), and ferns (Phylum Pl-7). All reproduce with spores. Spore dispersal in liverworts, hornworts, and horsetails is aided by elaters (Greek, "driver"), elongated structures that form within sporangia. The sporangium is an organ in which cells undergo meiosis and produce (haploid) spores. A spore (see Phylum Pl-3, Figure B) is a reproductive cell capable of developing into a mature plant without fusing with another cell; in comparison, an egg or sperm fuses with its complementary reproductive cell to produce a new plant. A spore usually consists of a single cell and is produced in a sporangium. In club mosses and horsetails, sporangia are borne on modified leaves clustered in cones called strobili (see Phylum Pl-6, Figure B). Club mosses and ferns also reproduce asexually by means of plantlets. These new plantlets are shed from the parent plant and, unlike moss gemmae, are diploid. Nonseed-bearing vascular plants have evolved a wide array of leaf structures. Leaves of Phyla Pl-7 through Pl-12 are megaphylls (see Figure 5-2); that is, comparatively large leaves with a web of veins or parallel veins and a gap above the junction of leaf with stem. These veins connect through several strands of vascular tissue to the vascular tissues of the stem. Leaves may be simple, as is a tulip leaf, with undivided blade, or compound, as is a walnut leaf, with a leaf blade composed of many leaflets. The club moss leaf, called a microphyll, has only a single vascular strand and lacks the leaf gap characteristic of megaphylls. Horsetail leaves are very small and scalelike; psilophytes (Phylum Pl-5) lack leaves altogether.

We call the three phyla of nonvascular plants (Pl-1 through Pl-3) Bryata—mosses, liverworts, and hornworts. Opinion varies regarding *Takakia*, considered a moss by some, including us, and a separate phylum (or division) by others. All nonvascular plants have a thallus—a plant body without true leaves, stem, or roots (see Phylum Pl-2, Figure B). Lacking true roots and vascular systems, they obtain moisture and nutrients from the environment by diffusion directly into their tissues. Within their bodies, diffusion, capillary action, and cytoplasmic streaming conduct fluids. Mosses also have conducting cells called leptoids and hydroids, but these are nonlignified. Hydroids are elongated cells that lack living cytoplasm at maturity. Their thin end walls are very permeable to solutes and water. Nutrient-conducting leptoid cells surround the hydroids in some mosses. Delicate uni- or multicellular filaments (rhizoids) anchor nonvascular plants to soil, rock, or tree bark (see Phylum Pl-1, Figure A). Most of the Bryata flourish in moisture-saturated habitats such as acidic bogs. Given these similarities among the nonvascular plants, significant differences—such as presence or absence of cuticle—remain such that the three nonvascular plant phyla may have diverged independently of one another from green algal ancestors.

All plant cells at all stages harbor plastids, usually many. Minimal plastids are 1 mm in diameter, membrane-bounded, colorless organelles (as in roots, colorless sprouts, and parasitic plants). Exposure to sunlight may transform colorless plastids into the chlorophyll-containing green form called chloroplasts. Fully developed plastids are so similar to those of green algae that biologists agree that these chlorophytes (Phylum Pr-28) were ancestral to plants. Other support for this hypothesis is that green algae and plants have similar cell-to-cell connections called plasmodesmata (singular:

plasmodesma). Some chlorophytes, such as *Klebsormidium,* even have cellulosic walls and patterns of mitotic cell division identical with those of plants. In these green algae, as in plants, a cell-wall structure called a cell plate (phragmoplast) develops perpendicularly to the mitotic spindle and separates the two daughter cells at the completion of mitosis.

Plants, like all other extant organisms, have aquatic ancestors, with land plants having evolved from only a small group of green algae. Plantlike fossils first appear in rocks of the Silurian period (430 to 408 million years ago) as rootless, leafless, but upright seaweedlike organisms. The earliest plants, for which the fossil record is abundant, were ancestral tracheophytes of two major types, represented during the Devonian period by the extinct *Zosterophyllum* and *Rhynia* (see phylogeny). Because nonvascular plants lack vascular tissue, they are presumed to have evolved before the appearance of vascular tissue—before tracheophytes. In apparent contradiction to this time sequence, though, the earliest bryophyte fossil found so far is only 350 million years old, which is later than the first tracheophytes in the fossil record. Better fossilization of lignified tissues of vascular plants compared with nonlignified tissue of nonvascular plants may explain this discrepancy.

The *Zosterophyllum* types gave rise to or share a common ancestor with lycopods (Phylum Pl-4), which have a fossil record as definite lycopods extending 400 million years into geological history. This tracheophyte group speciated extensively and included tree lycopods at the end of the Paleozoic era but is now reduced to a few genera of club mosses and their kin. Ancestral groups for psilophytes (Pl-5) and horsetails (Pl-6) are unknown at present.

Lycophytes and psilophytes have each been put forward as extant representatives of the first split in early lineages of vascular land plants. Chloroplast DNA studies tend to confirm the geological evidence that lycopods are more closely related to nonvascular plants (Phyla Pl-1 through Pl-3), whereas psilophytes are more closely related to vascular plants (Phyla Pl-6 through Pl-12) other than club mosses. The chloroplast gene order in modern lycopods is shared with that of *Marchantia,* a liverwort (Phylum Pl-2). Although psilophytes seem ancestral ("primitive")—they lack roots and have shoot protrusions that are probably branchlets rather than being homologous to leaves—their chloroplast DNA resembles that of ferns (Phylum Pl-7), gymnosperms (Phyla Pl-8 through Pl-11), and angiosperms (Phylum Pl-12). Psilophytes probably evolved directly from *Rhynia* types.

The *Rhynia* types of extinct tracheophytes were the ancestors of all the vascular land plants except club mosses. Many groups, such as the extinct phylum of seed ferns (Cycadofilicales or pteridosperms) and the phylum of horsetails (Sphenophyta, Phylum Pl-6), were far larger and more important in the past than they are now. We do not know if the ancestors of psilophytes (Phylum Pl-5) have any modern representatives other than *Psilotum* still living, though relationships among extant groups are now being sought with elegant molecular methods of inquiry. The *Psilophyton* types were ancestral to progymnosperms.

The details of seed, flower, fruit, and endosperm origins are under investigation, but we do know that these evolutionary innovations of flowering plants (Phylum Pl-12) changed the living world forever. Endosperm is a tissue, unique to flowering plants, that is neither sporophytic nor gametophytic. Endosperm develops from the union of sperm with polar nuclei of the central cell (female). Stored nutrients in endosperm are digested by the embryo. The remarkable innovation of the seed had evolved by at least 360 million years ago (in the late Devonian) and more than once. At least one lineage of seed plants—progymnosperms, which had seeds but no flowers or fruit—gave rise to the great Mesozoic forests of cone-bearing plants: cycads, ginkgos, conifers, and other gymnosperms. In what may be the most primitive animal pollination system, cycad cones (Phylum Pl-8) produce odors and heat that attract pollinating insects. The anthophytes—the angiosperms, or flowering plants (Phylum Pl-12)—by their production of nectar (sugary liquid produced in flowers that serves to attract and reward pollinating animals), flowers, and fruit created an environment in which we and so many other animals could thrive. Flowering plants are an enormous group and are relatively young, having appeared on the scene only about 130 million years ago, the newest plant phylum. Like the cycads, flowering plants were considered by some biologists to have descended from seed ferns. Now there is considerable evidence that flowering plants evolved from an *Ephedra*-like gnetophyte (Phylum Pl-11); double fertilization occurs in both.

If it seems that there are far fewer plant than animal groups, it is partly because plant and animal taxa are defined by morphological criteria, and the diversity of internal and external anatomy is more extensive in animals than in plants. The differences between many plants are subtle, often involving chemical distinctions. Plants produce many chemical compounds that are secondary metabolites used in the plant's defenses against fungi, animals, and other plants and are therefore only indirectly required for survival and reproduction. Secondary compounds include toxins, psychoactive compounds such as the marijuana alkaloids, and respiratory poisons such as cyanide—all of which deter predators from eating the plant. For example, black walnut trees leak compounds into the soil that prevent plants of other species from growing nearby. These poisons and other secondary metabolites, even gaseous compounds, are important in determining the distribution, growth rate, and abundance of plants in natural communities. Many of these compounds directly affect survival: the diterpenes, gibberellic acids, and—in reproduction, pollen and seed dispersal—flavonoids. Some activate genes, resulting, for example, in nodulation (swellings on roots) in *Rhizobium* in legumes and recognition of fungi in the establishment of mycorrhizae. Thousands of secondary metabolites are known, and many are used by the pharmaceutical industry as starting materials for the manufacture of drugs.

All plants develop from embryos, diploid multicellular young organisms supported by sterile or nonreproductive tissue; in conifers, gnetophytes, and flowering plants (Phyla Pl-10 through Pl-12), the cotyledon (embryonic seed leaf) provides nutrients to the young embryo (see Figure 5-1). The cotyledon of monocots absorbs food, whereas cotyledons of dicots store food. Unlike animal embryos, the plant embryo is not a blastula (see Figure 3-1. Unlike fungi, in which cells are either haploid (monokaryotic) or diploid (dikaryotic), except for the transient diploidy of the zygote during sexual reproduction, plants alternate haploid and diploid generations in their life cycle. Haploid plants (n) are called gametophytes; diploid plants ($2n$) are called sporophytes.

The life cycles of all nonvascular plants (Phyla Pl-1 through Pl-3) are dominated by the conspicuous green gametophyte (haploid), exemplified by a green mat of moss. To follow the life cycle of the moss (Figure 5-3), in broad outline an example of the general plant life cycle, we begin with sexual reproduction. As its name implies, the gametophyte produces gametes. The male reproductive organ—the antheridium—produces sperm having a pair of forward-pointing undulipodia (see Figure I-2 and Phylum Pl-1, Figure D). The female reproductive organ—the archegonium—produces an egg cell. Sperm of mosses are dispersed from the male mature gametophyte by splashing raindrops. In a magnified view of the archegonium, sperm can be seen swimming toward the egg. Sperm of cycads, ginkgos, gnetophytes, conifers, and flowering plants are carried to the egg in a pollen tube, formed after germination of a pollen grain. The tube nucleus of the mature pollen grain directs growth of the tube to the ovule. In flowering plants, the tube grows through stigma and style (see Phylum Pl-12, Figure B). The pollen tube provides a moist environment for sperm; thus its evolution eliminated the requirement for environmental water during fertilization. There—within the archegonium and still on the female gametophyte—fertilization of the egg by the sperm takes place. Fertilization restores diploidy (the $2n$ condition) and initiates development of the zygote into the embryo. The embryo develops into the diploid sporophyte that emerges from the archegonium. The young sporophyte derives nutrients from the female gametophyte, on which it permanently perches. Early in its growth, the moss sporophyte becomes green and photosynthesizes. As its name implies, the sporophyte produces spores. Meiosis within the capsule of the mature sporophyte results in haploid spores. The nonvascular plant sporophyte is small and often brown by the time it releases spores, like the stalked spore capsule of the moss shown in the life cycle. Spores that land in a favorable site germinate, beginning the (haploid) gametophyte generation and developing a strand of photosynthetic cells (protonema) that resembles a green algal filament. This young gametophyte forms a little outgrowth of cells

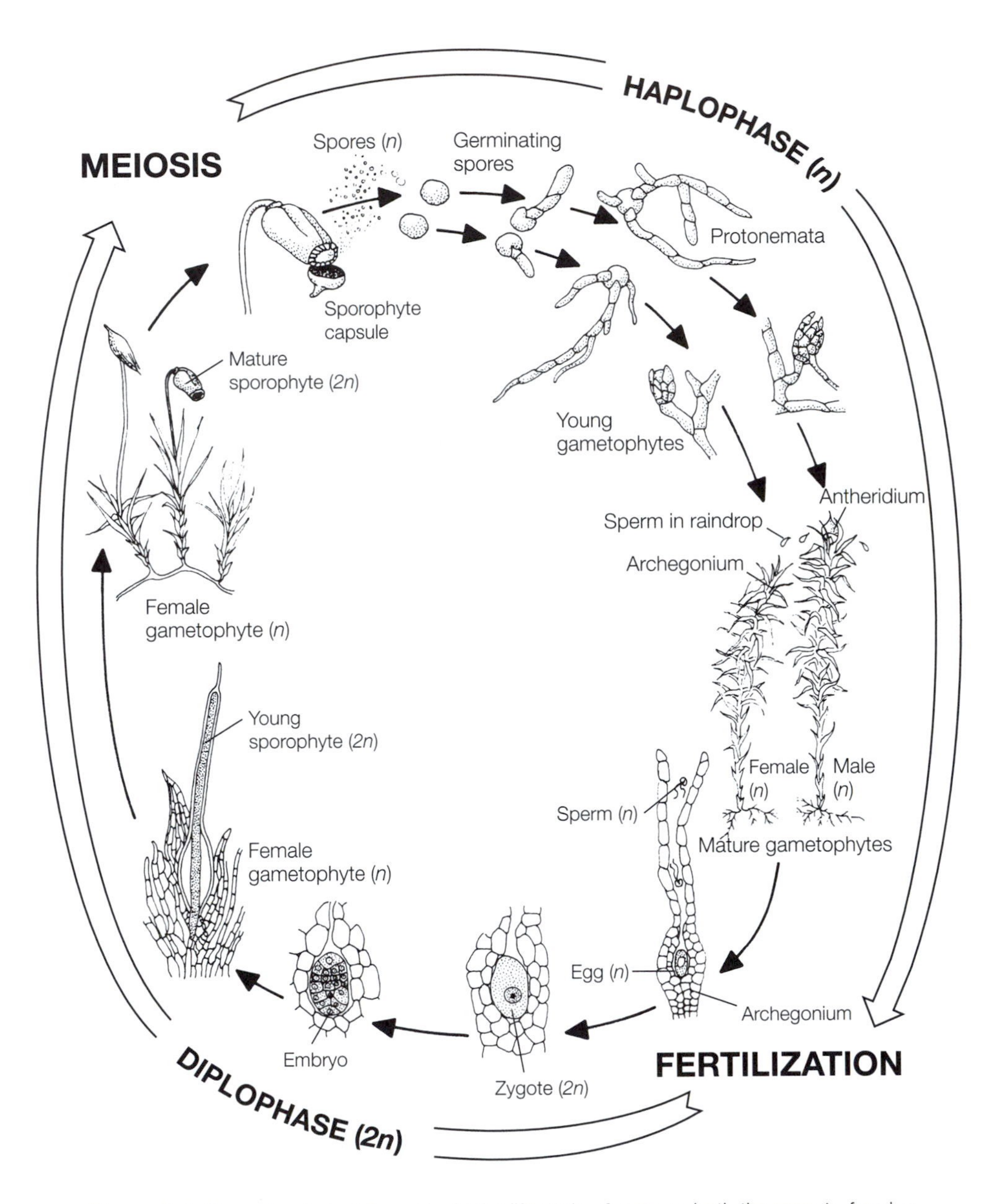

Figure 5-3 Generalized plant life cycle. In the life cycle of a moss, both the gamete-forming haploid phase and the spore-forming diploid phase are conspicuous. This moss, *Polytrichum,* is described in Phylum Pl-1. [Drawing by K. Delisle.]

called a bud, which grows into a mature gametophyte. The sporophyte of nonvascular plants is generally dependent on the gametophyte. In the vascular plants, on the other hand, the sporophyte is greener, larger, and more conspicuous than the gametophyte. For example, the oak tree is a sporophyte. The sporophyte generation dominates the life cycle in the most recently evolved extant phyla [horsetails (Phylum Pl-6), ferns (Phylum Pl-7), and the seed bearers (Phyla Pl-8 through Pl-12)], and the gametophyte is reduced in size. In the flowering plants, both the male and the female gametophyte, instead of being a separate plant as in mosses, is only a small group of cells entirely dependent on the sporophyte. The oak female gametophyte, as is the case for all flowers, is hidden within the flower; the oak male gametophyte is hidden with a grain of pollen. Extant seed plants, which produce nonmotile sperm in combination with pollen tubes, constitute one end of a series that extends from mosses, liverworts, hornworts, club mosses, horsetails, psilophytes, and ferns—all having swimming sperm—through ginkgos and cycads, which have pollen tubes as well as motile sperm, to conifers, gnetophytes, and flowering plants, with pollen tubes as conduits for nonmotile male gametes. Plant life histories are as elegantly diverse as their forms and colors.

Pl-1 Bryophyta

(Mosses)

Greek *bryon,* moss; *phyton,* plant

EXAMPLES OF GENERA

Andreaea
Bryum
Buxbaumia
Fontinalis
Funaria
Geothallus
Hypnum
Physcomitrium
Polytrichum
Sphagnum
Takakia
Tetraphis
Tortula

The nonvascular plants informally called "bryophytes" are the liverworts, hornworts, *Takakia,* and mosses, but only the mosses and *Takakia* are now included in the phylum Bryophyta. The presence of conducting cells and the lack of elaters—coils that facilitate spore dispersal—distinguish mosses from liverworts and hornworts. Mosses have leafy gametophytes with a stalk (stem) and multicellular rhizoids but lack the true leaves, stems, and roots present in vascular plants. The rhizoids, which are rootlike multicellular filaments, anchor the moss gametophyte to the substrate.

The moss life cycle alternates gameto- and sporophyte—universal in plants (see Figure 5-2). Fusing in pairs, the egg and sperm form a zygote. Embedded in the female, the embryo develops into a (diploid) sporophyte. Spores (haploid), produced by the sporophyte, germinate and give rise to the next generation of young gametophytes.

Mosses are low-growing plants that flourish in moist habitats including fresh water. Some grow on oceanside rocks, but none are marine. Most of the nearly 10,000 moss species live in moist tropical environments, though mosses are also more conspicuous than other nonvascular plants in temperate North America. The three classes of this phylum are Sphagnopsida (peat mosses), Andreaeopsida (granite mosses), and, containing the majority of species, Bryopsida, or true mosses, such as the *Polytrichum* shown in Figure A. *Takakia* (Figure B), a mosslike low-growing plant, is regarded by some workers as a moss similar to Andreaeopsida, by others as deserving separate phylum status. Many mosses are well adapted to withstand desiccation and survive as quiescent spores or dry gametophytes throughout the dry season. Several moss species are found in warm deserts, and mosses (for example, *Andreaea*), dominate the cold deserts of the Arctic tundra. Since its discovery in 1951, *Takakia* has been found in a great arc around the Pacific from Borneo to the Himalayas (Nepal, Sikkim, and China), Japan, the Aleutian Islands, the Alaskan panhandle, and British Columbia.

The conspicuous and familiar generation of mosses, including *Takakia,* is the green, leafy (not a true leaf) gametophyte—a haploid organism. Instead of the rootlike rhizoid of mosses, *Takakia*

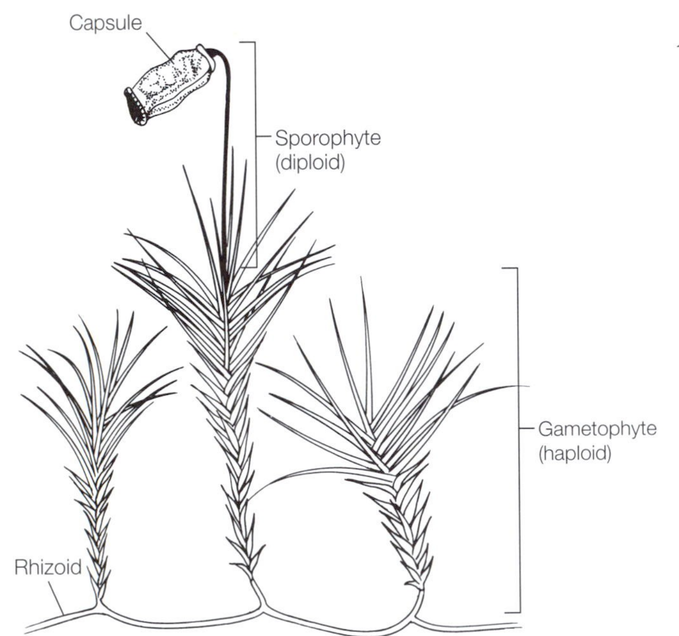

A *Polytrichum juniperinum,* a common ground cover in the mixed coniferous and deciduous forest of New England. Leafy gametophyte (haploid) with mature sporophyte capsule (diploid). Bar = 3 cm. [Photograph courtesy of J. G. Schaadt; drawing by L. Meszoly.]

B *Takakia* habit: female gametophyte plants bearing sporophytes with tapered capsules. Male plants nestled among female plants. [Photograph by permission of A. S. Heilman, D. K. Smith, and K. D. McFarland. From D. K. Smith and P. G. Davison, *Journal of the Hattori Botanical Laboratory,* No. 73:263–271 (1993).]

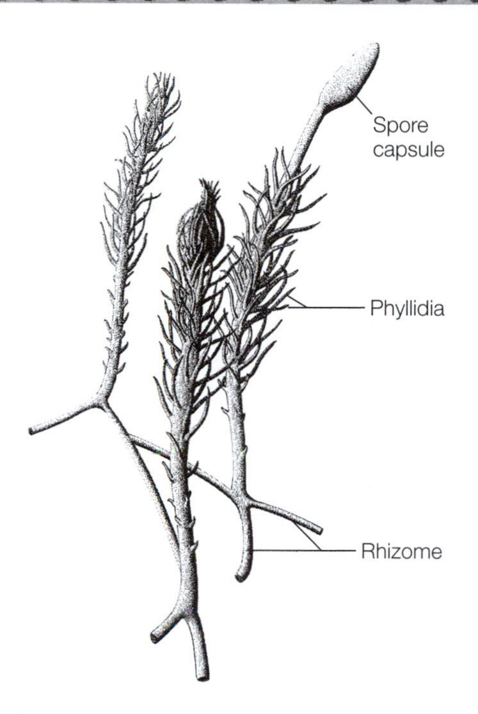

C *Takakia* gametophyte upright axis with rhizome, phyllidia, and spore capsule. [Drawing by C. Lyons.]

arises from an underground, branching rhizome. The leaves of mosses are one-cell thick with a midrib in some species, arrayed in a spiral around the stem. *Takakia,* in comparison, has phyllidia—solid, cylindrical appendages from three- to five-cells thick that arise singly or in twos, threes, or fours in an irregular spiral from an erect shoot of the gametophyte (Figure C). Phyllidia are unique to *Takakia.* Gametophytes and sporophytes of mosses, including *Takakia,* lack the lignified xylem and phloem of vascular plants. Conducting tissues of mosses are water-conducting hydroids and (rarely) leptoids. All mosses have stomata—minute openings in the epidermis through which gases move (like those of hornworts but of a different form from pores of liverworts)—on their gametophytes, and some have stomata on their sporophytes. *Takakia* lacks stomata. The stoma of some mosses is surrounded by a single bagel-shaped cell, in comparison with the paired dog bone–shaped guard cells surrounding stomata of vascular plants. Like hornworts, some mosses have a cuticle on their leaves; *Takakia* lacks a cuticle. *Polytrichum,* a common moss in temperate woodlands, typifies moss species in which the entire plant is either male or female. *Takakia* gametophytes are male, female, or sterile, resembling one another in size. In some other mosses, two sexes, with their reproductive organs, are on the same plant. Fragmentation of the stem or leaf or production of minute green spheres called gemmae also can give rise to new individual mosses, which is also true of liverworts. Each individual plant produced asexually bears the haploid genetic information of its parent plant. Asexual reproduction is not known in *Takakia.*

The moss gametophyte produces gametes in multicellular gametangia. These reproductive organs are either archegonia, which produce eggs, or antheridia, which produce sperm. After antheridia

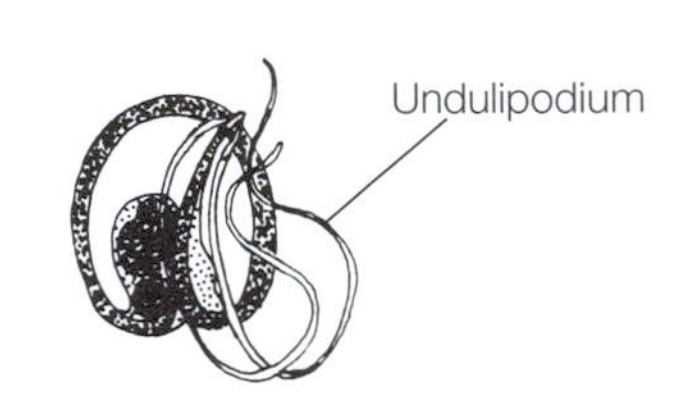

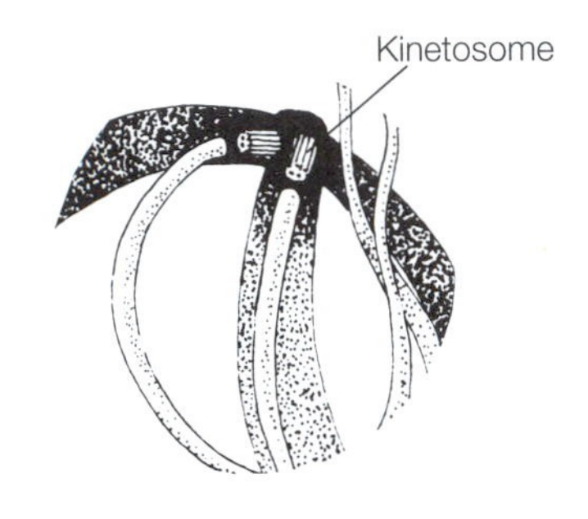

D *Polytrichum juniperinum* sperm. Moss sperm swim with undulipodia. [Drawing by K. Delisle.]

are produced, the shoot tip of *Takakia* may resume growth, unlike other mosses. Moss sperm have two forward-directed undulipodia, as do the gametes and swimming cells of liverworts and hornworts (Figure D). Sperm of *Polytrichum* and other mosses are dispersed from a splash cup: the head of the antheridium collects rain into which sperm are released. Water droplets splash out along with sperm, ejecting sperm long distances and enabling them to reach archegonia. Insects also disperse sperm-loaded water. At closer range, chemotaxis directs sperm in the water droplet to swim toward the moss egg.

Fertilization initiates development of the sporophyte embryo, which is retained in the archegonium (see Figure 5-3). The diploid sporophyte obtains nutrients from the female gametophyte, to which it attaches permanently by a foot. Later, the moss sporophyte becomes nutritionally independent and green; it photosynthesizes at least early in its growth, like hornwort and unlike liverwort sporophytes. Mature sporophytes of mosses release haploid spores in a sequence of events that includes the formation of spores by meiosis, drying and browning of the sporangium (capsule), opening of the capsule lid (operculum), and wind dispersal of spores (Figure E). In mosses of the class Bryopsida, tissue beneath the operculum splits into a circlet of teeth called the peristome. As they dry, the teeth curl, releasing spores. The other moss classes lack peristomata. A single capsule may release as many as 50 million spores. Spore germination initiates the haploid gametophyte generation, growing the protonema, a strand of photosynthetic cells that resembles a green algal filament (see Figure 5-3). Protonemata form budlike structures that give rise to the leafy upright gametophyte. Some spores give rise to male gametophytes, other spores to female gametophytes. In mosses with male and female reproductive organs in the same individual plant, one spore produces both. The sporophyte development of *Takakia* suggests evidence for classifying *Takakia* as a moss. *Takakia* sporophytes are erect, about 2 mm tall, developing like a moss with a tapered capsule atop the gametophyte on a slender stalk (Figure F). The sporangium eventually breaks open, releasing ripe spores (Figure G).

The spongy *Sphagnum* (peat moss) grows in acid bogs of northern Eurasia and in North America and contributes to the natural development of new soils. In intense sunlight, sphagnum develops red pigmentation, anthocyanin. In North Carolina and Virginia, the Great Dismal Swamp is a peat bog that originally encompassed 2200 square miles, now reduced to 750 square miles. Gardeners and florists use sphagnum to increase the water-holding capacity of soil.

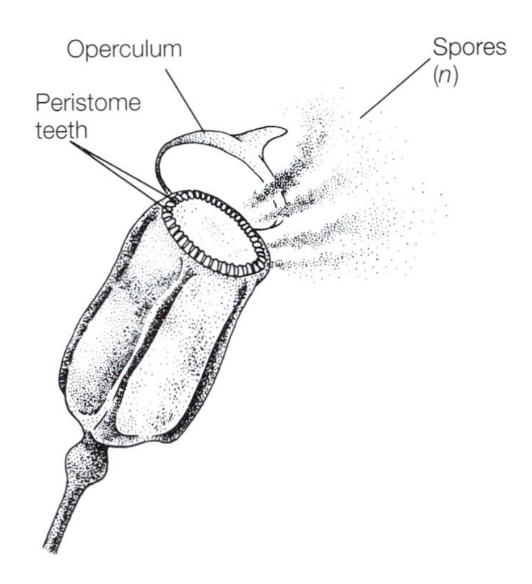

E *Polytrichum juniperinum* capsule, the sporangium. Spores, produced by meiosis, develop into an adult without fusing with another cell. [Drawing by L. Meszoly.]

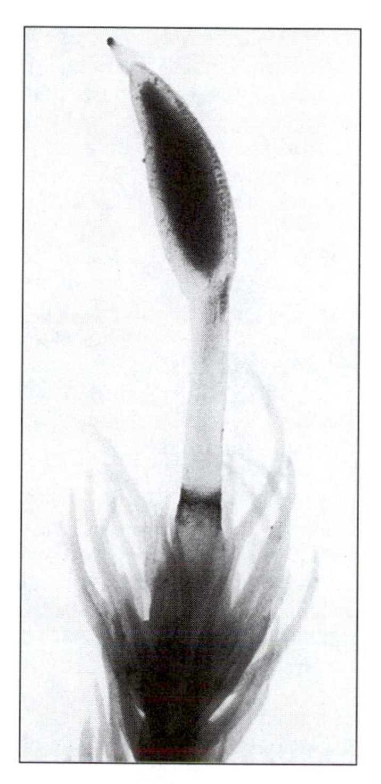

F *Takakia ceratophylla* sporophyte perched on gametophyte. [Photograph by permission of A. S. Heilman, D. K. Smith, and K. D. McFarland.]

Evidence put forward by scientists who place *Takakia* in its own phylum include the presence and patterns of development of phyllidia, growth by multiple meristematic cells (cells that continue to divide throughout the life of the organism), and the branching rhizome. In common with mosses, *Takakia* has stalked photosynthesizing archegonia and spiral leaf (recall that neither *Takakia* phyllidia nor moss appendages are true leaves) array. In common with some liverworts, *Takakia* has underground rhizomes and pitted hydroids (conducting vessels). In common with some lycopod sporophytes, *Takakia* has a multicellular meristem. In common with hornworts, *Takakia* has fewer plastids in mitotically active cells than in mitotically inactive cells. As finer details of development and physiology of the tiny, fascinating *Takakia* become better known, the puzzling relationship between *Takakia* and other nonvascular plants may be clarified.

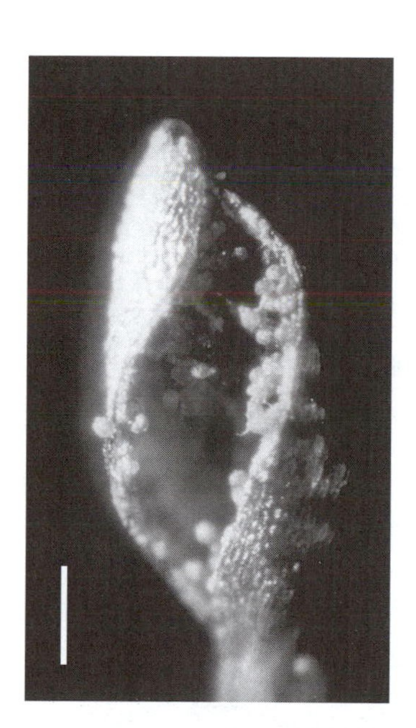

G *Takakia* capsule breaking open and releasing mature spores. Bar = 0.1 mm. [Photograph by permission of A. S. Heilman, D. K. Smith, and K. D. McFarland. From D. K. Smith and P. G. Davison, *Journal of the Hattori Botanical Laboratory,* No. 73:263–271 (1993).]

Compressed decayed sphagnum becomes carbonized in the absence of air, and is dug, dried, and burned as peat fuel. Sphagnum has proved an effective wound dressing, particularly in Europe, where it was utilized between 1880 and World War I; sphagnum absorbs many times its weight in moisture and contains iodine.

Chloroplasts and pigments of mosses resemble those of chlorophytes (Phylum Pr-28)—they contain chlorophylls *a* and *b* and carotenoids such as beta-carotene. Moss cells store starch as food reserve within chloroplasts, further evidence that mosses evolved from ancestral green algae. Mosses have a fossil record dating from the late Paleozoic, about 395 million years ago, but probably were never the dominant land plant form. They do not seem to be the ancestors of vascular plants (Phyla Pl-4 through Pl-12) or of hornworts or liverworts.

Pl-2 Hepatophyta

(Liverworts)

Greek *hepat,* liver; *phyton,* plant

EXAMPLES OF GENERA

Calypogeia
Carrpos
Conocephalum
Fossombronia
Frullania
Haplomitrium
Hymenophytum
Lejeunea
Lepidozia
Mannia
Marchantia
Marsupella
Monoclea
Neohodgsonia
Pellia
Plectocolea
Porella
Riccardia
Riccia
Ricciocarpus
Riella
Scapania
Sphaerocarpos
Targionia

Hepatophyta are commonly called liverworts—a term derived from the liver-shaped outline of their gametophyte. With the exception of *Takakia* (Phylum Pl-1), liverworts are the simplest of all extant plants. They thrive in moist habitats and are less well known than the mosses (Phylum Pl-1). Liverworts may be distinguished from mosses by their lobe shape and less-complex sporophyte.

The flattened green lobes of the gametophyte bear stalked reproductive organs (archegonia or antheridia) on the upper surface, and fine hairlike rhizoids—usually unicellular—project from the lower surface. The gametophyte is called a thallus and is undifferentiated into leaves, stem, and root. This thallus takes two forms—either thin as in "thallose" liverworts (Figure A) or thick as in "leafy" liverworts (Figure B). All liverwort thalli lack the mucus-filled cavity present in hornworts (Phylum Pl-3). The thallose liverwort *Marchantia* grows on stream banks, among mosses on rocks, and in wet ashes after fires. All liverworts lack a cuticle (the waxy, water-resistant layer present in mosses and hornworts) and exchange gases through barrel-shaped pores that open into air chambers within the thallus. Liverwort pores differ in form from the stomata of vascular plants. Only the gametophytes, not the sporophytes, of liverworts have air pores. Liverwort rhizoids are single celled in comparison with moss rhizoids, which are always multicellular. Rhizoids both anchor the thallus and absorb moisture and minerals. Many liverworts have a midrib, a thickened region that runs down the center of each thallus lobe. The thallus lacks vascular tissue, although some species have specialized tissue of uncertain function. In height, liverworts seldom exceed 5 cm.

Liverworts reproduce sexually (with gametes) and asexually (by spores, by fragmentation, and by gemmae), in broad outline like mosses. The liverwort egg is produced in the archegonium of the gametophyte (Figure C) by mitotic division. On a separate thallus (male gametophyte), antheridia produce motile, biundulipodiated sperm. Sperm transported by raindrops fertilize the egg. The liverwort embryo develops a sporophyte from the resulting diploid zygote. The liverwort sporophyte is permanently attached by a minute stalk to the female gametophyte. The sporophyte consists of a capsule (sporangium), stalk, and foot.

Meiotic cell division takes place at the sporophyte tip, leading to the production of haploid spores. After the capsule opens, spores are discharged by elaters, helical coils that twist as they dry and then snap suddenly, releasing spores. Hornworts and horsetails also have elaters, but mosses do not. Wind, animals, and water aid in spore dispersal. A spore germinates directly into a young thallus or, in a few genera, a filament of cells precedes the thallus. This leafy haploid gametophyte differentiates gametangia and the life cycle begins again. The haploid-dominated life cycle characterizes all mosses, liverworts, and hornworts.

In asexual reproduction by gemmae, liverworts reproduce haploid organisms that are genetically identical with the parent plant. On the upper surface of the thallus are small cup-shaped organs called cupules (Latin, "little cups"; see Figure A). Within the cupules, little green spheres called gemmae grow. When gemmae are dispersed by raindrops to suitable damp soil, they grow into new haploid liverworts.

Most of the 6000 liverwort species live in tropical regions throughout the world, on rock, shaded trees, fallen logs, and soil. Liverworts are often found in waterfalls and other rapidly running fresh water and as epiphytes, organisms that grow on other organ-

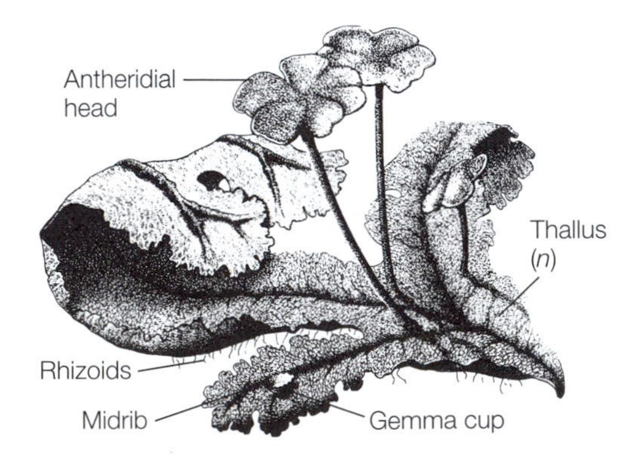

A The most common genus of liverwort, *Marchantia.* The gametophyte thallus with lobed, stalked reproductive structures bears antheridia on antheridial heads. Rhizoids differentiate on the lower surface of the thallus. [Photograph by K. V. Schwartz; drawing by C. Lyons.]

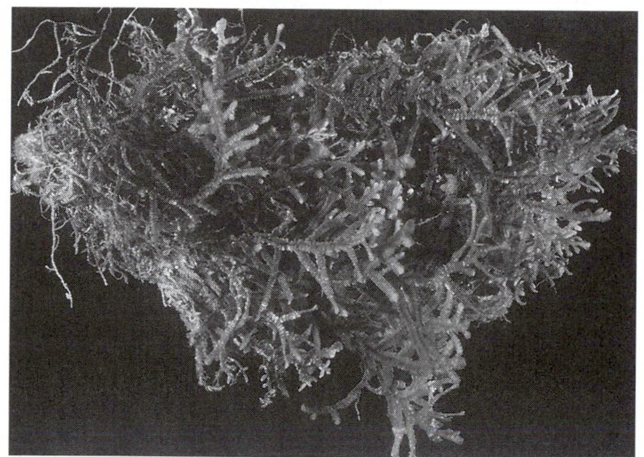

B *Porella,* a leafy liverwort collected in northern California. Two rows of minute "leaves" — not visible — grow along the stem. [Photograph by L. Graham.]

isms but are not parasitic. A number of species are known in Antarctica, where they may survive harsh environmental conditions by production of "antifreeze."

Liverworts were believed to be useful in treating liver ailments during the Middle Ages. At that time, plants that looked like an organ were used to treat medical conditions affecting that organ. Liverworts are not currently credited with therapeutic value and are not eaten. Their value lies in their function as pioneer plants in burned areas and other inhospitable habitats.

Combined morphological and molecular evidence indicates that liverworts likely evolved from green algal ancestors but independently of either hornworts or mosses. Another way of saying this is that the three groups of nonvascular plants appear to be paraphyletic. Like hornworts and mosses, liverworts gave rise to no other plant lineages.

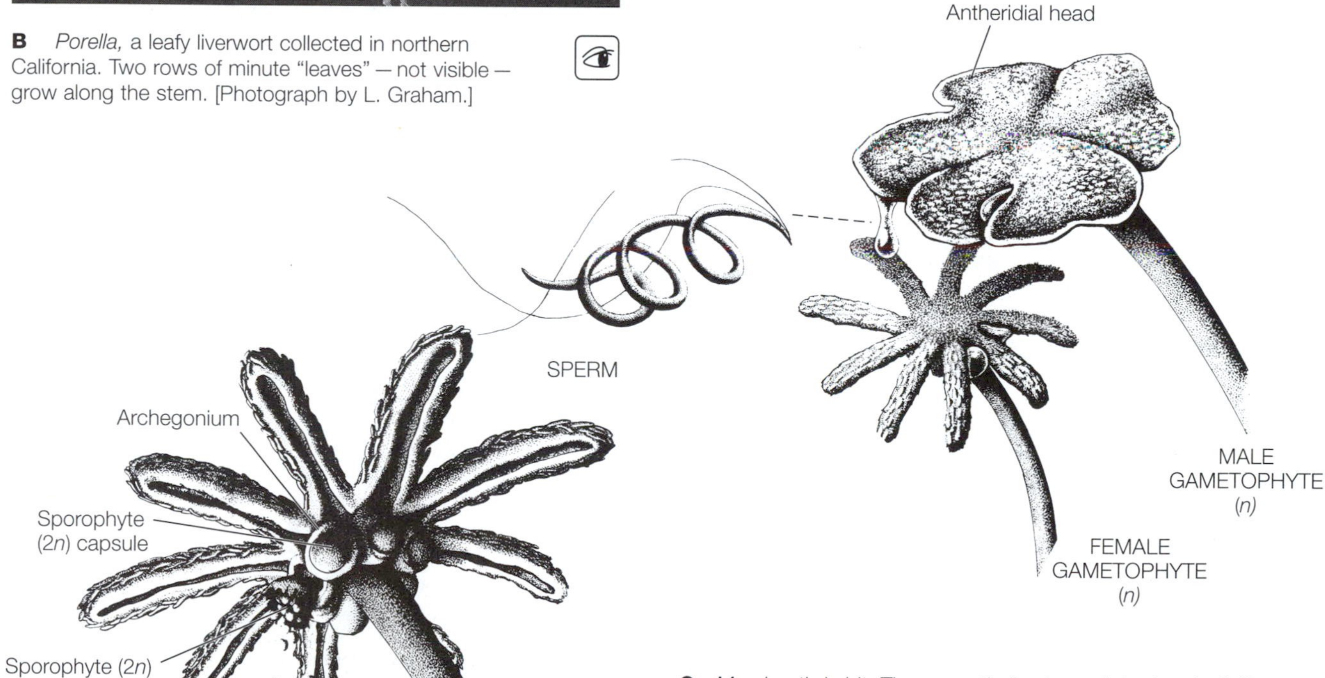

C *Marchantia* habit. The green thallus bears lobed umbrellalike structures that produce archegonia—the female reproductive organs—which produce eggs. Antheridia—the male organs, which contain sperm—differentiate on a separate thallus on the upper surface of stalked discs. A raindrop carries sperm from the male to the female, egg-bearing gametophyte. [Drawing by C. Lyons.]

Pl-3 Anthocerophyta

(Hornworts)

Greek *anthos,* flower; *keros,* wax; *phyta,* plant

EXAMPLES OF GENERA
Anthoceros
Dendroceros
Folioceros
Megaceros
Notothylas

Anthocerophyta, called horned liverworts or simply hornworts, derive their common name from the horn-shaped, elongate sporophyte that is embedded in the gametophyte by a foot. About 100 species live worldwide in temperate and tropical regions, on tree trunks, cliffs, and streamside water-splashed banks.

Members of this phylum superficially resemble mosses and liverworts—the gametophyte is a green dorsoventrally flattened thallus. Within the hornwort thallus is a mucus-filled cavity. A nitrogen-fixing cyanobacterium, *Nostoc* (Phylum B-5), lives inside the hornwort *Anthoceros.* Nitrogen-fixing organisms incorporate nitrogen from air into inorganic nitrogen-containing compounds available to plants. Thus, it is no surprise that hornworts are among plants that pioneer sterile substrate such as bare rock. Rhizoids connect the thallus to the substrate. Like liverworts (Phylum Pl-2), hornworts absorb moisture and inorganic ions across the flat thallus. Hornworts lack vascular tissue, stems, leaves, and roots. Some hornworts produce gemmae, small vegetative balls of cells that ultimately produce new thalli. Certain species of hornworts have both male and female sexual organs on the same thallus; others are unisexual. Sexual reproduction with swimming sperm is similar to the process in mosses (see Figure 5-3).

Sporophytes of hornworts bear stomata and are covered with cuticle, like moss sporophytes but unlike liverwort sporophytes. Hornwort sporophytes are green and photosynthesize (Figure A); like young sporophytes of mosses, they are nutritionally independent of the gametophyte. Unlike those of mosses and liverworts, which stop growing when they reach the height—as much as 4 cm—characteristic of each genus, the hornwort sporophyte keeps growing from a meristem at its base located between its foot and the sporangium. Meristem is tissue consisting of undifferentiated cells that give rise to new plant cells. The mature sporangium eventually splits from tip to base, ejecting spores that are often multicellular (Figure B). Most hornworts resemble liverworts in the way that spores are discharged. Packed among the spores in the sporophyte are elongate, helical, sterile hygroscopic (moisture-absorbing) cells called elaters that, when exposed to air, dry and expand to help disperse the spores. These spores initiate young gametophytes directly, without forming protonemata.

Each hornwort cell has a single large chloroplast and each chloroplast contains a pyrenoid, a feature unique to this phylum of the plant kingdom. Pyrenoids, which are common among algae such as *Spirogyra* (Phylum Pr-26), are morphologically defined regions of the chloroplast that are associated with photosynthate (sugars and starches) and probably function in food storage.

Hornworts, mosses, and liverworts probably evolved independently of one another. Fossilized hornwort spores from the Cretaceous (144 million to 66 million years before the present) are younger than the oldest moss fossils, which date from the early Carboniferous. So far, no fossil that links mosses, liverworts, and hornworts has been recognized. The origin of hornworts cannot be deduced by examining the fossil record; therefore workers must test hypotheses of ancestry based on clues in living hornworts.

A *Anthoceros.* This hornwort commonly grows on damp soil. Female and male reproductive organs (not visible) are embedded in the rosettelike thallus. Bar = 1 cm. [Photograph courtesy of E. Kozloff. From *Plants and Animals of the Pacific Northwest* (University of Washington Press, Seattle, WA, 1978), plate 33.]

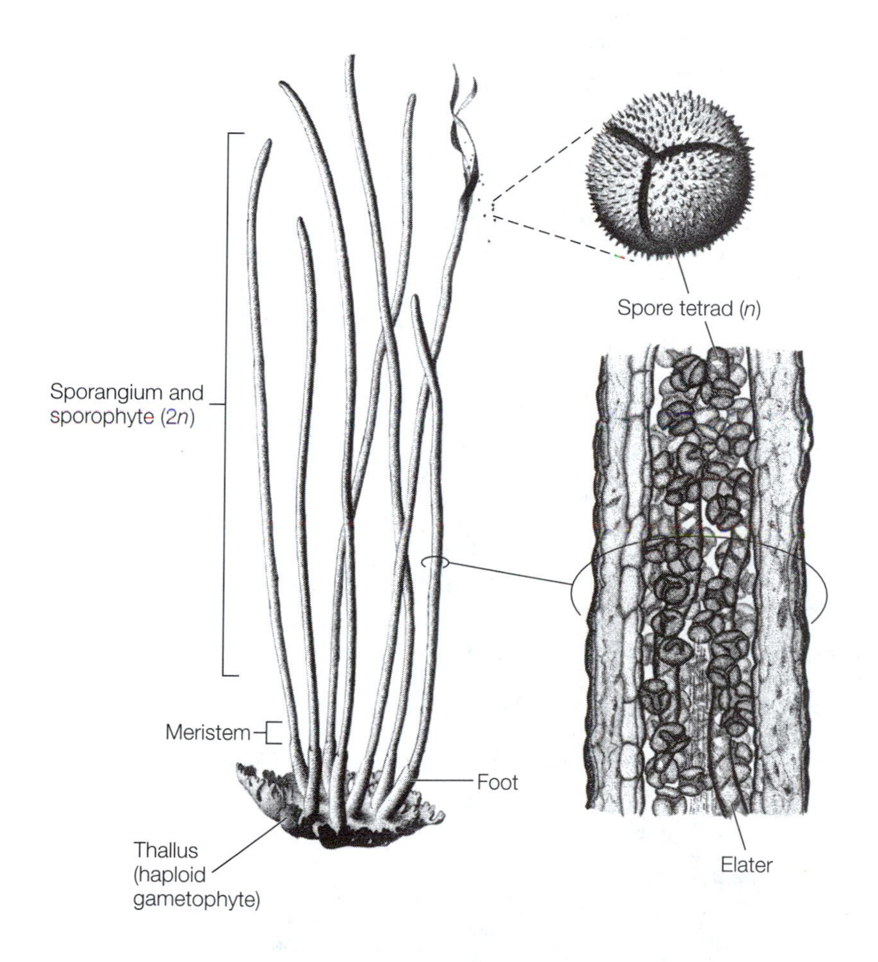

B *Anthoceros,* a gametophyte with horn-shaped sporangia. Left: The mature sporangia split in two, releasing spores. Upper right: tetrad of haploid spores. Lower right: longitudinal section of segment of sporophyte. [Drawings by C. Lyons.]

Pl-4 Lycophyta

(Club mosses, lycophytes, lycopods)

Greek *lykos,* wolf; *phyton,* plant

EXAMPLES OF GENERA
Isoetes
Lycopodium
Phylloglossum
Selaginella
Stylites

Lycophytes—club mosses, spike mosses, and quillworts—are relicts of a glorious 400-million-year-old past. The current name of the phylum, Lycophyta, is a contraction of the earlier phylum name, Lycopodophyta. The derivation of Lycopodophyta is from the Greek *lykos,* wolf, and *pous,* foot—based on a resemblance between the pattern of a wolf foot and the branching form of lycophytes.

Both treelike and herbaceous lycopods are found in the fossil record. Only from 10 to 15 genera comprising perhaps 1000 species are still living; many more that lived in the Devonian period are extinct. All living genera are herbaceous. However, *Isoetes, Selaginella,* and *Stylites* share characteristics [all are heterosporous (having two kinds of spores) and have ligules (projections from the modified leaves that bear sporangia)] with woody, ancient lycopods. The treelike lycopods—woody (fibrous) lepidodendrids—grew to heights of 40 m; they dominated the swampy Carboniferous coal forests long before the evolution of flowering trees until they died out some 280 million years ago. Giant lycopods are depicted in their Carboniferous community in the coal forest diorama at the Milwaukee Public Museum.

Lycophytes are evergreen vascular plants that bear neither seeds nor flowers. Most of the tropical species are epiphytes, depending on hosts for support. *Lycopodium* and *Selaginella* are two genera in temperate regions. *Lycopodium*—common club moss—consists of 200 species and is the most familiar lycophyte in the United States. A species of *Lycopodium* is used in winter decorations as a miniature conifer and is called ground pine or ground cedar by some and club moss by others. But these names are misleading—these plants are related neither to pines and cedar (Phylum Pl-10) nor to mosses (Phylum Pl-1).

The other well-known genus, *Selaginella* (spike moss), comprises about 700 species and flourishes in moist habitats such as Olympic National Park in Washington state. Paradoxically, the resurrection plant (*Selaginella lepidophylla*) is native to dry regions of Mexico and the southwestern United States. A curious feature of the resurrection plant is that it revives upon contact with water even after having been dry and dormant for months. Repeated cycles of desiccation and revival lead to no apparent loss of vigor.

Like all other plants, lycophytes alternate haploid and diploid generations. In lycophytes, the sporophyte (diploid) is more conspicuous than the gametophyte (haploid), as in other vascular plants. This is in contrast with the nonvascular plants (Phyla Pl-1 through Pl-3), in which the gametophyte is the more conspicuous form. The *Lycopodium* sporophyte consists of short, upright, branched stems with leaves attached, and creeping, branching rhizomes (underground stems) that lack leaves. Sparse adventitious roots attach to the rhizome.

The glossy leaves of *Lycopodium* are arranged in spirals or whorls, usually held close to the branches. The leaves characteristic of lycophytes and unique to them are called microphylls. These leaves probably evolved as outgrowths of the main photosynthetic axis of the plant. Eventually the outgrowths differentiated to form leaves with a single cylinder of vascular tissue that conducts water and nutrients. In contrast, the leaves called megaphylls have multiple strands of vascular tissue and probably originated by a different mechanism (see Phylum Pl-7). Megaphylls are characteristic of ferns and seed plants (Phyla Pl-7 through Pl-12). Some micro-

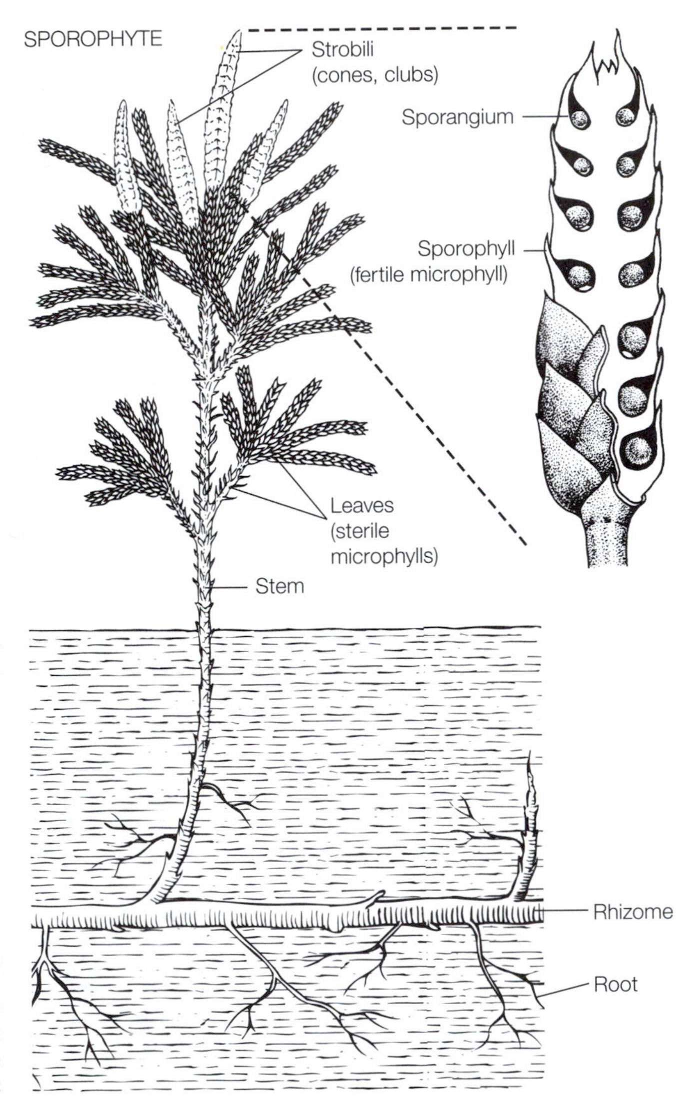

The club moss *Lycopodium obscurum* (shown here is a sporophyte) is widespread in the central and northeastern United States, in wooded areas under maples, pines, and oaks. The inset on the facing page exposes the sporangia. Meiosis occurring in cells within the sporangia produces spores. Bar = 6 cm. [Photograph courtesy of W. Ormerod; drawing by R. Golder.]

phylls are fertile; they bear sporangia. In some species, fertile microphylls—called sporophylls—and sterile microphylls (leaves) are interspersed; both are photosynthetic. The glossy leaves referred to earlier are sterile microphylls. In other species, such as *Lycopodium obscurum,* the fertile microphylls are nonphotosynthetic, scalelike structures grouped into cones (strobili). These cones form at the tips of top branches; cones are the "clubs" for which club moss is named.

Some lycophytes, such as *Lycopodium,* are homosporous, producing only one kind of (haploid) spore. Others of the phylum—*Selaginella* and *Phylloglossum,* for example—are heterosporous, forming two kinds of haploid spores on different sporophylls of the same plant: megaspores and microspores. Spores, growing by mitosis, germinate into haploid gametophyte plants that produce haploid gametes (eggs or sperm) by mitosis. Megaspores germinate into female gametophyte plants, forming archegonia containing eggs. Microspores germinate into male gametophytes, which produce sperm in male reproductive organs (antheridia). Or the microspore may simply release sperm, as in *Selaginella.* After the parent plant sheds both microspores and megaspores, the sperm swim to and fertilize eggs close by. The young sporophyte eventually sprouts root, stem, and microphylls. In homosporous lycophytes, the spores germinate into gametophytes that produce antheridia as well as archegonia on the same gametophyte. The gametophytes of homosporous lycophytes may be white subterranean tissue harboring symbiotic, mycorrhizal fungi in their tissues or they may be green and photosynthetic, living on the soil surface. These tiny gametophytes live inconspicuously for years. In all cases, fertilization of the egg by sperm requires at least a thin film of water so that the biundulipodiated sperm can swim into the nearby archegonium and fertilize the egg. As the resulting zygote develops into a green sporophyte, it may remain attached to the gametophyte on which it is nutritionally dependent, completing the life cycle.

Some club mosses—*Lycopodium lucidulum* and *L. selago,* for example—also reproduce by means of plantlets. Plantlets grow at the bases of the upper leaves. These small plants are produced asexually, are shed, and begin new diploid plants on their own. In comparison, mosses produce gemmae asexually, but gemmae are haploid.

Smooth-surfaced club moss spores—called lycopodium powder—have been used to coat pills and condoms. Ignited spores generated the flash for early photography and "pink lights," a type of fireworks.

Pl-5 Psilophyta
(Psilophytes, whisk fern)

GENERA
Psilotum
Tmesipteris

Greek *psilo,* bare, smooth; *phyton,* plant

The psilophytes, *Psilotum* and *Tmesipteris,* are unique among vascular seedless plants. They constitute the only phylum of vascular plants that—like the nonvascular liverworts, hornworts, and mosses—lack both roots and leaves. The dichotomously branched green stem has vascular tissue and alternate, minute outgrowths. These outgrowths—scalelike in *Psilotum* and leaflike in *Tmesipteris*—lack vascular tissue and are considered branchlets rather than microphylls or true leaves. *Psilotum's* distinctive three-part synangia (fused sporangia) produce spores and are supported in the axil (crotch between stem and scale) by the scalelike outgrowths. A rhizome from which rhizoids arise anchors the psilophyte sporophyte.

Psilotum and *Tmesipteris* (pronounced mezip'teris) are the only two living genera in this phylum. A plant buff can recognize both species of *Psilotum* (*P. nudum* and *P. complanatum*) in the subtropics and can maintain them in the temperate zone in a greenhouse. *Psilotum nudum*, the whisk fern (Figure A), grows in the Florida woods. In Hawaii, *Psilotum nudum,* known locally as moa, perches on tree trunks (in bits of soil), rock crevices, and soil. Interested naturalists can see *Tmesipteris* in Australia, New Zealand, and other South Pacific islands, growing as an epiphyte. A more likely opportunity to view *Tmesipteris* is in a world-class botanical garden such as the Royal Botanic Gardens, Kew, in London.

A casual glance at these herbaceous, leafless plants evokes images of a landscape rich in bacteria and protoctists some 400 million years ago. Then, in the late Silurian and early Devonian periods, Earth was barren except for early simple rootless, leafless, seedless, flowerless plants (along with bacteria and protoctists). Are the living psilophytes direct descendants of *Rhynia,* one of the first land plants? How do we know about *Rhynia*? In the quarry of the

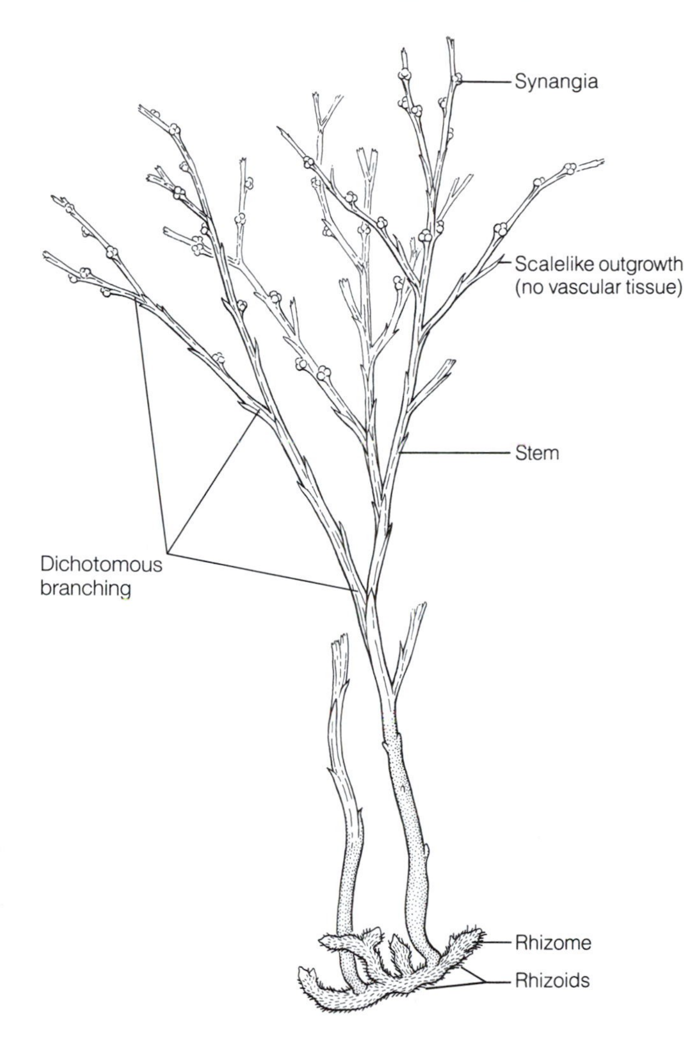

A *Psilotum nudum,* whisk fern, showing dichotomous branching, scalelike outgrowths, and synangia. This specimen (in photographs on the facing page), from ancestors in the Florida bush, has spent its life in a Boston greenhouse. [Photographs courtesy of W. Ormerod; drawing by L. Meszoly.]

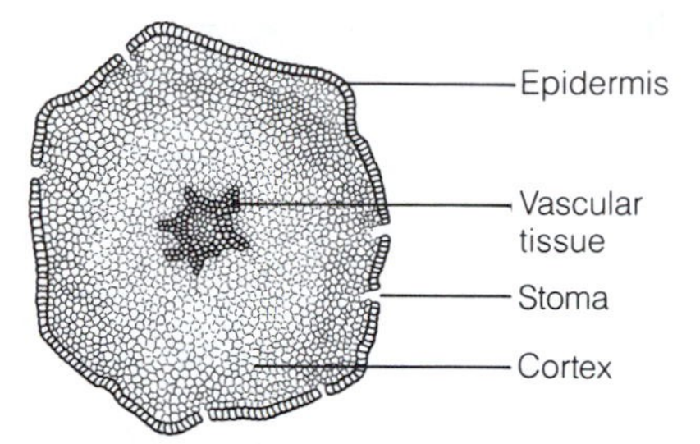

B *Psilotum nudum* stem cross section showing vascular tissue. [Drawing by L. Meszoly.]

Scottish town of Rhynie, black, smooth silica rocks have been known since the nineteenth century. Geologists tell us that these rocks, called cherts, probably precipitated in fresh water on the shores of an ancient lake. When cherts are cut and polished for microscopic study, some preserve ancient material so well that a multimillion-year-old covering of epidermal cells on the plants can still be distinguished. Fossil rhyniophytes have a leafless, dichotomously branching stem arising from a rhizome (an underground stem) with rhizoids like that of extant psilophytes. These beautifully preserved plant fossils, like the living psilophytes (Figure B), have vascular tissue in their stems, are cuticle covered, and have stomata. However, *Rhynia* sporangia were borne singly at the tips of the stems rather than in the axils of outgrowths as in present-day psilophytes. No intermediate fossils have been found that link modern psilophytes to ancient *Rhynia;* it is uncertain whether modern psilophytes are direct descendants of rhyniophytes.

Consistent with the idea of a direct relationship between the ancient *Rhynia* and modern *Psilotum* is the spectacular preservation of endomycorrhizae (fungus within the root) in 400-million-year-old fossils of the Rhynie chert. In the rhizoids of these rootless rhyniophytes, one can see spherical fungal reproductive structures (Figure C) that are remains of the ancient plant-fungus partnership. The spherical structure in this rhizome is interpreted as a fungal sporangium of an *Endogone*-like zygomycote (Phylum F-1). Rhizoids of the living *Psilotum* sporophyte also harbor mycorrhizal

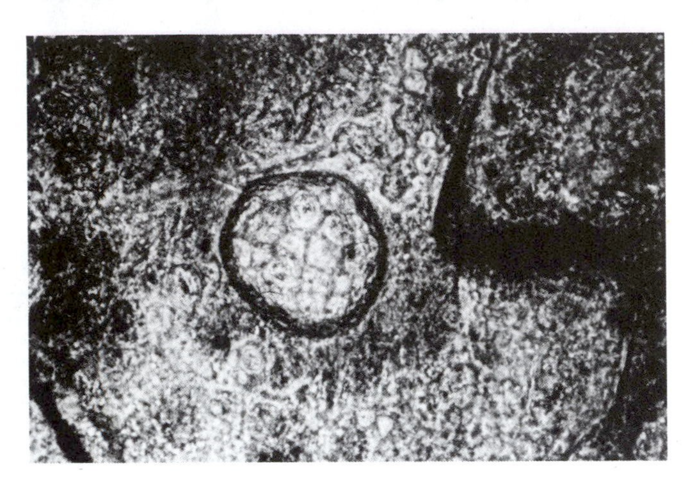

C Fossil *Rhynia* tissue section of rhizome, showing 400-million-year-old plant-fungus relationship. From Rhynie chert. [Photograph courtesy of Laurie Read.]

fungal hyphae that increase the flow of nitrate, phosphate, and organic compounds from soil to the nonphotosynthetic plant cells.

Chloroplast DNA comparisons suggest that psilophytes' closest relatives are nonlycophyte vascular plants such as ferns (Phylum Pl-7). And some botanists contend that today's psilophytes evolved directly from true ferns by simplification and loss of structure (rather than directly descending from ancient *Rhynia*). Psilophytes, as well as most ferns, horsetails (Phylum Pl-6), and some club mosses (Phylum Pl-4), have a single type of spore. Plants in these four phyla also have similar life cycles. The final word is not in regarding relationships of ancient rhyniophytes to the modern psilophytes, *Tmesipteris* and *Psilotum.*

Botanists search for clues that point to closest relative(s) of modern psilophytes by chemical comparisons—modern ferns and living psilophytes both produce secondary compounds. These biochemicals are not absolutely necessary for plant development but often play a crucial role in plant development and ecology. However, secondary compounds of modern ferns differ distinctly from those of *Psilotum.* This chemical evidence—in contrast with chloroplast DNA evidence—fails to support a strong evolutionary relation between the psilophytes and the ferns.

Additional similarities between modern psilophytes and some modern ferns are subterranean gametophytes and endophytic fungi both in gametophytes and in rhizomes.

Within each of the three chambers of the yellow-brown synangia on the sporophyte (see Figure A), one kind of haploid spore (homospore) is produced by meiotic cell divisions. Mature spores are released into the air, germinate in soil, and produce a bisexual haploid gametophyte called a prothallus (Figure D). Careful inspection reveals that the prothallus has fuzzy threads toward its center. These threads are endomycorrhizae.

Examination of the prothallus reveals two types of external sex organs (Figure E). The male sex organs, called antheridia, are microscopic bumps ringed with a layer of surface cells. A few cells away, on the same gametophyte (prothallus), are smaller female sex organs, archegonia. Each archegonium is composed of several ranks of cells with an opening that forms between them when the middle layer breaks down. The bisexual prothallus produces several archegonia, each with a mitotically produced egg at the base of the opening, as well as antheridia. Curled sperm with many undulipodia form inside the antheridia by mitosis. The sperms' undulipodia have the [9(2) + 2] organization of microtubules that reveals the protoctist ancestry of these plants. Sperm fertilize eggs within the archegonia. Because the sperm that are released into the soil must swim, moisture must be present for fertilization to occur. The resulting zygote develops into the multicellular diploid sporophyte embryo characteristic of all plants. At first, the young sporophyte is nourished through a foot anchored in the gametophyte. Later, the sporophyte takes up independent, photosynthetic life above ground.

Hawaiian men once used *Psilotum* spores as powder to prevent groin irritation from loin cloths. By boiling the moa plant, Hawaiians made laxative tea and a medicine to treat thrush (a yeast infection).

Tmesipteris and *Psilotum* fire the imagination—we envisage a past reign of dichotomously branched land plants that early in their phylogenic history already had established symbioses with members of the Kingdom Fungi. Some scientists hypothesize that this mycorrhizal association with fungi was a prerequisite to the coming ashore of all land plants.

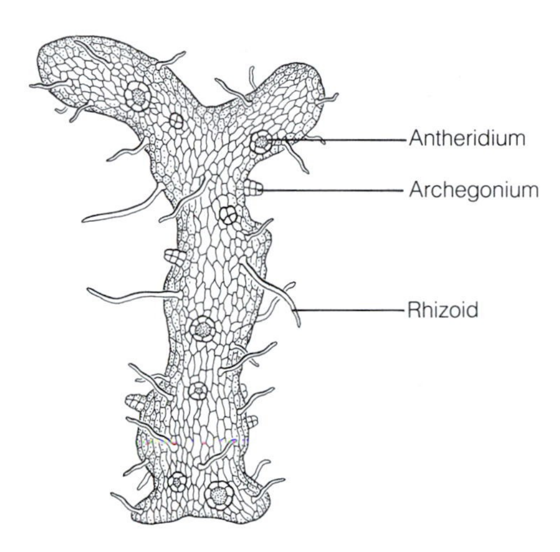

D A prothallus. This subterranean, bisexual, independent gametophyte of *Psilotum nudum* bears antheridia and archegonia, the reproductive organs. [Drawing by L. Meszoly.]

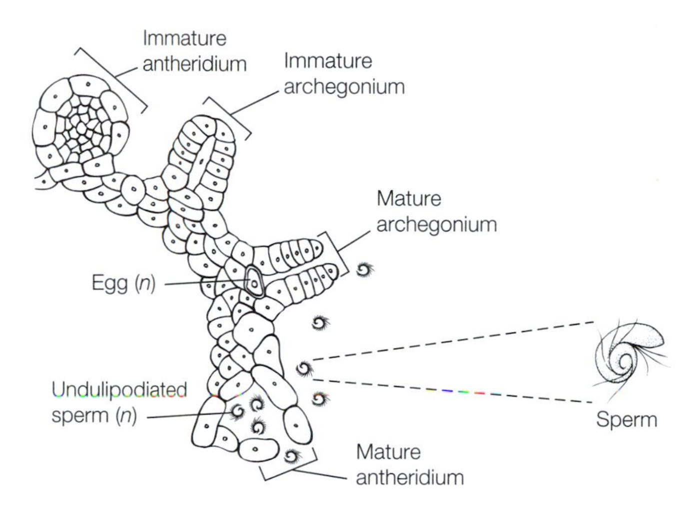

E *Psilotum nudum* prothallus cross section. Mature antheridia release spirally coiled, undulipodiated, sperm (n) that swim to mature archegonia. Each archegonium contains an egg (n), which is fertilized by a sperm.

Pl-6 Sphenophyta

(Sphenophytes, Equisetophyta, horsetails)

GENUS
Equisetum

Greek *sphen,* wedge; *phyton,* plant

The common horsetail belongs to the sphenophytes, a phylum of vascular, seedless plants, easily recognized by their jointed hollow stems with rough ribs. The abrasiveness is caused by silica within the epidermal cells of their stems, accounting for the plant's common name, scouring rush. Indeed, horsetails can be used to scour pots.

Like bryophytes (Phyla Pl-1 through Pl-3), lycophytes (Phylum Pl-4), and probably psilophytes (Phylum Pl-5), sphenophytes are relicts of a far more glorious past. In Devonian and Carboniferous forests, these plants dominated the plantscape; 300 million years ago, some horsetails were woody treelike plants reaching 0.5 m in diameter and about 15 m in height. Reconstructions of the ancient Carboniferous sphenophytes can be seen in a diorama at the California Academy of Science in San Francisco. Many horsetails had wedge-shaped leaves, hence their phylum name. (Some prefer the "jointed plants" name, Arthrophyta, but this name is too easily confused with the flowering plants, Anthophyta.) Whorls of branches may grow at the stem nodes, giving rise to the vernacular name horsetail. Today all 15 species belong to the single herbaceous genus *Equisetum.* They thrive on salt flats, along roadsides and stream banks, and in moist woods.

The familiar horsetail is the diploid, sporophyte generation; it may produce two kinds of shoots (aboveground part of the plant): green, photosynthetic shoots and pale, fertile shoots, each of the latter ending in a cone that produces one kind of spore (homosporous). The cone, called a strobilus, bears about 50 projections called sporangiophores (Figures A and B). Inside each sporangiophore are several sporangia, in which meiosis takes place, reducing diploid cells to haploid spore cells. The outer spore wall differentiates, forming coiled bands called elaters. Elaters uncoil when they dry out and thus help disperse the spore. If a spore settles in a sufficiently moist place, the spore germinates to form a dot-sized, free-living, green photosynthesizing gametophyte plant.

Gametophytes have many lobes of tissue emerging from rhizoids that anchor them to the soil. Gametophytes may be either bisexual or male—both have upper lobes that produce multicellular antheridia that give rise to sperm. On the sides of the bisexual gametophyte are female multicellular organs, archegonia, that produce eggs. The sperm, which bear many undulipodia, fertilize eggs on either the same gametophyte or other gametophytes. Sperm from male-only gametophytes swim through water to find a bisexual gametophyte bearing mature eggs. Several sperm, even from different plants, can fertilize the eggs on the same small gametophyte. The resulting zygotes then develop into independent diploid sporophytes as the parent gametophyte dies. Adult horsetails often grow in clusters, manifesting their development from a common gametophyte. Horsetails also propagate vegetatively from underground stems (rhizomes), forming clones.

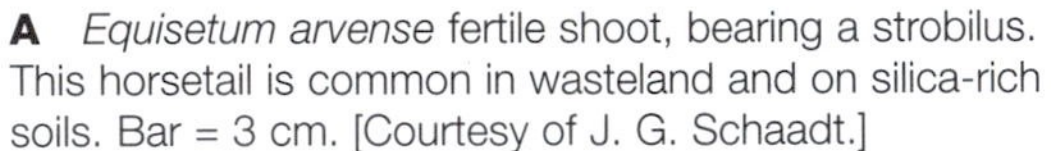

A *Equisetum arvense* fertile shoot, bearing a strobilus. This horsetail is common in wasteland and on silica-rich soils. Bar = 3 cm. [Courtesy of J. G. Schaadt.]

Native Americans, English, Tuscans, and Romans once consumed horsetails. Some contemporary references list them as edible. However, horsetails are known to be poisonous to livestock, especially cattle and horses. The toxicity is due to production of the enzyme thiaminase, which destroys the vitamin thiamine.

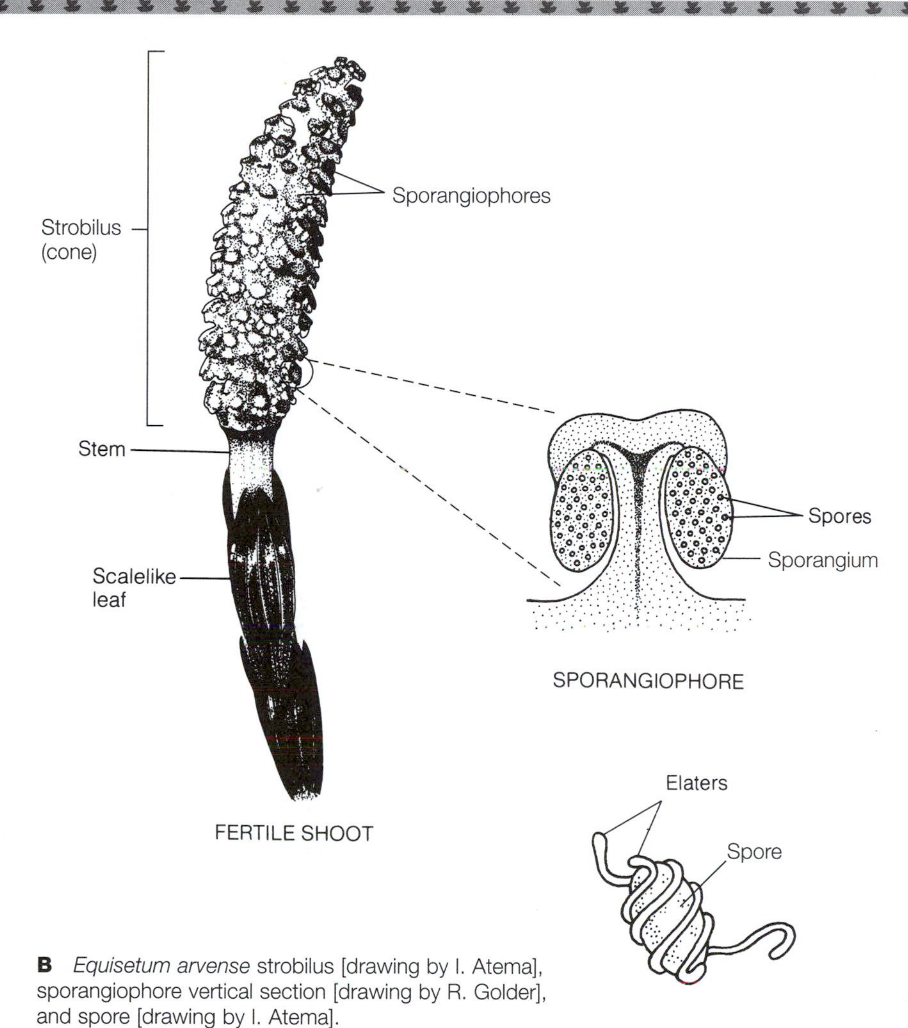

B *Equisetum arvense* strobilus [drawing by I. Atema], sporangiophore vertical section [drawing by R. Golder], and spore [drawing by I. Atema].

C *Equisetum hiemale,* common even in urban areas. Jointed stems with conspicuous nodes are evident. Bar = 15 cm [Courtesy of W. Ormerod.]

Horsetails' jointed, rough stem (Figure C) distinguishes them from other vascular, seedless plants (psilophytes, club mosses, and ferns). *Ephedra* (Phylum Pl-11) has jointed stems, but *Ephedra's* shrubby form and cones distinguish it from seedless, herbaceous horsetails.

Pl-7 Filicinophyta

(Pterophyta, Pterodatina, Pteridophyta, ferns)

Latin *felix,* fern; Greek *phyton,* plant; Greek *pteridion,* little wing, feather

EXAMPLES OF GENERA

Adiantum	*Dicksonia*	*Osmunda*
Anemia	*Dryopteris*	*Platyzoma*
Asplenium	*Hymenophyllum*	*Polypodium*
Azolla	*Marattia*	*Polystichum*
Botrychium	*Marsilea*	*Pteridium*
Cyathea	*Matteucia*	*Pteris*
Dennstaedtia	*Ophioglossum*	*Salvinia*

Ferns are seedless vascular plants that, like bryophytes, psilophytes, lycophytes, and sphenophytes (Phyla Pl-1 through Pl-6), reproduce and disperse by means of spores. But, unlike plants in these other phyla, fern sporophytes have megaphylls (Greek for "large leaf"), formerly called fronds in reference to ferns, that consist of the blade (expanded leaf part) and leaf stalk (stipe), which attaches to the rhizome (Figure A). The megaphyll is a relatively large leaf with a web of veins, in comparison with the single-veined microphylls of lycophytes. The fern megaphyll is usually compound, divided into leaflets called pinnae. Fern megaphylls may be fertile, bearing sporangia on the undersurface of modified leaves (as in *Polypodium,* Figures A and B) or on specialized stalks that emanate from the rhizome (as in *Osmunda,* Figure C), or nonreproductive (sterile). Fern sporangia tend to develop in clusters called sori (singular: sorus). In certain species, sori are bare. In many species, sori are covered with the indusium, a tissue that shrivels and folds back to expose the ripe sporangia. Many sporangia have an annulus, a strip of cells having a thin-walled outer surface. When mature sporangia dehydrate, annulus cells contract along their outer surface, ripping open the sporangium. When the water-surface tension in the annulus wall breaks, air penetrates the cell wall and the annulus snaps back, expelling the spores.

Most ferns form only one kind of spore and thus are homosporous; a few are heterosporous, producing both small and large spores. The sporophyte commences sexual reproduction with meiosis, producing haploid spores in sporangia. Spores store nutrients for future use, including some similar to proteins found in angiosperm seeds. Spores also accumulate the hormone abscisic acid, which may bring about dormancy in partly dehydrated spores. The waxy wall secreted by a spore prevents deadly dehydration. Water and usually light stimulate germination of a wind-borne spore. The young gametophyte that develops from the spore grows as a green photosynthesizing filament (protonema) toward the light. Blue light from sunlight switches development of the filament to lateral growth, forming a flat, heart-shaped gametophyte, called the prothallus. On the lower surface of the prothallus are numerous rhizoids that anchor it to the substrate and, in some ferns, form species-specific symbioses with fungi.

Depending on the species, fern gametophytes bear only antheridia (containing sperm) or only archegonia (containing eggs) or both (that is, they are bisexual, or hermaphroditic). Homosporous ferns usually produce bisexual gametophytes. In heterosporous ferns, the smaller male spores form thin gametophytes that develop only antheridia, whereas the larger female spores form rounded gametophytes that develop only archegonia. In some cases, environmental factors such as crowding can induce changes in the proportion of gametophytes with both sorts of sexual organs; a hormone secreted by gametophytes that stimulates antheridial development is responsible.

A film of water around each gametophyte (female and male) is required for fertilization. Water enters the antheridium and pops open the antheridium cap, enabling sperm to swim free. From several to thousands of motile undulipodia facilitate a sperm to move through the neck of the archegonium toward the egg. As the neck cells of the archegonium swell by water uptake, they part, creating a canal. Water dissolves mucus secreted by the neck cells. These processes probably release sperm-attracting molecules as by-products of cell respiration. Swimming fern sperm enter the archegonia, attracted by chemicals such as malic acid. A sperm (haploid) fertilizes an egg (haploid), beginning the diploid (sporophyte) generation. The embryo sporophyte may retain its connection to the gametophyte (prothallus), but soon organ development begins; the first leaf and rhizome grow out from under the gametophyte. This sporophyte grows into the familiar independent fern plant as the gametophyte dies, after having provided physical support, nutrients, and possibly hormones. Its rhizomes (subterranean or creep-

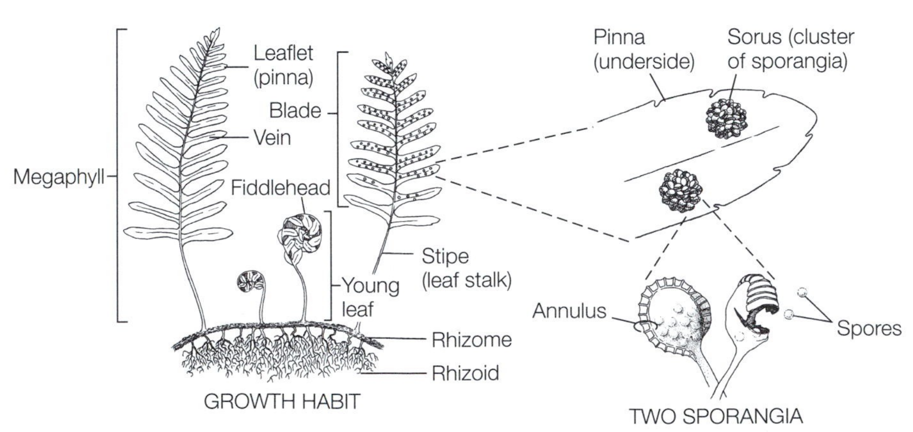

A Growth habit and reproductive structures of the sporophyte polypody fern. The name "polypody" is derived from Greek *poly* (many) and *pous* (foot), alluding to the branching rhizoids. [Drawing by R. Golder.]

B *Polypodium virginianum,* the rock polypody, showing clusters of sporangia on the underside of a fertile megaphyll (leaf). *Polypodium* is extensively distributed in North American and Eurasian woods. Bar = 15 cm. [Photograph by K. V. Schwartz.]

C *Osmunda cinnamomea,* the cinnamon fern, a species widespread in moist, shady areas, especially along the edges of ponds and streams. The sterile (nonreproductive) lateral leaves are easily distinguished from the upright fertile leaves. Bar = 50 cm. [Photograph by K. V. Schwartz.]

ing stems) send out aerial shoots (young leaves) along their lengths or near their tips. Adventitious roots originate on the rhizome, anchoring the rhizome and absorbing moisture and nutrients.

The developmental pathway is flexible. Environmental conditions can sometimes induce development of either gametophyte or sporophyte from cells other than spore and zygote. For example, ferns sometimes reproduce asexually—cells of the megaphyll tips divide by mitosis and form new diploid plants that fall to the ground. These vegetatively produced offspring are genetically identical with their parent, like club moss plantlets.

Fern spores and leaf impressions first appeared in the Devonian period (408 million to 360 million years ago). Fossilized ferns abound in the fossil record from the Carboniferous through the present. Of plants in phyla that do not form seeds, cones, or flowers, ferns are the most diverse. Their sperm swim to the egg, limiting ferns to habitats that are at least occasionally moist. About 12,000 living species are known, two-thirds of them in tropical regions. The genus *Polypodium* comprises nearly 1200 species, mostly tropical. Many tropical rain forest ferns grow high in the canopy, where they are watered by mist and rain and obtain nutrients from dust and decomposing organisms that land among their leaves. A few species live north of the Arctic Circle; like many ferns of cold areas, they produce new leaves from their rhizomes each growing season. Species are distinguished by spore morphology and details of their life histories. For example, although heterospory—production of two, differing spore types—is uncommon in ferns, water ferns such as *Marsilea, Salvinia,* and *Azolla,* as well as in *Platyzoma,* native to northern Australia, are heterosporous. Fern species are also distinguished by the nature of their sporangia. Like other seedless vascular plants, some fern species—such as *Ophioglossum* and *Marattia*—form eusporangia. Multicellular in origin, the eusporangium has a wall consisting of several cell layers. Other fern species (water ferns and *Polypodium,* for example) form leptosporangia, which develop from single cells and have single-cell-thick walls. Among the smallest ferns is the aquatic *Azolla.* These floating ferns harbor, in cavities on the undersides of their leaves, the nitrogen-fixing cyanobacterium *Anabaena* (Phylum B-5). The symbiotic complex provides nitrogen to rice paddies. *Azolla* leaves are typically 1 cm long, whereas the leaves of tree ferns may be 500 times as large. The largest ferns are thick-trunked tree ferns, with leaves as long as 5 m, stems 30 cm in diameter, and heights more than 25 m. "Tree" is a misnomer though; tree ferns lack the bark and fibrous woody tissue of trees. When moist, warm Carboniferous forest inhabitated what is now temperate North America and Europe, tree ferns flourished.

Fern fiddleheads—coiled, young sporophytes—are a delightful spring vegetable (see Figure A). Ostrich fern, *Matteucia struthiopteris* is commonly eaten and commercially grown in the United States. Ferns of all sizes provide texture and shape in landscaping. Thatch (from megaphylls), emergency food (starch from rhizomes and tree-fern pith), tea (from leaves), dye, and medicines are all products derived from ferns. Medicine that expels parasitic worms is prepared from the rhizome and root of *Polypodium aureum* in Puerto Rico. Hawaiians used fluff of tree-fern fiddleheads to stuff pillows.

Pl-8 Cycadophyta
(Cycads)

Greek *kykos*, a palm; *phyton*, plant

EXAMPLES OF GENERA
Bowenia
Ceratozamia
Chigua
Cycas
Dioon
Encephalartos
Lepidozamia
Marcrozamia
Microcycas
Stangeria
Zamia

Some cycads are small shrubs, such as *Zamia*, which is about 0.3 m in height, whereas others, such as *Cycas* and *Microcycas*, are palmlike trees more than 18 m tall. Members of the genus *Cycas* are sometimes called sago palms. Although they resemble palms, these cycads are not true palms, which flower and fruit and belong to Anthophyta (Phylum Pl-12). Cycads bear seeds (a structure formed after fertilization by maturation of the ovule in seed plants). Because their seeds are naked, cycads are classified as gymnosperms, along with conifers (Phylum Pl-10), ginkgos (Phylum Pl-9), and gnetophytes (Phylum Pl-11). As in other gymnosperm phyla, cycads lack flowers, and their seeds are exposed on female cones, instead of being enclosed in a fruit. In *Cycas* and *Dioon*, the petioles (leaf stalks) of shed leaves cover the trunk (the principal axis of the cycad, also called the stem). Shiny palmlike or fernlike leaves cluster at the apex of the stem. Like all vascular plants, cycads have megaphylls, each attached by a petiole to the stem, or trunk (Figure A). The leaves of some cycad species are subdivided into pinnules. Cycads have coralloid roots—named for their corallike appearance, which is unique to this phylum—that provide nitrogen to the cycad.

Like other vascular seed plants, cycads are heterosporous. Among all vascular seed-bearing plants, sexual reproduction in cycads is extraordinarily unusual owing to properties of the male gamete. Cycad sperm are motile, like sperm of ginkgo; sperm of conifers and gnetophytes lack undulipodia and are not motile. Cycad sperm are conveyed to the cycad egg in a pollen tube, as are sperm of conifers, ginkgo, and flowering plants (Phylum Pl-12). The combination of motile sperm and pollen tubes is characteristic of cycads and ginkgos and unique to these phyla; it is believed to be an evolutionary link between, on the one hand, ferns (Phylum Pl-7) and mosses (Phylum Pl-1), which have swimming sperm, and, on the other hand, extant seed plants (Phyla Pl-8 through Pl-12), with pollen tubes.

Separate male cycad plants bear male cones (Figure B). Cycad and conifer cones are analogous; they probably evolved independently rather than having derived from a common ancestor. Paleobotanical investigations were once believed to indicate that cycads were among the closest living relatives of flowering plants, related through their common ancestors, the seed ferns, now extinct. However, seed ferns and living cycads are no longer believed to be direct ancestors of flowering plants.

About 185 living species of cycads are grouped into eleven genera, all living in the tropics and subtropics. All cycads are listed as endangered species—vulnerable to habitat destruction and overcollection for gardens, clinging precariously to life in rain forests,

deserts, grasslands, and even mangrove swamps. At Foster Garden (Honolulu, Hawaii) and Fairchild Garden (Coral Gables, Florida), we may observe these fine plants, whose ancestors shared the early Earth with dinosaurs. In temperate zones, cycads are occasionally grown in greenhouses. *Zamia*, the only genus native to the continental United States (Georgia and Florida), is also found in the West Indies, Mexico, Central America, and northern South America. *Zamia* can be seen in Everglades National Park.

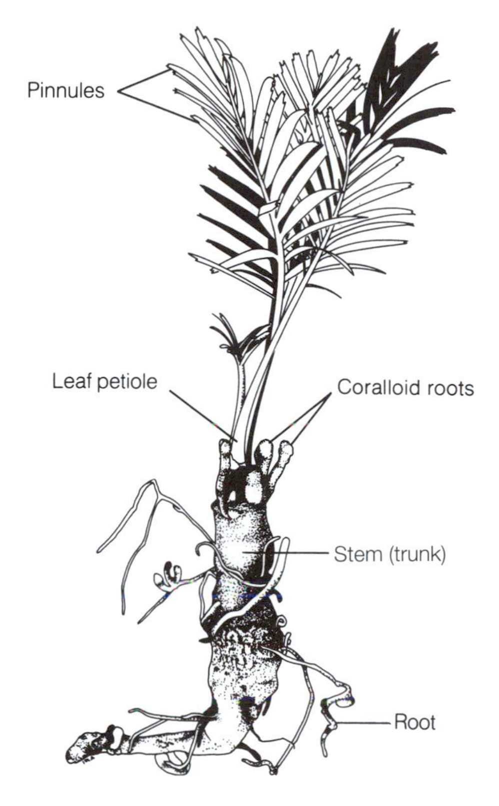

A *Macrozamia communis,* a very young sporophyte tree from sandy soil near Melbourne, Australia. Bar = 10 cm. [Photograph courtesy of C. P. Nathaniels and I. A. Staff; drawing by I. Atema.]

B A male cone of *Ceratozamia purpusii,* a cycad native to Mexico. Bar = 50 cm. [Photograph by K. V. Schwartz.]

Cycads tend to have unbranched trunks, under or above ground, with pith but little wood. The layer of cambial cells, the source of new woody tissue, divides sluggishly throughout the life of the cycad. As a result of limited cambial growth, cycads form little wood. Cycads have unique contractile trunks (stems) and roots that provide protection against adverse environments. When subject to drought or fire, subterranean stems of cycads contract as much as 30 percent in length, drawing the plant down into the protection of the soil. The contraction is due to the collapse of cells in the cortex—tissue of a root or stem bounded externally by epidermis and internally by vascular tissue—and pith, which reduces the root and stem in length.

Most cycads have taproots, some as long as 12 m, which reach deep into sand. In addition, the coralloid roots grow on or even above the soil surface (see Figure A). All cycads harbor nitrogen-fixing symbiotic cyanobacteria, generally *Anabaena* or *Nostoc*

(Phylum B-5), in these coralloid roots (Figure C). The cyanobacteria lie as a layer of single green cells just under the surface of cycad roots; when free living, they live as filaments of cells. The fixation of atmospheric nitrogen by the symbiotic bacteria probably permits cycads to populate areas where soils are depleted of nitrates. Other nitrogen-fixing bacteria, *Pseudomonas radicicola* (Phylum B-3) and *Azotobacter* (Phylum B-3) also are associated with coralloid roots.

Cycads bear their reproductive structures in cones that, in some species, are brilliant orange or velvety brown. Cycad sporophytes are either male or female (dioecious)—different individual mature plants bear reproductive structures of only one sex, either male, pollen-producing cones called microsporangiate cones (see Figure B) or female, seed-producing cones called megasporangiate cones. Female cones tend to be shorter and plumper than male cones of the same species. Male cones develop microsporangia on the lower surface of microsporophylls. The microsporophylls are packed into male cones. The female cones have megasporangia, borne on the surface of megasporophylls—modified leaves (cone scales); the megasporophylls are either packed into female cones or more loosely arranged in a leafy crown called a pseudocone. The cycad gametophytes are greatly reduced in size to only a multicellular structure in the ovule in the female and to pollen in the male. Within the ovule (the structure containing the gametophyte with its egg cell), meiosis results in megaspores that produce a haploid female gametophyte that produces haploid egg cells. Within the microsporangium, meiosis results in haploid microspores that produce pollen grains—the immature male gametophyte.

Ceratozamia mature cones of both sexes give off musty odors; these odors probably attract insects. Beetles, particularly weevils, lay their eggs in the male cycad cones. For example, the weevil *Tranes lyterioides* (Phylum A-20) is associated with *Macrozamia communis.* Both pollen and beetle larvae mature inside male cycad cones. The beetles feed on tissues of the cone but not on pollen. When adult beetles exit from the male cone, they chew through the microsporophylls and are dusted with pollen. (Some beetles remain behind, laying eggs and feeding on the male cone, continuing the beetle life cycle.) Beetles, as well as wind, transport pollen from male to female cones.

Heat, odor, and possibly sugar and amino acid–containing fluid produced by mature ovules attract insects to cycad cones at pollination time. The adult pollen-dusted beetles enter female cones through cracks between the scalelike megasporophylls. A pollen grain enters through a canal called a micropyle and is pulled into the ovule by a pollination droplet (also called micropylar fluid) secreted by the female gametophyte. The pollen (immature male gametophyte) germinates in the ovule, growing a haustorial pollen tube. Pollen that forms a tube that penetrates and absorbs is called haustorial pollen. The haustorial pollen tube transports the motile sperm to the neighborhood of the eggs, so water is not required to convey the sperm to the eggs, as it is for mosses and ferns. The mature male gametophyte consists of the germinated pollen grain, which produces sperm and a pollen tube. In a fluid-filled fertilization chamber inside the ovule beside the eggs, the pollen tube releases two large sperm, nearly 0.5 mm in diameter—the largest in the plant kingdom—each with some 40,000 undulipodia. The sperm swim in the fertilization chamber before fertilization; the entire process, from pollination to fertilization, takes as long as

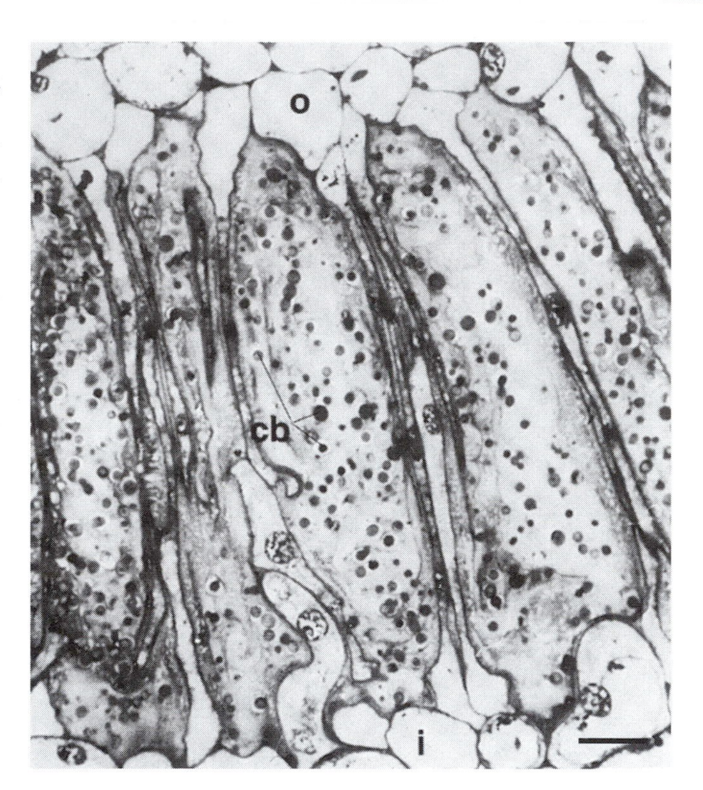

C Transverse section through a coralloid root of *Macrozamia communis,* showing the inner (i), outer (o) and cyanobacterial (cb) layers of the cortex. LM, bar = 10 μm.

5 months in cycads. (In comparison, angiosperm fertilization is often completed in a few hours.) Cumbersome, lengthy fertilization, like the unusual combination of motile sperm and pollen tubes, is unique to cycads and ginkgos. Fertilization results in a diploid embryo. Even when more than one egg produced by one female gametophyte is fertilized, generally only a single embryo survives. The embryonic sporophyte matures within a seed coat, nourished by the surrounding female gametophyte. The outer layer of the cycad seed coat is fleshy, is brightly colored, and contains starch; starch functions as a food reward for animals, from parrots to elephants. Birds, mammals, and water disperse cycad seed.

Similarities in the pollination of cycads in Australia and Africa by beetles suggest that this beetle-plant relation originated when Australia and Africa were still joined, in the Paleozoic era before the breakup of Gondwana. Pollinating beetles are attracted to a mature female cycad by an attractant, probably a scent, in what may be the most primitive animal pollination system. Thus, cycads provide clues to the origins of insect pollination.

Starch from cycad stem pith or seed kernels is a potential source of industrial alcohol by fermentation. The starchy seed inner kernel (female gametophyte) is roasted or made into food starch. The outer fleshy layer of the seed also is eaten, as is oil from this layer. Uncooked seeds are toxic. In certain species, the toxin is confined to the kernel. Edible starch leached from subterranean stems and roots of *Zamia* once was cooked as the staple of Florida's Seminoles. Kaffirs of Mozambique and Bantus of South Africa ate fermented stem pith of *Encephalartos,* called Kaffir bread. Cycad starch prepared from stem pith and seeds has furnished flour and bread to people wherever cycads grow from Asia to the Americas, Australia, and Africa. These cultures developed techniques including drying, fermentation, and leaching with water followed by cooking, which serve to remove a toxic glycoside called cycasin (macrozamin) in cycad starch that they prepared for consumption. The glycoside is a neurotoxin; failure to detoxify cycad causes illness and even death. Cycad toxicity may be a causal factor of the enigmatic neurodegeneration called "lytico-bodig" in Guam. Livestock that feed on raw leaves develop a paralysis called "zamia staggers."

Cycad leaves are used as thatch; they are dried for rituals and window displays. Gum exuding from cuts in cones, stems, and leaves is used as adhesive. Gum, pounded cycad seeds, and crushed buds may be used to dress ulcers, boils, and wounds. Cycads are extensively collected from the wild for interior and exterior landscaping.

Pl-9 Ginkgophyta

GENUS
Ginkgo

Japanese *ginkyo,* silver apricot; Greek *phyton,* plant

The ginkgo tree, *Ginkgo biloba,* is the only genus and species of phylum Ginkgophyta, a phylum of vascular seed plants. In number of species (one), this is the smallest plant phylum. Features that characterize the ginkgo include leaf veins that each branch into two smaller veins, active cambium (cells that produce wood), and fleshy, exposed ovules. Ginkgos, like all other gymnosperms—conifers (Phylum Pl-10), gnetophytes (Phylum Pl-11), and cycads (Phylum Pl-8)—are naked-seed plants; their ovules are enclosed only in the integument (outer layer of the ovule), giving rise to seeds not enclosed by fruit (a ripe, mature ovary—protective, seed-enclosing tissue). Like ferns (Phylum Pl-7), ginkgos have megaphylls, roots, motile sperm, and two spore types (heterospory). In contrast with the small, but independent, fern gametophyte, that of ginkgo is microscopic and totally dependent on the sporophyte, as is the case with cycads. The ginkgo sporophyte is a tree. As in all woody trees, the trunk, branches, and root of ginkgo increase in girth by division of cambium cells. Each growing season, a cylinder of cambium adds new xylem cells to the cylinder of xylem within it. These vascular tissues, living xylem, are called sapwood; they conduct water and dissolved minerals from the soil to the leaves. As the ginkgo ages, older xylem dies and becomes heartwood, a fibrous support tissue. Tissues external to the cambium, collectively called bark, surround the trunk and branches. Additional features found in ginkgo but not in Phyla Pl-1 through Pl-7 include pollen, sperm-conveying pollen tubes, wind pollination, ovules, and seeds. Among gymnosperms, ginkgos and conifers are wind pollinated; cycads and gnetophytes are pollinated by wind and insects. The combination of motile sperm and pollen—present only in cycads and ginkgo—is a transition between motile sperm (as in seedless plants such as ferns) and the pollen tubes combined with non-motile sperm (as in conifers, gnetophytes, and flowering plants).

Ginkgo biloba is the only living descendant of a great group of plants known from the fossil record to have been more extensive during the age of dinosaurs. The Ginkgo family originated in the Permian period of the Paleozoic era along with cycads and conifers. Petrified stumps of these great Mesozoic era trees still stand on the northwest coast of North America. The ginkgo tree is native to temperate forests of China. On steep slopes in southern China, a few semiwild ginkgo populations can still be found. *Ginkgo biloba* has been cultivated on temple and garden grounds in Asia for centuries. Photographs of living ginkgo trees are usually encumbered with houses and telephone wires because ginkgos, resistant to pollution, have been widely planted in urban settings (Figure A).

Common names of the living *Ginkgo* are ginkgo and maidenhair tree, the latter derived from the resemblance between the bilobed leaves of ginkgo and the leaves of the maidenhair fern (*Adiantum,* Phylum Pl-7). Ginkgo leaves with a bilobed, notched outline grow at tips of long shoots (branches) and seedlings. Leaves with a differing, fan-shaped outline are borne close to the stems on spur shoots (Figure B) that are shorter than long shoots, giving ginkgos a characteristic silhouette clearly recognizable even from a distance.

Ginkgo is dioecious—female and male sexes are on separate plants—and heterosporous. On male trees in hanging inflorescences, haploid microspores develop in microsporangia that are grouped into microstrobili (cones; Figure C). On female trees, haploid megaspores develop in megasporangia. The megasporangium, megaspore, and its protective integument constitute a single structure called the ovule (see Figure B). Cells within the mega- and microsporangia give rise by meiosis to megaspores and microspores, respectively.

The haploid megaspore develops into a haploid female gametophyte. This gametophyte develops within an ovule composed of tissue of the parent sporophyte (female tree), on which it depends

A *Ginkgo biloba,* the ginkgo tree, in an urban setting in the northeastern United States. Bar = 1 m. [Photograph by K. V. Schwartz.]

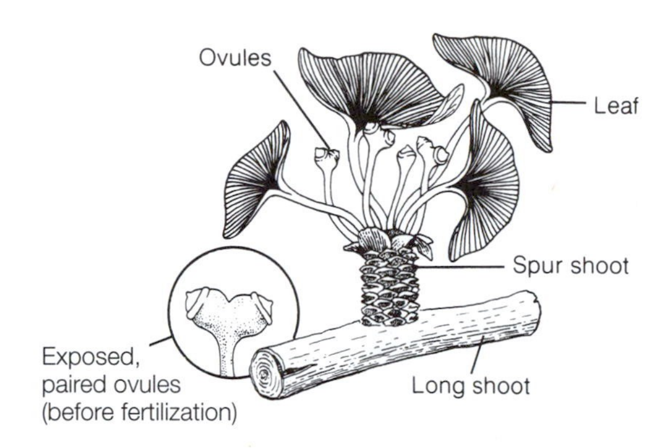

B *Ginkgo biloba* female branch, showing immature ovules. [Drawing by R. Golder.]

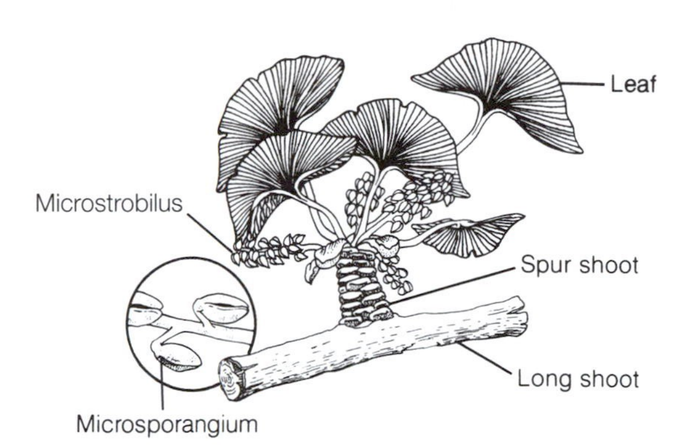

C *Ginkgo biloba* male branch, showing a microstrobilus comprising many microsporangia. [Drawing by R. Golder.]

entirely. Within the ovule, the female gametophyte develops a gamete—the egg.

In the microsporangium, each microspore develops into an immature male gametophyte, a grain of pollen. Wind transports pollen to female trees, where some pollen will land on exposed ovules. The female gametophyte secretes a liquid pollination droplet near the ovule. A pollination droplet pulls each pollen grain through a canal (micropyle) in the integument into a pollen chamber. There the haustorial pollen grain germinates, forming a much-branched pollen tube. Two sperm in each pollen grain develop only after the branched pollen tube has entered the micropyle. After about 5 months of growth nourished by the ovule, the pollen tube releases motile, helically coiled sperm, each having hundreds of undulipodia; one sperm fertilizes the egg. Because the ginkgo pollen tube delivers sperm within a liquid-filled pollen chamber, ginkgo fertilization needs no environmental water (as do ferns, for example). Embryonic development occurs after the seed has been shed from the parent tree. If pollination occurs in early April, fertilization takes place in September, embryo development is completed during winter, and seed germinates the following May. The embryo and the female gametophyte that nourishes it develop within a true but naked seed. This means that, unlike seeds of flowering plants (Phylum Pl-12), ginkgo seeds are not enclosed in a fruit. Instead, the outer skin (integument) of the ginkgo ovule develops a fleshy seed covering (seed coat) after fertilization (Figure D). A single female tree in one productive day drops thousands of these odoriferous seeds, which were probably eagerly sought and dispersed by now-extinct Mesozoic mammals. Contemporary diurnal Asian and North American squirrels as well as Asian wild cats eat and disperse ginkgo seeds. However, people are nauseated by the butyric acid and other foul-smelling substances of the ripening seed covering. If seed falls in a favorable site, it germinates and develops into a sporophyte seedling, completing the life history.

Ginkgo's major use today is ornamental—understandably, city dwellers prefer to cultivate odor-free male trees. The leaves of ginkgo turn golden before they are shed in autumn. Ginkgo is a traditional source of food in China and Japan—the outer fleshy seed covering (mistakenly called "fruit") is discarded and the inner seed kernel is roasted. Ginkgo leaf extracts have been used in Asian medicine for 5000 years to increase mental alertness and to ameliorate asthma, allergies, and heart ailments. Recently, the ginkgo leaf's medicinally active compounds, called ginkgolides, have been synthesized in the laboratory, and their therapeutic properties are now under investigation.

D *Ginkgo biloba* fleshy seeds (mature ovules) and fan-shaped leaves with distinctive venation—on a branch. Bar = 5 cm [Courtesy of W. Ormerod.]

Pl-10 Coniferophyta

(Conifers)

Latin *conus,* cone; *ferre,* to bear; Greek *phyton,* plant

EXAMPLES OF GENERA

Abies	*Larix*	*Sequoia*
Araucaria	*Metasequoia*	*Sequoiadendron*
Cedrus	*Picea*	*Taxodium*
Cryptomeria	*Pinus*	*Taxus*
Cupressus	*Podocarpus*	*Thuja*
Juniperus	*Pseudotsuga*	*Tsuga*

Most of these cone-bearing gymnosperms are trees, although some are shrubs and creeping, prostrate conifers. As in all gymnosperms (Phyla Pl-8 through Pl-11), the conifer sporophyte is the conspicuous generation, whereas the gametophyte is smaller and nutritionally dependent on the sporophyte. Like all seed plants, conifers are heterosporous and have megaphylls. Of the naked-seed plants (gymnosperms), conifers, with 550 living species, are the most familiar. Conifers are grouped into some 50 genera, including *Pinus* (pine, Figure A), *Taxus* (yew), *Abies* (fir), *Pseudotsuga* (Douglas fir), *Picea* (spruce), and *Larix* (larch). Two genera in Phylum Coniferophyta are the largest living plants: *Sequoiadendron giganteum,* the giant sequoias growing to 100 m high and 8 m wide in the Sierra Nevada mountains of northern California; and *Sequoia,* the redwoods of the California and Oregon coasts that exceed 115 m in height and are the tallest trees on the planet. *Araucaria,* the curly-branched monkey puzzle tree, originated in the southern hemisphere, where it covers vast tracts of mountainous terrain; it now also thrives in the congenial northern climate of California. Monkey puzzle tree is named for its twisted branches, said to deter climbing animals.

Conifers dominate many northern temperate forests; they are also common in the tropics and southern temperate forests. In mountainous habitats or in the far north, associations of symbiotic fungi with roots (called mycorrhizae) of conifers are especially characteristic. Contemporary conifers with a variety of mycorrhizal partners ring the cold northern temperate zone; this cold desert is arid in the sense that water frozen into ice is unavailable to plants for much of the year. The fungus sheaths but does not penetrate the conifer roots, forming ectomycorrhizal associations. By enhancing water and nutrient uptake from soil, thus promoting growth of the conifer, ectomycorrhizae make the trees more tolerant of drought.

Most conifers have leaves that are needle shaped, except *Podocarpus,* which has flat, narrow strap-shaped needles. All conifer leaves are simple—undivided into leaflets (Figure B); in comparison, compound leaves, such as those of walnut, are divided into leaflets. Conifer needles are arranged in a fascicle, a bundle comprising from one to eight needles with abbreviated leaves at their bases (Figure C). Needles are borne on a spur shoot, borne in turn on a long shoot. A heavy, transparent wax cuticle retards water loss from a needle yet allows light to enter cells in the interior for photosynthesis. Ducts, through which resin flows, penetrate the compact interior cells of the needles. Resin is a viscous, yellow brown, organic substance secreted by conifers, thought to protect wounds in the conifer from infection. Resin ducts and cuticle are particularly conspicuous in pines that inhabit arid places. Members of most genera shed their needles gradually—remaining evergreen and photosynthesizing even in winter. Some genera are deciduous—shed their needles; *Larix* (larch) and *Taxodium* (bald cypress) needles turn gold and are shed each autumn. Beneath the bark, cambium cells divide throughout the life of the plant, producing wood.

A *Pinus rigida,* a pitch pine on a sandy hillside in the northeastern United States. Bar = 1 m. [Photograph by K. V. Schwartz.]

Most conifers are monoecious—female and male reproductive structures, the cones, are borne on the same plant. Resin ducts are present in the cones. Cones contain cone scales that bear sporangia, the spore-producing structures. There are two types of sporangia and spores: male spores in microsporangia in small, male cones; female spores in megasporangia in the larger, more familiar, female pine cones (see Figure C). In the megasporangium, meiosis produces a megaspore cell that divides several times to produce a female gametophyte (haploid), which forms two haploid eggs. Similarly, in the microsporangium, meiosis produces microspores that develop into the immature male gametophyte, the pollen grain. Pollen, generally yellow and dustlike, is carried passively by wind from the male odorless cone. Some pollen may land on a female cone: this process is called pollination. A drop of sugary liquid secreted by the ovary dries, pulling pollen grains through an opening called the micropyle toward the ovule. When pollen reaches the ovule, the male gametophyte completes maturation, producing two immotile sperm and one pollen tube. One of the two sperm produced by each pollen grain degenerates. The pollen tube conveys the other sperm to the vicinity of the female gametophyte. Because conifer sperm are not motile and are transported by pollen tubes, fertilization is not dependent on environmental water. After pollination, fertilization may be delayed; in the genus *Pinus,* fusion of the male and female nuclei may be delayed for a year or more after pollination while the female gametophyte forms eggs. Eventually the haploid egg and a sperm nucleus fuse, and the resulting diploid zygote develops into an embryo, the diploid sporophyte. (Even when more than a single egg is fertilized, only one embryo develops in each ovule.)

Conifer seeds are naked and borne on compound cones. Embryos are embedded in the nutrient-providing female gameto-

B *Pinus rigida* branch, showing bundles of needle-shaped leaves and a mature female cone. Bar = 10 cm. [Photograph by K. V. Schwartz.]

phyte, enveloped by an integument that becomes the seed coat. Each ovule-bearing scale carries two ovules; these together with a sterile projection called a bract make up a seed-scale complex, present in female conifer cones (see Figure C). Male cones, in comparison, are simple, bearing the microsporangia directly on modified leaves called sporophylls. No fruit is produced, because the embryo is not covered by an ovary wall, such as develops into fruit in flowering plants (Phylum Pl-12). Conifer embryos are visible as a pair of raised areas on the underside of each female seed scale, two embryos on each scale. The embryos are dormant, young sporophytes. Conifer seeds have multiple cotyledons, or seed leaves, (usually eight) and, like all seeds, contain stored nutrients (gametophyte tissue). The winged, wind-dispersed seeds usually separate from female cones at maturity. If the seed reaches a suitable location, it germinates, and the seedling sporophyte commences to grow. In some pines, ripe seeds fail to separate from the cones until fire scorches the parent tree. Because their seeds germinate only after having been subjected to extreme heat, these pines repopulate forests after fires.

We value conifers immensely, as sources of lumber and paper pulp, and as horticultural trees and shrubs. Turpentine, pitch, tar, amber, rosin, and resin are products of conifer metabolism. Vanillin—a fragrant compound made by vanilla orchid seedpods and used to flavor perfumes and foods—can also be synthesized from conifers. Old English crossbows were made of the wood of yew trees. A drug used to combat ovarian cancer, Taxol, is derived from the Pacific yew, *Taxus brevifolia,* and has been partially synthesized. Taxol is also produced by cell culture of *Taxus media.* Thousands of conifers are cut each year in Europe and the Americas as Christmas trees. North Americans, Italians, Russians, and others eat edible seeds (pine nuts) of several pines. *Pinus sabiniana* (Digger pine), *P. coulteri* (Coulter pine), *P. lambertiana* (sugar pine), and several species of pinyon pine (*P. edulis*) in the United States are sources of protein-rich pine nuts.

Conifers likely descended from the progymnosperms, which may have given rise independently to the various phyla of gymnosperms (Phyla Pl-8 through Pl-11). (See plant phylogeny.) Conifers gave rise to no other plant phyla.

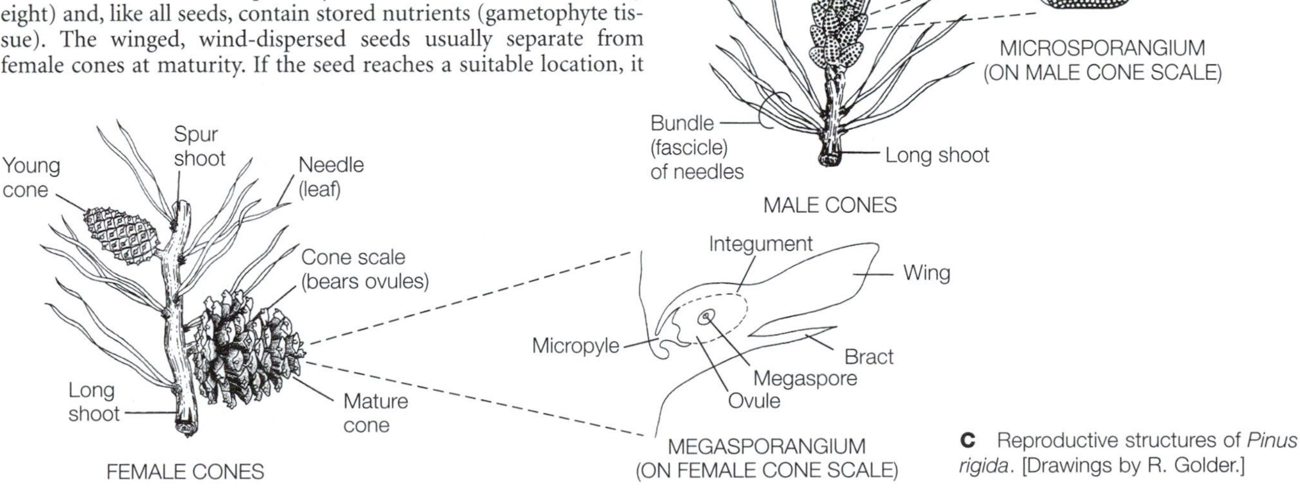

C Reproductive structures of *Pinus rigida*. [Drawings by R. Golder.]

Pl-11 Gnetophyta
(Gnetophytes)

Latin *gnetum* from Moluccan Malay *ganemu,* a gnetophyte species found on the island of Ternate; Greek *phyton,* plant

EXAMPLES OF GENERA
Ephedra
Gnetum
Welwitschia

Gnetophytes living today number about 70 species in three vastly different genera: *Welwitschia, Gnetum,* and *Ephedra.* Gnetophytes are vascular seed plants, distinct from other gymnosperms; gnetophytes have long, water-conducting tubes called vessels. Vessels form when cells join end to end in the xylem tissue and are common in flowering plants. Most gnetophytes produce naked (exposed) seeds, like the other gymnosperms—ginkgo, cycads, and conifers. *Gnetum leyboldii,* unlike most gymnosperms, produces seed enclosed in a juicy, fruit-mimicking layer. Motile sperm are absent from gnetophytes, as in conifers.

Like other seed plants, gnetophytes are heterosporous. Reproduction of gnetophytes resembles conifer reproduction in several ways. Microsporangiate (male) cones and megasporangiate (female) cones produce micro- and megaspores, respectively. Both the female and male gnetophyte cones are compound, like female (but not male) conifer cones. Gnetophyte cones and leaves lack the resin ducts present in conifer cones and leaves.

Most gnetophytes are dioecious, bearing male and female cones on different plants. *Welwitschia* cones (strobili) are borne on small branches that arise from the outer rim of the stem (Figure A). The male cone contains sterile ovules (female parts), suggesting that functioning male and female reproductive structures once resided on the same plant, as is found in conifers and many flowering trees (see Phylum Pl-12, Figure B). *Ephedra*'s and *Welwitschia*'s male and female reproductive structures are borne on bracts or cone scales in minute cones that superficially resemble cones of the conifer hemlock (Figure B). In *Gnetum* and *Ephedra,* the cones are attached along the stem at the nodes.

In the microsporangia of male cones, meiosis produces microspores that mature into pollen. After the pollen has been shed from the male cone, it germinates, producing immotile sperm. In the megasporangia of female cones, meiosis results in megaspores that produce eggs. One *Ephedra* (haploid) sperm fertilizes an egg cell (haploid), producing a diploid embryo. A second sperm nucleus fuses with a female gametophyte nucleus, producing another embryo. After this process—called double fertilization—both resulting zygotes initiate development. An integument—outer layer of the ovule—covers *Ephedra, Gnetum,* and *Welwitschia* seeds, but no gnetophyte embryo is enclosed in a fruit or nourished by endosperm. The gnetophyte seed resulting from fertilization of the egg stores its food reserves in a pair of cotyledons, which are embryonic. *Gnetum gnemon* also undergoes double fertilization, although it is unresolved whether double fertilization occurs in other species of *Gnetum. Gnetum gnemon* (and perhaps *Welwitschia*) does not form egg cells; instead, as a product of meiosis, it forms egg nuclei that are free in the female gametophyte. Each ovule forms a pollination droplet that retracts, pulling pollen into the ovule. In *Gnetum,* each of two nuclei in a binucleate sperm from one pollen tube fertilizes a separate female (haploid) nucleus; two separate fertilizations form two (diploid) embryos. Ultimately, just one of the two embryos reaches maturity in each seed. Subsequent to fertilization, several female *Gnetum* nuclei fuse, developing polyploid, embryo-nourishing tissue that is the functional equivalent of endosperm but of differing origin. Double fertilization in *Gnetum* and *Ephedra*—whether it takes place in *Welwitschia* is unknown—results in two zygotes; in comparison, double fertilization in many flowering plants results in the production of one zygote and polyploid endosperm. Researchers suggest that the gnetophyte pattern of rudimentary double fertilization with formation of two embryos evolved in a common ancestor of gnetophytes and anthophytes. After divergence of gnetophytes from their common ancestry, the flowering plants may have modified one of the two zygotes result-

A *Welwitschia mirabilis* (male), growing in the desert in southwestern Africa. Bar = 25 cm. [Photograph courtesy of E. S. Barghoorn; drawing of plant by I. Atema; drawings of cones by R. Golder.]

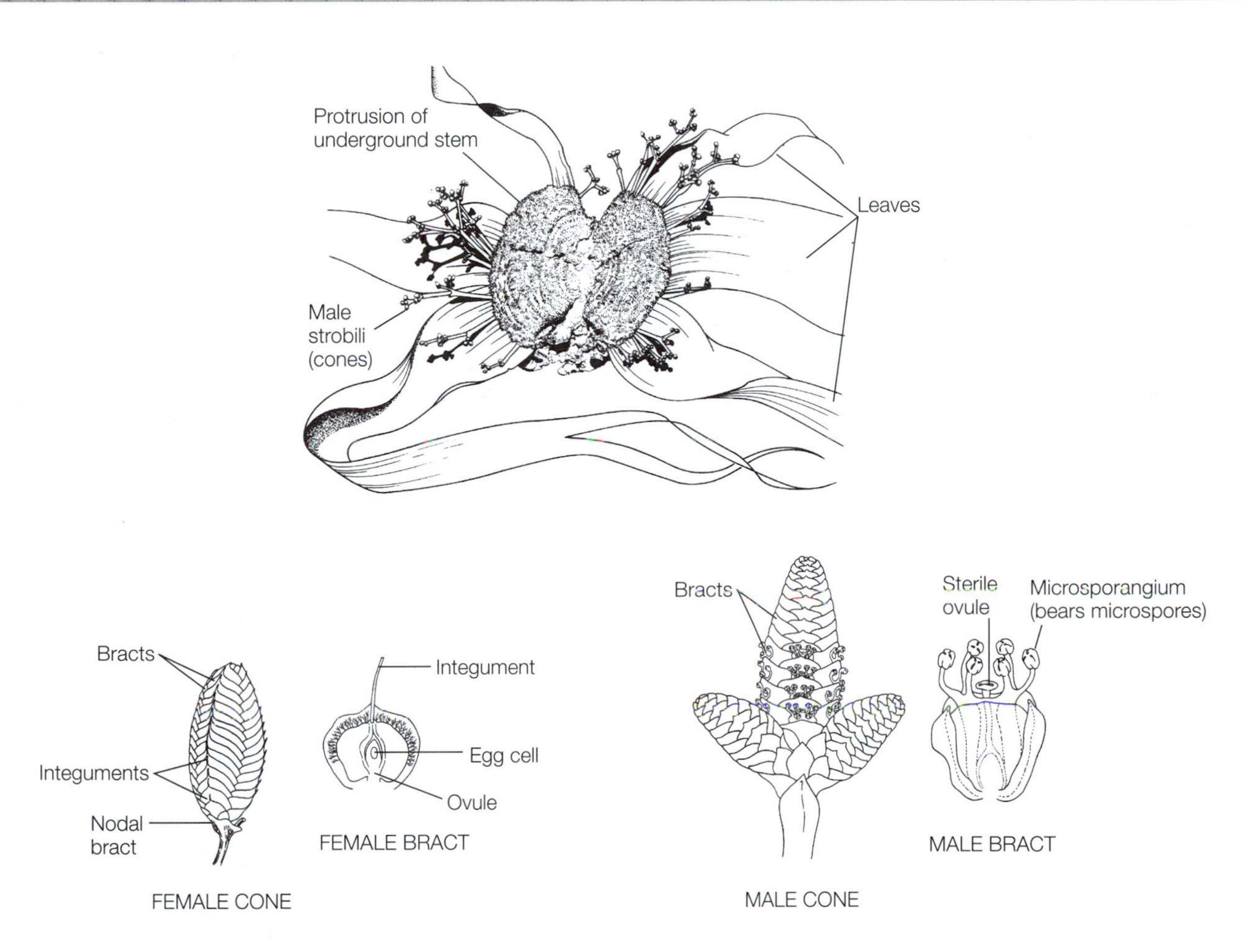

ing from double fertilization to form (triploid) endosperm, unique to flowering plants.

Gnetum survives in the angiosperm-dominated tropical rain forest even though individual plants are widely spaced, as are individual plants of angiosperm species. In several ways, dioecious *Gnetum* plants resemble many angiosperms: insect pollination and seed dispersal by fruit-eating birds. At least some species of *Gnetum, Ephedra,* and *Welwitschia* produce nectar, which enhances the efficiency of pollination. Insects pollinate *Gnetum*; wind and insects probably pollinate the other gnetophytes. *Gnetum leyboldii* seed is covered by a fleshy integument layer called a pseudofruit. Its sweet flesh entices the chestnut-mandible toucan to feed on the pseudofruit. Later the toucan regurgitates the seed, sometimes dispersing *Gnetum* seed.

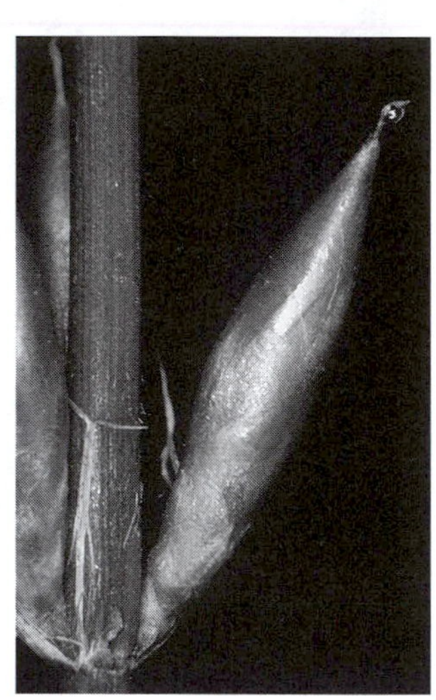

B Reproductive structures of *Ephedra trifurca,* the long-leaf ephedra. This desert shrub ranges from Texas to Baja California, Mexico. Left: Pollen-producing, microsporangiate cones—male cones. Each pollen grain produces a single pollen tube and two sperm. Right: Megasporangiate cones—female cones. A glistening pollination droplet at the tip of each 0.5-cm-long cone draws pollen into the ovule. Upon fertilization, two sperm fertilize a binucleate egg cell, producing two embryos. Ultimately, one embryo will mature and become a seed. [Courtesy of Karl Niklas, Cornell University.]

Welwitschia was discovered in 1859 by an Austrian, Friedrich Welwitsch, in arid southwestern Africa. This genus of extraordinary plants contains but one extant species, *Welwitschia mirabilis.* Some *Welwitschia* plants may be 2000 years old. Each plant grows two strap-shaped leaves as much as several meters long. These leaves grow at their attachment points during the entire life of the plant and tatter with age, lying on the sand. The leaves are attached to the outer margin of a woody, top-shaped stem that protrudes a meter or more above ground. Depending on the plant's age, the stem may be as much as a meter wide. Below ground, the stem tapers to a root that branches after a meter or so. Morning fog forms perhaps 100 days per year in *Welwitschia*'s coastal Namib Desert habitat, equivalent to about 50 mm of water per year. Although water absorption by the leaves on foggy mornings has been proposed as the main method of water uptake, the taproot may take up water as well.

The genus *Gnetum* comprises about 30 species of woody, large-leafed trees, shrubs, and lianas (woody vines) that are native to the tropical deserts, rain forests, and mountains of Asia, Africa, and Central and South America. *Gnetum gnemon* is a small tree that grows on islands in the Pacific. About 7 m high, it has glossy leaves and scarlet seeds. This tree of the forest is also cultivated for its edible seeds. Wood is produced by old stems of *Gnetum* vines as well as by *Gnetum* trees.

Ephedra, called joint fir, is native to southwestern deserts and uplands of North America, to the Mediterranean, and across the Himalayas to Mongolia. The 40 or so *Ephedra* species are evergreen perennial shrubs as tall as 3 m. Greenish, slender, jointed branches are longitudinally ribbed. Scalelike leaves sheath the nodes, where two stem joints meet. Reduction in leaf surface is probably a response to water loss in semiarid regions. At first glance, *Ephedra* resembles joint-stemmed horsetails (Phylum Pl-6), but *Ephedra*'s shrubby habit and cones (see Figure B) distinguish it from horsetails.

The oldest gnetophyte fossil dates from the Triassic period, which began 245 million years ago (see Figure I-4). Gnetophytes are believed not to have given rise to any other plant lineage. Fossil pollen evidence dating back 280 million years suggests that *Welwitschia* originated more than 300 million years ago from progymnosperm-derived cone-bearing plants that are common ancestors not only of gnetophytes, but also of modern conifers.

Double fertilization, which was once considered exclusive to angiosperms, either has evolved independently twice or else *Ephedra* and *Gnetum* are closely related to angiosperms through an extinct common ancestor. Molecular evidence suggests that gnetophytes are the phylum closest to flowering plants; data from ribosomal RNA are supported by ribulose-1,5-bisphosphate carboxylase (*rbcL*) gene sequence information. Other similarities between gnetophytes and flowering plants (especially dicots) include vessels in the wood, seeds with two cotyledons, and leaves with netlike veins (in *Gnetum*). Vessels are present in groups of plants only distantly related to each other—gnetophytes, angiosperms (Phylum Pl-12), several fern species (Phylum Pl-7), *Equise-*

tum (Phylum Pl-6), and some *Selaginella* (Phylum Pl-4). Vessels and pseudofruit exemplify convergent evolution—that is, the independent development of similar structures by distantly related organisms.

Native Americans and settlers brewed teas from *Ephedra*'s yellow male cones and dried stems, giving rise to its common names Mormon tea, Brigham tea, and Mexican tea. *Ephedra* and *Gnetum* seeds can be roasted and eaten. *Gnetum* seeds are also edible raw. Native Americans ground *Ephedra* seeds into meal. Some *Ephedra* species native to Asia and the Mediterranean are harvested for the alkaloids ephedrine and pseudoephedrine. These compounds have diverse therapeutic uses: they dilate bronchioles and are thus used to treat nasal congestion, hay fever, emphysema, and colds; as vasoconstrictors, they staunch nosebleeds and alleviate hypotension during anesthesia. In Chinese medicine, *Ephedra* has been used for at least 5000 years as remedies for colds, malaria, headaches, and cough. Although extracts from the North American *Ephedra* species were used in folk treatment of venereal disease (leading to the name "whorehouse tea"), this therapeutic use is unproved. Gum from *Gnetum nodiflorum* is used as a medicine to reduce swelling caused by muscle damage. On islands of Southeast Asia, young leaves of *Gnetum* are eaten as a cooked vegetable.

Pl-12 Anthophyta

(Angiospermophyta, Magnoliophyta, flowering plants)

Greek *anthos*, flower; *phyton*, plant

EXAMPLES OF GENERA (m, monocot; d, dicot)

Acer (d)
Agave (m)
Allium (m)
Artocarpus (d)
Aster (d)
Atropa (d)
Avena (d)
Beta (d)
Brassica (d)
Camellia (d)
Capsella (d)
Chondodendron (d)
Chrysanthemum (d)
Cinchona (d)
Cinnamomum (d)
Cocos (m)
Coffea (d)
Colchicum(m)
Cucurbita (d)
Cuscuta (d)
(continued)

Flowering plants, the angiosperms, are superstars of diversity and abundance. More than 230,000 angiosperm species are grouped into about 350 families. If we had more botanist explorers to identify them, the number of described angiosperm species would probably be closer to a million. The flower (Figure A), the reproductive organ common to all flowering plants, reveals their common ancestry. The gametophyte is barely visible within the flower; so, in this phylum, the sporophyte is the more familiar generation. Flowers and fruits uniquely distinguish this phylum (Figure B). Flowering plants have female reproductive structures (ovules enclosed in a carpel) as well as male reproductive structures (stamens), which may be on the same flowers and plants or on different ones. Fertilized eggs in the ovules become seeds, and fruits develop around the seeds as flowering plants mature. In corn, for example, the tassel is the male reproductive structure, and the corn ear is the female structure. Fertilized eggs become corn kernels and, when mature, corn seed. All familiar trees, shrubs, garden plants, crops, and wildflowers that produce flowers and fruits are members of this phylum. Flowering plants, like gnetophytes, ginkgo, cycads, and conifers (Phyla Pl-8 through Pl-11), are seed plants.

The detail of this vast diversity of plants is unknowable by a single person. How, then, are flowering plants organized? The major subdivisions established by Antoine-Laurent de Jussieu (1748–1836) that divide plants lacking seed leaves, such as mosses, and those bearing one (monocot) or two (dicot) seed leaves is still valid. Monocotyledones and Dicotyledones are subphyla within phylum Anthophyta. A seed leaf, also called a cotyledon, is a leaflike structure of the seed-plant embryo; the seed leaf often contains stored food (in dicots), absorbs food (in monocots), and provides nutrients used during seed germination (see Figure 5-1).

Among the 65,000 monocot species are bananas (*Musa*), cattails (*Typha*), coconut palm (*Cocos*), grasses such as maize (*Zea*), and crocus (*Colchicum*). Monocots are easily distinguished from dicots: in addition to the defining characteristic of one seed leaf (cotyledon), monocots display a complex array of primary vascular bundles in their stems, their leaf veins run in parallel through the leaf, and their petals and other flower parts often grow in threes. Primary growth increases plant length by means of growth at tips of shoots and roots. Monocots lack secondary cambium, the tissue that secondarily increases stem girth. Most monocots are herbaceous rather than woody shrubs or woody trees: for example, the banana plant is herbaceous; it lacks secondary (woody) growth and thus is not a tree. Palms, monocots that appear to have woody trunks, have trunks stiffened by strands of fiber rather than the woody growth characteristic of dicots.

By far the largest group of flowering plants, the dicots comprise some 170,000 species. Dicots include roses (*Rosa*), sunflowers (*Helianthus*), maples (*Acer*), pumpkins (*Cucurbita*), grapes (*Vitis*),

A *Liriodendron tulipifera* (tulip tree) flower. Bar = 10 cm. [Photograph by K. V. Schwartz.]

peas (*Pisum*), tulip tree (*Liriodendron*, Figure C), and asters (*Aster*, Figure D). Dicot characteristics include two seed leaves (two cotyledons), branching leaf veins, primary vascular bundles in a ring within the stem, flower parts in fours or fives, and the presence of secondary (woody) growth.

Our major food plants are just a few species of flowering plants. Our cereals—rice, maize (corn), wheat, oats, barley, millet, rye—are all seeds of monocots. We consume the monocot seeds of maize and coconut (coconut meat) and press edible oils from them. Sugar cane (a monocot) is crushed to release its stem juices; sugar is crystallized from the resulting liquid. Coffee, soybeans, potatoes, tomatoes, beans, lentils, buckwheat, plantains, and apples are dicots. Cotton is fiber surrounding the cotton seeds. Linen and hemp are fiber from plant stems. Most pharmaceuticals are of plant origin. Digitalis, a medicine that slows the heartbeat, is extracted from leaves of the foxglove, dicot *Digitalis*. Atropine—an antisecretory drug—is extracted from belladonna, *Atropa belladonna*. Bark of the cinchona tree *Cinchona calisaya* is the source of quinine, an antimalarial. Morphine and codeine, prescribed to relieve pain, are extracted from *Papaver somniferum*, the opium poppy.

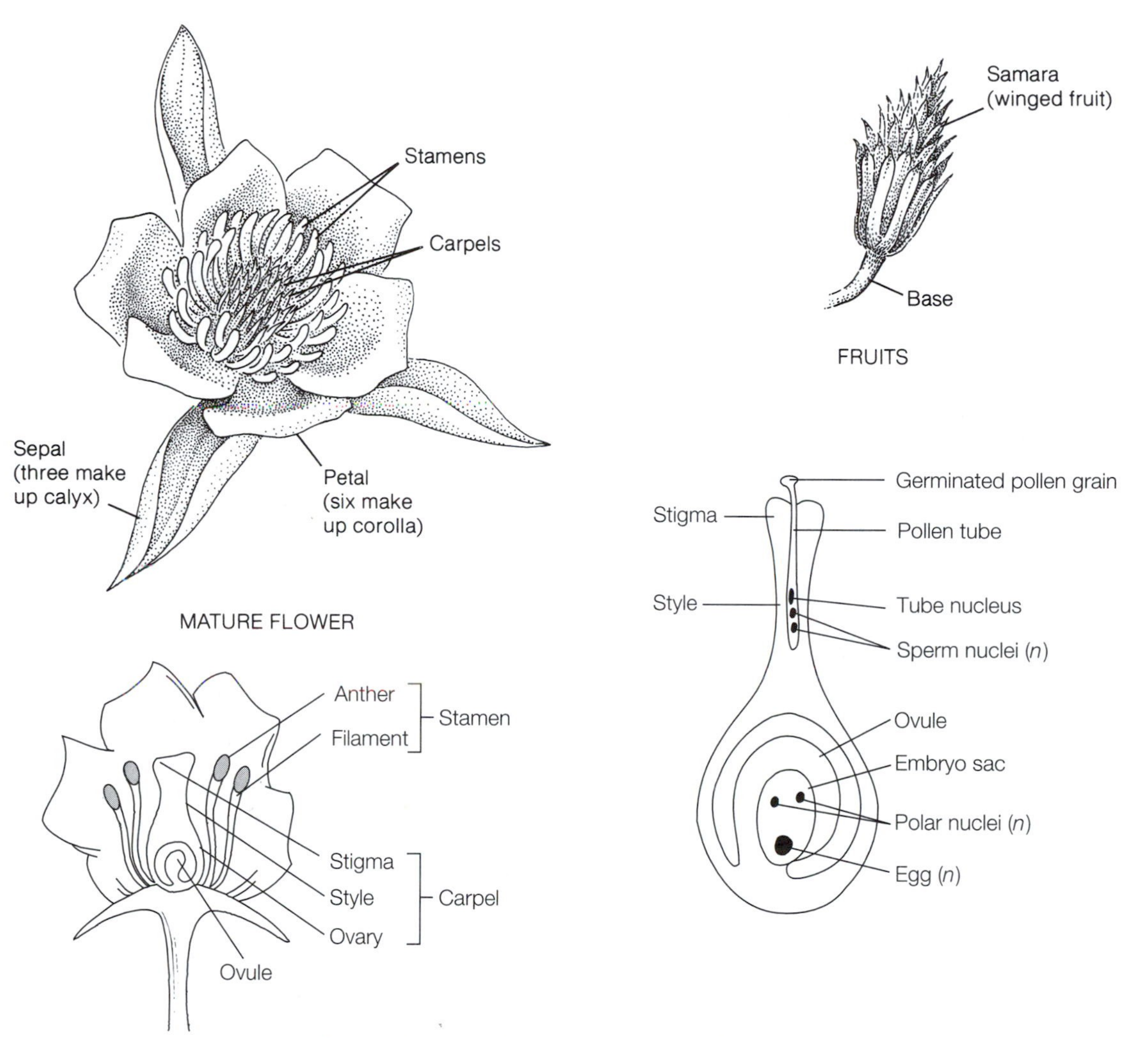

B *Liriodendron tulipifera,* fruit and flower. Female reproductive structures constitute a carpel — a leaflike floral structure including stigma, style, and ovary. Within the female gametophyte, which is called the embryo sac, are two polar nuclei and one egg nucleus. Male reproductive structures constitute the stamen — composed of anthers and the filaments that carry them. Within the male gametophyte — the germinated pollen grain — is a tube nucleus and a pair of sperm nuclei. Only a single carpel and four stamens are illustrated; however, many flowers, including that of the tulip tree, have multiple carpels and stamens. [Drawings of mature flower and fruits by L. Meszoly.]

C *Liriodendron tulipifera,* the tulip tree, in summer in Illinois. Bar = 5 m. [Courtesy of Arnold Arboretum, Harvard University.]

Both female and male flower structures evolved from the leaf as a modification of a whorl of leaves into a shoot specialized for reproduction. Clues for this line of reasoning come from plant development: leaf and flower development share many similarities. The female flower structure, the carpel (composed of stigma, style, and ovary; see Figure B), began as a folded leaf blade. The ovules of the earliest angiosperms probably formed in rows along the inner surface of the carpels. One or more carpels make up the ovary, enclosing the maturing seeds in a protective layer. Several carpels probably fused in the evolution of the more complex flowers; tomato—a fruit that grows from a single flower that has fused carpels—is an example. The stamen, the male flower part (consisting of a stalk, or filament, bearing an anther) also is a modified leaf. The leaf shoots that form the various floral parts—sepals, petals, carpels, stamens—shortened and fused; so the whorled arrangement of the evolved leaves is undetectable in most species.

The flower bears both sporophylls (modified leaves that bear sporangia) and gametophytes. Female and male reproductive parts can be present in the same flower, as in magnolia and apple, in different flowers of the same plant, as in corn, or in different plants, as in date palm. In corn, tassels bear male flowers, whereas female reproductive parts are borne within the young ear. Megaspores (female) form by meiosis in the ovule; microspores (male) form by meiosis in the anther. From these haploid spores, gametophytes germinate; a gametophyte consists of fewer than a dozen specialized cells hidden inside the flower and is physically dependent on the sporophyte plant. The haploid gametophytes develop haploid gametes, sperm in the male and eggs in the female. The archegonium and antheridium of mosses and other nonvascular plants have thus been replaced in angiosperms as in all other seed plants—cycads, ginkgoes, conifers, and gnetophytes (Phyla Pl-8 through Pl-11)—with male and female gametophytes greatly reduced in size.

Male flower parts of a typical angiosperm consist of a stalk carrying anthers, which form microsporangia called pollen sacs. Microsporogenesis takes place in the microsporangia, generating haploid microspores, which develop into pollen grains, the male gametophytes. A pollen grain consists of a tough outer coat enclosing two or three haploid nuclei within a larger tube cell. When the mature anther releases the pollen, wind or animals carry the pollen grains to the female part of the same flower or of another flower, a necessary precondition for fertilization.

The female flower part, the carpel, consists of the stigma, style, and ovary. Inside the ovary are the ovules, the female sporophyte flower tissue in which megasporogenesis produces megaspores. In the ovule, the megasporocyte (a diploid cell) divides meiotically to form four haploid megaspores. Three of these haploid megaspores usually degenerate and one divides mitotically to form several haploid nuclei. One nucleus becomes the egg, several others are short-lived, and two nuclei form a binucleate central cell ($n + n$), separate from the egg cell. Growth and development of the nuclei and associated cytoplasm leads to the mature female gametophyte (megagametophyte), also called the embryo sac. The egg—the female gamete—lies within the embryo sac along with the central

Musa (m)
Oreodoxa (m)
Oryza (m)
Papaver (d)
Phaseolus (d)
Pisum (d)
Primula (d)
Prunus (d)
Quercus (d)
Rauwolfia (d)
Rosa (d)
Saccharum (m)
Solanum (d)
Spartina (m)
Spiradela (m)
Taraxacum (d)
Theobroma (d)
Tradescantia (m)
Trifolium (d)
Tritucum (m)
Tulipa (m)
Typha (m)
Viburnum (d)
Vitis (d)
Wolffia (m)
Yucca (m)
Zea (m)
Zostera (m)

D *Aster novae-angliae.* The New England aster blooms purple in meadows and along roadsides. This wildflower belongs to Compositae, the largest of the plant families. Bar = 1.5 cm. [Photograph by K. V. Schwartz.]

cell. Production of female gametes is complex and yet remarkably uniform in thousands of flowering plants.

If the pollen grain reaches a receptive female reproductive part (the stigma), it germinates and produces a pollen tube. The germinated pollen grain is the mature male gametophyte (microgametophyte), consisting of three haploid nuclei—two sperm nuclei and one tube nucleus—with surrounding cytoplasm all in a single cell. Directed by the tube nucleus, the pollen tube grows down through the style into the embryo sac, usually conveying two sperm nuclei. One (haploid) sperm fertilizes the (haploid) egg to form the zygote (diploid); the second sperm nucleus (haploid) usually fertilizes two female nuclei (of the binucleate central cell) of the embryo sac, forming a cell—often triploid—that becomes the endosperm. This is "double fertilization," which also occurs only in *Ephedra* and *Gnetum* (gnetophytes), indicating that gnetophytes and flowering plants may have a common ancestor. The zygote sporophyte embryo is surrounded by the triploid endosperm tissue; endosperm is destined never to reproduce but to grow as the nutritive support tissue of the seed and, in some species, the seedling. Maternal diploid sporophyte tissue (integument) of the preceding generation forms a protective seed coat around both embryo and endosperm. As zygote and endosperm develop into a mature seed, the ovary matures into fruit. Reserve nutrients are mobilized from the endosperm and stored in the cotyledons that emerge from the embryo as it grows within the seed coat; many but not all plant embryos do this—for example, the dicot peas. Many dicots and some monocots use all or most of the endosperm before dormancy—arrested growth. In dicots, such as the peanut, the embryo nestles between two cotyledons, the kernels of the peanut. Many dicots develop nutrient-storing, fleshy cotyledons that themselves provide food for both the developing embryo and the young seedling. A plump acorn is such a dicot seed; if you dig up an oak seedling, you will see the acorn. In monocots such as wheat, the embryo is tucked toward one end of the single cotyledon. Peas absorb most of the stored nutrients before the pea seed becomes dormant, well before the seedling grows forth. Corn finally absorbs the food reserves of the endosperm at germination. Endosperm is present only in angiosperms, whereas gymnosperm seeds are nourished by stored nutrients provided by the female gametophyte. Other seed plants—conifers, gnetophytes, cycads, and ginkgos—also have cotyledons. In comparison with the embryo of flowering plants, the conifer embryo absorbs food stored not in the cotyledons or endosperm but in the female gametophyte tissues within the conifer seed.

Eventually, the fruit sheds its seed. Humans consume the fruit but not the seed (such as an apple), only the seed (such as a lima bean), or the fruit including the seed (such as green beans). Seeds are dispersed by ejection (jewelweed), wind (milkweeds), water (coconut), and animals (apples). Birds consume wild cherries, later depositing the seeds with their feces sometimes distant from the parent cherry tree. The cherry seed itself is both bitter and toxic, discouraging animals that eat the cherry fruit from breaking the seed itself up and digesting it. Eventually, the seed may sprout and develop into a seedling, the diploid sporophyte.

These fertile machinations led to one of the greatest of all evolutionary innovations—the angiosperm seed (see Figure 5-1). Seeds are commonly genetically programmed to remain dormant until conditions are favorable for resumption of growth. Seed coats are prepared for many contingencies, depending on the species; many contain chemical inhibitors that maintain dormancy until the seed passes through an animal's digestive tract or the inhibitor is otherwise removed. Some seed coats require burning or scarring (cutting or abrasion of the seed coat by rain and weathering) to germinate; others must be chilled or frozen; still others must be exposed to certain wavelengths of light. Chilling followed by warming and light signal Spring. Environmental cues signal suitable conditions for germination of the seed, involving hormone production and mobilization of food reserves, and subsequent growth of the young sporophyte.

Strawberry plants and a number of other flowering plants also reproduce asexually by sending out runners on which new tiny plants grow, by rhizomes (potato), and from unfertilized (haploid) seed (dandelion). These mechanisms are somewhat analogous to asexual reproduction by fragmentation or production of gemmae and plantlets in mosses, liverworts, club mosses, and ferns.

We find the earliest angiosperm fossils—angiosperm pollen grains—in the early Cretaceous period, near the end of the Mesozoic about 140 million years ago. At that time, flowering plants became the dominant plant phylum worldwide, although they undoubtedly evolved earlier in the Permian. The earliest well-preserved flowering plant fossil (120 million years old) is the extinct *Koonwarra* collected in southeastern Australia. Some early fossil angiosperms resemble large-flowered magnolias; others, like the herbaceous *Koonwarra,* are smaller. The presence of double fertilization, net-veined leaves, seeds with two cotyledons, and vessels (water-conducting cells) in the wood in both gnetophytes and angiosperms suggests that they may share a common ancestor (see plant phylogeny). *Welwitschia*—a gnetophyte—has strobili that contain both male (stamens) and female (ovules, though the ovules are aborted) structures, somewhat like flowers that contain both male and female structures. Seventy-five million years ago, great Mesozoic forests of gymnosperms—seed ferns, conifers, cycads, and ginkgos—were displaced by flowering plants. By the opening of the Cenozoic era, 65 million years ago, many modern families and even genera (for example, *Viburnum*) of angiosperms appear in the fossil record. Flowers reveal certain evolutionary trends with time. Older, earlier-evolving flowers, such as those of tulip tree (see Figures A and B) have many parts, indefinite in number, and tend to be radially symmetrical. Younger, later-evolving flowers have fewer parts, definite in number, and tend to be bilaterally symmetrical (such as orchids).

Angiosperm dominance in the plant kingdom is related, in part, to characteristics that afforded angiosperms greater success in completing their life cycle than that of gymnosperms. Such a characteristic is the coevolution of plants with land animals, especially with insects (Phylum A-20) and terrestrial chordates (Classes Reptilia, Aves, and Mammalia of Phylum A-37). Flower color and the animal species that transfer pollen for that flower evolved together, with the result that color often advertises to a certain animal pollinator. Yellow crocuses, for example, are attractive to insect pollinators, whereas white night-blooming cactus is more visible in low light and therefore attracts nocturnal moths and bats. Wind pollination of conifers is much less efficient than animal pollination; animal pollination permits production of a smaller quantity of pollen. In angiosperms, fertilization rapidly follows pollination, in contrast with the delay between pollination and fertilization in conifers. Flowers are pollinated and seeds enclosed in fruit are dispersed by beetles, butterflies, moths, mice, bats, birds, and hundreds of other animal species. Chipmunks stash beechnuts and acorns for later consumption; dispersed, forgotten seeds may sprout, dispersing the tree seeds. The flowers of the mango tree are pollinated by bats, blueberry blossoms and sunflowers by insects, columbine by hummingbirds, and sage and alfalfa by honey bees. Water-dispersed seed, such as coconut, tend to be buoyant. These angiosperm innovations—efficient pollination and seed dispersal, reduced water loss resulting from seasonal shedding of leaves (also characteristic of the gymnosperm larches and bald cypress), short life cycle compared with conifers, tough seed coat, chemical diversity, and others—led to the dominance of angiosperms in the tropic and temperate regions of Earth.

Bibliography: Plantae

General

Attsatt, P. R., "Fungi and the origin of land plants." In L. Margulis and R. Fester, eds., *Symbiosis as a source of evolutionary innovation*. MIT Press; Cambridge, MA; 1991.

Balick, M. J., and P. S. Cox, *Plants, people, and culture: The science of ethnobotany*. Scientific American Library, W. H. Freeman and Company; New York; 1996.

Bazzaz, F. A., and E. D. Fajer, "Plant life in a CO_2-rich world." *Scientific American* 266(1):68–74; January 1992.

Britton, N. L., and A. Brown, *An illustrated flora of the northern United States and Canada*, 3 vols. Dover; New York; 1970.

Carpenter, P. L., and T. D. Walker, *Plants in the landscape*, 2d ed. W. H. Freeman and Company; New York; 1989.

Chaloner, W. G., and P. MacDonald, *Plants invade the land.* Royal Scottish Museum; Edinburgh; 1980.

Cox, P. A., and M. J. Balick, "The ethnobotanical approach to drug discovery." *Scientific American* 270 (6):82–87; June 1994.

Fernald, M. L., *Gray's manual of botany: A handbook of the flowering plants and ferns of the central and northeast United States and adjacent Canada.* Dioscorides (Timber Press); Portland, OR; 1987.

Friedman, W. E., "The evolutionary history of the seed plant male gametophyte." *Trends in Ecology and Evolution* 8:15–20; 1993.

Gifford, E. M., and A. S. Foster, *Morphology and evolution of vascular plants,* 3d ed. W. H. Freeman and Company; New York; 1989.

Goulding, M., "Flooded forests of the Amazon." *Scientific American* 266(3):114–120; March 1993. Plant and animal biodiversity and adaptations.

Govindjee and W. C. Coleman, "How plants make oxygen." *Scientific American* 262(2):50–58; February 1990. Green plant photosynthesis compared with bacterial.

Graham, L. E., *Origin of land plants.* Wiley; New York; 1993.

Handel, S. N., and A. J. Beattie, "Seed dispersal by ants." *Scientific American* 263(2):76–83. August 1990.

Harlan, J. R., "Plants and animals that nourish man." *Scientific American* 235(3):88–97; September 1976.

May, R. M., "How many species inhabit the earth?" *Scientific American* 267 (4):42–48; October 1992.

"Origin and relationships of the major plant groups." *Annals of the Missouri Botanical Garden* 81(3); 1994. Special edition.

Pirozynski, K., *Symbiosis as source of evolutionary innovation.* MIT Press; Cambridge, MA; 1989.

Raven, P. H., R. F. Evert, and S. Eichhorn, *Biology of plants,* 5th ed. Worth; New York; 1992.

Rosenthal, G. A. "Chemical defenses of higher plants." *Scientific American* 254(1):94–99; January 1986.

Scagel, R. F., R. J. Bandoni, G. E. Rouse, W. B. Schofield, J. R. Stein, and T. M. C. Taylor, *Plants: An evolutionary survey.* Wadsworth; Belmont, CA; 1984.

Scientific American 235(3); September 1976. (Special issue: Food and Agriculture.)

Scientific American 271(4); October 1994. (Special issue: Life in the Universe.)

Stebbins, G. C., Jr., *Variation and evolution in plants.* Books on Demand; Ann Arbor, MI; 1950.

Stix, G., "Back to roots." *Scientific American* 268(1):142–143; January 1993. Pharmaeuticals derived from plants.

Taylor, T. N., and E. L. Taylor, *Biology and evolution of fossil plants.* Prentice-Hall; Englewood Cliffs, NJ; 1993.

Van De Graaff, K. M., S. R. Rushforth, and J. L. Crawley, *A photographic atlas for the botany laboratory.* Morton; Englewood, CO; 1995.

Pl-1 Bryophyta

Conard, H. S., and P. L. Redfearn, Jr., *How to know the mosses and liverworts,* 2d ed. Brown; Dubuque, IA; 1980.

Crandall-Stotler, B., "Morphogenesis, developmental anatomy and bryophyte phylogenetics: Contraindications of monophyly." *Journal of Bryology* 14:1–23; 1986.

Mishler, B. D., L. A. Lewis. M. A. Buchheim, K. S. Renzaglia, D. J. Garbary, C. F. Delwiche, F. W. Zechman, T. S. Kantz, and R. L. Chapman, "Phylogenetic relationships of the 'green algae' and 'bryophytes.'" *Annals of the Missouri Botanical Garden* 81:451–483; 1994.

Richardson, D. H. S., *The biology of mosses.* Wiley; New York; 1981.

Schofield, W. B., *Introduction to bryology.* Macmillan; New York; 1985.

Smith, D. K., and P. G. Davison, "Antheridia and sporophytes in *Takakia ceratophylla* (Mitt.) Grolle: Evidence for reclassification among the mosses." *Journal of the Hattori Botanical Laboratory* No. 73:263–271; 1993.

Pl-2 Hepatophyta

Schofield, W. B., *Introduction to bryology*. Macmillan; New York; 1985.

Schuster, R. M., *The Hepaticae and Anthocerotae of North America east of the hundredth meridian,* 6 vols. Columbia University Press; New York; 1966–1992.

Pl-3 Anthocerophyta

Paolillo, D. J., Jr., "The swimming sperms of land plants." *BioScience* 31:367–373; 1981.

Schofield, W. B., *Introduction to bryology*. Macmillan; New York; 1985.

Pl-4 Lycophyta

Cobb, B., *A field guide to ferns,* 2d ed. Houghton Mifflin; Boston; 1984.

DiMichele, W. A., and J. E. Skog, "The Lycopsida: A symposium." *Annals of the Missouri Botanical Garden* 79:447–449; 1992.

Hoshizaki, B. J., *Fern growers' manual.* Knopf; New York; 1975.

Mickel, J. T., *How to know the ferns and fern allies.* Brown; Dubuque, IA; 1979.

Smith, G. M., *Cryptogamic botany,* 2d ed., Vol. 2, *Bryophytes and pteridophytes.* McGraw-Hill; New York; 1955.

Wherry, E. T., *The fern guide.* Doubleday; Garden City, NY; 1961.

Pl-5 Psilophyta

Banks, H. P., *Evolution and plants of the past.* Wadsworth; Belmont, CA; 1970.

Bierhorst, D. W., *Morphology of vascular plants.* Macmillan; New York; 1971.

Cooper-Driver, G., "Chemical evidence for separating the Psilotaceae from the Filicales." *Science* 198:1260–1262; 1977.

Hoshizaki, B. J., *Fern growers' manual.* Knopf; New York; 1975.

Raven, P., R. Evert, and S. Eichhorn, *Biology of plants,* 5th ed. Worth; New York; 1992.

Sporne, K. R., *Morphology of pteridophytes,* 4th ed. Hutchinson University Library; London; 1975.

Pl-6 Sphenophyta

Cobb, B., *A field guide to ferns,* 2d ed. Houghton Mifflin; Boston; 1984.

Hoshizaki, B. J., *Fern growers' manual.* Knopf; New York; 1975.

Smith, G. M., *Cryptogamic botany,* 2d ed., Vol. 2, *Bryophytes and pteridophytes.* McGraw-Hill; New York; 1955.

Wherry, E. T., *The fern guide,* Doubleday; Garden City, NY; 1961.

Pl-7 Filicinophyta

Brooklyn Botanical Garden. *Handbook on ferns,* a special printing of *Plants and Gardens* 25 (1); Brooklyn Botanical Garden; Brooklyn, NY; 1969.

Camus, J. M., A. C. Jermy, and B. A. Thomas, *A world of ferns.* Natural History Museum Publications; London; 1991.

Cobb, B., *A field guide to ferns,* 2d ed. Houghton Mifflin; Boston; 1984.

Foster, G. *Ferns to know and grow.* Timber Press; Portland, OR; 1993.

Hoshizaki, B. J., *Fern growers' manual.* Knopf; New York; 1975.

Lellinger, D. B., *A field manual of the ferns and fern-allies of the United States and Canada.* Smithsonian Institution Press; Washington, DC; 1985.

Pryer, K. M., A. R. Smith, and J. E. Skog, "Phylogenetic relationships of extant ferns based on evidence from morphology and the *rbcL* sequences." *American Fern Journal* 85(4):205–282; 1995.

Sporne, K. R., *Morphology of pteridophytes,* 4th ed. Hutchinson University Library; London; 1975.

Tryon, R. M., and A. F. Tryon, *Ferns and allied plants with special reference to tropical America.* Springer-Verlag; New York; 1982.

Pl-8 Cycadophyta

Bold, H. C., C. J. Alexopoulos, and T. Delevoryas, *Morphology of plants and fungi,* 5th ed. Harper College; New York; 1990.

Chamberlain, C. J., *Gymnosperms: Structure and evolution.* Dover; New York; 1966. (Reprint of 1935 edition.)

Dallimore, W., and A. B. Jackson, *A handbook of Coniferae and Ginkgoaceae,* 4th ed., rev. S. G. Harrison. St. Martin's; New York; 1966.

Jones, D. L., *Cycads of the world.* Smithsonian Institution Press; Washington, DC; 1993.

Norstog, K., "Cycads and the origin of insect pollination." *American Scientist* 75(3):270–279; May 1987.

Stevenson, D., ed., "Biology, structure, and systematics of the Cycadales." *Memoirs of the New York Botanical Garden,* Vol. 57. Symposium Cycad '87, Beaulieu-sur-Mer, France, April 17–22. New York Botanical Gardens; New York; 1987.

Pl-9 Ginkgophyta

Chamberlain, C. J., *Gymnosperms: Structure and evolution.* Dover; New York; 1966. (Reprint of 1935 edition.)

Dallimore, W., and A. B. Jackson, *A handbook of Coniferae and Ginkgoaceae,* 4th ed., rev. S. G. Harrison. St. Martin's; New York; 1966.

Dirr, M. A., *Manual of woody landscape plants,* 4th ed. Stipes; Champaign IL;1990.

Gifford, E. M., and A. S. Foster, *Morphology and evolution of vascular plants,* 3rd ed., W. H. Freeman and Company; New York; 1988.

Hori, T., ed., "Ginkgo biloba—a global treasure—from biology to medicine." *Journal of Plant Research.* Springer-Verlag Tokyo; Tokyo; 1997. Special issue.

Sargent, C. S., *Manual of the trees of North America,* Vol. 1. Dover; New York; 1961. (Reprint of 1922 edition; Houghton Mifflin; Boston.)

Pl-10 Coniferophyta

Chamberlain, C. J., *Gymnosperms: Structure and evolution.* Dover; New York; 1966. (Reprint of 1935 edition.)

Dallimore, W., and A. B. Jackson, *A handbook of Coniferae and Ginkgoaceae,* 4th ed., rev. S. G. Harrison. St. Martin's; New York; 1966.

Denison, W. D., "Life in tall trees." *Scientific American* 228(6):74–80; June 1973. High forest canopy community of old-growth Douglas fir: animals, lichens, and nitrogen pathways.

Hartzell, H., *The yew tree.* Hulogosi Communications; Eugene, OR; 1991.

Nemecek, S., "Rescuing an endangered tree." *Scientific American* 274(3):22; March 1996. A Florida conifer, *Torreya taxifolia.*

Nicolaou, K. C., R. K. Guy, and P. Potier, "Taxoids: New weapon against cancer." *Scientific American* 274(6):94–98; June 1996.

Rushforth, K. D., *Conifers.* Facts on File; New York; 1987.

Pl-11 Gnetophyta

Beck, C. B., ed., *Origin and evolution of Gymnosperms.* Columbia University Press; New York; 1988.

Bold, H. C., C. J. Alexopoulos, and T. Delevoryas, *Morphology of plants and fungi,* 4th ed. Harper and Row; New York; 1980.

Bornman, C. H., *Welwitschia: Paradox of a parched paradise.* Struik; Cape Town, South Africa; 1978.

Carmichael, J. S., and W. E. Friedman. "Double fertilization in *Gnetum gnemon* (Gnetaceae): Its bearing on the evolution of sexual reproduction within the Gnetales and the anthrophyte clade." *American Journal of Botany* 83(6): 767–780; 1996.

Carmichael, J. S., and W. E. Friedman. "Double fertilization in *Gnetum gnemon*: The relationship between the cell cycle and sexual reproduction." *Plant Cell* 7:1975–1988; 1995.

Chamberlain, C. J., *Gymnosperms: Structure and evolution.* Dover; New York; 1966. (Reprint of 1935 edition.)

International Journal of Plant Sciences 157(6); November 1996. (Supplement: Biology and Evolution of the Gnetales: A Symposium.)

Stewart, L., *Guide to palms and cycads.* Angus and Robertson; Sydney; 1994.

Pl-12 Anthophyta

Bray, F., "Agriculture for developing nations." *Scientific American* 271(1):30–37; July 1994. Rice polyculture compared with monoculture.

Brockman, C. F., *Trees of North America.* Golden; New York; 1968.

Doyle, J. A., M. J. Donoghue, and E. A. Zimmer, "Integration of morphological and ribosomal RNA data on the origin of the angiosperms." *Annals of the Missouri Botanical Garden* 81:419–450; 1994.

Friedman, W. E., "Double fertilization in *Ephedra,* a nonflowering plant: Its bearing on the origin of angiosperms." *Science* 247:951–954; February 23, 1990.

Friedman, W. E., "Evidence of a pre-angiosperm origin of endosperm: Implications for the evolution of flowering plants." *Science* 255:336–339; January 17, 1992.

Friis, E. M., W. G. Chaloner, and P. R. Crane, *Origin of angiosperms and their biological consequences.* Cambridge University Press; Cambridge, UK; 1987.

Gleason, H. A., and A. Cronquist; *Manual of vascular plants of northeastern United States and adjacent Canada,* 2d ed., New York Botanical Garden; Bronx, NY; 1991.

Perry, D. R., "The canopy of the tropical rain forest." *Scientific American* 251(5):138–147; November 1984.

Rickett, H. W., ed., *Wildflowers of the United States,* 6 vols. New York Botanical Garden; Bronx, NY; 1970–1990.

Robacker, D. C., B. J. D. Meeuse, and E. H. Erickson, "Floral aroma." *BioScience* 38:390–396. 1988.

Swaminathan, M. S., "Rice." *Scientific American* 250(1): 80–93. January 1984. Compare with Bray, F., "Agriculture for developing nations." *Scientific American* 271(1): 30–37; July 1994.

Symonds, G. W., *Shrub identification book.* Morrow; New York; 1973.

Symonds, G. W., *Tree identification book.* Morrow; New York; 1973.

APPENDIX

List of Phyla

SUPERKINGDOM PROKARYA
(Prokaryotes)

KINGDOM BACTERIA (PROKARYOTAE, PROCARYOTAE, MONERA)

Subkingdom ARCHAEA

MENDOSICUTES (deficient walls)

Phylum B-1 **Euryarchaeota**—methanogens (methane producers) and halophils (salt lovers)

Phylum B-2 **Crenarchaeota**—thermoacidophils (hot-acid lovers)

Subkingdom EUBACTERIA

GRACILICUTES (Gram-negative walls)

Phylum B-3 **Proteobacteria**—purple bacteria

Phylum B-4 **Spirochaetae**—spirochetes (corkscrew bacteria with internal flagella)

Phylum B-5 **Cyanobacteria**—oxygenic photosynthesizers

Phylum B-6 **Saprospirae**—fermenting gliders

Phylum B-7 **Chloroflexa**—green nonsulfur phototrophs

Phylum B-8 **Chlorobia**—green sulfur phototrophs

TENERICUTES (no walls)

Phylum B-9 **Aphragmabacteria**—wall-less bacteria

FIRMICUTES (Gram-positive walls)

Phylum B-10 **Endospora**—bacilli and related formers of internal spores (endospores)

Phylum B-11 **Pirellulae**—stalked, protein-walled bacteria

Phylum B-12 **Actinobacteria**—mycelial (fungus-like) bacteria

Phylum B-13 **Deinococci**—radioresistent micrococci (tiny spheroidal bacteria)

Phylum B-14 **Thermotogae**—thermophilic fermenters

SUPERKINGDOM EUKARYA
(Eucaryotae, Eukaryotes)

KINGDOM PROTOCTISTA

AMITOCHONDRIATES (cells lack mitochondria)

Phylum Pr-1 **Archaeprotista**—amitochondriates (amebas, mastigotes without mitochondria)

Phylum Pr-2 **Microspora**—microsporans

AMEBOIDS (ameba stages in life history)

Phylum Pr-3 **Rhizopoda**—amastigote amebas

Phylum Pr-4 **Granuloreticulosa**—foraminifera (marine ameboid protists with granular pseudopodial network) and relatives

Phylum Pr-5 **Xenophyophora**—large deep-sea benthic protoctists

Phylum Pr-6 **Myxomycota**—plasmodial slime molds

ALVEOLATES (alveolar sacs at cell surface)

Phylum Pr-7 **Dinomastigota**—dinoflagellata (whirling biundulipodiated cells)

Phylum Pr-8 **Ciliophora**—ciliates (macro- and micronucleated unicells)

Phylum Pr-9 **Apicomplexa**—symbiotrophs with apical complex

SWIMMING MASTIGOTES

Phylum Pr-10 **Haptomonada**—prymnesiophytes (Coccolithophorids and other yellow-brown planktic unicells with haptoneme filaments)

Phylum Pr-11 **Cryptomonada**—cryptophytes (flattened asymmetric unicells, usually photosynthetic and motile)

Phylum Pr-12 **Discomitochondria**—unicells with discoid cristae; euglenoids, kinetoplastids, mastigote amebas

HETEROKONTS (Stramenopiles; bearers of mastigonemes)

Phylum Pr-13 **Chrysomonada**—chrysophytes; golden brown algae

Phylum Pr-14 **Xanthophyta**—yellow green algae

Phylum Pr-15 **Eustigmatophyta**—green eyespot mastigote algae

Phylum Pr-16 **Diatoms**—bacillariophytes

Phylum Pr-17 **Phaeophyta**—brown algae

Phylum Pr-18 **Labyrinthulata**—slime nets and thraustochytrids

Phylum Pr-19 **Plasmodiophora**—symbiotrophs with multinucleate protoplasts

Phylum Pr-20 **Oomycota**—egg molds; zoosporic conjugators

Phylum Pr-21 **Hyphochytriomycota**—zoosporic fresh water osmotrophs with single anteriorly directed undulipodium

PROPAGULE-FORMING SYMBIOTROPHS

Phylum Pr-22 **Haplospora**—unicells with haplosporosomes in tissues of marine animals

Phylum Pr-23 **Paramyxa**—nesting cells by endogenous budding in tissues of marine animals

Phylum Pr-24 **Myxospora**—multicellular polar filament-formers in tissues of fish, sipuculans, and annelids

CONJUGATING ALGAE

Phylum Pr-25 **Rhodophyta**—red algae

Phylum Pr-26 **Gamophyta**—Conjugaphyta (conjugating green algae)

RADIOLARIATES

Phylum Pr-27 **Actinopoda**—"sun animalcules"; marine planktic axopod-formers

ANCESTRAL PHYLA

Phylum Pr-28 **Chlorophyta**—green algae (plant ancestors)

Phylum Pr-29 **Chytridiomycota**—chytrid water molds (fungal ancestors)

Phylum Pr-30 **Zoomastigota**—zoomastigotes, opalinids, choanomonads (animal ancestors)

KINGDOM ANIMALIA

Subkingdom PARAZOA

Phylum A-1 **Placozoa**—trichoplaxes

Phylum A-2 **Porifera**—poriferans (sponges)

Subkingdom EUMETAZOA

Phylum A-3 **Cnidaria**—hydroids (hydras) and medusas (jellyfish)

Phylum A-4 **Ctenophora**—comb jellies

Phylum A-5 **Platyhelminthes**—flatworms

Phylum A-6 **Gnathostomulida**—gnathostomulids (jaw worms)

Phylum A-7 **Rhombozoa**—dicyemids and heterocyemids

Phylum A-8 **Orthonectida**—orthonectids

Phylum A-9 **Nemertina**—nemertines, Rhynchocoela, Nemertea (ribbon worms)

Phylum A-10 **Nematoda**—nematodes (thread worms, round worms)

Phylum A-11 **Nematomorpha**—nematomorphs (Gordian worms, horsehair worms)

Phylum A-12 **Acanthocephala**—thorny-headed worms

Phylum A-13 **Rotifera**—rotifers

Phylum A-14 **Kinorhyncha**—kinorhynchs

Phylum A-15 **Priapulida**—priapulids

Phylum A-16 **Gastrotricha**—gastrotrichs

Phylum A-17 **Loricifera**—loriciferans

Phylum A-18 **Entoprocta**—entoprocts

ARTHROPODS (Phyla A-19 through A-21)

Phylum A-19 **Chelicerata**—chelicerates (horseshoe crabs, spiders, and sea spiders)

Phylum A-20 **Mandibulata**—Uniramia (insects, centipedes, and millipedes)

Phylum A-21 **Crustacea**—crustaceans and pentastomes

Phylum A-22 **Annelida**—annelid worms

Phylum A-23 **Sipuncula**—sipunculids, sipunculans (peanut worms)

Phylum A-24 **Echiura**—echiurans, echiurids (spoon-worms)

Phylum A-25 **Pogonophora**—beard worms, tube worms

Phylum A-26 **Mollusca**—molluscs

Phylum A-27 **Tardigrada**—tardigrades (water bears)

Phylum A-28 **Onychophora**—velvet worms, peripatuses

LOPHOPHORATES (Phyla A-29 through A-31)

Phylum A-29 **Bryozoa**—Ectoprocta, bryozoans (moss animals)

Phylum A-30 **Brachiopoda**—brachiopods (lampshells)

Phylum A-31 **Phoronida**—phoronids

Phylum A-32 **Chaetognatha**—arrow worms

Phylum A-33 **Hemichordata**—enteropneusts, and pterobranchs (acorn worms)

Phylum A-34 **Echinodermata**—echinoderms (sea urchins, sea stars, or starfish, and sea cucumbers)

ACRANIATE CHORDATES (Phyla A-35 and A-36)

Phylum A-35 **Urochordata** (tunicates)—ascidians (sea squirts), larvaceans (appendicularians), and thaliaceans (salps, doliolids, and chain tunicates)

Phylum A-36 **Cephalochordata**—lancelets

Phylum A-37 **Craniata**—craniate chordates (fishes, reptiles, amphibians, mammals, and birds)

Kingdom FUNGI

Phylum F-1 **Zygomycota**—zygomycotes, zygomycetes (molds)

Phylum F-2 **Basidiomycota**—basidiomycotes, basidiomycetes (mushrooms, puffballs, rusts, smuts, and jelly fungi)

Phylum F-3 **Ascomycota**—ascomycotes, ascomycetes (molds, yeasts, and lichens)

Kingdom PLANTAE

BRYATA (nonvascular plants: Phyla Pl-1 through Pl-3)

Phylum Pl-1 **Bryophyta**—mosses

Phylum Pl-2 **Hepatophyta**—liverworts

Phylum Pl-3 **Anthocerophyta**—hornworts (horned liverworts)

VASCULAR PLANTS (Phyla Pl-4 through Pl-12)

Phylum Pl-4 **Lycophyta**—lycophytes (lycopods, club mosses, and spike mosses)

Phylum Pl-5 **Psilophyta**—whisk ferns, psilopsids

Phylum Pl-6 **Sphenophyta**—sphenophytes, Equisetophyta, Arthrophyta (horsetails)

Phylum Pl-7 **Filicinophyta**—Pterophyta, Pteridophyta, Pterodatina (ferns)

GYMNOSPERMS (naked-seed plants: Phyla Pl-8 through Pl-11)

Phylum Pl-8 **Cycadophyta**—cycads

Phylum Pl-9 **Ginkgophyta**—ginkgo (maidenhair tree)

Phylum Pl-10 **Coniferophyta**—conifers

Phylum Pl-11 **Gnetophyta**—gnetophytes (joint fir, ephedra, and *Welwitschia*)

Phylum Pl-12 **Anthophyta**—Angiospermophyta, Magnoliophyta (flowering plants, or monocots and dicots)

Genera Assigned to Phyla

This list includes all the genera mentioned in this book, and many others as well. Page numbers for genera mentioned in this book are given in the index. The list names the phylum of each genus and, for many genera, gives the common names by which their members are known. No common names are available for most genera.

GENUS	PHYLUM	COMMON NAME
Abies	Coniferophyta	Fir
Acanthamoeba	Rhizopoda	
Acantharia	Actinopoda	
Acanthocephalus	Acanthocephala	
Acanthocystis	Actinopoda	
Acanthodasys	Gastrotricha	
Acanthogyrus	Acanthocephala	
Acanthometra	Actinopoda	
Acanthopriapulus	Priapulida	
Acanthoscurria	Chelicerata	Tarantula
Acer	Anthophyta	Japanese maple, Norwegian maple, sugar maple, mountain maple, red maple, silver maple, black maple
Acetabularia	Chlorophyta	Mermaid's wine glass
Acetobacter	Proteobacteria	
Achlya	Oomycota	
Achnanthes	Diatoms	
Acholeplasma	Aphragmabacteria	
Achromobacter	Proteobacteria	
Acidaminococcus	Saprospirae	
Acidianus	Crenarcheota	
Acrasia	Rhizopoda	
Acronema	Zoomastigota	
Acropora	Cnidaria	
Actinobacillus	Proteobacteria	
Actinomyces	Actinobacteria	
Actinophrys	Actinopoda	
Actinoplanes	Actinobacteria	
Actinosphaera	Actinopoda	
Actinosphaerium (*Echinosphaerium*)	Actinopoda	
Acytostelium	Rhizopoda	
Adiantum	Filicinophyta	Maidenhair fern
Aerobacter	Proteobacteria	
Aerococcus	Deinococci	
Aeromonas	Proteobacteria	
Agardhiella	Rhodophyta	Coulter's seaweed
Agaricus	Basidiomycota	Field mushroom, commercial mushroom
Agelas	Porifera	
Agrobacterium	Proteobacteria	
Agave	Anthophyta	Century plant, sisal, henequen, maguey
Alaria	Phaeophyta	Winged kelp, dapper-locks, honey ware, mirlin
Alatospora	Myxospora	
Albertia	Rotifera	
Albugo	Oomycota	
Alcaligenes	Proteobacteria	
Alcyonidium	Bryozoa	

GENUS	PHYLUM	COMMON NAME
Alcyonium	Cnidaria	
Alginobacter	Proteobacteria	
Alligator	Craniata	American and Chinese alligators
Allium	Anthophyta	Chive, leek, onion, garlic, ramps
Allogromia	Granuloreticulosa	
Allomyces	Chytridiomycota	
Alternaria	Ascomycota	Potato blight
Althornia	Labyrinthulata	
Alysiella	Saprospirae	
Amanita	Basidiomycota	Caesar's amanita, fly agaric, death cap, destroying angel
Ambystoma	Craniata	Tiger salamander, marbled salamander
Amia	Craniata	Bowfin
Amoeba	Rhizopoda	
Amoebobacter	Proteobacteria	
Amorphomyces	Ascomycota	
Amphidinium	Dinomastigota	
Amphioxus	Cephalochordata	
Amphipleura	Diatoms	
Amphiporus	Nemertina	
Amphiroa	Rhodophyta	
Anabaena	Cyanobacteria	
Anacystis	Cyanobacteria	
Anaplasma	Aphragmabacteria	
Anas	Craniata	Mallard duck, black mallard, gadwall, pintail duck, teal
Ancalochloris	Chlorobia	
Ancalomicrobium	Proteobacteria	
Ancyclostoma	Nematoda	
Andreaea	Bryophyta	Granite moss, black moss
Anemia	Filicinophyta	Pine fern
Angiococcus	Actinobacteria	
Anisolpidium	Hyphochytrio-mycota	
Anopheles	Mandibulata	
Anser	Craniata	Greylag goose, snow goose, white-fronted goose
Anthoceros	Anthocerophyta	Horned liverwort
Antipathes	Cnidaria	Black coral
Aphanomyces	Oomycota	
Aphrodite	Annelida	Sea mouse
Apis	Mandibulata	Honey bee
Aplanochytrium	Labyrinthulata	
Aplysia	Mollusca	Sea hare
Apodachlya	Oomycota	
Aptenodytes	Craniata	Emperor penguin
Apteryx	Craniata	Common kiwi, little spotted kiwi, great spotted kiwi
Aquaspirillum	Proteobacteria	
Aquifex	Thermotogae	
Arachnoidiscus	Diatoms	
Araucaria	Coniferophyta	Monkey puzzle tree, bunya pine, Moreton Bay pine, Norfolk Island pine, Paraná pine
Arbacia	Echinodermata	Purple sea urchin
Arca	Mollusca	Ark shells
Arcella	Rhizopoda	Shelled ameba
Archangium	Proteobacteria	
Archechiniscus	Tardigrada	Water bear
Archilochus	Craniata	Ruby-throated hummingbird
Architeuthis	Mollusca	Giant squid
Arcyria	Myxomycota	

GENUS	PHYLUM	COMMON NAME
Ardea	Craniata	Great blue heron, gray heron, purple heron, goliath heron
Argiope	Chelicerata	Orb weaver spider
Argonauta	Mollusca	Paper nautilus, argonaut
Argyrotheca	Brachiopoda	
Armadillium	Crustacea	Pill bug
Armillifera	Crustacea	Pentastomid
Artemia	Crustacea	Brine shrimp
Arthrobacter	Actinobacteria	
Arthromitus (*Coleomitus*)	Endospora	
Artocarpus	Anthophyta	Breadfruit, jackfruit
Asbestopluma	Porifera	
Ascaris	Nematoda	Roundworm of pigs and human beings
Aschemonella	Xenophyophora	
Ascidia	Urochordata	
Ascophyllum	Phaeophyta	Knotted wrack, yellow tang, sea whistle
Aspergillus	Ascomycota	Black mold
Aspidosiphon	Sipuncula	
Asplanchna	Rotifera	
Asplenium	Filicinophyta	Lady fern
Astasia	Discomitochondria	
Aster	Anthophyta	Aster
Asterias	Echinodermata	Common starfish (seastar; of North Atlantic), northern (purple) starfish
Asterionella	Diatoms	
Asticcacaulis	Proteobacteria	
Atolla	Cnidaria	Crown jellyfish
Atropa	Anthophyta	Belladonna
Atubaria	Hemichordata	
Auerbachia	Myxospora	
Aulacantha	Actinopoda	
Aurantiactinomyxon	Myxospora	
Aurelia	Cnidaria	Moon jelly
Austrognatharia	Gnathostomulida	
Avena	Anthophyta	Oat
Azolla	Filicinophyta	Water fern, pond fern, mosquito fern
Azomonas	Proteobacteria	
Azotobacter	Proteobacteria	
Babesia	Apicomplexa	
Bacillaria	Diatoms	
Bacillus	Endospora	
Bacteroides	Saprospirae	
Balaenoptera	Craniata	Fin whale, sei whale, Bryde's whale, rorqual
Balanoglossus	Hemichordata	
Balantidium	Ciliophora	
Balanus	Crustacea	
Bambusina	Gamophyta	
Bangia	Rhodophyta	
Barbeyella	Myxomycota	
Barbulanympha	Archaeprotista	
Barentsia	Entoprocta	
Bartonella	Aphragmabacteria	
Basidiobolus	Zygomycota	
Bathybelos	Chaetognatha	
Bathycordius	Urochordata	
Bathycrinus	Echinodermata	
Bathyctena	Ctenophora	
Bathyspadella	Chaetognatha	
Batillipes	Tardigrada	Water bear

GENUS	PHYLUM	COMMON NAME
Batrachospermum	Rhodophyta	
Bdellovibrio	Proteobacteria	
Beggiatoa	Proteobacteria	
Beijerinckia	Proteobacteria	
Beneckea	Proteobacteria	
Beroë	Ctenophora	Thimble jelly, sea mitre
Beta	Anthophyta	Sugar beet, Swiss chard, red beet, mangel, spinach
Biddulphia	Diatoms	
Blakeslea	Zygomycota	
Blastobacter	Pirellulae	
Blastocladiella	Chytridiomycota	
Blastocrithidia	Discomitochondria	
Blatta	Mandibulata	Oriental cockroach
Blepharisma	Ciliophora	
Bodo	Discomitochondria	
Boletus	Basidiomycota	Edible bolete, lurid bolete, yellow cracked bolete
Bolinopsis	Ctenophora	Lobed comb jelly
Bombyx	Mandibulata	Silkworm moth
Bonellia	Echiura	Spoon worm
Borrelia	Spirochaetae	
Bos	Craniata	Cattle, cow, zebu, yak, guar, banteng, kouprey
Bothrioplana	Platyhelminthes	
Botrychium	Filicinophyta	Moonwort, grape fern
Botrydiopsis	Xanthophyta	
Botrydium	Xanthophyta	
Botryllus	Urochordata	
Botryococcus	Xanthophyta	
Bowenia	Cycadophyta	

GENUS	PHYLUM	COMMON NAME
Brachionus	Rotifera	
Bradypus	Craniata	Three-toed sloth, maned sloth
Bradyrhizobium	Proteobacteria	
Branchioceranthus	Cnidaria	
Branchiostoma	Cephalochordata	Lancelet, amphioxus
Brassica	Anthophyta	Cabbage, broccoli, turnip, black mustard, rutabaga, kohlrabi
Brucella	Proteobacteria	
Bryopsis	Chlorophyta	Sea fern
Bryum	Bryophyta	Silver moss
Bubalus	Craniata	Water buffalo
Bubo	Craniata	Great horned owl
Bufo	Craniata	American toad
Bugula	Bryozoa	
Busycon	Mollusca	Knobbed whelk, lightning conch
Buteo	Craniata	Red-tailed hawk
Buxbaumia	Bryophyta	Bug-on-a-stick, elf-cap moss
Caedibacter	Proteobacteria	
Caenorhabditis	Nematoda	
Cafeteria	Zoomastigota	
Calanus	Crustacea	Copepod
Calcidiscus	Haptomonada	
Calciosolenia	Haptomonada	
Calicium	Ascomycota	Pear-shaped lichen, stubble lichen
Callophyllis	Rhodophyta	Red sea fan
Calonympha	Archaeprotista	
Calvatia	Basidiomycota	Giant puffball
Calymmatobacterium	Proteobacteria	
Calypogeia	Hepatophyta	
Calyptrosphaera	Haptomonada	

GENUS	PHYLUM	COMMON NAME
Cambarus	Crustacea	
Camellia	Anthophyta	Tea, camellia
Camelus	Craniata	Arabian (dromedary) camel, two-humped (bactrian) camel
Campyloderes	Kinorhyncha	
Cancer	Crustacea	
Candida	Ascomycota	Monilia, yeast agent of thrush
Canis	Craniata	Dog, gray wolf, coyote, dingo, jackal, red wolf
Canteriomyces	Hyphochytrio-mycota	
Cantharellus	Basidiomycota	Chanterelle
Capnocytophaga	Saprospirae	
Capsella	Anthophyta	Shepherd's purse
Carchesium	Ciliophora	
Carcinonemertes	Nemertina	
Carcinoscorpinus	Chelicerata	
Cardiobacterium	Proteobacteria	
Carinina	Nemertina	
Carrpos	Hepatophyta	
Casuarius	Craniata	Cassowary
Cateria	Kinorhyncha	
Caulerpa	Chlorophyta	
Caulobacter	Proteobacteria	
Cavia	Craniata	Guinea pig
Cavostelium	Myxomycota	
Cedrus	Coniferophyta	Atlas cedar, Cyprus cedar, deodar, cedar of Lebanon
Cellulomonas	Actinobacteria	
Centritractus	Xanothophyta	
Centroderes	Kinorhyncha	
Centropyxis	Rhizopoda	
Cepedea	Zoomastigota	
Cephalobaena	Crustacea	Pentastomid
Cephalodiscus	Hemichordata	
Cephalofovea	Onychophora	
Cephalosporium	Ascomycota	
Cephalothamnium	Discomitochondria	
Cephalothrix	Nemertina	
Ceratiomyxa	Myxomycota	
Ceratium	Dinomastigota	
Ceratomyxa	Myxospora	
Ceratozamia	Cycadophyta	Mexican horncone
Cercomonas	Myxomycota	
Cerebratulus	Nemertina	Ribbon worm
Cerelasma	Xenophyophora	
Cerelpemma	Xenophyophora	
Cervus	Craniata	Whitetail deer
Cestum	Ctenophora	Venus's girdle
Chaetocladium	Zygomycota	
Chaetoderma	Mollusca	
Chaetomium	Ascomycota	
Chaetomorpha	Chlorophyta	
Chaetonotus	Gastrotricha	
Chaetopterus	Annelida	Parchment worm
Challengeron	Actinopoda	
Chamaesiphon	Cyanobacteria	
Chantransia	Rhodophyta	
Chara	Chlorophyta	Stonewort
Characiopsis	Xanthophyta	
Charonia	Mollusca	Triton
Chattonella	Chrysomonada	
Chelifer	Chelicerata	Pseudoscorpion
Chigua	Cycadophyta	
Chilomastix	Archaeprotista	
Chilomonas	Cryptomonada	
Chironex	Cnidaria	Sea wasp
Chironomus	Mandibulata	Midge
Chlamydia	Pirellulae	

GENUS	PHYLUM	COMMON NAME
Chlamydomonas	Chlorophyta	
Chlamydomyxa	Xanthophyta	
Chloramoeba	Xanthophyta	
Chlorarachnion	Rhizopoda	
Chlorella	Chlorophyta	
Chloridella	Xanthophyta	
Chlorobium	Chlorobia	
Chlorobotrys	Eustigmatophyta	
Chlorochromatium	Chlorobia	
Chlorococcum	Chlorophyta	
Chlorodesmis	Chlorophyta	
Chloroflexus	Chloroflexa	
Chloroherpeton	Chlorobia	
Chloromyxum	Myxospora	
Chloronema	Chlorobia	
Chloropseudomonas	Proteobacteria	
Chondrococcus	Actinobacteria	
Chondrodendron	Anthophyta	
Chondromyces	Proteobacteria	
Chondrus	Rhodophyta	Irish moss, carrageen
Chorda	Phaeophyta	
Chordaria	Phaeophyta	
Chordeiles	Craniata	Common nighthawk
Chordodes	Nematomorpha	
Chordodiolus	Nematomorpha	
Chromatium	Proteobacteria	
Chromobacterium	Proteobacteria	
Chromogaster	Rotifera	
Chromulina	Chrysomonada	
Chroococcus	Cyanobacteria	
Chroomonas	Cryptomonada	
Chrysanthemum	Anthophyta	Costmary, ox-eye daisy, pyrethrum
Chrysarachnion	Rhizopoda	
Chrysemys	Craniata	Painted turtle
Chrysobotrys	Chrysomonada	
Chrysocapsa	Chrysomonada	
Chrysochromulina	Haptomonada	
Chrysosphaerella	Chrysomonada	
Chytria	Chytridiomycota	
Ciliocincta	Orthonectida	
Ciliophrys	Actinopoda	
Cinchona	Anthophyta	Fever-bark tree
Cinnamomum	Anthophyta	Cinnamon, cassia
Ciona	Urochordata	Sea squirt
Citrobacter	Proteobacteria	
Cladochytrium	Chytridiomycota	
Cladonia	Ascomycota	Reindeer lichen, spoon lichen, pixie cup, ladder lichen, British soldiers
Cladophora	Chlorophyta	
Clastoderma	Myxomycota	
Clathrulina	Actinopoda	
Clavaria	Basidiomycota	Coral mushrooms
Claviceps	Ascomycota	Ergot fungus
Clibanarius	Crustacea	Hermit crab
Clevelandina	Spirochaetae	
Closterium	Gamophyta	
Clostridium	Endospora	Lockjaw and botulism agents
Clypeoseptoria	Ascomycota	
Coccidia	Apicomplexa	
Coccolithus	Haptomonada	
Coccomyxa	Myxospora	
Cochlonema	Zygomycota	
Cocos	Anthophyta	Coconut
Codium	Chlorophyta	Sea staghorn, oyster thief, spongy cushion, dead-man's fingers
Coelomomyces	Chytridiomycota	
Coeloplana	Ctenophora	

GENUS	PHYLUM	COMMON NAME
Coelospora	Apicomplexa	
Coenonia	Rhizopoda	
Coffea	Anthophyta	Coffee
Colacium	Discomitochondria	
Colaptes	Craniata	Flicker
Colchicum	Anthophyta	Autumn crocus
Coleomitus	Endospora	
Colius	Craniata	Mouse bird
Collema	Ascomycota	Pulpy, thousand-fruited, tee, blistered jelly, bat's wing, tapioca, crowded pulp, black pepper, dusky jelly, pimpled jelly, granular jelly, lichen
Collozoum	Actinopoda	
Colpoda	Ciliophora	
Columba	Craniata	Band-tailed pigeon, some other doves and pigeons
Comatricha	Myxomycota	
Conidiobolus	Zygomycota	
Coniocybe	Ascomycota	Mealy stubble lichen
Conocephalum	Hepatophyta	Great scented liverwort
Conochilus	Rotifera	
Conocyema	Rhombozoa	
Conus	Mollusca	Cone shells
Convoluta	Platyhelminthes	
Copromonas	Cryptomonada	
Corallina	Rhodophyta	
Corallium	Cnidaria	Red coral
Coronarctus	Tardigrada	Water bear
Coronula	Crustacea	
Coronympha	Archaeprotista	
Corvus	Craniata	Crow, raven, rook, jackdaw
Corynebacterium	Actinobacteria	
Coscinodiscus	Diatoms	
Cosmarium	Gamophyta	
Cowdria	Aphragmabacteria	
Coxiella	Proteobacteria	
Crania	Brachiopoda	
Craspedacusta	Cnidaria	
Crepidula	Mollusca	Slipper shells, boat shells, slipper limpets
Cristatella	Bryozoa	
Cristispira	Spirochaetae	
Crithidia	Discomitochondria	
Cronartium	Basidiomycota	White pine blister rust
Crotalus	Craniata	Diamondback rattlesnake, sidewinder
Crucibulum	Basidiomycota	Common bird's-nest fungus
Cryptocercus	Mandibulata	Wood-eating cockroach
Cryptochiton	Mollusca	Chiton
Cryptococcus	Ascomycota	
Cryptohydra	Cnidaria	
Cryptomeria	Coniferophyta	Japanese cedar
Cryptomonas	Cryptomonada	
Cryptosporium	Ascomycota	
Ctenoplana	Ctenophora	
Cucumaria	Echinodermata	Tar-spotted sea cucumber, other sea cucumbers
Cucurbita	Anthophyta	Pumpkin, squashes, gourds
Cupelopagis	Rotifera	

GENUS	PHYLUM	COMMON NAME
Cupressus	Coniferophyta	Cypress
Curvularia	Ascomycota	
Cuscuta	Anthophyta	Dodder
Cutleria	Phaeophyta	
Cyanea	Cnidaria	Sea blubber, lion's mane, red winter jellyfish, pink jellyfish
Cyanomonas	Cryptomonada	
Cyanophora	Cryptomonada	
Cyathea	Filicinophyta	Tree fern
Cyathomonas	Cryptomonada	
Cyathus	Basidiomycota	Striate bird's-nest fungus
Cycas	Cycadophyta	Australian nut palm, sago palm
Cyclosalpa	Urochordata	Tunicate
Cyclotella	Diatoms	
Cygnus	Craniata	Mute swan, black swan, whistling swan
Cylindrocapsa	Chlorophyta	
Cylindrocystis	Gamophyta	
Cymbella	Diatoms	
Cynocephalus	Craniata	Flying lemur
Cystodinium	Dinomastigota	
Cytophaga	Saprospirae	
Cytoseira	Phaeophyta	
Dactylopodola	Gastrotricha	
Dallinella	Brachiopoda	
Daphnia	Crustacea	Water flea
Daptobacter	Proteobacteria	
Dasya	Rhodophyta	Chenille weed
Datura	Anthophyta	Jimsonweed, thornapple
Daucus	Anthophyta	Carrot, Queen Anne's lace
Deinococcus	Deinococci	
Deltotrichonympha	Archaeprotista	
Dendroceros	Anthocerophyta	Hornwort
Dendrohyrax	Craniata	Tree hyrax, bush hyrax
Dendronephthya	Cnidaria	
Dendrostomum	Sipuncula	
Dennstaedtia	Filicinophyta	Hay-scented fern
Dentalium	Mollusca	Tooth shells, tusk shells
Derbesia	Chlorophyta	
Dermacentor	Chelicerata	Tick that transmits Rocky Mountain spotted fever
Dermatophilus	Actinobacteria	
Dermocarpa	Cyanobacteria	
Derxia	Proteobacteria	
Desmarella	Zoomastigota	
Desmidium	Gamophyta	
Desulfacinum	Proteobacteria	
Desulfobacter	Proteobacteria	
Desulfonema	Proteobacteria	
Desulfotomaculum	Proteobacteria	
Desulfovibrio	Proteobacteria	
Desulfurococcus	Crenarchaeota	
Desulfuromonas	Proteobacteria	
Devescovina	Archaeprotista	
Diatoma	Diatoms	
Dicarpa	Urochordata	
Dicksonia	Filicinophyta	Tree fern
Dictydium	Myxomycota	
Dictyosiphon	Phaeophyta	
Dictyostelium	Rhizopoda	
Dictyota	Phaeophyta	
Dictyuchus	Oomycota	
Dicyema	Rhombozoa	
Dicyemennea	Rhombozoa	
Didemnum	Urochordata	Tunicate
Diderma	Myxomycota	

GENUS	PHYLUM	COMMON NAME
Didinium	Ciliophora	
Didymium	Myxomycota	
Difflugia	Rhizopoda	
Digitalis	Anthophyta	Foxglove
Dileptus	Ciliophora	
Dinema	Discomitochondria	
Dinenympha	Archaeprotista	
Dinobryon	Chrysomonada	
Dinothrix	Dinomastigota	
Dioctophyme	Nematoda	
Dioon	Cycadophyta	Dioon
Diplocalyx	Spirochaetae	
Diplococcus	Deinococci	
Diplolaimella	Nematoda	
Diploneis	Diatoms	
Diplosoma	Urochordata	Tunicate
Dipylidium	Platyhelminthes	Double-pored tapeworm
Dirofilaria	Nematoda	Heartworm
Discorbis	Granuloreticulosa	
Discosphaera	Haptomonada	
Distigma	Discomitochondria	
Doliolum	Urochordata	
Doris	Mollusca	Sea slug, nudibranch
Dracunculus	Nematoda	Guinea worm
Dreissena	Mollusca	Zebra mussel
Drosophila	Mandibulata	Fruit fly
Dryopteris	Filicinophyta	New York fern, shield fern
Dugesia	Platyhelminthes	Laboratory planarian
Dunaliella	Chlorophyta	
Durvillaea	Phaeophyta	
Echinarachnius	Echinodermata	Sand dollar
Echiniscoides	Tardigrada	Water bear
Echiniscus	Tardigrada	Water bear
Echinocactus	Anthophyta	Barrel cactus
Echinochrysis	Chrysomonada	
Echinococcus	Platyhelminthes	Dog tapeworm, hydatid tapeworm
Echinoderes	Kinorhyncha	
Echinorhynchus	Acanthocephala	
Echinospaerium (*Actinosphaerium*)	Actinopoda	
Echinostelium	Myxomycota	
Echiurus	Echiura	Spoon worm
Ectocarpus	Phaeophyta	
Ediacara	Cnidaria	
Edwardsiella	Proteobacteria	
Ehrlichia	Aphragmabacteria	
Eimeria	Apicomplexa	
Elephas	Craniata	Indian elephant
Ellipsoidion	Eustigmatophyta	
Elodea	Anthophyta	Waterweed
Elphidium	Granuloreticulosa	
Elsinoe	Ascomycota	
Elysia	Mollusca	
Embata	Rotifera	
Emiliania	Haptomonada	
Emplectonema	Nemertina	
Encephalartos	Cycadophyta	Kaffir bread (bread "palm")
Encephalitozoon	Microspora	
Endocochlus	Zygomycota	
Endogone	Zygomycota	Endomycorrhizal fungus
Endothia	Ascomycota	
Entamoeba	Rhizopoda	Dysentery ameba
Enterobacter	Proteobacteria	
Enterobius	Nematoda	Pinworm
Enteromorpha	Chlorophyta	
Entophysalis	Cyanobacteria	
Ephedra	Gnetophyta	Mormon tea, long-leaf ephedra, joint fir

GENUS	PHYLUM	COMMON NAME
Ephelota	Ciliophora	
Epifagus	Anthophyta	Beech drops
Epigonichthys	Cephalochordata	
Epiperipatus	Onychophora	
Epipyxis	Chrysomonada	
Equisetum	Sphenophyta	Horsetail, scouring rush
Equus	Craniata	Horse, ass, zebra, donkey, burro, mule
Erwinia	Proteobacteria	
Erythrocladia	Rhodophyta	
Erythropsidinium	Dinomastigota	
Erythrotrichia	Rhodophyta	
Escarpia	Pogonophora	
Escherichia	Proteobacteria	Colon bacterium
Eucalyptus	Anthophyta	Gum tree, ironbark, Australian mountain ash
Eubacterium	Endospora	
Euchlanis	Rotifera	
Euchlora	Ctenophora	
Euglena	Discomitochondria	
Eukrohnia	Chaetognatha	
Eunice	Annelida	Palolo worm
Eunotia	Diatoms	
Euphausia	Crustacea	Krill
Euplectella	Porifera	Venus's flower basket
Euplotes	Ciliophora	
Eurhamphea	Ctenophora	Comb jelly
Eustigmatos	Eustigmatophyta	
Fabespora	Myxospora	
Fagus	Anthophyta	Copper beech, American beech, European beech
Fasciola	Platyhelminthes	Sheep liver fluke, cattle liver fluke
Felis	Craniata	Cat, mountain lion, jaguarundi, ocelot, marbled cat, serval, golden cat, Pallas's cat, European wild cat, jungle cat, African black-footed cat, little spotted cat, African wild cat, margay cat, pampas cat, leopard cat
Ferrobacillus	Proteobacteria	
Fervidobacterium	Thermotogae	
Fischerella	Cyanobacteria	
Flavobacterium	Proteobacteria	
Flexibacter	Saprospirae	
Flexithrix	Saprospirae	
Floscularia	Rotifera	
Foaina	Archaeprotista	
Folioceros	Anthocerophyta	
Fomes	Basidiomycota	Rusty-hoof fomes, bracket fungus
Fontinalis	Bryophyta	Fountain moss
Fossombronia	Hepatophyta	
Fragilaria	Diatoms	
Frankia	Actinobacteria	Root-nodule bacterium
Fredericella	Bryozoa	
Fregata	Craniata	Frigate bird
Fritschiella	Chlorophyta	
Frullania	Hepatophyta	Leafy liverwort
Fucus	Phaeophyta	Rockweed, wrack
Fulica	Craniata	Coot
Fuligo	Myxomycota	
Funaria	Bryophyta	Cord moss

GENUS	PHYLUM	COMMON NAME
Fundulus	Craniata	Killifish
Fusarium	Ascomycota	
Fusobacterium	Saprospirae	
Fusulina	Granuloreticulosa	
Gadus	Craniata	Atlantic cod
Gaffkya	Deinococci	
Galatheammina	Xenophyophora	
Gallus	Craniata	Chicken, jungle fowl
Gastrodes	Ctenophora	
Gastrostyla	Ciliophora	
Gavia	Craniata	Loon
Geastrum	Basidiomycota	Triplex earthstar, crowned earthstar
Gelidium	Rhodophyta	
Gelliodes	Porifera	
Gemmata	Pirellulae	
Genicularia	Gamophyta	
Geococcyx	Craniata	Roadrunner
Geonemertes	Nemertina	
Geothallus	Bryophyta	
Geotrichum	Ascomycota	
Gephyrocapsa	Haptomonada	
Giardia	Archaeprotista	
Gigantocypris	Crustacea	Ostracod
Gigantorhynchus	Acanthocephala	
Ginkgo	Ginkgophyta	Ginkgo, maidenhair tree
Glabratella	Granuloreticulosa	
Glandiceps	Hemichordata	
Globigerina	Granuloreticulosa	
Gloeocapsa	Cyanobacteria	
Gloeothece	Cyanobacteria	
Glomus	Zygomycota	
Glossobalanus	Hemichordata	
Glottidia	Brachiopoda	
Gluconobacter	Proteobacteria	
Glugea	Microspora	
Glycine	Anthophyta	Soybean
Gnathostomaria	Gnathostomulida	
Gnathostomula	Gnathostomulida	
Gnetum	Gnetophyta	Gnetum, bawale, longonizia, bulso
Golfingia	Sipuncula	
Gonatozygon	Gamophyta	
Goniotrichum	Rhodophyta	
Gonium	Chlorophyta	
Gononemertes	Nemertina	
Gonyaulax	Dinomastigota	
Gonyostomum	Chrysomonada	
Gordionus	Nematomorpha	Gordian worm
Gordius	Nematomorpha	Gordian worm
Gorilla	Craniata	Gorilla
Gracilaria	Rhodophyta	
Grantia	Porifera	Purse sponge
Gregarina	Apicomplexa	Gregarine
Guttulina	Rhizopoda	
Guttulinopsis	Rhizopoda	
Guyenotia	Myxospora	
Gymnodinium	Dinomastigota	
Haemophilus	Proteobacteria	
Haemoproteus	Apicomplexa	
Hafnia	Proteobacteria	
Haliclystus	Cnidaria	
Halicryptus	Priapulida	
Halimeda	Chlorophyta	
Haliotis	Mollusca	Abalone
Halisarca	Porifera	Jelly sponge
Haloarcula	Euryarchaeota	
Halobacterium	Euryarchaeota	
Halobiotus	Tardigrada	
Halococcus	Euryarchaeota	
Halocynthia	Urochordata	Sea peach
Haloferax	Euryarchaeota	
Halteria	Ciliophora	
Hantschia	Diatoms	
Haplognathia	Gnathostomulida	

GENUS	PHYLUM	COMMON NAME
Haplomitrium	Hepatophyta	
Haplopappus	Anthophyta	Wild aster
Haplosporidium	Haplospora	
Hartmannella	Rhizopoda	
Helianthus	Anthophyta	Sunflower, Jerusalem artichoke
Helicosporidium	Myxospora	
Heliobacterium	Endospora	
Heliopora	Cnidaria	Blue coral
Heliothrix	Chloroflexa	
Helix	Mollusca	Edible land snail
Hemiselmis	Cryptomonada	
Hemithyris	Brachiopoda	
Hemitrichia	Myxomycota	
Henneguya	Myxospora	
Heptabrachia	Pogonophora	
Herdmania	Urochordata	
Herpetomonas	Discomitochondria	
Herpetosiphon	Saprospirae	
Herpothallon	Ascomycota	Red blanket lichen
Heterodendron	Xanthophyta	
Heterokrohnia	Chaetognatha	Arrow worm
Heteronema	Discomitochondria	
Heterophrys	Actinopoda	
Hevea	Anthophyta	Rubber plant
Hexacapsula	Myxospora	
Hexamastix	Archaeprotista	
Hexamita	Archaeprotista	
Hildebrandia	Rhodophyta	
Hillea	Cryptomonada	
Himanthalia	Phaeophyta	
Hippospongia	Porifera	
Hirudo	Annelida	European medicinal leech
Histiona	Zoomastigota	
Histomonas	Archaeprotista	
Histoplasma	Ascomycota	

GENUS	PHYLUM	COMMON NAME
Hollandina	Spirochaetae	
Holomastigotoides	Archaeprotista	
Holothuria	Echinodermata	
Homarus	Crustacea	Lobster
Homo	Craniata	Human being
Hordeum	Anthophyta	Barley
Hyalodiscus	Rhizopoda	
Hyalotheca	Gamophyta	
Hydra	Cnidaria	Hydra, green hydra
Hydrogenomonas	Proteobacteria	
Hydrurus	Chrysomonada	
Hyella	Cyanobacteria	
Hymenolepis	Platyhelminthes	Dwarf tapeworm of human beings and rodents, tapeworm of chickens
Hymenomonas	Haptomonada	
Hymenophyllum	Filicinophyta	Filmy fern
Hymenophytum	Hepatophyta	
Hyphochytrium	Hyphochytrio-mycota	
Hyphomicrobium	Proteobacteria	
Hypnum	Bryophyta	Sheet moss, carpet moss
Hypsibius	Tardigrada	
Ichthyosporidium	Microspora	
Ikeda	Echiura	
Ikedosoma	Echiura	
Ilex	Anthophyta	Yaupon, holly
Incisitermes (*Kalotermes*)	Mandibulata	Dry-wood termite
Ipomoea	Anthophyta	Sweet potato, morning glory
Iridia	Granuloreticulosa	
Isaacsicalanus	Crustacea	
Ishige	Phaeophyta	
Isoachlya	Oomycota	

GENUS	PHYLUM	COMMON NAME
Isoetes	Lycophyta	Quillwort
Isophaera	Pirellulae	
Isospora	Apicomplexa	
Ixodes	Chelicerata	Tick that transmits Lyme disease
Jakoba	Zoomastigota	
Japonochytrium	Labyrinthulata	
Joenia	Archaeprotista	
Juniperus	Coniferophyta	Juniper, red cedar
Kalanchoë	Anthophyta	
Kalotermes (*Incisitermes*)	Mandibulata	Dry-wood termite
Karotomorpha	Zoomastigota	
Kickxella	Zygomycota	
Kinorhynchus	Kinorhyncha	
Klebsiella	Proteobacteria	
Klebsormidium	Chlorophyta	
Krohnitta	Chaetognatha	
Krohnittella	Chaetognatha	Arrow worm
Kudoa	Myxospora	
Labyrinthorhiza	Labyrinthulata	Slime net
Labyrinthula	Labyrinthulata	Slime net
Labyrinthuloides	Labyrinthulata	
Lacazella	Brachiopoda	
Lactobacillus	Endospora	
Lagenidium	Oomycota	
Lamellibrachia	Pogonophora	
Lamellisabella	Pogonophora	
Laminaria	Phaeophyta	Oarwood, sea girdle, fingered kelp, tangle, sea belt, kelp
Lampea (*Gastrodes*)	Ctenophora	
Lamprothamnium	Chlorophyta	Stonewort
Larix	Coniferophyta	Tamarack, larch
Latimeria	Craniata	Lobe-finned fish
Latrodectus	Chelicerata	Black widow spider
Latrostium	Hyphochytriomycota	
Legionella	Proteobacteria	
Leiobunum	Chelicerata	Harvestmen, daddy longlegs
Leishmania	Discomitochondria	
Lejeunea	Hepatophyta	
Lemanea	Rhodophyta	
Lemna	Anthophyta	Duckweed
Leocarpus	Myxomycota	
Lepidodermella	Gastrotricha	
Lepidozamia	Cycadophyta	
Lepidozia	Hepatophyta	
Lepiota	Basidiomycota	Smooth lepiota, yellow lepiota, parasol mushroom, shaggy parasol
Lepraria	Ascomycota	
Leptomonas	Discomitochondria	
Leptonema	Spirochaetae	
Leptorhynchoides	Acanthocephala	
Leptosomatum	Nematoda	
Leptospira	Spirochaetae	
Leptothrix	Proteobacteria	
Leptotrichia	Proteobacteria	
Lepus	Craniata	Arctic hare, varying hare, European hare, jack "rabbit"
Leuconostoc	Endospora	
Leucosolenia	Porifera	
Leucothrix	Proteobacteria	
Licea	Myxomycota	
Lichenopora	Bryozoa	
Lichenothrix	Ascomycota	
Lichina	Ascomycota	
Ligniera	Plasmodiophora	
Lilium	Anthophyta	Turk's-cap lily, tiger lily, wood lily, some other lilies

GENUS	PHYLUM	COMMON NAME
Limenitis	Mandibulata	Viceroy butterfly, white admiral butterfly, some other butterflies
Limnodrilus	Annelida	
Limulus	Chelicerata	Horseshoe crab
Lineola	Endospora	
Lineus	Nemertina	Bootlace worm
Linguatula	Crustacea	Pentastomid, pentastome worm
Lingula	Brachiopoda	
Liriodendron	Anthophyta	Tulip tree, tulip poplar
Lissoclinum	Urochordata	Tunicate
Lissomyema	Echiura	
Listriolobus	Echiura	Spoon-worm
Lithacrosiphon	Sipuncula	
Lithothamnion	Rhodophyta	
Littorina	Mollusca	Periwinkle
Llama	Craniata	Guanaco, llama, alpaca
Lobaria	Ascomycota	Speckled lichen
Locusta	Mandibulata	Locust
Loligo	Mollusca	Common squid
Loxosceles	Chelicerata	Brown recluse spider
Loxosoma	Entoprocta	
Loxosomella	Entoprocta	
Lumbricus	Annelida	Earthworm
Lycogala	Myxomycota	Slime mold
Lycopodium	Lycophyta	Club moss
Lycosa	Chelicerata	Wolf spider
Lyngbya	Cyanobacteria	
Maccabeus	Priapulida	
Macracanthorhynchus	Acanthocephala	
Macrobdella	Annelida	American medicinal leech
Macrobiotus	Tardigrada	Water bear
Macrocystis	Phaeophyta	Kelp
Macrodasys	Gastrotricha	
Macromonas	Proteobacteria	
Macroperipatus	Onychophora	
Macrozamia	Cycadophyta	
Magnolia	Anthophyta	Magnolia, cucumber tree, sweet bay
Malacobdella	Nemertina	
Mallomonas	Chrysomonada	
Manihot	Anthophyta	Yucca, cassava, manioc, tapioca
Mannia	Hepatophyta	
Manus	Craniata	Pangolin, scaly anteater
Marattia	Filicinophyta	King fern, tree fern
Marchantia	Hepatophyta	Common liverwort
Marsilea	Filicinophyta	Waterclover
Marsupella	Hepatophyta	
Marteilia	Paramyxa	
Mastigamoeba	Archaeprotista	
Mastigina	Archaeprotista	
Matteucia	Filicinophyta	Ostrich fern
Maudammina	Xenophyophora	
Mayorella	Rhizopoda	
Megaceros	Anthocerophyta	
Megaceryle	Craniata	Belted kingfisher
Megascolides	Annelida	Giant earthworm
Megasphaera	Saprospirae	
Megathyris	Brachiopoda	
Meiopriapulus	Priapulida	
Melanopus	Mandibulata	Grasshopper
Meleagris	Craniata	Turkey
Melosira	Diatoms	
Membranipora	Bryozoa	Sea mat
Membranosorus	Plasmodiophora	
Mercenaria	Mollusca	Quahog, cherrystone, hard-shell clam

GENUS	PHYLUM	COMMON NAME
Meridion	Diatoms	
Merotricha	Chrysomonada	
Mertensia	Ctenophora	Sea walnut
Mesocricetus	Craniata	Golden hamster
Mesognatharia	Gnathostomulida	
Mesoperipatus	Onychophora	
Mesotaenium	Gamophyta	
Metabonellia	Echiura	Spoon-worm
Metacrinus	Echinodermata	Sea lily
Metadevescovina	Archaeprotista	
Metasequoia	Coniferophyta	Dawn redwood
Metatrichia	Myxomycota	
Methanobacillus	Euryarchaeota	
Methanobacterium	Euryarchaeota	
Methanobrevibacter	Euryarchaeota	
Methanococcoides	Euryarchaeota	
Methanococcus	Euryarchaeota	
Methanocorpusculum	Euryarchaeota	
Methanoculleus	Euryarchaeota	
Methanogenium	Euryarchaeota	
Methanohalophilus	Euryarchaeota	
Methanolobus	Euryarchaeota	
Methanomicrobium	Euryarchaeota	
Methanoplanus	Euryarchaeota	
Methanopyrus	Euryarchaeota	
Methanosaeta	Euryarchaeota	
Methanosarcina	Euryarchaeota	
Methanosphaera	Euryarchaeota	
Methanospirillum	Euryarchaeota	
Methanothermus	Euryarchaeota	
Methanothrix	Euryarchaeota	
Methylococcus	Proteobacteria	
Methylomonas	Proteobacteria	
Methylosinus	Proteobacteria	
Metridium	Cnidaria	Plumose anemone
Micrasterias	Gamophyta	
Microciona	Porifera	Redbeard sponge

GENUS	PHYLUM	COMMON NAME
Micrococcus	Deinococci	
Microcycas	Cycadophyta	Corcho
Microcyema	Rhombozoa	
Microcystis	Cyanobacteria	
Micromonospora	Actinobacteria	
Micropterus	Craniata	Large-mouth black bass
Microscilla	Saprospirae	
Microspora	Chlorophyta	
Microthopalodia	Archaeprotista	
Microtus	Craniata	Vole
Miliola	Granuloreticulosa	
Millepora	Cnidaria	Fire coral
Milnesium	Tardigrada	
Minakatella	Rhizopoda	
Minchinia	Haplospora	
Mischococcus	Xanthophyta	
Mnemiopsis	Ctenophora	Lobed comb jelly
Mobilifilum	Spirochaetae	
Molgula	Urochordata	
Monas	Chrysomonada	
Moniliformis	Acanthocephala	
Monoblepharella	Chytridiomycota	
Monoblepharis	Chytridiomycota	
Monocercomonas	Archaeprotista	
Monoclea	Hepatophyta	
Monodopsis	Eustigmatophyta	
Monosiga	Zoomastigota	
Monotropa	Anthophyta	Indian pipe
Morchella	Ascomycota	Morel
Mortierella	Zygomycota	
Mougeotia	Gamophyta	
Mucor	Zygomycota	Bread mold
Murex	Mollusca	Murex or rock shell
Mus	Craniata	House mouse
Musa	Anthophyta	Banana, plantain, abaca
Musca	Mandibulata	House fly

GENUS	PHYLUM	COMMON NAME
Mya	Mollusca	Soft-shell clam, long-neck or sand clam, short clam
Mycobacterium	Actinobacteria	
Mycococcus	Actinobacteria	
Mycoplasma	Aphragmabacteria	
Mycosphaerella	Ascomycota	
Myotis	Craniata	Little brown bat, some other bats
Mytilus	Mollusca	Common edible mussel, scorched mussel
Myxidium	Myxospora	
Myxobolus	Myxospora	
Myxochloris	Xanthophyta	
Myxococcus	Proteobacteria	
Myxostoma	Myxospora	Salmon twist-disease parasite
Myzocytium	Oomycota	
Myzostoma	Annelida	
Naegleria	Discomitochondria	
Nanaloricus	Loricifera	
Nanochloropsis	Eustigmatophyta	
Nanognathia	Gnathostomulida	
Natronobacterium	Euryarchaeota	
Natronococcus	Euryarchaeota	
Nautilus	Mollusca	Chambered nautilus
Navicula	Diatoms	
Nebalia	Crustacea	
Necator	Nematoda	Hookworm
Nectonema	Nematomorpha	Gordian worm, horsehair worm
Nectonemertes	Nemertina	
Neisseria	Proteobacteria	
Nemalion	Rhodophyta	Threadweed
Nematochrysis	Chrysomonada	
Nematodinium	Dinomastigota	
Neocallimastix	Chytridiomycota	
Neoechinorhynchus	Acanthocephala	
Neohodgsonia	Hepatophyta	
Neomenia	Mollusca	
Neopilina	Mollusca	
Neorickettsia	Proteobacteria	
Nephasoma	Sipuncula	
Nephroselmis	Cryptomonada	
Nephthys	Annelida	Paddle-footed worm, bristle worm
Nereis	Annelida	Clamworm
Nereocystis	Phaeophyta	Bull or bladder kelp
Netrium	Gamophyta	
Neurospora	Ascomycota	Pink bread mold
Nitella	Chlorophyta	Stonewort
Nitellopsis	Chlorophyta	Stonewort
Nitrobacter	Proteobacteria	
Nitrococcus	Proteobacteria	
Nitrocystis	Proteobacteria	
Nitrosococcus	Proteobacteria	
Nitrosogloea	Proteobacteria	
Nitrosolobus	Proteobacteria	
Nitrosomonas	Proteobacteria	
Nitrosospira	Proteobacteria	
Nitrospira	Proteobacteria	
Nitzschia	Diatoms	
Nocardia	Actinobacteria	
Noctiluca	Dinomastigota	
Nodosaria	Granuloreticulosa	
Noguchia	Spirochaetae	
Nosema	Microspora	
Nostoc	Cyanobacteria	
Notheia	Phaeophyta	
Notila	Archaeprotista	
Notommata	Rotifera	
Notosaria	Brachiopoda	
Notothylas	Anthocerophyta	

GENUS	PHYLUM	COMMON NAME
Numida	Craniata	Guinea fowl
Nyctotherus	Ciliophora	
Nymphon	Chelicerata	Sea spider
Oasisia	Pogonophora	
Obelia	Cnidaria	
Oceanospirillum	Proteobacteria	
Ochetostoma	Echiura	
Ochrolechia	Ascomycota	Cudbear lichen
Ochromonas	Chrysomonada	
Octomyxa	Plasmodiophora	
Ocultammina	Xenophyophora	
Ocyropsis	Ctenophora	Comb jelly
Oedogonium	Chlorophyta	
Oenothera	Anthophyta	Evening primrose
Oikomonas	Zoomastigota	
Oikopleura	Urochordata	Tunicate
Oligobrachia	Pogonophora	
Olpidium	Chytridiomycota	
Onchnesoma	Sipuncula	
Onychognathia	Gnathostomulida	
Oocystis	Chlorophyta	
Ooperipatus	Onychophora	
Opalina	Zoomastigota	
Ophiocytium	Xanthophyta	
Ophioglossum	Filicinophyta	Adder's-tongue fern
Ophiostoma	Ascomycota	Dutch elm disease
Ophiura	Echinodermata	Brittle star
Opisthocomus	Craniata	Hoatzin
Opisthopatus	Onychophora	
Opisthorchis	Platyhelminthes	Chinese liver fluke
Oreodoxa	Anthophyta	Cabbage palm
Ormieractinomyxon	Myxospora	
Ornithorhynchus	Craniata	Duck-billed platypus
Ortholinea	Myxospora	
Orycteropus	Craniata	Aardvark
Oryctolagus	Craniata	Domestic rabbit, Old World gray rabbit
Oryza	Anthophyta	Cultivated rice
Oscillatoria	Cyanobacteria	
Oscillochloris	Chloroflexa	
Osmunda	Filicinophyta	Interrupted fern, royal fern, cinnamon fern
Ovis	Craniata	Domestic sheep, bighorn sheep, mouflon, Dall mountain sheep, Marco Polo sheep, Laristan sheep
Oxymonas	Archaeprotista	
Oxytricha	Ciliophora	
Pagurus	Crustacea	
Pan	Craniata	Chimpanzee
Pandorina	Chlorophyta	
Papaver	Anthophyta	Opium poppy
Parachordodes	Nematomorpha	
Paracoccus	Proteobacteria	
Paragordius	Nematomorpha	
Paramarteilia	Paramyxa	
Paramecium	Ciliophora	
Paramoeba	Rhizopoda	
Paramyxa	Paramyxa	
Paranema	Discomitochondria	
Paranemertes	Nemertina	
Paratetramitus	Discomitochondria	
Parmelia	Ascomycota	Boulder lichen, crottle, shield lichen
Parvicapsula	Myxospora	
Pasteurella	Proteobacteria	
Pauropus	Mandibulata	
Pavo	Craniata	Peafowl (peacock and peahen)
Pecten	Mollusca	Scallop
Pectinatella	Bryozoa	

GENUS	PHYLUM	COMMON NAME
Pedicellina	Entoprocta	
Pedicellinopsis	Entoprocta	
Pediculus	Mandibulata	Human body louse
Pelagophycus	Phaeophyta	
Pellia	Hepatophyta	
Pellicularia	Basidiomycota	
Pelodera	Nematoda	
Pelodictyon	Chlorobia	
Pelomyxa	Archaeprotista	
Pelvetia	Phaeophyta	
Penicillium	Ascomycota	Blue mold, green mold
Penicillus	Chlorophyta	Neptune's shaving brush
Penium	Gamophyta	
Peptococcus	Endospora	
Peptostreptococcus	Endospora	
Peranema	Discomitochondria	
Perichaena	Myxomycota	
Peridinium	Dinomastigota	
Peripatoides	Onychophora	
Peripatopsis	Onychophora	
Peripatus	Onychophora	Velvet worm
Periplaneta	Mandibulata	American cockroach
Peromyscus	Craniata	Deer mouse
Peronospora	Oomycota	
Petromyzon	Craniata	Sea lamprey
Pfeisteria	Dinomastigota	
Phacus	Discomitochondria	
Phaeoceros	Anthocerophyta	Horned liverwort
Phaeocystis	Haptomonada	
Phaeothamnion	Chrysomonada	
Phallus	Basidiomycota	Common stinkhorn
Phascolarctus	Craniata	Koala
Phascolion	Sipuncula	
Phascolopsis	Sipuncula	
Phaseolus	Anthophyta	String (snap) bean, scarlet runner bean, wax bean, shell bean, mung bean, tepary bean, lima bean
Phasianus	Craniata	Ring-necked pheasant
Philodina	Rotifera	
Phoca	Craniata	Harbor seal, hair seal
Phoronis	Phoronida	
Phoronopsis	Phoronida	
Photobacterium	Proteobacteria	Luminous bacterium
Phycomyces	Zygomycota	
Phylloglossum	Lycophyta	
Physalia	Cnidaria	Portuguese man-of-war
Physarella	Myxomycota	
Physarum	Myxomycota	
Physcomitrium	Bryophyta	Urn moss
Physoderma	Chytridiomycota	
Phytomonas	Discomitochondria	
Phytophthora	Oomycota	Potato blight
Picea	Coniferophyta	White spruce, black spruce, Norway spruce
Pillotina	Spirochaetae	
Pilobolus	Zygomycota	Dung fungus, cap thrower
Pinnularia	Diatoms	
Pinus	Coniferophyta	Pitch pine, yellow pine, long-leaf pine, white pine
Pirellula (formerly *Pirella*)	Pirellulae	
Pipetta	Actinopoda	
Pisaster	Echinodermata	Ochre seastar, purple seastar, common seastar

GENUS	PHYLUM	COMMON NAME
Pisum	Anthophyta	Peas
Plagiorhynchus	Acanthocephala	
Planaria	Platyhelminthes	Freshwater planarian
Planctomyces	Pirellulae	
Planktoniella	Diatoms	
Planococcus	Deinococci	
Plasmodiophora	Plasmodiophora	Club root disease
Plasmodium	Apicomplexa	Malarial parasite
Plasmopara	Oomycota	Grape mildew
Platymonas	Chlorophyta	
Platyzoma	Filicinophyta	
Plectocolea	Hepatophyta	
Pleisiomonas	Proteobacteria	
Pleurobrachia	Ctenophora	Sea gooseberry, cat's eye
Pleurochloris	Eustigmatophyta	
Pleurotricha	Ciliophora	
Pliciloricus	Loricifera	
Plumatella	Bryozoa	
Pneumocystis	Ascomycota	Agent of cystic pneumonia
Pocheina	Rhizopoda	
Podangium	Actinobacteria	
Podarke	Annelida	
Podiceps	Craniata	Horned grebe, other grebes
Podocarpus	Coniferophyta	Yellowwood
Polyangium	Proteobacteria	
Polybrachia	Pogonophora	
Polyedriella	Eustigmatophyta	
Polykrikos	Dinomastigota	
Polymorphus	Acanthocephala	
Polymyxa	Plasmodiophora	
Polyphagus	Chytridiomycota	
Polypodium	Filicinophyta	Rock polypody, resurrection fern
Polyporus	Basidiomycota	Sulfur polyporus

GENUS	PHYLUM	COMMON NAME
Polysiphonia	Rhodophyta	
Polysphondylium	Rhizopoda	
Polystichum	Filicinophyta	Holly fern, Christmas fern
Polytrichum	Bryophyta	Haircap moss, pigeon wheat
Polyxenus	Mandibulata	Millipede
Pongo	Craniata	Orangutan
Pontosphaera	Haptomonada	
Porella	Hepatophyta	
Porocephalus	Crustacea	Pentastomid
Porphyra	Rhodophyta	Purple laver, nori
Porphyridium	Rhodophyta	
Postelsia	Phaeophyta	Sea palm
Prasinociadus	Chlorophyta	
Priapulopsis	Priapulida	
Priapulus	Priapulida	
Primula	Anthophyta	Primrose
Proales	Rotifera	
Problognathia	Gnathostomulida	
Prochlorococcus	Cyanobacteria	
Prochloron	Cyanobacteria	
Prochlorothrix	Cyanobacteria	
Procotyla	Platyhelminthes	Freshwater planarian
Prometor	Echiura	
Propionibacterium	Actinobacteria	
Prorocentrum	Dinomastigota	
Prorodon	Ciliophora	
Prosthecochloris	Chlorobia	
Prostoma	Nemertina	Freshwater nemertine
Proteromonas	Zoomastigota	
Proteus	Proteobacteria	
Protococcus	Chlorophyta	
Protoopalina	Zoomastigota	
Protopsis	Dinomastigota	

GENUS	PHYLUM	COMMON NAME
Protostelium	Myxomycota	
Prunus	Anthophyta	Peach, plum, beach plum, bitter cherry, pin cherry, black cherry, choke cherry, sweet cherry
Prymnesium	Haptomonada	
Psammetta	Xenophyophora	
Psammina	Xenophyophora	
Psammohydra	Cnidaria	
Pseudechiniscus	Tardigrada	Water bear
Pseudicyema	Rhombozoa	
Pseudobryopsis	Chlorophyta	
Pseudocharaciopsis	Eustigmatophyta	
Pseudomonas	Proteobacteria	
Pseudotrebouxia	Chlorophyta	Lichen alga
Pseudotricho-nympha	Archaeprotista	
Pseudotsuga	Coniferophyta	Douglas fir
Psilocybe	Basidiomycota	
Psilotum	Psilophyta	Whisk fern
Psoroptes	Chelicerata	Mange mite
Pteridium	Filicinophyta	Brake, bracken fern
Pteris	Filicinophyta	
Pterognathia	Gnathostomulida	
Pterosagitta	Chaetognatha	
Pterotermes	Mandibulata	Sonoran desert termite
Ptychodera	Hemichordata	Acorn worm
Puccinia	Basidiomycota	Wheat rust
Puffinus	Craniata	Shearwater
Pycnophyes	Kinorhyncha	
Pyramimonas	Chlorophyta	
Pyrodictium	Crenarchaeota	
Pyrolobus	Crenarchaeota	
Pyrosoma	Urochordata	
Pyrsonympha	Archaeprotista	

GENUS	PHYLUM	COMMON NAME
Pythium	Oomycota	
Quercus	Anthophyta	White oak, cork oak, red oak, post oak, live oak, yellow oak
Raillietiella	Crustacea	Pentastomid
Raja	Craniata	Skate
Rangifer	Craniata	Reindeer, caribou
Raphidomonas	Chrysomonada	
Rauwolfia	Anthophyta	Snakeroot
Reckertia	Chrysomonada	
Reclinomonas	Zoomastigota	
Reighardia	Crustacea	Pentastomid
Renilla	Cnidaria	Sea pansy
Reticulammina	Xenophyophora	
Reticulitermes	Mandibulata	Subterranean termite
Reticulomyxa	Granuloreticulosa	
Retortamonas	Archaeprotista	
Rhabdias	Nematoda	Roundworm of amphibians
Rhabdopleura	Hemichordata	
Rhabdosphaera	Haptomonada	
Rhea	Craniata	Greater (common) rhea
Rhipidium	Oomycota	
Rhizidiomyces	Hyphochytrio-mycota	
Rhizobium	Proteobacteria	Nodule-forming bacteria
Rhizochloris	Xanthophyta	
Rhizochrysis	Chrysomonada	
Rhizoctonia	Basidiomycota	
Rhizomyces	Ascomycota	
Rhizophydium	Chytridiomycota	
Rhizopus	Zygomycota	Black bread mold
Rhodobacter	Proteobacteria	
Rhodoferax	Proteobacteria	

GENUS	PHYLUM	COMMON NAME
Rhodomicrobium	Proteobacteria	Budding bacterium
Rhodopseudomonas	Proteobacteria	
Rhodospirillum	Proteobacteria	
Rhodymenia	Rhodophyta	Dulce or dulse
Rhopalodia	Diatoms	
Rhopalura	Orthonectida	
Riccardia	Hepatophyta	
Riccia	Hepatophyta	Slender riccia
Ricciocarpus	Hepatophyta	Purple-fringed riccia
Rickettsia	Proteobacteria	
Rickettsiella	Proteobacteria	
Ridgeia	Pogonophora	
Riella	Hepatophyta	
Riftia	Pogonophora	Vestimentiferan
Rosa	Anthophyta	Rose
Rotaliella	Granuloreticulosa	
Rothschildia	Mandibulata	
Rubrivivax	Proteobacteria	
Rugiloricus	Loricifera	
Ruminobacter	Proteobacteria	
Ruminococcus	Endospora	
Sabella	Annelida	
Saccharomyces	Ascomycota	Bakers' yeast, brewers' yeast
Saccharum	Anthophyta	Sugarcane
Saccinobaculus	Archaeprotista	
Saccoglossus	Hemichordata	
Sagitta	Chaetognatha	Arrow worm
Salamandra	Craniata	European fire salamander
Salmonella	Proteobacteria	
Salpa	Urochordata	Salp (tunicate)
Salvinia	Filicinophyta	Salvinia, water spangle
Sappinia	Myxomycota	
Saprolegnia	Oomycota	
Saprospira	Saprospirae	
Sarcina	Deinococci	
Sarcinochrysis	Chrysomonada	
Sarcoscypha	Ascomycota	Some cup fungi
Sargassum	Phaeophyta	Kelp
Scapania	Hepatophyta	Leafy liverwort
Schistosoma	Platyhelminthes	Blood fluke
Schizochytrium	Labyrinthulata	
Schizocystis	Apicomplexa	
Schizophyllum	Basidiomycota	
Scolopendra	Mandibulata	
Scolymastra	Porifera	
Scorpio	Chelicerata	Scorpion
Scytosiphon	Phaeophyta	Whip tube
Seirococcus	Phaeophyta	
Seison	Rotifera	
Selaginella	Lycophyta	Spike mosses
Selenaria	Bryozoa	
Selenidium	Apicomplexa	
Semaeognathia	Gnathostomulida	
Semipsammina	Xenophyophora	
Semnoderes	Kinorhyncha	
Sepia	Mollusca	Common sepia, cuttlefish
Septemcapsula	Myxospora	
Sequoia	Coniferophyta	California redwood
Sequoiadendron	Coniferophyta	Giant sequoia
Serratia	Proteobacteria	Blood-of-Christ bacterium
Shigella	Proteobacteria	
Siboglinum	Pogonophora	
Siliqua	Mollusca	Pacific razor clam
Simonsiella	Saprospirae	
Sinuolinea	Myxospora	
Siphonosoma	Sipuncula	
Sipunculus	Sipuncula	
Snyderella	Archaeprotista	
Solanum	Anthophyta	White potato, Jerusalem cherry, nightshade, eggplant

GENUS	PHYLUM	COMMON NAME
Solaster	Echinodermata	Purple sun star, eleven-armed sun star
Solentia	Cyanobacteria	
Sordaria	Ascomycota	
Sorex	Craniata	Masked shrew, long-tailed shrew
Sorodiscus	Plasmodiophora	
Sorogena	Ciliophora	
Sorosphaera	Plasmodiophora	
Spadella	Chaetognatha	Arrow worm
Spartina	Anthophyta	Marsh grass, salt marsh grass
Speleoperipatus	Onychophora	Velvet worm
Spengelia	Hemichordata	
Sphacelaria	Phaeophyta	
Sphaeractinomyxon	Myxospora	
Sphaerocarpos	Hepatophyta	
Sphaerotilus	Proteobacteria	Sheathed iron bacterium
Sphaeromyxa	Myxospora	
Sphaerospora	Myxospora	
Sphagnum	Bryophyta	Peat moss, sphagnum
Speciospongia	Porifera	
Sphenodon	Craniata	Tuatara
Spiradela	Anthophyta	Duckweed
Spirillum	Proteobacteria	
Spirobolus	Mandibulata	Millipede
Spirobrachia	Pogonophora	
Spirochaeta	Spirochaetae	Mud spirochete
Spirogyra	Gamophyta	
Spironympha	Archaeprotista	
Spiroplasma	Aphragmabacteria	Corn stunt disease agent
Spirostomum	Ciliophora	
Spirosymplokos	Spirochaetae	
Spirotrichonympha	Archaeprotista	
Spirulina	Cyanobacteria	
Spongia	Porifera	Bath sponge
Spongilla	Porifera	Freshwater sponge
Spongomorpha	Chlorophyta	
Spongospora	Plasmodiophora	Powdery-scab disease agent
Sporocytophaga	Saprospirae	
Sporolactobacillus	Endospora	
Sporomusa	Endospora	
Sporosarcina	Endospora	
Squalus	Craniata	Spiny dogfish shark
Stangeria	Cycadophyta	Hottentot's head
Stannoma	Xenophyophora	
Stannophyllum	Xenophyophora	
Staphylococcus	Deinococci	"Staph" infection toxic shock agent
Staurastrum	Gamophyta	
Staurojoenina	Archaeprotista	
Stelangium	Proteobacteria	
Stemonitis	Myxomycota	
Stentor	Ciliophora	
Stephanoceros	Rotifera	
Stephanonympha	Archaeprotista	
Stephanopogon	Discomitochondria	
Sterna	Craniata	Common tern, roseate tern
Sticholonche	Actinopoda	
Stigeoclonium	Chlorophyta	
Stigmatella	Proteobacteria	
Stigonema	Cyanobacteria	
Stoecharthrum	Orthonectida	
Stomatopora	Bryozoa	
Streptobacillus	Proteobacteria	
Streptococcus	Deinococci	
Streptomyces	Actinobacteria	
Stromatospongia	Porifera	
Strongylocentrotus	Echinodermata	Western purple sea urchin, giant red sea urchin, green sea urchin

GENUS	PHYLUM	COMMON NAME
Struthio	Craniata	Ostrich
Stylites	Lycophyta	
Stylonychia	Ciliophora	
Stylopage	Zygomycota	
Sulfolobus	Crenarchaeota	
Surirella	Diatoms	
Sus	Craniata	Pig, wild boar
Symbiodinium	Dinomastigota	Zooxanthella
Symperipatus	Onychophora	
Synchaeta	Rotifera	
Synchytrium	Chytridiomycota	
Synechococcus	Cyanobacteria	
Synechocystis	Cyanobacteria	
Synura	Chrysomonada	
Syracosphaera	Haptomonada	
Syringammina	Xenophyophora	
Tabellaria	Diatoms	
Tachyglossus	Craniata	
Taenia	Platyhelminthes	Tapeworm of beef and pork
Takakia	Bryophyta	
Talaromyces	Ascomycota	
Taraxacum	Anthophyta	Dandelion
Targionia	Hepatophyta	
Tatjanellia	Echiura	
Taxodium	Coniferophyta	Bald cypress
Taxus	Coniferophyta	Yew
Telomyxa	Myxospora	
Temnogyra	Gamophyta	
Tenebrio	Mandibulata	Darkling beetle (larva: mealworm)
Terebratula	Brachiopoda	
Terebratulina	Brachiopoda	
Teredo	Mollusca	Shipworm
Tetrahymena	Ciliophora	
Tetrakentron	Tardigrada	
Tetramyxa	Plasmodiophora	
Tetranchyroderma	Gastrotricha	

GENUS	PHYLUM	COMMON NAME
Tetraphis	Bryophyta	Four-tooth moss
Tetraspora	Chlorophyta	
Tevnia	Pogonophora	
Textularia	Granuloreticulosa	
Thalassema	Echiura	Spoon worm
Thalassicola	Actinopoda	
Thalassiosira	Diatoms	
Thalassocalyce	Ctenophora	Comb jelly
Thallochrysis	Chrysomonada	
Thaumatomastix	Chrysomonada	
Thecamoeba	Rhizopoda	
Themiste	Sipuncula	
Theobroma	Anthophyta	Cacao
Thermoactinomyces	Actinobacteria	
Thermococcus	Crenarchaeota	
Thermodesulfo-rhabditis	Proteobacteria	
Thermofilum	Crenarchaeota	
Thermomonospora	Actinobacteria	
Thermoplasma	Crenarchaeota	
Thermoproteus	Crenarchaeota	
Thermosipho	Thermotogae	
Thermotoga	Thermotogae	
Thermozodium	Tardigrada	
Thermus	Deinococci	
Thiobacillus	Proteobacteria	
Thiobacterium	Proteobacteria	
Thiocapsa	Proteobacteria	
Thiocystis	Proteobacteria	
Thiodictyon	Proteobacteria	
Thiomicrospira	Proteobacteria	
Thiopedia	Proteobacteria	
Thiosarcina	Proteobacteria	
Thiospirillum	Proteobacteria	
Thiospira	Proteobacteria	
Thiothece	Proteobacteria	
Thiovulum	Proteobacteria	
Thraustochytrium	Labyrinthulata	

GENUS	PHYLUM	COMMON NAME
Thuja	Coniferophyta	Arbor vitae
Thyone	Echinodermata	Sea cucumber
Tilopteris	Phaeophyta	
Tinamus	Craniata	Gray tinamou, solitary tinamou, black tinamou, great tinamou, white-throated tinamou
Tjalfiella	Ctenophora	
Tmesipteris	Psilophyta	
Tokophrya	Ciliophora	Suctorian
Tolypella	Chlorophyta	Stonewort
Tortula	Bryophyta	Wall moss, twisted moss
Torulopsis	Ascomycota	
Toxoplasma	Apicomplexa	
Trachelomonas	Discomitochondria	
Trachypheus	Chelicerata	
Tradescantia	Anthophyta	Spiderwort
Tranes	Mandibulata	Weevil
Trebouxia	Chlorophyta	Lichen alga
Tremella	Basidiomycota	Orange jelly fungus
Trentonia	Chrysomonada	
Treponema	Spirochaetae	Yaws, syphilis agent
Triactinomyxon	Myxospora	
Tribonema	Xanthophyta	
Trichechus	Craniata	Manatee
Trichinella	Nematoda	Trichina worm
Trichodorus	Nematoda	
Trichomitus	Archaeprotista	
Trichomonas	Archaeprotista	Trich
Trichonympha	Archaeprotista	Wood-eating hypermastigote
Trichophyton	Ascomycota	Athlete's foot, cattle ringworm, tinea cruris
Trichoplax	Placozoa	Trichoplax
Tricoma	Nematoda	
Tridacna	Mollusca	Giant clam
Trididemnum	Urochordata	Tunicate
Trifolium	Anthophyta	White clover, red clover, yellow clover
Trilospora	Myxospora	
Tripedalia	Cnidaria	Sea wasp
Triticum	Anthophyta	Bread wheat, durum (pasta) wheat
Trypanochloris	Xanthophyta	
Trypanosoma	Discomitochondria	Trypanosome
Tsuga	Coniferophyta	Canadian hemlock, Carolina hemlock
Tuber	Ascomycota	Truffle
Tubifex	Annelida	"Red worm" sold as aquarium fish food
Tubiluchus	Priapulida	
Tubipora	Cnidaria	Organ-pipe coral
Tubulanus	Nemertina	
Tubularia	Cnidaria	Oaten-pipes hydroid
Tubulipora	Bryozoa	
Tulipa	Anthophyta	Tulip
Turbanella	Gastrotricha	
Typha	Anthophyta	Cattail
Uca	Crustacea	Fiddler crab
Udotea	Chlorophyta	
Ulkenia	Labyrinthulata	
Ulothrix	Chlorophyta	
Ulva	Chlorophyta	Sea lettuce
Umbilicaria	Ascomycota	Rock tripe, toad-skin lichen
Unicapsula	Myxospora	
Unicauda	Myxospora	
Urechis	Echiura	Fat innkeeper

GENUS	PHYLUM	COMMON NAME
Urnatella	Entoprocta	
Urodasys	Gastrotricha	
Uroleptus	Ciliophora	
Urospora	Chlorophyta	
Urosporidium	Haplospora	
Usnea	Ascomycota	Old-man's beard
Ustilago	Basidiomycota	Corn smut, corn mushroom
Vacuolaria	Chrysomonada	
Vahlkampfia	Discomitochondria	
Vairamorpha	Microspora	
Vaucheria	Xanthophyta	
Velamen	Ctenophora	Venus's girdle
Velella	Cnidaria	By-the-wind sailor
Vema	Mollusca	
Verrucaria	Ascomycota	Marine lichen
Verticillium	Ascomycota	
Vibrio	Proteobacteria	
Viburnum	Anthophyta	Arrow-wood, wild raisin
Vischeria	Eustigmatophyta	
Vitis	Anthophyta	Fox grape, muscadine grape, frost grape, other grapes
Volvox	Chlorophyta	
Vorticella	Ciliophora	
Vulpes	Craniata	Kit fox, red fox
Waddycephalus	Crustacea	Pentastomid
Wardia	Myxospora	
Warnowia	Dinomastigota	
Welwitschia	Gnetophyta	
Wolbachia	Aphragmabacteria	
Wolffia	Anthophyta	Watermeal, duckweed
Woronina	Plasmodiophora	
Wuchereria	Nematoda	
Xanthomonas	Proteobacteria	
Xenopus	Craniata	African clawed frog
Xenorhabditis	Proteobacteria	Luminescent nematode bacterium
Xyloplax	Echinodermata	Sea daisy
Yersinia	Proteobacteria	Plague bacterium, agent of pestilence
Yucca	Anthophyta	Beargrass, Spanish bayonet, yucca, Joshua tree
Zamia	Cycadophyta	Coontie, Seminole bread
Zanardinia	Phaeophyta	
Zea	Anthophyta	Maize, Indian corn, teosinte
Zelleriella	Zoomastigota	
Zenkevitchiana	Pogonophora	
Zonaria	Phaeophyta	
Zostera	Anthophyta	Eel grass
Zygacanthidium	Actinopoda	
Zygnema	Gamophyta	
Zygogonium	Gamophyta	
Zymomonas	Proteobacteria	Palm wine bacterium, pulque bacterium, tequila bacterium

GLOSSARY

Terms generally restricted to certain kingdoms or phyla are indicated by abbreviations in parentheses. However, the terms are not always restricted to those kingdoms and phyla and do not necessarily apply to all members of the kingdoms and phyla.

aboral Away from the mouth

abyss Oceanic habitat at greatest depths

acanthella (pl. **acanthellae**) Larval stage of acanthocephalan that follows the acanthor larva (A-12)

acanthor Larval stage of acanthocephalan that follows the embryo (A-12)

acervulus (pl. **acervuli**) Mat of hyphae giving rise to conidiophores closely packed together to form a bedlike mass (F-3)

aciculum (pl. **acicula**) Chitinous support rod of polychaete annelid parapodia (A-22)

acoel Flatworm that lacks a gut (A-5)

acoelomate Animal lacking a coelom (A-5 through A-9, A-16)

acraniate Chordate that lacks a cranium or skull (A-35, A-36)

acritarch Spherical microfossil, possibly the test of a shelled ameba (Pr-3) or chrysomonad (Pr-13)

actin One of the two major classes of muscle protein (A); also found in filaments that participate in motility of protoctists (Pr)

actinopod Pseudopod containing filaments and microtubules (Pr-27)

actinospore Spore of actinobacteria (actinomycetes, B-5)

actinotroch Free-swimming feeding larva of phoronids (A-31)

adhesive disc Attachment structure of certain rotifers (A-13)

adhesive organ Organ for attachment, for example, in some platyhelminthes (A-5)

adhesive sac Structure that glues ectoproct larva to substrate (A-29)

adhesive tube Secretory locomotory structure of kinorhynch (A-14)

adhesorium Adhesive organelle of plasmodiophoran plant parasites (Pr-19)

adventitious root Root growing from an atypical location, as on a stem or leaf (Pl-7 through Pl-12)

aerobe Organism requiring free (molecular) oxygen (A, F)

aerobic Metabolic mode requiring gaseous oxygen

aerobiosis Living in the presence of oxygen

agamete Asexual reproductive cell (Pr-4)

agamogony Series of nuclear or cell divisions giving rise to individuals (agamonts) that are not gametes or are not capable of forming gametes (Pr-4)

agamont Life-cycle stage that does not produce gametes (Pr-4)

agar Material extracted from seaweed and used as base for bacterium and fungus culture medium (B, F) and in prepared foods and pharmaceuticals

alimentary tract Digestive or gastrointestinal tract; a tubular passage that extends from mouth to anus and whose function is to digest and absorb nutrients (A)

alkaloid Organic, nitrogen-containing ring compounds produced by fungi and plants that taste bitter; some alkaloids such as nicotine and amanita mushroom alkaloids are poisonous to humans (F, Pl)

allelochemical Defensive secretion of certain plants (for example, *Artemisia,* sagebrush) that is toxic to other plant species, clearing the immediate environment of potential competitors (Pl)

allophycocyanin Type of phycobiliprotein; blue water-soluble extract; found in cyanobacteria (B-5), cryptomonads (Pr-11), and rhodophytes (Pr-25)

alternation of generations Reproductive cycle in which haploid (N) phase alternates with diploid (2N) phase (Pr, Pl)

alveolus (pl. **alveoli**) Small sac or vesicle, characteristic of alveolate protoctists (Pr-7 through Pr-9)

amastigote Single-celled organism that lacks undulipodia [for example, foram (Pr-4), haptomonad (Pr-10)]

ambulacrum (pl. **ambulacra**) In echinoderms (A-34), a tube-foot lined, ciliated groove leading down the center of each arm (starfish) or over the test (sea urchin); conducts food to the mouth

amebocyte Cell having ameboid locomotion; in sponges, a cell that differentiates into eggs, spongin-secreting cells, others—see *archaeocyte;* in chelicerates, a cell involved in blood clotting (A-2, A-15, A-19, A-28)

ameboid Shaped like an ameba; having a cell form with ever-changing cytoplasmic protrusions, or pseudopods (Pr, A)

amebomastigote Undulipodiated ameba [for example, *Paratetramitus jugosus* (Pr-30)]

amphiblastula (pl. **amphiblastulae**) Hollow, ciliated larva of certain sponges (A-2)

amphiesmal vesicle Membranous sac underlying test; thought to be responsible for test production in dinomastigotes (Pr-7)

ampulla (pl. **ampullae**) Contractile bulb that moves seawater into echinoderm tube foot (A-34)

anaerobic Metabolic mode requiring the absence of gaseous oxygen

anal sac Structure that moves fluid from coelom to gut of echiuran (A-24)

analogous Of structures or behaviors that have evolved convergently; similar in function but different in evolutionary origin

anastomosis (pl. **anastomoses**) The process of linking branches, filaments, or tubes to form networks

ancestrula (pl. **ancestrulae**) Ancestral zooid of ectoproct colony (A-29)

aneuploid Deviation from the normal haploid (N) or diploid (2N) number of chromosomes (for example, 2N + 1, 2N – 2)

angiosperm One species of a plant group having seeds borne in a fruit; also called flowering plants (Pl-12)

anisogamete Gamete that differs in form or size from others of the species (A)

anisogamy Formation of gametes that differ in size or morphology

annulus (pl. **annuli**) Ring, for example, on the stems of certain mushroom species (F-2); specialized cells on fern sporangia (Pl-7)

anoxic Without molecular oxygen

antenna (pl. **antennae**) Sensory appendage on the head for detecting sound, touch, etc. (A-20, A-21, A-28)

antennal gland Gland of water excretion in crustaceans—see *green gland* (A-21)

anterior Toward the front or head end of a cell or organism

anther Pollen-bearing part of the stamen (pollen is the immature male gametophyte of angiosperms) (Pl-12)

antheridium (pl. **antheridia**) Multicellular male sexual organ; the sperm-producing gametangium of plants other than seed plants (Pl-1 through Pl-7, Pr-17, Pr-25)

antherozoid Motile male gamete of the Monoblepharidales (Pr-29)

anthocyanin Red or blue pigment in plant-cell sap and in certain algae (Pr, Pl)

Anthozoa Class of cnidarians (A-3)

antibiotic Substance produced by organisms, typically fungi or bacteria, that injures or kills other organisms, typically bacteria, or prevents their growth (B, F)

aphanoplasmodium (pl. **aphanoplasmodia**) One of three types of plasmodium produced by myxomycotes (Pr-6), intermediate between phaneroplasmodia and protoplasmodia in size and complexity; formed by most members of the class Stemonitomycetidae

apical cell Cell at top; for example, the two unciliated cells of the rhombozoan infusoriform larva (A-7)

apical pore Opening at the apex of a structure or organism

apical tuft Tuft of cilia on a trochophore larva, for example, of annelids (A-22)

apophysis Swollen region

apothecium (pl. **apothecia**) Open ascoma, a base for the spore-bearing structures called asci (F-3)

aragonite A needle crystalline form of the mineral calcium carbonate; component of shells, algal cell walls, molluscs, sponge spicules, etc. (B, Pr, A, Pl)

archaeocyte Amebocyte that digests food, eliminates waste, secretes spongin or spicules, and produces gametes in sponges (A-2)

archegonium (pl. **archegonia**) Multicellular female sexual organ; egg-producing gametangium of a plant other than a seed plant (Pl-1 through Pl-7, Pr-17, Pr-25)

Arthrophyta Alternate phylum name for Sphenophyta (horsetails; Pl-6)

articulation Hinge connecting valves (shells) of articulate brachiopod (A-30)

ascogenous Giving rise to ascospores, in ascomycetes (F-3)

ascoma (pl. **ascomata**) Multihyphal structure of ascomycetes that bears asci; formerly ascocarp (F-3)

ascospore Spore formed by karyogamy and meiosis and contained in an ascus (F-3)

ascus (pl. **asci**) A saclike structure generally containing a definite number of ascospores (F-3)

asexual Reproductive process that does not include union of gametes (for example, reproduction, such as fission or budding, in which offspring has a single parent); parthenogenesis; term best restricted to fungi, plants, animals

astropyle Nipplelike aperture projecting from the central capsule of some actinopods (radiolarians), Phaeodorina (Pr-27)

ATP Adenosine triphosphate; molecule that is the primary energy carrier for cell metabolism and motility

atriopore Pore through which seawater moves from the atrium of a cephalochordate into the sea (A-36)

atrium (pl. **atria**) In entoprocts (A-18), the space within the tentacles; in tunicates (A-35), the chamber between tunic and pharynx; in cephalochordates (A-36), the space around the pharyngeal gill slits into which the respiratory current flows; in craniate chordates (A-37), the vestibule into which blood enters the heart

auricle In certain comb jellies, ciliated lobes (A-4). In molluscs, heart chambers (A-26)

autogamy Union of two nuclei, both derived from a single parent nucleus (Pr)

autotroph An organism that grows and synthesizes organic compounds from inorganic compounds by using energy from sunlight or from oxidation of inorganic compounds

auxospore A diatom cell, that has been released from its rigid test, often zygotic products of fertilization (Pr-16)

axenic Growth in pure culture—that is, culture in the complete absence of members of other species

axial cell Cylindrical cell that surrounds axoblast cells of rhombozoans (A-7)

axial filament Solid long thin structure that is aligned longitudinally and more or less centrally in a cell or organelle

axil In seed plants (Pl-8 through Pl-12), the point at which the leaf stalk (petiole) joins the stem; origin of lateral buds

axoblast Reproductive cell of rhombozoan (A-7)

axon Portion of nerve cell that transmits nerve impulse away from nerve cell body (A-25)

axoneme A tubule or shaft of tubules extending the length of an undulipodium, pseudopod, or axopod (Pr)

axoplast Central granule; in actinopods (Pr-27), microtubule organizing center from which axonemes of axopods arise; devoid of inner differentiation

axopod Permanent pseudopod stiffened by a microtubular axoneme (Pr-27)

axostyle Axial motile structure of members of the Zoomastigina (Pr-30); composed of a patterned array of microtubules and their crossbridges

backbone Dorsal spine of vertebrates (A-37)

bacteriophage Virus that reproduces in bacteria

bacteroid Transformed bacterium in a root nodule; site of nitrogen fixation

balancer In aboral sense organ, cilia tuft that supports statolith of comb jelly (A-4)

bark Tissues external to the vascular cambium, including phloem, in a woody shrub or tree (Pl-9 through Pl-12)

basal apparatus, basal body (1) Kinetosome (Pr, A, Pl); (2) thin cylindrical plates found at the base of bacterial flagella (B); see Figure I-3

basal disc Attachment organ of some entoprocts (A-18)

basal plate Part of feeding structure of gnathostomulid worm (A-6)

basidiogenesis In basidiomycetes, formation of basidia that bear basidiospores, as in mushroom or puffball (F-2)

basidioma (pl. **basidiomata**) Reproductive structure of basidiomycotes that consists of packed hyphae; formerly basidiocarp (F-2)

basidiospore Spore resulting from karyogamy and meiosis and borne on the outside of a basidium (F-2)

basidium (pl. **basidia**) In basidiomycote fungi, reproductive cell in which nuclear fusion, meiosis, and spore production take place, as in mushrooms (F-2)

benthic Dwelling in or on sea substrate (A-4, A-5, A-7, A-9, A-13, A-25, A-26, A-32, A-35)

bilateral symmetry Anatomical arrangement in which the right and left halves of an organism or part are approximately mirror images

bilharzia Schistosomiasis; see *bilharziasis* (A-26)

bilharziasis Schistosomiasis, a severe disease of human beings; it is caused by infection with blood flukes (trematodes, A-5) and transmitted by snails (A-26)

binary fission Process of cell division in which a single parent cell divides to form two identical offspring cells; see *fission*

biocoral Substitute bone material derived from coral; biomaterial used for bone graft (A-3)

biogenic Produced by living organisms or their remains

bioluminescence Light generated biochemically and emitted by organisms [for example, *Photobacterium* (B-3), dinomastigotes (Pr-7), fireflies (A-20), and some fish (A-37)]

biomass Total mass of live organisms in a given area

biomaterial Material that originates from natural organism, such as biocoral (A-2)

biosphere The part of Earth's volume that is occupied by living organisms; Earth's surface

biosynthesis Multienzyme-catalyzed chemical reactions that form chemical compounds in live organisms

biota The sum total of all organisms alive today (B, Pr, A, F, Pl)

bioturbation Reworking of soil or other sediment by a living organism such as a worm (A-22, A-23)

bioturbator Organism, such as polychaete annelid, that reworks sediment (A-22)

biradial Organism with bilateral as well as radial symmetry (A-4)

biramous Having two branches (for example, appendage of crustacean) (A-21)

bisexual Having male and female reproductive organs; see *hermaphroditic* (A, Pl-5, Pl-6, Pl-7)

bitunicate In ascomycetes, ascus having a double wall (F-3)

black smoker Submarine vent that emits black smoke; same as hydrothermal vent (A)

bladder Hollow structure that collects liquid (for example, urinary bladder) (A-13, A-19, A-21, A-28, A-37)

blastocoel Cavity of the blastula (A)

blastopore Opening connecting the cavity of the gastrula stage of an embryo with the outside (A); represents the future mouth of some animals (protostomes, A-19 through A-28) and the anus of others (deuterostomes, A-32 through A-37)

blastula (pl. **blastulae**) Animal embryo after cleavage and before gastrulation; usually a hollow, liquid-filled sphere, the walls of which are composed of a single layer of cells

blastulation Formation of animal blastula embryo (A)

book gill In chelicerate, gas exchange organ shaped like pages of book (A-19)

book lung In arachnids, internal book gills (A-19)

bothrosome Structure at the labyrinthulid or thraustochytrid membrane that produces new membrane, sequesters calcium, and filters cytoplasm for the production of the proteinaceous extracellular slime net matrix (Pr-18); also called a sagenogenosome

bract Modified, sometimes reduced, leaf on gnetophyte

(Pl-11) or conifer (Pl-10) cone; modified, often colored, leaf beneath a flower or flower cluster (Pl-12)

branchial plume Tentacle crown of vestimentiferans (A-25)

bridle Pair of oblique ridges of thickened cuticle on the mesosome of pogonophorans (A-25)

brood To protect offspring on body of parent (A-18, A-21, A-22, A-25, A-26, A-29, A-30, A-31, A-34)

bryophyte Mosses, hornworts, and liverworts (Pl-1 through Pl-3)

buccal canal In loriciferans, anterior extrusible part of digestive tract into which the mouth opens (A-17)

buccal pouch Diverticulum from the mouth into the proboscis of hemichordate (A-33)

buccal tentacle In cephalochordates, tentacle beside the mouth; also called oral cirrus (A-36)

buccal tube In tardigrades, the anteriormost part of the foregut, between mouth and pharynx (A-27)

bud Projection that develops into a new plant, flower, fungus, or animal (A-1, A-3, A-22, A-29, A-31, A-33, A-35, F-3, Pl-1, Pl-8, Pl-12)

budding Asexual reproduction by outgrowth of a bud from a parent cell or body (A, F, Pl)

Burgess shale Middle Cambrian rocks in British Columbia, Canada; especially rich in fossils of soft-bodied animals from approximately 530 million years ago

bursa (pl. **bursae**) Saclike cavity; surface slit (A-34)

caecum (pl. **caeca**) Cavity, open at one end, usually part of the gastrointestinal tract (A)

calcareous Composed of calcium carbonate (A-2)

calcite A stable form of the mineral calcium carbonate; component of brachiopod shells (A-30), skeleton of echinoderms (A-34), dinomastigote cysts, shells of some amebas, etc. (B, Pr, A, Pl)

calyx Flower part; the sepals, the cup-shaped outer series of two series of floral leaves (Pl-12); the cup-shaped structure of a crinoid or of an entoproct (A-18, A-34)

cambial Composed of cambium, a meristem tissue that gives rise to new cells (Pl-4, Pl-6 through Pl-12)

cambium (pl. **cambia**) Cylindrical sheath of dividing cells that produce phloem and xylem (Pl-4, Pl-6 through Pl-12)

canopy Tree crowns that shade the forest floor (Pl)

capillary Minute blood vessel that interconnects veins and arteries (A-26)

capillary action Combined adhesion, cohesion, and surface tension that draws liquid along a tube such as a plant vessel (Pl)

capillitium System of sterile threads in the spore-bearing structure of true slime molds (Pr-6)

capsule Sporangium of moss, hornwort, or liverwort (Pl-1 through Pl-3)

carapace Hard covering on the back of turtles and arthropods (for example, crustaceans) (A)

carbohydrate Organic compound that consists of long chain of carbon atoms with hydrogen and oxygen attached in a 2:1 ratio (for example, cellulose, sugars)

carboxysome Organelle inside plastids; thought to harbor the CO_2-fixing enzyme ribulosebisphosphocarboxylase

cardiac stomach The stomach closer to the mouth [for example, in echinoderms (A-34)]

carnivore An organism that obtains food by eating live animals (A, F, Pl); compare *herbivore*

carnivorous Flesh-eating plant, fungus, or animal (A-2, A-15, F, Pl)

carotenoid Member of a group of red, orange, and yellow hydrocarbon pigments found in plastids (Pr, Pl)

carpel Flower part; structure enclosing an ovule of an angiosperm; typically divided into ovary, style, and stigma (Pl-12)

carpospore Meiotically produced spore of rhodophytes (Pr-25)

carposporophyte Diploid red algal organism (Pr-25) produced after fertilization; a phase characterized by the presence of carposporangia

cartilage Elastic, tough form of connective tissue (A-35, A-36, A-37)

cartilaginous Composed of cartilage, as in skeleton of cartilagenous fish (A-37)

casting Excrement of worms (A-22)

catabolism Metabolic breakdown of organic compounds to release energy and endproducts

catalase Enzyme that catalyzes the breakdown of hydrogen peroxide to water and oxygen

catkin In many flowering plants (for example, willows, poplars, birches; Pl-12), a dangling type of inflorescence bearing male (pollen-producing) flowers

caudal Pertaining to the tail region; compare *oral* (A)

caudal lobe Lobe in the tail region (A-11)

cell plate Phragmoplast; membranous structure that forms at the equator of the spindle in cell division and is the site of formation of the new cell wall between the daughter cells (Pr, Pl)

cell wall Structure external to the plasma membrane produced by cells; generally rigid and composed primarily of cellulose and lignin in plants (Pl), chitin in fungi (F), and peptidoglycans (diaminopimelic acid, muramic acid, and peptides) in bacteria (B); it is absent (A, Pr) or of various composition in protoctists (Pr)

cellulose Polysaccharide made of glucose units; chief constituent of the cell wall in all plants (Pl) and chlorophytes (Pr-28)

cement gland In acanthocephalans (A-12), a gland that secretes a cement into the vagina to close it off; in gastrotrichs (A-16) and rotifers (A-13), it secretes a substance by which the animals temporarily attach themselves to objects

central capsule Spherical structure that encloses the central part of cells of actinopods (Pr-27)

central cell In angiosperms, the binucleate cell that is formed in female flower tissue and contributes two of the three nuclei that form endosperm tissue (Pl-12)

centriole Small barrel-shaped organelle seen at each pole of the spindle, formed during cell division (A, some Pr), homologous to kinetosome; see Figure I-3

centromere Kinetochore; a proteinaceous structure at a constricted region of a chromosome; holds sister chromatids together and is the site of attachment of the microtubules forming the spindle fibers in cell division; when a distinction is made, kinetochore refers to visible structure whereas centromere is inferred from genetic behavior (Pr, A, Pl)

centroplast Single, central microtubule-organizing center from which axonemes of axopods arise in certain actinopods (Pr-27); a tripartite disc consisting of an electron-lucent exclusion zone and interaxonemal substance sandwiched between two caps of electron-dense material

cephalic hood Body wall that can be drawn over the head of arrow worm (A-32)

cephalic lobe Anterior part of the protosome region of Pogonophora (A-25)

cephalic pit Depression on the head of onychophoran and used in reproduction (A-28)

cephalic shield In pterobranch hemichordates, a muscular structure used in locomotion (A-33)

cephalic slit On the head of nemertine, slit of unknown function (A-9)

Cephalorhyncha Phylum that includes priapulids (A-15), nematomorphs (A-11), kinorhynchs (A-14), and loriciferans (A-17)

cephalothorax The fused head and thorax of some arthropods (A-19)

cephalum Head (A)

cercaria One of several types of trematode (fluke) larvae (A-5)

cerebral organ Sense organ of unknown function in nemertines (A-9)

chelicera (pl. **chelicerae**) In some arthropods, the first pair of appendages (for example, those used by arachnids for seizing and crushing prey) (A-19)

chemoautotrophic Requiring carbon dioxide as well as electrons from inorganic donor for biosynthesis (A-25)

chemoheterotroph Organism requiring compounds from the environment as sources of energy and electrons

chemolithoautotrophic Requiring carbon dioxide and electrons from inorganic donor for biosynthesis (A-25)

chemoorganotroph Organism requiring organic compounds as a source of both energy and carbon

chemosynthesis Production of organic compounds by using inorganic ones as the source of both carbon and energy

chert Fine-grained silica rock that forms in sediment

chitin Tough, resistant, nitrogen-containing complex polysaccharide, a polymer of glucosamine; component of exoskeletons (A) and cell walls (Pr, F)

chlamydospore Asexual spherical structure of fungi or funguslike protoctists originating by differentiation of a hyphal segment (or segments); used primarily for perennation (overwintering), not for dissemination (Pr, F)

chlorocruorin Iron-containing, green respiratory pigment (A-22)

chlorophyll Green photosynthetic pigment that is receptor of light energy for photosynthesis and that includes magnesium held in a porphyrin ring; in plants, algae, and photosynthetic bacteria (B, Pr, Pl)

chloroplast Green plastid, a membrane-bounded photosynthetic organelle containing chlorophylls *a* and *b* (Pr-10, Pr-12, Pr-26, Pr-28, Pl)

choanocyte Collared cell having undulipodia (Pr-8, A-2)

choaste Ribbed lorica (Pr-30)

cholesterol Organic molecule in cell membrane; biosynthetic precursor of steroids; in animal fat, bile, blood

chordate Animal having a notochord; dorsal, hollow nerve cord; pharyngeal gill slits (A-35, A-36, A-37)

chromatid Longitudinal half of a chromosome; seen when chromosomes appear during prophase of mitosis in eukaryotes

chromatin Material of which chromosomes are composed; made of nucleic acids and protein, it stains deep red with Feulgen reagent

chromatophore In photosynthetic prokaryotes, a vesicle containing photosynthetic pigment (B)

chromoneme DNA strands (B, Pr-7)

chromosome Intranuclear organelle made of chromatin; visible during cell division; chromosomes contain most of a cell's genetic material (Pr, A, F, Pl)

chrysolaminarin vesicle Membranous intracellular sac filled with oil (Pr-13, Pr-14, Pr-16, Pr-17)

chrysoplast Yellow plastid, the membrane-bounded photosynthetic organelle of chrysomonads (Pr-13)

cilium (pl. **cilia**) Short undulipodium, intracellular but protruding organelle of motility (Pr, A, Pl); see *flagellum* and Figure I-3

cingulum Plate of the equatorial groove of a dinomastigote test (Pr-7)

cirrus (pl. **cirri**) In lancelets (A-36), buccal tentacles; in annelid worms (A-22), modified parapodia (feet); tuft-shaped organelle formed from fused cilia (Pr-8)

clamp connection In basidiomycetes, connection in which hyphae of complementary mating types fuse, their nuclei divide and are shuffled so that each cell of the hypha then contains two dissimilar nuclei (F-2)

cleavage Successive cell divisions of a zygote that form the blastula (A)

cloaca Exit chamber from gastrointestinal tract; it also serves as the exit for the reproductive and urinary systems (A)

clone Genetically identical individuals, as a clone of horsetail plants (A, Pl-6)

closed circulation Circulatory system in which blood flows through blood vessels rather than open sinuses (A-22, A-26, A-36, A-37)

closed vascular system Closed circulatory system (A-23, A-24)

clypeus In some insects, a median plate that connects the labrum (upper lip) to the head (A-20)

cnidoblast Cell within which nematocyst forms (A-3)

cnidocil Exposed filament that triggers discharge of nematocyst (A-3)

coccolithophorid Haptomonad (Pr-10) bearing coccoliths, small, numerous surface plates made of calcium carbonate; often abundant fossils in chalk

coccus (pl. **cocci**) Spherical organism (B, Pr)

cocoon Sheltering wrap formed by developmental stage of flatworm, annelid, or insect (A-5, A-20, A-22)

coelenterate Cnidarians and comb jellies (A-3, A-4)
coelom Body cavity lined on all surfaces by mesoderm (A-15, A-19 through A-37)
coelomate Animal having a coelom (A-9 [?], A-15, A-19–A-37)
coelomic Having a body cavity lined with peritoneum (coelom) between the outer body wall and the gut (A-25)
coelomopore Pore through which eggs exit from ectoproct (A-29)
coenocytic Of a mass of cytoplasm containing nuclei but lacking cell membranes or walls (Pr, F, Pl); see *syncytium*
cofactor Nonprotein substance that enzymes require to function; some cofactors are called coenzymes, others are metal ions
cold seep Submarine vent that emits cold water (A)
collagen Fibrous connective tissue in bone and cartilage (A-11, A-23, A-34)
collar Structure that encircles anterior end of hemichordate (A-33)
colloblast Adhesive cell of comb jelly; also called lasso cell (A-4)
colonial Refers to genetically identical organisms or cells that live in a permanent association (A-2, A-3, A-29, A-33, A-35)
colony Permanent association of organisms or cells that are genetically identical (A-18); cells or organisms of the same species living together in permanent but loose association
columella The central column of a gastropod shell, around which the spiral architecture develops
comb plate Ciliated body plate used by ctenophores for swimming; also called a ctene (A-4)
commensal Refers to species that live closely associated, one of which benefits from the relation and the other not harmed by it (A)
companion cell Small, specialized cell found next to conducting sieve tube in the phloem of vascular plants (Pl-4, Pl-6 through Pl-12)
compound cone Type of cone in which sporangium attaches indirectly by a sterile bract to the cone axis of female conifer (Pl-10) and of both sexes of gnetophyte (Pl-11) cones; compare *simple cone*
compound eye Visual organ composed of a number of simple eyes (for example, insect eye) (A-20, A-21)
compound leaf Leaf having a blade divided into leaflets, as black walnut leaf; compare *simple leaf* (Pl)
conceptacle Small saclike structure bearing gametangia, gamete-producing organs (Pr-17)
conducting cell Cell that transports fluid within a plant body [for example, water-conducting cells of moss (hydroid; Pl-1) and phloem (Pl-4 through Pl-12)]
cone Reproductive structure composed of modified leaves (bracts, scales) that produces spores, ovules, or pollen; ovule-bearing modified leaves grouped into a cone, often on a stem tip; see *strobilus* (Pl-4, Pl-6, Pl-8 through Pl-11)
cone scale Part of a cone that is a modified leaf bearing ovule or microspore (Pl-10, Pl-11)
conicopseudopod Conical pseudopod, characteristic of acanthamebas (Pr-3)
conidiophore Structure that bears conidia (F-1, F-3)
conidiospore Spore formed asexually, usually at the tip or side of a hypha (F)
conidium (pl. **conidia**) Fungal spore—usually multinucleate, always produced in absence of sexual fusion—not contained in a sporangium but borne on a hypha (F-1 and F-3)
conifer Cone-bearing tree or shrub with needle or blade-like leaves and with cones having exposed seeds (Pl-10)
coniferous forest Forest principally of cone-bearing trees and shrubs (Pl-10)
conjugation The transmission of genetic material from a donor to a recipient cell (B); fusion of nonundulipodiated gametes or gamete nuclei (Pr-25, Pr-26, F)
connecting canal Canal leading from the radial canal to the tube feet (A-34)
conodont Paleozoic fossil once interpreted as extinct cyclostome (for example, lamprey) tooth (A-37) or invertebrate remains (A-6); now classified as (fossil) animal phylum Conodonta

contractile vessel In sipunculans, structure in which tentacular fluid is stored (A-23)
convergence, convergent evolution Independent evolution of similar structures for similar functions in taxa that are not directly related
copulation Mating of animals (A-6)
copulatory spicule Secreted cuticular rod used in mating (A)
coralloid root In cycads (Pl-8), specialized root harboring green tissue layers composed of cyanobacteria (B-7)
corolla Flower part; the petals, the inner of two series of floral leaves (Pl-12)
corona Crown or crown-shaped structure
coronate larva Crown-shaped larva of ectoproct (A-29)
cortex Outer layer of stem or root internal to the epidermis and external to vascular tissue (Pl); outer layer of organ (for example, kidney cortex) (A); protein complexes (Pr-8); outer layer of fungus in a lichen (F-3); see *medulla*
cortical layer Cortex or outer layer
costa Highly motile intracellular rod, nonmicrotubular (Pr-30)
cotyledon Leaflike structure of seeds; stores food for dicot embryo (Pl-11, Pl-12) and absorbs food for monocot embryo
craniate Having a skull or cranium (for example, craniate chordate; A-37)
cranium (pl. **crania**) Skull, either bony or cartilaginous, that encloses the brain (A-37)
crop Food-storage region of the digestive tract (A-22, A-24)
cross fertilize Fertilization of one individual organism by another (A-22)
crustose Of low-lying and crusty lichens (F-3)
cryptobiosis Temporary suspension of metabolic activity induced by adverse environmental condition such as desiccation (for example, developmental arrest of rotifer) (A-13, A-27)
crystalline style Enzyme-releasing organ of the digestive system of bivalve molluscs (A-26); rotated by undulipodia, it grinds algal food
ctene Comb plate, organ of locomotion of comb jelly (A-4)
ctenidium (pl. **ctenidia**) Gill of a mollusc (A-26)
ctenoid scale In bony fish, thin scale that forms internally in the skin and has a toothed rim (A-37)
cuitlacoche Commercial name for spore masses of corn smut, a basidiomycete infection of corn consumed by humans (F-2); also called corn mushrooms
cupule Little cup on the thallus in which liverworts reproduce gemmae (Pl-2)
cuticle Outer layer or covering, usually composed of metabolic products (wax) rather than of cells (A, Pl)
cyanobacterium (pl. **cyanobacteria**) Formerly blue-green alga (B)
cycloid scale In bony fish, thin scale that forms internally in the skin and has a smooth outer margin (A-37); compare *ctenoid scale*
cyclosis The streaming of cytoplasm within a cell
cydippid Free-swimming larva of ctenophore (A-4)
cyst Encapsulated form (often a dormant stage) of one or several organisms; formed in response to extreme environmental conditions (Pr, A); in *Pneumocystis carinii,* "cysts" are fungus spores (F-3)
cystacanth Resting stage of acanthocephalan (A-12)
cystocarp Carposporophyte, the asexual generation of Floridean rhodophytes (Pr-25); it grows parasitically on the female gametophyte
cystosorus (pl. **cystosori**) Structures into which cysts may be united, the presence and morphology of which are of taxonomic significance (for example, in plasmodiophorids; Pr-19)
cytochrome Small protein that contains iron heme and acts as electron carrier in respiration and photosynthesis
cytogamy Fertilization by fusion of gametes (sex cells) (A, F, Pl); compare *karyogamy* (fusion of nuclei)
cytokinesis Division of cell cytoplasm; distinguished from karyokinesis, division of a cell nucleus (Pr, A, F, Pl)
cytoplasm In a cell, the fluid, ribosome-filled part exterior to the nucleus or nucleoid
cytoplasmic streaming Streaming movement of cytoplasm, fluid exterior to the nucleus of a cell (Pl)

cytostome Ingestive opening or mouth of a protoctist (Pr)

Dauer larva Nematode juvenile in developmental arrest (A-10) [Dauer (German), duration]

daughter cells See *offspring cells*

deciduous Shed or sloughed off with season (examples are leaves of a temperate zone tree and anthers)

deciduous forest Forest principally of trees that seasonally shed leaves (Pl)

desmosome Intercellular membranous junction fastening cells together in tissues (A)

Desor larva Postgastrula nonswimming larva characteristic of heteronemertean worms; it remains inside the egg membrane and its development resembles that of pilidium larvae; named for its discoverer, Pierre Jean Edouard Desor (1811–1882), Swiss geologist and archaeologist who saw this larva develop from *Lineus* egg obtained from the New England coast (A-18)

detrital Composed of detritus (A-2)

detritus Loose natural material, such as rock fragments or organic particles, that results directly from disintegration of rocks or organisms

deuteromycote Fungus that reproduces only by mitotic production of spores and that has lost all sexual processes; member of the ascomycetes (F-3)

deuterostomes Includes all coelomate animals in which the blastopore becomes the anus (A-32 through A-37) series of Grade Coelomata

dichotomous Branching or forking (for example, veins on wings, leaves, or fronds) (A, Pl)

dicotyledon Member of a subphylum of angiosperms (Pl-12); a plant having two cotyledons in the seed; also called dicot; compare *monocotyledon*

dictyosome Golgi apparatus or Golgi body (A); parabasal body (Pr-30); a layered, cuplike organelle composed of modified endoplasmic reticulum; plays a part in the storage or secretion of metabolic products

diffusion Movement of gas or liquid from a region of higher concentration to one of lower concentration

digenean Trematode having two or more hosts (A-5)

dikaryon Organism composed of mycelium containing pairs of closely associated nuclei, each derived typically from a different mating type (F-2, F-3)

dikaryosis Dikaryotic condition

dikaryotic Diploid; 2N

dimorphism Morphological or genetic difference between two individual members of the same species—for example, between males and females or between winter and summer forms

dioecious Having male and female structures on different individuals of the same species (Pr, A, Pl)

diploid Of cells in which the nucleus contains two sets of chromosomes; 2N (Pr, A, F, Pl)

direct development Development directly from the zygote to the adult without a larval stage (A-6, A-28, A-32); in fungi, development of spores by mitosis (F-1)

divergence Evolutionary term; accumulation of dissimilar characters in organisms related by descent but exposed to unlike environments for many generations

diverticulum (pl. **diverticuli**) Outpocket of the gut or other internal organ (A-27, A-33)

DNA Deoxyribonucleic acid; a long molecule composed of nucleotides in a linear order that constitutes the genetic information of cells; capable of replicating itself and of synthesizing RNA

dorsal Toward the back side (Pr, A); compare *ventral*

double fertilization Egg and sperm fuse with concurrent fusion of polar nuclei and a second sperm nucleus (resulting in triploid endosperm in angiosperms; resulting in polyploid, embryo-nourishing tissue in gnetophytes; Pl-11, Pl-12)

duogland Gland having two functions (for example, adhesion and release) (A-5)

dynein Protein component of undulipodium; forms cross-bridges between 9(2)+2 microtubules and cleaves ATP during movement

ecdysis Process of shedding an exoskeleton (A-21)

ectobiont Organism that lives on the outside of another (A-17)

ectoderm Outermost layer of body tissue (A, Pl); in particular, the outermost layer of tissue formed in gastrulation (A)

ectomycorrhiza (pl. **ectomycorrhizae**) Symbiosis between roots of plant and soil fungus that forms a mantle around plant roots (F-2)

ectomycorrhizal Characterized by ectomycorrhizae (F)

ectoparasite A species that benefits from and lives on the outside of another at the second organism's expense (A-13, A-17)

ectoplasm Outer, more or less rigid, and granule-free layer of cytoplasm (Pr)

egg Female gamete or oocyte (A-1 through A-37, Pl-1 through Pl-12)

elater Hygroscopic cell or band usually attached to the spore—for example, of a horsetail (Pl-6) or the spore of a moss, hornwort, or liverwort (Pl-1 through Pl-3); aids in dispersing spores

electron Subatomic particle that orbits the nucleus of an atom and determines that atom's chemical properties; an electron carries a negative charge

embryo Early developmental stage of multicellular organisms produced from a zygote, or fertilized egg (A, Pl)

embryo sac Female gametophyte of angiosperm; consists, at maturity, of an egg nucleus and accessory nuclei (Pl-12)

endoderm Innermost layer of body tissue formed in the gastrulation of an embryo (A)

endomycorrhiza (pl. **endomycorrhizae**) Symbiosis between roots of plant and soil fungi that penetrate the roots, typical of herbaceous plants and zygomycotes (F-1, Pl-5)

endoparasite Relation between species in which one species lives inside and benefits from another at the expense of the second (A-8, A-11, A-13)

endophyte Ecological term referring to the topology of symbiotic associations with plants; refers to fungi, protoctists, or bacteria living within the tissue of plants or other photosynthetic organisms; because "-phyte" may refer to fungi, protoctists, or bacteria, none of which are plants, the term should be replaced with endobiont, endosymbiotic bacteria, or another specific name (B, Pr, F, Pl)

endoplasm Inner, relatively fluid central part of cytoplasm (Pr)

endoplasmic reticulum Extensive system of membranes inside eukaryotic cells; called rough if coated with ribosomes, called smooth if not (Pr, A, Pl)

endoskeleton Internal support (for example, calcareous ossicles of seastar; A-34)

endosome Nucleolus, intranuclear body, or karyosome, composed of precursors of ribosomes (Pr, A)

endosperm Tissue surrounding angiosperm plant embryo; contains stored food; develops from the union of haploid (N) male nucleus and two haploid (2N) polar nuclei of female and is therefore triploid (3N) (Pl-12)

endospore Desiccation- and heat-resistant spore produced inside bacteria (B-3, B-6, B-10)

endostyle In tunicates and lancelets, structure that secretes mucus (A-35, A-36)

endosymbiont Organism living inside an organism of a different species; may be intracellular or intercellular

enzyme Molecule that accelerates rate of reaction without being consumed in that reaction; biological catalyst (A, F, Pl)

ephyra (pl. **ephyrae**) Larval stage of scyphozoan cnidarians, reproduced asexually by sessile polyps; ephyrae (minute mobile medusae) develop into adult medusae (A-3)

epibiont Relation in which one organism lives on the surface of another (A-21, F)

epibiotic Living on the surface of another organism

epicone Upper surface (hemisphere) of a dinomastigote test (Pr-7)

epidermal Of the epidermis (A-35, Pl)

epidermis Outer layer of leaf, young root, young stem, or skin (A-3, A-4, A-6, A-9, A-11, A-13, A-21, A-24, A-28, A-33, A-34, Pl-1, Pl-5, Pl-8)

epiphyte Plant or fungus that lives on body of another plant with no metabolic exchange (for example, bromeliad and some lichens) (Pl-12, F-3)

epitheliomuscular cell In cnidarians, part of the contractile system (A-3)

epithelium (pl. **epithelia**) Tissue covering the outer or inner surface of an organ or structure (A, Pl)

epitoke A gamete-bearing, sexually mature, swimming individual or bud of polychaete annelid (A-22)

epizoic Commensal relation in which one animal lives on the surface of another animal species, as in rotifers (A-13)

ergosterol A sterol present in animal and plant tissues that, when irradiated with ultraviolet rays, becomes vitamin D2 (A, Pl)

ergot Disease caused in plants by infection with *Claviceps purpurea,* the ergot fungus (F-3)

ergotism Disease caused in human beings and stock animals if they consume ergot-infected grain (F-3)

esophagus The part of the digestive tract between the pharynx and the stomach (A)

estuary Passage, as at the mouth of a lake or a river, where ocean tide meets fresh water

eucarpic Of the mode of development in which reproductive structures form on certain parts of the thallus while the thallus itself continues to perform its somatic functions (Pr, F)

Eufungi Kingdom equivalent to Kingdom Fungi (F)

eukaryote A cell having a membrane-bounded nucleus, organelles such as mitochondria and plastids, and several chromosomes in which the DNA is coated with histone proteins (Pr, A, F, Pl)

Eumetazoa True metazoa or eumetazoans; all heterotrophic, diploid, multicellular animals that develop from blastulae, excepting Parazoa [placozoans (A-1) and poriferans (A-2)], rhombozoans (A-7), and orthonectids (A-8)

eusporangium (pl. **eusporangia**) In certain ferns, sporangium that has a wall consisting of several cell layers (Pl-7)

Eustacian tube Tube that connects the middle ear with the pharynx and equalizes pressure on the eardrum (A-37)

evergreen Plant, such as club moss and conifer, that does not shed its leaves seasonally; see *deciduous* (Pl-4, Pl-10, Pl-12)

excretory pore Opening through which excretory products leave the organism (A-14)

excurrent siphon In tunicates, the opening through which the feeding current leaves the animal (A-35)

exocytosis Extrusion of entire cell (for example, in entoproct waste discharge) (A-18)

exoskeleton Protective secreted support structure (for example, a crayfish shell and a clam valve) against which muscles act (A-3, A-19 through A-21)

exospore Externally borne reproductive structure (cell) not necessarily heat or desiccation resistant (B-5)

extant Refers to organism living in the present, distinguished from extinct

extrusome Extrusive organelle; membrane-bounded structure, the contents of which are extruded by protoctists in response to a variety of stimuli (for example, predators, prey, changes in acidity); generalized term referring to various, probably nonhomologous, structures (such as cnidocyst, discobolocyst, ejectosome, and kinetocyst)

facultative anaerobe Optional anaerobe; distinguished from obligate anaerobe (B)

facultative autotroph Organism that, depending on conditions, can grow either by autotrophy (photo- or chemosynthesis) or by heterotrophy

false branching Growth habit of a filamentous structure in which apparent branching is caused by slippage between two rows of cells; only two growth points are simultaneously active on any single cell (B, Pr)

falx Structure of opalinids (Pr-30); composed of rows of closely set kinetosomes and their undulipodia

fascicle A bundle of needle leaves of gymnosperms (Pl-10)

fauna Animal life

femur Thighbone of vertebrates (A-37)

fermentation Anaerobic respiration; degradation of organic compounds in the absence of oxygen, yielding energy and organic end products (the ferment); organic compounds are the terminal electron acceptors in all fermentations

fertilization Fusion of two haploid cells, gametes, or gamete nuclei to form a diploid zygote (Pr, A, F, Pl)

fertilization chamber In cycads, liquid-filled chamber inside the ovule into which the pollen tube releases sperm (Pl-8)

fertilize To form zygote, diploid cell, or diploid nucleus by fusion of two haploid cells or gametes (A, F, Pl)

fibula Outer bone between ankle and knee of the lower leg of a tetrapod (A-37)

fiddlehead Emergent young fern frond (Pl-7)

filament Threadlike structure (A-2, A-27, Pl-1, Pl-7, Pl-12)

filopod Pseudopod composed entirely of ectoplasm; typically thin and pointed, with microfibrillar ultrastructure (Pr-3, Pr-4); also called filopodium

filter feeder Animal that filters plankton and detritus from a water current; a form of suspension feeding (A-15)

fin Winglike organ of an arrow worm (A-32), fish, or aquatic mammal (A-37) used for stabilizing or swimming

fission Asexual reproduction by division of cells or organisms into two or more parts of equal or nearly equal size

flagellate Prokaryote motile by means of flagella; see *mastigote*

flagellin Member of a class of proteins that are structural components of bacterial flagella

flagellum (pl. **flagella**) (1) Long, thin, solid extracellular organelle of bacterial motility; composed of a protein, belonging to the class of proteins called flagellins; (2) undulipodium; long, fine, intrinsically motile intracellular structure used for locomotion or feeding; covered by plasma membrane and underlain by regular array of nine doublet microtubules and two central microtubules composed of tubulin, dynein, and other proteins, not flagellin (Pr, A, Pl); eukaryotic "flagella" tend to be longer than cilia but have the same internal structure; the term undulipodium refers to both the "flagella" and the cilia of eukaryotes

flame cell Ciliated cell of excretory tubule associated with excretion, as in flatworms (A-5), polychaete annelids (A-22), rotifers (A-13)

flora Plant life

flosculus (pl. **flosculi**) Rosette of tiny microvilli on lorica of loriciferans (A-15, A-17)

flowering plant See *angiosperm*

foliose Of lichens having leafy form (F-3)

food web Relations in a community regarding feeding in which an organism obtains nutrients from another organism and in turn provides nutrients to yet another (A-21, A-35, F, Pl)

foot In molluscs and ectoprocts, muscular organ of locomotion (A-26, A-29); in mosses, liverworts, and hornworts, a structure that is embedded in the female gametophyte and supports the sporophyte (Pl-1 through Pl-3)

foramen (pl. **foramina**) Fenestra: small opening, orifice, or perforation; in foraminifera, an opening in a septum separating two chambers (Pr)

form-taxon In fungi, organism distinguished from others by form or shape (F-3)

fossil Evidence of ancient life as preserved in the geological record (for example, leaf imprint, dinosaur footprint, and fossil fungus spore) (A, F, Pl, B, Pr)

free living Refers to organism not attached to substrate (A, Pl)

frond Leafy part, as of a fern (Pl-7) or seaweed (Pr-12, Pr-13); see *megaphyll*

fruit Characteristic of angiosperms; ripe and mature ovary (or ovaries) that contains seeds (Pl-12)

fruiting body Structure that contains or bears cysts, spores, or other generative structures (B, Pr, F, Pl)

fruticose Of lichens that bear upright reproductive bodies (F-3)

Fungi Imperfecti Deuteromycotes reclassified as ascomycetes, formerly believed to lack sexual stages (F-3)

funiculus (pl. **funiculi**) Body structure resembling a cord (A)

fusion Union of two cells (Pr, A, Pl) or hyphae (F)

fusule Aperture through which an axopod extends (Pr-27)

gall Secretions of the liver that are stored in the gall bladder (A-37)

gametangium (pl. **gametangia**) Organ or cell in which gametes are formed (Pr, F, Pl)

gamete Mature haploid reproductive cell whose nucleus fuses with that of another gamete of an opposite sex to form a zygote (Pr, A, Pl)

gametic meiosis Meiosis that takes place immediately before gametogenesis

gametogenesis Production of gametes (Pr, A, Pl)

gametophyte Haploid (1N) gamete-producing generation, as in plants having alternation of gametophyte and sporophyte generations (Pl)

gametophytic Of the gametophyte (Pl)

gamogony Series of cell or nuclear divisions leading eventually to gamonts, individuals that produce gamete nuclei or gametes capable of fertilization (Pr-4)

gamont Gamete-producing form of an organism; sexually differentiated individual (Pr-4)

ganglion (pl. **ganglia**) Aggregate of bodies of nerve cells (A)

gap junction Intercellular, membranous, discontinuous junction fastening cells together in tissues; thought to regulate the flow of ions between cells (A)

gastric gland Multicellular epithelial organ that secretes enzymes for extracellular digestion (A)

gastrodermis In cnidarians and comb jellies, the inner cell layer that lines the gastrovascular cavity (A-3, A-4)

gastrovascular cavity Cavity in which both digestion and circulation take place (A)

gastrula (pl. **gastrulae**) Embryo in the process of gastrulation, during which the blastula with its single layer of cells becomes a three-layered embryo (A)

gastrulation Developmental stage during which the blastula embryo is transformed into the gastrula (A)

G–C ratio Proportion of guanine and cytosine base pairs in the total number of base pairs in any DNA; a wide difference in G–C ratio is taken to preclude a phylogenetic relationship between organisms (B, F)

gemma (pl. **gemmae**) Asexual reproductive structure, a small mass of vegetative tissue that can develop into a new individual (for example, moss gemmae) (Pr-20, Pl-1)

gemmule Asexual reproductive structure; in sponges, a small mass of vegetative tissue that can be released and develop into a new individual (A-2)

generative nuclei Nuclei capable of further division or of fertilization followed by further division (Pr, Pl)

genital pore Exit aperture for fertilized eggs (A)

genome Complete set of genetic material of an organism

genophore Genetic material of prokaryotes, not coated with protein (B)

genotype Genetic basis of a trait in an organism; compare *phenotype*

germ Microorganism that causes disease

germ cell Ovum (egg), spermatozoon (sperm); gamete; cell requiring fertilization; sex cell

germ layer Any of the three primary cell layers (ectoderm, mesoderm, endoderm) formed from a blastula during gastrulation (A)

germinal cell Precursor to gamete or to asexual reproductive cell (Pr, A, Pl)

germinal nucleus Precursor of gamete (A-14); also called germinal cell

germinarium In rotifers, the ovary (A-13)

germinate Seed begins to grow (Pl-8 through Pl-12); spore commences to form gametophyte (for example, a horsetail; Pl-6) or fungus (F)

germinating Process of releasing spores or converting into a growing structure (F, Pl)

germination Process of germinating

germovitellarium In rotifers, the combined germinarium (ovary) and vitellarium (organ in which embryo develops) (A-13)

gill Respiratory organ used for uptake of oxygen, release of carbon dioxide, and regulation of diffusible ions by aquatic animals (A); in fungi, flat platelike structure on underside of mushroom cap on which spores form, in basidiomycetes (F-2)

gill bailer In crustaceans, a modified mouthpart that generates the water current through the gill chamber (A-21)

gill flap In horseshoe crabs, flattened abdominal appendage for gas exchange (A-19)

girdle Equatorial groove or other structure (Pr, A); cingulum of a dinomastigote (Pr-7); one of a pair of ridges dividing the trunk of a pogonophoran (A-25)

girdle lamella Centrally or equatorially located flat intracellular structure (Pr)

gizzard Digestive organ in which food is ground (A-22, A-37)

glycogen A long-chain carbohydrate composed of glucose units (Pr, A)

glycolysis Metabolic pathway in which glucose is broken down into organic acids and CO_2, releasing energy

Golgi body See *dictyosome*

gonad Animal organ composed of tissues that produce gametes; ovary (produces eggs) or testis (produces sperm)

Gondwana Ancient land mass that separated into present-day South America, Antarctica, Africa, India, and Australia in the Cretaceous

gonoduct Reproductive duct that provides transport of sperm and eggs (A-14, A-28)

gonopore Opening from reproductive tract to the exterior (A-4, A-10, A-12, A-18, A-28, A-32)

Gram-negative Failing to retain the purple stain (crystal violet) when subjected to the Gram staining method; indicates the presence of certain component layers in the cell wall (B)

Gram-positive Retaining the purple stain (crystal violet) when subjected to the Gram staining method; indicates absence of a certain component layer in the cell wall (B)

granellare The plasma body ("protoplasm") of a xenophyophoran and its surrounding tubes, which are yellowish and branched in varying degrees (Pr)

granum Inclusion in a plastid; seen as a minute grain under the light microscope, but known to consist of stacked thylakoids (Pr, Pl)

green gland Antennal gland of crustaceans; in freshwater species, this gland resorbs salt and excretes water (A-21)

gymnomastigote Eukaryotic organism motile by means of undulipodia, with the cell body naked—that is, test or lorica absent (Pr)

gymnosperm Seed plant that bears exposed seed, not enclosed in fruit (Pl-8 through Pl-11)

gynospore See *megaspore*

habitat Natural environment in which an organism normally lives (A, Pl, F)

haploid Of cells in which the nucleus contains only one set of chromosomes; 1N (Pr, A, F, Pl)

haplosporosome Electron-dense membrane-bounded organelle with unknown function, generally spherical but sometimes having profiles of various shapes; another unit membrane (which distinguishes the organelle from other membrane-bounded, electron-dense inclusions in other eukaryotic cells) is found internally in various configurations free of the delimiting membrane (for example, of haplosporidians and possibly myxozoans and paramyxeans) (Pr)

haptoneme Coiled, threadlike microtubular structure often used as a holdfast (Pr-10)

haustorial pollen tube Pollen tube that branches, absorbs, and penetrates (Pl-8, Pl-9)

haustorium (pl. **haustoria**) Growth of a hypha, or other heterotrophic tube, generally into plant or algal tissue; it penetrates between or into cells to absorb nutrients (F, Pl-9)

heartwood Nonliving wood in the center of a woody plant and surrounded by living, water-conducting tissue (Pl-9, Pl-10, Pl-12)

hemal System of uncertain function (perhaps excretory or providing nutrients to gonads) consisting of canals, oral and aboral rings, and axial organ that circulates fluid in echinoderms (A-34)

hemal channel In onychophorans, channel in which colorless blood circulates (A-28)

heme Organic ring compound bound to iron and attached to protein (such as in hemoglobin)

hemerythrin Iron-containing respiratory pigment used for storage of oxygen in the blood of various invertebrates (A)

hemocoel Fluid-filled body cavity that functions as part of the circulatory system (A); also called hemocoelom

hemocyanin Copper-containing, blue respiratory pigment of most pulmonate molluscs, all prosobranch and cephalopod molluscs, and crustaceans (A-19, A-26)

hemoglobin Iron-containing protein consisting of a heme part and a protein part and used for the transport or storage of oxygen; found in the blood of animals (A), the root nodules of some legumes (Pl) that have symbiotic bacteria (B), and in some strains of *Paramecium* and *Tetrahymena* (Pr)

hepatic Refers to the liver; in hemichordates, the abdominal region (A-33)

hepatic caecum In cephalochordates, the liver (A-36)

hepatopancreas Organ that combines liver and pancreas (A-26)

herbaceous Nonwoody plant or herb meaning nonwoody seed plant (Pl)

herbivore Organism that obtains nutrition and energy by eating plants or algae; compare *carnivore* (A)

herbivorous Of organism that eats plants or phytoplankton (A)

hermaphroditic Having both female and male reproductive systems; simultaneous hermaphrodite: both sexes present and functioning at the same time; protandric: male system functions first, followed by female system (A, Pl)

heterocyst A differentiated cyanobacterial cell that fixes nitrogen (B)

heterokaryon Organism composed of mycelium containing pairs of closely associated nuclei, the members of each pair known to be derived from different mating types (differing in genotype) (F)

heterokont Biundulipodiated cell in which the two undulipodia are unequal in length (Pr)

heterosporous Individual that simultaneously forms spores of two kinds; typically, microspores and megaspores (Pl); see *homosporous*

heterospory Bearing spores of two different kinds, as in lycopods (Pl-4), certain ferns (Pl-7), and seed plants (Pl-9 through Pl-12)

heterotrichous Hairy or filamentous with hairs or filaments of more than one type (Pr-8, Pl)

heterotroph Organism that obtains carbon and energy from organic compounds ultimately produced by autotrophs: osmotrophs, parasites, carnivores, and others (B, Pr, A, F)

Higgins' larva Larval stage of loriciferans; named in honor of Dr. Robert Higgins (A-17)

histology Study of tissue (F, Pl, A)

histones Class of positively charged chromosomal proteins that bind to DNA, tend to be lysine and arginine rich, and stain with fast green

holdfast Organ or organellar structure for attachment (Pr, A, F)

holocarpic Of the mode of development in which the thallus is entirely converted into one or more reproductive structures (Pr, F)

homeotherm Warm-blooded organism capable of regulating its body temperature around a set point fairly independently of external temperatures (A-37)

homokaryon Pair of genetically identical haploid nuclei (F)

homokaryotic Having nuclei of a single genotype, characteristic of a homokaryon (F-2)

homologous Of structures or behaviors that have evolved from common ancestors, even if the structures or behaviors have diverged in form and function

homospore One kind of spore, as in some (homosporous) lycopods, psilophytes, and most ferns (Pl-4, Pl-5, Pl-7)

homosporous Individual that forms only one kind of spore (Pl-4, Pl-5, Pl-7)

homothallism Condition in which mating hyphae have grown from the same basidiospore; self-fertilizing fungi (F-2)

hormogonium (pl. **hormogonia**) Cyanobacterial life-history stage; motile gliding filaments of cylindrical cells (B)

hormone Molecule synthesized in one body region, secreted by ductless glands (in animals) into body fluids, and carried to its target organ or cell in another region, where the hormone (for example, estrogen) elicits a specific physiological response (A, F-2, Pl)

horned liverwort Same as hornwort (Pl-3)

host Organism that provides nutrition or lodging for symbionts or parasites

humerus In tetrapods, the bone of the upper forelimb, between shoulder and elbow (A-37)

hydra Large genus of freshwater cnidarians that form colonies (A-3)

hydrogenase Enzyme that catalyzes the breakdown of organic compounds and the simultaneous release of hydrogen (H_2)

hydroid Cnidarians with polyp form (A-3). In mosses, nonlignified, water conducting cells (Pl-1)

hydrostatic pressure Pressure exerted by a liquid on the walls of a container

hydrothermal Refers to hot water containing dissolved minerals that is emitted from new ocean-bottom crust (A)

hydrothermal vent Vent in earth crust from which hot water is emitted (for example, hot sulfide-rich seawater habitat of pogonophorans, molluscs) (A-21, A-25, A-26)

Hydrozoa Class of cnidarian coelenterates (A-3)

hygroscopic Describes any substance that absorbs water from the atmosphere

hymenium Fertile layer of fungal tissue consisting of asci or basidia (F-2, F-3)

hypermastigote Motile heterotroph that can possess many thousands of undulipodia (Pr-30)

hypersaline Saltier than sea water (> 3.4 percent NaCl)

hypertrophy Overgrowth; unusual increase in size or number

hypha (pl. **hyphae**) Threadlike tubular filament, a component of a mycelium (B-10, Pr, F)

hypobranchial gland Mucus-producing gland in molluscs (A-26)

hypocone Lower surface (hemisphere) of a dinomastigote test (Pr-7)

hypodermic copulation Mating mode during which male injects sperm through the female body wall (A-13)

hypothecium Thin layer of interwoven hyphae between the hymenium and the apothecium (F-3)

ilium (pl. **ilia**) Uppermost part of three sections (ilium, ischium, and pubis) of the innominate bone, which makes up the lateral half of the pelvis (A-37)

indusium (pl. **indusia**) Tissue outgrowth on a fern megaphyll under which sori develop (Pl-7)

inflorescence Flower cluster (Pl-12)

infusoriform larva Ciliated immature form (A-7) produced by the rhombogen germ cell; a swarm larva

infusorigen In rhombozoan, hermaphroditic gonad that produces both sperm and eggs (A-7)

inhalent canal Incurrent opening through which water enters sponge (A-2)

integument Skin or enveloping layer of an organism (Pr, A, F, Pl)

intermediate host Host organism that is one of a series that provide shelter or nutrients for parasite or symbiont (A-12)

internode Region between hard or enlarged junction of stem or branch segments (Pl)

interstitial Living between sediment particles (A-3, A-13, A-21, A-27)

intertidal Refers to zone between high and low tidemarks

introvert Slender anterior body part that can be turned inside out and completely withdrawn into the trunk of the body (A-23)

invertebrate Animal lacking a backbone; compare *vertebrate* (A-1 through A-36)

isogamy The formation of gametes of unlike mating types that are alike in size and morphology (Pr)

isokont Undulipodiated cell in which the undulipodia are of equal length (Pr)

isoprenoid The C_5H_{10} unit of a class of organic compounds; they are synthesized from multiples of isopentenyl pyrophosphate; phytol, carotenoids, and steroids are isoprenoids

Iwata larva One of several larval types of nemertine worm, named for Fumio Iwata, a biologist at Hokkaido University, Japan (A-9)

jacket cell Ciliated cell that makes up outermost layer of rhombozoans and orthonectids (A-7, A-8)

karyogamy Fusion of nuclei, often in zygote formation (Pr, A, F, Pl); compare *cytogamy*

karyokinesis Nuclear division (Pr, A, F, Pl)

karyomastigont system Mastigont system with associated nucleus (or nuclei—for example, diplomonads); see *mastigont system* (Pr)

keratin Class of tough, fibrous proteins that are components of hair, skin, claws, and horn (A)

kernel Seed, as of corn; inner part of nut (Pl-12)

kinesin Microtubule-associated protein (MAP) that, in the presence of ATP, moves vesicles along microtubules in an anterograde direction

kinetid Unit structure that includes at least one kinetosome and characteristic of all undulipodiated cells; typically composed of microtubule sheets or ribbons, striated fibers, and kinetosomes; see Figure I-3; older literature used the terms basal apparatus or flagellar insertion apparatus

kinetochore See *centromere*

kinetoplast Intracellular DNA-containing structure, a modified mitochondrion with an associated kinetosome characteristic of the class Kinetoplastida (Pr-12; for example, *Trypanosoma*)

kinetosome Organelle at the base of all undulipodia and responsible for their formation; like a centriole, its cross section shows a characteristic circle of nine triplets of microtubules (see Figure I-3); centriole that has an associated axoneme (Pr, A, Pl)

labrum (pl. **labra**) Animal mouthpart; upper lip of insects (A-20)

lamella (pl. **lamellae**) Flattened saclike structure (Pr, A, F)

larva (pl. **larvae**) Immature form of an animal, distinguishable morphologically from the adult (A)

lasso cell Adhesive cell of comb jelly; also called colloblast (A-4)

leaf Organ of photosynthesis and transpiration in plants (Pl)

leaf blade Flat part of leaf, excluding the stalk (petiole) (Pl-7 through Pl-12)

leaf impression Fossil consisting of imprint of a leaf (Pl-7)

leaf primordium Lateral outgrowth that will become a leaf (Pl)

leaf stalk Same as stipe (Pl-7) and petiole (Pl-7 through Pl-12)

leaflet One of several parts of compound leaf (Pl-7, Pl-12)

leafy liverwort Liverwort having thicker thallus; compare *thallose liverwort* (Pl-2)

lemniscus (pl. **lemnisci**) Fluid-filled reservoir used by an acanthocephalan worm to evert the proboscis (A-12)

lepidodendrid Treelike lycopod with scaly bark and big, spore-bearing cones that flourished in Carboniferous (Coal Age) coal forests, now extinct

leptoid In certain mosses, refers to elongate, food-conducting cells having degenerate nuclei at maturity (Pl-1)

leptosporangium (pl. **leptosporangia**) In certain ferns, sporangium that has a single-cell-thick wall and that develops from a single cell; compare *eusporangium* (Pl-7)

leucosin Storage oil usually found in the membranous vesicles of chrysophytes (Pr-13)

liana Climbing vine of woody plant (Pl-11, Pl-12)

lichen Symbiotic association between a fungus (F-3) and a photosynthetic alga

life cycle Developmental sequence in an organism, from gametes onward to death

lignin Polymer that is similar to cellulose and contributes stiffness and strength to woody plants (Pl-9, Pl-10, P-12)

ligule Tiny appendage at base of leaf of some lycopods (Pl-4) and grasses (Pl-12)

linella (pl. **linellae**) Long, thin threads composed of a cementlike matter found outside the granellare of xenophyophorans and regarded as an organic part of the test (Pr-5)

lipid One of a class of organic compounds soluble in organic and not aqueous solvents; includes fats, waxes, steroids, phospholipids, carotenoids, and xanthophylls

lobate Having lobes, as in certain comb jellies (A-4)

Lobopodia Arthropod subphylum proposed by some to contain tardigrades (A-27) and onychophorans (A-28)

long shoot Longer of two types of shoots in ginkgo and conifers; compare *spur shoot* (Pl-9, Pl-10)

lophophorate Animal bearing a lophophore: ectoprocts, brachiopods, phoronids (A-29 through A-31)

lophophore Food-trapping ridge surrounding the mouth and bearing hollow, ciliated tentacles (A-29 through A-31)

lophotrichous Bearing polar bundles of flagella (B) or undulipodia (Pr-30)

lorica Secreted protective covering: test, shell, valve, or sheath (Pr, A)

lumen In animals, a hollow cavity within an organ (for example, stomach lumen); in plants, a channel inside a tube or organelle (for example, space inside thylakoid) (A-31, Pl)

luminesce To become luminescent (A-2)

luminescence See *bioluminescence*

luminescent Capable of giving off light (A-21, A-22)

lycopod Herbaceous plant having scalelike leaves and spore-bearing cones, as in club mosses and relatives (Pl-4)

lycopodium powder Spores of club mosses; used commercially in fireworks and pharmaceuticals (Pl-4)

macrobenthic Refers to animal longer than 0.5 mm that dwells on the seafloor (A-15)

macrogamete Large gamete; in most cases, female (Pr, Pl)

macronucleus Larger of the two types of nuclei in ciliates (Pr-8); site of mRNA synthesis; contains many copies of each gene; required for asexual growth and division

macroscopic Visible to the naked eye; compare *microscopic*

madreporite External aboral opening of the echinoderm water vascular system; connects to the stone canal (A-34)

Malpighian tubule Excretory organ of insects and eutardigrades (A-20, A-27)

mammary gland Milk-secreting gland of mammals (A-37)

mandible Jaw (A)

mantle Covering or coat; body wall that secretes a shell (A-26)

manubrium Protrusion that bears the mouth (for example, in hydrozoans) (A-3)

marine Oceanic (A, F, Pl)

mastax Horny, toothed chewing apparatus in the pharynx of rotifers (A-13)

mastigoneme Lateral appendage of undulipodia; also called flimmer, tinsel, and scale

mastigont system Intracellular organellar complex; minimally kinetids with kinetosomes, undulipodia, and associated fibers; when nucleus is attached (by rhizoplast) organellar complex is called *karyomastigont* (Pr)

mastigote Eukaryotic microorganism motile by means of undulipodia; eukaryotic "flagellate"

mating type Designation of an individual organism to distinguish it from others capable of mating or fusing with it; individuals of the same mating type cannot mate with each other; in many species, individuals of different mating types are indistinguishable except by their willingness to mate with each other; some species have hundreds of mating types (B, Pr, F, Pl)

maxilla (pl. **maxillae**) One of several appendages associated with the head of a crustacean (A-21)

medulla Inner part of gland or other structure surrounded by cortex (A); fungal part of lichen (F-3) that consists of loose hyphae

medusa (pl. **medusae**) Free-swimming bell- , box- , or umbrella-shaped stage (diplophase) in the life cycle of many cnidarians (A-3) and all comb jellies (A-4); jellyfish

megalops Final larval stage characterized by large eyes of certain crustaceans (crabs and some shrimp) (A-21)

megaphyll Large leaf, having a web of veins or parallel veins and having a leaf gap; compare *microphyll* (Pl-7 through Pl-12)

megasporangiate cone Cone that bears megasporangia (Pl)

megasporangium (pl. **megasporangia**) Structure in which female meiotic products form; usually produces from one to four megaspores (Pl)

megaspore Haploid spore that develops from megaspore mother cell; develops into the female gametophyte (Pl-7 through Pl-12); also called gynospore

megasporophyll Leaf or leaflike appendage that bears megasporangia (Pl)

meiobenthic Refers to animal less than 0.5 mm in length that dwells on the seafloor; members of the meiobenthos (A-15)

meiofauna Animals that fall between two mesh-size openings (0.042 mm and 1 mm) of a collecting screen (A)

meiosis One or two nuclear divisions in which the number of chromosomes is reduced by half (Pr, A, F, Pl)

meiosporangium (pl. **meiosporangia**) In yeast, ascus-formed diploid zygote that undergoes meiosis (F-3)

meiotic Concerning meiosis

membranelle Organelle formed by the fusion of rows of adjacent cilia (Pr-8)

Mendelian genetics Genetics of visible heritable traits and their genetic control, discovered by Gregor Mendel (1822–1884)

meristem Plant tissue composed of undifferentiated cells capable of division (Pl)

merozoite Life-cycle stage; agamete produced by multiple mitoses of a trophozoite (Pr-9)

mesenchyme Gelatinous material containing amebocytes or other cells (A-1, A-2); unspecialized embryonic tissue that gives rise to circulatory and connective tissue (A)

mesentery In coelomate animals, thin sheet of peritonium from which the gut is suspended (A-15, A-28, A-30, A-31, A-37)

mesocoel Middle part of the coelom of many deuterostomes (A)

mesoderm Middle layer of body tissue (between the ectoderm and endoderm) formed in gastrulation (A)

mesoglea Gelatinous noncellular layer formed between ectoderm and endoderm (A-2, A-3)

mesohyl In sponges, middle, gelatinous layer containing living cells (A-2)

Mesozoa Former phylum consisting of rhombozoans and orthonectids (A-7, A-8)

messenger RNA (**mRNA**) RNA produced from DNA-directed polymerization in pieces just large enough to carry the information for the synthesis of one or several proteins

metabolism Sum of enzyme-catalyzed chemical reaction sequences taking place in cells and organisms

metacercaria One of several types of trematode (fluke) larvae (A-5)

metacoel Trunk part of the coelom of an ectoproct (A-29)

metamorphose To discontinuously transform from immature into intermediate or adult body form, as in tadpole to frog (A-3, A-4, A-20)

metamorphosis An abrupt transition from the immature to an intermediate or adult form (for example, tadpole to frog)

metanephridium (pl. **metanephridia**) Type of nephridium (kidney) (A-23, A-26, A-31); see *protonephridium*

metatroch Middle band of cilia on a trochophore larva (A-22, A-26)

Metazoa Kingdom Animalia; excludes protoctists (protozoa)

methanogenic Describes any organism or process that produces methane (B-1)

microaerophil Aerobic organism requiring an oxygen concentration less than normal (less than 20 percent by volume)

microbe Microscopic organism (B, Pr, A, F)

microbial mat Matlike community of microorganisms; living precursor of a stromatolite

microcyst Desiccation-resistant sporelike structure that is released from bacterial fruiting bodies and may develop into a new bacterium (B-3)

microfibril Any solid, thin, fibrous proteinaceous structure, generally those in the cytoplasm of eukaryotic cells; some participate in movement (Pr, A)

microgamete Small gamete, typically male (Pr, Pl)

micronucleus Smaller of the two types of nuclei in ciliates (Pr-8); does not synthesize messenger RNA; most are diploid and required for meiosis and autogamy but not for asexual growth and division

micropaleontology Study of fossil microbes and microscopic parts of fossil organisms

microphyll Small leaf that is characteristic of mosses, hornworts, liverworts, and lycopods, has a single vein, and is not associated with a leaf gap (characteristic of a megaphyll) (Pl-1 through Pl-4)

micropyle The opening in the ovule integument through which the pollen tube grows (Pl-8 through Pl-12)

microscopic Invisible to the naked eye; see *macroscopic*

microsporangiate cone Cone that bears microsporangia (Pl-8, Pl-11)

microsporangium (pl. **microsporangia**) Structure in which male meiotic products—microspores—form (Pl)

microspore Haploid spore that develops from microspore mother cell; develops into a male gametophyte (Pl-7 through Pl-12); in a flowering plant (Pl-12), it becomes a pollen grain; also called an androspore

microsporophyll Leaf bearing microsporangia (Pl)

microstrobilus (pl. **microstrobili**) Male cone (Pl)

microtubule Slender, hollow proteinaceous intracellular structure; most are 24 nm in diameter; found in axopods, axonemes, mitotic spindles, undulipodia, haptonemes, nerve cell processes (axons and dendrites), and other intracellular structures; often, their formation can be inhibited by colchicine (Pr, A, F, Pl)

microvillus (pl. **microvilli**) Minute fingerlike projections of the plasma membrane of a cell (A)

midrib Center vein of leaf (Pl)

miracidium (pl. **miracidia**) One of several types of trematode (fluke) larvae (A-5)

mitochondrion (pl. **mitochondria**) Organelle in which the chemical energy in reduced organic compounds (food molecules) is transferred to ATP molecules by oxygen-requiring respiration (Pr, A, F, Pl)

mitosis Nuclear division in which attached pairs of duplicate chromosomes move to the equatorial plane of the nucleus, separate at their centromeres, and form two separate, identical groups; subsequent division of the cell will thus produce two identical offspring cells (Pr, A, F, Pl)

mitotic Pertaining to mitosis (A, F, Pl)

mitotic apparatus Mitotic spindle

mitotic spindle Microtubular structure formed during mitosis and responsible for the poleward movement of the chromosomes (Pr, A, F, Pl)

mole Quantity of a substance containing a standard number (Avogadro's number, 6.023×10^{23}) of atoms or molecules; quantity of a substance whose mass in grams is equal to the atomic or molecular weight of the substance

molecular systematics Classification on the basis of molecular criteria

molt To shed all or part of an outer covering, such as cuticle, skin, or feathers (A)

monocotyledon Member of a subphylum of angiosperms (Pl-12); includes plants having a single cotyledon in the seed; also called monocot; compare *dicotyledon*

monoecious Hermaphroditic; botanical term describing plants with male parts (for example, anthers, pollen cones) separate from female parts (for example, carpels, ovulate cones) on different structures of the same organism (Pl-10 through Pl-12)

monogenean Trematode flatworm having a single host (A-5)

monokaryon Mycelial organism all of whose nuclei are descended from a single mating type (F)

monokaryotic Having only single, unpaired nuclei; see *haploid* (F-2, Pl-I)

monomorphic Having a single form [for example, in cnidarians, having a polyp (monomorphic) compared with having both a polyp and a medusa (polymorphic)] (A-3, A-4)

monophyletic Of a trait or a group of organisms derived directly from a common ancestor

monopodial Having one main axis of growth (Pl) or one pseudopod (Pr)

monosaccharide Simple sugar, such as glucose or ribose

morphogenesis Development of the size, form, or other structural features of an organism

morphological Of the form and structure of an organism or its parts (A, F, Pl)

morphology Form and structure, study of form

morula cell Cluster of cells that is shaped like a mulberry, as in a tunicate (A-35)

mouth cone Protrusible snout that contains the mouth of a kinorhynch (A-14) or loriciferan (A-17)

mucilage Secreted slime or extracellular cementing material composed in part of nitrogen-containing polysaccharides

mucopolysaccaride Member of a class of polymers composed of nitrogen-substituted simple sugars

Müller's larva Larval form of some marine turbellarian flatworms, named for a biologist (A-5)

mutation Inheritable change in genetic information in which one form of an allele is changed to a different form

mycelium (pl. **mycelia**) Mass of hyphae constituting the body of a fungus or funguslike protoctist (Pr-19 through Pr-21, Pr-29, F)

Mychota Kingdom equivalent to Kingdom Fungi (F)

mycobiont Fungus partner in a symbiotic association such as a lichen; see *photobiont* (F-3)

mycorrhiza (pl. **mycorrhizae**) Association between the hyphae of a fungus and the roots of a plant; see *ecto-* and *endomycorrhiza* (F, Pl)

myelin Fatty substance that ensheaths most nerve cells, except cnidarian nerves (A-3)

myoneme Fibrillar organelle that has a contractile function (Pr-8)

myosin One of the two major proteins of muscle, in which it makes up the thick filaments (A); also found in filaments that participate in the motility of protoctists (Pr); see *actin*

mysis One of several types of decapod crustacean larvae (A-21)

myxospore A spore or other desiccation-resistant stage of myxobacteria (B-3)

naked seed Seed that is not enclosed in a fruit, as in all seed plants except for Pl-12 (Pl-8 through Pl-11)

nauplius Larva having three segments, characteristic of crustaceans (A-21)

necrotrophy Nutritional mode in which a symbiotroph damages or kills its host; parasitism or pathogenesis

nectar Sugar-rich fluid secreted by gland of angiosperms and some gnetophytes (Pl-11, Pl-12)

needle Slender, modified leaf shape of conifer (Pl-10)

nematocyst Capsule containing a threadlike stinger used for anchoring, defense, or capturing prey; some—for example, the cnidoblasts of coelenterates (A-3)—contain poisonous or paralyzing substances

nematogen Life-cycle stage of rhombozoan (A-7); the ciliated wormlike stage of a rhombozoan living in a host cephalopod (A-26)

nephridiopore Exit pore for wastes secreted by a nephridium (A)

nephridium (pl. **nephridia**) Excretory organ of many invertebrates (A)

nephrostome Ciliated funnel-shaped inner opening of the nephridium of coelomate animals (A)

nerve net Interconnecting web of nerve cells characteristic of cnidarians and comb jellies (A-3, A-4)

nerve ring Circular network of nerve fibers; a component of the central nervous system (A)

neurotoxin Toxin that destroys nerve tissue (A-32)

nexin In the shaft of an undulipodium, one of a set of axonemal proteins that connect one set to another of the pairs (2s) in the 9(2)+2 arrangement

niche Role performed by members of a species in a biological community

nitrogen fixation Incorporation of atmospheric nitrogen into organic nitrogen compounds; requires nitrogenase (B, Pl)

nitrogenase Enzyme complex containing iron and molybdenum; converts atmospheric nitrogen into organic nitrogen (B-3, B-5, B-6)

node Typically hard or swollen junction of two internodes of a stem or branch; may bear a leaf or leaves (Pl)

nodulation Formation of swellings in roots of legumes upon infection by bacteria (Pl-12)

nonsexual Refers to structure or process that lacks a sexual component (for example, fission)

nonvascular Lacking vascular tissue, as in bryophytes (Pl-1 through Pl-3)

notochord Long elastic rod that serves as an internal skeleton in the embryos of chordates; replaced by the vertebral column in most adult chordates (A-35 through A-37)

notum (pl. **nota**) Part of the dorsal thoracic exoskeleton of an insect (A-20)

nuclear cap Crescent-shaped sac surrounding a third or more of the zoospore nucleus of some chytrids (Pr-29); it contains virtually all the ribosomes of the cell

nuclear envelope Nuclear membrane; membrane surrounding the nucleus of a cell

nucleocytoplasm That part of a eukaryotic cell that includes the nucleus and the cytoplasm with its membranes and inclusions but excludes organelles such as plastids and mitochondria (Pr, A, F, Pl)

nucleoid Genophore; DNA-containing structure of prokaryotic cells; not bounded by membrane (B)

nucleolus (pl. **nucleoli**) Structure in the cell nucleus; contains DNA, RNA, protein, precursors of ribosomes (Pr, A, F, Pl)

nucleotide Single unit of nucleic acid; composed of an organic nitrogenous base, deoxyribose or ribose sugar, and phosphate

nucleus (pl. **nuclei**) Large membrane-bounded organelle that contains most of a cell's genetic information in the form of DNA (Pr, A, F, Pl)

nummulite Large Cenozoic fossil foraminiferan (Pr-29) having a coin-shaped shell

nutritive-muscular cell In cnidarians, a cell capable of contracting that also takes up nutrients from the gastrovascular cavity (A-3)

obligate anaerobe Organism that can survive and grow only in the absence of gaseous oxygen (B)

obligate parasite Organism unable to survive and grow when removed from its host

obturaculum In vestimentiferans, structure that plugs the tube when the worm withdraws inside (A-25)

ocellus (pl. **ocelli**) Eye or eyespot (Pr, A)

offspring cells Two genetically and morphologically similar products of equal cell division (B) or mitosis; also called daughter cells (Pr, A, F, Pl)

oidium Thin-walled cell that functions as a spore or in sexual processes (F)

oocyst Desiccation-resistant thick-walled structure in which sporozoans are transferred from host to host (Pr-30)

oocyte Cell that eventually develops into an egg (ovum) (A)

oogamy Mode of sexual reproduction; anisogamy in which one of the gametes (the egg) is large and nonmotile, whereas the other gamete (the sperm) is smaller and motile (Pr, A)

oogonium (pl. **oogonia**) Unicellular female sex organ that contains one or several eggs (Pr); a female gametangium

oosphere Large, naked, nonmotile female gamete (Pr)

oospore Thick-walled spherical structure developing from an oosphere after fertilization in oomycotes (Pr-20)

opaline Shiny and whitish; resembling amorphous silica (opal); opal is component of diatom frustule and skeletons of some arthropods and phytoliths (Pr, A, Pl)

open circulation Circulatory system in which blood flows through open sinuses rather than blood vessels; compare *closed circulation* (A-28)

operculum (pl. **opercula**) Lid; covering flap (for example, the horny structure that seals a mollusc into its shell) (Pr, A-26, A-29, F, Pl-1)

opisthe Posterior cell formed by binary fission of ciliates (Pr-8)

opisthosoma In chelicerates, the abdomen, as distinct from the cephalothorax (A-19); in pogonophorans, the hind body region (A-25)

oral Refers to mouth region

oral papilla (pl. **papillae**) In onychophorans, paired openings beside the mouth through which adhesive is squirted (A-28)

oral siphon Canal through which seawater enters tunicate (A-35)

oral stylet Needlelike structure beside the mouth of a kinorhynch; also called style (A-14)

organ Structure of organism that has a specific function(s) and is composed of specialized tissue (A, F, Pl)

organelle "Little organ"; distinct intracellular structure composed of a complex of macromolecules and small molecules—for example, nuclei, mitochondria, undulipodia, and lysosomes (B, Pr, A, F, Pl)

organoheterochemotroph Heterotrophic organism that metabolizes organic molecules; nonphotosynthesizer, such as animal or fungus

organoheterotrophy Nutritional mode of any organism for which the form of carbon, energy, and electrons is organic compounds (B, Pr, A, F, and, rarely, Pl)

osculum (pl. **oscula**) Excurrent opening of a sponge (A-2)

osmoregulatory Regulating the concentration of dissolved substances in a cell, organ, or organism, despite changes of the concentrations in the surrounding medium

osmotroph Organism that obtains nutrients by absorption (F)

osmotrophic Obtaining nutrients by absorption, direct uptake of molecule-sized food compounds across membranes (F)

osphradium (pl. **osphradia**) In molluscs, chemical and touch receptor that samples incurrent seawater (A-26)

ossicle Calcareous structure that, in numbers, forms the endoskeleton of echinoderms (A-34)

osteoblast Cell that produces bone (A-37)

ostium (pl. **ostia**) Incurrent opening of a sponge (A-2); perforation (A-19)

ovarian ball In acanthocephalans, fragments of ovary within the female pseudocoelom (A-12)

ovary Multicellular female reproductive organ surrounding egg (A, Pl)

ovicell Modified ectoproct zooids that brood embryos (A-29)

oviduct Reproductive tract that transports eggs (A-21, A-28, A-32, A-37)

oviger Legs on which male sea spiders brood their fertilized eggs (A-19)

oviparous Of a mode of sexual reproduction in which shelled fertilized eggs are laid that develop outside the female's body (for example, bird eggs) (A-28)

ovoviviparous Of a mode of sexual reproduction in which eggs or embryos develop inside the maternal body but do not receive nutritive or other metabolic aids from the parent; offspring are released as miniature adults (A, Pl)

ovule Megasporangium enclosed in integument in seed plants; contains specialized tissues and an egg cell, which will become a seed after fertilization (Pl-8 through Pl-12)

ovuliferous scale In conifer cones, plant tissue on which female reproductive structures develop (Pl-10); a female cone scale (Pl-11)

ovum (pl. **ova**) Egg (Pr, A, Pl)

palintomy Rapid sequence of binary fissions, typically within a cyst and with little or no intervening growth, resulting in production of numerous small offspring cells; common in various parasitic protists (for example, ciliates and dinomastigotes; Pr-7, Pr-8)

palp In molluscs, a ciliated flap that sorts food particles (A-26)

pansporoblast Membrane surrounding multinucleate structure; syncytium that undergoes cytokinesis to yield parasites; gives rise to two sporoblasts contained within a single membrane in some apicomplexans (Pr-9); in actinosporean myxozoans (Pr-24), two- to four-celled envelope containing groups of eight spores and sometimes called a pansporocyst

papilla (pl. **papillae**) A small bump (A-28)
parabasal body Organelle located near kinetosomes; resembles dictyosome and Golgi apparatus in structure and probably in function (Pr-30)
parabasal folds Creases in the parabasal body (Pr-30)
paralarva (pl. **paralarvae**) Juvenile stage of certain cephalopod molluscs (A-26)
paramylon Carbohydrate composed of glucose units; forms the reserve foodstuff of some protoctists (Pr-10, Pr-12)
paraphyletic Refers to more inclusive taxa that include ancestral taxon as well as some but not all descendants [for example, gymnosperms are paraphyletic because flowering plants (Pl-12), descendants of certain gymnosperms, are excluded]
paraphysis One of several cellular trichomes, threadlike structures that surround the egg in the phaeophyte oogonium (Pr-17), in mosses (Pl-1), and in ascomycotes (F-3)
parapodium (pl. **parapodia**) Fleshy, segmental appendage of polychaete annelid worms (A-22)
paraproct One of a pair of ventrolateral lobes bordering the anus of insects (A-20)
parapyle Tubular aperture of the central capsule of radiolarians of the suborder Phaeodorina (Pr-27)
parasexual cycle Production of individuals having more than one parent without meiosis and fertilization (B, Pr, F)
parasite Organism that lives on or in an organism of a different species and obtains nutrients from it
parasomal sac Structure associated with each kinetid of the cortex of ciliates (Pr-8)
Parazoa Subkingdom comprising animals (parazoans) that lack tissues organized into organs; placozoans, and sponges (A-1, A-2)
parenchyma Tissue made of unspecialized cells capable of growth in three dimensions (Pr, A, F, Pl); in nemertines and nematomorphs, mesodermally-derived tissue loosely packed between gut and body wall (A-9, A-11)
parenchymula A solid, ciliated larval type of sponge (A-2)
parthenogenesis Development of an unfertilized egg into an organism (A)
pathogen Organism that causes disease (B, Pr, A, F)
peat Partly decayed, compressed plant material composed of peat mosses and other plants (Pl-1)
pebrine Disease of silkworm larvae (A-20)
pectin Complex carbohydrate found between plant cells and in their cell walls (Pl)
pedal disk Structure that attaches sedentary cnidarian such as sea anemone to substrate (A-3)
pedal gland Gland that secretes adhesive for attaching an animal to a substrate (A-10 through A-14, A-16 through A-18); some pedal glands are also called cement glands or adhesive tubes
pedal laceration Mode of asexual reproduction of sea anemone by splitting off part of pedal disk (A-3)
pedicel Attachment stalk, holdfast [for example, in some chonotrich ciliates (Pr-8) and choanomastigotes (Pr-30)]; elongated protrusion from the posterior end of a cell
pedicle Stalk that attaches a bryozoan to substrate (A-30); see *pedicel*
pedipalp One of several paired, jointed head appendages of chelicerates such as horseshoe crab (A-19)
peduncle Holdfast, base, stalk, or stemlike structure (Pr, A)
pelagic Dwelling in the open ocean
pelagosphera Posttrochophore, feeding larval stage of certain sipunculans (A-23)
pellicle Thin, typically proteinaceous outer layer of a cell or organism, outside plasma membrane
pen Skeleton or internal horny shell of cephalopods (A-26)
penis Male organ of copulation (A-5, A-12, A-21, A-26, A-37)
perfect Denoting complete set of sexual reproductive structures in a fungus or plant (F-3, Pl)
periatrial Pertaining to region around atria (A)
periaxostylar bacteria Endosymbiotic bacteria found around the axostyle (Pr-30)
periaxostylar ring Morphologically distinctive ring of material surrounding the axostyle (Pr-30)

pericardial Pertaining to region around the heart (A-26)

perihemal One of several coelomic systems of echinoderms (A-34)

perinuclear Surrounding the nucleus

periostracum Proteinaceous outer layer of a shell (A-26, A-30)

periplasm Peripheral cytoplasm; in prokaryotes, the space between the inner plasma membrane and the peptidoglycan layer of the cell wall (B)

peristalsis Successive contraction and relaxation of transverse muscles of tubular organs, resulting in the movement of contained fluid and solids in one direction (A)

peristome Circle of teeth under the capsule lid of moss sporangia (Pl-1)

perithecium (pl. **perithecia**) Closed ascocarp with a pore, or ostiole, at the top and a wall of its own (F-3)

peritoneum Mesodermal membrane lining the body cavity (true coelom) and forming the external covering of the visceral organs (A)

peritrichous Bearing flagella uniformly distributed over the body surface (B)

peritrophic membrane In onychophorans, membrane that forms around ingested food in the digestive tract (A-28)

permafrost Permanently frozen soil

petal Flower part; one of the (usually colored) leaves of a corolla (Pl-12)

petiole Stalk of a leaf (Pl-8 through Pl-12)

petrify To replace original substance of an organism by mineral, as in petrified plant fossil (Pl-9)

pH Scale for measuring the acidity of aqueous solutions; pure water has a pH of 7 (neutral); solutions having a pH greater than 7 are alkaline; less than 7 are acid

phaeoplast Brown plastid, the membrane-bounded photosynthetic structure of brown algal cells (Pr-17)

phagocytosis Ingestion, by a cell, of solid particles by flowing over and engulfing them whole (Pr, A)

phagotroph Organism with mode of heterotrophic nutrition in which particulate food, bacteria sized or larger, is ingested

phaneroplasmodium Largest and most conspicuous of the three types of plasmodia formed by myxomycotes, primarily of the order Physarales (Pr-6); plasmodium consisting of thin, fanlike advancing regions and a branching network of veins; veins consist of an outer gel zone of protoplasm and an inner fluid zone, in which protoplasm streams

pharyngeal basket Basket-shaped pharynx that is respiratory and feeding structure of tunicates (A-35)

pharynx Throat; part of the alimentary tract between the mouth cavity and the esophagus (A)

phasmid Sense organ of parasitic nematodes (A-10)

phenotype Physical manifestation of a trait in an organism; determined by genotype plus environment

pheromone Hormone transported through air to its target organism; if the hormone is produced by one sex and responded to by the other sex, that hormone is a sex hormone (A)

phloem A plant vascular tissue that transports nutrients; in trees, the inner bark (Pl-4 through Pl-12)

phosphorescence Luminescence caused by reemission of absorbed radiation as light that continues after the impinging radiation has ceased

photic zone Ecological term referring to the illuminated part of a water column, soil profile, microbial mat, etc.; the layer in which, because of the penetration of light, photosynthesis can take place; also called euphotic zone

photoautotroph Organism able to produce all nutrient and energy requirements by using inorganic compounds and visible light (B-3, B-5, Pl)

photobiont Photosynthetic partner in a symbiotic association, such as algal partner in a lichen; also called phycobiont (F-3)

photoheterotroph Photosynthetic organism that can also obtain nourishment heterotrophically, such as *Euglena* (Pr-12)

photolithoautotroph Photosynthetic organism that obtains metabolic energy from light, electrons from inorganic species such as water or hydrogen sulfide, and carbon from CO_2 (for example, green plants and algae)

photoreceptor Specialized accumulation of pigment that reacts to light stimuli (B, Pr, A, F)

photosynthate Chemical products of photosynthesis (B, Pr, Pl)

photosynthesis Production of organic compounds from carbon dioxide and water by using light energy captured by chlorophyll (B, Pr, Pl)

phototactic Moving in response to light (A-29)

phototaxis Movement toward a light source (A, F)

phototrophy The process of obtaining metabolic energy from light, as in photosynthesis

phragmoplast See *cell plate*

phycobilin One of various blue-green or reddish pigments that take part in photosynthesis in cells of cyanobacteria and rhodophytes; see *phycocyanin* and *phycoerythrin* (B-5, Pr-12)

phycobiont Phototrophic component of lichen partnership (B-5, Pr-28, F-3)

phycocyanin Type of phycobiliprotein; water-soluble extract is blue; found in cyanobacteria (B-5), rhodophytes (Pr-25), and cryptomonads (Pr-11)

phycoerythrin Type of phycobiliprotein; water-soluble extract is red; found in cyanobacteria (B-5), rhodophytes (Pr-25), and cryptomonads (Pr-11)

phycology Study of algae; algology

phyllidium (pl. **phyllidia**) Solid, cylindrical appendage on the shoot of the gametophyte of *Takakia;* unique to *Takakia* (Pl-1)

phyllosoma One of several types of decapod crustacean larvae (A-21)

phylogeny Evolution of a genetically related group of organisms; also, a schematic diagram representing that evolution

phytolith Plant stone; opaline or calcareous stones produced by some plants that tend to preserve well in the fossil record (Pl); stony material part of a living plant or alga that secretes mineral matter (Pr, Pl)

phytoplankton Free-floating microscopic photosynthetic organisms (Pr)

pileus (pl. **pilei**) Cap or upper part of certain ascocarps and basidiocarps (F-2, F-3); the cap of a mushroom

pilidium Free-swimming larval stage of a nemertine (A-9)

pilus (pl. **pili**) Hairlike structure on bacterial cells that functions in adhesion to surfaces; also may take part in conjugation

pinacocyte Epithelial cell of sponges (A-2)

pinna (pl. **pinnae**) Primary subdivision of a megaphyll (fern frond or leaf); may be further subdivided into pinnules (Pl-7 through Pl-12)

pinnule Protrusion of tentacles of pogonophoran worm (A-25); division of pinna of fern leaf (Pl-7)

pinocytosis Ingestion, by a cell, of liquid droplets by engulfing them whole (Pr, A)

pioneer plant One of the first plants or lichens to colonize an area devoid of vegetation (F-3, Pl)

pit connection Protoplasmic connection joining cells into tissues (Pr-25, F-2, Pl)

pith Tissue, usually parenchyma, occupying the center of the stem within the vascular cylinder of a plant (Pl-4, Pl-6 through Pl-12)

placenta Temporary structure formed in part from the inner lining of the maternal uterus and in part from the embryonic membranes; facilitates the exchange of nutrients and wastes between embryo and mother (A-28, A-37)

placoid scale Cartilaginous fish scale that is covered by enamel and composed of dentine (A-37)

plankton Free-floating microscopic or small aquatic organisms, both photosynthetic and heterotrophic (Pr, A)

planogamete Motile gamete having undulipodia (Pr)

plantlet Small plant that is produced asexually and shed, forming new individual club moss, fern, cycad, or flowering plant (Pl-4, Pl-7, Pl-8, Pl-12)

planula (pl. **planulae**) Ciliated, free-swimming larva of cnidarian coelenterates (A-3)

plasmalemma Cytoplasmic membrane; cell membrane; cell envelope; plasma membrane

plasmid Bacterial DNA fragment that is not part of the genophore; circular in organization; not always essential for growth

plasmodesma (pl. **plasmodesmata**) Strands of living cytoplasm that interconnect plant cells through pores in cell wall (Pr, Pl)

plasmodium Multinucleate mass of cytoplasm lacking internal cell boundaries (membranes, walls), cells, syncytium, coenocyte (Pr-6, Pr-19)

plasmogamy Fusion of two cells or protoplasts without fusion of nuclei (Pr, F)

plastid Cytoplasmic, photosynthetic pigmented organelle (such as a chloroplast) or its nonphotosynthetic derivative (such as a leukoplast or etioplast) (Pr, Pl)

pleural region Membranous sidewall of an insect thorax (A-20)

pleurite Part of insect exoskeleton that covers the thorax (A-20)

ploidy level The number of complete chromosome sets present in a cell [for example, haploid (one set) versus diploid (two sets) versus polyploid (more than two 2 sets); all human genes are present on the 23 chromosomes in a haploid sperm or egg cell; a diploid body cell contains 46 chromosomes (23 pairs)]

podium (pl. **podia**) Food-gathering or locomotory extension of the cytoplasm of a cell; tube foot of an echinoderm (A-34)

poikilotherm Organism unable to internally regulate its body temperature (A)

polar nucleus Nucleus derived from one of the two poles of the embryo sac; usually two in number (one from each pole), the polar nuclei travel to the middle of the embryo sac and join a male nucleus to form (triploid) endosperm (Pl-12)

polaroplast Structure consisting of a series of flattened sacs and vesicles, thought to take part in polar tube extrusion in microsporan spores (Pr)

pollen cone Male pollen-producing cone (strobilus) (Pl)

pollen sac Flower structure consisting of hollow interior of anther in which pollen is produced (Pl-12)

pollen tube Nucleated tube formed by germination of the pollen grain; it grows down the style of the carpel and carries the male nuclei to the female nuclei in the ovary (Pl-9 through Pl-11) or is haustorial (Pl-8 through Pl-12) or both

pollination Transfer of pollen grains from male to female reproductive parts (Pl-8 through Pl-12)

pollination droplet Drop of liquid that pulls pollen into the ovule (Pl-8, Pl-9, Pl-11)

polymastigote Motile heterotroph with few (less than 16) undulipodia (Pr-30)

polymorphism Morphological or genetic differences between individuals that are members of the same species; see *monomorphic*

polyp Coelenterate body form; cylindrical tube closed and attached at one end, open at the other by a central mouth typically surrounded by tentacles (A-3)

polyphyletic Of a trait or group of organisms derived by convergent evolution from different ancestors

polyploidization The production of a cell whose nucleus contains more than the diploid number of chromosomes (Pr, A, F, Pl)

polysaccharide Carbohydrate composed of many monosaccharide units joined to form long chains (for example, starch, cellulose, glycogen, and paramylon)

pore Opening in an organism to exterior, as in sponges and basidiomycote fungi (A-2, F-2, Pl-2)

porocyte Tubular sponge cell through which water moves into the cavity of the sponge (A-2)

porphyrin Nitrogen-containing heterocyclic organic compound; derivatives include chlorophyll and heme

postannular papilla Small bump that probably secretes chitinous tube material or adhesive; part of a pogonophoran (A-25)

posterior Pertaining to the rear end

predation Mode of capturing and consuming animals, as in predatory insects (A-22)

predator Animal that captures and consumes other animals (A-9, A-26, A-32, A-34, A-36)

prehensile Capable of grasping (for example, prehensile arms of octopus) (A-26)

presoma Barrel-shaped, retractile body region anterior to the trunk of priapulids (A-15)

primary growth Growth that arises from apical meristems of root and shoot and gives rise to increase in length in plants; compare *secondary growth* (Pl)

proboscis Tubular protrusion or prolongation of the head or snout (Pr, A)

progymnosperm Large Devonian trees, now extinct, that had fernlike leaves and the first true wood; likely ancestors of extant gymnosperms (Pl-10)

prokaryote Cell or organism composed of cells lacking a membrane-bounded nucleus, membrane-bounded organelles, and DNA coated with histone proteins (B); members of Bacteria (Prokaryotae, Monera) kingdom

proloculus Initial test formed by free-swimming foraminiferan; precedes mature, multichambered test (Pr-4)

propagule Uni- or multicellular structure produced by an organism and capable of survival, dissemination, and growth; generative structure (for example, cyst, spore, overwintering egg, seed) (A, F-1 through F-3, Pl)

prosoma Anterior body region (fused cephalothorax) of chelicerates (A-19)

prostomium (pl. **prostomia**) Preoral lobe of an annelid (A-22)

protandria Type of hermaphroditism in which individual organism is first male and then female (A-18)

protandric Refers to hermaphrodite that is first male and then female (A-26, A-29)

protandry Prevention of self-fertilization in hermaphroditic animals by production of sperm and eggs at different times (sperm first)

protein Macromolecule composed of amino acid subunits (B, Pr, A, F , Pl)

proter New anterior cell formed by binary fission of ciliates (Pr-8)

prothallus (pl. **prothalli**) In ferns, the photosynthetic gametophyte; also called prothallium (Pl-7)

protocoel Anterior part of coelom in freshwater ectoprocts (A-29)

protogynia Type of hermaphroditism in which individual organism is first female and then male (A-18); see *protandria*

protonema (pl. **protonemata**) Filamentous or platelike structure produced by germinating spore (Pl-1 through Pl-7)

protonephridium (pl. **protonephridia**) Rudimentary tubular excretory organ; has a blind inner end, typically occupied by a ciliated flame cell, and an open outer end, or external pore, through which it discharges collected fluid wastes (A)

protoplasmic cylinder Long thin cytoplasm and nucleoid of a spirochete; the part of a spirochete bounded by the plasma membrane (B-4)

protoplast Nucleus and cytoplasm of a cell after cell wall is removed

protostome Includes all animals in which the blastopore becomes the mouth (A-19 through A-28); series of Grade Coelomata

prototroch Most apical girdle of cilia on a trochophore larva (A-22, A-26)

protozoan Informal name of a member of the animal phylum Protozoa in the traditional two-kingdom (animal and plant) classification system; consists mostly of heterotrophic, microscopic eukaryotes (small protoctists and their photosynthetic relatives); obsolete term

psammophile Organism that lives in sand, especially in the spaces between sand grains (Pr, A)

pseudergate Worker insect (A-20)

pseudocoelom Internal body cavity lying between outer body wall and digestive tract; not lined with mesoderm (A-10 through A-14, A-17, A-18); also called pseudocoel

pseudocoelomate Animal having a pseudocoelom (A-10 through A-14, A-17, A-18)
pseudocone In certain cycads, loose array of megasporangia in a leafy crown (Pl-8)
pseudofruit In gnetophytes, seed enclosed in fleshy integument that is not an angiosperm fruit (Pl-11)
pseudopod Temporary cytoplasmic protrusion of an ameboid cell; used for locomotion or phagocytotic feeding (Pr, A)
psittacosis Disease of birds caused by rickettsia (B-3); can be transmitted to people
psychrophil Organism whose optimum environmental temperature is well below 15°C (B, Pr)
Pteridophyta Alternative phylum name for Filicinophyta (Pl-7)
pteridosperm Extinct group of gymnosperm that bore seed rather than spores and had fernlike leaves (Pl)
Pterodatina Alternative phylum name for Filicinophyta (Pl-7)
Pterophyta Alternative phylum name for Filicinophyta (Pl-7)
pusule Fluid-filled intracellular sac responsive to pressure changes (Pr-7)
pycnidium (pl. **pycnidia**) Hollow reproductive structure of fungi that is lined with conidiophores (F-3)
pycnotic Darkly staining, as of nongerminative nuclei or those in moribund cells (Pr, A, F, Pl)
pyloric stomach More aboral of the two stomachs of echinoderms (A-34)
pyrenoid Proteinaceous structure inside some plastids; serves as a center of starch formation (Pr, Pl-3)
pyriform Having the form of a tear or pear

rachis Axis from which the pinnae or leaflets of a frond or a compound leaf arise (Pl-7 through Pl-12)
radial canal (1) Part of the digestive system of medusae (A-3); (2) tube of the water vascular system that passes into each ray and to the tube feet of echinoderms (A-34); (3) canals of some sponges (A-2)
radial symmetry Regular arrangement of parts around one longitudinal axis; any plane through this axis will divide the object or organism into similar halves
radius (pl. **radii**) Inner bone of forearm (on the side of the thumb) between elbow and wrist of vertebrates (A-37)
radula (pl. **radulae**) Horny toothed organ of molluscs (A-26); used to rasp food and carry it into the mouth
raphe In most pennate diatoms, slit, elongate cleft, groove, or pair of grooves through the valve that facilitates gliding cell motility (Pr-16)
ray Stiff projection, as in fin ray (A-32)
receptacle Fertile area at tip of gametophyte (Pr-17)
recombinant Refers to progeny organisms, cells, or DNA molecules bearing gene combinations that differ from the combinations in the parents
recombination The appearance, in progeny, of gene combinations that differ from the combinations in the parents
redia One of several types of trematode (fluke) larvae (A-5)
reduction potential Electric potential (voltage) required to reduce (donate electrons to) a chemical species
reef Underwater ridge of coral and other organisms (A-3, A-23, A-29, A-34)
regenerate To grow again (for example, to regrow a lost tentacle (A-1, A-3, A-9, A-15, A-22, A-31)
regeneration Developmental process of regrowing a lost body part, for example, regeneration of amputated limb or plant part (A-4, A-5, A-23, A-24, A-32, A-34, Pl)
regolith Loose material consisting of soil, sediments, or broken rock overlying solid rock
renal pore Pore through which mollusc kidney discharges waste into mantle cavity (A-26)
renal sac Structure in which some tunicates accumulate excretory waste (A-35)
replicon Product of DNA replication
reproduction Increase in number of organisms or of cells
reservoir Vestibule or holding structure; deep part of the oral structure of some protoctists (Pr-8, Pr-12)
resin Viscous organic substance secreted by conifers; protective against fungus and insect attack (Pl-10)
resin duct Intercellular space, as in needles and wood of

conifers, that is lined with resin-secreting cells and contains resins (Pl-10)

respiration Oxidative breakdown of food molecules and release of energy from them; the terminal electron acceptor is inorganic—it may be oxygen (B, Pr, A, F, Pl) or, in anaerobic organisms, methane, nitrate, nitrite, or sulfate (B)

respiratory membrane Intracellular membrane in which enzymes for oxygen or nitrogen respiration are embedded (for example, in mitochondria and bacteria)

respiratory tree Gas-exchange organ consisting of diverticuli of cloaca in sea cucumbers (A-34)

reticulate Netlike or covered with netlike ridges

reticulipod Pseudopod that is part of a network of cross-connected pseudopods (Pr-4)

rhizoid Rootlike structure that anchors and absorbs (Pr, A-29, F-1, Pl-1 through Pl-3, Pl-5 through Pl-7)

rhizome Subterranean or creeping stem; differing histologically from a root, it sends out shoots from near its tip or at nodes along its length (Pl-1, Pl-4 through Pl-12)

rhizomycelium (pl. **rhizomycelia**) Rhizoidal system extensive enough to resemble mycelium superficially (Pr-29)

rhizoplast Basal rootlet, proteinaceous striated strand connecting the nucleus and a kinetosome; some contractile and calcium sensitive (Pr)

rhodoplast Red plastid, the membrane-bounded photosynthetic structure of red algal cells (Pr-25)

rhombogen Sexually mature life-cycle stage of rhombozoans (A-7), attained only when the population is dense

rhynchocoel In nemertines, cavity into which proboscis can be retracted (A-9)

rhyniophyte Extinct plant having a rhizome with rhizoids and a leafless stem that branched dichotomously; ancestral to Pl-5 through Pl-12

ribosome A spherical organelle composed of protein and ribonucleic acid; the site of protein synthesis

ring canal Ring-shaped tube of the water vascular system of echinoderms (A-34)

RNA Ribonucleic acid; a molecule composed of a linear sequence of nucleotides; can store genetic information; takes part in protein synthesis; see *messenger RNA* and *rRNA*

RNA core Inner, genetic part of an RNA-type virus

roe Eggs of fish (A-37) or sea urchins (A-34)

root Plant organ that absorbs water and soil nutrients (Pl-4 through Pl-12)

root nodule Spherical growth on the roots of leguminous plants; contains nitrogen-fixing bacteroids

rosette Star-shaped structure composed of radiating parts (A-17, A-34)

rostrum In certain gnathostomulid worms, the anterior-most body region (A-6); term describing apical end of cell when it is beak shaped or has a protuberance (especially ciliates or mastigotes; Pr); head; usually less conspicuous than a proboscis

rRNA Ribosomal RNA, the class of RNA that, together with proteins, makes up the substance of ribosomes; approximately 85 percent of all RNA in the cell is ribosomal RNA

rumposome Intracellular structure; honeycomb-like organelle of unknown function consisting of regularly fenestrated membrane vesicles in chytrid zoospores (Pr-29)

runner Horizontal offshoot of plant (for example, strawberry runner) that produces plantlet (Pl)

Saefftigen's pouch Fluid-filled sac—seminal vesicle—used during mating by an acanthocephalan male (A-12)

sagenogen Limited to Phylum Labyrinthulata, organelle on cell surface from which the ectoplasmic network arises (in the sagenogen, an electron-dense plug separates the cytoplasm from the matrix of ectoplasmic network); also called bothrosome or sagenogenosome (Pr-18)

salivary gland Organ that secretes saliva (A-28)

samara Dry, one-seeded, indehiscent winged fruit (Pl-12)

saprobe Osmotroph, organism that excretes extracellular digestive enzymes and absorbs dead organic matter (B, Pr, F)

saprobic Refers to nutritional mode of being a saprobe (F-1 through F-3)

saprophyte See *saprobe*

sapwood Part of the wood that is living xylem in which water is conducted and which is external to heartwood in woody plants (Pl-9, Pl-10, Pl-12)

scale Thin, hard surface plate that overlaps with others to coat a surface (Pr, A); small rudimentary or vestigial leaf (for example, those that protect the buds of a plant) (Pl)

scalid In kinorhynchs, of spines on the head (A-14); in priapulids, of spines on the introvert (A-15); in loricifer-ans, of spines on the head and neck (A-17)

scapula (pl. **scapulae**) Shoulder blade of vertebrates (A-37)

scavenger Animal that consumes decaying organic material (for example, some sea stars) (A-5, A-20, A-22, A-34)

schistosomiasis See *bilharziasis*

schizogony Multiple mitoses without increase in cell size; gives rise to schizonts (Pr-9)

schizont Organism that is the product of schizogony or that will give rise to more schizonts by schizogony (Pr-9)

sclerocyte Cell that secretes spicules of sponges (A-2)

Scyphozoa Class of cnidarian coelenterates (A-3); jellyfish

seasonal gonad Temporary reproductive organ present only at some times of year (A-22 through A-24)

secondary compound Molecule that may act in plant development and ecology but is not absolutely necessary (Pl)

secondary growth In plants, growth that results in increase in diameter and is derived from vascular and cork cambium; compare *primary growth* (Pl-8 through Pl-12)

sedentary Refers to bottom-dwelling animal capable of limited locomotion; compare *sessile* (A-22, A-31, A-33)

seed Propagule of seed plant (Pl-8 through Pl-12)

seed coat In seed, outer layer that develops from integuments of the ovule (Pl-8 through Pl-12)

seed cone Cone that bears seeds (Pl-8, Pl-10)

seed fern Extinct Carboniferous gymnosperm that had fernlike fronds and bore microsporangia and seeds (for example, *Medullosa*); also called pteridosperm (Pl)

seed leaf Cotyledon (Pl-11, Pl-12)

seed plant Any plant that bears seed (Pl-8 through Pl-12)

seedless plant Plant species lacking seeds (Pl-1 through Pl-7)

seedling Young sporophyte that sprouts from a plant seed (Pl-8 through Pl-12)

segment In arthropods, part of appendage between joints (A-19 through A-21); in annelids, one of a number of repeat units of the body (A-22)

semen Complex nutrient fluid in which male gametes (sperm) are transferred to a female (A)

seminal fluid Liquid that transports sperm (A-12)

seminal receptacle Organ or vessel that receives semen (A)

seminal vesicle Saclike structure containing semen (A-33)

sensorium Entire sensory apparatus (A-6)

sensory plate Apical mass of nerve tissue in trochophore larvae (A-22)

sepal Flower part; one of the leaves of a calyx (Pl-12)

septum (pl. **septa**) Cross wall in a plasmodium, trichome, or hypha (B, Pr, F)

sessile Attached, not free to move about

seta (pl. **setae**) Bristle

sexual apex Sexually differentiated anterior or top part of a thallus (Pr-17, Pr-25)

sexual dimorphism Pertaining to differences in size or shape (morphology) between females and males of a species (A-21, A-24)

sexual reproduction Reproduction that requires more than a single parent and leads to individual offspring

sheath External layer or coat of bacteria; typically composed of mucopolysaccharides (B)

shoot Aerial part, such as stem and leaves, of plant (Pl)

shrub Woody perennial plant that has multiple stems; compare to tree (Pl-8, Pl-10 through Pl-12)

siliceous Containing silica, as in sponge spicules (A-2)

simple cone Type of cone in which sporangium attaches directly to the cone axis; characteristic of male conifer cones (Pl-10); compare *compound cone*

simple eye Visual organ composed of a single eye; compare *compound eye* (A-20)

simple leaf Leaf having an undivided leaf blade; compare *compound leaf* (Pl)

sinus Space or hollow, sometimes fluid filled (for example, blood sinus of horseshoe crab and forehead sinus of mammal) (A-19, A-26, A-33, A-37)

siphon Tubular organ used for drawing in or ejecting fluids (Pr, A)

skeleton Hardened biogenic scaffolding, structural material often composed of calcium carbonate, silica, or calcium phosphate (Pr, A)

skull Cranium: bony box enclosing the brain of craniate chordates (A-37)

slurp gun Device for collecting aquatic organism by slurping it into a cylinder (A-4)

solenocyte Undulipodiated cell at the inner end of a protonephridial tube; distinguished from a flame cell by having only a single long undulipodium (A)

soma The parts of an organism's body that lack genetic continuity, as distinguished from its gametes (eggs, sperm) or reproductive structures (gemma, spores), which have genetic continuity through time

somatic Refers to parts of a body (soma) other than gametes or reproductive structures (A-13)

somatic cell Differentiated cell composing the tissues of the soma (body); any body cell except the germ cells (Pr, A, Pl)

somatic nucleus Nucleus incapable of further division or of fertilization followed by further division (Pr-4, Pr-8)

soredium (pl. **soredia**) Asexual reproductive structure of lichens; fragment containing both fungal hyphae and algal cells that disseminates lichen (F-3)

sorocarp Fruiting body (F, Pr, Pl)

sorus (pl. **sori**) Cluster of sporangia (Pl, F); found, for example, on the underside of fern megaphylls (Pl-7)

sperm Male gamete, of animal or plant (A, Pl)

sperm nucleus The nucleus of the pollen that fertilizes the egg or the two polar nuclei to form the zygote (diploid) or the endosperm nucleus (triploid); see *tube nucleus* (Pl)

spermary In nematomorph (A-11) and cnidarians (A-3), testes or male organ producing sperm

spermatium (pl. **spermatia**) Nonmotile male gamete of red algae (Pr-25); male gamete of certain ascomycete fungi (F-3)

spermatophore Packet in which sperm are wrapped and passed to a second individual in mating (A-14, A-22, A-25, A-26, A-28, A-32)

sphincter Ring-shaped muscle capable of closing a tubular opening by constriction (A)

spicule Slender, typically needle shaped biogenic crystal (for example, those secreted by sponge cells for skeletal support) (A-2)

spindle pole body Nucleus-associated organelle, granulofibrosal and microtubular material found at the poles of mitotic spindles; centriole (Pr, A, F)

spinneret Abdominal appendage through which spider extrudes silk (A-19)

spiracle Opening of gas-exchange organ to the environment (for example, spiracle of chelicerate book lung) (A-19, A-20, A-28)

spirillum (pl. **spirilla**) Spiral or helical filamentous cell (B-3, B-4, B-6)

spongin Horny, sulfur-containing protein that makes up the flexible skeleton of some sponges (A-2); related to the keratin of vertebrate hair and nails

spongocoel Central body cavity of a sponge (A-2)

sporangiophore Branch or hypha bearing one or more sporangia (F, Pl-6)

sporangiospore Spore borne within a sporangium (F)

sporangium (pl. **sporangia**) Hollow unicellular or multicellular structure in which spores are produced and from which they are released (Pr, F, Pl)

spore Small or microscopic propagule containing at least one genome and capable of maturation, often desiccation and heat resistant (B, Pr, F, Pl)

spore wall Outer tough resistant complex layer of spore; composed of peptidoglycan impregnated with dipicolinic acid (B-1, B-3, B-10, Pl) or sporopollenin (Pl)

sporoblast Structure giving rise to spores (for example, myxozoans, Pr-24); elliptical, nucleated structure pointed at the end, the result of a process of protoplasmic segmentation in apicomplexans (Pr-9)

sporocarp Spore-bearing structure often of protoctists; fruiting structure erect with spores or cysts at apex

sporocyst One of several types of trematode (fluke) larvae (A-5)

sporogony Multiple mitoses of a spore or zygote without increase in cell size; produces sporozoites (Pr-9)

sporont Stage in the life cycle that will form sporocysts [for example, in coccidians (Pr-9), zygote within the oocyst wall], sporoblasts (haplosporidians, Pr-22), or spores (paramyxeans, Pr-23)

sporophore Structure that bears spores (Pr, F, Pl)

sporophyll Leaf that bears sporangia (Pl)

sporophyte Spore-producing diploid plant (Pl)

sporophytic Of the sporophyte generation (Pl)

sporoplasm Infective body (Pr-9); ameboid organism within a spore (Pr-2)

sporozoite Life-cycle stage of apicomplexans (Pr-9); motile product of multiple mitoses (sporogony) of zygote or spore; usually infective

sporulation Development of spores (F-1)

spur Stubby projection (for example, nonadhesive toe of rotifer, A-13)

spur shoot Short stubby branch (Pl-9, Pl-10); compare *long shoot*

stachel Bulletlike structure of plasmodiophorids (Pr-19) contained in the rohr, whose pointed end is oriented toward the approsorium (germ tube specialized for nutrient uptake; penetrates cell wall) and the host cell wall; German, meaning stinger or spine

stalk In plant, main axis or stem (Pl); in fungi, stalk of mushroom; also called stipe (F-2)

stamen Male flower part composed of a filament (stalk) and pollen-bearing anthers (microsporangia) (Pl-12)

star cell Contractile cell that causes nutrients to move inside an entoproct (A-18)

statoblast Overwintering armored cell ball in ectoprocts; asexually produced propagule produced under severe conditions, disseminated, and grows into new individual (A-29)

statocyst (1) Organ of balance; a vesicle containing granules of sand or some other material that stimulate sensory cells when an animal moves (A-3, A-4); (2) silicified chrysomonad cell (Pr-13) that overwinters or resists desiccation

statolith In comb jellies, calcareous grain in statocyst, the organ of balance that responds to gravity (A-4)

stem In plants, main aerial axis or stalk as well as underground parts that have similar tissue, such as rhizome (subterranean stem) (Pl-5 through Pl-12); in mushrooms, stipe or stalk (F-2)

stercomare Mass, usually formed as strings of stercomes (xenophyophore fecal pellets) lumped together in large numbers and covered by a thin membrane, produced by xenophyophores (Pr-5)

stereoblastula One of several patterns of blastulae (A)

sterile (1) Unable to reproduce; (2) devoid of all life

sternite Ventral part or shield of a cuticle of an arthropod (A-20); also called sternum

sternum (pl. **sterna**) Breastbone (A-37)

steroid Class of certain organic compounds (such as cholesterol and sex hormones) synthesized by animal, fungal, and plant tissue (A, F, Pl)

stigma (1) Female flower part; receptive surface of carpel on which pollen germinates (Pl-12); (2) eyespot (Pr-15)

stipe Stalk of an organ or organism (F-2, Pl-7)

stolon Horizontal stalk near the base of a plant or a colonial animal; produces new individuals by budding (A, Pl)

stoma (pl. **stomata**) One of many minute openings bordered by guard cells in the epidermis of leaves and stems; route of gas exchange between plant and air (Pl)

stone canal Calcareous tube leading from the madreporite to the ring canal of the water vascular system of echinoderms (A-34)

stramenopiles Member of a group of taxa defined by the presence of cells with dissimilar undulipodia (heterokonts) and tubular mastigonemes (Pr-13 through Pr-21); Greek, meaning straw bearer

stratigraphy Field of geology that deals with the origin, composition, distribution, and succession of sediments (strata)

strobilus (pl. **strobili**) Cone; modified ovule-bearing leaves or scales grouped together on an axis (Pl-4, Pl-6)

stroma Support tissue formed by ascomata (F-3)

stromatolite Laminated carbonate or silicate rocks;

organosedimentary structures produced by growth, metabolism, trapping, binding, and precipitating of sediment by communities of microorganisms, principally cyanobacteria (B-5)

style Female flower part; slender column of tissue that arises from the top of the ovary and down through which the pollen tube grows (Pl-12); in animals, crystalline style (A)

stylet Rigid elongated organ or appendage (A)

substrate (1) Foundation to which an organism is attached; for example, a rock or carapace; (2) compound acted upon by an enzyme

subtidal Below the low-tide level

sulcal groove In dinomastigote hypocone (Pr-7), groove in which longitudinal undulipodium lies

sulfuretum Marine habitat in which the sand smells of rotten eggs because of bacterial reduction of sulfate to hydrogen sulfide; community in sulfide-rich sediment (A-6)

supratidal Normally just above the reach of high tide but inundated during storms

swimmeret Appendage of crustacean located beneath the abdomen and used in swimming (A-21)

symbiogenesis Process by which organelles of the eukaryotic cell (for example, mitochondria, chloroplasts, and kinetosomes) evolved from ancient bacterial endosymbionts

symbiont Member of a symbiosis—that is, organism that has protracted and intimate association with one or more organisms of a different species (A, F, Pl)

symbiosis Intimate and protracted association between two or more organisms of different species (B, Pr, A, F, Pl)

symbiotic Engaging in symbiosis (A, F, Pl)

symbiotroph Heterotrophic member of a symbiotic association that derives carbon and energy from a living partner (A-4, A-7, A-8, A-11, A-14, F-1 through F-3)

symbiotrophic Engaging in mode of nutrition that involves symbiotrophs (A-10, A-13, A-21, A-24, F)

synangium (pl. **synangia**) Two or more sporangia joined together (Pl-5)

syncytium (pl. **syncytia**) Multinucleate mass of cytoplasm lacking internal cell boundaries (membranes, walls) (Pr, A); see *plasmodium*

syngamy Fertilization; process by which two haploid cells fuse to form a diploid zygote

tactile Refers to sensory organ receptive to touch (A-6)

taproot Stout, tapering main root from which arise smaller, lateral branches (Pl-9, Pl-12)

taxon (pl. **taxa**) Unit that classifies all extant organisms in hierarchy of inclusiveness (for example, kingdom, phylum, class, order, family, genus, species) (B, Pr, A, F-1, Pl)

teliospore In basidiomycete fungi, thick-walled spore from which basidia develop; propagule formed by rust or smut (F-2)

telotroch Most posterior circlet of cilia on a trochophore larva (A-22, A-26)

tentacle Appendage or "arm" of coelenterate (A-3, A-4)

tergite Body plate on the dorsal surface of arthropod (A-20)

terminal electron acceptor In a metabolic pathway, the compound finally reduced by the acceptance of electrons and converted into the compound to be excreted

terrestrial Dwelling on land (A-37, F, Pl)

test Shell, hard covering, valve, or theca (Pr, A)

testis (pl. **testes**) Male gonad or male gamete-producing organ (A-4, A-6, A-8, A-10, A-12, A-14, A-17, A-19, A-27, A-28, A-31, A-32, A-37)

tetrapod Animal having four legs—amphibians, reptiles, mammals (A-37)

tetrasporangium Structure in which tetraspores are formed (Pr-25, Pr-28)

tetraspore One of a set of four spores, products of meiosis, that will develop into haploid thalli (Pr-25, Pr-28, Pl)

tetrasporophyte Diploid thallus in rhodophytes that produces tetrasporangia (Pr-25)

thallose liverwort Liverwort having a thin thallus; compare *leafy liverwort* (Pl-2)

thallus (pl. **thalli**) Simple flat leaflike body undifferentiated into organs such as leaves or roots (Pr, F, Pl-1 through Pl-3)

theca Test, shell, valve, coat, hard covering, enveloping sheath or case

thermophil Organism whose optimum environmental temperature is well above 30°C

thorax In chordates, the chest, the part of the body that contains the heart and the lungs (A-37); in arthropods, the leg-bearing membranous segments between head and abdomen (A-20 through A-21)

thylakoid Photosynthetic membrane, saclike membranous lamella that, stacked in numbers, constitutes the grana of chloroplasts

tibia Inner bone of the lower leg (between knee and ankle) in vertebrates (A-37)

tinsel See *mastigoneme*

tissue Aggregation of similar cells organized into a structural and functional unit; component of organs (Pl, A)

toe In rotifers, a posterior organ of attachment to substrate (A-13); in loriciferans, a paddlelike posterior appendage for swimming (A-17)

tornaria larva Immature, ciliated form of certain enteropneust hemichordates (A-33)

trachea Air-conducting tube (A-20, A-37); windpipe (A-37)

Tracheata Vascular plants (Pl-4 through Pl-12)

tracheophyte Vascular plant; plant that contains xylem and phloem—vascular tissues (Pl-4 through Pl-12)

transduction Transfer of host genes from one bacterial cell to another by a virus

transport host Host organism that carries a parasite between successive hosts; while in the transport host, the parasite does not develop further (A-12)

tree Woody perennial plant having a single stem, compared with shrubs, which have multiple stems (Pl-8 through Pl-12)

trepang Dried sea cucumber used as human food; also called bêche-de-mer (A-34)

trichinosis Disease caused by infection with the parasitic nematode *Trichinella* (A-10)

trichocyst Extrusome underlying the surface of many ciliates and some mastigotes; capable of sudden discharge to sting prey; probably nonhomologous structures present in disparate groups (Pr)

trichocyst pore Opening through which contents of a trichocyst are released (Pr)

trichogyne In rhodophytes (Pr-25) and many fungi, receptive protuberance or threadlike elongation of a female gametangium to which male gametes become attached

trichome Filament consisting of a string of connected cells (B, Pr, F, Pl)

trilobite Extinct Paleozoic arthropod having jointed legs and body divided in three lobes, sometimes placed in own phylum, sometimes grouped with arthropods; trilobite larva of horseshoe crab named for resemblance to ancient trilobites (A-19, A-20)

triploblastic Of developing embryos having three distinct tissue layers (A-5, A-6, A-10)

triploid 3N, as in endosperm (Pl-11, Pl-12)

trochophore larva Free-swimming, ciliated marine larva (A-22 through A-24, A-26)

troglodyte Cave-dwelling animal (A)

trophi Specialized circlet of cuticle-covered cilia (undulipodia) forming the jaws of rotifers (A-13)

trophosome Midgut of vestimentiferan packed with sulfide-oxidizing bacteria in symbiotrophic association with the vestimentiferan (A-25)

trophozoite Life cycle stage; growing, vegetative stage of Apicomplexa (Pr-9)

true branching Growth habit of a filamentous structure produced by the presence of three growth points on a single cell (B, Pr, F); see *false branching*

trunk Upright stem of a tree (Pl)

trypanosomiasis Disease caused by trypanosome infection (Pr-30); borne by the tse-tse fly (A-20)

tube Tubule on underside of certain (polypore) basidiomycete caps; basidiospores form within the tubes and are released to the exterior through tube pores (F-2)

tube foot Hydraulically controlled foot, part of the gas-exchange and water vascular system of echinoderms (A-34)

tube nucleus The nucleus of the pollen that controls pollen tube development (Pl-12); see *sperm nucleus*

tubercle Small projection or raised bump, in priapulids

presumed to be sensory (A-15); nodule or small eminence, especially those produced in host tissue by *Mycobacterium tuberculosis* (B-12); a translucent mass containing giant cells and surrounded by layer of spindle-shaped connective tissue cells

tubulin Protein component of microtubules

tun Name given to cryptobiotic tardigrade because of resemblence to wine cask (A-27)

tunic In tunicates (Urochordata), external carbohydrate test, sometimes enclosing many individual zooids (A-35)

tunicin Polysaccharide related to cellulose of which tunicate tunic is composed (in ascidians and thaliaceans but not in larvaceans) (A-35)

ulna Outer of two bones of the forearm between elbow and wrist of vertebrates (A-37)

ultrastructure Detailed structure of cells and organs; structure visible by transmission electron microscopy

undulipodium (pl. **undulipodia**) A cilium or eukaryote "flagellum"; used primarily for locomotion and feeding (Pr, A, Pl); see Figure I-3

uniramous Single (that is, unbranched), refers to appendages of certain arthropods (A-20)

unitunicate In ascomycetes, ascus having a homogenous wall (F-3)

urn cell Cell taking part in waste removal in sipunculans (A-23)

urogenital pore Opening of priapulid reproductive tract to the exterior (A-15)

uterine bell Funnel that connects the acanthocephalan body cavity to the uterus (A-12)

uterus (pl. **uteri**) In mammals, the chamber of the female reproductive tract in which the offspring develops and is nourished before birth (A-28, A-37)

vacuole Cavity that contains fluid or air and is surrounded by a membrane in cell cytoplasm

vagina In female reproductive tract, region leading from the exterior to the uterus; also called birth canal (A-12, A-21, A-37)

valve Test, shell, hard covering (Pr, A)

vanadium Rare metallic chemical element, concentrated in tissues of urochordate tunic (A-35)

vascular Concerning vessels that conduct fluids (water, sap, blood) in a body (Pl-4 through Pl-12, A)

vascular bundle See *vascular cylinder*

vascular cylinder Column of tissue containing primary xylem and primary phloem (Pl)

vascular plant Member of plant kingdom having vascular tissue (Pl-4 through Pl-12)

vascular system Blood vessel system (A-9, A-18, A-25)

vector Organism that transmits a disease-producing organism from one host to another (A-19)

vegetative apex Sexually undifferentiated anterior or top part of a thallus or shoot

vegetative cell Somatic cell, produced asexually by mitosis (distinguished from germ cells)

vein (1) Rib supporting an insect wing (A-20); (2) vessel carrying blood from capillaries toward the heart (A-37); (3) conduit for fluid in a leaf (Pl)

veliger larva Posttrochophoral stage of many molluscs (A-26)

velum (pl. **vela**) A membranous part resembling a veil or curtain (B, A)

venom Poison or toxin (A-9, A-19)

ventral On or toward the belly or undersurface; in human beings, toward the front (A)

ventricle Chamber(s) from which blood leaves the heart (A-26, A-37)

vermiform Wormlike in form (A)

vermiform larva Worm-shaped larva of orthonectid (A-7)

vernal Present in spring, as in vernal pools—a seasonal temporary habitat (A-21)

vertebra (pl. **vertebrae**) Bones that make up the spinal column in vertebrates (A-37)

vertebral column Backbone of craniate chordates (A-37)

vertebrate Animal having vertebrae (A-37)

vessel In angiosperms and gnetophytes, water-conducting tubes of xylem made up of stacks of long cells having perforated ends (Pl-11, Pl-12)

vestimentum In vestimentiferans, body wall folds (A-25)
virus Tiny (15–300 μm) infectious but nonliving particle consisting of genetic material (RNA or DNA, but not both) surrounded by a protein coat (capsid); unable to replicate on its own, but must infect specific host cells; after infection, the host enzymes may be redirected to produce new virus particles (lytic phase) or the virus may be incorporated into the host genome (latent, temporary phase)
viscera Collective term for the internal organs (A)
vitamin Organic substance found in some foods; some vitamins are synthesized in the body; essential in small amounts in metabolism (F-1, Pl-6, A)
vitellarium (pl. **vitellaria**) Part of the ovary; produces yolk-filled cells that nourish the eggs (A)
viviparous Of a mode of sexual reproduction in which offspring develop inside the maternal parent, from which they receive nutritional and other metabolic aids; offspring are released either immature or as miniature adults (A)

wall See *cell wall*
water vascular system Ambulacral system; set of hydraulic canals derived from the coelom and equipped with tube feet (podia) used for gas exchange, movement, food handling, and sensory reception (A-34)
winter egg Rotifer egg in arrested development that overwinters (A-13)
Wolffian duct Tubule that transports sperm from testes in vertebrates (A-37)
wood Secondary xylem (Pl-8 through Pl-12)

xanthin Member of a class of oxidized isoprenoid derivatives; typically yellow or orange
xanthoplast Plastid (photosynthetic organelle) of members of Phylum Xanthophyta (Pr-14)
xenophya (pl. **xenophyae**) Foreign bodies of which the inorganic part of xenophyophoran tests is composed (Pr-5)
xylem Plant vascular tissue through which most of the water and minerals are conducted from the root to other parts of the plant; constitutes the wood of trees and shrubs (Pl-4 through Pl-12)

yolk Fatty substances deposited in egg for nourishment of the developing embryo (A-18, A-23, A-29, A-37)

zoea One of several types of decapod crustacean larvae (A-21)
zoochlorella (pl. **zoochlorellae**) Green alga living symbiotically within protists or animals (B-5, Pr-28)
zooid Colonial individual that resembles, but is not, a separate organism (A)
zooplankton Free-floating heterotrophic microorganisms (Pr, A)
zoosporangium Sporangium that produces zoospores
zoospore Undulipodiated motile cell capable of germinating into a different developmental stage without being fertilized (Pr)
zooxanthella Yellow alga, often dinomastigotes of the genus *Gymnodinium* (Pr-7), living symbiotically within protoctists or animals
zygosporangium (pl. **zygosporangia**) Multinucleated, resistant structure (resting sporangium) that results from the fusion of two gametangia; formerly zygospore (F-1)
zygospore Resistant structure formed by conjugation; thick-walled zygote of the conjugating green algae (Pr-26); large, multinucleate resting stage
zygote Diploid nucleus or cell produced by the fusion of two haploid cells and destined to develop into a new organism (Pr, A, F, Pl)
zygotic meiosis Meiosis that takes place in a zygote immediately after it has been formed

INDEX

Page references followed by the letter "f" refer to a figure; by the letter "t" to a table, and by "phy" to the phylogenetic diagrams at the beginning of each chapter. Letters and numbers in parentheses after an entry refer to the kingdom and phylum. See the Table of Contents or the Appendix for the entire list. Examples of genera from the heading of each phylum do not appear in the Index but are listed in the Appendix.